PIPELINES IN THE CONSTRUCTED ENVIRONMENT

PROCEEDINGS OF THE 1998 PIPELINE DIVISION CONFERENCE

August 23–27, 1998
San Diego, California

EDITED BY
Joseph P. Castronovo
James A. Clark

1801 ALEXANDER BELL DRIVE
RESTON, VIRGINIA 20191–4400

Abstract: This proceedings, *Pipelines in the Constructed Environment*, consists of papers presented at the 1998 Pipeline Division Conference, held in conjunction with the Prestressed Concrete Cylinder Pipe (PCCP) Users Forum, which was held in San Diego, California, August 23-27, 1998. The fact that new pipeline construction, the maintenance of existing pipeline, or the rehabilitation of a deteriorating pipeline often takes place where there is development over or adjacent to it can place a variety of constraints on a project. These papers explore new construction technology, such as horizontal directional drilling, that allows this construction or rehabilitation to take place with minimal inconvenience to the public. In addition, various papers on existing technology, such as CAD systems, explain how they can be used to help engineers construct new facilities with longer service life.

Library of Congress Cataloging-in-Publication Data

Pipelines in the constructed environment / edited by Joseph P. Castronovo and James A. Clark.
    p.   cm.
Includes bibliographical references and index.
ISBN 0-7844-0372-4
1. Pipelines–Congresses. 2. Pipelines–Maintenance and repair–Congresses. 3. Pipe, Concrete –Congresses. I. Castronovo, Joseph P. II. Clark, James Anthony, 1907- . III. American Society of Civil Engineers. Pipeline Division. Conference (1998: San Diego, Calif.) IV. Prestressed Concrete Cylinder Pipe Users Forum (4th: 1998: San Diego, Calif.)
TA660.P55P59   1998
621.8'672–dc21   98-35557
                                                                                                    CIP

Any statements expressed in these materials are those of the individual authors and do not necessarily represent the views of ASCE, which takes no responsibility for any statement made herein. No reference made in this publication to any specific method, product, process or service constitutes or implies an endorsement, recommendation, or warranty thereof by ASCE. The materials are for general information only and do not represent a standard of ASCE, nor are they intended as a reference in purchase specifications, contracts, regulations, statutes, or any other legal document.

ASCE makes no representation or warranty of any kind, whether express or implied, concerning the accuracy, completeness, suitability, or utility of any information, apparatus, product, or process discussed in this publication, and assumes no liability therefore. This information should not be used without first securing competent advice with respect to its suitability for any general or specific application. Anyone utilizing this information assumes all liability arising from such use, including but not limited to infringement of any patent or patents.

Photocopies. Authorization to photocopy material for internal or personal use under circumstances not falling within the fair use provisions of the Copyright Act is granted by ASCE to libraries and other users registered with the Copyright Clearance Center (CCC) Transactional Reporting Service, provided that the base fee of $8.00 per chapter plus $.50 per page is paid directly to CCC, 222 Rosewood Drive, Danvers, MA 01923. The identification for ASCE Books is 0-7844-0372-4/98/ $8.00 + $.50 per page. Requests for special permission or bulk copying should be addressed to Permissions & Copyright Dept., ASCE.

Copyright © 1998 by the American Society of Civil Engineers,
All Rights Reserved.
Library of Congress Catalog Card No.: 98-35557
ISBN 0-7844-0372-4
Manufactured in the United States of America.

FOREWORD

The 1998 Pipeline Division Conference, in conjunction with the Prestressed Concrete Cylinder Pipe (PCCP) User's Forum, was held in San Diego, California, on August 23-27, 1998. The theme for the Conference and Forum was "Pipelines in the Constructed Environment." The Conference Committee is pleased to present these proceedings on behalf of the Pipeline Division and the San Diego Section of the American Society of Civil Engineers.

New pipeline construction, the maintenance of existing pipelines, and the rehabilitation or replacement of deteriorating pipelines is, often times, taking place with constraints imposed by development over or adjacent to these facilities. New construction technology is allowing the construction and rehabilitation to take place with minimal inconvenience to the public. In addition, the technology exists to construct new facilities with longer service lives to reduce future replacement costs. These topics are of increasing importance in the pipeline industry and the Committee wishes to thank all of the Conference participants for making this an excellent opportunity to share and learn more about these issues.

A special feature of this conference was the fourth meeting of the PCCP User's Forum. This group evolved in 1990 at a meeting in Denver, Colorado, where several owners of PCCP met to discuss common issues with maintaining their PCCP pipelines in corrosive environments. Each of the previous meetings focused on discussing and developing inspection and rehabilitation techniques. The Committee is extremely pleased that this year's Pipeline Conference included the PCCP User's Forum.

Each paper included in the proceedings has received two peer reviews. We want to thank everyone who contributed to this peer review process for their time, expertise, and patience. All papers are eligible for discussion in the ASCE Journal of Transportation Engineering and are eligible for ASCE awards.

Finally, the Committee wishes to thank all of those who participated in every way to make this conference a success. Your energy and enthusiasm were the primary reason we were able to accomplish our task. The conference and these proceedings are the result of your efforts.

On behalf of the Committee,

Michael T. Stift, P.E., *Conference Co-Chairman*

W. Jeffery Moncrief, P.E., *PCCP User's Forum Co-Chairman*

Conference Organizing Committee

Michael T. Stift, P.E.	*Conference Co-Chairman*
W. Jeffery Moncrief, P.E.	*PCCP User's Forum Co-Chairman*
Joseph P. Castronovo, P.E.	*Technical Program Co-Coordinator*
James A. Clark, P.E.	*Technical Program Co-Coordinator*
Charles R. Spinks, P.E.	*ASCE San Diego Section Liaison*
Michael S. Hewitt	*ASCE Conference Liaison*
Michael S. Tucker, P.E.	*Exhibits Coordinator*
Randall C. Conner	*Exhibits & 1999 ASCE Pipeline Conference Chairman*
Richard D. Brady, P.E.	*Local Arrangements Coordinator*
Lawrence F. Catalano, P.E.	*Pipeline Division Chair and Committee Liaison*
Lawrence E. Shaw	*Publicity/Marketing Coordinator*

Contents

Session 2A: Pipeline Design 1
Kathleen Haynes, Moderator

Integrated Design Procedure for Flexible Pipe 1
 Robert W. Miles and B. Jay Schrock

The Suitability of Spangler's Iowa Formula for Predicting Deflection in All Flexible Pipes 14
 James C. Schluter and Theodore A. Capossela

Replacing E' with the Constrained Modulus in Flexible Pipe Design 28
 Timothy J. McGrath

Session 2B: Pipeline Crossings
Tom Cooper, Moderator

Willow Street/Lower Sweetwater River Pipeline Crossings 41
 Michael Marks, Hector Martinez, and James Smyth

Tijuana River Crossing Using Horizontal Directional Drilling: A Case History 47
 Moi Arzamendi and Terry Smith

Ma Wan Submarine Crossing Using Horizontal Directional Drilling 57
 Steven R. Kramer and Stephen V.L. Barrett

Session 2C: Reclaimed/Wastewater Planning & Design
Larry Catalano, Moderator

The Successful Design of a Large Diameter Gravity Sewer 66
 Jamal Batta, John Harris, and Mark Giandoni

Sacramento County Northwest Interceptor: Effective Planning Ensures Project Success and Saves Money 71
 Thomas Kalkman and Michael Watson

Using GIS-Based Models To Plan Regional Reclaimed Water Pipeline Networks 80
 Scott Lynch, Kim Martin, and Dave Bramwell

Session 3A: Transients
Howard Arnold, Moderator

Analysis of Surge Pressures in the Inland Feeder and Eastside Pipeline 88
 R. Scott Foster

Avoiding Common Thrust Restraint Mistakes 97
 Andrew E. Romer

Surge Protection Design for the City and County of San Francisco Water Transmission System 103
 R. Scott Foster

Session 3B: Microtunneling 1
Dan Badaluco, Moderator

The Influence of Geologic Setting on Microtunneling 113
David C. Mathy and Dru R. Nielson

Recipe for Successful Microtunneling 124
John Goodwin, Jesse Gill, and David Mathy

Microtunneling in Downtown Honolulu 133
James Kwong, Steve Klein, Galen Nagle, Glen Okita, and Steve Duke

Session 3C: Bidding, Construction and Mitigation
Curt Lehnhoff, Moderator

Development of a Coastal Sage Scrub Revegetation Program Following Pipeline Construction, San Diego County, California 144
Tim Cass and Dave Chamberlain

Bidding Projects with Alternative Pipeline Materials 154
Karen Larson Henry, Dave Zoumaras, Timothy Pflum, and Mark Giandoni

Horizontal Directionally Drilled River Crossing Meets Regional Water Needs 162
Daniel J. Nichols, Thomas J. Meinhart, and Steven R. Kramer

Session 4A: Corrosion
George Ruchti, Moderator

Electromagnetic Conductivity Survey of Pipeline Alignments for Soil Corrosivity and Corrosion Control Design for Large Diameter Water Pipelines 172
G.E.C. Bell, R.A. Pannell, and J.J. Galleher, Jr.

Polyethylene Encasement of Buried Conduit 180
George L. Ash

Cathodic Protection of Cast Iron Pipe 187
Nicholas J. Irias and Marilyn L. Miller

Session 4B: Pipeline Design/Construction
James Ashcraft, Moderator

*Las Vegas Pump Station No. 2 Underground Construction
Stephen J. Navin

Value Engineering Savings on Pipeline Project 195
Paul W. Johnson, Antonio J. Perez, Burt K. Yu, and Carlos de Leon

CAD Design Expedites Environmentally Friendly Project for CCWA 203
Dan Masnada, Ken Ferguson, and Lyndel Melton

Session 4C: Community Relations
Rick Fornelli, Moderator

Public Outreach and Construction Strategies for the Mission Trails Pipeline and Flow Regulatory Structure 210
Zachary Ahinga and Janice Collins

*Not available at time of publication

Rancho Penasquitos Pipeline Public Affairs Support 221
James R. Melton

City of San Diego Mid-City Pipeline Project: Integrating a Public Outreach
Program into the Planning and Alignment Selection Process 232
Marc R. Weinberger

Session 4D: Inspection 1
Matt Tebbetts, Moderator

Diagnosis of Buried Sewers: Tools and Methods, The French Experience 241
Youssef Georges Diab

Impact Echo Testing of In Situ Precast Concrete Cylinder Pipe 250
Dennis A. Sack and Larry D. Olson

Inspection, The Key to Successful Rehabilitation 260
Paul J. Williams

Session 5A: Pipeline Design 2
Anka Fabian, Moderator

Environmental Design Factors for Concrete Pressure Pipe 268
Wayne R. Brunzell

Pipeline Construction for Designers 277
Deon T. Fowles

Rethinking the Approach to Engineering Pipelines 284
Andrew E. Romer

Session 5B: Case Histories—Wastewater
Timothy Stanton, Moderator

Pipeline Planning and Design: City of San Diego Water Department CIP 291
Frank X. Collins and Steven W. Wallace

Balancing Environment and Reliability: East Mission Gorge Trunk Sewer
Rehabilitation Project Case Study 298
Michael E. Conner, Marnel Hale, Stephen L. Deering, and D. Michael Metts

Construction of the East Mission Gorge Trunk Sewer Rehabilitation Project,
San Diego, California 308
Denis M. O'Malley, Peter J. Barden, Michael E. Conner, and Marnell L. Hale

Session 5C: Rehabilitation
Lynn Osborn, Moderator

Evaluation of 70-Year-Old Non-Reinforced Concrete Sewers 318
James Biery, Alison Ratliff, and Sylvia Hall

Water Line Splitting in Gainesville, Florida 328
John S. Gifford and Larry J. Ruffin

Wastewater Force Mains Problems and a Solution 334
Roger M. Cimbora, Sr.

Session 5D: ASCE New Manuals of Practice
Larry Catalano, Moderator

ASCE Water Resource Division Pipeline Manual (Presentation Only)
Roger Beieler

MOP #46, Pipeline Route Selection (Presentation Only)
Nick Day

MOP, Pipeline Installation, Inspection and Acceptance Testing (Presentation Only)
Malcom Stephens

Session 6A: PCCP Corrosion Evaluations 1
Nathan Jones, Moderator

Evaluating a Proposed Right-of-Way for PCCP 345
Risque L. Benedict

Corrosion Control of Prestressed Concrete Cylinder Pipe 356
Sylvia C. Hall

Extending The Life of Prestressed Concrete Cylinder Pipe with Pulse Cathodic Protection 367
Ted Doniguian, Harry Kipps, and John Barnes

Session 6B: Inspection 2
Paul Klein, Moderator

External Corrosion Control of Water Mains To Maximize Operating Life 377
Bryan M. Bradish and Michael J. Szeliga

Internal Inspection of 20-Inch Natural Gas Pipeline 387
Barbara B. Ostrander

The Use of Submersible Remotely Operated Vehicles for the Inspection of Water-Filled Pipelines and Tunnels 397
Ronald E. Heffron

Session 6C: Seismic/Land-Slide
Michael Metts, Moderator

Seismically Upgrading the Mokelumne Aqueducts 405
David L. Pratt, Christopher F. Dodge, Frederick N. Brovold, and Howard O. Wilson

Seismic Design of Puerto Rico's North Coast Superaqueduct 413
Mehdi S. Zarghamee, Rajesh S. Rao, Michael A. Yako, Edward M. Motley, Felix Garcia, and Anibal Camacho

Session 6D: Pipeline Design 3
Kathy Schuler, Moderator

Case Study: Crossing of Existing PCCP Aqueducts A and B, Southern Nevada Water Authority, Las Vegas, Nevada 423
Philip K. Ryan and R. Ted Davis

Pipe Joint Failure Caused by an Inadequately Specified Constructed
Environment 433
   Kenneth K. Kienow

Unified Design Methodology for Most Pipeline Materials 451
   Jey K. Jeyapalan and Sri K. Rajah

Session 7A: PCCP Monitoring & Rehabilitation
Jay Schrock, Moderator

Acoustic Monitoring of Prestressed Concrete Cylinder Pipe..468
   Mark Holley and Doug Buchanan

An Update on Acoustic Emission Testing of PCCP 477
   Will Worthington

Pipeline Rehabilitation and Repair 485
   Tim Gwaltney

Session 7B: Microtunneling 2
Allison Ratliff, Moderator

Horizontal Directional Drilling with Ductile Iron Pipes...494
   Randall C. Conner

Municipal Infrastructure: Innovative Trenchless Replacement Method Utilizing
Bell-Less Ductile Iron Pipe, Case Studies 506
   Al Tenbusch and Ralph Carpenter

Geotechnical Investigations for Tunneling & Pipe Jacking..516
   Gregory L. Raines

Session 7C: PCCP & Concrete Pipeline Design
James Rasmus, Moderator

Role of Surge and Possible Mitigation in PCCP Design 528
   Sri K. Rajah and Jey K. Jeyapalan

Large Diameter Reinforced Concrete Pipe Bedding Design and Installation
in Adverse Soil Conditions 536
   Mark Giandoni

Evaluation of New Installations for Concrete Pipe 546
   James J. Hill, John M. Kurdziel, Charles R. Nelson, and James A. Nystrom

Session 7D: Pipe Materials 1
Rick Nelson, Moderator

TRWD Experience with Prestressed Concrete Pipe 556
   David H. Marshall

Soap Lake Siphon Receives Rehabilitation...566
   James E. Wolfe and M. Wayne Cardwell

PCCP Research at the National Research Council of Canada (Presentation Only)
   John Makor

Session 8A: PCCP Corrosion Evaluations 2
Tim McGrath, Moderator

When Should PCCP with Interplace Class IV Wire Be Replaced? (Presentation Only)
Bryan M. Bradish

Evaluation of Prestressed Concrete Cylinder Pipe in a High Chloride Environment after 19 Years of Service 575
Jose L. Villalobos

Effects of Environment on the Durability of Prestressed Concrete Cylinder Pipe 584
Robert E. Price, Richard A. Lewis, and Bernard Erlin

Session 8B: PCCP Rehabilitation 2
Christine Waters, Moderator

Rehabilitation of a 183 cm PCCP with Steel Plate Liners 594
Gary P. Stine and Michael T. Stift

Overcoming the Challenges of Replacing 20 Km of Defective 1524 mm Diameter PCCP 602
Terry L. Walsh and David S. Hodge

Design and Construction Aspects of the PCCP Water Main Rehabilitation 612
Sufian A. Khondker and John R. Mitchell, Jr.

Session 8C: Case Histories—Water
Gerald Copeland, Moderator

The Lake Gaston Pipeline: 76 Miles of Controversy 622
Joe Bivins, Tom Leahy, and Jim Richards

Design and Construction of the Valley Center Pipeline 632
Mark Butier, W. Jeffery Moncrief, Gary P. Stine, and Richard Trembath

Designing and Building a Major Transmission Main through a Constructed Environment, The North County Distribution Pipeline 638
Steve Tedesco and Edward Stewart

Session 9A: PCCP Design
Lisa Jackson, Moderator

PCCP Designs Check Using C304-92 in a Spreadsheet 646
Jey K. Jeyapalan and Sri K. Rajah

Assuring Top Quality Prestressing Wire in PCCP 656
Ralph T. Rundle, John Olden, and Will Worthington

Advancements in Design and Installation of Prestressed Concrete Cylinder Pipe 664
David P. Prosser

Session 9B: Case Histories—Tunnels
John Bomba, Moderator

North City Tunnel Connects to New Water Reclamation Plant 674
Greg Arakaki, Duane Larson, John Kinneen, and Alan Redmon

Construction Challenges for Soft Ground Tunneling 681
Gregory W. McBain, Luciano Meiorin, Jon Y. Kaneshiro, Stephen J. Navin,
and Rolf H. Lee

San Diego's Conveyance Tunnels: A Historical Perspective 692
Gregory L. Raines and Rick Wright

Session 9C: PCCP Inspection
Raz Konyalian, Moderator

Condition Assessment and Repair of Prestressed Concrete Pipeline 702
Mehdi S. Zarghamee, Rasko P. Ojdrovic, and Roger Fongemie

In-Line Electromagnetic Inspection of PCCP 714
Brian J. Mergelas and David L. Atherton

Internal Inspection and Database Development of PCCP 721
John J. Galleher, Jr., and Michael T. Stift

Session 10A: PCCP Rehabilitation 3
Jaime Moreno, Moderator

Sources of Funding for Replacement and/or Repair of Defective PCCP 731
Geoffrey Johnson

Repair of PCCP with Fiber Reinforced Composites (Presentation Only)
V. M. Karbhari, F. Seible, and A. Mullen

Practical Repair Procedures for Concrete Pressure Pipe 742
Richard I. Mueller

Session 10B: Pipe Materials 2
Hans Torabi, Moderator

Lessons Learned about Cured-in-Place Pipe During Construction 752
Mark W. Hutchinson

Pipeline Market—20 Billion Dollars for 1998: Would You Like Some
of This Work? 763
Jey K. Jeyapalan

Compliance Audits of Concrete Pressure Pipe Manufacturers (Presentation Only)
Robert Orr

Session 10C: I/I Reduction & Drainage
Doug Gillingham, Moderator

Miami Beach Infiltration/Inflow Reduction Program: A Project That
Pays for Itself 782
Russell Barnes

Pipeline Drainage Discharge Analysis 787
Timothy M. Smith

Design and Construction of the Sweetwater Reservoir Urban Runoff Diversion
System 48-Inch Pipeline 797
Richard Bottcher, Claud Seal, Tucker, and Jim Smyth

Subject Index 806
Author Index 810

Integrated Design Procedure for Flexible Pipe

Robert W. Miles[a], B. Jay Schrock[b]

ABSTRACT

Engineers are presented with a formidable, often conflicting, amount of technical information concerning design of flexible pipes. When preparing design documents for projects, selection of pipe materials often becomes a time-consuming task that can result in wide variations in design approach between engineers in the same agency or firm. These inconsistencies reduce the productivity of the design process and can result in avoidable problems during construction and service. This paper presents an approach to design of flexible pipe that integrates recent research, design procedures, and construction practices. The approach allows design engineers to determine early in a project if a particular type of flexible pipe is suitable.

TECHNICAL BACKGROUND

Basis for Design

Most types of thermoplastic and composite steel and concrete pipes are designed by comparing calculated values for vertical deflections with acceptable limits. Excessive calculated deflections require the engineer to modify the design until the deflections become less than the limits. The modifications involve improvement of strength of embedment soil that surrounds the pipe or strengthening of the pipe structural design to generate additional resistance to vertical earth and surface loads. For thermoset composite pipe materials, a limitation on the strain of the pipe wall is used as the basis of design instead of vertical deflection. However, for design purposes, strain limitations are converted to equivalent deflection values to simplify calculations.

Basic Deflection Equation

Beginning in 1913 research and experimental work was done at Iowa State College on the performance of flexible pipes. Spangler and Marston developed the early pipe deflection formula, which became known as the Iowa Formula. Spangler continued this work in the 1920's and 1930's. In the 1950's the work of Spangler and Watkins resulted in the Modified Iowa Formula shown below.

---

[a] Principal, Robert W. Miles, Consulting Civil Engineer, P.O. Box 627, Brentwood CA 94513
[b] Principal, JSC International Engineering, 1313 Gary Way, Carmichael CA 95608

$$\Delta X = \frac{D_L K_b W_c r^3}{EI + 0.061 E' r^3}$$

In the equation:

ΔX	Average horizontal deflection	K_b	Bedding factor
D_L	Deflection lag factor	W_c	Earth load on pipe
r	Mean radius of pipe	E'	Modulus of soil reaction
E	Modulus of tensile elasticity of the pipe material	I	Moment of inertia of the pipe wall

Simplified Deflection Equation

The Modified Iowa Formula may be written in a very simplified form as an expression for vertical deflection[1]:

$$\delta_v = \frac{W}{S_p + S_s}$$

In this equation the Modified Iowa Formula has been rewritten to show δ_v as the vertical deflection due to load, W as the vertical load on the pipe, S_p as the pipe stiffness term, and S_s as the soil stiffness term. For flexible pipes the soil stiffness term S_s is much larger than the pipe stiffness term S_p, often amounting to more than 90 percent of the total stiffness. As an example, the following deflection computation has been performed for a 15 inch PVC pipe with 20 feet of cover:

$$\delta_v = \frac{W}{S_p + S_s} = \frac{1.6 psi}{6.9 psi + 55 psi} = 2.6\%$$

In this example, the soil stiffness term S_s at 55 psi is 89 percent of the total stiffness. The example illustrates a reality that has been borne out by experience and research; that the deflection performance of flexible types of pipe depends much more on the stiffness, or strength, of the soil envelope around the pipe than on the stiffness of the pipe itself. To the design engineer, this underscores the importance of determining the insitu and embedment soil strengths and making certain that the design assumptions are implemented during construction.

Important research has been done under the direction of Howard[2,3] while at the Bureau of Reclamation on the modulus of soil reaction E' and the factors that can be used to estimate soil stiffness. And research and analysis in Germany by Leonhardt[4] and others, and in the United States by Greenwood, Lang[1], Selig[5], and others since 1979 has allowed engineers to more fully understand relationships between insitu soils and pipe embedment soils, pipe stiffness and soil stiffness interaction, and other complex issues.

GUIDELINES

Development

The above information can be used to construct guidelines for design and installation of flexible pipes with greater levels of confidence than in the past. The guidelines can take the form of design criteria and tables that illustrate limitations on use of various types of flexible pipe. After completion, review, and approval of the guidelines, agency or consulting firm master specifications, field testing, and monitoring procedures would be revised to implement the guidelines.

Building on the simplified formula for deflection presented above, the following formulae allow development of a refined deflection equation that reflects recent research and experience with flexible plastic pipe.

Vertical Loads on Pipe

The following formula is recommended to reflect the long-term vertical design load on the pipe:

$$W = K_V((C_L \gamma H + w_L)$$

Pipe Stiffness

The pipe stiffness term, S_p, can be expressed as the following equation:

$$S_p = \frac{8EI}{D^3}$$

The pipe stiffness factor EI is usually calculated for custom designed pipe such as mortar-lined and coated steel pipe. It is established from tests for mass-produced pipe products such as polyvinyl chloride plastic pipe using the procedures in ASTM D 2412 and the following equation:

$$EI = \frac{F}{\Delta Y} 0.149 r^3$$

Soil Stiffness

The soil stiffness term can be expressed as the following equation:

$$S_s = 0.061 \xi C_m' C_l E_b$$

The term C_m' is used to represent the influence of construction testing on the soil modulus, in accordance with the following table. The lowest value for C_m' is selected based upon the condition that applies to the project under consideration.

Construction Testing	C_m' Value
Embedment soils testing	
No testing	0.3
At least one test/lift per 1,000 feet	1.0
Deflection testing	
No testing	0.5
Testing at one month	0.8
Testing at 11 months	1.0

The term C_t has been adapted from Greenwood and Lang[1] to represent a long-term soil modulus retention factor.

Non-Ellipsoidal Deflection

The deflection ratio dy/dx is based upon the pipe to soil stiffness ratio[6] and the tendency of installations with a high ratio to develop non-elliptical distortion with total deflections greater than predicted by the Modified Iowa Formula. The minimum value for dy/dx is 1.1, the ratio of vertical to horizontal deflection for a pipe that has deflected into an ellipsoidal shape without distortion. It is recommended that this factor be calculated using the embedment soil stiffness E_b.

Influence of Trench Width

The factor ξ is the trench width factor, as defined by Leonhardt and summarized in the ATV design methods[4]. The factor is a function of trench width, pipe diameter, native soil modulus, and embedment soil modulus. The factor is applied to the embedment soil modulus to produce a composite modulus that considers differences between the embedment backfill and native soil that surrounds the pipe installation. The factor ξ is defined as follows:

$$\xi = \frac{1.662 + 0.639(B/D - 1)}{(B/D - 1) + [1.662 - 0.361(B/D - 1)]\frac{E_b}{E_s}}$$

The native soil modulus E_s is assumed to vary by soil type and depth[5,7]. Stronger soils and greater depths of cover produce larger values for E_s. Weaker soils and lesser depths produce lower values for E_s. Table A shows estimated variations of E_s with soil type and depth for installations without groundwater. Presence of groundwater requires reduction in values of E_s. The proposed correlation of insitu soil characteristics and relative strengths to the soil moduli in Table A need to be reviewed before adoption as guidelines. The embedment soil modulus E_b is recommended to be based on the work by Howard[2].

Deflection Variations from Average

The soil moduli and other factors are average values, and therefore the resulting calculations will result in average values for deflection. However, it is the maximum values for deflection that are the most important in design. Typically, deflection measurements vary considerably along a pipeline in response to a number of factors. Research by Greenwood and Lang[1] offers some guidance for application of a deflection variability deflection of δ_c. Application of this factor needs evaluation since it may have some relationship with the deflection ratio dy/dx. Addition of δ_c to the average deflection results in an estimate for the maximum deflection.

Refined Deflection Equation

Substitution and addition of the above terms results in a deflection equation that can be used for design purposes:

$$\delta_v = \left(\frac{dy}{dx}\right) \frac{K_x(C_L \gamma H + w_L)}{S_p + 0.061 \xi C_m C_t E_b} + \delta_c$$

In the above equations:

δ_v Vertical deflection due to loads, percent of pipe diameter
δ_c Vertical deflection due to variability
K_x Bedding factor, dependent on apparent bedding angle, usually taken as 0.08
$\dfrac{dy}{dx}$ Deflection ratio due to pipe/soil stiffness ratio. Minimum value is 1.1
C_L Soil arching factor. For long term, value is 1.0
γ Unit weight of trench soil
H Depth of pipe cover
w_L Surface live load
S_p Long-term pipe stiffness term
ξ Leonhardt trench width factor
C_t Soil modulus retention factor
C_m Construction testing factor
E_b Embedment zone soil modulus
E_s Insitu soil modulus
B Excavation trench width
D Pipe diameter

DEFLECTION CRITERIA

Hydraulic and Structural Performance Limits

Pipe materials have their respective performance limits. For flexible pipes the performance limits are related to flow capacity and structural considerations.

Analysis of pipeline hydraulic performance indicates that a flow reduction of approximately 2 percent results from a uniform pipe deflection of 10 percent[8] Therefore, an average deflection of less than 10 percent will limit loss of flow to relative insignificance considering the overall accuracy of hydraulic calculations.

Structural performance of flexible pipe is measured by the amount of deflection the pipe/soil system assumes, state of strain of the pipe material, resistance to buckling of the pipe wall, and integrity of the joints. As an example, a brief discussion of some structural performance limitations for several types of commonly used pipe is presented below.

Generally, flexible pipe made of high density polyethylene (HDPE) material can assume significant long-term deflections without failure because the material flows plastically over time to reduce initial stresses caused by deflection. An elliptical deflection as high as 30 percent can be sustained in short-term laboratory tests by HDPE profile pipe without onset of structural failure by reverse curvature of the crown, buckling, rupture, cracking, rib failure, or other structural problems.

Polyvinyl chloride material (PVC) does not exhibit the plastic flow capacity of HDPE but nonetheless also can deflect approximately 30 percent under short-term laboratory conditions without reverse curvature and collapse. For these plastic pipes the consensus standards have used material structural performance factors of up to 4.0 to ensure that pipes will maintain structural integrity for extended time periods. Use of these structural performance factors results in a structural performance limit of 7.5 percent for deflection. Many engineers will specify pressure stress-rated plastic materials to ensure that long-term performance is realized.

Although HDPE and PVC pipes can experience large deflections under controlled conditions, long-term field performance is another matter. The tendency of these pipes to develop flat spots and skewed, non-elliptical shapes at large deflections prompts some engineers to consider a deflections in the range of 12 to 16 percent and 10 to 12 percent as structural performance limits under field conditions for HDPE and PVC pipes, respectively.

Mortar-lined and coated steel pipe is a composite material that has deflection limitations based upon onset of objectionable cracking of the mortar lining or coating. To prevent cracking of the mortar coating the deflection must be limited to about 2.0 percent. Steel pipe that is mortar-lined only can assume greater deflection before cracking of the lining, up to about 3.0 percent.

Other types of pipe should be analyzed for their performance limitations as indicated above. Careful consideration of available objective research, field performance, and testing data is required in these analyses.

<u>Design Factor</u>

Inclusion of a design factor F_d into the design requires that performance limits be divided by the factor as shown below, where D_l is the deflection limit criteria and P_l is the deflection performance limit for hydraulic or structural considerations:

$$D_l = \frac{P_l}{F_d}$$

The deflection limit criterion becomes the lowest of those calculated from the hydraulic or structural performance limits.

Any condition that has potential to compromise the reliability or capacity of a pipeline system should be respected in design. For instance, a pipeline interceptor system may be essential for operation of a municipal sewage system because there is no way to bypass flow and it must be in operation 24 hours per day, every day. In addition, interceptor facilities should have some reserve protection that respects the potential cost of failure and potential unplanned loads from sources such as earth surcharge, crossings, and excavations for connections. To obtain the required reliability, a design factor F_d of at least 1.25 should be applied to the long-term pipe deflection performance limits itemized above. Less critical facilities, or pipelines with a short design service life, should be considered for a lower design factor, but not less than 1.0.

Deflection Limitation Criteria

The above discussions on performance limits and design factors can be used to calculate a deflection limit criterion for each type of pipe as it applies to a particular project. For the examples of HDPE profile wall pipe, PVC sewer pipe, and steel pipe used above, the following table shows resulting long-term deflection limitation criteria.

Pipe Type	Lowest Deflection Performance Limit P_l, %	Deflection Limitation Criteria D_l, %
HDPE profile wall pipe	10.0[a]	8.0[a]
PVC sewer pipe, solid/profile wall	10.0[a]	8.0[a]
Steel pipe, mortar lined/coated	2.0[b]	1.6

Notes:
a. Hydraulic considerations govern. If a structural performance factor of 4.0 is used, the performance limit becomes 7.5 percent, and deflection limitation criteria becomes 6.0 percent.
b. Cracking of mortar coating governs
c. Design factor used above is 1.25.

ADDITIONAL CONSIDERATIONS

Some Factors Not Considered Here

Other factors should include considerations for internal pressure and pipe wall buckling. Also, many engineers include considerations for minimum pipe stiffness for handling and installation of the various types of materials. These issues are beyond the scope of this paper but are very important.

Selection and Calibration of Design Criteria

The project team's task is to establish design criteria using the above information as a guide. Many agencies and firms may want to establish criteria that are specific for regional soils, prevalent pipe materials, local construction methods, and local pipe embedment materials. Over time, field studies of local pipe installations will yield deflection data for refinements and calibration of criteria.

CHARTS FOR PRELIMINARY SELECTION OF FLEXIBLE PIPE

Using the above information, Tables B, C, and D have been constructed as examples of the type of design charts useful for preliminary selections of pipe. These examples include 48-inch mortar lined and coated steel pipe, SDR35 PVC sewer pipe, and 36-inch HDPE profile wall pipe. The design information is presented as values of deflection. Shaded portions of the charts indicate where deflections exceed the selected deflection limitation criteria as presented in the above table. The benefit of these charts is that they allow quick visual determination of acceptable depths of cover for the indicated insitu soil types.

PRELIMINARY DESIGN OF FLEXIBLE PIPE

Preliminary Geotechnical and Surveying Programs

During preliminary design either existing or extrapolated geotechnical information can be used or an abbreviated exploration and testing program can be undertaken to provide the required information. The minimum information required for preliminary design includes the soil classification at the anticipated elevation of the pipeline and an estimate of the groundwater surface elevation. The surveying information is necessary to determine pipe cover depths along the proposed pipeline alignment, and for these purposes, existing information is often sufficient.

Use of Charts as Design Guides

The charts presented as Tables B, C, and D have been made for pipe materials that have application in various soil types. The charts graphically indicate that the example pipe types have limitations of insitu soil type, pipe size, and depth of cover. Often, the pipe or embedment soils must be strengthened to accommodate the insitu soils or cover depth. In some cases semi-flexible or rigid pipe should be selected to ensure a successful installation. Whatever the situation, the charts will facilitate selection of the proper materials for each project.

FINAL DESIGN OF PROJECTS USING FLEXIBLE PIPE

Verification of Geotechnical Conditions

During final design the geotechnical program should be extended to provide confirmation of the pipe selection criteria. Often, the more detailed program can generate information related to issues of constructibility. An example of this is the use of test trenches, which can yield information about soil bedding and any tendency of the soils to slough and soften during or following excavation. Soils that exhibit sloughing or weakening after excavation should have discounted values for the insitu soil modulus.

Concurrence on Construction Requirements

An essential element of design is to develop the construction methods, frequency of soil testing, and methods used for deflection testing. This information can only be developed during discussions with the construction management staff because often adjustments to field inspection procedures, staff training programs, or specifications are necessary. The key issue is to develop genuine concurrence on the field installation requirements, even if it requires adjustments to design criteria to accommodate reasonable deviations from optimum construction procedures.

Final Design Calculations

Final pipe design calculations, or final versions of the charts, should be prepared to reflect the refined design criteria and the physical conditions of the project. The design calculations will serve as a record of design criteria and as a source of information that can be used in calibration of criteria after the pipeline has been installed and deflections measured. Comparison of measured field deflections with design deflections is valuable feedback for management of infrastructure quality.

Construction Contract Documents

Successful project performance will depend heavily on clear and effective specifications for pipeline installation. Specifications must implement project design criteria but not be so restrictive that they cannot be applied by the contractor or the construction management team. The specifications must have as their highest priority effective provisions for the properties, installation, and compaction of the pipe embedment zone backfill.

REFERENCES

1. Greenwood, Mark E. and Lang, Dennis C., Vertical Deflection of Buried Flexible Pipes, ASTM STP 1093, October 1990.
2. Howard, Amster K., Modulus of Soil Reaction Values for Buried Flexible Pipe, ASCE Journal Geotechnical Division, January 1977.
3. Howard, Amster K., Prediction of Flexible Pipe Deflection, Table 1, Draft, Bureau of Reclamation, October 1991.
4. Wastewater Engineering Society, Inc., Guidelines for Static Calculation of Drainage Conduits and Pipelines, Second Edition, translation from German, December 1988.
5. Selig, E. T., Soil Properties for Plastic Pipe Installation, ASTM STP 1093, October 1990.
6. Jeyapalen, Jey K., and Abdelmagid, A. M., Importance of Pipe-Soil Stiffness Ratio in Plastic Pipe Design, ASCE Conference, October 1984.
7. Hartley, James D., and Duncan, James M., E' and Its Variation with Depth, ASCE Journal of Transportation Engineering, September 1987.
8. Walker, Robert, The Effects of Deflection on Sewer Hydraulics, UniBell PVC Pipe News, date unknown.

TABLE A
IN-SITU SOIL MODULUS VALUES BASED ON DEPTH OF COVER AND SOIL TYPE

Depth of Cover		In-Situ Soil Characterization and Relative Strength					
		Rock, Very Dense Granular Soils, Very Hard Cohesive Soils spt>50 uct>6.0 Relative Strength A	Dense Granular Soils, Hard Cohesive Soils spt=31-50 uct=4.1-6.0 Relative Strength B	Very Firm Granular Soils, Very Stiff Cohesive Soils spt=21-30 uct=2.1-4.0 Relative Strength C	Firm Granular Soils, Stiff Cohesive Soils spt=11-20 uct=1.1-2.0 Relative Strength D	Loose Granular Soils, Medium Cohesive Soils spt=5-10 uct=0.6-1.0 Relative Strength E	Very Loose Granular Soils, Soft Cohesive Soils spt=2-4 uct=0.25-0.5 Relative Strength F	Very Loose Granular Soils, Very Soft Cohesive Soils spt<2 uct<0.25 Relative Strength G
From	To							
3	3	2200	1600	400	300	140	100	50
3	5	2400	1800	700	450	220	150	70
6	10	2600	2000	1000	600	300	200	100
11	15	2800	2200	1300	750	380	260	160
16	20	3000	2400	1600	900	460	320	220
21	25	3200	2600	1900	1050	540	380	280
26	30	3400	2800	2200	1200	620	440	340
31	35	3600	3000	2500	1350	700	500	400
36	40	3800	3200	2800	1500	780	560	460

spt - Standard penetration test, blows per foot, for granular materials only
uct - Unconfined compression test, tons per square foot

TABLE B
SELECTION OF 48" DR 160 STEEL PIPE BASED ON DEPTH OF COVER AND SOIL TYPE

Depth of Cover		In-Situ Soil Characterization and Relative Strength						
		Rock, Very Dense Granular Soils, Very Hard Cohesive Soils spt>50 uct>6.0 Relative Strength A	Dense Granular Soils, Hard Cohesive Soils spt=31-50 uct=4.1-6.0 Relative Strength B	Very Firm Granular Soils, Very Stiff Cohesive Soils spt=21-30 uct=2.1-4.0 Relative Strength C	Firm Granular Soils, Stiff Cohesive Soils spt=11-20 uct=1.1-2.0 Relative Strength D	Loose Granular Soils, Medium Cohesive Soils spt=5-10 uct=0.6-1.0 Relative Strength E	Very Loose Granular Soils, Soft Cohesive Soils spt=2-4 uct=0.25-0.5 Relative Strength F	Very Loose Granular Soils, Very Soft Cohesive Soils spt<2 uct<0.25 Relative Strength G
From	To							
3	3	1.5	1.7	3.0	3.7	4.8	5.4	5.7
3	5	1.1	1.2	1.7	2.2	3.0	3.5	3.7
6	10	1.1	1.2	1.4	1.9	2.5	3.0	3.2
11	15	1.2	1.3	1.5	2.1	2.8	3.5	3.7
16	20	1.4	1.5	1.6	2.3	3.2	4.1	4.3
21	25	1.6	1.6	1.7	2.5	3.6	4.7	4.9
26	30	1.8	1.8	1.9	2.8	4.0	5.2	5.6
31	35	1.9	2.0	2.0	3.0	4.3	5.8	6.1
36	40	2.1	2.1	2.1	3.1	4.6	6.3	6.7

B/D - 2.0
Embedment zone backfill - crushed rock
Design factor - 1.25
spt - Standard penetration test, blows/foot, for granular materials only
uct - Unconfined compression test, tons/square foot

TABLE C
SELECTION OF PVC SDR35 SEWER PIPE BASED ON DEPTH OF COVER AND SOIL TYPE

Depth of Cover		In-Situ Soil Characterization and Relative Strength					
		Rock, Very Dense Granular Soils, Very Hard Cohesive Soils spt>50 uct>6.0 Relative Strength A	Dense Granular Soils, Hard Cohesive Soils spt=31-50 uct=4.1-6.0 Relative Strength B	Very Firm Granular Soils, Very Stiff Cohesive Soils spt=21-30 uct=2.1-4.0 Relative Strength C	Firm Granular Soils, Stiff Cohesive Soils spt=11-20 uct=1.1-2.0 Relative Strength D	Loose Granular Soils, Medium Cohesive Soils spt=5-10 uct=0.6-1.0 Relative Strength E	Very Loose Granular Soils, Soft Cohesive Soils spt=2-4 uct=0.25-0.5 Relative Strength F	Very Loose Granular Soils, Very Soft Cohesive Soils spt<2 uct<0.25 Relative Strength G
From	To							
3	5	3.0	3.3	6.0	8.2	15.1	21.5	28.8
3	10	2.4	2.5	3.3	4.5	7.4	11.1	15.8
6	10	2.3	2.4	2.7	3.6	5.5	8.4	11.7
11	15	2.5	2.5	2.8	3.8	5.8	8.9	11.9
16	20	2.7	2.8	2.9	4.0	6.3	9.7	12.7
21	25	3.0	3.0	3.1	4.3	6.7	10.6	13.6
26	30	3.2	3.2	3.3	4.6	7.1	11.3	14.4
31	35	3.4	3.4	3.5	4.8	7.5	12.0	15.1
36	40	3.7	3.7	3.6	5.1	7.9	12.7	15.8

B/D - 2.5
Embedment zone backfill - crushed rock
Design factor - 1.25
spt - Standard penetration test, blows/foot, for granular materials only
uct - Unconfined compression test, tons/square foot

TABLE D
SELECTION OF 36" HDPE SEWER PIPE BASED ON DEPTH OF COVER AND SOIL TYPE

Depth of Cover		In-Situ Soil Characterization and Relative Strength						
From	To	Rock, Very Dense Granular Soils, Very Hard Cohesive Soils spt>50 uct>6.0 Relative Strength A	Dense Granular Soils, Hard Cohesive Soils spt=31-50 uct=4.1-6.0 Relative Strength B	Very Firm Granular Soils, Very Stiff Cohesive Soils spt=21-30 uct=2.1-4.0 Relative Strength C	Firm Granular Soils, Stiff Cohesive Soils spt=11-20 uct=1.1-2.0 Relative Strength D	Loose Granular Soils, Medium Cohesive Soils spt=5-10 uct=0.6-1.0 Relative Strength E	Very Loose Granular Soils, Soft Cohesive Soils spt=2-4 uct=0.25-0.5 Relative Strength F	Very Loose Granular Soils, Very Soft Cohesive Soils spt<2 uct<0.25 Relative Strength G
3	3	3.6	4.0	7.5	10.3	18.7	26.1	34.1
3	5	2.9	3.1	4.1	5.6	9.3	13.7	18.9
6	10	2.8	2.9	3.4	4.5	7.0	10.4	14.2
11	15	3.0	3.1	3.4	4.7	7.2	11.0	14.5
16	20	3.2	3.3	3.6	4.9	7.8	12.0	15.6
21	25	3.5	3.5	3.7	5.2	8.3	13.0	16.6
26	30	3.7	3.8	3.9	5.5	8.7	13.9	17.6
31	35	3.9	4.0	4.0	5.7	9.1	14.6	18.4
36	40	4.1	4.2	4.2	5.9	9.5	15.3	19.2

B/D - 2.0
Embedment zone backfill - crushed rock
Design factor - 1.25
spt - Standard penetration test, blows/foot, for granular materials only
uct - Unconfined compression test, tons/square foot

THE SUITABILITY OF SPANGLER'S IOWA FORMULA FOR PREDICTING DEFLECTION IN ALL FLEXIBLE PIPES

James C. Schluter, P.E.[1]
Theodore A. Capossela, P.E.[2]

ABSTRACT

Since it was first published in 1941, the Spangler Iowa Formula has become the primary means of predicting deflection in flexible pipes. While the original research used corrugated steel pipe and sandy loam as a backfill material, the Iowa Formula is now applied to all flexible pipes. With Amster Howard's work, published in 1976, E' values are available for a wide variety of backfill conditions.

In Spangler's work, the pipe stiffness of the pipes ranged from 25 to 60 psi. Backfill, even in its uncompacted form provided more than half of the combined soil and pipe stiffness as they are expressed in the denominator of the formula. The accuracy of the Iowa Formula used with Howard's E' values varies substantially. A simple review of the original data and actual field performance of very stiff (semi-rigid) and very flexible pipes demonstrates that this approach excessively overstates the deflections on one end (very stiff) of the scale while it understates them on the other end.

An alternative soil strain methodology is suggested to more accurately determine deflections in these areas.

INTRODUCTION

The Modified Iowa Formula is widely used for predicting deflections in flexible pipe. In its basic form (below), deflection is caused by soil loads and controlled by a combination of pipe and soil stiffness.

---

[1] Chief Engineer, CONTECH CONSTRUCTION PRODUCTS INC., 1001 Grove St., Middletown, OH 45044
[2] Manager, CONTECH CONSTRUCTION PRODUCTS INC., 1001 Grove St., Middletown, OH 45044

$$\Delta = \frac{Kp}{\text{Pipe Stiffness} + \text{Soil Stiffness}} = \frac{Kp}{0.149\, PS + .061\, E'}$$

Where:

Δ = Deflection $\quad K$ = bedding coefficient $\quad p$ = Soil pressure (psi)
PS = Pipe stiffness $\quad\quad\quad\quad\quad\quad\quad\quad\quad\quad\quad\; E'$ = Modulus of soil reaction

Dr. Spangler's work in developing his empirical Iowa formulas was based on measured deflections of corrugated steel pipe (CSP). Burial tests used 36 to 60 inch, 2 2/3 x 1/2 CSP with a sandy clay loam backfill that was either tamped or dumped and shovel placed.

The pipe stiffness of the CSP involved ranged from 61 lbs/in/in for the 36 inch size down to 25 lbs/in/in for 60 inch. Using the reported E' value for the untamped backfill indicates that soil stiffness (as opposed to pipe stiffness) provided at least 60% of the total resistance to deflection as measured by the denominator of the equation. Through the entire testing range, the soil was stiffer than the pipe, and provided the majority of the deflection control.

Any "built in" deflections, due to installation loads would be expected to be minor. The pipes exceeded today's minimum installation stiffness required for their diameter. The installations were done under controlled, test conditions rather than typical, uninspected construction conditions.

In 1976 Amster Howard published his modulus of soil reaction (E') estimates. Howard estimated E' values for various soils and soil densities. He then demonstrated the accuracy of these values by using them in the Iowa formula. He compared calculated deflections against those measured and reported in various field and laboratory studies. In all, this effort reports correlation's with 113 installations.

The Modified Iowa Formula is widely used with Amster Howard's E' values. The accuracy of the predicted deflections are normally adequate for design within the range of the pipe and soil stiffness relationships covered by the research.

FIELD DATA SOURCE

Substantial field performance data is available for pipes beyond the pipe stiffness range of Spangler's work. In the late 1960's - early 1970's, 175 installations of semi-rigid pipe (200 lbs/in/in minimum pipe stiffness) were measured for actual deflections. While the Howard work covered only 7 installations in the 150 to 1000 lbs/in/in range, on the low stiffness end Howard's data base included 30 installations of pipe with very low (20 lbs/in/in or less) stiffnesses.

COLLECTION OF FIELD DATA - HIGH STIFFNESS PIPE

Actual deflections were measured in 175 different installations of high stiffness (semi-rigid) plastic pipe. These high stiffness pipes were composite pipe (ASTM D2680) with a measured pipe stiffness (ASTM D2412) in excess of 200 lbs/in/in. The investigation was initiated to determine if excessive deflection is a governing factor for the structural design of high stiffness/semi-rigid pipe. The data came from a cross section of installations from 30 different states with a wide variety of embedment materials and schemes, cover or loading conditions, construction techniques, inspection practices, etc. Typically, testing was conducted at least six months after installation and after 1 year for majority of sites.

The deflection data was gathered through the use of electro-mechanical deflectometer equipment. Strain gages, mounted on two independent vertical and horizontal probes, monitored changes in the pipe shape as the deflectometer moved through the pipe. Movement of the probe arms changes their strain gage's electrical resistance which equates to a linear change in the pipe's dimensions. By moving the deflectometer throughout the sewer length, readings were accumulated at each station for each probe and then converted to compute percent deflection at that station.

EVALUATION OF FIELD PERFORMANCE DATA - HIGH STIFFNESS PIPE

When measured deflections for these high stiffness composite pipes are compared with the estimated deflections from Spangler Iowa Formula, it becomes obvious that Spangler substantially over predicts deflections. Table 1 compares the average actual deflection vs. the Spangler Iowa Formula predicted values. A review of Table 1 data indicates:

1. In 96% (168 of 175) of the installations, Spangler's Iowa Formula over predicted the deflection vs. those actually measured for the high stiffness, composite pipes.

2. In 82% (138 of 168) of the installations that Spangler's Iowa Formula over predicted the deflection, the error was more than 100%.

3. In 80% (111) of those 138 installations, Spangler's Iowa Formula over predicted the deflection by a margin of error over 200%.

4. Typically, the margin of error increases as the loading (depth of cover) increased and/or the soil modulus (E') decreased. This is consistent with Amster Howard's findings.

5. There were only 7 installations (4% of the total) where Spangler was essentially accurate.

In comparison to the field data summarized here for 175 composite pipe installations, only 15 of Howard's 113 pipes had pipe stiffness in excess of 100 lbs/in/in (see Table 2). Five (5) of these were essentially rigid (with stiffness in excess of 1000 lbs/in/in) and did not exhibit significant deflection levels.

The 10 remaining stiff pipes evaluated in Howard's work exhibit stiffness ranging from 111 to 440 lbs/in/in. Within this stiffness range, a review of the Howard data indicates:

- The Iowa formula is accurate for stiff pipes only when coupled with a stiff soil so that the soil stiffness remains as the greatest portion of the total resistance to deflection.

- When the soil provides less than 60% of the resistance, the Iowa formula over predicts deflections in more than 70% of the cases.

- When the soil provides 60% or more of the resistance, the combination of stiff pipe and stiff backfill result in little actual deflection, leaving the accuracy of the analysis moot.

EVALUATION OF THE FIELD PERFORMANCE DATA - LOW STIFFNESS PIPE

The Howard data base for pipes with a 20 lbs/in/in or lower pipe stiffness is summarized in Table 3. The 21 of these installations using AASHTO type backfill materials (those with E' values of 1000 psi or higher) are shown in bold print.

A review of this data indicates that:

1. Over all, the estimated deflection level accurately predicts the actual (measured) average deflection in 3 of 21 cases. The formula over estimates the actual, average deflection 11 times and under estimates it 7 times.

2. If only cases with an estimated deflection of 1% or more (i.e. cases with a significant deflection level) are considered, this methodology under predicts the actual, average deflection half the time. In the worst case it under predicts by 160%. However, for all the data points exhibiting 1% or larger deflections, the Iowa formula under predicts actual average deflections by an average of 40%.

3. As a tool for estimating the actual, maximum deflection, this methodology under predicts maximum deflections in 11 of 17 cases. In the worst instance, it under predicts by 425%.

4. Looking only at cases with predicted deflections of 1% or more, the approach under predicts the actual, maximum deflections by as much as

200%. For all the data points with predicted deflection of 1% or more, actual, maximum deflection levels are under predicted by an average of 62%

Accuracy considerations become more meaningful when uses of these deflection predictions are considered. Where these estimates are of interest only to provide a general indication of deflection, the Iowa formula may tend to serve us well. However, where accuracy is needed - such as in the evaluation of stains in a structural check - the Iowa formula may be inadequate within this low stiffness range. This becomes even more significant when one considers:

1. Many of the test installations evaluated, were just that - test installations installed under the watchful eye of the researcher. They could not be expected to display a normal amount of installation deflection.

2. The low stiffness pipes evaluated where E' was 1000 psi or greater, were primarily RPM pipes. None were thermoplastic. Thus they did not display the expected temperature, strain rate or hoop compression deflection considerations associated with thermoplastics.

3. In this low pipe stiffness range, the effects of non elliptical deflections (i.e. squaring) must also be accounted for when the deflection estimate is to be used for structural design.

SUMMARY AND CONCLUSIONS

While Spangler's Iowa formula coupled with Amster Howard's E' values is a useful design tool for many flexible pipe applications, it clearly needs to be applied within the range of its data base. A review of the work indicates this design approach is appropriate only for conditions that have both:

1. Pipe in the 25 - 60 lbs/in/in pipe stiffness range
2. Soil stiffness that provide at least 60% of the combined resistance to deflection

When stiffer pipes are installed in the recommended manner, the pipe typically provides more than 40% of Spangler's combined resistance to deflection. Substantial field experience with pipes in the 200 lbs/in/in pipe stiffness range indicates that using Spangler with Howard's E' values substantially over predicts deflections. The stiffness of these pipes is well outside of the range of the Spangler work. The typical installation of these pipes relies more heavily on pipe stiffness (as opposed to soil stiffness) to resist deflection.

Using the Iowa formula to predict pipe deflections, in low stiffness pipes also appears to require additional considerations. When deflection results will be used for structural design, suitable adjustment factors, load factors, etc. must be provided for the

Iowa equation unless a more accurate approach is found for estimating ring bending deflections.

ALTERNATIVE METHOD

Generally speaking, a flexible pipe will deflect about the same amount as the embedment soil around it strains. Watkins soil strain theory, expanded by Hoechst in Germany to address low stiffness polyethylene pipes they worked with, provides a more accurate method of predicting deflection in pipes that do not fit the data base of the Iowa formula.

This methodology is summarized in figure 1. Table 4 summarizes estimated pipe deflections for several of the composite pipe installations from Table 1 that used low stiffness backfill. When Howard's E' is used as an estimate of soil stiffness, this approach provides more accurate deflection estimates for these stiff pipes in installations with relatively low stiffness embedment soils.

The use of E' is suggested as a measure for soil stiffness because it approximates the constrained soil modulus reasonably well and is a familiar term. Unfortunately, predicted deflections based on soil strain theory are no less sensitive to the accuracy of the assumptions used for soil stiffness than are Spangler Iowa formula results.

A review of Table 4 demonstrates that the accuracy suffers as burial depths increase. While estimated deflections remain better than those provided by the Iowa Formula, accuracy would be much improved by recognizing the increase in the soil stiffness with depth. Work by Selig and others provides background for this approach. Further analysis is needed to properly predict deflection for high stiffness (semi-rigid) pipe in burial depths exceeding 10 feet.

REFERENCES

1 Howard, Amster K, "Modulus of Soil Reaction (E') Values for Buried Flexible Pipe", Transportation Research Board, Session 57, 1976.
2 Selig, Ernest T., "Soil Parameter for the Design of Buried Pipe Lines", ASCE Conference on Pipe Line Infrastructure; NY, 188.
3 Spangler M.G., "The Structural Design of Flexible Pipe Culverts" Iowa State Bulletin 153, 1941
4 Watkins, R. K. and A. B. Smith, "Ring Deflection of Buried Pipe", Journal AWWA, Volume 59, No. 3; March 1967.

Table 1

Location	Diam. (in)	Cover (ft)	Bedding Class	Soil Mod. E'	Avg. Actual Deflection	Calc. Deflec. DLF = 1.0	Difference
Little Rock, Ark.							
Coachlight Dr.	8	8	III	400	1.06	2.13	1.07
Coachlight Dr.	8	6	III	400	0.97	1.60	0.63
Mesa, Arizona							
Keating Dr.	15	8	III	400	0.62	2.13	1.51
Fresno, Cal.							
Maroa St.	8	9.5	III	400	1.23	2.53	1.30
Santa Ana, Cal.							
Forward-Corta Dr.	8	10	II	400	1.61	2.67	1.06
Backward-Corta Dr.	8	10	II	400	1.38	2.67	1.29
Santa Margarita							
Rocio St.	8	11	III	400	0.86	2.93	2.07
Oshawa, Canada							
Ormand Dr.	8	10.5	I	1000	0.62	1.29	0.67
Sarasota St.	8	9	I	1000	1.08	1.11	0.03
Tampa Crescent	8	9	I	1000	0.44	1.11	0.67
Orlando Ct.	8	9	I	1000	1.04	1.11	0.07
Glover Rd.	8	9	I	1000	1.38	1.11	(0.27)
Colorado Springs, Colo.							
Ferber St.	8	9	II	400	1.02	2.40	1.38
Del Rey Dr.	8	20	II	400	1	5.33	4.33
Dawson Dr.	8	10	II	400	1.36	2.67	1.31
Jacksonville, Fla.	8	8	II	400	0.77	2.13	1.36
College St. MH 7-8	8	6	III	400	0.05	1.60	1.55
College St. MH 7-6	8	8	III	400	0.49	2.13	1.64
College St. MH 5-4	10	10	III	400	0.87	2.67	1.80
College St. MH 4-3	10	10	III	400	0.41	2.67	2.26
Merritt Island, Fla.	8	10	II	400	1.08	2.67	1.59
Orlando, Fla.	8	10	II	400	1.09	2.67	1.58
Atlanta, Ga.							
Milan Estates	8	35	II	400	1.28	9.33	8.05
Glen Castle	8	12	I	1000	0.79	1.47	0.68
Savannah, Ga.	10	12	III	400	1.49	3.20	1.71
St. Simons Is., Ga	8	2	II	400	0.65	0.53	(0.12)
Addison, Ill.							
Kingspoint West Sub.	10	28	I	1000	1.2	3.44	2.24
Kingspoint West Sub.	10	20	I	1000	2.6	2.46	(0.14)
Arlington Heights, Ill.	8	5	IV	200	0.73	2.19	1.46
Champaign, Ill	8	4	III	400	0.24	1.07	0.83
Decatur, Ill.							
Cantrell St.	8	17	II	400	1.18	4.53	3.35
Divernon, Ill.							
Second St.	8	12	II	400	0.7	3.20	2.50
Forest View, Ill.							
Oil Terminal	15	6	I	1000	1.9	0.74	(1.16)
Green Valley, Ill.							
Geraldina St.	15	20	II	400	1.43	5.33	3.90
Morrison, Ill.							
Anthony St.	8	9.5	III	400	0.83	2.53	1.70
Morton, Ill.							
Burton Industrial Park (Morn.)	8	10	IV	200	0.44	4.37	3.93
Burton Industrial Park (After.)	8	10	IV	200	0.65	4.37	3.72
Normal, Ill.	8	8	II	400	0.81	2.13	1.32
Springfield, Ill.							
Scarborough St.	8	10	III	400	0.08	2.67	2.59
Scarborough St.	8	5	III	400	0.73	1.33	0.60

Table 1

Location	Diam. (in)	Cover (ft)	Bedding Class	Soil Mod. E'	Avg. Actual Deflection	Calc. Deflec. DLF = 1.0	Difference
Oakland (Pontiac), Mich.							
Club House Area	10	27	II	400	1.12	7.20	6.08
Oakdale	10	18	II	400	1.1	4.80	3.70
Waverly	10	17	I	1000	0.93	2.09	1.16
Woodmont	8	14	I	1000	0.79	1.72	0.93
Parma, Mich.							
John St.	8	9	IV	200	0.94	3.94	3.00
Union St.	8	10	IV	200	1.14	4.37	3.23
Wyoming, Mich.							
Cricklewood St.	8	8	II	2000	0.19	0.52	0.33
Cricklewood St.	8	20	II	2000	0.86	1.29	0.43
Porter St.	10	14	II	2000	0.01	0.91	0.90
Ypsilanti, Mich.	8	11	II	400	0.22	2.93	2.71
Bloomington, Minn.	10	24	II	400	1.22	6.40	5.18
Circle Pines, Minn.	15	28	II	400	1.12	7.47	6.35
Orono, Minn.	8	30	IV	200	1.3	13.12	11.82
	8	30	IV	200	0.73	13.12	12.39
Ste. Genevieve, Mo.	8	17	III	400	0.93	4.53	3.60
St. Louis, Mo.							
MH 3E to 3F	8	10	I	1000	0.66	1.23	0.57
MH 3F to 3G	8	14	I	1000	0.68	1.72	1.04
MH 3N to 3P	8	7	I	1000	0.45	0.86	0.41
MH 3W to 3L	8	10	I	1000	0.28	1.23	0.95
Middletown, N.J.							
Herb Rd.	8	8	II	400	0.78	2.13	1.35
Stephane Rd.	8	6	I	1000	0.84	0.74	(0.10)
Coram, N.Y.							
MH 1-4 to 1-3	8	5	II	400	0.27	1.33	1.06
MH 1-1 to 1-2	8	2	II	400	0.6	0.53	(0.07)
MH 1-6 to 1-5	8	5	II	400	0.29	1.33	1.04
Greece, N.Y.							
Athena Dr.	8	10	II	400	1.23	2.67	1.44
Fargo, N.D.							
Prairiewood Dr.	10	24	II	400	0.13	6.40	6.27
32nd St.	12	25	II	400	0.4	6.67	6.27
Butler County, Ohio							
Manor Dr.	15	28	IV	200	2.26	12.25	9.99
Marion, Ohio	15	22	II	400	0.92	5.87	4.95
Mentor, Ohio							
Truman Ct.	8	14	I	1000	0.59	1.72	1.13
Truman Ct.	8	10	I	1000	0.64	1.23	0.59
Sidney, Ohio							
Doorly Run #1	12	8	II	400	0.17	2.13	1.96
Doorly Run #2	8	15	II	400	1.23	4.00	2.77
Strongsville, Ohio							
Whitney Rd.	15	28	I	1000	0.18	3.44	3.26
Webster Rd.	8	30	I	1000	0.15	3.68	3.53
Trenton, Ohio							
Kay Dr.	10	12	II	400	0.42	3.20	2.78
Kay Dr.	10	15	II	400	0.81	4.00	3.19
Kay Dr.	10	12	II	400	0.86	3.20	2.34
Tulsa, Okla.							
94th E. Ave.	8	12	II	400	-0.7	3.20	3.90
94th E. Ave.	8	12	II	400	0.26	3.20	2.94
Henfield Twp., Pa.	8	15	IV	200	1.07	6.56	5.49
Vernon Twp., Pa.	8	13	II	400	0.4	3.47	3.07

PIPELINES IN THE CONSTRUCTED ENVIRONMENT

Table 1

Location	Diam. (in)	Cover (ft)	Bedding Class	Soil Mod. E'	Avg. Actual Deflection	Calc. Deflec. DLF = 1.0	Difference
Carmel, Ind.							
Brookshire	8	12	IV	200	1.64	5.25	3.61
Carmel Dr.	8	12	I	1000	0.85	1.47	0.62
Mohawk Hills	8	8	IV	200	0.81	3.50	2.69
Fort Wayne, Ind.							
(Soap Box Derby Hill Area)							
MH-3 -- MH-4	12	28	II	400	1.62	7.47	5.85
MH-3 -- MH-4	12	28	II	400	1.89	7.47	5.58
MH-3 -- MH-4	12	28	II	400	2.03	7.47	5.44
MH-2 -- MH-3	12	22	II	400	1.43	5.87	4.44
MH-2 -- MH-3	12	22	II	400	0.65	5.87	5.22
Hampton Rd.	8	13	II	400	1.02	3.47	2.45
Morrow Way	10	10	II	400	1.23	2.67	1.44
Morrow Way	10	10	II	400	1.71	2.67	0.96
Morrow Way	10	10	II	400	1.73	2.67	0.94
Muncie, Ind.	10	25	II	400	0.12	6.67	6.55
Topeka, Kan.							
Seward Ave.	8	7	IV	200	0.43	3.06	2.63
Seward Ave.	8	4	IV	200	0.08	1.75	1.67
Louisville, Ky.							
Washington State Ct.	8	8	I	1000	0.67	0.98	0.31
Washington State Ct.	10	8	I	1000	-0.61	0.98	1.59
Houma, La.	8	5	IV	200	1.33	2.19	0.86
New Orleans, La.							
Downman Rd.	8	15	I	1000	1.11	1.84	0.73
Downman Rd.	8	15	I	1000	1.04	1.84	0.80
Downman Rd.	8	15	I	1000	0.84	1.84	1.00
Downman Rd.	8	15	I	1000	0.76	1.84	1.08
Lourdes St.	8	7	I	1000	0.66	0.86	0.20
Sandle Ct.	8	9	I	1000	0.35	1.11	0.76
Mansfield St.	12	13	I	1000	0.89	1.60	0.71
Dorchester St.	12	16	II	400	0.93	4.27	3.34
New Castle St.	15	16	II	400	0.7	4.27	3.57
Brewer, Maine							
Nottingham Way	8	9	II	400	1.32	2.40	1.08
Rotherdale St.	8	9	II	400	3.04	2.40	(0.64)
Portland, Maine							
Tucker Ave. MH 3+10 to 1.+ 95	12	15	II	400	0.58	4.00	3.42
MH 1.+95 to 1.+00	12	15	II	400	-0.78	4.00	4.78
Tucker Ave. MH 5+56 to 3+10	10	13	II	400	0.74	3.47	2.73
Beal St. MH 9+07 to 7+57	8	13	II	400	0.98	3.47	2.49
Beal St. MH 10+98 to 9+07	8	13	II	400	0.92	3.47	2.55
Bailey St.	8	10	II	400	1.14	2.67	1.53
Beal St. MH 0+00 to 2+00	8	10	II	400	0.33	2.67	2.34
Beal St. MH 5+50 to 2+75	8	14	II	400	0.9	3.73	2.83
Beal St. MH 2+75 to 0+00	8	12	II	400	0.89	3.20	2.31
Avon Twp. (Detroit), Mich.	8	12	II	400	0.69	3.20	2.51
Devil's Lake, Mich.							
Walnut Hill	8	20	II	400	0.69	5.33	4.64
Devil Lake Hwy.	8	20	II	400	0.68	5.33	4.65
Marysville, Mich.							
Range Rd.	8	10	II	400	1.35	2.67	1.32
Range Rd.	12	17	II	400	0.94	4.53	3.59
Range Rd.	15	18	II	400	1.28	4.80	3.52
Monroe, Mich.	12	12	II	400	0.13	3.20	3.07

Location	Diam. (in)	Cover (ft)	Bedding Class	Soil Mod. E'	Avg. Actual Deflection	Calc. Deflec. DLF = 1.0	Difference
Sioux Falls, S.D.							
58th St. MH 17-16	8	23	II	400	0.15	6.13	5.98
58th St. MH 16-8	8	23	II	400	-0.05	6.13	6.18
Batcheller Lane MH 2-3	8	20	II	400	0.47	5.33	4.86
Batcheller Lane MH 3-4	8	20	II	400	0.96	5.33	4.37
Memphis, Tenn.	8	18	II	400	0.89	4.80	3.91
Ft. Worth, Texas	8	8	II	400	0.63	2.13	1.50
Chesapeake, Va.							
Border Rd.	8	11	III	400	1.83	2.93	1.10
Tacoma, Wash.							
MH E1-A5	8	8	II	400	0.1	2.13	2.03
MH E3-E4	8	4.5	II	400	0.48	1.20	0.72
Brookfield, Wisc.	8	8	II	400	1.26	2.13	0.87
Green Bay, Wisc.							
Wentworth Ave.	8	11	II	400	1.02	2.93	1.91
Wentworth Ave.	8	11	II	400	1.28	2.93	1.65
Wentworth Ave.	8	11	II	400	0.95	2.93	1.98
Wentworth Ave.	8	11	II	400	1.08	2.93	1.85
Mequon, Wisc.	8	24	II	400	0.36	6.40	6.04
Appleton, Wisc.							
Clara St.	8	15	II	400	0.71	4.00	3.29
Newberry St.	8	14	II	400	1.2	3.73	2.53
Albuquerque, NM							
Somervell St. MH 916-MH 915	8	14	III	400	0.9	3.73	2.83
Somervell St. MH 705-MH 812	8	10	III	400	0.14	2.67	2.53
Granite St.	8	14	III	400	0.14	3.73	3.59
Ft. Collins, Colorado							
Bryan St.	8	10	IV	200	0.66	4.37	3.71
Jefferson City, MO							
Algoa Rd.	8	12	III	400	1.15	3.20	2.05
Omaha, Neb.							
52nd & Ida MH 27 - MH 28	8	9	IV	200	1.14	3.94	2.80
52nd & Ida MH 28 - MH 29	8	8.5	IV	200	1.25	3.72	2.47
96th & Burt MH 38 - MH 25	8	22	III	400	1.43	5.87	4.44
96th & Burt MH 25 - NMH	8	22	III	400	1.46	5.87	4.41
96th & Burt NMH - MH 28	8	15	III	400	0.91	4.00	3.09
54th & Ida	8	11	IV	200	1.54	4.81	3.27
54th & Ida	8	9	IV	200	1.11	3.94	2.83
Jacksonville, FL							
Fort Carolina Club Estates	8	8	II	400	0.77	2.13	1.36
Springdale, Ohio	12	10	II	400	0.94	2.67	1.73
Woodville, NY							
South Court St.	10	14	I	1000	0.08	1.72	1.64
Bailey Ave.	12	10	I	1000	0.67	1.23	0.56
Ashland, Illinois							
Progress St. MH 112- MH 110	8	16	IV	200	1.36	7.00	5.64
Progress St. MH 113- MH 112	8	14	IV	200	0.98	6.12	5.14
Franklin St. MH 132 - MH 92	8	14	IV	200	0.75	6.12	5.37
Franklin St. MH 134 - MH 132	8	12	IV	200	0.9	5.25	4.35
Franklin St. MH 136 - MH 134	8	10	IV	200	0.88	4.37	3.49
Buchanan St. MH 10 - MH 9	8	16	II	400	0.58	4.27	3.69
Virginia Beach, VA							
Torngat Ct. MH 8 - MH 9	8	11	I	1000	0.02	1.35	1.33
DaVinci Dr. MH 16 - MH 22	8	10	I	1000	0.06	1.23	1.17
Shelbyville Lake, IL	8	5	II	400	0.92	1.33	0.41
	8	17	II	400	0.51	4.53	4.02
LombardoTrailer Park	8	6	III	400	0.81	1.60	0.79
	8	6	III	400	0.68	1.60	0.92

Table 2

Howard Data For Pipes
Stiffer Than 100 lbs/in/in
(Summary)

Test a* No. Resistance	Deflection: Inch (%) Estimated		Measured		Pipe Stiffness (lbs/in/in)	Soil Stiffness E'	Soil as % of
1	.94	(3.9)	.77	(3.2)	251	400	39
3	.54	(1.3)	.25	(0.6)	1,074	400	13
7	.30	(1.0)	.30	(1.0)	182	1,000	69
8	.30	(1.0)	.24	(0.8)	219	1,000	65
9	.09	(0.3)	.09	(0.3)	1,540	1,000	21
11	.14	(0.7)	.20	(1.0)	440	1,000	48
12	1.30	(3.1)	1.34	(3.2)	111	400	60
25	.36	(1.0)	.22	(0.6)	195	50	9
27	.10	(0.3)	.10	(0.3)	234	2,000	78
28	.14	(0.4)	.14	(0.4)	234	2,000	78
48	.07	(0.2)	.24	(0.7)	1,248	2,000	40
49	.07	(0.2)	-.31	(-.9)	1,248	2,000	40
50	.08	(0.2)	.04	(0.1)	1,148	3,000	22
82	.37	(.23)	.29	(1.8)	141	100	41
83	.06	(0.4)	.10	(0.6)	141	2,000	93

*$0.061 E'/(.149PS + .061E')$

Table 3

Howard Data For Pipes
With Stiffness Less Than 20 lbs/in/in

Test No.	Pipe Stiffness lbs/in/in	E' (psi)	Deflections (%)		
			Theoretical	Actual Ave	Actual Low/High
30	13.4	50	7.1	7.8	6.1/7.9
31	13.4	200	2.5	3.5	3.0/4.2
32	13.4	1,000	0.6	0.1	-0.3/3.0
33	13.4	100	4.4	5.1	3.6/6.5
34	13.4	1,000	0.6	0.6	0.6/0.6
51	14.1	2,000	0.7	0.6	0.5/0.7
52	14.1	1,000	1.7	1.6	0.9/2.4
63	17.5	1,000	0.9	0.7	0.4/1.1
65	17.5	200	4.0	3.2	
67	17.5	2,000	1.0	0.7	0.6/0.7
68	17.5	200	8.5	7.0	5.8/3.2
70	10.7	1,000	0.8	0.7	-0.4/4.2
71	13.4	2,000	1.0	1.1	
72	13.4	3,000	0.7	0.8	
73	8	2,000	1.0	1.1	
74	8	3,000	0.7	0.7	
91	2	50	3.4	2.1	1.8/2.5
92	2	50	3.9	7.6	5.3/9.4
102	14.1	2,000	0.5	0.5	0.4/0.6
103	14.1	3,000	0.3	-0.1	-0.3/0.1
104	14.1	2,000	0.5	0.2	0.1/0.3
105	14.1	400	2.3	2.8	1.4/3.6
106	14.1	1,000	1.5	3.6	2.3/4.5
107	14.1	1,000	1.5	1.0	0.4/1.5
108	14.1	2,000	0.7	0.3	-0.2/0.9
109	14.1	2,000	0.7	0.9	0.8/1.0
110	14.1	2,000	0.7	0.2	1.0/1.4
111	14.1	2,000	0.7	0.8	0.6/1.2
112	14.1	1,000	1.5	3.9	3.1/4.5
113	14.1	100	11.3	2.9	2.5/3.3

Figure 1: Estimating Deflectons on Soil Strain Theory
(from Watkins/Hoechst)

Extended Watkins graph

a Region for steel pipe according to Watkins
b Region for P Thermoplastics pipe according to our test results
c Extrapolated region
m Mean curve

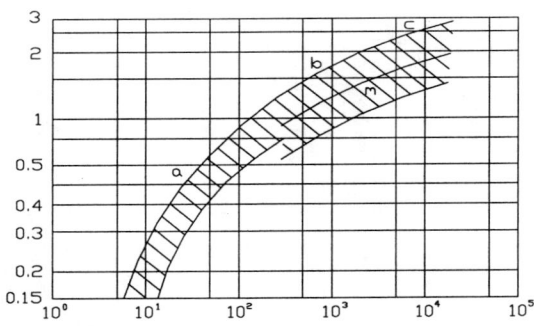

R_s-value

from Hoechst A. G., Frankfurt

$$R_s = \frac{\text{Soil Stiffness}}{\text{PipeStiffness}} = \frac{E_s}{E I / D^3}$$

$$\varepsilon_s = \text{Soil Strain} = \frac{\sigma}{E_s}$$

$$\Delta = \frac{\delta v}{\varepsilon_s} (\varepsilon_s)$$

Where:

σ = Soil Stress
ε_s = Soil Modulus
E = Pipe Material Modulus
I = Pipe Wall Moment of Inertia
Δ = Pipe Deflection
D = Pipe Diameter

Table 4. Deflections Estimated on Soil Strain Theory
(Assuming $E_s = E'$)

Location	Diam. (in)	Cover (ft)	Bedding Class	Soil Mod. E'	Avg. Actual Deflection	Estimated Deflection $E_s = E'$
Little Rock, Ark.						
Coachlight Dr.	8	8	III	400	1.06	1.1
Coachlight Dr.	8	6	III	400	0.97	0.8
Mesa, Arizona						
Keating Dr.	15	8	III	400	0.62	1.1
Fresno, Cal.						
Maroa St.	8	9.5	III	400	1.23	1.3
Santa Ana, Cal.						
Forward-Corta Dr.	8	10	II	400	1.61	1.4
Backward-Corta Dr.	8	10	II	400	1.38	1.4
Santa Margarita						
Rocio St.	8	11	III	400	0.86	1.5
Colorado Springs, Colo.						
Ferber St.	8	9	II	400	1.02	1.2
Del Rey Dr.	8	20	II	400	1	2.7
Dawson Dr.	8	10	II	400	1.36	1.4
Jacksonville, Fla.	8	8	II	400	0.77	1.5
College St. MH 7-8	8	6	III	400	0.05	1.1
College St. MH 7-6	8	8	III	400	0.49	1.5
College St. MH 5-4	10	10	III	400	0.87	1.4
College St. MH 4-3	10	10	III	400	0.41	1.4
Merritt Island, Fla.	8	10	II	400	1.08	1.4
Orlando, Fla.	8	10	II	400	1.09	1.4
Atlanta, Ga.						
Milan Estates	8	35	II	400	1.28	4.7
Savannah, Ga.	10	12	III	400	1.49	1.6
St. Simons Is., Ga	8	2	II	400	0.65	0.3
Butler County, Ohio						
Manor Dr.	15	28	IV	200	2.26	5.3
Parma, Mich.						
John St.	8	9	IV	200	0.94	1.7
Union St.	8	10	IV	200	1.14	1.9
Ypsilanti, Mich.	8	11	II	400	0.22	1.5
Bloomington, Minn.	10	24	II	400	1.22	3.2
Circle Pines, Minn.	15	28	II	400	1.12	3.8
Orono, Minn.	8	30	IV	200	1.3	5.6
	8	30	IV	200	0.73	5.6
Ste. Genevieve, Mo.	8	17	III	400	0.93	2.3

Replacing E' with the Constrained Modulus in Flexible Pipe Design

Timothy J. McGrath[1]

Abstract

In design of buried flexible pipe the soil stiffness has traditionally been modeled using the modulus of soil reaction, E'. This is a semi-empirical parameter required as input to the Iowa formula for predicting deflection of buried pipe. Unfortunately, E' is not a true soil parameter and efforts to develop tests to evaluate it have been largely unsuccessful. Also, the commonly used design values for E', proposed by Howard, do not reflect the change in soil stiffness that occurs with increasing depth of fill. An alternate to use of the modulus of soil reaction is the true soil property, the constrained or one-dimensional modulus, M_s. The relationship between E' and M_s has been discussed often in the literature with several researchers concluding that the two parameters are interchangeable. Design values for M_s are derived using the hyperbolic model for Young's modulus developed by Duncan and the hyperbolic model for bulk modulus developed by Selig. The hyperbolic model is currently incorporated in the finite element buried culvert computer programs CANDE and SPIDA, and was the basis for the development of the SIDD installations for concrete pipe recently accepted by AASHTO. The model demonstrates an increase in soil stiffness with increasing depth of fill and the values for M_s of soils under low heights of fill compare well with the Howard values for E'.

Introduction

Flexible pipe provide good service by interacting with the surrounding embedment soil to create a pipe-soil structural system. Generally the soil parameter of most interest is the stiffness of the soil. This concept was first understood by Spangler (1) when he developed the Iowa equation to predict deflection of buried pipe due to earth load. Spangler proposed the use of two values for soil stiffness, one for compacted soil and one for uncompacted soil. Howard (2, 3) later proposed a table of about 15 values of soil stiffness based on soil classification and soil density. This table has come into common use; however, the soil stiffness parameter, the modulus of soil reaction, E', is empirical, and has not been successfully related to true soil properties that can be

---

[1] Principal, Simpson Gumpertz & Heger Inc., Consulting Engineers, 297 Broadway, Arlington, MA 02174

evaluated by test. Some researchers have investigated a relationship between E' and the constrained or one-dimensional modulus, M_s, and have proposed various relationships between the two but there has been no consensus on the matter. The development of the hyperbolic soil model (4, 5, 6) provides a non-linear soil model that has been used successfully in finite element analyses of buried pipe and culverts and in the development of the SIDD installations for reinforced concrete pipe recently adopted by AASHTO (7, 8). This paper investigates development of design values of the constrained modulus as predicted by the hyperbolic soil model and its relationship to published values of the modulus of soil reaction.

The Iowa Formula and the Modulus of Soil Reaction

The Iowa deflection formula as developed by Spangler (1) and modified by Watkins and Spangler (9), is a simple pipe-soil interaction formula that predicts change in horizontal diameter in a pipe based on a load term, a pipe stiffness term, and a soil stiffness term. In simple format the Iowa formula can be stated as:

$$\text{Deflection} = \frac{\text{Load}}{\text{Pipe Stiffness} + \text{Soil Stiffness}} \quad (1)$$

While the Iowa formula has been regularly criticized, it remains the best known simplified method for computing deflections. The formula was developed based on the pressure distribution shown in Figure 1 where the vertical load is distributed uniformly over the top of the pipe, the vertical reaction is uniformly distributed over a width of the bottom of the pipe selected by the designer, and the lateral pressure has a parabolic distribution over 100 degrees at the side of the pipe. The magnitude of the pressure at the side is a function of the outward deflection of the pipe. The Iowa formula, as proposed by Spangler predicted the change in the horizontal diameter of the pipe due to soil placed over the top of the pipe. The soil stiffness parameter proposed by Spangler was an empirical spring constant, the modulus of passive resistance, with units of force per length cubed. Later, Watkins and Spangler (9) proposed the use of the modulus of soil reaction, E', with units of force per length squared. E' is computed as the modulus of passive resistance times the pipe radius.

In 1977 Howard reported on measurements of deflection of many flexible pipe installations and proposed a table of design values of E' based on soil classification and soil unit weight. Howard altered the use of E' somewhat in that he correlated vertical deflection with E', rather than horizontal deflection as originally conceived by Spangler. The values proposed by Howard are presented in Table 1. Howard found no correlation between E' and depth of fill. He limited the use of the proposed values of E' to depths of fill less than 50 ft based on the depths of fill over the pipe in his data set.

Thus, the Iowa formula, as most often used today, is:

$$\Delta_v = \frac{D_1 K W_p}{\frac{EI}{R^3} + 0.061 E'} \quad (2)$$

where:

Δ_v = Change in vertical diameter, mm, (in.)
D_l = Deflection lag factor to account for time related increase in deflection after installation is complete
K = A bedding constant to account for the width of soil support at the bottom of the pipe
W_p = Load on the pipe, MN/m, (lb/in.)
E = Pipe material modulus of elasticity, MPa, (psi)
I = Moment of inertia of the pipe wall, mm^4/mm, (in.^{4}/in.)
R = Radius to centroid of the pipe wall, mm, (in.)
E' = Modulus of soil reaction, MPa, (psi)

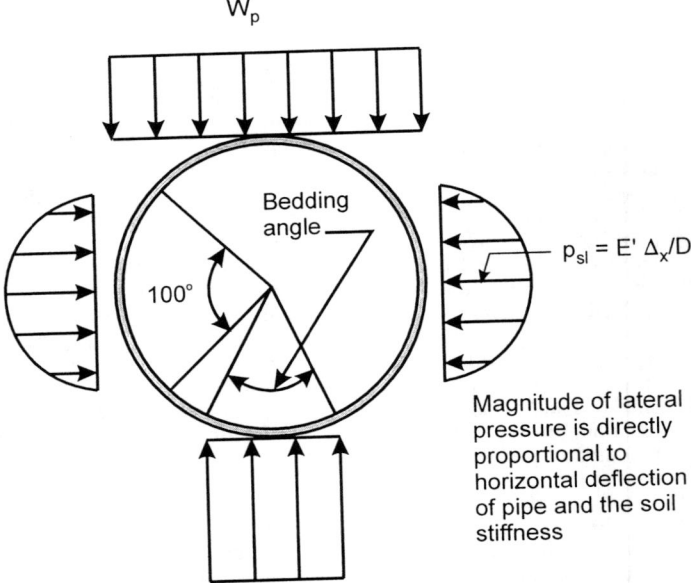

Figure 1. Spangler Pressure Distribution

The key elements in the use of the Iowa formula are the load terms, D_l, K, and W_p, and the soil stiffness, E'. The pipe stiffness is generally low and does not affect the result of the calculation significantly. The modulus of soil reaction is empirical in nature and, despite the efforts of a number of researchers (10) has not been related to a true soil property, nor has a standardized test been developed to evaluate values for it for a specific soil.

Table 1. E' Values Proposed by Howard (2)

Backfill type (ASTM D2487)	E' for degree of compaction of bedding (MPa)			
	Dumped	Slight <85% Proctor <40% relative density	Moderate 85-95% Proctor 40-70% relative density	High >95% Proctor >70% relative density
Fine-grained soils (liquid limit > 50) CH, MH, CH-MH	No data available; consult a competent soils engineer; otherwise use E' = 0			
Fine-grained soils (liquid limit < 50) CL, ML, ML-CL, with less than 25% coarse-grained particles	0.3	1.5	3	7
Fine-grained soils (liquid limit < 50) CL, ML, ML-CL, with more than 25% coarse-grained particles *Coarse-grained soils with fines*, GM, GC, SM, SC[1] with more than 12% fines	0.7	3	7	14
Coarse-grained soils with little or no fines GW, GP, SW, SP with less than 12% fines	1.5	7	14	21
Crushed rock	7		21	
Accuracy in terms of percent deflection[2]	±2%	±2%	±1%	±0.5%

[1] Or any borderline soil beginning with one of these symbols (i.e., GM-GC, GC-SC).
[2] For ±1% accuracy and predicted deflection of 3%, actual deflection would be between 2% and 4%.

Note: A. Values applicable only for fills less than 50 ft.
B. Table does not include any safety factor.
C. For use in predicting initial deflections only, appropriate deflection lag factor must be applied for long-term deflections.
D. If bedding falls on the borderline between two compaction categories select lower E' value or average the two values.
E. Percent Proctor based on laboratory maximum dry density from ASTM D-698, AASHTO T-99.
F. 1 MPa = 145 psi.

Finite Element Analysis and the Hyperbolic Soil Model

Finite element analysis has been developed as a numerical method to solve complex problems in continuum mechanics, and permits a very precise treatment of the pipe-soil interaction problem. The soil mass around a pipe is set up as a mesh of soil elements, such as shown in Figure 2, and each element can be assigned different properties. Katona (13) developed the finite element program CANDE which is the most commonly used program for analyzing buried pipe. Heger et al. (14) developed the finite element program SPIDA, which is proprietary, but was used to develop the SIDD installations (15) recently adopted by AASHTO (7, 8) for reinforced concrete pipe. Both CANDE and SPIDA incorporate the non-linear hyperbolic soil model developed by Duncan et al. (4) and modified by Selig (5, 6). The model is based on the hyperbolic nature of a soil stress-strain curve in a triaxial compression test. Nine parameters are required to completely define a soil with this model, including both strength and stiffness parameters. The model correctly represents true soil behavior by softening when compressed with constant confining stress (such as a triaxial compression test) and hardening when compressed under confined conditions (such as a one-dimensional compression test).

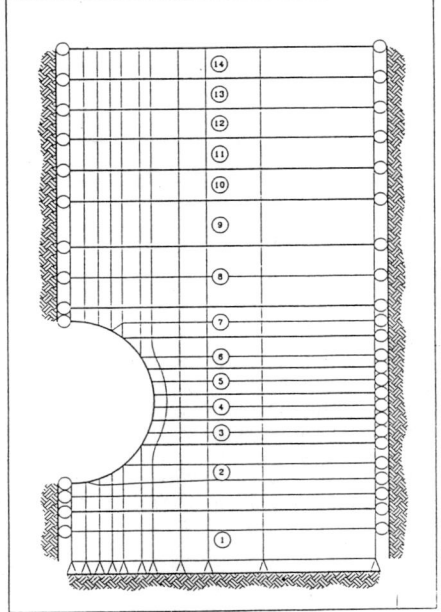

Figure 2. Finite Element Model of Buried Pipe Installation

E' vs M_x

As noted, many researchers have attempted to correlate the modulus of soil reaction, E', with other true soil properties that can be evaluated by test. The most common parameter used in these efforts is the constrained modulus, M_s, which is the soil stiffness under uniaxial strain conditions (Figure 3). M_s is related to the more common Young's modulus by the relationship:

$$M_s = \frac{E_s(1 - v)}{(1 + v)(1 - 2v)} \tag{3}$$

M_s = Constrained soil modulus, Mpa, psi
E = Young's modulus of soil, Mpa, psi
v = Poisson's ratio of soil

Typically, values for M_s are computed as the slope of the secant from the origin of the stress-strain curve to the stress level on the curve that represents the free field soil stress at the side of the pipe (the average modulus in Figure 3).

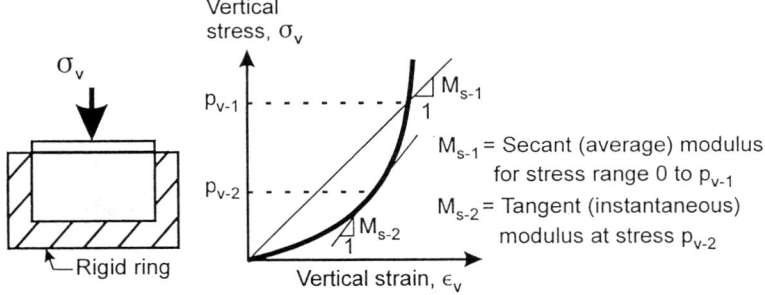

Figure 3. One-Dimensional Compression Test

Chambers et al. (12) presented a comparison of E' and M_s values. Hartley and Duncan (11) proposed a table of E' values backcalculated from finite element analyses. This table incorporated depth of fill as a variable to model the increased stiffness of soil with increased confinement. Krizek et al. (10) reported that M_s could vary from 0.7 to 1.5 times E'. Hartley and Duncan, and McGrath (15) proposed a direct substitution, i.e. E' = M_s. Although due to their empirical and theoretical nature, there can be no truly correct relationship between the two parameters, it is reasonable for design purposes to assume a direct substitution as acceptable for two reasons:

1. Researchers agree that the factor relating E' and M_s is close to 1.0, and

2. For purposes of buried pipe installations the precision of the design models, such as the Iowa formula, is sufficiently low that an approximate relationship is acceptable.

In developing an elasticity model for a pipe embedded in uniform soil mass Burns and Richard (16) used the constrained modulus as the soil property most representative of soil behavior in the ground.

M_s from the Hyperbolic Soil Model

The constrained modulus can be derived directly from the hyperbolic soil model.

Two constants are required to define behavior of an elastic material. The hyperbolic model uses Young's modulus and the bulk modulus as the parameters. These parameters are both affected by the soil strength and state of stress. The basic equations for stress-vertical strain, and volumetric strain, as presented in Selig (5), are:

$$(\sigma_1 - \sigma_3) = \frac{\epsilon_v}{\dfrac{1}{E_i} + \dfrac{\epsilon_v}{(\sigma_1 - \sigma_3)_u}} \qquad (4)$$

where

σ_1 = major principal stress, kPa, (psi),
σ_3 = minor principal stress, kPa, (psi),
$(\sigma_1 - \sigma_3)$ = deviator stress, kPa, (psi),
ϵ_v = vertical strain, mm/mm (in./in.),
E_i = the initial Young's modulus, kPa, (psi),
$(\sigma_1 - \sigma_3)_u$ = ultimate deviator stress, kPa, (psi).

and

$$\sigma_m = \frac{B_i \, \epsilon_{vol}}{1 - \dfrac{\epsilon_{vol}}{\epsilon_u}} \qquad (5)$$

where

σ_m = mean stress = $(\sigma_1 + 2\sigma_3)/3$, kPa, (psi), (6)
B_i = initial bulk modulus, kPa, (psi),
ϵ_{vol} = volumetric strain,
ϵ_u = ultimate volumetric strain,

The one-dimensional compression test imposes the additional restriction that the volumetric strain is equal to the vertical strain:

$$\epsilon_{vol} = \epsilon_v \qquad (7)$$

Substituting and rearranging, σ_1 can be expressed in terms of ϵ_v and σ_m:

$$\sigma_1 = \frac{0.667 \, \epsilon_v}{\dfrac{1}{E_i} + \dfrac{\epsilon_v}{(\sigma_1 - \sigma_3)_u}} + \sigma_m \qquad (8)$$

The initial Young's modulus is a function of the soil parameters, K and n, and the confining stress, σ_3:

$$E_i = K\, P_a (\sigma_3/P_a)^n \tag{9}$$

Rearranging and substituting Eq. (6) into Eq. (9):

$$E_i = K\, P_a \left(\frac{3\sigma_m - \sigma_1}{2P_a}\right)^n \tag{10}$$

The ultimate deviator stress is a model parameter that is a function of the actual deviator stress at failure and the model parameter, R:

$$(\sigma_1 - \sigma_3)_u = \frac{(\sigma_1 - \sigma_3)_f}{R_f} \tag{11}$$

where the deviator stress at failure is a function of the soil friction angle, ϕ, the cohesion intercept, C, and the confining stress, σ_3:

$$(\sigma_1 - \sigma_3)_f = \frac{2C(\cos\phi) + 2\sigma_3(\sin\phi)}{1 - \sin\phi} \tag{12}$$

Rearranging and substituting Eq. (6) into Eq. (12), and the result into Eq. (11) gives the expression:

$$(\sigma_1 - \sigma_3)_u = \frac{2C(\cos\phi) + 2\left(\dfrac{3\sigma_m - \sigma_1}{2}\right)\sin\phi}{(1 - \sin\phi)R_f} \tag{13}$$

Finally, the major principal stress, σ_1, can be expressed in terms of the vertical strain (which by definition of the one-dimensional compression test is the volumetric strain), by substituting Eqs. (13) and (10) into Eq. (8):

$$\sigma_1 = \frac{0.667\,\epsilon_v}{\dfrac{1}{K\,P_a\left(\dfrac{3\sigma_m - \sigma_1}{2P_a}\right)^n} + \dfrac{\epsilon_v}{\left(\dfrac{2C(\cos\phi) + 3\sigma_m\sin\phi - \sigma_1\sin\phi}{(1-\sin\phi)R_f}\right)}} + \sigma_m \tag{14}$$

This is the expression for the one-dimensional stress-strain curve and can be used to compute the constrained modulus, M_s.

Eq. (14) is based on the assumption of a linear failure envelope (constant soil friction angle, ϕ, at all stress levels). To incorporate the effect of a curved failure envelope, the expression for ϕ may be corrected by introducing a stress sensitive model parameter, $\Delta\phi$:

$$\phi = \phi_o - \Delta\phi\,\log_{10}(\sigma_3/P_a) \tag{15}$$

Eq. (6) may be rearranged and substituted into Eq. (15):

$$\phi = \phi_o - \Delta\phi \, \log_{10}\left(\frac{3\,\sigma_m - \sigma_1}{2\,P_a}\right) \qquad (16)$$

Substituting Eq. (16) into Eq. (14) produces a complete expression that can be solved for the stress-strain curve under confined conditions. The complete expression is complex but is solved by publicly available mathematics software packages such as MathCad.

The complete expression for soil stress-strain curve under one-dimensional strain can be used with the hyperbolic model parameters to develop values for the secant constrained modulus, M_s, at various stress levels. This was done for two sets of soils and then compared to two sets of E' values. The soil properties include:

- the "SIDD" soil properties proposed by Selig (5),
- the "CANDE" properties proposed by Selig (6),
- the "Duncan" E' values proposed by Hartley and Duncan (11), and
- the "Howard" E' values (2).

The E' and secant M_s values are compared directly for three types of soil at different densities in Figure 4, which shows:

- The CANDE properties produce values of M_s that are consistently about twice the values of the SIDD properties.
- At stress levels of about 70 kPa (10 psi), the SIDD properties are consistently similar to the E' values backcalculated by Howard based on actual installations.
- The Duncan properties are somewhat erratic relative to the other sets of properties.

The comparison in Figure 4 suggests that for design purposes E' can be assumed equal to M_s and that the SIDD properties are roughly equivalent to the Howard values which represent a substantial amount of field data. This association could mean that soil models for simplified flexible pipe design and for reinforced concrete pipe design would be the same, which is a significant positive step in bringing together the currently diverse design methods used by different industries. Based on this conclusion suggested design values for M_s, for use as a soil stiffness parameter in the Iowa deflection formula, are proposed in Table 2. The proposed values are secant moduli.

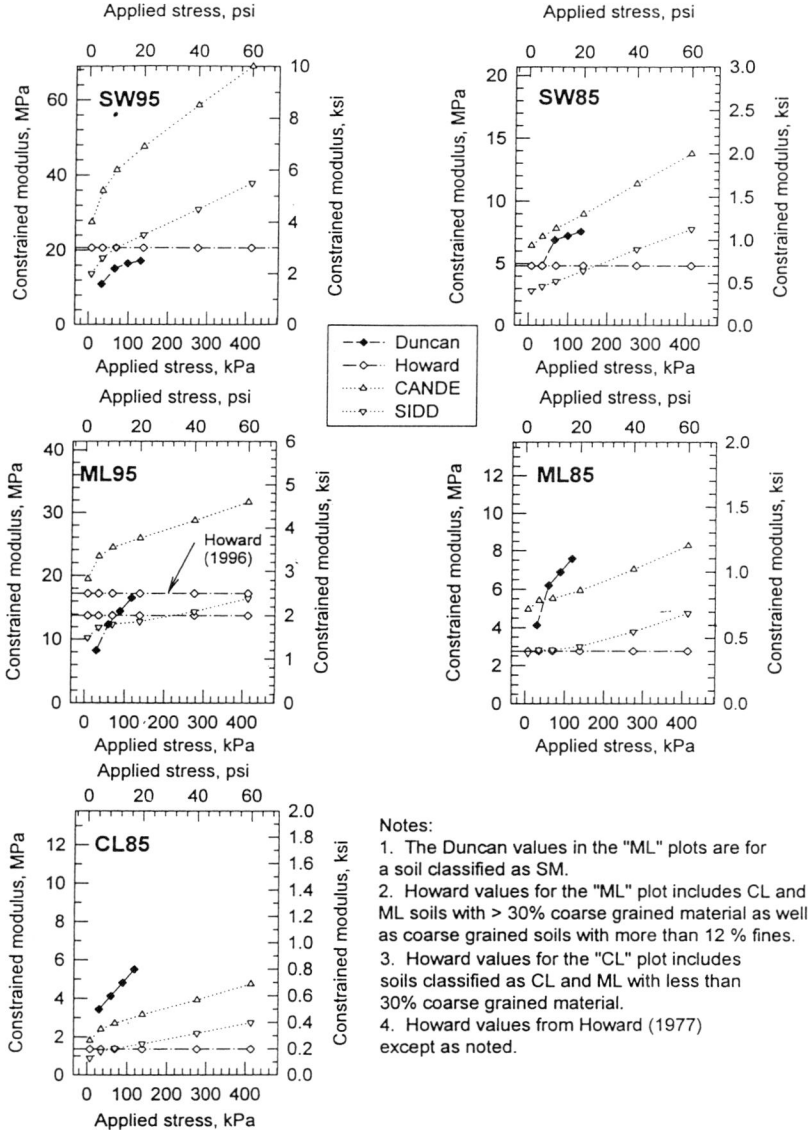

Figure 4. Comparison of E' and M_s Values

Table 2. Suggested Design Values for Constrained Soil Modulus, M_s

Stress level (kPa)	Soil type and Compaction Condition (MPa)								
	SW95	SW90	SW85	ML95	ML90	ML85	CL95	CL90	CL85
7	13.8	8.8	3.2	9.8	4.6	2.5	3.7	1.8	0.9
35	17.9	10.3	3.6	11.5	5.1	2.7	4.3	2.2	1.2
69	20.7	11.2	3.9	12.2	5.2	2.8	4.8	2.4	1.4
138	23.8	12.4	4.5	13.0	5.4	3.0	5.1	2.7	1.6
275	29.3	14.5	5.7	14.4	6.2	3.5	5.6	3.2	2.0
413	34.5	17.2	6.9	15.9	7.1	4.1	6.2	3.6	2.4

1 MPa = 145 psi

The values of M_s from the CANDE properties were proposed by Selig more recently than the SIDD properties; however, they are substantially higher than past practice based on Howard's E' values. They may be appropriate for research or analysis purposes, but are not selected here for use in routine design.

Conclusions

Various sets of soil stiffness parameters used for the design of flexible and rigid pipe were investigated and compared. Analysis indicates that:

- values for the modulus of soil reaction, E', and the constrained soil modulus, M_s, can be assumed interchangeable for design purposes; and

- soil properties used to develop the SIDD installations for reinforced concrete pipe compare very well when compared to Howard's E' values at low depths of fill.

Suggested values for the secant constrained soil modulus are proposed for use in the Iowa formula for deflection of buried pipe and other design equations that had adopted the use of E'.

References

1   Spangler, M.G. The Structural Design of Flexible Pipe Culverts. *Iowa Engineering Experiment Station, Bulletin No. 153*, Ames, IA, 1941.

2   Howard, A.K. Modulus of Soil Reaction Values for Buried Flexible Pipe. *Journal of the Geotechnical Engineering*, ASCE, Vol 103, No. GT1, New York, NY, USA, 1977.

3   Howard, A.K. *Pipeline Installation*. Relativity Publishing, Lakewood, CO, 1996.

4   Duncan, J.M., P. Byrne, K.S. Wong, and P. Mabry. Strength, Stress-Strain and Bulk Modulus Parameters for Finite Element Analyses of Stresses and Movements in Soil Masses. *Department of Civil Engineering Report No. UCB/GT/80-01*, University of California, Berkeley, CA, 1980.

5   Selig, E.T. Soil Parameters for Design of Buried Pipelines. *Pipeline Infrastructure – Proceedings of the Conference*, American Society of Civil Engineers, New York, NY, 1988, pp. 99-116.

6   Selig, E.T. Soil Properties for Plastic Pipe Installations. *Buried Plastic Pipe Technology, ASTM STP 1093*, George S. Buczala and Michael J. Cassady, Eds., American Society for Testing and Materials, Philadelphia, PA, 1990.

7   AASHTO. *Standard Specifications for Highway Bridges*. 16th Edition, American Association of State Highway and Transportation Officials, Washington, D.C., 1996.

8   AASHTO. *LRFD Bridge Design Specifications*. 1st Edition, American Association of State Highway and Transportation Officials, Washington, D.C., 1994.

9   Watkins, R.K., and M.G. Spangler. Some Characteristics of the Modulus of Passive Resistance of Soil. A Study in Similitude. *Proceedings HRB*, Vol. 37, 1958, pp. 576-583.

10 Krizek, R.J., R.A. Parmelee, N.J. Kay, and H.A. Elnaggar. Structural Analysis and Design of Buried Culverts. *National Cooperative Highway Research Program Report 116*, National Research Council, Washington, D.C., 1971.

11 Hartley, J.P., and J.M. Duncan. E' and its Variations with Depth. *Journal of Transportation Engineering*, ASCE, Vol. 113, No. 5, 1987, pp. 538-553.

12 Chambers, R.E., T.J. McGrath, and F.J. Heger. Plastic Pipe for Subsurface Drainage of Transportation Facilities. National Cooperative Highway Research Program Report 225. Transportation Research Board, National Research Council, October 1980.

13 Katona, M.G., et al. CANDE: Engineering Manual - A Modern Approach for the Structural Design of Buried Culverts. *Report No. FHWA/RD-77*, NCEL, Port Hueneme, CA, 1976.

14 Heger, F.J., A.A. Liepins, and E.T. Selig. SPIDA: An Analysis and Design System for Buried Concrete Pipe. *Advances in Underground Pipeline Engineering – Proceedings of the International Conference*, American Society of Civil Engineers, 1985, pp. 143-154.

15 McGrath, T.J., R.E. Chambers, and P.A. Sharff. Recent Trends in Installation Standards for Plastic Pipe. *Buried Plastic Pipe Technology, ASTM STP 1093*, George S. Buczala and Michael J. Cassady, Eds., American Society for Testing and Materials, Philadelphia, 1990.

16   Burns, J. Q. And Richard, R. M., (1964), "Attenuation of Stresses for Buried Cylinders," *Proceedings of the Symposium on Soil Structure Interaction,* University of Arizona, Tucson, Arizona, pp.378-392.

Willow Street/Lower Sweetwater River Pipeline Crossings

Michael Marks[1], Hector Martinez[2], James Smyth[3]

Abstract

The Lower Sweetwater River traverses through portions of the City of Chula Vista and County of San Diego, Ca. The City of San Diego Engineering and Capital Projects Department (City of San Diego) and the Sweetwater Authority (Authority), a public water agency, each has a potable water transmission main crossing perpendicular to the lower Sweetwater River. The City's main is a 36-inch diameter welded steel pipeline, and the Sweetwater Authority's is a 32-inch diameter riveted steel pipeline. The two lines are parallel to each other with a 20-foot horizontal separation and have been in service for over 60 years. Over the years, the scour of the riverbed above these pipelines has resulted in exposing each of these lines. The City of Chula Vista (Chula Vista) and County of San Diego (County) contend that each of these pipelines impeded the flow of the river and created backwater conditions that has flooded the upstream golf course and nearby streets via drainage pipes to this river. Both Chula Vista and the County received continued public criticism for not improving the flow in the river to minimize the flooding potential.
After a number of years of considerable negotiations, it was decided by all four agencies to jointly participate in the cost of lowering each of these lines in place. The Authority and the City of San Diego each performed the design of lowering its perspective water line. However, the Authority was named the responsible party to develop the construction specifications and construction management to lower both water mains. The project area is within sensitive riparian woodlands requiring a permit from the U.S. Fish and Wildlife Service. The County took the lead in acquiring the permits. The project dealt with constraints due to aforementioned

---

1 Project Manger, City of San Diego Water Department, CIP Program Management Division, 600 B Street, Suite 700, San Diego, CA 92101-4502, Phone: (619) 533-4271, Fax: (619) 533-5176
2 Principal Engineer, Sweetwater Authority, P.O. Box 2328, Chula Vista, CA 91912-2328, Phone: (619) 420-1413, Fax: (619)425-7469
3 Chief Engineer, Sweetwater Authority, P.O. Box 2328, Chula Vista, CA 91912-2328, Phone: (619) 420-1413, Fax: (619)425-7469

permits, operational shut down conditions by the City of San Diego and the Authority, impending weather and potential flooding, coordination with a medical facility adjacent to the construction site, and providing a design for future channel improvements by Chula Vista. Initial estimates for construction were $500,000 based on 200 linear feet of lowering for each pipe and a three-week construction time. An innovative design by the Authority's construction manager and an experienced contractor resulted in completing the installation in five days and a completed cost of $440,000, of which $380,000 was for construction and $60,000 was for permit processing with the U.S. Fish and Wildlife Service and other. This paper will include more detailed discussions that relate to the challenges of having four public agencies seeking resolution of this problem, and planning, design, environmental and construction issues relating to these pipeline lowerings.

Introduction

At one time or another we have heard that dealing with bureaucracies is challenging, if not an impossible task. It takes a great deal of patience to achieve the completion of any project. What would happen if you had four bureaucracies dealing with one another? You would think that it would lead to a paralytical disaster. This manuscript tells the story of how four public agencies got together to solve a problem that affected the community of Bonita, the City of Chula Vista and the County of San Diego.

History

The City of Chula Vista and the County of San Diego had expressed its concern with the continual flooding of local streets due to poor drainage characteristics of the lower Sweetwater River. The main basis of this concern was that the two pipelines were creating a dam across the Sweetwater River. Chula Vista and the County asked the City of San Diego and the Authority to have the pipe lowered. Not one of the four agencies wanted to accept full responsibility for the problem. The City of San Diego and the Authority stated that their pipe had been installed with adequate cover and had become exposed due to upstream improvements within the vicinity of the river causing the streambed centerline to change its course. Further, each of these pipelines was located within an easement giving prior rights. Therefore, these pipe-owning agencies were not responsible for any costs to relocate. Chula Vista and the County stated that the pipes were subject to normal conditions of a river; therefore, should be lowered or relocated. In 1993, the County of San Diego area was designated a disaster area due to the heavy rainfall locally known as the "Miracle March" rains. Attempts were made by all agencies to obtain FEMA funding for lowering the pipelines. This was unsuccessful because FEMA only reimburses for damages and not prevention of disaster. Compounding this problem was the sudden extreme growth of arundo cane within the river in the vicinity of the two pipelines, which also contributed to the damming effect on flow in the river. Budget constraints limited the flexibility of all agencies. Finally, both the City of San Diego and the Authority had future pipe replacement and relocation plans within a ten-year window. This made it difficult for the pipe owners to spend their limited funds for a temporary pipe. The replacement plans called for relocating the pipes to a new location approximately 100 feet upstream of its present location. Each of these agencies faced a decision to either spend funds to lower the pipelines in-place and consider it a temporary solution, or spend additional funds to relocate them to a permanent location.

Agreement

A series of meetings between the General Manager for Sweetwater Authority, the Director of Public Works for the County of San Diego, the Director of Public Works for the City of Chula Vista, and the City of San Diego Engineering Department were held. The General Manager finally "broke the ice" by proposing a four way deal in which all four agencies would share in costs and responsibilities. The general concept agreed to was to lower each of these pipelines in-place with an offset fitting. The City of San Diego would design its own pipeline lowering. The Authority would design its own pipeline, hire a contractor to lower the pipeline and perform construction management. The County would obtain all the permits and Chula Vista would perform the scour analysis in that section of the river to determine how low the pipe needed to be and design the future relocation of the river at this location which would improve flow characteristics. The total cost of the project construction would be divided one-third by the County, one-third by Chula Vista, one-sixth by the City of San Diego and one-sixth by the Authority. Also, the agreement stipulated that Chula Vista would provide an easement at no cost for future pipe relocation to the Authority and the City of San Diego.

Planning

Since the construction site is located in the river, it was imperative that when work took place there was minimal runoff. However, no construction could take place in the warmer months because of operational concerns by Sweetwater Authority and the City of San Diego. High system demands in the summer and early fall limited the construction period between November 1 and March 30. Unfortunately, these are typically the wettest months of the year. For this reason, it was important that construction commences on November 1st, and once construction started, the project be completed promptly to minimize the potential for river flooding damaging the construction site. Also, preceding the construction was the relocation of several willow trees within the potential pipe excavation area. The Authority had experience with one contractor who could remove, relocate, and replant mature willow trees. This contractor was hired to remove approximately 30 trees. Due to potential nesting of endangered California gnatcatcher and least Bell's vireos, the trees could be removed only between the months of October and March.

Design

As stated earlier, there are two steel pipes across the Sweetwater River at this location. One belongs to the City of San Diego and was installed in the 1950's. The Authority's pipe was installed in 1929 and is made of riveted steel. The initial approach was to replace one pipeline at a time. The pipeline would be taken out of service, removed, and a new section of pipe would be placed in the same location but deeper. The Authority's Construction Manager expressed concerns regarding the non-cohesive, sandy material within the trench that may cause the pipe remaining in service to be jeopardized and overall safety for the construction crews. Therefore, the design was Figure 1. Pipeline Profile modified to remove and install both pipes and decrease the pipe separation. This would reduce the size of the pit and concrete encasement. Other savings due to the pit size reduction included less dewatering, less revegetation, less reconstruction, etc. Finally, Chula Vista provided the proposed river realignment, which provided the points of pipe removals and connections for the new pipelines. Figure 1 shows the horizontal and vertical alignments.

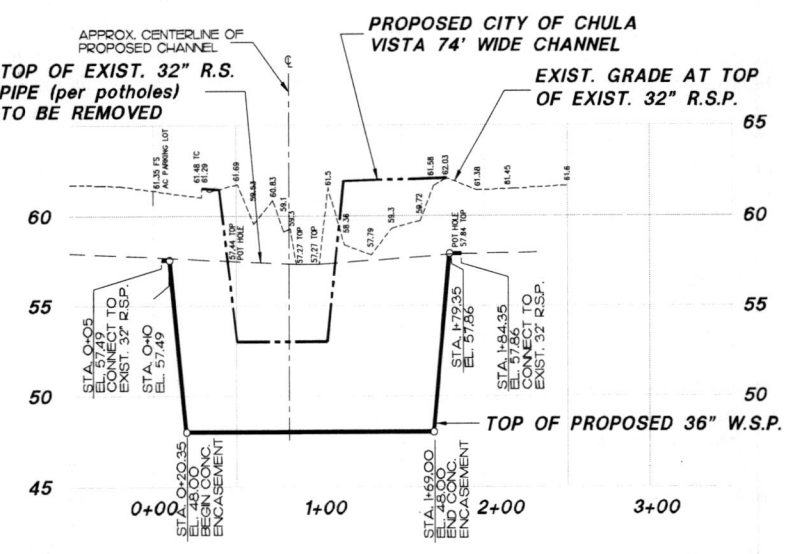

Environmental

As stated earlier, it was agreed that the County of San Diego would be responsible for obtaining the permits and environmental clearance for the project. Coordination was required with the U.S. Army Corps of Engineers and U.S. Fish and Wildlife Service. The County, in a very expeditious manner, obtained all necessary permits including environmental compliance, allowing construction to start within the designated window starting in October and ending in April. Mitigation for the project was the pre-construction transplantation of willow trees within the construction area to an adjacent bare area 100 feet west of the site. It was also required that arundo cane, which is considered an invasive non-native species, be removed with all the roots. After the project was completed, small willows were planted within the impacted area and a temporary irrigation plan was implemented for the transplanted trees and the new willows.

Construction

Construction was originally estimated to last three weeks and it was to occur at the beginning or tail end of the construction window. It actually only took one week to complete the entire project. This was mainly due to field design changes, excellent construction coordination, right size and number of equipment for the environmentally restricted site. A serious delay almost resulted when coordinating this work with an adjacent medical facility. The work required that several parking spaces be temporarily eliminated which posed a problem, as this was the highest annual use of this facility due to patients receiving flu shots. The construction manager assisted in restriping the parking lot to maintain the number of original stalls. This also required working with the local fire department because some of these temporary stalls were within a fire lane.

Payment

The Authority handled the payment of funds to the contractor. However, some of the agencies requested payment for their activities not noted on the agreement. This surprised everyone and after reviewing the contract it was discovered that not all-actual expenses were included. Once again, all of the agencies came together to determine how to resolve this. This was further exasperated when a leak developed on the City of San Diego's pipeline adjacent to one of the connection points that caused extensive damage to the parking lot adjacent to the construction site. Sweetwater contended that the leak was not related to the construction, but the City of San Diego disagreed. At the end, all of the agencies agreed to consider the additional costs including the cost of the leak repairs. Listed below is a breakdown of project costs.

Description	Amount
Initital Budget	$500,000
Construction Cost	$380,000
Subtotal	$120,000
Unexpected Costs	$60,000
Balance	$60,000

Table 1 - Project Costs

The lesson learned from this phase of the project and the agreement development was the necessity to be more detailed on potential costs (i.e., construction, environmental, design, and permitting), or that each agency needed to cover its own pre-construction costs. Since the construction cost was $120,000 below budget, there was funding available to cover the additional project costs resulting in avoiding a request from its Boards and Council.

Conclusion

All of the agencies involved were amazed as to the short time it took to construct the pipeline lowering. It was apparent that some compromise was needed to consider proceeding regardless of how strong one or all of the agencies felt its position was on this matter. There was a genuine care to solve a problem involving public safety. Following the completion of this project, several winter "El Niño" storms hit this area with no major flooding as had occurred in the past.

Also, the agreement could have been more detailed. Everyone was eager and motivated to develop, sign and get the work completed once the concept was approved. There was a fear that momentum would slow down this process if every detail had to be discussed, which in turn would rekindle thoughts of "we're not responsible for that" attitudes. It was fortunate that each agency continued to work together at the end to resolve all of the problems encountered after the construction was completed. It was, of course, a feeling of relief that the overall costs did not exceed the original budgeted amount. Public agencies can work together effectively.

Tijuana River Crossing
Using Horizontal Directional Drilling
A Case History

Moi Arzamendi[1], P.E. and Terry Smith[2], P.E.

Abstract

The South Bay International Wastewater Treatment Plant (SBIWTP) has been constructed by the International Boundary and Water Commission (IBWC) to eliminate major transboundary discharges of untreated wastewater from Tijuana, Mexico into the United States (Figure 1). The SBIWTP facilities require the use of potable and nonpotable water for sewage treatment processes. Therefore, a new water forcemain pipeline was required to reach a remote parcel of land located between the U.S./Mexico border and the Tijuana River in the southern-most part of the City of San Diego. Existing access roads and bridges are susceptible to flooding and erosion and do not provide a reliable means of extending a pipeline across the Tijuana River. A pipeline was installed using horizontal directional drilling (HDD) to cross the 1,500-foot wide alluvial floodplain (Figure 2). A 20-inch O.D. HDPE pipe was selected for the river crossing (Figure 3). Analyses and design for the HDD installation included a geotechnical investigation, river scour evaluation, permitting, constructibility analyses, and value engineering. The use of HDD minimized environmental impacts, costs, and significantly decreased construction time in order to meet the accelerated schedule of the SBIWTP project.

---

[1] Moi Arzamendi, Senior Project Engineer, USRGreiner/Woodward-Clyde, 1615 Murray Canyon Road, Suite 1000, San Diego, CA 92108, (619) 294-9400, mxarzam0@wcc.com

[2] Terry Smith, Senior Project Engineer, Malcolm Pirnie, Inc., 703 Palomar Airport Road, Suite 150, Carlsbad, CA 92009, (760) 431-0500, tsmith@pirnie.com

Project Background

The first phase of the SBIWTP project included construction of a 25 MGD advanced primary treatment plant expandable to 100 MGD with secondary treatment. The Tijuana River pipeline crossing was required to meet the process demands of the new facility. The waterline is a critical infrastructure component which was designed and constructed on a fast-track schedule in order to ensure water availability for testing and operation of the SBIWTP. Providing a reliable water source was an engineering challenge since the SBIWTP is remotely located in a zone vulnerable to flooding and erosion.

The Tijuana River typically swells laterally well beyond its active channel during periods of high rainfall and flooding. The SBIWTP site was designed to be protected from the 333-year flood event (135,000 cfs). Typically, scour during peak flows is concentrated along the river channel thalweg. Preliminary scour analyses indicated the scour potential was about 30 to 40 feet. Hence, the waterline would need to be below this scour elevation or above the flood elevation.

Various alternatives for providing potable water were considered such as 1) the use of groundwater wells with onsite storage, 2) construction of a new bridge or a dedicated pile supported above-ground pipeline crossing, 3) obtaining water from Mexico, 4) cut-and-cover construction, 5) microtunnelling, and 6) HDD. Value engineering determined that a HDD river crossing utilizing a high density polyethylene (HDPE) carrier pipe would provide the most value.

HDD is an outgrowth of oil drilling technology. In recent years, HDD has become generally accepted for river crossings for water, sewer, gas, electricity, telephone, and other specialty conduits. The use of HDD provided a means to expedite both design and construction. Lengthy environmental review and permitting were unnecessary since excavation was not required in the floodplain. A U.S. Army Corps of Engineers (USACE) Section 404 permit and a NPDES construction dewatering discharge permit were not required.

Geotechnical Investigation

The entire 5,300-foot pipeline project extended from Interstate 5 to the south side of the Tijuana River (Figure 1). The 1,500-foot river crossing alignment is along a washed-out segment of Old Dairy Mart Road (Figure 2). Woodward-Clyde performed a geotechnical investigation to evaluate geologic conditions and seismic hazards along the forcemain route and their effects of on design, construction, and serviceability. The geotechnical investigation addressed requirements for both cut-and-cover and HDD construction methods.

Subsurface explorations within the river crossing area consisted of 5 borings and 12 cone penetrometer test (CPT) soundings to depths up to 50 feet. Geotechnical laboratory testing was performed to evaluate moisture, density, grain size, strength, and compressibility of the encountered materials. The CPT load cell measures tip bearing resistance, skin friction, and porewater pressure as a function of penetration depth. The CPTs, borings, and laboratory testing allows for determination of subsurface material classification, strength, and groundwater conditions. The CPT was effectively used to evaluate soil liquefaction potential and settlement.

Geologic Conditions

The Tijuana River Valley is a large active drainage which crosses the international border and flows west to the Pacific Ocean (Figure 1). The thickly-alluviated valley is bordered on the north by a broad lowland area underlain by late Pleistocene-age marine and near-shore terrace deposits. Conversely, the southern edge of the valley is flanked by the sharp relief of the Border Highlands. The Border Highlands area consists primarily of Tertiary-age marine and nonmarine sandstones and conglomerates which are locally capped by early Pleistocene-age terrace sediments. Late Pleistocene-age terrace deposits of the Bay Point Formation have variable amounts of clay, silts, sand, gravel, cobble, and boulders which underlie a low lying narrow terrace along the northeast edge of the Border Highlands area.

Surface and Subsurface Conditions

Selected areas of the alluvial floodplain had been extensively mined for construction quality sand and gravel in the past. Channelization projects by the USACE have redirected river flows through southerly portions of the subject floodplain. At the time of design, the river crossing area had surface elevations ranging from +27 to +39 feet MSL (Figure 2). Groundwater ranged from approximately +23 feet to +27 feet MSL over the entire project alignment.

The river crossing traverses the deep alluvial soils. Explorations indicate that the upper 40 to 50 feet of the alluvial deposits are typically composed of silty to poorly-graded sands with localized zones of scattered small gravels which increase in density and strength with depth (Figure 2). Alluvial deposits above an elevation of +20 feet MSL are generally in a loose to medium dense state. Between elevations of +20 to -10 feet MSL the sandy deposits grade from dense to very dense with a few gravels. These materials also have discontinuous zones of silts and clays. A layer of very dense coarse sands with gravels, cobbles, and boulders is believed to exist below an elevation of about -5 to -10 feet MSL. These coarse materials are believed to be greater than 10 to 20 feet thick. The depth of alluvium in the Tijuana River valley is anticipated vary from about 100 to 200 feet.

Based on interpretation of subsurface conditions, it has been estimated that alluvial deposits above elevation +20 feet MSL may be Holocene age (<11,000 years old). In general, sandy alluvial deposits below elevation +20 feet MSL have CPT tip resistance greater than 200 tsf and standard penetration test (SPT) sampler blow counts in excess of 40 blows per foot (Figure 2).

The ideal subsurface conditions for HDD consist of moderately overconsolidated, cohesive, fine-grained materials which can easily maintain an open hole structure after being drilled. However, dense cohesionless sandy materials can be drilled with little difficulty if appropriate drilling muds are used to keep the hole open. Drilling and maintaining an open hole and carrier pipe pullback are more difficult when gravels and cobbles are present.

Faulting and Ground Shaking

Several northerly trending faults were observed in quarry cut slopes along the international border during geologic mapping of the region for the SBIWTP project. Detailed studies of these faults indicated that they have not displaced alluvial sediments or the Bay Point Formation. As such, these faults would not be considered "active" by State of California guidelines. However, the orientation and character of these faults appear similar to the complex system of "potentially active" faults which do displace other Tertiary-age deposits in areas of the Border Highlands to the west. The potential for fault ground rupture is considered low along the pipeline alignment. Maximum probable peak ground acceleration (475-year event) for the project area was estimated to be on the order of 0.3g. The maximum credible earthquake has been estimated to produce a maximum peak ground acceleration on the order of 0.4g.

Liquefaction Potential and Ground Deformation

Seismically induced soil liquefaction is a phenomena in which loose to medium dense, saturated granular materials undergo matrix rearrangement, develop high pore water pressure, and lose shear strength due to cyclic ground motions generated by earthquakes or other means. This rearrangement and strength loss may be followed by a reduction in bulk volume. Manifestations of soil liquefaction can include loss of pipe bearing capacity below the invert, settlement, hyrdostatic uplift, and instabilities in areas of sloping ground. Soil liquefaction potential was evaluated based on the results of CPT soundings, SPT blow counts, soil grain size distribution analyses, assumed ground motions, and published analytical procedures. Saturated loose to medium dense sandy soils below the groundwater table and above an elevation of about +20 feet MSL were identified to be susceptible to soil liquefaction.

Liquefaction induced ground surface settlements on the order of 2 to 4 inches were estimated. Differential settlements less than 1 to 2 inches over a horizontal distance of 50 feet were estimated. In addition, large temporary lateral shear strains (progressing toward liquefaction) of up to 20 percent could induce significant short-term axial strain in the pipeline as it enters and exits the upper alluvial soils. Pipes with an average inclination of 20 degrees (measured from horizontal) through the deformed soil mass could undergo axial strain of about 6 percent. However, the same pipe at 10 degrees could be subject to an axial strain of about 10 percent. Fortunately, HDPE carrier pipe should be able to resist temporal axial strains of up to 10 percent.

Pipeline Design and Material Selection

Malcolm Pirnie, Inc. performed alignment studies, pipeline engineering, secured right-of-way acquisition documents, prepared construction documents, provided permitting assistance, and construction management. Pipe materials utilized were AWWA C905 PVC for cut-and-cover construction north of the river, ductile iron pipe for a 50-foot bridge crossing in a backwater area, and HDPE for the HDD river crossing. A 20-inch

O.D. SDR-11 HDPE pipe was selected for the river crossing in order to eliminate the need for costly coatings and cathodic protection.

Engineering analyses for the pipeline included bending stresses, axial pull-back stresses, internal water service and surge pressures, external earth loading, buckling, and earthquake-induce deformations. In addition, hydrokinetic forces due to drilling fluid displacement around the annular zone during the carrier pipe pullback was evaluated. The design operating pressure is 100 psi. The minimum specified radius of curvature for the 20-inch O.D. SDR-11 (1.8-inch wall thickness) HDPE pipe was 1,400 feet (Figure 3).

Cherrington Corporation, pioneers of HDD trenchless technology, performed a HDD project analysis during the project design phase. The project analysis provided information pertaining to budgetary costs, pricing breakdown, project feasibility, schedule, execution planning, and HDD phase descriptions. Having a HDD contractor assist during the design phase proved to be instrumental to the education of all parties involved with the project.

Horizontal Directional Drilling

The contractor awarded the horizontal directional drilling work was Environmental Crossings, Inc. (ECI) from Traverse City, Michigan. Prior to mobilization of the drilling equipment, approximately 1,500 feet of the HDPE pipe was laid out on the north side of the Tijuana River where it was butt fused together with a fusion machine and then hydrostatically pressure tested. The HDD rig was set up on the south side of the river on a 20-foot thick compacted fill pad at the SBIWTP site. ECI chose to use an American Auger DD-90 drill rig with a 100,000 lb push/pull and 22,000 ft-lb torque rated capacities. The guidance and survey package consisted of a magnetic steering system located behind the lead assembly. The Tru-Tracker™ telemetry system was used as a secondary method of pipeline alignment verification.

Pilot Drill Hole

The first phase of HDD consists of drilling a small diameter pilot drill hole along the desired profile using a computer controlled steerable lead end and drill string. The inclination and azimuth of the lead end is continuously monitored. If intolerable deviations from the designed alignment occur, then the drill string is withdrawn an appropriate distance followed by an additional attempt at forward advancement.

The contractor proceeded with drilling of the 8.75-inch diameter pilot hole using a high pressure spud jet to erode material and back circulate sediment loaded drilling fluid to an automated desanding system. The drilling fluid used by the contractor was a bentonite/polymer/water mixture. The bentonite acts as a lubricant and hydrostatically maintains an open drill hole during the reaming process. Progress of the pilot drill hole proceeded as anticipated except that some gravels and cobbles were encountered in a relatively short initial segment of the alignment as drilling proceeded downward to the planned springline elevation of -3 feet MSL. Although the production rate slowed somewhat through the gravel and cobble zones, the production rate increased substantially through the remaining very dense alluvial sand along the remainder of the HDD alignment.

Halfway through the drilling operation the contractor received inconsistent depth readings from the down-hole steering tool and the Tru-Tracker™ system. It appeared that the drill bit had deflected downward as indicated by the down-hole readings even though the Tru-Tracker™ readings remained relatively constant. The discrepancy in the readings was about 2 to 3 feet. The contractor maintained the current elevation with the belief that the drill path was below the planned elevation and that the Tru-Tracker™ depth readings were incorrect. In order to punch out at the planned location, the contractor started the upward ascent sooner to account for the greater depth. However, the actual depth of the pilot drill hole was per plan. This resulted in the pilot hole punching out approximately 30-feet short of the exit location. Opportunely, the area north of the river was an abandoned road. However, an existing 30-inch sewer line located north of the planned exit location was narrowly missed. The incorrect vertical profile provided an unacceptable separation beneath the sewer line which necessitated that the contractor re-drill the last portion of the alignment. Once it was determined that the Tru-Tracker™ readings were correct and the down-hole readings were erroneous, the contractor pulled back the drill string to the bottom of the vertical curve and drilled along a corrected path. The final punch out of the pilot hole was within 6 inches of the planned location.

As planned, the pilot hole drilling operation was completed within 2 days. Although some gravel zones were encountered as expected, cobbles did not appear to be present along the alignment during the drilling operation of the pilot hole. Thus, the pilot hole was deemed ready for prereaming

Prereaming

The second HDD phase consists of prereaming (overdrilling) the pilot hole to a diameter large enough to accommodate the carrier pipe. The contractor chose to forward ream the hole and use a 30-inch diameter fly cutter. Soon after proceeding with the reaming operation, the gravel zone was encountered which had slowed the pilot hole operation. The gravel zone encountered was less than 200 feet from the entry point. Since production rates were slow, the contractor decided to pull the reamer back out of the hole. Observation of the fly cutter revealed that a number of the teeth had been broken off. The damage was attributed to the presence of cobbles.

Therefore, the contractor chose to perform the reaming operation in 2 passes. An 18-inch diameter reamer with a shape that was more suitable for breaking up or removing the cobbles was installed. Although production rates were very slow in two areas along the alignment, the reconfigured reamer was able to penetrate the suspected cobble zones. Upon completion of the first reaming pass, the hole was swabbed with a barrel reamer and a reconfigured 30-inch diameter reamer was installed to complete the process of opening the hole to the desired diameter. A 26-inch diameter barrel reamer was pulled back through to swab the hole and verify that the hole was completely open and free of obstructions. This operation was completed satisfactorily such that pullback of the carrier pipe was ready to begin. The reaming process required approximately 6 days more time than the contractor had scheduled. However, part of the extended construction period was attributed to unrelated mechanical problems of the drill rig.

Carrier Pipe Pullback

The last HDD phase performed is the carrier pipe pull-back through the prereamed drill hole. The planning for the pullback operation took into account the presence of existing overhead utilities, trees, and other objects. The 20-inch diameter HDPE pipe was fused to a nose piece with a swivel and was prepared for pullback. With the pipe placed on rollers, the pullback operation began with the pipe trailing a barrel reamer and was performed in a continuous operation until pullback was complete. Water was added to the pipeline during pullback in order to reduce the buoyancy of the pipeline and therefore reduce the friction of the pipe against the wall of the drilled hole. The bentonite displaced by the pipe during pullback was pumped out of the overflow pit which had been constructed next to the entry pit.

The pullback operation was performed smoothly without significant resistance and was completed within 3 hours. The carrier pipe was pressure tested upon completion of the pullback phase. Subsequent to the horizontal directional drilling, the general contractor completed testing, CCTV inspection, disinfection and connection of the pipeline to the distribution system. The project was successfully completed in June of 1996.

Differing Site Conditions

As a result of the cobble zones encountered, the contractor submitted a request for equitable adjustment due to differing site conditions. Although the geotechnical report indicated that some cobbles may exist due to the nature of broad alluvial sedimentation, the borings logs performed along the alignment did not indicate any material larger than 3 inches in size. Due to previous experience in the area, all engineers involved in the SBIWTP project knew that it was highly likely that cobbles would be encountered on the south side of the river. Therefore, it was subsequently determined by the owner that the language in the text of the geotechnical report was not specific enough to warrant an increase in the contractor's bid to compensate for the potential that cobbles may exist. This was substantiated by the fact that the 4 lowest bidders were within 3.5 percent of each other and therefore they likely made similar assumptions in their cost estimates. A negotiated settlement was reached between the owner and contractor on this issue. The resulting adjustment to the contract amount brought the total cost of the project closer to the original engineer's cost estimate.

The total cost for the 1,500-foot river crossing portions of the project was on the order of $475,000 including mobilization, drilling, installation, and testing. The planned HDD installation period was 17 working days. The actual installation lasted 23 days. The carrier pipe pullback phase was completed in 3 hours. Overall, the project was viewed by all parties involved as very successful. The project was completed on time and within budget. Environmental impacts were avoided as the result of using horizontal directional drilling. The experience of the HDD superintendent enabled appropriate adjustments to be made during drilling operations as needed.

Lessons and Conclusions

The three important lessons and conclusions that can be drawn from this project. First, geotechnical baseline reports are essential. These reports should closely follow recommended guidelines presented in "Geotechnical Baseline Reports for Underground Construction" (ASCE Special Publication, 1997). Particularly close attention should be paid to the evaluation and understanding of subsurface conditions that are expected to be encountered and what potential differing site conditions may be possible. Second, the contractor should utilize a verifiable method of determining the location of the pilot drill head. This is especially true where restrictive site conditions exist. Third, a minimum drill rig thrust and torque rated capacity should be specified in order to minimize the potential equipment damage or failure.

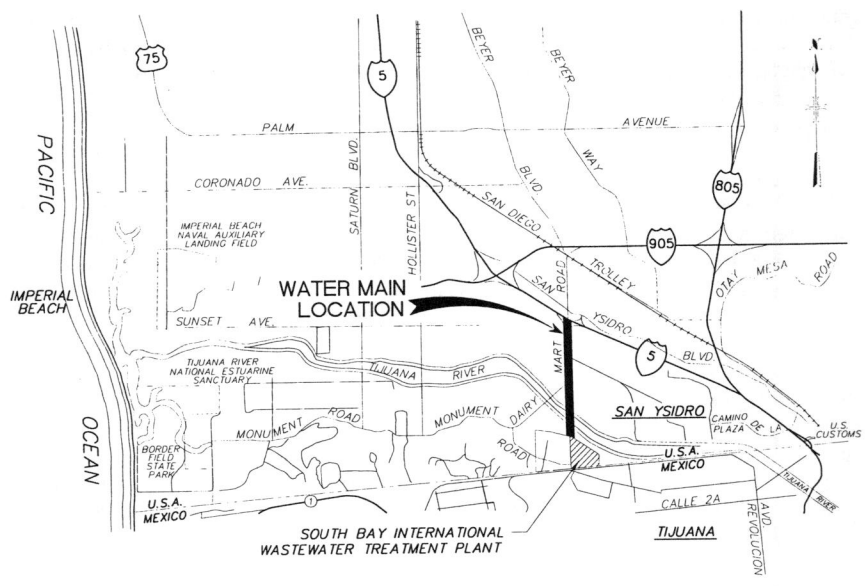

Figure 1. Vicinity Map

PIPELINES IN THE CONSTRUCTED ENVIRONMENT

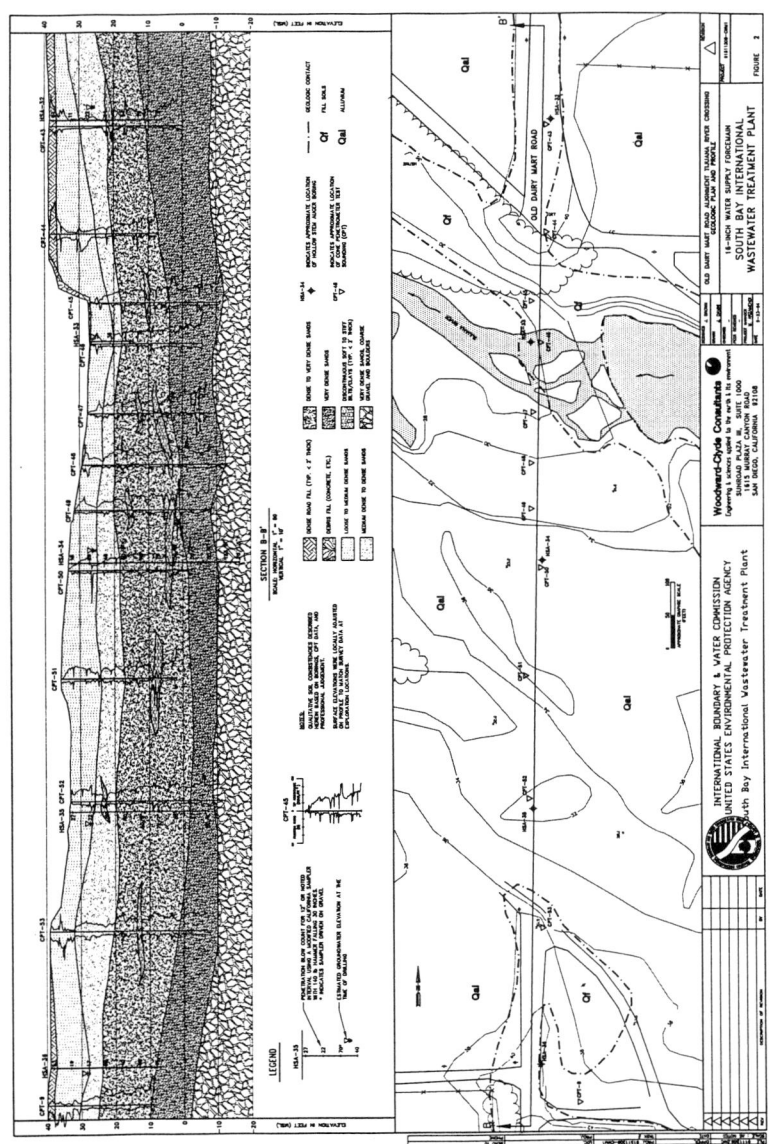

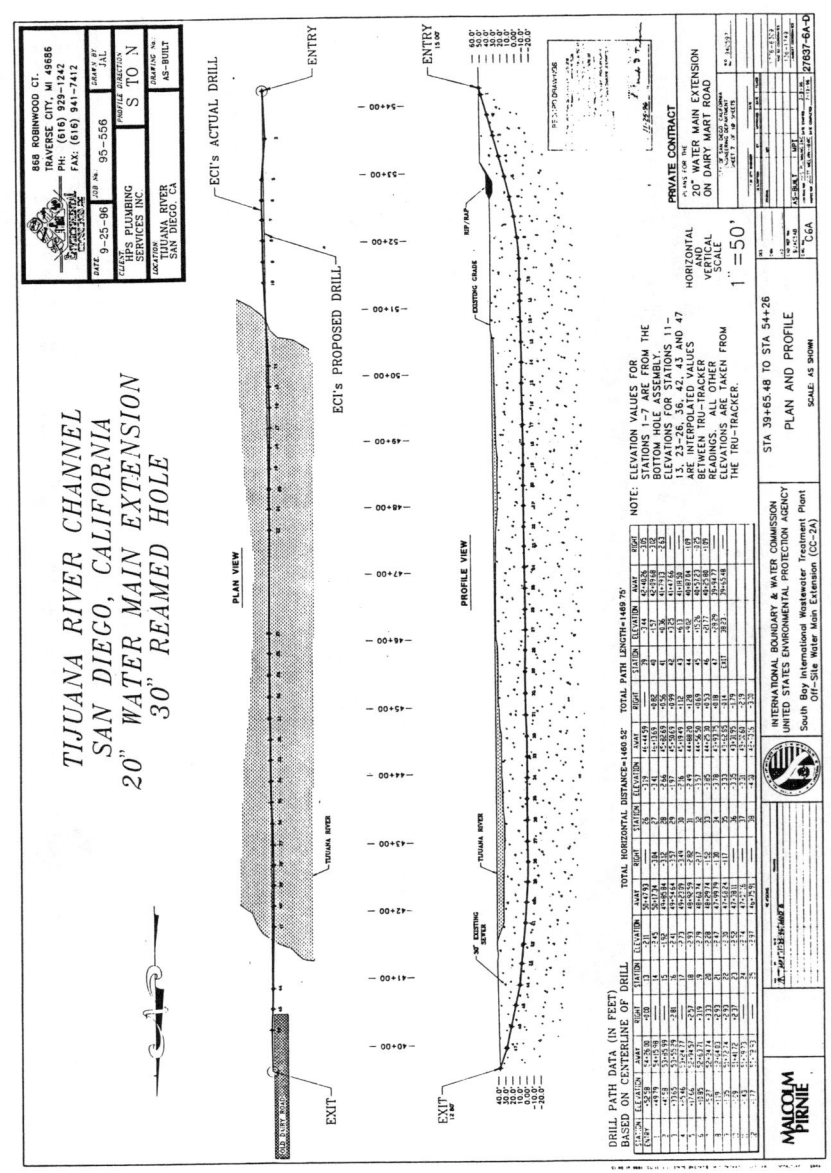

Ma Wan Submarine Crossing Using Horizontal Directional Drilling

Steven R. Kramer, P.E., M. ASCE[1] and Stephen V.L. Barrett, P. Eng.[2].

Abstract

This paper describes the underground investigation and design of a state-of-the-art crossing of the Ma Wan Channel in Hong Kong. The Ma Wan Channel is a major shipping channel next to the New Territories Peninsula and Ma Wan Island. The project includes the installation of twin 609.6mm (24 in.) diameter pipelines over a distance of 1,369m (about 4,500 ft.) in very complex ground conditions. The geotechnical investigation indicated the proposed drillpath would encounter granitic and volcanic formations, and inactive regional fault structures. The pipelines were designed to deliver potable water to a new island development which will eventually inhabit 15,000 people. The Ma Wan development is a $1 billion project to construct a resort style community with residential housing, commercial centers and a theme park.

Project Background

One of Hong Kong's major property developers is currently constructing a new residential and commercial community on Ma Wan Island in the Western New Territories, Hong Kong. The existing fresh water supply to Ma Wan consists of a 200mm (8 in.) diameter water main resting on the bottom of the Ma Wan Channel. This water main is inadequate to provide the volumes of water necessary to support the new development and two new 300mm (11.8 in.) I.D. water mains are required. The second main is required to provide system redundancy in the event one of the water mains ever needs to be replaced.

----

[1]Principal, Jason Consultants International, Inc. 2000 Massachusetts Ave., NW, Washington, DC 20036
[2]Project Manager, Golder Associates (HK) Ltd., 2/F Jockey Club Environmental Building, 77 Tat Chee Ave., Kowloon Tong, Hong Kong

The construction of the existing water main during 1987/88 was reported to be difficult. Heavy marine traffic, strong currents, irregular bottom topography and stringent Marine Department requirements in the Ma Wan Channel were the primary causes of the problems. After considering conventional sea floor trenching, directional drilling and tunneling, directional drilling was selected as the optimum means for constructing the new water mains on the basis of no marine interference, minimal impact to the environment, system protection from hazards such as dragged anchors, maintenance considerations and cost.

Pipeline Requirements

The requirements of the Water Supplies Department of the Hong Kong Government were as follows:

i) Each water main should be replaceable and each end should be easily accessible for maintenance purposes.
ii) The water mains should be capable of withstanding installation loads, normal operating pressures and anticipated dynamic pressure surges.
iii) The water mains should have a design life of at least 50 years.

Based on these requirements, a 457.2mm (18.3 in.) O.D., SDR 6.8 High Density Polyethylene (HDPE) water pipe was selected which would be installed inside an internally and externally Fusion Bonded Epoxy (FBE) coated 609.6mm (24 in.) O.D., 12.7mm (0.5 in.) mm wall thickness, API-5L Grade B steel casing pipe. To provide protection to the FBE coating during installation of the steel casings inside the completed bores, an exterior coating of polymer concrete was also specified.

Site Conditions and Constraints

<u>Physical Setting</u>

The Ma Wan Channel lies in an east-west orientation between the south shore of the Western New Territories and the north shore of Ma Wan Island (Figure 1). The Ma Wan Channel is approximately 1,100m (3,609 ft.) across between the pier at Angler's Beach and the recently constructed reclamation at Pak Wan on the north shore of the island. The alignment of the proposed crossing lies just to the west of the pier at Angler's Beach and tracks south southeast across the channel to intersect the island to the east of the reclamation area.

Subsurface Conditions

Based on the results of a marine seismic refraction and hydrographic survey which was conducted along the proposed alignment, the sea floor and bedrock topography of the channel is highly irregular and characterized by a ridge and valley topography. The base of the deepest trough is over 46m (150.9 ft.) deep and is bound by steep sidewalls. The troughs are believed to be the surface expression of fault structures and/or areas of lithological contrast. Available geotechnical borehole information indicates the northern portion of the channel is underlain by granitic rocks, while the southern portion of the channel and Ma Wa Island is underlain by volcanic tuffs and rhyolitic and granitic intrusions. The depth of weathering in the fault zones cannot be stated with certainty, but deep tropical weathering, which occurred during ancient periods of sea level lows, have been encountered at depths of 130m (426.5 ft.) or more elsewhere in offshore locations within Hong Kong.

The results of the geophysical investigation and existing marine borehole information indicate that in the foreshores, where the tidal currents are much reduced, marine and alluvial deposits overlie rock. In the central deeper portions of the channel the fast tidal currents have essentially removed all the soft sediment and bedrock is exposed on the channel floor. The subsurface geology therefore dictated that this would be a hard rock crossing, which also was likely to contain multiple fault zones and highly fractured and weathered rock.

Site Constraints

The Ma Wan Channel is a heavily traveled inland waterway. Bulk carriers, towed barges, small ships, lighters, high-speed ferries and fishing craft all use the channel. These vessels travel day and night at speeds that make marine construction activities in the channel dangerous.

The New Territories shoreline is also very busy. Castle Peak Road runs parallels to it and carries large volumes of traffic, which restricts available access for an entry staging area to a small 80m by 15m (262 ft. by 49 ft.) rectangular parking lot to the north of the roadway at Angler's Beach. The seawall to the south of Castle Peak Road and the utilities beneath the roadway dictated that the entry points be located in a 4m (13.1 ft.) deep, 18m long by 12m wide (59.1 ft. by 39.4 ft.) entry pit within the parking lot. Normally it would be desirable to launch the rig at an angle varying from about 8 to 15 degrees, which is compatible with existing rig configurations and generally reduces pulling loads. However, in this case it was necessary to increase the entry angle to 20 degrees to keep the pit depth at 4m (13.1 ft.) and still be able to drill the holes safely beneath the toe of the sea wall. Because of the lack of cover beneath Castle Peak Road and the beach beyond the sea wall, it was necessary to specify that the drill holes be cased through the overburden deposits into rock. This will assist in containing the drilling fluids.

Available sites for exit on Ma Wan Island were also restricted due to the island's topography, the layout of the proposed development and governmental restrictions on the alignment of the water work's reserve. After evaluating a series of alternatives, an exit site was finally chosen which met most of the key selection criteria. However, the site's steep topography approaching the exit area required that the exit angle be 15 degrees, which is steeper than the 8 to 10 degrees normally desired to facilitate pipe handling during pullback. Because a steeper angle has been selected, it is anticipated that the pipe will need to be elevated behind the exit point and special handling equipment will be required to prevent kinking of the pipes prior to their entry into the drill holes. The space available for pipe fabrication and stringing behind the exit site is also restricted, with only 230m (754.6 ft.) of open space available. This will require that the pipes be pulled back in stages.

Crossing Design

Proper design, planning and preparations are required to successfully complete a job of this magnitude. Throughout the detailed design of the crossing, it was important to thoroughly analyze different construction scenarios for the installation of the pipelines. This included the usual variables for a horizontal directional drilling project plus additional factors deemed unique and important for this crossing.

The Ma Wan project is near the technical envelope in terms of length and diameter for hard rock crossings utilizing horizontal directional drilling. Most of the crossings using directional drilling at distances over 1,000m (3,280 ft.) have occurred in alluvial materials. Hard rock drilling is relatively new and most experience has occurred in the last five years.

The Ma Wan crossing posed three major challenges: 1) drilling parallel 1,369m (4,492 ft.) long holes beneath a 46m (151 ft.) deep channel; 2) drilling the holes in complex underground conditions; and 3) executing the job within the confinement of a tight work site.

The key design issues for the project are discussed below.

Route Selection

As described earlier, the subsurface conditions along the proposed alignment are highly complex and likely to consist of multiple rock types of varying strength and hardness, as well as fault zones where the ground may be more like soil than rock. This dictated that the crossings must be deep in order to maximize the cover beneath the channel floor and to minimize the amount of drilling through weathered rock within the fault zones. Based on the bedrock profile indicated by the seismic refraction survey, a crossing elevation of 76m (249.4 ft.) below Hong Kong Principal Datum (which is approximately equivalent to sea level) was chosen for both drill paths.

The alignments in plan were restricted by the location and size of the available entry and exit areas and the requirement to keep the alignments as straight as possible. A straight alignment would reduce both pipe installation loads during pullback and drilled length of the crossings. As the interpreted faults were linear features and therefore difficult to avoid, the geological selection criteria was to try and avoid areas where two or more faults may intersect, as the depth and extent of weathering would likely be greater in these areas than in other areas of the channel.

Directional Tolerances

Due to the restricted size of the entry and exit areas, the alignment of the pipelines was designed to be approximately parallel with 5m (16.4 ft.) separating the entry points and 10m (32.8 ft.) separating the exit points. This necessitated very tight drilling tolerances of +/- 1m (3.3 ft.) horizontally and vertically from the centreline of the bore. By designing the drill paths to diverge slightly from the entry to the exit points, the potential for bore interference due to hole deviation was also reduced.

Hole Size

The final size of the reamed bore has been left to the Contractor to select. However it is anticipated that it will range from 800mm to 900mm (32 to 36 in.), i.e. a 33% to 50% overcut for the 609.6mm (24 in.) O.D. steel casing. The actual size will be dependent on the quality of the pilot holes, encountered ground conditions and the Contractor's selected equipment and method of operation. Because of the size, depth and length of the bores, it is anticipated that special measures will be required to facilitate cutting removal and reduce the volume of cuttings at any given time in the hole. Extra attention will also need to be paid to mud cleaning so as to minimize cutter wear.

Installation/Construction Loads

The installation load on the pipe is a function of the following five elements:

1) Buoyancy force
2) Total angle of turns/Installation alignment
3) Roughness of the drill hole
4) Pipe weight
5) Installation depth and length

These forces will increase as the pipe is installed in the hole and typically these loads are more severe than the operating loads.

The industry guideline of 1,200 times the outside pipe diameter was used to define the turning radius to be used when drilling. For a 609.6mm (24 inch) diameter casing pipe, the minimum turning radius would then be 730m (2,400 ft.). This information was used to develop the crossing design, along with the selected entry and exit angles and the alignment and crossing depth defined by the geology.

DrillPath™ software was used to perform the installation load analysis. DrillPath allows the users to perform multiple sensitivity analyses to find an optimum design based on the varying the drill path and pipe type. The final drillpath design for the Ma Wan crossing is shown on Figure 2. The total drilled length of the proposed installation is approximately 1,369m (4,492 ft.). In the analysis it was assumed that the drill rig would be situated in the New Territories with the pipe pulled back from Ma Wan Island.

The results of the analyses showed that the installation stresses in the pipelines were below the allowable levels and that a 500,000 lb rig would have adequate capacity to complete the installation. While the maximum pulling loads were calculated to be less than this value, it is normally considered to be prudent practice to provide extra rig capacity as the designed drill path is an ideal case, which does not account for any additional pulling load required due to hole deviation.

Contract Options

There are several contract types that have been used for horizontal directional drilling (HDD) projects. These include: lump sum or fixed price, daily rate, footage rate, mixed price and target price contracts. In most cases, the lump sum contract has been used. However, there are situations where an alternative contract may be advantageous for both owner and contractor. According to one of the largest directional drilling contractors, there have been numerous directionally drilled crossings let on a lump sum basis which have failed. Our past experience supports this statement that traditional contract types have not always been appropriate as contractors have expanded the envelope for HDD projects. There is a trend to utilize alternative contract types for the more challenging HDD projects.

The particular choice of which contract type to utilize is dependent upon the technical feasibility, economic feasibility and risk of the project. The technical feasibility is defined by the following project parameters: installation length and depth, hole diameter, ground conditions and potential for damage to nearby or surrounding facilities. As crossing design is largely experience-based, designers and contractors need to use past projects to assess the feasibility of future proposals. If the installation is near the limit or outside of the operating envelope, a standard contract type may not be suitable.

Contract Type for Ma Wan Channel Crossing

For the Ma Wan crossing, the owner preferred to utilize a standard lump sum contract with pre-qualification of the HDD contractors based on their previous rock drilling experience. At the time of tender, it was recognized that the tenderers would likely be reluctant to submit a lump sum bid or need to include a large contingency built into their price. It was therefore necessary to develop a contractual solution that balanced the owner's desires and contractor's need not to take on undue risk. Balancing these factors, a solution was found by splitting the project into two lump sum contracts.

The first contract covers the drilling of the first of the two pilot holes to allow all parties to gain a better understanding of the likely drilling rates and ground conditions along the proposed alignment. By investing in the pilot hole, the Owner is protecting himself against large contingencies associated with unanticipated ground conditions, which would need to be included in the contract price if the work were to be let as a single contract. The Owner also minimizes the potential for large claims, in the event the ground conditions are so adverse that the crossing proves not to be feasible.

The second contract will cover the remainder of the work and will only be awarded after successful completion of the pilot hole. During the second contract's tender period, the as-built records and reports will be available to all tenders to highlight potential problems and allow the Contractor to make appropriate contingencies.

Conclusions

The Ma Wan Submarine crossing is a state-of-the-art directional drilling project which is approaching or is at the limits of the envelope of HDD crossings in rock. By careful design, planning and preparation, the Owner has been given a vehicle to achieve a potentially cost effective solution to a difficult construction problem. By taking a risk sharing approach with respect to the contract type, the Owner has substantially reduced the risk exposure to both himself and the Contractor. Should the pilot hole therefore prove successful, the Owner will then be able to achieve a more competitive total project price than if the project had been let as a single, traditional, lump sum job.

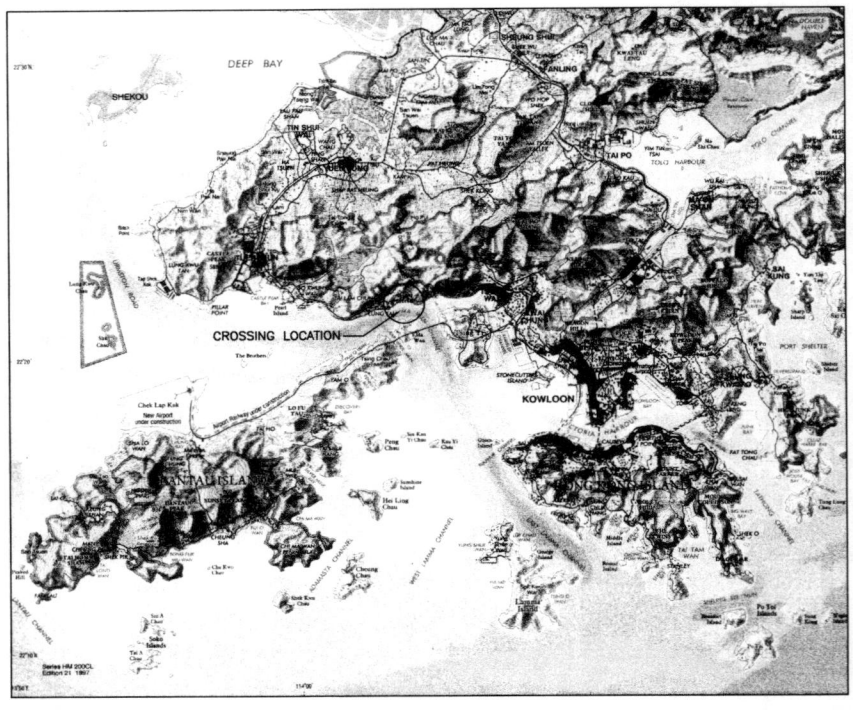

Figure 1. Project location within the Hong Kong Special Administrative Region

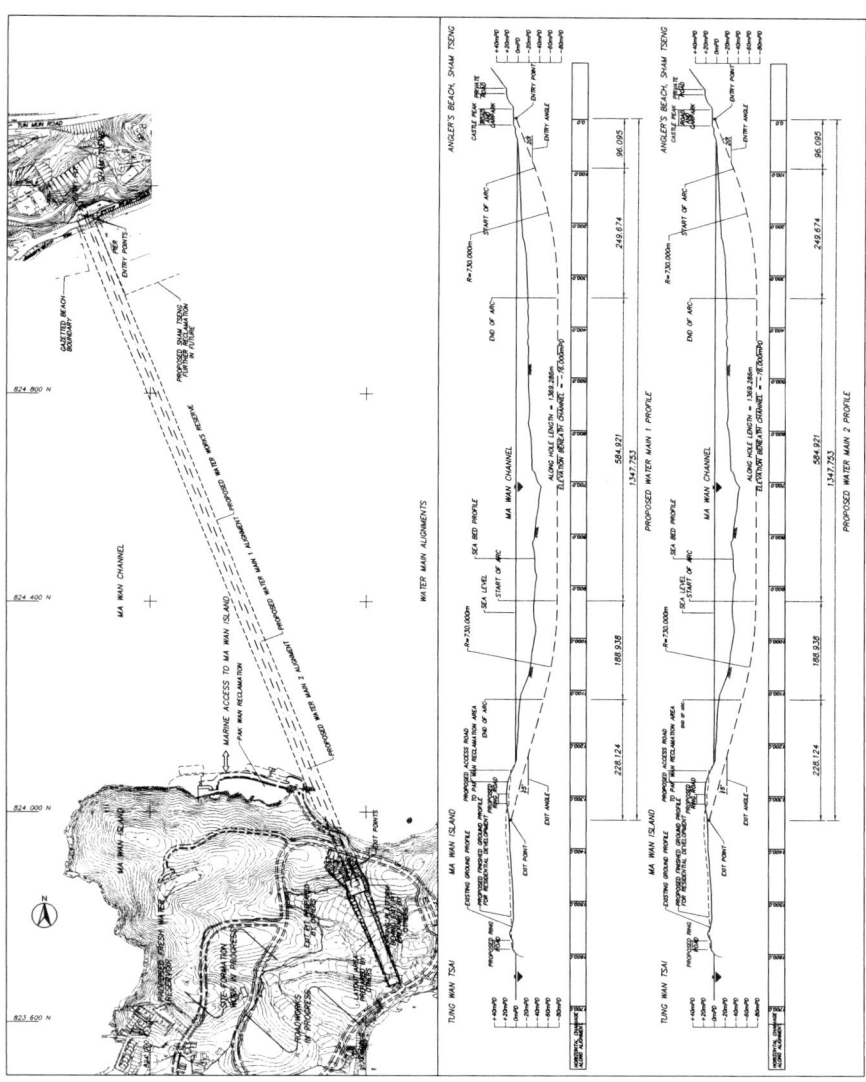

Figure 2. Alignment and profile of the drillpath as designed.

The Successful Design of a Large Diameter Gravity Sewer

Jamal Batta, P.E.[1]
John Harris, P.E., MASCE[2]
Mark Giandoni, P.E., MASCE[3]

Abstract

This paper presents the case study of the successful design and installation for the City of San Diego's North Mission Valley Interceptor Sewer, Phase 2 (NMVIS2). The NMVIS2 project included over 3,660 meters (12,000 feet) of 1,950 mm (78-inch) plastic lined reinforced concrete pipe interceptor sewer and over 2,135 meters (7,000 feet) of 300 mm (12-inch) to 900 mm (36-inch) collector sewer pipeline. The NMVIS2 project total construction cost was $20,000,000. This project received the ASCE 1997 Outstanding Civil Engineering Project Award for Wastewater Systems, San Diego, the 1997 Trenchless Technology Project of the Year Award, and the City of San Diego William Earl Hayden Outstanding

Introduction

To successfully complete this project the engineering design consultant and City of San Diego planned and designed for the use of new and alternative construction technologies and materials. The alternative construction technologies included the use of four large diameter utility tunnels for the installation of over 610 meters (2,000') of 1,950 mm (78-inch) pipe. The new construction techniques included the use of microtunnel installation techniques for over 1,525 meters (5,000') of collector sewers. The alternative material allowed was fiberglass reinforced plastic mortar (FRPM, Hobas). These alternative materials and construction methods provided additional competition during the bidding process, which allowed the City of San Diego to realize a more cost effective pipeline installation.

The City of San Diego's North Mission Valley Interceptor Sewer (NMVIS) is the main collector sewer between the San Diego Metropolitan Wastewater System and

---

[1] Associate Engineer, City of San Diego, 600 B Street, San Diego, CA 92101
[2] Vice President, Hirsch & Company Consulting Engineers, 3550 Camino Del Rio North, San Diego, CA 92108
[3] Senior Project Engineer, Hirsch & Company Consulting Engineers

six contributing wastewater agencies to the east of the City of San Diego. Within the City of San Diego there are eight communities that contribute sewage flows to the NMVIS. The original NMVIS pipeline was constructed in 1960 and was reaching it's maximum capacity in the mid-1980s. In a 1986 study it was determined that the NMVIS needed to be replaced due to lack of capacity and deteriorating conditions on several reaches. A new NMVIS would be designed and constructed to convey flows for the ultimate buildout of the sewage basin with peak flows of 51,097 m^3/day (135 MGD).

The NMIVS pipeline is located in the Mission Valley area of San Diego (See Location Map). The alignment passes through heavily urbanized commercial developments, riparian habitats of the San Diego River and through relatively undisturbed areas in two golf courses. The project coordination and permitting requirements required numerous complex negotiations with property owners and resource agencies.

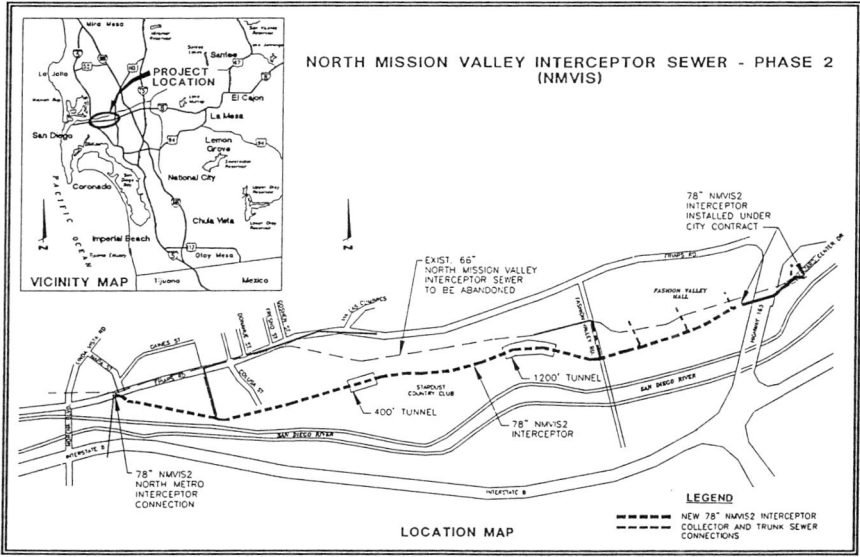

Tunnels

The design for the 1,950 mm (78-inch) diameter pipeline included four tunnels that were required for several different reasons. Two of the tunnels, including the longest tunnel on the project, passed underneath areas that were known to have cultural resources from ancient Native American settlements. The other tunnels were required to cross underneath a state highway right of way and large storm drain structures.

The tunnel construction underneath the cultural resource areas provided an efficient method to mitigate potential impacts to Native American resources due to construction. The additional costs for the tunneling were justified by eliminating the requirement for a cultural resource data recovery program. The data recovery program was estimated to cost $1,000,000 and take six to twelve months to complete. The additional costs to construct the tunnels were approximately the same as the cost to complete the data recovery program, however the City of San Diego saved the six months that the additional cultural resources studies would have required.

The contract documents for the tunnels on this project allowed the contractor several different options for the tunnel construction. The longest tunnel at 365 meters (1,200-feet) had alternatives for constructing a two pass tunnel with steel rib and steel liner plates, or steel rib with wood lagging as the primary liner and support. The contractor chose the steel rib and wood lagging as the most efficient alternative for the 365 meter (1,200-foot) tunnel. The second tunnel underneath the cultural resource site was 122 meter (400-feet) long. The contract documents allowed the contractor to choose either direct jacking, two pass tunneling with steel rib and wood lagging or steel rib with steel liner plates for this tunnel. The contractor chose to use direct jacking of the concrete pipe for this tunnel.

The tunnel passing underneath the state highway was 92 meter (300-feet) long and was constructed below an overpass that had only six feet of vertical clearance. The state highway department had concerns regarding traditional pipeline construction techniques required for the installation of a 1,950 mm (78-inch) pipeline. The highway department's concerns that large construction equipment moving in and around the overpass could damage the facility, impact traffic and impact the quality of the installed pipeline. The tunnel construction addressed these concerns and the highway department's requirement of a double containment structure for the pipeline. The double containment requirement was addressed by including the options for construction of the tunnel with steel rib and steel liner plates or a jacked steel casing. Secondary containment was achieved by pressure grouting the annular space between the primary liner and pipe O.D. The contractor chose to use a steel rib and liner plate primary tunneling support system.

The fourth large diameter tunnel was designed to cross underneath a large storm drainage facility. The crossing was underneath four parallel 1,950 mm (96-inch) arched pipe storm drains. The contractor had the option to remove the storm drains during construction or tunnel underneath the facilities. They chose to tunnel underneath the drains. The tunneling was faster than disassembling and then reassembling the drains and was safer since the storm drains would be in continuous operation during construction. This tunnel was 27 meter (90-feet) long and provided an efficient alternative to open trench construction.

In all of the tunnels constructed for the North Mission Valley Interceptor Sewer, the contract documents allowed the contractor to choose between at least two appropriate alternative construction techniques for the installation of the tunneled pipelines. This allowed the contractors bidding the project to use their skills and expertise to determine and use the most appropriate and efficient construction methods. By allowing the contractor to choose the appropriate alternative construction methods the City of San Diego was able to get a cost effective and successful tunnel project.

Collector Sewers

The collector sewer pipelines for the North Mission Valley Interceptor Sewer involved the installation of over 213 meters (7,000 feet) of collector sewers in sizes of 300 mm (12") to 900 mm (36"). Over 1,524 meters (5,000 feet) of these sewers were installed using microtunneling construction. The installation of the collector sewers had numerous challenges during the design process. These challenges included pipelines at depths from 4.6 meters (15 feet) to 9.2 meters (30 feet) underneath busy urban streets. Ground water well above the pipelines and surrounding soils that were made up of weak alluviums or a well cemented clay cobble conglomerate.

The original collector sewer design had microtunneling as an alternative to open trench construction. However, after discussing the project with community groups and major business owners in the project area, the City of San Diego decided to require microtunnel construction to minimize the disruptions to the community.

The designs for these microtunneled pipelines required that the engineer address the high groundwater and mixed tunneling conditions and allow the contractor to use their specialized skills in the constructed project. The tunneling design and hydraulic design were coordinated so that the contractor could use one pipe size for all of the microtunneled pipelines. The project plans noted the minimum pipe diameter allowed and the specifications provided the maximum pipe size. The design of the oversized pipes included additional hydraulic calculations that ensured minimum cleansing velocities in the larger pipes. Much of the pipeline needed to be only 300 mm (12-inches) in diameter, however the design allowed the use of up to 600 mm (24-inch) pipe. The use of one pipe size for the microtunneled pipe allowed the contractor to construct the pipelines using only one tunneling machine. The contractor chose to address the high groundwater pressures and the potential for cobbles by using a pressure balanced microtunnel machine that had rock crushing capabilities and by using the 600 mm (24-inch) clay pipe. The larger sized pipe that the contractor chose allowed the use of the larger tunnel machine that was capable of crushing cobbles up to the expected size of 300 mm (12-inches).

The contract document provided different appropriate alternatives for the microtunnel installation of the North Mission Valley Interceptor Sewer collector sewers that allowed the contractor to use the most efficient materials and equipment for the project.

Alternative Materials

As part of the design studies for the North Mission Valley Interceptor Sewer, the City of San Diego required the design consultant to complete an alternative pipeline materials analysis. This analysis investigated pipeline materials that would be suitable as an alternative to City of San Diego's standard for large diameter sewers, plastic lined reinforced concrete pipe (PLRCP). The designer reviewed numerous pipeline materials including, high density polyethylene (HDPE) pipes, both profile wall and smooth wall, fiberglass reinforced plastic mortar (FRPM), profile wall PVC pipe materials, steel pipe and pre-stressed concrete pipe. From the studies completed it was determined that the only other material that would be suitable for use in the NMVIS project was FRPM. The selection process compared the pipe materials ability to meet the challenging installation conditions that were expected in the NMVIS project. These conditions included changing ground water levels, installation depths of over 12.2 meters (forty feet), embankment fills and low strength soils.

The fiberglass reinforced plastic mortar pipe (FRPM), manufactured by Hobas was capable of meeting the structural requirements expected on this project. The contract documents for the purchase of the pipe materials included plastic lined reinforced concrete pipe and the FRPM pipe. Including the FRPM pipe as an alternative to PLRCP in the contract documents encouraged the competitive bidding process. The estimated cost savings to the City of San Diego for the purchase of the 1,950 mm (78-inch) pipe was over $1,000,000 (15% of the pipe materials cost).

A Successful Project

The determination of the success of the North Mission Valley Interceptor Sewer Phase 2 project is based on the numerous accolades given to the design team by regional and national engineering and construction organizations. These awards included the 1997 San Diego Section ASCE Project of the Year Award for Wastewater Projects, the 1997 Trenchless Technology Project of the Year and the City of San Diego William Earl Hayden Award for Pipelines.

This engineering and pipeline construction project addressed the project challenges during design and used appropriate alternative technologies to address the project challenges and maximize the competitive bidding process to allow the City of San Diego to realize cost savings and efficiencies above and beyond those of the typical pipeline project.

Sacramento County Northwest Interceptor: Effective Planning Ensures Project Success and Saves Money

Thomas Kalkman[1]
Michael Watson[2]

Abstract

The Northwest Interceptor is being constructed by the Sacramento Regional County Sanitation District (SRCSD) to provide relief for the existing sewer system within the northeast part of Sacramento County, California and provide sewerage capacity to areas along the northern portion of the County. This paper focuses on planning for the interceptor and how effective preliminary design saves money and time relative to future projects.

Introduction

The Sacramento Regional County Sanitation District (District) is expanding its sewage collection system to accommodate new growth and expand its service area. A major part of this expansion is the Northwest Interceptor Project. This $120 million 29-kilometer interceptor system ranges in size from 1-meter to 2-meter diameter and includes a 1.5 m^3/sec pump station.

The Northwest Interceptor Project will relieve the northeast part of Sacramento County and provide sewerage capacity to areas along the Northern portion of the County. The District has embarked on a three-phase planning process to bring the project from concept to completion. The first phase, master planning, determined the sewerage needs and proposed a project to satisfy these needs. The second phase is a design report,

---

[1]Associate, Carollo Engineers, 2700 Ygnacio Valley Road, Suite 300, Walnut Creek, California 94598.

[2]Senior Engineer, Sacramento Regional County Sanitation District, 9660 Ecology Lane, Sacramento, California 95827.

which fine-tunes the recommendations of the master plan and documents the preliminary design issues to provide continuity to the third phase, final design. In final design, the entire project is broken into manageable segments for construction contracts.

The Northwest Interceptor Design Report evaluated the recommendations of the comprehensive master plan in detail. This additional evaluation, particularly review of flow transfer and pipeline routing alternatives, saved the District approximately $10 million over the master plan recommendations.

Background

Sewerage services within the Sacramento metropolitan area are provided by SRCSD and the three Contributing Agencies that provide sewage collection services within the SRCSD service area: Sacramento County Sanitation District No. 1 (CSD-1), the City of Sacramento, and the City of Folsom. SRCSD was formed in 1973 to comply with the basin plan objectives of the California State Water Quality Control Board, Central Valley Region. Under the Master Interagency Agreement (MIA) that defines the operational, financial, and administrative responsibilities of SRCSD, the Contributing Agencies, and Sacramento County; SRCSD is responsible for financing, construction, and operation of new interceptor sewers designed to carry 10 mgd or more of sewage from new development in its service area.

Much of the funding used to build the existing regional interceptor system was provided by the Clean Water Grant Program. A condition of this funding limited interceptor capacity to existing flows plus 20 years of anticipated growth. Since the majority of the regional interceptor system was designed in the mid 1970's and therefore was approaching design capacity, SRCSD embarked on a system-wide evaluation of the existing interceptor system. The result of this system-wide evaluation was a document entitled the Sacramento Sewerage Expansion Study, completed in 1993.

In addition to identifying capacity deficiencies within the existing trunk system, the 1993 Sacramento Sewerage Expansion Study identified interceptor sewer facilities (including the Northwest Interceptor) required to serve the areas proposed for urban development in the County's Draft General Plan Update prepared in 1990 and the land uses adopted in the General Plans of the Cities of Sacramento and Folsom.

Project Objectives

The primary objective of this project is to provide the continuity necessary for successful design and construction of the Northwest Interceptor System under a phased program utilizing multiple design and construction contracts.

Other goals of this project were to:

- Establish the most cost effective solution to relieving wastewater flows in the Northwest Area.

- Provide sufficient design documentation to allow preparation of an environmental document for the entire Northwest Interceptor System.

- Provide sufficient route and alignment identification to allow SRCSD staff to acquire necessary rights-of-way.

- Provide plan and profile drawings in a format which may be readily used by other consulting firms in the future to create final design and construction drawings.

- Provide a standard reference document and basis of understanding for addressing specific questions and concerns over the course of the development of the Northwest Interceptor System.

Planning Analysis

A number of factors are involved in planning a large project like this.

Flow Transfer: An evaluation of flow transfer between existing sewers and the proposed Northwest Interceptor was conducted. This evaluation included location, method, and quantity of flow to be transferred. The result of this evaluation leads to a major change in the Northwest Interceptor route proposed in the original 1993 Expansion Study and significant cost savings.

Route Selection: Various pipeline route alternatives were considered including consideration of pumped versus gravity alternatives where applicable. The route of the Northwest Interceptor was ultimately changed from Antelope Road to Elkhorn Boulevard as part of this route selection process.

Utilities: A search of existing major utilities was performed and their potential impact on horizontal and vertical alignments was identified.

Alignment Selection: Once the final route was selected and major utilities identified, alternative pipeline alignments within the route were evaluated and selected.

Hydraulics: Hydraulic analysis of the recommended interceptor system including: depth of flow and velocities for both initial average dry weather, and build-out peak wet weather flow conditions was conducted.

Environmental and Hazardous Substance Review: Both preliminary environmental and hazardous substance surveys were conducted to identify potential existing conditions which might have a significant impact on construction through a portion of the interceptor route.

Geotechnical Investigation: Geotechnical borings were performed approximately every 305 meters. The borings were analyzed to determine the subsurface conditions in which the interceptor is to be constructed.

Pump Station Evaluation: Review of pump station siting alternatives and configurations was conducted for a pump station which will pump flow from the Northeast Area into the Northwest Interceptor.

Route Selection

Route selection is a key aspect in the planning process. The objectives of route selection are to:

- Identify alternative interceptor routes within these reaches.

- Conduct a comparison analysis of the selected alternatives based upon selection criteria.

- Recommend a preferred interceptor route based upon the conclusion of the comparison analysis.

The potential routes were reviewed to identify fatal flaws which would render a particular route unfeasible. Routes not exhibiting fatal flaws were subjected to preliminary screening using a matrix evaluation. The two most desirable alternatives, based on the matrix evaluation, were then subjected to conceptual level cost comparison. The recommended route would be the best apparent alternative based on the matrix and cost evaluations.

Each alternative route is generally set within public right-of-way as much as possible. Where possible, existing roadways were used to establish the alternatives to avoid encroachment upon private property. In certain areas where roadways have not been built out to their ultimate widths, some acquisition of private rights-of-way may be necessary.

The evaluation of the alternative interceptor routes follows three major steps: a search for "fatal flaws," a preliminary screening analysis, and a cost comparison.

Fatal flaws are conditions which would render an alternative infeasible to construct. Such conditions include lack of adequate cover at stream beds, extraordinarily narrow rights-of-way, conflicts with large underground utilities, unmitigable major environmental issues, and similar unsurmountable conditions. No apparent fatal flaws were found in any of the alternative routes identified.

- **Flexibility** - A route alternative is considered advantageous with respect to flexibility when existing and future trunk sewers can connect to the interceptor without pumping or low-flow problems in the connecting sewers.

- **Environmental Concerns** - Environmental concerns include impacts to cultural and biological resources and the potential to encounter hazardous materials during construction.

- **Constructability** - A measure of physical constraints such as crossing of waterways, major utilities, bridges and other interferences such as adverse soil or groundwater conditions. Also included is acceptability by CalTrans, the local Reclamation District or other "non-landowner" groups or agencies of locating the interceptor in their right-of-ways.

- **Easements/Right-of-Ways Acquisition** - A measure of the potential difficulty or costs associated with obtaining the necessary temporary and permanent right-of-ways and/or easements for the interceptor. Alternative routes requiring easement acquisition in areas not zoned for urban development are least favored. Alternatives requiring easements in otherwise developed areas are second least favored, while designated buffer zones or utility corridors are most favored.

- **Accessibility** - Once the interceptor is completed and in service, routes with improved access for maintenance or repairs are favored.

- **Comparative Cost** - Comparative construction costs based on cost criteria. The lower comparative construction cost ranks higher.

The matrix evaluation was developed as a method of comparing the route alternatives. This approach evaluates each alternative route by assigning a value between 1 and 5 for each of the screening criteria. An alternative with a higher score is considered preferable to one with a lower score for a particular criterion. The alternative routes with the highest total scores are preferred over alternative routes with lower total scores.

Table 1 presents the preliminary screening matrix for the route Alternatives. The total points in the evaluation ranged from a low of 18 to a high of 27. The three routes with the lowest scores (Alternatives 2, 3 and 4) are eliminated from further evaluation. The routes with the two highest scores (Alternatives 4 and 5) are retained for the final evaluation process.

Each of the highest scoring alternative interceptor routes are more critically evaluated in the final screening process. The basis of this evaluation is the final screening construction cost, since each retained alternative has been identified as being capable to serve as the Northwest Interceptor route. Construction cost comparison provides a non-subjective basis to assess the remaining alternatives with respect to the ease of

Table 1	Screening Matrix - Route Analysis Northwest Interceptor					
		Alternatives				
Criteria		1	2	3	4	5
Flexibility		5	4	2	2	2
Environmental Concerns		3	4	5	5	5
Constructability		4	4	4	4	5
Easements/Right-of-Way Acquisition		3	1	2	4	5
Accessibility		4	4	5	5	5
Comparative Cost		2	1	3	4	5
Total		21	18	21	24	27

NOTE: Scoring is on a scale of 1 to 5 (1 = poor, 5 = excellent).

construction, connection to existing and future sewers, and other considerations which differ from one route to the next.

Alignment Selection

Following completion of the routing alternative analysis, horizontal and vertical positioning of the pipeline along the selected route was established based on a number of design considerations. These considerations included hydraulics, topography, existing trunk sewer connections, avoidance of conflicts with major utilities, accommodation of right-of-way and construction constraints.

Major utilities were plotted along the selected interceptor route along with topographic information generated by the aerial photogrammetrist. Major utilities were defined as any utility 15-cm in diameter or larger. This included sewer, water, drainage, electrical, communication, and gas utilities. Based on this information, alternative interceptor alignments within the chosen route were examined and evaluated.

Project Segments

To facilitate better management of the overall design and construction phase effort, the Northwest Interceptor System was divided into 14 project segments. Thirteen of the 14 project segments are interceptor reaches and one is a lift station. The interceptor segments provide the District with moderate sized projects ranging in length from one to two miles. The actual length of each segment varies depending on factors such as:

locations of manholes and junction structures, physical land features (i.e. roads, canals, rail lines, etc), manageable construction distances, and anticipated project phasing. Figure A illustrates the extent of individual project segments based on these criteria. Table 2 lists the 14 project segments along with key technical data and scheduling information. Segments are grouped into Phase I, Phase II, and Phase III projects.

Table 2	Proposed Project Segments for the Northwest Interceptor System Northwest Interceptor Design Report				
Project Segment	Station Start	Station End	Description	Length (m)	Construction Phase
NW1	10+00	117+50	Natomas Pump Station to North of Del Paso Blvd.	3,278	2
NW2	117+50	195+00	North of Del Paso Blvd. to South of Elkhorn Blvd.	2,363	2
NW3	195+00	283+70	South of Elkhorn Blvd. to Natomas East Main Drainage Canal	2,705	2
NW4	283+70	339+00	Natomas East Main Drainage Canal to W. 2nd St.	1,686	3
NW5	339+00	393+72	W. 2nd St. to Cherry Lane	1,668	3
NW6	393+72	471+00	From Cherry Lane to West of 20th St.	2,357	1
NW7	471+00	530+00	West of 20th St. to 28th St.	1,799	1
NW8	530+00	620+00	From 28th St. to West of Sloan Dr.	2,745	1
NW9	620+00	684+00	West of Sloan Dr. to West Side of Union Pacific RR	1,952	1
NW10	684+00	748+00	West Side of Union Pacific RR to East Side of Verner Ave.	2,516	1
NW11	748+00	830+50	East Side of Verner Ave. to Lift Station at Auburn and Van Maren	8,250	1
NW12	--	--	Northwest Interceptor Lift Station	--	1
NW13	830+50	905+00	Lift Station at Auburn and Van Maren to Auburn and Carriage	2,272	1
NW14	905+00	1009+53	Auburn and Carriage to Fair Oaks and Old Auburn	3,188	1

The Phase I projects include those segments (Northwest 6 through 14) that need to be designed and constructed by 1999.

The Phase II projects include those segments (Northwest 1, 2, and 3) that need to be designed and constructed by 2004.

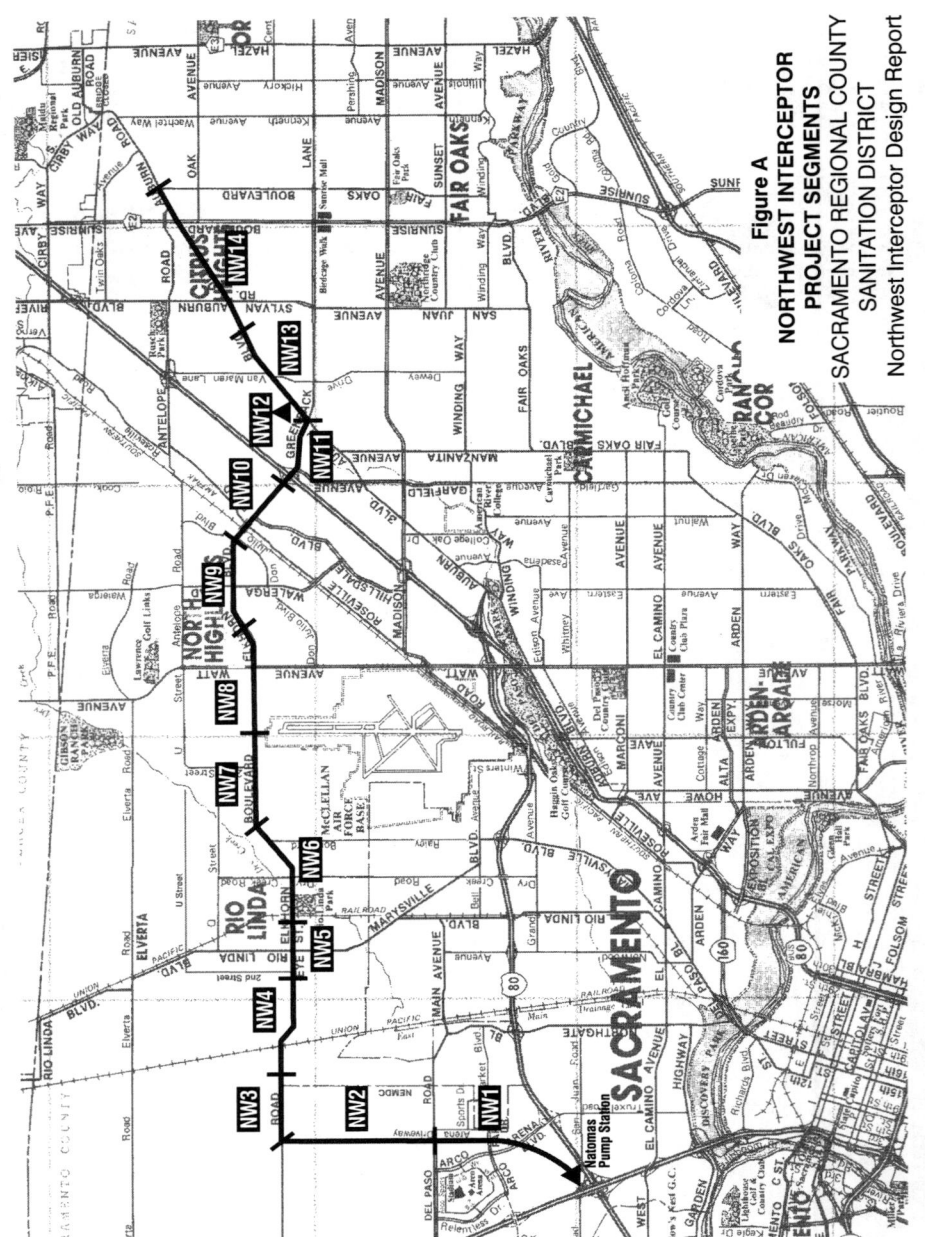

Figure A
NORTHWEST INTERCEPTOR
PROJECT SEGMENTS
SACRAMENTO REGIONAL COUNTY SANITATION DISTRICT
Northwest Interceptor Design Report

The Phase III projects include those segments (Northwest 4 and 5) that need to be designed and constructed by 2009.

Conclusion

On a large project such as this, effective planning at an early stage can save significant amounts of money and ensure project success. Preliminary design is often a stage where too little effort is given. Often, due to the critical nature of most projects, once a project is defined there is a great deal of pressure to get it constructed as soon as possible.

In 1993, a route was recommended for the Northwest Interceptor. Three years later when preliminary design began, this route was reevaluated and based on a new route, the District benefited by a $10 million cost savings. In addition to the cost savings, a detailed planning effort which included 8 different elements insured that this project will be successful and effective when it moves into the construction phase next year.

Using GIS Based Models to Plan Regional Reclaimed Water Pipeline Networks

Scott Lynch, P.E. (Associate)[1]
Co-Authors: Kim Martin and Dave Bramwell

Abstract

This paper is a summary of the first phase of a study initiated by the United States Bureau of Reclamation (USBR), seven Southern California water agencies, and the California Department of Water Resources. The objective of this study, known as the Southern California Comprehensive Water Reclamation and Reuse Study (SCCWRRS), is to identify opportunities and constraints for maximizing water reuse in Southern California using two time horizons, 2010 and 2040. Maximizing water reuse will help to meet the increasing demands while facing limited supplies of fresh water in the future.

Introduction

In 1993, the participating agencies adopted a Plan of Study for the Southern California Comprehensive Water Reclamation and Reuse Study (SCCWRRS) to evaluate the feasibility of a regional water reclamation plan. The Plan of Study called for a three-part, six-year comprehensive effort to identify a regional reclamation system and develop potential capital projects. Major objectives of the project include the following:

- Develop a relational database containing information through the year 2040 on total reclaimed water supply and demand, detailed information on existing and planned Southern California wastewater treatment and water reclamation facilities, and existing and potential reclaimed water demands; including irrigation, landscaping, industry, groundwater basins, surface storage reservoirs, and environmental enhancement projects

- Examine the potential for maximizing reclaimed water use through groundwater recharge projects, indirect potable reuse via existing surface water reservoirs, and environmental enhancement projects.

---

[1] Project Engineer, CH2M HILL, 3 Hutton Centre Dr., Suite 200, Santa Ana, CA 92707

- Develop reconnaissance level reclaimed water system designs and costs.

A Geographic Information Systems (GIS) database was developed from public and private data sources for the Southern California region. The database includes base-map data, transportation and hydrologic features, land use and terrain models, and information on reclaimed water supplies and demands.

The GIS software, ARC/INFO by Environmental Systems Research Institute (ESRI), was used in developing the model. This ARC/INFO based optimization model uses GIS GRID modeling to allocate reclaimed water supplies to identified demands by the most cost effective pipeline route. Cost curves were created and included construction and operation and maintenance costs for treatment plants, pump stations, and pipelines. Pipeline and pumping costs took into consideration elevation, existing land uses, freeways and highways, rivers and water bodies, open terrain, and existing pipelines. Treatment costs included the costs to upgrade all wastewater facilities to full tertiary treatment and additional treatment costs to meet the users' water quality needs.

The model created a comprehensive GIS cost surface using GRID modeling for the study area by applying the cost curves to the underlying GIS database. The model generated a cost surface in order to establish a least cost distribution system in connecting the reclaimed water demands with potential supplies or wastewater treatment plants. The model selected the most cost effective reclaimed water supply, taking into consideration treatment costs, distance to the facility, costs of the associated pipe route, and the water quality needs of the demand. The output of the model is a distribution network showing sizes and costs for each segment of pipe, maps of the pipe network, and total costs (treatment, pipeline, and pumping) for each evaluated alternative. The power of the model is its ability to quickly process large quantities of information and to help in developing solutions for a wide range of conditions.

To optimize reclaimed water use and identify constraints, the current paradigm of reclaimed water use was challenged. Expanded reclaimed water use in groundwater recharge and indirect potable recharge of surface storage reservoirs constitute potentially tremendous opportunities; however, many institutional, regulatory, and public acceptance issues surround these types of projects. These issues were considered in different analyses. For each analysis, the conditions varied under which the water was distributed, from the currently constrained climate to an aggressive climate, reflecting the regulatory requirements and public acceptability regarding the use of reclaimed water.

Defining the Analysis

The SCCWRRS Project focuses on the Southern California area and is comprised of the South Coast and Colorado River hydrologic basins, covering 6 counties. The study area is shown in Figure 1. The Study's database includes the following:

- 117 water reclamation and wastewater treatment plants.

- Approximately 7,400 individual reclaimed water demand points.
- Cataloging of groundwater recharge, surface storage reservoirs, and environmental enhancement opportunities for utilizing reclaimed water.

Figure 1. Project Study Area (Southern California)

The Study attempted to answer the major questions posed during the Study by dividing the alternative analyses several ways. The Study examines two time horizons in seeking to maximize reuse. A planning horizon of 2010 is used in the short-term and 2040 is used for the long-term planning horizon. The Study also examined subregional and regional solutions for the two planning horizons.

In addition to studying the issues of a planning horizon and subregional versus regional optimizations, the regulatory conditions under which the water is distributed was varied; from a highly constrained climate to an aggressive climate for reclaimed water use. These climates reflect regulatory requirements and public acceptability regarding the use of reclaimed water. The following criteria were assumed for the three regulatory conditions:

- Condition 1 (Existing Regulations)   - 20% Groundwater Dilution
                                        - No Indirect Potable Reuse (IPR)

- Condition 2 (Moderate Regulations)   - 50% Groundwater Dilution
                                        - 50% Blend for limited IPR sites

- Condition 3 (Relaxed Regulations)   - 100% Groundwater Dilution
                                        - 50% Blend for all IPR sites

Table 1 shows a summary of the various conditions used in this study. As indicated in Table 1, the regional analyses were built upon the subregional results by connecting areas of surplus supply with demand deficit areas. Note that no analysis was conducted for the 2010 time frame for Condition 3 as this more relaxed level of regulations was foreseen as too aggressive for the near future.

Condition	2010		2040	
	Subreg.	Regional	Subreg.	Regional
Existing Regulations, 20% G/W Dilution, No IPR	X	X	X	X
50% G/W Dilution, 50% Blend on Selected IPR's	X	X	X	X
100% G/W Dilution, All IPR's (50% Blend)			X	X
Analysis Building Blocks	■■■→■■■		■■■→■■■	

Table 1. Conditions for Analysis

The output for each of the these analyses consisted of reclaimed water distribution systems and the associated costs for these systems. All cost estimates generated for the Study included the costs to upgrade existing wastewater treatment plants to produce reclaimed water of suitable quality and to distribute the water to numerous users. The costs did not include already planned expansions or upgrades of existing facilities, nor did they include the costs for constructing, operating, and maintaining the existing facilities. The costs generated for each analysis were used to identify opportunities and constraints for maximizing reuse, and to help identify a list of candidate projects for further evaluation.

Data Input

The model requires two basic categories of data. The first is information pertaining to the supply sources and the potential reclaimed water users. Data collected for the supply sources includes the name of the treatment facility, location, elevation, extent of wastewater treatment (primary, secondary, tertiary, and advanced), capacity and anticipated production, and influent and effluent water qualities as measured in Total Dissolved Solids (TDS). Demand data collected includes the name of the demand, location, elevation, type of user, projected demand (in acre-feet per year), and required water quality in TDS.

The second type of data needed for the optimization model is the base map information stored in GIS. The GIS base map data provides the land uses for the entire study area. Land use data was obtained from the United States Geological Society's (USGS) GIRAS Land Use Data. Roadway, highway, railroad, and hydrography coverages were obtained from the USGS's DLG data and converted to ARC/INFO coverages. Other base map data was collected for displaying and geographic reference such as county lines and other political boundaries.

Role of the Model

The model was developed in Arc Macro Language (AML) and consists of several GIS programs which are used to process, analyze, and display the data. The model uses the Grid, Network, and TIN modules from the ARC/INFO software to analyze and store the data. This analysis process includes creating cost surfaces and performing a least cost routing routine to connect the demands to the supplies. Output information is stored in GIS based coverages and databases. Much of this output data is then used for further analyses using the model.

A GIS based model was chosen as the best way to help in analyzing the vast amounts of data in order to optimize the use of reclaimed water. Figure 2 shows an example display of the pipeline routing model.

Figure 2. Display of Pipeline Distribution Network

Optimization in allocating the reclaimed water in the model is based on connecting users in order of their cost effectiveness; from most cost effective to least cost effective. The model creates a distribution network for each analysis by connecting individual users to their most cost effective supply sources. In addition, demands are allowed to be partially satisfied or served reclaimed water by more than one treatment plant. Unit costs for treatment, pipelines, and pumping were developed for use by the model to determine the optimal route and supply source as a function of total cost to the user. The model also utilizes the established distribution network to connect subsequent demands. Pipelines are increased in size as more reclaimed water is conveyed through the pipe segments to additional users. Grid modeling techniques are used to create these distribution systems and to determine the least cost routes from the users to the supplies.

Cost Surfaces

In order for the model to connect potential users to the supply points using the most cost effective pipeline route, cost surfaces are generated using some built-in procedures of the ARC/INFO software. These cost surfaces are simply GIS GRIDs (raster data) which represent the cost to the user for treating and distributing reclaimed water from a treatment plant to that user. These costs include the treatment, pipeline, and pumping costs that would be required to connect the user to any of the available supply sources in the study area. Each of these component costs is computed in a separate coverage layer and then combined into one final total cost surface. Figure 3 shows a graphical depiction of typical cost layers used when creating the costs for a given user. Both capital and operation and maintenance costs were used in the analysis. Figure 4 shows a sample GIS plot of a final cost surface.

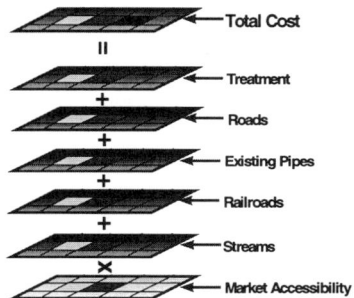
Figure 3. Typical Cost Layers

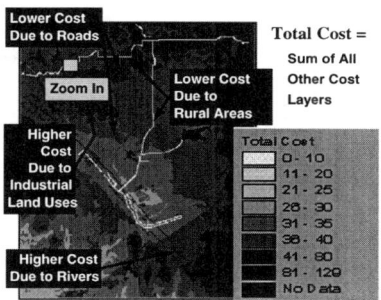
Figure 4. Final Cost Surface

Demands are connected to supplies using a built-in least cost routing routine in the ARC/INFO software. This function determines the cheapest cost from a beginning point to an endpoint which may be one of several possible alternatives. For this model, the beginning point is the actual demand's location, and all the potential supply sources or treatment plants are the potential end points for the least cost route. This least cost routine is best understood when compared to the path a rain drop would take in traveling from the point of origin or the highest point to the lowest point or endpoint. Instead of elevations, the model uses costs to determine how to travel from the beginning point to the endpoint. Figure 5 illustrates this least cost routing process by showing the least cost route a rain drop would take in traversing downward through a cost surface or elevation surface. The model connects all of the users in this manner in generating the most cost effective reclaimed water distribution networks.

Conclusion

The total costs and utilization of reclaimed water for each analysis is computed by the model and is then used to compare and contrast the assumed regulatory conditions for the different planning horizons. By analyzing the study area the under varying conditions, the potential for maximizing reuse was evaluated for the entire study area. Reconnaissance level designs and costs were also generated for potentially feasible future projects. This model has provided an effective way to optimize the use of reclaimed water and has helped to identify constraints in the use of reclaimed water. This model demonstrates the powerful capabilities of GIS in aiding large scale studies or planning projects. Other applications for which the model could be utilized include brine line optimization studies, water and sewer system planning, and pipeline or electrical transmission line routing studies.

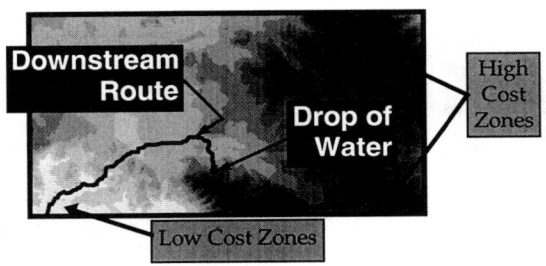

Figure 5 - Demonstration of Least Cost Routing

Acknowledgements

The authors wish to acknowledge the participating member agencies for their roles in this study; United States Bureau of Reclamation, California Department of Water Resources, Central and West Basin Municipal Water Districts, City of San Diego, Los Angeles Department of Water and Power, Metropolitan Water District, San Diego County Water Authority, Santa Ana Watershed Project Authority, and the South Orange County Reclamation Authority. In addition, the authors appreciate the assistance of Don Marske, Scott Goldman (Water 3 Engineering), Mike Savage, Jim McKibben, and Brent Fulgham for their contributions to the Study and this paper, as well as a special thanks to Rick Martin and Kris Mills of the USBR.

References

City of Los Angeles, *Technical Memorandum No. 6c—Cost Estimating Methodology for Preliminary Planning*, Wastewater Program Management Division Clean Water Program Advanced Planning Project, September 18, 1989.

City of San Diego, *Reclaimed Water Distribution System Planning Criteria*, Program Manager for Clean Water Program, September, 1991.

Metropolitan Water District of Southern California, *Cost Guide for the Local Resources Program*, HYA Consulting Engineers, October, 1995.

United State Environmental Protection Agency (US EPA), *Construction Cost for Municipal Wastewater Conveyance Systems 1973-1979*, February, 1982.

United States Geological Society, USGS GIRAS Land Use Data.

United States Geological Society, USGS DLG Data.

United States Bureau of Reclamation (USBR), *Administrative Draft Report—Central California Regional Water Recycling Project Step 1 Feasibility Study*, USBR and Bay Area Water and Wastewater Agencies, June 30, 1995.

ANALYSIS OF SURGE PRESSURES IN THE INLAND FEEDER AND EASTSIDE PIPELINE

R. Scott Foster[1], M. ASCE

ABSTRACT

The Metropolitan Water District of Southern California (MWD) is in the process of designing and constructing the massive Eastside Reservoir and all its appurtenances in the California counties of San Bernardino and Riverside. With all the different modes of operation for a project of this size, there are many possibilities for the creation of surges in the system. There are two large pressure control structures, a 53,690 kW (72,000 HP) pump station, in-line flow control valves, a reservoir inlet/outlet tower, a secondary inlet structure, and approximately 84 km (52 miles) of 3660-mm (12-ft) diameter tunnel and pipeline. In addition, the 71 km (44.2 mile) long section of pipeline and tunnel known as the Inland Feeder crosses numerous earthquake faults.

The interaction of all these facilities has the potential to create damaging surges if their operations are not properly coordinated. The goal of the analysis is to maintain the maximum HGL elevation below the allowable limits as outlined by MWD. The surges generated would be a result of the operation of the numerous valves throughout the system and the operation of the pump station. This paper deals with the transients resulting from the operation of the twelve pumps and their corresponding pump control butterfly valves.

Numerous simulations were run to determine the proper valve characteristics and operation times required to minimize the magnitude of the pressure surges in the system resulting from the operation of the P-1 pumping plant. These values had to be checked against many possible flow conditions and system configurations under that MWD had designed the system to operate.

---

[1] Project Engineer, Flow Science Incorporated, 599 North Fair Oaks Avenue, Pasadena, CA 91103

INTRODUCTION

The Eastside Reservoir Project, currently under construction in Riverside County, California, just outside the town of Hemet, will provide Southern California with water in the event of a water shortage resulting from drought or earthquake. The completed reservoir will have a capacity of around 987,000,000 m^3 (800,000 acre-ft), nearly doubling Southern California's surface water storage capacity. As part of this project, the Metropolitan Water District of Southern California (MWD) is constructing a pump station capable of delivering water from the Eastside Reservoir Forebay (ERF), located at the base of the Eastside Reservoir (ER), to either the ER or to the Colorado River Aqueduct (CRA). The operation of the pumps and their corresponding pump control valves will create pressure surges in the system that if not addressed could drop the pressure low enough to collapse the pipelines or raise the pressure high enough to overpressurize the pipelines.

The project consists of many facilities that are used to deliver water in the flexible manner that is vital to the overall delivery scheme of MWD. This paper addresses the pressure transients that can occur as a result of the operation of the pump station located at the P-1 Pump Station Facility when delivering flow to the ER. Figure 1 gives a schematic view of the system, showing the relative locations of these facilities. Design changes have been made by MWD since these analyses were originally completed. However, the basic concepts and results still hold true.

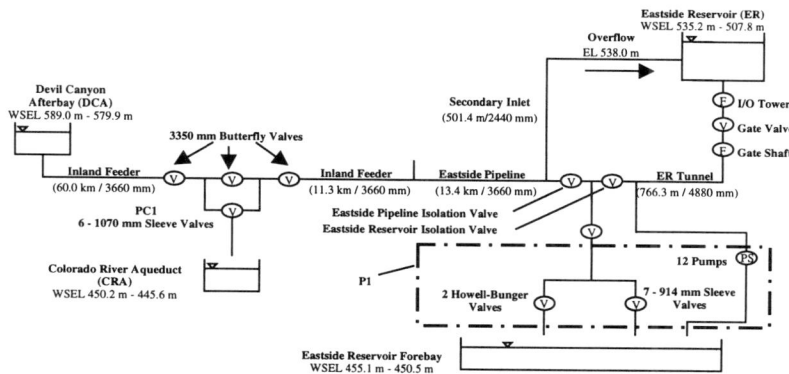

Figure 1 – Schematic of MWD Eastside Reservoir System

P-1 PUMP STATION FACILITY

The P-1 Pump Station Facility contains a pump station, a pressure control structure, and a low-level outlet. They are intended to give maximum flexibility to the entire project by allowing flow to be delivered to or from the ER, the CRA, and the ERF.

The proposed pump station contains twelve vertical turbine pumps, each rated to deliver 4957 m^3/s (175 cfs) at a total dynamic head of 71.6 m (235 ft). The motors are each 4475 kW (6000 HP). The pump/motor combinations have a rated speed of 450 rpm and a total polar moment of inertia, WR^2, of 6250 kg-m^2 (148,300 lb-ft^2). Each pump is controlled by a 1370-mm (54-inch) Pratt butterfly valve. There are no check valves on the pump discharge and the pumps are not ratcheted to prevent reverse rotation. During a power failure event, flow will reverse through the pumps into the forebay until the pump control valves close. The forebay supplies flow to the pumps at a WSEL varying between a high of 455.1 m (1493 ft) and a low of 450.5 m (1478 ft). The pumps can also be used in turbine mode to generate power when delivering flow to the ERF from either the Devil Canyon Afterbay (DCA) via the Inland Feeder (IF) and Eastside Pipeline (EPL) or the ER via the Eastside Reservoir Tunnel (ERT).

Adjacent to the pump station is the P-1 pressure control structure consisting of seven, 914-mm (36-inch) Bailey sleeve valves and two 1980-mm (78-inch) Howell-Bunger valves at the end of the ERT. They can both be used when flow is being delivered to the ERF from either the Devil Canyon Afterbay (DCA) via the Inland Feeder (IF) and Eastside Pipeline (EPL), or the ER via the Eastside Reservoir Tunnel (ERT). Neither the sleeve valves or the Howell-Bunger valves are in use while the pumps are operating.

When the pumps are operating to deliver flow to either the ER or the CRA, the flow is delivered to the ERT where the flow can split. The 3660-mm (12-ft) diameter EPL isolation ball valve needs to be closed to isolate the EPL and the IF from potential transients generated by the P-1 pump station. Another 3660-mm (12-ft) diameter ball valve is used to isolate the ERT when delivering flow to the CRA. For the analysis discussed in this paper, both valves are open.

PC-1 PRESSURE CONTROL FACILITY

The location where water can be delivered to the CRA is the PC-1 Pressure Control Facility (PC1). This facility consists of six, 1070-mm (42-inch) sleeve valves and three, 3350-mm (11-ft) diameter, in-line butterfly valves. During delivery to the ER, flow can be prevented from entering the CRA at this location by any one of three ways. The first is by closing the 3660-mm (12-ft) diameter EPL isolation ball valve previously mentioned. This would prevent any flow from entering the EPL. Another way is to close the six sleeve valves that comprise PC1. The third way, and the method used in this analysis, is by closing the most downstream of the three 3350-mm (11-ft) diameter butterfly valves at PC1.

EASTSIDE RESERVOIR

The 4880-mm (16-ft) diameter, 766 m (2,514-ft) long ERT delivers flow to the ER Inlet/Outlet Tower (I/O Tower) where the flow is delivered into the ER through a series of nine pairs of 1370-mm (54-inch) butterfly valves set at different levels within the I/O Tower. The I/O Tower can be isolated from the ERT by a 4880-mm (16-ft) sluice gate. On the upstream side of the sluice gate is a shaft that allows water to be rejected out into the ER in the event that the sluice gate is dropped. When water is pumped to the ER, the sluice gate is open and at least one pair of butterfly valves in the I/O Tower is open. The ER has a maximum operating WSEL of 537.9 m (1765 ft) and a minimum operating WSEL of 507.8 m (1666 ft).

INLAND FEEDER AND EASTSIDE PIPELINE

The 3660-mm (12-ft) diameter IF connects the DCA to PC1 and the EPL. It consists of a series of alternating tunnels and pipelines for the 60 km (37.2 miles) from the DCA to PC1. Beyond PC1, the IF continues for an additional 11.3 km (7 miles) to the EPL. The EPL is a 13.4 km (8.3 mile) long, 3660-mm (12-ft) diameter, cement mortar lined and coated steel pipeline. For this analysis, only the portion of the IF downstream of PC1 and the entire EPL is of concern.

SECONDARY INLET

Approximately 1300 m (4,260-ft) upstream of the connection of the EPL and ERT is the Secondary Inlet (SI). The SI is a 2440-mm (8-ft) diameter, 501 m (1645 ft) long pipeline that will be used as an alternative to the I/O Tower to deliver flow into the ER. The SI will always be open and will have the ability to intercept transients generated elsewhere in the system and minimize their effects. The SI has an overtopping elevation of 535.2 m (1756 ft).

MWD CONCERNS

MWD has two main concerns with respect to the operation of this system. First, there is the possibility that upon the loss of power to the pumps at the P1 Pump Station, the resulting downsurge will be sufficient to drop the internal pressure low enough that the pipeline lining and coating are damaged or even more disastrous, the pipeline collapses. The second concern is that the ensuing return upsurges will overpressurize the system, again possibly damaging the system.

For the purpose of the analysis, the collapse strength of the IF and EPL were calculated. For a pipe with a diameter of 3660-mm (144-inches), a thickness of 15.9-mm (0.625-inches), and yield strength of 241,290 kPa (35,000 psi), the pipe would be able to withstand a vacuum of 17.24 kPa (2.5 psi) including a factor of safety of 2.0. The equation used to calculate the collapse strength of the pipeline is the Timoshenko Elastic Formula with Eccentricity [1], which is then divided by a factor of safety of 2.0.

It should be noted that even if the pipeline were thick enough to withstand collapse due to the drop in pressure, any type of mortar lining may be damaged by the low pressures, which could over time lead to the corrosive failure of the pipeline where the lining was damaged.

For this analysis, the maximum allowable HGL on the downstream side of PC-1 was 606.5 m (1990 ft) and decreased linearly to the start of the Eastside Pipeline (EPL) where the maximum allowable HGL elevation was 593.2 m (1946.5 ft). From this point the maximum allowable HGL was assumed to be equal to 551.7 m (1810 ft) plus 50 percent of the steady state pressure along the pipeline to the 3,660-mm (12-ft) diameter EPL isolation ball valve. For the P1 Pump Station Facility the maximum allowable HGL is 576.1 m (1890 ft), and decreases linearly to 536.4 m (1760 ft) at the I/O Tower. On the downstream side of the 3660-mm (12-ft) ball valve on the EPL, at the junction of the EPL with the ERT the allowable HGL is approximately 570.0 m (1870 ft). These values were given by MWD for use in calculating the potential for damages resulting from transients in the system and were used in the analysis. They have since been changed.

TRANSIENT ANALYSIS
Pump Control Valve Operation

The first concern with the operation of the system was how fast to close the pump control valves upon the loss of power to the pumps. Initially, a closing time of 40 seconds was proposed by MWD. However, calculations show that for wavespeeds, a, in the EPL and IF of approximately 914 m/s (3000 ft/s), the total round trip wave time from the pump station at P1 to PC1 and back, $2L/a$, is slightly more than 60 seconds. Thus, if the EPL and IF were not isolated by closing the 3660-mm (12-ft) EPL ball valve discussed earlier, closing the pump control valves in 60 seconds or less would result in a maximum pressure rise equivalent to the instantaneous closing of the pump control valves. To prevent this, the closing time of the pump control valves should be increased to a time greater than 60 seconds. The table below shows the predicted HGL elevations at various locations along the system with the pump control valves closing in 40, 60, 90, and 120 seconds after the failure of all 12 pumps operating at a flow rate of 44.5 m^3/s (1572 cfs).

Location within System	Max. HGL for Pump Control Valve Closure Time			
	40 sec	60 sec	90 sec	120 sec
PS	573.6 m (1882 ft)	559.0 m (1834 ft)	551.7 m (1810 ft)	547.1 m (1795 ft)
SI, m (ft)	540.1 m (1772 ft)	537.4 m (1763 ft)	539.2 m (1769 ft)	538.9 m (1768 ft)
Upstream End EPL	579.1 m (1900 ft)	579.1 m (1900 ft)	567.8 m (1863 ft)	566.0 m (1857 ft)
PC1	616.9 m (2024 ft)	616.9 m (2024 ft)	595.0 m (1952 ft)	591.0 m (1939 ft)

The results shown in this table demonstrate the dramatic impact that the closing time of the pump control valves have on the maximum predicted HGL elevations within the system. Figures 2 and 3 show the resulting predicted pressure record at PC1 for the four pump control valve closing times listed in the table above.

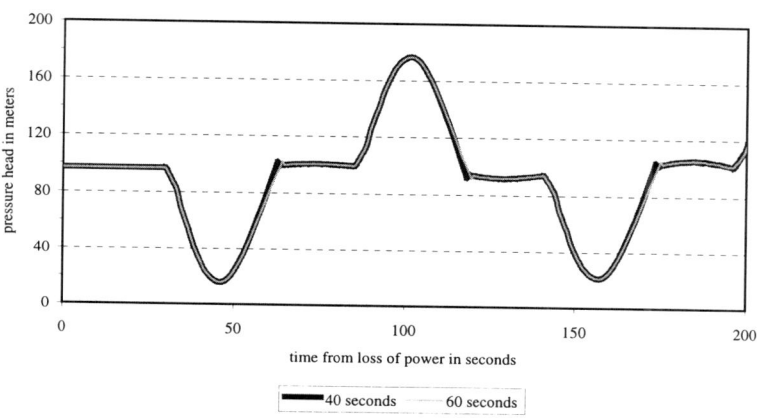

Figure 2 – Pressure record comparison of pump control valve closure times, 40 sec & 60 sec.

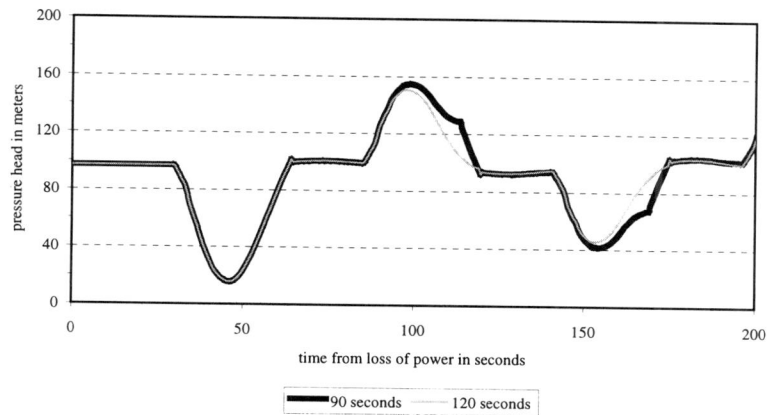

Figure 3 – Pressure record comparison of pump control valve closure times, 90 sec & 120 sec.

Due to the results of this analysis, MWD increased the closing time for the pump control valves to 120 seconds to limit the impact that closing the pump control valves would have on the system.

With the operating times of the pump control valves determined, analysis of the system under differing flow scenarios could now be completed. The operating scenarios examined in the original analysis involving the pump station are listed below. Flow scenario 2 will be discussed in the following text.

1. Pumping to the ER with the EPL isolation valve closed.
2. Pumping to the ER with the EPL isolation valve open.
3. Pumping back up the EPL and IF to the CRA.

Case 2
Pumping to the ER with the EPL isolation valve open.

For this case, with all twelve pumps operating, the pump station is delivering 44,530 m^3/s (1572 cfs) at a speed of 450 rpm and the pump control valve is set to close in 120 seconds upon the loss of power to the pump station. The ERF is at WSEL 450.5 m (1478 ft) and the ER is at WSEL 535.2 m (1756 ft). The 3660-mm (12-ft) ball valve EPL isolation ball valve is open. The opening of this valve allows the transients generated by the pump station to propagate not only up the ERT to the ER, but also up the EPL and IF to PC1. When the pumps lose power, the resulting downsurge wave drops the pressure in the system. The wave splits when it reaches the ERT. Part of the wave travels the short distance to the reservoir where it is repressurized, and the other part travels down the EPL and the Inland Feeder to PC-1, which is closed. At this point, the downsurge, upon reaching the closed PC1 facility, doubles in magnitude and propagates back up the IF and EPL to the pump station. Figure 4 below shows the results of this simulation.

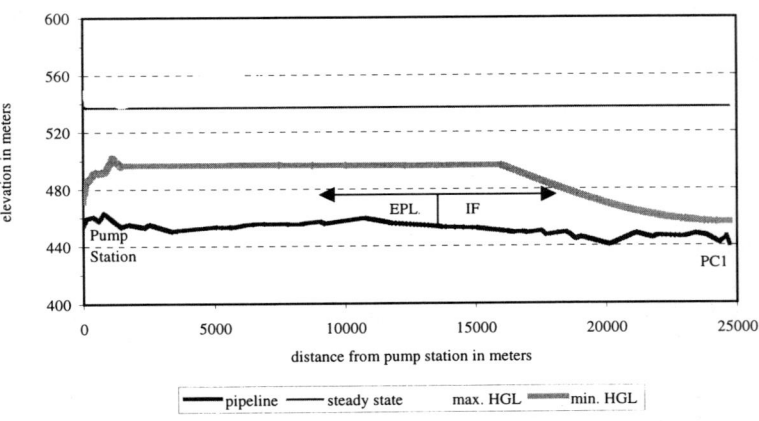

Figure 4 - Predicted HGL elevations along the Eastside Pipeline and Inland Feeder after pump power failure.

In this figure the predicted maximum, minimum, and steady state HGL elevations, are shown with the pipeline profile. It can be seen that the minimum HGL elevation in the IF and EPL is not predicted to drop low enough to result in the opening of vacuum relief valves along these pipelines. While the pump control valves are closing, flow reverses back through the pumps causing them to turbine. When the pump control valves come to a close, they shut off a significant amount of flow, creating an upsurge in the system. This upsurge increases the predicted pressure in the ERT, increasing the WSEL in both the gate shaft and the I/O Tower. Figure 4 shows the effect of this upsurge. Contributing to the magnitude of the downsurge and the upsurge is the SI. When the initial pressure drop wave passes the SI, water begins flowing out into the EPL, helping to limit the magnitude of the drop. When the return upsurge passes the SI, water is forced into the SI, causing it to overtop but limiting the maximum pressure rise. These complex interactions are what create the shape of the HGL elevation plots shown in Figure 4. The predicted pressure record at PC1 for this scenario can be seen in Figure 3, which shows the predicted pressure records at PC1 for this case with a pump control valve closure time of 90 seconds and 120 seconds.

The predicted pressure record as shown in Figure 5 below, shows the magnitude of the initial pressure drop resulting from the loss of power to the pump station. The closure of the pump control valves is also evident on this plot.

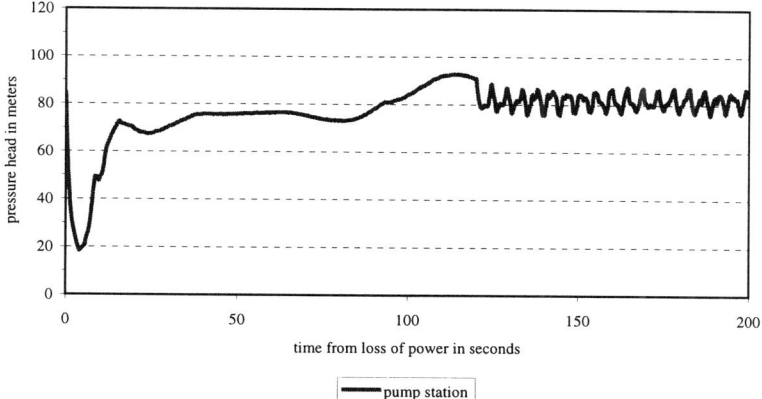

Figure 5 – Pressure record at the pump station following pump power failure.

It was mentioned above that the SI plays a significant role in the action of the pressure surges in the system. The SI acts as both a reflection point and a relief point for the surges. During steady state operations, the WSEL in the SI is only 2 ft below its overflow elevation. Any pressure rise in the system will result in the overflowing of the SI. Figure 6 shows the pressure record at the SI during the simulation. This figure shows the pressure drop wave and the corresponding rise in pressure that causes the SI to overflow into the ER.

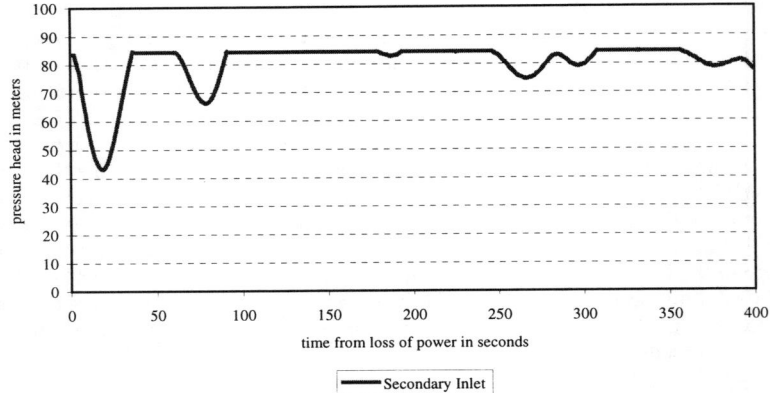

Figure 6 – Pressure record at the Secondary Inlet (SI) for pump power failure.

As was seen in the previous figures, the SI does not have the ability to limit the maximum HGL elevation in the EPL and the IF to 538.0 m (1765 ft), the overtopping elevation of the SI_t.

CONCLUSIONS

The transient analysis shows the effects that seemingly small changes in the timing of pump control valves can have on the pressures that can be created as a result of their operation. As was shown, the timing of the pump control valve is vital to ensuring that the system is not subjected to pressures that would raise the maximum predicted HGL elevation above the allowable pipeline pressure. In this case, the pump control valves should not be operated any faster than 120 seconds to avoid such an occurrence when delivering flow to the Eastside Reservoir.

Also of importance is the understanding of the effects that appurtenances can have on the transients within a pipeline system, such as the effect the Secondary Inlet has on the surges in the Eastside Pipeline and the Inland Feeder. In this scenario, the Secondary Inlet helps to limit the magnitude of the upsurge and the downsurge in the system.

Reference:

[1]   Roscoe Moss Company 1990 *Handbook of Groundwater Development*, John Wiley & Sons, New York, 183p.

AVOIDING COMMON THRUST RESTRAINT MISTAKES

Andrew E. Romer, PE[1]
Member, ASCE

Abstract

Unbalanced forces resulting from angular changes in pressure pipelines must be resolved through either restrained joints or externally applied reactions. Design equations for thrust restraint presented in references published by AWWA and others are not universally applicable. Misunderstanding or misapplication of those equations must be avoided. A check of the statics utilized for any angular change in pressure pipelines must indicate that each vector represent an actual applied force or reaction.

Introduction

The longitudinal force in any cross section of a pressure pipeline is the product of the pressure P, and the cross sectional area A, of the pipeline; PA. The vector sum of PA in each leg at a geometric discontinuity (such as an elbow) in a pressure pipe is easily calculated from statics. **Figure 1** shows a typical elbow of deflection angle Δ in a pipeline with unrestrained joints. The vector sum, T of the longitudinal force PA in each leg acting on the elbow of deflection angle Δ is equal to

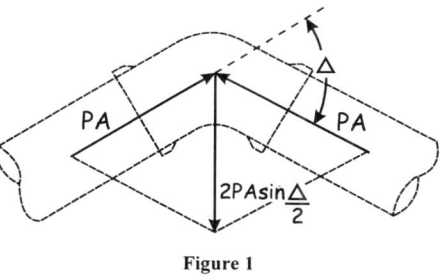

Figure 1

$$T = 2 \, PA \sin \Delta/2. \qquad [\text{Equation 1}]$$

When an externally applied reaction of magnitude T is applied at the elbow colinear with the angle bisector, the elbow remains in static equilibrium. The maximum value, and thus the design value of PA is at the joint closest to the

---

[1] Senior Engineer, Boyle Engineering Corporation, 1501 Quail Street, Newport Beach, California 92658-9020. Telephone 949-476-3546, Facsimile 949-721-7482.

discontinuity. The most common means of application of that external reaction is with a thrust block.

Another means of providing the resisting force is a drag force directed axially along each length of pipe away from the geometric discontinuity. This is accomplished by means of longitudinally restraining the pipe joints a sufficient length L away from the geometric discontinuity to mobilize the friction between the pipe and soil. The magnitude of force which must be dissipated through pipe-soil friction when using restrained joints is the product of the pressure and the cross sectional area of the pipe: PA. This is illustrated in **Figure 2**.

$$PA = PA \qquad [\text{Equation 2}]$$

The force is independent of the deflection angle, Δ. The force PA is transmitted through the shell of the pipe in tension. Because the resisting axial drag forces are colinear with the imposed forces, the sum of the forces and moments is zero, and static equilibrium exists. This is also the case at bulkheads, wyes, tees and bifurcations.

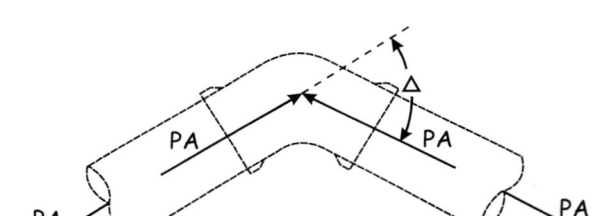

Figure 2

The magnitude of force PA is also independent of the length L available to mobilize the friction between the pipe and soil. That is, the reaction may be provided, much as a Coke™ bottle bottom provides the reaction opposite the Coke™ bottle top, provided that there is physical continuity of the pipe shell between the discontinuities.

PA (1 - cos Δ) Method

Some references including AWWA Manual M-11[2] indicate that the magnitude of force which must be dissipated through pipe-soil friction at each side of an elbow is PA (1 - cos Δ). Only one leg of the pipeline needs to be restrained away from the elbow in this case. The vector sum PA (1 - cos Δ) is the force which must be restrained only when the pipe on one side of the elbow has unrestrained joints, and an external restraint, such as a thrust block, is applied to the elbow centroid perpendicular to the restrained joint pipe. This is illustrated in **Figure 3**.

---

[2] Steel Pipe-A Guide for Design and Installation, AWWA Manual M-11, Second Edition, 1987, p. 148.

The magnitude of force which determines the sizing of the thrust block, in this case only, is PA sin (Δ). The most common application of this is the junction of a pipeline with restrained joints and a pipeline with unrestrained joints. If supplemental joint restraint cannot be added to the existing unrestrained-joint pipeline, such as welding the joints, then this is a practical method.

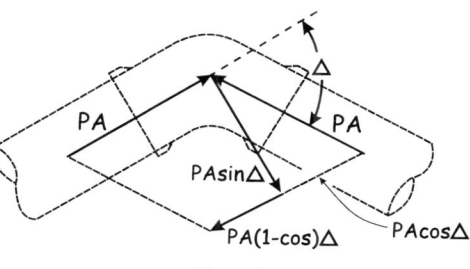

Figure 3

It is not practical to assume that the force PA sin (Δ) will be resisted by the passive resistance of the soil, except in very favorable soil conditions and very small diameter pipelines. If both legs away from an elbow are provided with restrained joints, then no thrust block is used, and the magnitude of the force to be restrained in each leg is PA, [Equation 2] as discussed previously.

PA sin (Δ/2) Method

Some references including AWWA Manual M-9[3] indicate that the magnitude of force which must be dissipated through pipe-soil friction at each side of an elbow is PA sin (Δ/2). That method, illustrated in **Figure 4**, is based upon the assumption that

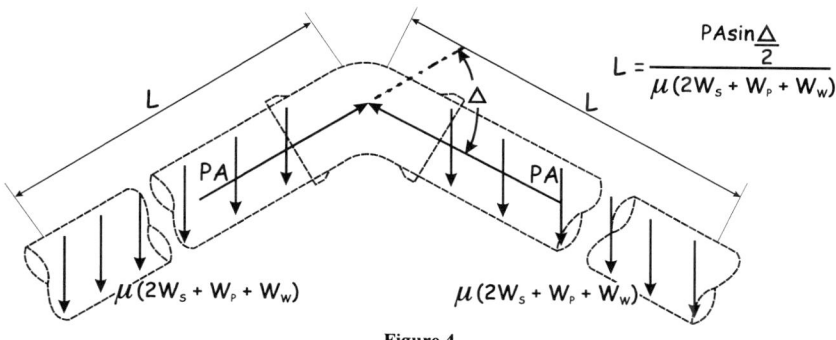

Figure 4

the resisting forces are uniformly mobilized along each pipe leg and directed parallel with the angle bisector. The magnitude of the assumed distributed force is one-half of the resultant reaction which would need to be applied at the elbow colinear with the

---

[3] Concrete Pressure Pipe, AWWA Manual M-9, Second Edition, 1995, p.125.

angle bisector. In other words, it is assumed that one half of the force which would be utilized to determine sizing of a thrust block [Equation 1] is redistributed along the pipe length. This cannot be considered a reasonable assumption. Recalling statics: Only moments are free vectors. Thus the vector sum which is appropriate for sizing an external reaction (see thrust block discussion above) cannot be arbitrarily redistributed. A check of the statics with any restrained joint length will show that the proposed distributed reactions would induce significant shears and bending moments in the pipeline.

Combination of Friction and Passive Resistance

Thrust blocks or anchors are commonly and economically sized on the basis of mobilizing the passive resistance of the soil. Passive resistance will be developed only upon the movement of the pipe towards the soil. Thus flexibility must be provided (such as a rubber-gasket jointed elbow) so that the movement does not impose shear and bending forces on the pipeline. Static friction is obviously exceeded before the passive friction is mobilized.

Thrust restraint methods have been proposed, principally by manufacturer's trade associations, to include allowance for passive resistance mobilized along each pipe leg and directed parallel with the angle bisector. These methods are based upon the assumption that the joint restraint allows some axial movement and rotation prior to engaging the axial force.

The joint restraint utilized may indeed allow some axial movement and rotation prior to engaging the axial force, but this should not be the basis for structural stability of a pipeline. Consideration of the potential for trench excavation parallel to any pipeline should be enough caution to avoid this method. As with the PA sin ($\Delta/2$) method discussed earlier, a check of the statics assumed along any restrained joint length will show that the proposed distributed reactions would induce significant shears and bending moments in the pipeline. It is recommended that these difficulties be avoided by utilizing the axial drag force method. Should there be concern that the joint restraint specified will allow axial movement (and rotation) prior to engaging the axial force, a means should be developed to fully engage the tensile capacity of the joint prior to pressurizing the pipeline.

Estimating Pipe-Soil Friction

The magnitude of the longitudinal resisting force F, available through pipe-soil friction is the product of the coefficient of friction, μ, and the weight of the soil Ws, the net weight of the pipe Wp, and the weight of the water Ww. In general,

$$F = \mu(Ws + Wp + Ww). \quad [\text{ Equation 3 }]$$

The pipeline will be stable when:

$F \geq PA$.   [Equation 4]

Some references[4] indicate that the friction can be mobilized on both the top and bottom of the pipe. The effective resistance of friction on both the top and bottom of the pipe can only be utilized if the frictional resistance between the column of soil above the pipe and the adjacent undisturbed earth can be demonstrated to be in excess of the frictional resistance mobilized between the pipe and the trench backfill.

The block of soil above the pipe will be stable if the at-rest pressure of the soil multiplied by the tangent of the soil internal friction angle is greater than the frictional resistance offered by the soil. This is illustrated in **Figure 5**. For the soil block to be stable:

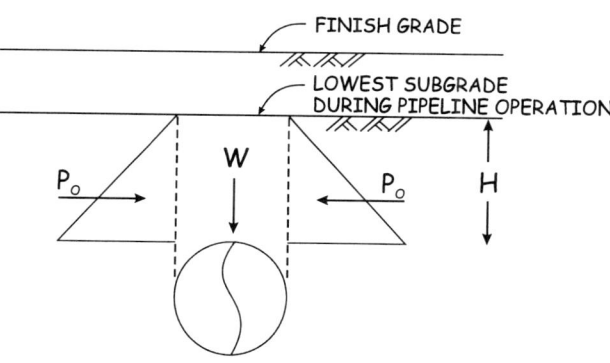

Figure 5

$2 P_o \tan \phi \geq W \mu$ [Equation 5]

Where: $P_o = (\gamma / 2) H^2 k_o$
γ = unit soil weight; buoyant if groundwater is above pipe,
H = minimum depth to top of pipe,
k_o = coefficient of passive soil pressure at rest,
W = weight of soil prism above pipe,
μ = coefficient of friction between pipe and backfill,

When this is satisfied, the pipeline is deep enough to keep the soil above from moving with the pipe. In that case and only then can friction be assumed to be effective at the top of the pipe, and then is:

$F = \mu(2Ws + Wp + Ww)$.   [Equation 3A]

---

[4] ibid, p. 126.

It should also be noted that because the soil prism is assumed to be fully mobilized to resist sliding, that no consideration can be given to mobilization of any passive soil pressure against the pipe bells.

Summary

Unbalanced forces resulting from angular changes in small diameter pressure pipelines can be economically resolved through use of externally applied reactions provided by thrust blocks. When sizing thrust blocks on the basis of passive resistance the land which provides the passive resistance is usually encumbered to ensure the permanent stability of the pipeline. For very large pipelines, the size of the thrust blocks and the encumbered land become economically unfeasible. The most economical and straightforward means of providing the resisting force is a drag force directed axially along each length of pipe away from the geometric discontinuity. That force is transmitted by axial restraint of the pipe joints.

The magnitude of force which must be dissipated through pipe-soil friction when using restrained joints is the product of the pressure and the cross sectional area of the pipe: PA. The longitudinal force PA resisted by pipe-soil friction is the product of the coefficient of friction, μ, and the weight of the soil W_s, the net weight of the pipe W_p, and the weight of the water W_w. When the pipeline is buried deep enough to keep the soil column above the pipe from moving under the influence of the above forces, significant economy can be achieved in reduction of the length of pipeline joints which must be welded.

Other means of distributing the restraining forces, although not readily apparent when applied to small diameter pipelines, may lead to failures of large diameter and high pressure pipelines. Application of statics to each force discontinuity, rather than cookbook solutions, will result in elimination of thrust restraint mistakes.

SURGE PROTECTION DESIGN FOR THE CITY AND COUNTY OF SAN FRANCISCO WATER TRANSMISSION SYSTEM

R. Scott Foster[1], M. ASCE

ABSTRACT

The City and County of San Francisco were concerned about the ability of their water transmission system's ability to withstand pressure surges resulting from the unplanned operation of their system flow control valves and their two primary pump stations, Lake Merced and Baden. Flow Science Incorporated was hired to determine the extent of the surges in the system and recommend alternatives to keep the maximum hydraulic gradelines within the system below 130 percent of the maximum static pressure within the system.

The transmission system consists of two main pump stations, (Lake Merced and Baden), the Tracy water treatment plant, flow control valves R60, T60, R20, and T20, and is constructed of many different pipeline materials of varying diameters. The system was constructed from around 1900 through to the 1980's and is designed to provide potable water to the City and County of San Francisco. The maximum total flows into the system can run as high as approximately 6308 l/s (100,000 gpm). There are five storage reservoirs to which the water from the pump stations and the WTP flow. The system is divided into two pressure zones: (1) the low pressure zone, which is served directly by the Upper Crystal Springs Reservoir and is located upstream of the two pump stations, and (2) the high pressure zone, which is served by the two pump stations and the Tracy Water Treatment Plant.

The analysis showed that the loss of power to the two pump stations created surges in the system that raised the maximum HGL elevations within some of the main transmission pipelines well in excess of 130 percent of the maximum static pressure that was used as the upper limit for allowable pressures. The existing surge protection that was installed at the Lake Merced pump station did little to lower these pressures due to the configuration of the surge tank connection. It did however change the problem from a waterhammer problem to a surge problem.

---

[1] Project Engineer, Flow Science Incorporated, 599 North Fair Oaks Avenue, Pasadena, CA 91103

Protection was to be designed to prevent the maximum HGL elevations from exceeding the 130 percent of maximum static pressure threshold. However, it was determined through a series of simulations with various protection measures that this would not be feasible. The best solution would be to limit the pressures as close as possible to the allowable level. The surge protection alternative that was determined to provide the most protection to the system was the installation of a 28.3 m^3 (1000 ft^3) surge tank on the discharge side of the Lake Merced pump station and a 610-mm (24-inch) surge relief valve on the suction side of the Lake Merced pump station. The surge relief valve is designed to work in combination with the existing surge tank at this location. At the Baden pump station a 127.4 m^3 (4500 ft^3) surge tank was required on the discharge side of the pump station and a 42.5 m^3 (1500 ft^3) surge tank was required on the suction side of the surge tank.

The installation of the surge protection measures described above were predicted to provide the system with protection that would keep the maximum pressures as near to the 130 percent of the maximum static pressures for the system as possible.

INTRODUCTION

The City and County of San Francisco's water transmission system is the primary source of potable water for the San Francisco region. The City and County of San Francisco Department of Public Works became concerned that in the event of a loss of power to the primary pump stations, Lake Merced and Baden, the resulting hydraulic transients would severely damage the pipelines in the system. This would put a great strain on their ability to deliver water to the city.

Analyses were performed for specific flow conditions that were most likely to be experienced during the operation of the system. These included (1) the maximum pumping rate scenario of three pumps operating at Lake Merced Pump Station and two pumps operating at Baden Pump Station, (2) the normal maximum pumping condition of three pumps operating at Lake Merced and Baden pump station idle, and (3) three gravity flow conditions with Baden not operating. This paper describes the results of the analysis for the worst case operating scenario of three pumps at Lake Merced and two pumps at Baden pump stations operating upon a system wide loss of power.

SYSTEM DESCRIPTION

The transmission system considered in this analysis consists of two main pump stations, Lake Merced and Baden, a water treatment plant, Tracy, flow control valves, R60, T60, R20, T20, and is constructed of many different pipeline materials of varying diameters. The system was constructed from around 1900 through to the 1980's and is designed to provide potable water to the City and County of San Francisco. The maximum total flows into the system can run as high as approximately 6308 l/s (100,000 gpm). There are five storage reservoirs to which the water from the pump stations and the WTP flow. The water transmission system, shown in the schematic Figure 1, is divided into two pressure zones: (1) the low pressure zone, which is served directly by the Crystal Springs Bypass Tunnel (CSBT) and is located upstream of the two pump stations, and (2) the high pressure zone, which is served by the pump stations and the Tracy Water Treatment Plant.

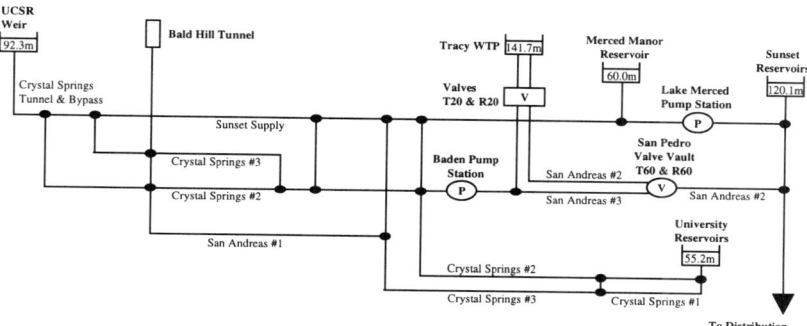

Figure 1 – Schematic of water transmission system

The low pressure zone is supplied from a weir with a crest elevation of 92.3 m (303 ft) at the Upper Crystal Springs Reservoir (UCSR). The high pressure zone can be served either by the Baden Pump Station (BPS) and the Lake Merced Pump Station (LMPS) that are supplied by the low pressure zone, and/or the Tracy WTP from a reservoir at WSEL 141.7 m (465 ft). In this paper, the high pressure zone is supplied by the two pump stations and the Tracy WTP is isolated from the system.

The Baden Pump Station (BPS) consists of two 746 kW (1000 hp) pumps and one 522 kW (700 hp) pump that was assumed to be on standby. The large pumps are rated to operate at 1185 rpm with an 88 percent efficiency for a flow rate of 852 l/s (13,500 gpm) at 68.6 m (225 ft) TDH. The total polar moment of inertia (WR^2) of the pump/motor combinations is 40.8 kg-m^2 (969 lb-ft^2). Each pump has a 510-mm (20-inch) ball pump control valve on the discharge set to close upon loss of power to the pumps in 6.6 seconds. There are no existing surge protection devices present at the pump station. The two large pumps were assumed to be operating.

The Lake Merced Pump Station (LMPS) is located adjacent to Lake Merced and contains four different sizes of pumps. There are two 149 kW (200 hp), one 373 kW (500 hp), two 522 kW (700 hp), and three 746 kW (1000 hp) pumps existing at the station. For the analysis, the 746 kW (1000 hp) pumps are of interest. The other pumps supply other systems and will have minimal effect on the main transmission system. The 746 kW (1000 hp) pumps deliver flow from the low pressure system and discharge the flow into the high pressure zone Sunset Reservoirs for delivery to the city. Each of these pumps is rated to deliver 725 l/s (11,500 gpm) at a TDH of 79.2 m (260 ft) and an efficiency of 89 percent and operate at 1200 rpm. The combined total WR^2 of the pump/motor combinations is 88.7 kg-m^2 (2110 lb-ft^2). Downstream of each of these pumps is a 610-mm (24-inch) swing-type check valve.

On the suction side of the LMPS is an existing surge tank. The tank is 26.2 m (86 ft) long, 2.7 m (9 ft) in diameter, and is sitting at a 24.5 degree angle from horizontal on a hill adjacent to the pump station. The surge tank is connected to the suction pipeline by a 76.2 m (250 ft) long, 1220-mm (48-inch) diameter pipeline. Based on information provided on the surge tank, when all three 746 kW (1000 hp) pumps are operating, the water level is in the 1220-mm (48-inch) connection pipeline. Therefore, the connecting pipeline is assumed to be part of the surge tank. Based on this, the total volume of water in the tank under steady state conditions with three pumps operating is 47.6 m^3 (1679 ft^3), or 19 percent of the total volume of the tank.

The pipelines throughout the distribution system vary in size from the 760-mm (30-inch) diameter Baden-Merced Pipeline to the 2900-mm (114-inch) diameter Crystal Springs Tunnel. The pipe materials are generally steel or prestressed concrete with some riveted steel segments in the older lines. Some of the steel is protected by a cement mortar lining and all the steel pipelines are of varying thickness and class. For the design of the surge protection for the system, the maximum permissible HGL elevation was set to provide a pressure limit that was 30 percent above the static pressure for the pipelines in the system. The following static HGL elevations were used as the basis for the static pressure for the listed locations. The pipelines may be able to withstand these pressures, but for the purposes of this analysis exceedence of this limit is regarded to be indicative of potential problems.

EL 92.3m - all pipelines in the low pressure zone (those on the upstream side of the pump stations).
EL 120.1m - all high pressure zone pipelines downstream of the LMPS and the SPVL.
EL 141.7m - all high pressure zone pipelines downstream of the Tracy WTP including the Sunset Branch upstream of the Capuchino valves.
EL 60.0m - the Baden Merced pipeline.

There are numerous flow scenarios under which the system can operate. This paper addresses both Baden and Lake Merced pump stations operating at full capacity with the valves R20 and T20 closed, thus allowing no flow contribution from the Tracy WTP.

Under steady state conditions for this flow scenario, the total computed flow rate from the two operating pumps at the BPS is 2252 l/s (35690 gpm) at a TDH of 57.3 m (188 ft). The three pumps at the LMPS will discharge a combined 2799 l/s (44,380 gpm) at a TDH of 69.2 m (227 ft). The total flow rate into the system at the upstream weir is 6308 l/s (100,000 gpm).

TRANSIENT ANALYSIS

Upon the simultaneous loss of power to the pump stations, an upsurge is created on the upstream side of both pump stations. The upsurge generated by the BPS travels out from the pump station as a waterhammer wave to the Crystal Springs Pipeline No. 2 (CS2). The wave splits numerous times as it travels upstream from the BPS. At the BPS, the predicted pressure records presented in Figure 2 show oscillations that slowly attenuate on both sides of the pump station.

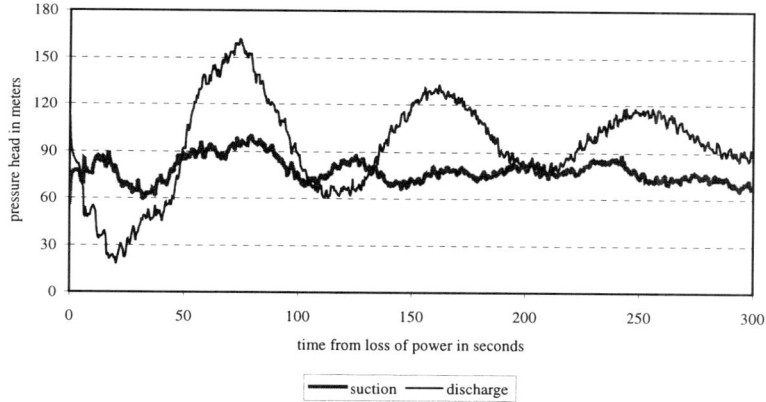

Figure 2 – Pressure records at the BPS following simultaneous loss of power.

After the initial downsurge, the return upsurge on the discharge side will reach a peak at around 70 seconds after the loss of power. On the suction side, the upsurge peaks at approximately the same time as the discharge side. The closing of the pump ball control valves in 6.6 seconds contributes to the magnitude of the upsurge on the discharge side of the BPS. These valves close while flow is still being delivered through the pumps, creating surges of greater magnitude than if check valves were installed. Check valves are predicted to close 52 seconds after loss of power, thus allowing much more flow through the pumps and lowering the magnitude of the surges on both the upstream and downstream sides of the BPS. Due to the numerous modes of operation in which the system could be operated, the timing of the ball valves cannot be adjusted for every case. Check valves at this location would close at such times to minimize potential pressure swings under all the possible flow scenarios and therefore, should be used here with the existing ball valves.

The initial rise in the pressure on the suction side of the LMPS is a consequence of the waterhammer wave generated by the failure of the BPS. This initial rise in pressure is then followed by the upsurge created by the inertial flow in the Sunset

Supply Pipeline (SS) that is still traveling toward the LMPS. This is shown graphically in Figures 3 and 4.

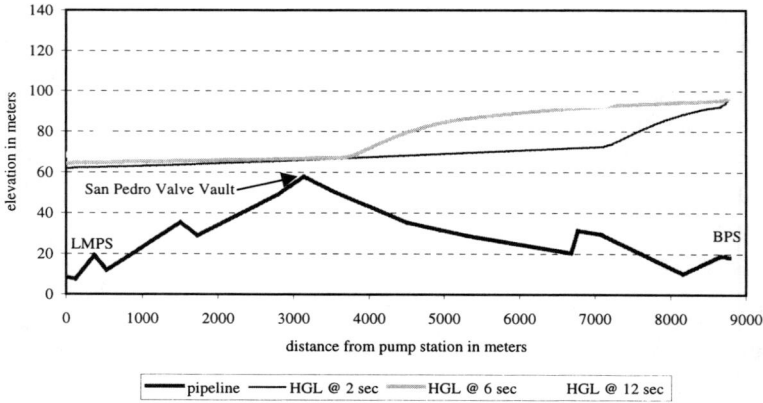

Figure 3 – Plot of HGL elevation along pipeline between the two pump stations following simultaneous loss of power at t = 2 sec, 6 sec, and 12 sec.

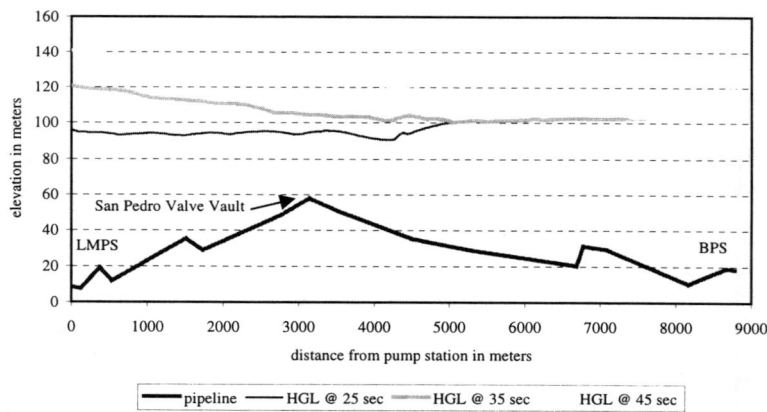

Figure 4 – Plot of HGL elevation along pipeline between the two pump stations following simultaneous loss of power at t = 25 sec, 35 sec, and 45 sec.

Figure 3 shows the initial waterhammer wave propagating from the BPS towards the LMPS at 2, 6, and 12 seconds after the loss of power. Figure 4 shows the increase in the pressure at the LMPS due to the inertial flow that is in the SS traveling toward the LMPS. The upsurge resulting from the inertial flow is what causes the excessive pressures on the suction side of the LMPS. The existing surge tank provides some protection to the low pressure system but is not adequate to keep the maximum predicted HGL elevations below the 130 percent of the static pressure level. The plot of the pressure vs. time records for both the suction and discharge sides of the LMPS are shown in Figure 5.

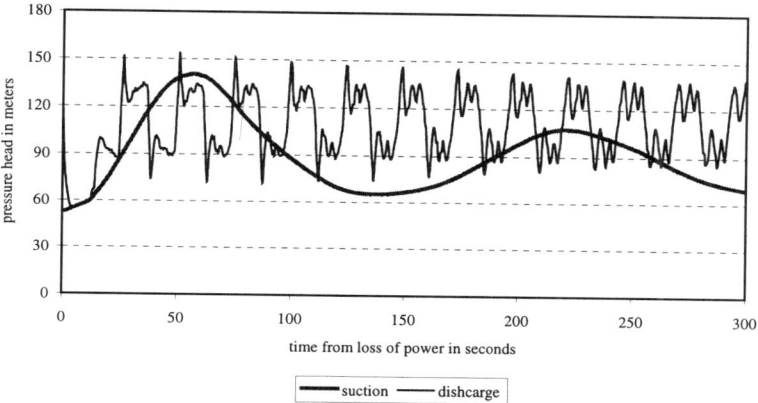

Figure 5 – Pressure records at the LMPS following simultaneous loss of power.

This figure shows the effect the surge tank has on the suction side pressure. Although it shows the typical pressure oscillation shape created by a surge tank, the pressure is allowed to get higher than the 130 percent of static criterion level discussed earlier.

Meanwhile, the downsurge created on the discharge side of the LMPS by the loss of power initially drops the pressure in the section of the SS between the LMPS and the Sunset Reservoirs. When the downsurge reaches the reservoirs, it is reflected as a repressurization wave, and travels back to the LMPS where it closes the check valves, compounding the upsurge. As a result of all these factors, the HGL elevation in much of the SS is predicted to be above the 130 percent of static pressure level.

On the discharge side of the LMPS, the maximum pressure is predicted to be approximately 152.4 m (500 ft). The surge waves slowly attenuate on the discharge side but the peaks remain above 137.2 m (450 ft), even 300 seconds after the loss of power. It should be noted that in Figure 3 where the suction pressure head exceeds that of the discharge pressure head, the check valves at the pump station will open allowing flow through from the suction to the discharge side of the pumps, until the gate valves downstream of the check valves have closed. In this analysis, it is assumed that once the check valves close, they remain closed. Thus flow was not allowed to continue through the pumps after loss of power even when the HGL on the suction side of the pump station exceeded the HGL on the discharge side of the pump station. This allowed for a worst case scenario. In the model, the closing of the gate valves would have no effect on the analysis, provided the check valves close before the gate valves.

The predicted maximum HGL elevations along most of the SS exceed 130 percent of the static pressure elevation, both upstream and downstream of the LMPS. At the downstream end of the Crystal Springs Bypass Tunnel, the minimum pressure is predicted to drop below atmospheric to a minimum of -4.9 m (-16 ft). This low pressure head could result in damage to the tunnel lining. Additionally, the maximum pressure at this location is predicted to exceed the allowable. Numerous other locations within the system were predicted to exceed the 130 percent of static criterion, drop to vapor pressure, or be near vapor pressure. The possibility of damage to the pipeline either from the high pressures or the low negative pressures warranted the installation of surge protection in an attempt to minimize the potential for damage.

RECOMMENDED SURGE PROTECTION MEASURES

Due to the complexity of the system, no practical way could be found to limit the predicted surges throughout the entire system to less than the 130 percent of the static pressure criterion for all possible operating scenarios. There is also the consideration of economic feasibility. In some cases, vastly increasing the size and scope of surge protection measures for the system provides relatively little additional protection for the cost of the added protection.

Baden Pump Station

In order to adequately protect the low and high pressure systems directly affected by the operation of the BPS, surge tanks should be installed on both sides of the pump station in addition to the installation of check valves on the discharge of the pump station.

In the discussion of the effects of loss of power to the pump stations with no additional surge protection, it was pointed out that the closing of the pump ball control valves were compounding problems within the system under certain flow conditions. From the series of computations, it was concluded that the installation of check valves upstream of each ball control valve is required to help in the protection of the system from high HGL elevations.

When the pumps lose power, the ensuing downsurge on the discharge side of the pump station is reflected and returns to the pump station, creating an upsurge. The installation of a 127.4 m^3 (4500 ft^3) surge tank, such as a 3.7 m (12 ft) diameter by 12.2 m (40 ft) long tank, is required to keep the maximum predicted HGL elevations in the high pressure zone below the 130 percent of static pressure level. The surge tank should be connected to the discharge manifold with a connecting pipeline that is equivalent to an 460-mm (18-inch) orifice producing losses of 10 velocity heads for flow both in and out of the tank. The tank should contain 55 percent air under steady state flow conditions.

On the suction side of the pump station, another surge tank is required to keep the maximum HGL elevation within the sections of the low pressure system from exceeding the 130 percent criterion. A 42.5 m^3 (1500 ft^3) surge tank containing 80 percent air under steady state conditions was modeled. This tank should be connected to the suction manifold by a connecting pipeline with losses equivalent to a 380-mm (15-inch) orifice producing losses of 10 velocity heads for flow both in and out of the tank. Figure 6 shows the pressure record at the pump station with the two surge tanks installed.

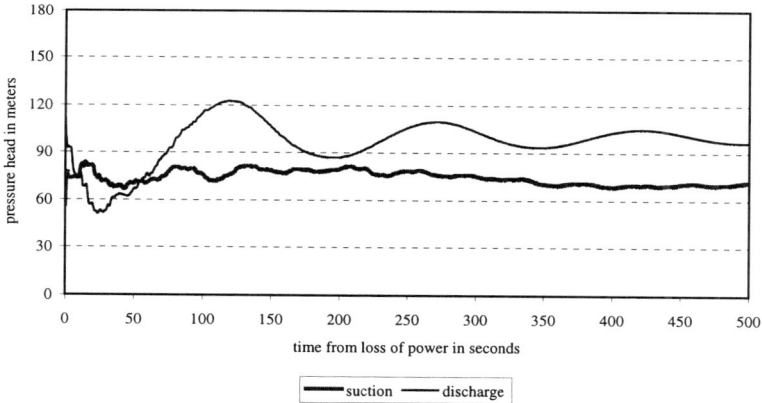

Figure 6 – Pressure records at the BPS following simultaneous loss of power with recommended surge protection installed.

Lake Merced Pump Station

As previously discussed, a considerable section of the SS on the upstream side of the LMPS is predicted to have maximum HGL elevations well in excess of the 130 percent of static pressure criterion. The existing surge tank appears to do little to fully mitigate the surges in this pipeline. In order to prevent the maximum HGL elevation from exceeding the 130 percent of static pressure criterion on the upstream side of the LMPS, a 610-mm (24-inch) surge relief valve set to open at an HGL elevation of 97.5 m (320 ft) in combination with the existing surge tank

To lower the pressure on the discharge side of the pump station below the 130 percent of static pressure criterion, a 2.4 m (8 ft) diameter by 6.1 m (20 ft) long, or equivalent volume (28.3 m^3), pressurized surge tank should be connected to the discharge manifold of the LMPS with a connecting pipeline that is equivalent to a 380-

mm (15-inch) diameter orifice producing losses of 10 velocity heads for flow both in and out of the tank. Under steady state conditions, the tank should contain 55 percent air. Figure 7 shows the predicted pressure record at the LMPS with the surge relief valve and discharge surge tank installed.

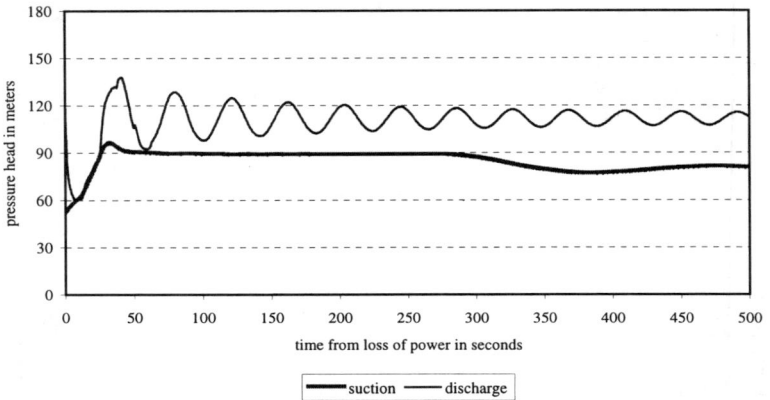

Figure 7 – Pressure records at the LMPS upon simultaneous loss of power with recommended surge protection installed.

This recommended surge protection is predicted to limit the HGL elevations along almost the entire SS to below the 130 percent of static pressure level with the primary exception being at the Crystal Springs Bypass Tunnel due to the low static head it is under.

CONCLUSIONS

The hydraulic transient analysis of the City and County of San Francisco's potable water system shows that without properly designed surge protection, large sections of the system may be subjected to pressures that may be sufficiently high to cause damage to the pipelines. The complexity of the system further complicates the issue due to the numerous possible methods of operation. The uncertainty of the conditions of some of the pipelines in the system add to the difficulty in determining the ability of the system to withstand high pressures.

The analysis performed made assumptions that were conservative in nature to assure that under the worst possible scenario, the system would be afforded the most protection. The analysis shows how transient events can lead to potentially damaging pressures that can disable large systems. Proper analysis of such systems can identify potential problems and lead to the design of mitigation measures that eliminate the danger that transients can create.

The Influence of Geologic
Setting on Microtunneling

David C. Mathy[1] & Dru R. Nielson[2]

Abstract

A thorough and accurate characterization of geologic setting and definition of geotechnical conditions is critical to the planning, design and construction of microtunneling projects. Geologic setting controls soil, bedrock and groundwater conditions. An understanding of where subsurface materials originated (provenance) and how and in what setting they were transported and deposited (paleoenvironment) and their subsequent burial, compaction, consolidation, cementation, erosion, and structural deformation are key elements in characterizing geologic setting. Specific unique aspects of microtunneling impacted by geologic setting and geotechnical conditions include: cutter-head configuration and tunnel face stability; cutter-head performance, steering corrections and line and grade control in mixed and changed-face conditions; cutter-head abrasion; crushing chamber configuration for oversize materials; spoil handling; slurry and slurry additives; mechanical slurry separation systems; hydrofracture and slurry loss; overcut stability; lubrication and pipe friction loads; jacking forces and drive lengths; intermediate jacking stations; thrust block and reaction wall capacity; pipe selection; long-term jacking and receiving pit stability; and ground settlement. Understanding the unique aspects of pipeline installation by microtunneling and how subsurface conditions are controlled by geologic setting is a critical step to successful planning, design and construction of all microtunnel projects.

Introduction

Geologic setting influences all surface and underground construction. The degree of this influence is, in part, a function of the type of project. The influence of geologic setting is significantly greater for pipelines involving thousands of

---
[1]Principal Engineer, M. ASCE, [2]Engineering Geologist, DCM/Joyal Engineering, 484 North Wiget Lane, Walnut Creek, California 94598

feet of continuous excavation than for buildings involving localized excavations within relatively small footprint areas. An understanding of geologic setting provides a framework upon which to characterize the range of subsurface conditions along a proposed pipeline alignment including changing soil and soil properties, changes from soil to bedrock, changing bedrock and bedrock properties and changing groundwater conditions. Traditional methods of open cut pipeline construction are somewhat forgiving (although not always so) in excavating through variable subsurface conditions. Microtunneling as a pipeline construction method is far more complex than open cut trenching and as a result is much more sensitive (and less forgiving) to subsurface conditions and variations in subsurface conditions along a pipeline alignment. In fact, there are certain combinations of geologic setting and subsurface conditions where microtunneling may not be feasible (e.g., small diameter tunnels through cobbles and boulders which are too large for the microtunnel machine's crushing chamber to ingest).

Microtunneling is most commonly described as a remotely controlled, guided, pipe jacking process that provides continuous support to the excavation face. The microtunneling process does not require personnel entry into the tunnel (Bennett, 1995). Microtunneling equipment has five independent systems (Iseley, 1997):

- microtunnel boring machine (MTBM),
- jacking or propulsion system,
- spoil removal system,
- laser guidance and remote control system, and
- pipe lubrication system.

Of these five components, only the laser guidance and remote control system is not directly impacted by geologic setting and subsurface conditions on every construction project. However, steering can be influenced (or required) in mixed and changed conditions at the tunnel face and, in that event, all five independent systems of microtunneling are directly impacted by geologic setting and subsurface conditions. Even the selection of pipe material is influenced by geologic setting and subsurface conditions based on face resistance and pipe friction resistance to jacking. Long drives and high anticipated jacking loads may limit the designers pipe choice to high strength materials.

Microtunneling is a hybrid of the tunneling industry (miniturization of tunnel boring machines in Japan in 1975) and the pipeline industry where pipe jacking has been used for more than 100 years. Detailed descriptions of microtunneling are contained in Stein, 1989 and Thomson, 1993. Many designers and most owners arrive at microtunneling through the pipeline industry (most commonly gravity sanitary sewers) rather than through the tunneling industry. As a result, designers and owners often specify geotechnical investigations which are typical for open cut pipeline design and construction rather than the more rigorous geotechnical investigations (with emphasis on geologic setting) typical for tunnel design and construction (Bickel, 1996). While microtunneling design may combine certain aspects of both construction methods, the geotechnical engineering investigation should be approached as a tunnel not as open cut trenching. Table No. 1 is a qualitative comparison of the geologic and geotechnical factors that influence open cut trenching pipeline design and construction and microtunnel pipeline design and construction. As illustrated on Table No. 1, microtunneling is far more sensitive to subsurface conditions than open cut trenching. While microtunneling offers distinct advantages in the areas of shoring, dewatering, bedding and backfill, and contamination, it poses new challenges for spoils handling, oversize material, pipe friction, mixed and changed face conditions, abrasion etc. Therefore, understanding geologic setting and providing a sound framework for interpreting geotechnical information along a pipeline alignment is critical to microtunneling. Without a thorough and accurate understanding of geologic setting, vital features of the subsurface that adversely impact microtunneling may be missed. Of course, the more complex the geologic setting, the more important this understanding becomes.

Geologic Setting

The evaluation of geologic setting is relatively straight forward and should be included in all geotechnical investigations for microtunneling projects. The typical design geotechnical investigation for microtunneling should be completed in the following three phases:

   Phase I - Geologic setting of alignment (and alignment alternatives).

   Phase II - Preliminary subsurface investigation (borings, test pits, large diameter borings, cone penetration tests, geophysical investigation etc., widely spaced along the pipeline alignment).

   Phase III - Final subsurface investigation (borings at all pits, mid-drive borings, large diameter borings, test pits, cone penetration tests, geophysical investigation, piezometers, etc.).

For smaller projects, the above three phases of investigation are sometimes combined into a single continuous effort. For larger projects each phase may be independent; however, all phases of the geotechnical investigation should be

TABLE NO. 1

Geologic/Geotechnical Factor	Open Cut Trenching	Microtunneling Pits	Tunnel
1. Shoring	●	●	○
2. Groundwater & dewatering	●	●	○
3. Unstable soils (raveling, running, flowing - no stand-up time)	●	●	◗
4. Contamination above pipe zone	●	●	○
5. Bedding & backfill	◗	◗	○
6. Soil gradation down to 2 microns	○	○	◗
7. Pipe friction and jacking resistance [1]	○	◗	●
8. Cobbles/boulders, size and distribution	○	○	●
9. Cobbles/boulders, compressive strength, abrasiveness	○	○	●
10. Swelling clays and claystones	○	○	●
11. Tree roots, wood, fill debris and metal obstructions	○	○	●
12. Face stability and voids generation [2]	○	○	●
13. Shallow cover, hydrofracture and slurry loss	○	○	◗
14. Systemic settlement	○	○	◗
15. Reaction wall bearing capacity	○	●	●
16. Recompression settlement	○	◗	○
17. Bedrock hardness/strength	◗	◗	●
18. Bedrock fracturing	◗	◗	◗
19. Bedrock abrasiveness	○	○	●
20. Bedrock slake/durability	○	○	◗
21. Variable cementation	○	○	◗
22. Mixed-face condition [3]	○	○	●
23. Changed-face condition [4]	○	○	●

● - critical   ◗ - important   ○ - not critical

(1) influences pit spacing, thrust wall design, pipe selection
(2) soil caused, operator influenced
(3) e.g., soil and bedrock in the tunnel face cross-section - impacts cutter head performance and line and grade control
(4) e.g., transition from soil to bedrock in the tunnel alignment - impacts cutter-head performance

completed by the same geologist and geotechnical engineer to assure continuity of information and analysis. A thorough and accurate characterization of geologic setting in Phase I will significantly improve the geotechnical team's, design team's and owner's ability to:

- evaluate alignment alternatives at an early stage of design;
- plan the details of the subsurface geotechnical investigation (e.g., type and size of drill rig, large diameter borings or test pits, geophysical testing, groundwater monitoring, etc.);
- plan the details of special laboratory testing (e.g., grain size distribution of cobble and boulder deposits, uniaxial compressive strength of cobbles and boulders, abrasion testing of cobbles and boulders, swell testing in over-consolidated clays and claystones, uniaxial compression testing in bedrock, abrasion testing in bedrock, slake/durability testing in bedrock);
- interpret geophysical testing and other indirect testing;
- interpolate subsurface conditions between borings;
- visualize the project alignment in three dimensions rather than the typical two dimension profile; and
- interpret subsurface conditions between jacking and receiving pits at crossings where mid-drive borings or testing are not possible (e.g., some river crossings, freeway undercrossings, runway undercrossings, wetlands undercrossings, contaminated sites, etc.).

Evaluation of geologic setting includes two separate tasks:

Task No. 1 - Collect and organize published information along the pipeline alignment pertaining to geology, seismicity, soils, groundwater, and possible contamination.

Task No. 2 - Reconnaissance of the pipeline alignment and surrounding area to field check the collected geologic information. The alignment reconnaissance should include an area substantially larger than the linear pipeline alignment to incorporate regional trends in geology and geomorphology (landforms).

Task No. 1 - Data Collection

The following are typical sources for collection of published information pertaining to geology, seismicity, soils, groundwater and possible contamination:

State and Federal Sources:
- U.S. Department of Agriculture, Soil Conservation Service
- U.S. Geologic Survey
- State Department of Mines and Geology
- Regional Water Quality Control Board
- State Department of Health Services
- State Department of Transportation (e.g., Caltrans)

Local Sources:
- City and County Geologists
- City and County building permit files
- County Health Services well permits files
- Sewer/Water Districts
- Local transportation agencies (e.g., BART)

Miscellaneous Sources:
- Historic aerial photograph libraries
- Local historical societies
- Local colleges and universities

On any individual project, some or all of these sources of information will be useful. The following paragraphs discuss in more detail three of the above sources of information which are not often used in pipeline investigations that we have found to be valuable in characterizing geologic setting for microtunneling.

Most microtunneling projects are located in densely populated and developed urban areas. There is often a wealth of subsurface information along the proposed pipeline alignment that can be obtained by reviewing geotechnical investigation reports for existing buildings near and along the alignment. These reports are available in the City or County building permit files. By reviewing subsurface boring logs near and along the alignment, a preliminary model of soil and bedrock type, bedrock elevations, and groundwater elevations can be developed. Another benefit of reviewing geotechnical reports for structures adjacent to the pipeline alignment is to check for possible obstructions (e.g., abandoned basements, old pile foundations, fill, tie-backs, etc.) that may exist in the proposed pipeline alignment.

Detailed groundwater elevation information is often available within environmental reports archived by the Regional Water Quality Control Board, State Department of Health Services and local Department of Health Services. This information often includes well logs that have been monitored over extended periods of time and can, therefore, give insight into seasonal and/or tidal fluctuations in groundwater elevation. Of course, these environmental reports will also provide information on soil and groundwater contamination and potential plumes of contaminated groundwater that may pass through the pipeline alignment that will have to be addressed in design and construction.

Historic aerial photographs of urban areas are typically available from public agencies (e.g., U.S. Geological Survey), university collections (e.g., Fairchild Collection at Whittier College in Southern California), and private aerial survey companies. Given the geographic coordinates of proposed project alignments, these sources can provide a list of dates (some as early as the 1930's) and altitudes (typically ranging from an equivalent 1:12000 to 1:36000 scale for proof prints) of aerial photographs available from their archives. Prints of selected historical aerial photograph sequences can be reviewed for recent geomorphic changes that may have occurred or which may be hidden by urban development. Reviewing an historical sequence of aerial photographs of a proposed pipeline alignment over a 50 to 60 year time span is a valuable tool in evaluating natural and manmade changes in the area that may influence microtunneling such as:

- natural geomorphic changes (e.g., streams, wetland sloughs, creeks, landslides);
- man-made alterations including cuts and fills; re-alignment of natural drainages; former structures and foundations; the existence of reclaimed land over wetlands, marshes sloughs, ponds, etc.;
- prior agricultural uses along the proposed pipeline alignment involving orchards and wind row trees (see Item 11 in Table No. 1);
- active and/or abandoned quarries and gravel pits (see Items 8 and 9 in Table No. 1); and
- prior industrial uses of properties (see Item 4 in Table No. 1).

Task No. 2 - Reconnaissance

Once all readily available archived data is collected and organized, it must be field checked during a reconnaissance of the pipeline alignment. This field reconnaissance should include a thorough walk-over of the pipeline alignment with special attention given and notations taken at road cuts and bedrock outcrops, and at creek, stream, river and/or flood control channel crossings. In addition, the location of potential sources of contamination (e.g., gasoline service stations) should be mapped. This reconnaissance should include a substantial area beyond the limits of the alignment (particularly uphill). Bedrock exposed in road cuts and outcrops should be sampled and described including rock type, color, structural attitude, particle size gradation, cementation, hardness, fracture spacing and character, durability, abrasiveness, etc. Provenance (bedrock source) of alluvium in nearby drainages should be determined along with particle size gradation of the exposed alluvium. In highly developed urban areas where geologic exposures are limited along the pipeline alignment, interpolation of regional geology is essential.

Application

The San Francisco Bay area has one of the most complex and interesting geologic settings of any densely populated urban area in the United States. Figure No. 1 is a general geologic map of the San Francisco Bay area (Helley, 1979). Even in this generalized map the wide diversity of geologic conditions around San Francisco Bay is apparent. Thorough and accurate characterization of geologic setting is particularly critical in the San Francisco Bay area due to the variety of bedrock (igneous, metamorphic and sedimentary) and soils (Bay Mud, clay, silt, sand [including dune sand], gravel, cobbles, and boulders) that are present in the region. Deformation of these materials through tectonic folding and landsliding has further complicated the geologic setting. A few brief examples of how of geologic setting has influenced microtunneling in the San Francisco Bay area are described in the following paragraphs.

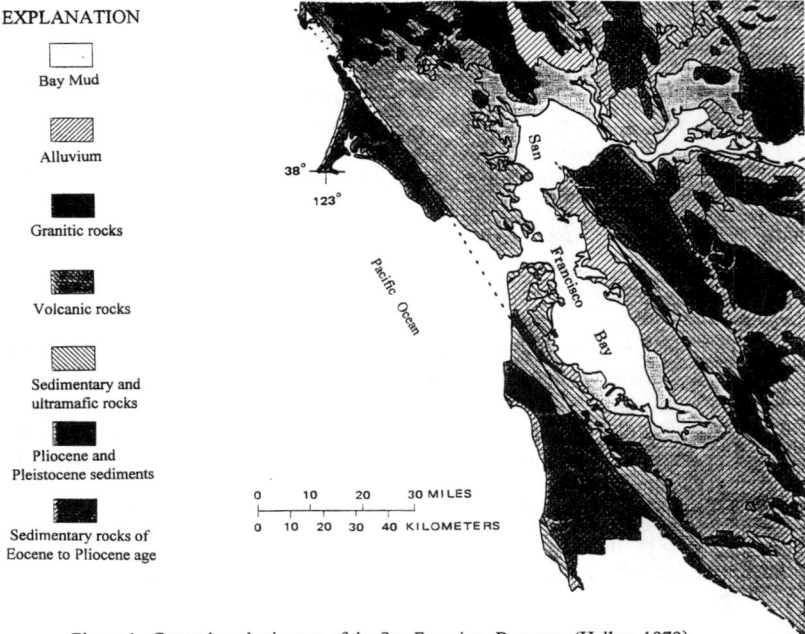

Figure 1. General geologic map of the San Francisco Bay area (Helley 1979).

Bay Mud forms a rim around the margin of San Francisco Bay (Helley, 1979; Lee, 1969) and consists predominantly of clay but can locally have significant fine loose sandy strata. In general, Bay Mud is saturated, normally consolidated, highly compressible and very weak. At the tunnel head, the soft clay behaves as squeezing ground and the loose sands behave as flowing ground. The construction of jacking and receiving pits in Bay Mud is very difficult and expensive and usually requires caisson or cofferdam type construction. It is usually not possible to develop sufficient reaction wall capacity for the pipe jacking system without some form of ground treatment (e.g., jet grouting) to strengthen the soil behind the jacking pit. Launching the microtunnel machine requires careful contractor consideration to minimize diving and steering corrections just outside the pit. Some form of ground treatment is usually required at the launch portal to ease the machine's transition from the rigid pit to soft soils. Consolidation settlement at manholes can occur if there is any net increase in load upon jacking pit and receiving pit backfilling and manhole construction.

Alluvial fan deposits originate from the coastal mountains which surround the Bay. These deposits are coarse-grained (including cobbles and boulders) near the fan heads and fine-grained at the Bay margins. At certain locations around the Bay the source rock for these alluvial fans includes high strength (hard) volcanic bedrock. This high strength bedrock weathers to cobbles and boulders which are then carried into creeks and streams and spread out near fan heads. Cobbles and boulders from 150 mm (6 in.) to 450mm (18 in.) are not uncommon within 3.4 km (2 mi.) of the foothills. These cobble and boulder sizes can and have stopped small diameter, 600 mm (24 in.), microtunnel machines.

Within the coastal mountains and valleys around the Bay, upland bedrock consisting of high plasticity claystone is common. These claystones are highly overconsolidated and can swell significantly upon pressure relief and the introduction of water. Evaluating adequate overcut and polymer application is critical to minimize claystone swelling and resultant high pipe friction and high jacking loads. The claystone is also weak in shear strength; therefore, landslides are relatively common. Ancient landslide deposits mapped by the U.S. Geological Survey often cover areas as large as several city blocks. Mixed-face and changed-face conditions are frequent in these deposits with often abrupt transitions from soil to bedrock and abrupt changes in groundwater conditions.

Closure

A thorough and accurate characterization of geologic setting provides the context and framework for:

- evaluating alignment alternatives;

- planning a detailed geotechnical investigation;
- interpolating geotechnical information between borings;
- interpreting geophysical testing; and
- viewing the pipeline alignment in three dimensions rather than a two dimension profile.

A typical geotechnical investigation for a 2,000m (6,558 ft.) long, 600mm (24 in.) diameter microtunnel project with borings at 100m (328 ft.) intervals and continuous sampling in the tunnel zone will only expose about 0.007% of the total tunnel excavation. Filling the gaps between borings with thoughtful interpretations of subsurface conditions based on geologic setting and the context it provides is vitally important. The contractor will excavate 100% of the soil/bedrock along the pipeline alignment and any significant differences from the subsurface conditions described in the contract documents may result in alignment difficulties and pipeline deviation from line and grade, emergency (911) excavations to remove obstructions, construction delays, potential claims by the contractor, and disruption to the community.

APPENDIX - References

1. Bennett, R.D., Guice, L.K., Khan S., Staheli, K., (1995), "Guidelines for Trenchless Technology: Cured-in Place (CIPP), Fold and Formed Pipe (FFP), Mini-Horizontal Directional Drilling (Mini-HDD), Microtunneling," TTC Technical Report #400.
2. Bickel, J.O., Kuesel, T.R., King, E.H., (1996), Tunnel Engineering Handbook, Chapman & Hall, New York.
3. Helley, E.J., Lajoie, K.R., Spangle, W.E., and Blair, M.C., (1979), Flatland Deposits of the San Francisco Bay Region, California - Their Geology and Engineering Properties and Their Importance to Comprehensive Planning, U.S. Geological Survey Professional Paper 943.
4. Iseley, T., Gokhale, S.B., (1997), "Trenchless Installation of Conduits Beneath Roadways", NCHRP, Synthesis 242, Transporation Research Board.
5. Lee, C.H., and Prazker, M., (1969), Bay Mud Developments and Related Structural Foundations, in. Goldman, H.B., ed., Geologic and Engineering Aspects of San Francisco Bay Fill, California Division of Mines and Geology, Special Report 97, p. 41-85.
6. Stein D., Mollers, K., Bielecki R., (1989) Microtunneling, Ernst & Sohn, Berlin.
7. Thomson, J., (1993), Pipe Jacking and Microtunneling, Blackie Academic & Professional, Glasgow.

The Influence of Geologic Setting on Microtunneling

David C. Mathy & Dru Nielson

Biographies

- David C. Mathy

Mr. Mathy is the founder and Principal Engineer of DCM/Joyal Engineering in Walnut Creek, California. Mr. Mathy has a bachelors degree in Civil Engineering from California State Polytechnic University in Pomona and a masters degree in Geotechnical Engineering from the University of California at Berkeley. Mr. Mathy is a registered civil engineer in California, Nevada and Oregon and a registered geotechnical engineer in California.

- Dru R. Nielson

Mr. Nielson is a Senior Geologist with DCM/Joyal Engineering in Walnut Creek, California. Mr. Nielson has a bachelors and masters degree in Geology from Brigham Young University and is a registered geologist and certified engineering geologist in the State of California. Mr. Nielson has completed geologic evaluations for over 20,000 linear feet of microtunnel pipeline alignments in the San Francisco Bay area.

RECIPE FOR SUCCESSFUL MICROTUNNELING

John Goodwin[1], Jesse Gill[2], David Mathy[3]

Abstract

This paper presents planning and design elements that have been developed to ensure successful microtunneling. The authors have been involved in five recent microtunneling projects for Union Sanitary District and have developed a successful approach to microtunneling in poor soil conditions, areas with high groundwater, soil and groundwater contamination, and high traffic areas. The paper presents thorough soil investigation procedures, specification requirements to ensure that jacking/receiving pits are watertight and stable, procedures to minimize settlement during construction, requirements to minimize impact on traffic, and requirements for microtunneling equipment and procedures.

Background Information

Union Sanitary District (District) is a public agency located approximately 56 kilometers (35 miles) south of the city of San Francisco, California. The District provides wastewater collection, treatment, and disposal services to the tri-city area of Fremont, Union City, and Newark. The District has completed seven microtunneling projects since 1992. These projects total approximately 6,700 meters (22,000 feet) in length and range in size from 305 to 915 millimeters (12 to 36 inches) in diameter.

Earlier Failures With Open Cut And Bore And Jack. In the early 1990s, the District employed traditional open cut and auger bore and jack on several gravity sewer projects that failed due to the contractor's inability to control groundwater and flowing sands and silts. The soil investigation reports for these early projects indicated layers of loose silts and sands below the water table at a

---

[1] Principal Engineer, Brown and Caldwell, 3480 Buskirk Avenue, Pleasant Hill, CA 94523-4342
[2] Senior Engineer, Union Sanitary District, 37532 Dusterberry Way, Fremont, CA 94536
[3] Principal Engineer, DCM/Joyal Engineering, 484 N. Wiget Lane, Walnut Creek, CA 94598

depth of 3.7 meters (12 feet). When the contractors started the auger bore operation, they could not control the soil at the face of the auger bore. The soil at the tunnel heading was not adequately supported and flowed through the open-end steel casing into the jacking pit (see Figure 1). This resulted in the formation of sinkholes above the tunnel heading and street pavement cracking and settling. In addition, the steel casing was installed 0.6 meters (2 feet) off line and 0.3 meters (1 foot) off grade. Another project experienced similar soil and groundwater conditions which resulted in voids/sink holes beneath major streets and adjacent to railroad tracks and poor line and grade.

Figure 1. Previous Problems with Flowing Silts and Sands

Our experience has shown that the most important factors for microtunneling project success in these subsurface conditions is a thorough geotechnical investigation and specifications with an emphasis on the construction of watertight and stable jacking and receiving pits. The following is a discussion of the design and specification requirements developed for the Upper Fremont Boulevard Sewer and the Baine Avenue Sewer Projects. Both of these projects were completed in 1997. The Upper Fremont Boulevard Sewer project consisted of 1,650 linear meters (5,400 feet) of 530-millimeter-diameter (21-inch) sewer installed by microtunneling. The Baine Avenue Sewer project consisted of 610 linear meters (2,000 feet) of 915-millimeter-diameter (36-inch) sewer installed by microtunneling.

Geotechnical Investigation

The investigations included research into geologic setting and definition of geologic formations along the alignment, followed by drilling, logging, and taking soil samples from locations as close as possible to the proposed jacking and receiving pits. We also installed several observation wells (piezometer) for each project to monitor groundwater levels. Test pits were excavated using a backhoe to document the behavior of soils in unsupported vertical cuts.

Geologic Setting. The project sites are located on flatlands near the southeast margin of San Francisco Bay. The flatland soil deposits are Holocene age, medium- to coarse-grained alluvium underlain by older alluvial fan deposits (e.g., late Pleistocene alluvium). The depth to bedrock around the Bay margins is variable but typically to hundreds of feet. The Holocene deposits are less than 7,000 years old and include fine sand, silt, and clays with occasional layers of coarse sand. Coarse sands are more abundant near alluvial fan heads. The coarse-grained deposits at the heads of alluvial fans transition into fine-grained alluvial fan and marsh deposits near the shore of San Francisco Bay. At the margins of the Bay, the alluvium is underlain by Bay Mud. Groundwater is relatively shallow and abundant in the coarse-grained alluvium with tidal influence near the Bay.

General Soils Profile. In general, both the Upper Fremont and Baine Avenue Sewer projects consist of interlayered clays, silts, and sands. The Baine Avenue project also contained occasional thin beds of coarse sand and gravel. Figure 2 is a general soil profile for the Baine Avenue Sewer Project. Based on the soil profile and previous experience, we expected to experience difficulties associated with flowing silts and sands, development of voids and sinkholes, dewatering, vibration-induced settlement, and recompression settlement. The following is a description of each of these conditions.

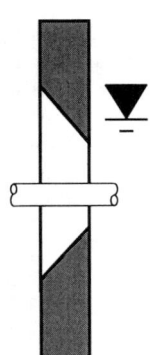

Clay layer approximately 1.5 to 3 meters (5 to 10 feet) thick
N = 7 to 30, Ave N=12, Stiff

Silt and sand granular sequence, relatively permeable, 3 to 4.6 meters (10 to 15 feet) thick
N = 5 to 20, Ave N=7, Loose

Clay and silt, relatively impermeable, depth from about 4.6 to 7.6 meters (15 to 25 feet)
N = 7 to 13, Ave N=10, Stiff

Figure 2. Typical Soil Profile

Voids and Sinkholes. The cofferdam style sheeting and shoring would have to be carefully installed to prevent uncontrolled inflow of soils and the formation of voids and sinkholes outside the sheeting and shoring system. Seepage of groundwater into the excavation would carry noncohesive, fine-grained soils into the excavation. After a period of time, significant voids can develop behind the sheeting and shoring which would likely develop into sinkholes.

Difficulty Dewatering. Based on previous experience in the project area and a groundwater dewatering pump test, we concluded that dewatering of the silt and sand layer using a dewatering well (gravity system) would be ineffective. The predominant grain size distribution of the silt and sand layer is too fine for effective gravity drainage. In addition, the silts and sands at the lower clay boundary cannot be completely dewatered. It was estimated that if soils were to be dewatered, an extensive and costly well point dewatering system (suction pumps) would be required with well points spaced less than 5 feet, center to center, along all sides of the pit excavation. An alternative to dewatering prior to excavation would be to install a sheeting and shoring system (cofferdam) that would exclude the groundwater from entering the pits.

Vibration-Induced Settlement. Vibration-induced settlement can occur as a result of localized liquefaction and densification of loose, saturated, uniformly graded, noncohesive soils. This could occur during vibratory sheet pile driving and extraction. On a previous District project, manhole settlement of approximately 150 to 230 millimeters (6 to 9 inches) was observed during vibratory extraction of sheet piles.

Recompression Settlement. Recompression settlement was anticipated for pipelines and manholes installed by open cut. Approximately 6 meters (20 feet) of pipe was installed by open cut at each jacking and receiving pit. Jacking and receiving pits were approximately 4.5 to 6.1 meters (15 to 20 feet) deep. Removal of the native soils resulted in rebound of the pit bottom due to overburden removal. Recompression settlement is caused by the weight of the backfill material recompressing the underlying soils. The recompression occurs immediately during backfill. It was anticipated that the maximum recompression of the pit bottom would be approximately 50 millimeters (2 inches) for undisturbed soil. Backfill loading upon soils weakened by disturbance, softening, or boiling effects can produce random settlements much greater than 50 millimeters (2 inches).

Sheeting And Shoring Specification Requirements

A performance-based specification was prepared to indicate the requirements of the sheeting and shoring system. Acceptable methods included the use of interlocked steel sheet piling (Figure 3), caissons, or other systems proposed by the Contractor and approved by the Engineer that would meet the conditions of the specifications. The following requirements were included in the specifications.

Figure 3. Interlocked Steel Sheet Piling

Watertight Sheeting & Shoring. The project team concluded that complete dewatering for the excavation of the jacking and receiving pits would be impractical. To prevent uncontrolled inflow of soils and the formation of voids and sinkholes, the specifications required that the sheeting and shoring systems be watertight including corners and launching and retrieving faces. This required the use of specially constructed interlocked corners (if steel sheet piles were utilized) and soil grouting at launching and retrieving faces to prevent groundwater infiltration. The base of the shoring should extend through any permeable soils and penetrate sufficiently into relatively impermeable clays to act as a groundwater cutoff and minimize groundwater inflow into the base of the excavation. Adequate toe embedment of the sheet piling also provided for a stable base of excavation with minimal softening and boiling of the bottom of the pits.

Abandon-in-Place Sheeting and Shoring. To reduce the likelihood of vibration-induced settlement, the specifications required that the sheeting and shoring system be abandoned in place instead of the common practice of vibratory extraction of the sheeting and shoring. The shoring was to be cut at a point 1.8 meters (6 feet) below ground and all sheet piles below that point abandoned in place. The Contractor's cost for used sheet piles was approximately $57 per square meter ($5.30 per square foot). The material cost for abandoning sheet piling for a 3.6-meter by 6.1-meter (12-foot by 20-foot) pit was approximately $7,000.

Manhole Recompression Settlement

To allow for recompression settlement of the manholes and pipe installed at the jacking and receiving pits, a detail to allow for the recompression settlement was included in the design. A simplified detail is shown on Figure 4. The pipe and manhole installed within the excavated jacking or receiving pit would experience approximately 25 to 50 millimeters (1 to 2 inches) of recompression settlement. By constructing the pipe and manhole slightly above the actual design grade, the manhole and pipe would settle to approximately design grade. The amount of the adjustment can be modified from pit to pit by the construction manager depending on observed settlement during construction.

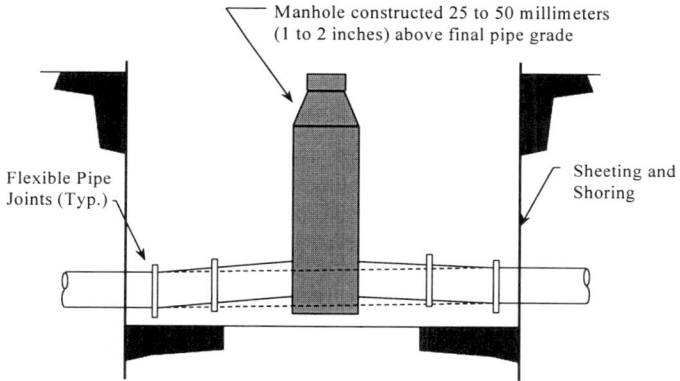

Figure 4. Recompression Settlement Allowance Detail

Alignment Tolerances

The majority of the District's service area lies in very flat terrain. Several of the District's microtunneling projects were designed with a slope of 0.0010 (feet per feet). The specifications required a vertical grade tolerance of plus or minus 13 millimeters (1/2 inch). This would theoretically allow for a maximum sag of 25 millieters (1 inch). The horizontal alignment tolerance was 76 millimeters (3 inches).

Contamination

Both of the projects are located in areas that were subjected to prior groundwater and soils contamination. The Upper Fremont Boulevard Project involved total petroleum hydrocarbon (TPH) contamination from underground fuel storage tanks at two locations. The Baine Avenue Sewer (and the earlier Newark Subbasin Relief Sewer Project) involved volatile organic compound (VOC) contamination. Subsurface investigations, which included review of existing documentation, field sampling, and testing, was performed to evaluate the occurrence and concentrations of chemicals in the soil and groundwater.

Based on the investigations, "potential contamination areas" were identified on the drawings. The specification requirements for each project were similar. Initial screening of the soil and groundwater was performed by the Engineer. For soil that was determined to be contaminated, the Contractor was required to load and haul the soil directly to a Class II landfill. Within the potential contamination area, the Contractor was required to pump the groundwater from the dewatering operation to storage tanks. The Engineer performed initial screening of the groundwater. If the groundwater was below a specific contamination concentration level, the Contractor was allowed to discharge the groundwater directly to the District's sanitary sewer system. For groundwater with a higher concentration level, the Contractor provided treatment by two-stage granular activated carbon mobile treatment units prior to discharge to the sanitary sewer system (Figure 5).

Unit price bid items were specified for the soil and groundwater handling. The Contractor was paid per cubic yard for transportation and disposal of soil to a Class II landfill. For contaminated groundwater, the Contractor was paid a lump sum for groundwater handling and unit price bid items for the mobile treatment units and storage tanks.

Figure 5. Contamination Treatment Unit

Microtunneling Through Existing Sewer

A portion of the preferred alignment for the Upper Fremont Boulevard Sewer was along the alignment of an existing 122-meter-long (400-foot), 380-millimeter-diameter (15-inch) VCP sewer. It was decided to microtunnel the new 533-millimeter-diameter (21-inch) sewer through the existing 380-millimeter-diameter (15-inch) sewer. The following steps were completed to ensure a successful installation:

1. Verify that the existing VCP did not contain any steel bands. The steel bands would damage the microtunneling equipment.
2. Break the existing 380-millimeter-diameter (15-inch) VCP by pipe bursting prior to microtunneling. This would help keep the tunneling machine on grade.
3. Fill the broken 380-millimeter-diameter (15-inch) VCP with low-strength grout to allow the microtunneling machine to maintain slurry balance.

Traffic Issues And Special Work Hours

The Upper Fremont Boulevard Sewer project included several restrictions for construction due to traffic (Figure 6). These included the following:

1. Contractor was allowed to work 7 days a week, 24 hours per day provided that there were no noise complaints from businesses or residents adjacent to the work area.
2. No construction equipment was allowed on the street between 3 p.m. and 7 p.m., Monday through Friday. This included slurry transport, pipe and material deliveries.
3. All jacking and receiving pits were required to be constructed on the weekend.

Figure 6. Traffic Conditions

Prequalification Of Microtunneling Subcontractors

The specifications require that the Contractor or the subcontractor performing the microtunneling work be prequalified. The list of prequalified tunneling contractors is developed prior to advertising for construction bid of the microtunneling project.

Following is a partial list of prequalification requirements:

1. A minimum of two projects using slurry microtunneling equipment. At least one project within the last 12 months.

2. One project in similar geologic conditions.

3. Project Superintendent and microtunneling machine operator shall have 10 and 5 years' experience on trenchless pipeline projects, respectively. Each person shall have worked on a minimum of two slurry microtunneling projects.

Project Partnering

Union Sanitary District strongly believes in the concept of partnering on their projects. The District's philosophy is to work closely with the local community and local agencies to minimize disruption and be sensitive and responsive to their needs, concerns, and suggestions, both during the design and construction of the projects. The District works closely with consultants and contractors and is actively involved both during the design and construction phases. This philosophy has helped the District to reduce conflicts on projects and has resulted in win-win situations for everyone involved. In addition, the District takes a very positive approach to issue resolution and decisions are made and implemented expeditiously. This philosophy and positive approach has helped the District to establish good relations with the community, local agencies, consultants, and contractors.

Summary

Many factors play a part in a successful microtunneling project, including good luck. Our experience has shown that the following aspects will help produce a successful project.

1. A comprehensive Geotechnical Investigation.

2. Careful consideration during design to achieve stable, water-tight jacking and receiving pits.

3. Consider traffic control, project access, and working hour requirements.

4. A positive partnering approach to issue resolution by the Owner, Engineer, Construction Manager, and Contractor.

5. Establish good relations with community and local agencies.

Microtunneling in Downtown Honolulu

by James Kwong[1], M.ASCE, Steve Klein[2], M.ASCE, Galen Nagle[2], M.ASCE, Glen Okita[3], M.ASCE, & Steve Duke[4]

Abstract

Highly congested urban utility corridors present some of the most difficult conditions for construction of new underground pipelines especially when subsurface conditions are also very challenging. The need to construct a new sewer pipeline approximately 2,287 m (7,500 ft) long and 915 mm (36 in) in diameter through one of these utility corridors in the busiest part of downtown Honolulu was successfully resolved by specifying microtunneling methods for construction of the pipeline.

Introduction

The Nimitz Highway Reconstructed Sewer Project will be constructed in downtown Honolulu, from Ala Moana Boulevard to Hotel Street, Honolulu, Hawaii (Figure 1). This project is being constructed by the Department of Wastewater Management, City and County of Honolulu (the City) to upgrade a portion of the sewer system in downtown Honolulu. This project involves the construction of a new sewer pipeline approximately 2,293 m (7,520 ft) long and 915 mm (36 in) in internal diameter (ID). Of the 2,293 m (7,520 ft) of new sewer to be constructed, approximately 2,232 m (7,320 ft) will be installed using microtunneling methods with the remainder to be installed using open cut trenching methods. The microtunneling will be completed in a series of 21 drives ranging in length from about 24.4 to 244 m (80 to 800 ft). The crown of the sewer line will be about 2.4 to 4.6m (8 to 15 ft) below the existing ground surface. The project also involves the construction of 36 manholes.

URS Griener Woodward-Clyde: [1]Honolulu, HI, [2]Oakland, CA, [4]Santa Ana, CA. [3]City and County of Honolulu, Department of Wastewater Management

Construction of this pipeline through a congested urban area like downtown Honolulu will be very difficult and is further complicated by the numerous existing underground utilities (see Figure 2, for example) and the challenging subsurface conditions. Several key issues that needed to be addressed in planning and design of the project include:

- Highly complex and difficult subsurface conditions;
- Close proximity to many existing utilities;
- Potential for encountering obstructions along the alignment;
- Minimizing construction noise and disruption to residents and businesses;
- Restrictions regarding work hours on City streets and State Highways and specific traffic control requirements;
- Presence of 1800's burial sites along a portion of the pipeline alignment;
- Contaminated soils and ground water issues.

This paper discusses some of these key design issues and describes how they were addressed in design of the project. Planning issues relating to this project were discussed in (Matsunaga, Kwong, and Okita, 1998).

Geotechnical Investigations

To supplement the available subsurface data and explore the subsurface conditions for design and construction of the new sewer line, a geotechnical investigation consisting of a total of twenty-one exploratory borings was completed (Woodward Clyde, 1997). The exploratory borings were generally completed to depths of about 3.4 to 24.4 m (11 to 80 ft) below the existing ground surface. An idealized geologic profile depicting the complex subsurface conditions shown on Figure 3.

In addition to the exploratory drilling program, a ground penetrating radar (GPR) investigation was performed to try to locate existing underground utilities, buried utility trenches or related structures that may be obstructions to microtunneling. The results were used to better identify pothole locations for evaluating actual subsurface conditions. The GPR investigation consisted of an initial calibration test and a total of fifty-eight (58) GPR transects. The GPR data was recorded by pulling a 500 MHz antenna along the transects. A calibration test was performed adjacent to an open excavation where subsurface conditions were known. The calibration tests indicated that the majority of the GPR energy is being reflected at the groundwater table interface; and therefore, targets below the groundwater table are not being imaged.

An example GPR transect is shown on Figures 4A and B. On this profile, the horizontal axis corresponds to the two-way travel time of the GPR signal, from 0 nanoseconds (ground surface) at the top of the profile to 50 nanoseconds (approximately 3.05 m or 10 ft) below the ground surface) at the bottom if the profile. On Figure 4B, the upper 0.6 to 0.8 m (2 to 2.5 ft) correspond to asphalt concrete pavement and road base material. Below this depth, various GPR reflectors and anomalies, corresponding to various utilities and related structures, can be seen down to the depth of the groundwater table (about 2.3 m or 7.5 ft) below the ground surface). At the saline ground water table, a prominent reflector is evident, and below this reflector there is no coherent GPR data. Based on the results of the GPR investigation, a total of ten areas were selected for potholing to locate and examine specific features.

Anticipated Subsurface Conditions

Based on the results of the geotechnical investigations, the deposits were divided into seven generalized geologic units; Fill, Alluvium, Volcanic Cinders, Lagoonal Deposits, Coralline Deposits, Coral Reef, and Estuarine Deposits. Except for Fills occurring above the pipeline interval, all of these deposits are expected to be encountered in microtunneling for the new sewer pipeline (Figure 3). Each of these units are briefly described below.

- **Fill** - Fill materials encountered consist mainly of loose to medium dense gravely silt, silty, and clayey sand, and sandy and clayey gravel to depths of approximately 3m (10 ft) or more below the existing ground surface. The fill also contains cobbles and boulders composed of hard coral, basalt, and other igneous rocks.

- **Alluvium-** Alluvial deposits consisting primarily of interbedded medium stiff to stiff silt and clay, dense silty sand, and dense silty and sandy gravel are present underlying the surficial fill deposits along portions of the new sewer alignment (Figure 3).

- **Lagoonal Deposits** - Lagoonal deposits generally consist of gravel-sized coral fragments embedded in a matrix of normally consolidated, highly compressible, very soft to soft clay, and very loose to loose sand. These deposits were encountered along the edge of the original shoreline along Ala Moana Boulevard and the southern 457 m (1,500 ft) the alignment along South Street (Figure 1 and 3).

- **Volcanic Cinders** - Episodic volcanic eruptions have deposited layers of volcanic cinders consisting of medium dense to very dense sand and silty sand along portions of the sewer alignment   The volcanic cinders are locally fused into relatively hard layers.

- **Coralline Deposits** - Coralline deposits consisting of medium dense to very dense silty sand, silty and clayey gravel, and well-graded gravel are the most widespread geologic unit in the area. Coarse gravels, and cemented cobble and boulder size coral reef detritus up to a maximum dimension of 457 mm (18 in) are also present in this unit. In some areas the Coralline deposits exhibit weak to strong cementation.

- **Coral Reef** - Coral reef formations were encountered in the subsurface, at depths ranging from 5.3 to 6.4 m (18 to 21 ft) ranging in thickness from 1.5 to 8.2 m (5 to 27 ft). The coral formations included hard, moderately fractured limestone cemented coral fragments and coral sand with zones of friable sands and gravel. Unconfined compression tests on core samples of the hard coral reef deposits exhibited strength values ranging from 1.8 MPa (260 psi) to 13.6 MPa (1,975 psi).

- **Estuarine Deposits** - Estuarine deposits consisting primarily of saturated, slightly underconsolidated to normally consolidated, very soft, highly compressible organic silts, sandy silts, and silty clays were encountered at or near the mouth of the original channels in the area such as along River Street and the Nimitz Highway (Figure 1). Timber debris was also encountered in the estuarine deposits, possibly resulting from old abandoned wharves, ships, or other previous waterfront developments. A thick sequence of estuarine deposits underlie surface fills to depths of over 24.2 m (80 ft) beneath Nimitz Highway and River Street (Figure 3).

Groundwater was encountered in all of the borings completed for the project at depths ranging from 1.4 to 2.7 m (4.5 to 9 ft) below the present ground surface. Numerous existing underground utilities are located within 1 m (3 ft) of the new sewer alignment, and some unknown utilities, sheetpiles, and concrete working slab remnants could encroach into the microtunneling interval in some areas. Portions of the sewer alignment may also encounter soil and groundwater contaminated with hydrocarbons.

Key Geotechnical Design Considerations

Based on the results of the geotechnical investigations, the most important geotechnical factors that needed to be considered in design included:

- The soft compressible lagoonal and estuarine deposits along portions of the pipeline alignment;
- The high groundwater levels and the permeable coralline sands and gravels along most of the alignment; and
- Localized areas of hard cemented coral, coralline deposits, cobble-and boulder-size rocks in fill, and the presence of other obstructions.

The following discussion briefly indicates how these various factors were addressed in design of the project.

Pipeline Support - Settlement resulting from consolidation of soft, saturated, highly compressible estuarine and lagoonal deposits could adversely impact the long-term performance of the new sewer line. Two methods were evaluated for providing long-term support for the new sewer line to minimize future settlement: stabilizing the compressible Lagoonal and Estuarine deposits using jet grouting methods; or supporting the new sewer line on piles.

Jet grout columns would extend down through the soft, compressible materials to firmer underlying materials providing stable support for the new sewer line. Microtunneling would then proceed through the jet grout treated soil. This method would be similar to the method used for construction of the nearby Nimitz Highway Relief Sewer (Honke, Lum & Raines, 1997). Concrete piles driven into firm bearing material, with the pipe constructed on a concrete cradle would be another method to support the new sewer line. This will require the use of open cut trenching methods in an area underlain by numerous utilities and contaminated soils and groundwater. Because this option would require mobilization of pile driving equipment onto Nimitz Highway, where State Department of Transportation (DOT) allows closure of only one traffic lane at any time, the driving of piles was not considered feasible for this project, therefore jet grouting methods were specified for support of the new sewer line where it is underlain by compressible soils.

Obstructions - Another important consideration for this project is the potential for alignment conflicts with either manmade or natural subsurface obstructions. Much of the alignment lies within downtown Honolulu, where there are many existing underground utilities. Furthermore, portions of the alignment traverse the original shoreline, and old buried timber structures such as wharves and piers may have been partially or completely demolished, or otherwise simply buried and constructed over. The potential for encountering obstructions is considered to be a major factor in the selection of microtunneling equipment for the project.

Construction of the existing utilities, particularly those located below the groundwater level, may have required the installation of sheetpiles and placement of concrete working slabs. Remnants of this construction work such as debris, sheetpiles, or concrete may extend below some of these utilities. A review of available building plans was conducted for most of the structures adjacent to the alignment to check if tiebacks were used for shoring basement excavations. The results of this review did not indicate the presence of tiebacks projecting into the streets along the planned sewer alignment.

Hard basalt cobbles may be encountered in fills, particularly in previous trench and manhole excavations that encroached into the pipe zone of the new sewer line.

Natural obstructions that may also be encountered during microtunneling and in shaft excavations are expected to include coral reef deposits, cemented coralline layers, and hard coralline to volcanic (basalt) cobbles and boulders.

Microtunneling - About 97 percent of the new sewer line will be constructed using microtunneling methods in 21 drives. Anticipated ground conditions for the geologic units to be encountered during microtunneling are summarized in Table 1. The entire alignment of the new sewer is expected to be .6 to 4.3 m (2 to 14 ft) below the groundwater level. The saturated very loose to medium dense sands and gravels in the lagoonal and coralline deposits that were encountered along the new sewer alignment will exhibit flowing ground conditions if not stabilized prior to excavation, or controlled with appropriate microtunneling equipment along the new sewer alignment.

In addition, the soft clays, silts, and organic clays and silts associated with the estuarine and lagoonal deposits indicate the potential for squeezing ground conditions. Flowing and squeezing ground conditions, if uncontrolled, can lead to significant loss of ground at the tunnel heading, resulting in surface settlement that could damage existing utilities, adjacent buildings, and other improvements. Improving the strength or stability of the ground or utilizing appropriate microtunneling equipment can prevent this loss of ground.

In order to avoid dewatering along the alignment, use of a closed-face microtunneling machine with the capability to provide a positive stabilizing pressure at the tunnel face is a specification requirement. Microtunneling equipment with this capability will maintain face stability by applying a stabilizing pressure at the tunnel face which will control groundwater inflows, and minimize loss of ground and surface settlement without the need for dewatering or ground improvement.

Table 1
Anticipated Ground Conditions

Geologic Unit	Anticipated Ground Conditions
Fill	Flowing, Fast Raveling (Running above the groundwater level)
Alluvium	Flowing, Slow to Fast Raveling
Lagoonal Deposits	Squeezing, Fast Raveling
Volcanic Cinders	Flowing, Fast Raveling
Coral Reef Deposits	Firm, Flowing
Coralline Deposits	Flowing, Fast Raveling
Estuarine Deposits	Squeezing

A slurry microtunneling machine is required in order to be able to positively counterbalance the hydrostatic pressures in the clean sands associated with the coralline deposits and the volcanic cinders. Microtunneling systems used for this project are required to be remotely controlled with an articulated, steerable, closed-faced machine; a laser guidance system; a lubrication injection system; and an automated slurry spoil transportation system capable of coordinating the volume of material excavated with the rate of pipe installation to avoid overexcavation and loss of ground.

The slurry microtunneling machine is required to have as large a jet as practical for this pipe size and the ability to backflush the slurry lines to avoid clogging up the cutterhead and slurry ports if buried timber debris or wood fragments are encountered such as old wharves, remnants of old ships, or other previous waterfront developments. Because of the relatively shallow ground cover along the sewer alignment, and particularly at River Street, it will be important to closely monitor and control the slurry pressure applied to the tunnel face to avoid ground heave or fracturing of the ground and resultant slurry discharge to the ground surface.

The microtunneling machine must also be able to excavate through coral reef deposits (i.e. limestone) and cemented coral layers that are 41 MPa (6,000 psi) in unconfined compressive strength (UCS), and be able to crush or excavate hard rock boulders up to 305 mm (12 in) in maximum dimension and 138 MPa (20,000 psi) in UCS. Based on strength test results of coral reef samples from other projects in Honolulu, coral reef rocks and cobbles or boulder size chunks/clasts have exhibited typical UCS values of up to 13.8 MPa (2,000 psi) and occasionally hard coral heads and ledges may have an UCS of over 41 MPa (6,000 psi). In order to cut rock of this strength, the machine is required to be equipped with a rock cutterhead fitted with disc cutters or roller bits (Strawberry cutters) because coral reef or cemented coral is too strong to be excavated with drag teeth, although, a combination of disc or roller cutters and drag teeth may also be appropriate. Mixed face conditions associated with the coral reef/cemented coral deposits will tend to deflect the machine off line and grade requiring careful steering procedures to achieve specified line and grade tolerances.

Some of the coralline deposits may have unusual properties relative to other soils such as low specific gravity. The properties of the various geologic units as indicated by the grain size tests, hydrometer tests, and specific gravity tests will have to be taken into account in selecting an appropriate slurry separation plant. In particular, materials with low specific gravity, may not settle out rapidly in a cascading tank type of slurry separation system, and a slurry separation plant will be necessary for efficient muck separation.

Shafts - Shafts excavations will be required for jacking and receiving pits and to construct manholes. At least 32 shaft excavations will be constructed. The entire alignment of the new sewer is heavily traversed by existing utilities that in some cases will have to be relocated, and must also be protected from damage during construction. Where compressible estuarine or lagoonal deposits underlie the new sewer line and planned manholes, jet grouting will be required as the groundwater control system and to support the sewer and manholes to minimize the potential for post-construction settlement.

Groundwater control will be a key factor in construction of shaft excavations and in determining appropriate support systems. To reduce potential dewatering effects on adjacent structures and to minimize handling of contaminated groundwater, a watertight support system is required. Feasible excavation support systems for this project depending on subsurface conditions include: continuous interlocking sheetpiles, or overlapping jet grout columns, both with appropriate internal bracing, systems. Due to the presence of strong, cemented coral reef deposits, and coralline sheetpiles to support the shaft and trench excavations. The pre-drilled holes will need to be overlapping (i.e., essentially continuous).

Localized ground treatment using jet or chemical grouting will be required for groundwater control when breaking out of jacking pits and breaking into receiving pits. Launching seals will be required at jacking pits to prevent leakage of pipe lubricants.

Specification Requirements

In view of the highly variable subsurface conditions and complex project requirements, prequalification of the bidders and contractor's work force was performed during the bidding period prior to bid opening. In order to compensate the contractor fairly for uncertain items of work a number of special bid items were included in the Contract:

- Specific utility relocations as indicated on the Plans;
- Additional potholing to locate other existing utilities as indicated on the Plans or as required based on field conditions;
- Removal of identified obstructions along the sewer alignment prior to microtunneling such as existing manholes, sheetpiles and other shoring left in place, construction debris, and abandoned existing utilities;
- Recovery shafts if unanticipated obstructions are encountered during microtunneling;
- Chemical testing of potentially contaminated soils and groundwater;
- Handling and disposal of contaminated soils; and
- Off-duty police officers to assist with traffic control and help maintain smooth traffic around the work areas.

There are a number of constraints that will also be very challenging for the contractor. City streets must be open for traffic between 7:00 and 8:30 a.m. and 3:30 and 6:30 p.m. (Matsunaga, Kwong & Okita, 1998). The contractor will have to re-open the streets to traffic during these time periods, which will require excavations to be covered, and equipment to be removed from the streets. City noise ordinances and the community will limit night work. The City has applied for a variance to allow noise levels of 85 dBA at adjacent property lines between the hours of 6:00 p.m. and 10:00 p.m., and 70 dBA between the hours of 10:00 p.m. and 7:00 a.m., Monday through Friday. Whether these limits will be acceptable to the community remains to be seen.

Status of Project

Design was completed in November of 1997 and the project was bid in December. The low bidder was Frank Coluccio Construction Company of Seattle, Washington for about $17.8 million. Notice to proceed is expected to be issued in August, 1998 and construction is expected to begin with potholing, jet grouting, and shaft excavation that same month.

Acknowledgments

The authors would like to thank the Department of Wastewater Management, City and County of Honolulu for their permission to publish this paper, and Woodward-Clyde's primary teaming partner for the project, M&E Pacific, Inc. of Honolulu.

References

Honke, J.K., Lum, G., and Raines, G.L., 1997, "Maintaining the Long Term Stability of a Sewer Microtunneling in Soft Compressible Soils: An Owner's Perspective", Water Environment Federation 70th Annual Conference and Exposition, October 18-22, Chicago, Illinois.

Matsunaga, R.M., Kwong, J.K.P., and Okita, G., 1998, "A New Perspective - Mitigation of Social Costs as a Best Management Practice", 16th International ISTT NO-DIG '98, Lansanne, Switzerland, June 8 to 11.

Woodward-Clyde International, Inc., 1997, "Geotechnical Investigation Report, Nimitz Highway Reconstructed Sewer (Auahi Street to Hotel Street) Honolulu, Oahu, Hawaii, October.

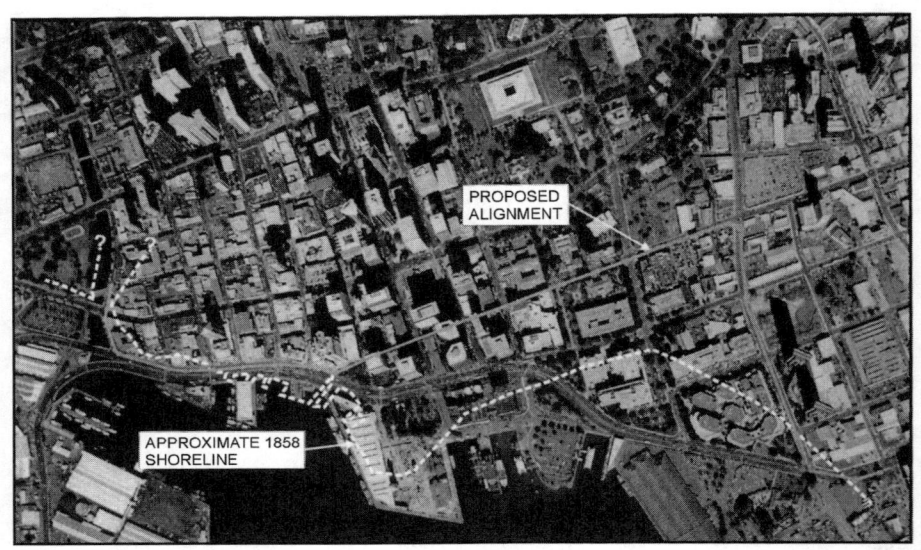

Figure 1 Project location map showing new sewer alignment and 1858 shoreline. Historic shorelines provided useful information on approximate extent of recent sediments and locations of former wharves and piers.

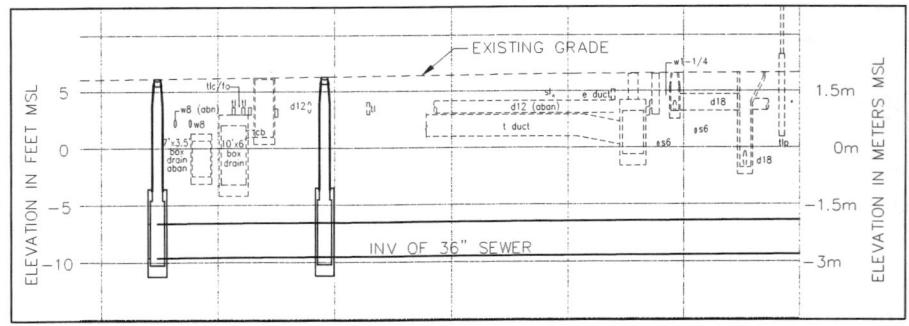

Figure 2 Example of underground utilities near new sewer line interval.

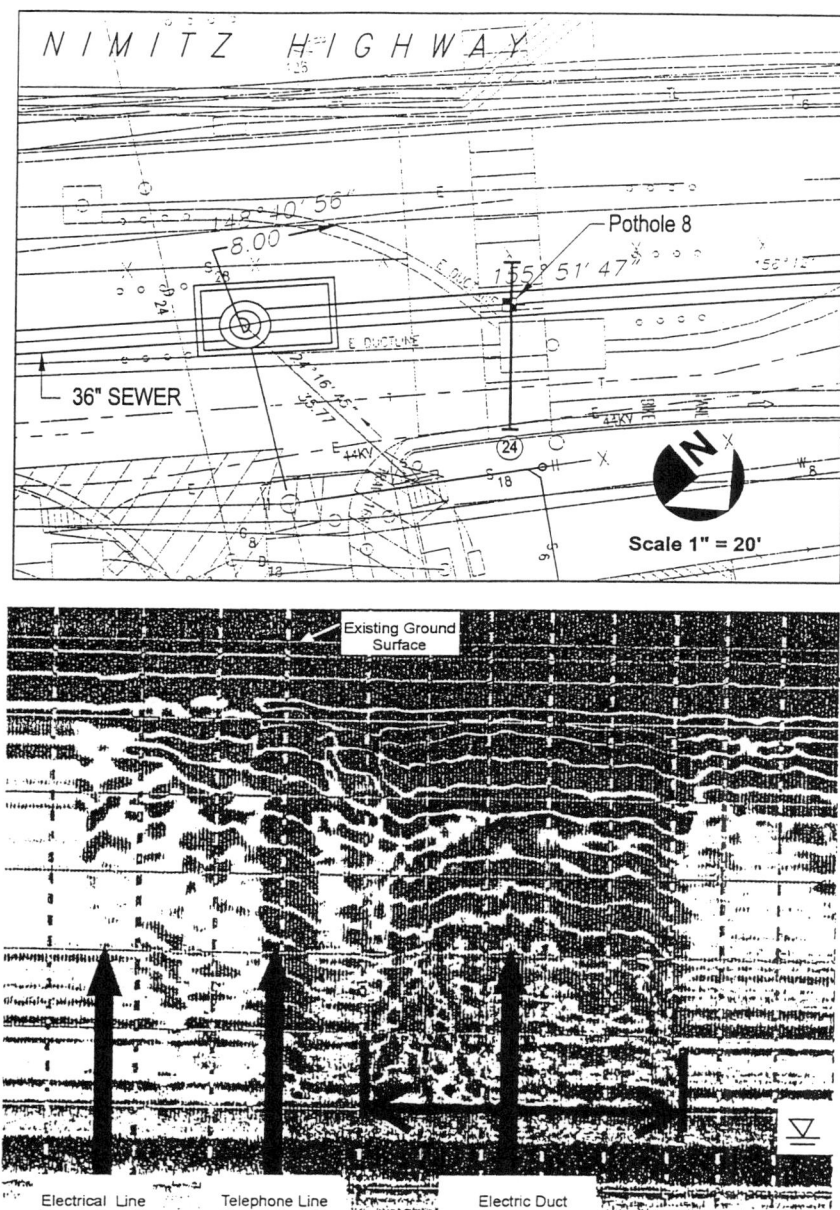

Figure 4 A. location of GPR transect.
B. Example of GPR results showing buried utility trench backfill and manhole.

DEVELOPMENT OF A COASTAL SAGE SCRUB REVEGETATION PROGRAM FOLLOWING PIPELINE CONSTRUCTION SAN DIEGO COUNTY, CALIFORNIA

Tim Cass[1] and Dave Chamberlain[2]

Abstract

The San Diego County Water Authority (Authority) has constructed several large diameter water pipelines in San Diego County. Much of the undeveloped portions of the pipeline right-of-way are vegetated with native coastal sage scrub, which is habitat for several threatened and endangered species. Federal and State resource agencies require timely and successful re-establishment of coastal sage scrub as an environmental permit condition. Most of the negative impacts to vegetation caused by pipeline construction are temporary, and revegetation can be accomplished within five to eight years. Successful revegetation programs for coastal sage scrub include the preparation of detailed revegetation specifications, close construction monitoring and a two-year post-construction maintenance period.

The three key elements of the Authority's revegetation program are topsoil salvage, supplemental seeding, and weed control. Methods of topsoil salvage, storage, and redistribution are detailed in the construction specifications. Salvaged topsoil often contains desirable native plant seed, but also contains undesirable non-native weed seed. Mulching vegetation into the topsoil prior to salvage is an effective means to incorporate seed and organic matter into the salvaged topsoil, which benefits revegetation. Supplemental native plant seed has been found necessary and can be applied by hydroseeding, drill seeding, or hand seeding. Control of selected non-native weed species is often necessary to reduce competition with native plant species.

Proper planning, specification preparation, and construction monitoring will improve revegetation success and result in lower post-construction maintenance costs. Construction site revegetation is not only required by resource agencies permit, but benefits the project and Authority by providing erosion control and visual enhancement. A successful revegetation project increases community and resource agency acceptance of pipeline construction projects.

---
[1] Water Resources Specialist, [2] Senior Civil Engineer, San Diego County Water Authority, 3211 Fifth Avenue, San Diego, CA 92103

Introduction

In 1989, the San Diego County Water Authority (Authority) embarked on an ambitious Capital Improvement Program (CIP) to improve system reliability and meet future imported water requirements for the rapidly growing San Diego region. The major CIP component is a new large-diameter regional water transmission pipeline (62.1 miles of 72 to 108-inch diameter pipeline) stretching north and south essentially the length of San Diego County. The CIP also includes three smaller-diameter east/west branch pipelines (7.9 miles of 54 to 66 inch diameter pipeline), two underground storage reservoirs and numerous flow metering facilities. All told, the Authority will be investing $835 million to implement these new facilities.

Routes selected for the new pipelines generally parallel existing Authority-owned pipelines constructed in the early 1950s to the middle 1970s. Much of the pipeline route is covered by coastal sage scrub vegetation which has naturally regrown since original pipeline construction. Coastal sage scrub is habitat to several Federal threatened and endangered species, including the California gnatcatcher. The gnatcatcher is a small songbird that was listed as a threatened species by the U.S. Fish and Wildlife Service (USFWS) on March 30, 1993. This listing had the potential to halt or significantly delay the progress of many of the Authority's ongoing CIP construction projects. Authority projects were allowed to continue pursuant to a Biological Opinion issued by the U. S. Fish and Wildlife Service (USFWS) on July 19, 1993. Mitigation requirements for temporary construction impacts included off-site purchase of California gnatcatcher coastal sage scrub habitat at a ratio of 1:1, and implementation of an on-site revegetation program for areas disturbed by construction.

To comply with the Biological Opinion, the Authority has implemented a mitigation and monitoring program. This program involves pre-construction biological surveys, design of project specific environmental mitigation measures, revegetation with native plant species, construction and post-construction monitoring, and reporting revegetation success to the resource agencies. By demonstrating to the resource agencies that construction sites can be successfully revegetated, the Authority has been able to negotiate reduced off-site mitigation requirements. Other benefits of this program include project streamlining and cost savings due to reduced mitigation requirements.

Biological Mitigation

The Biological Opinion provided conditions for preserving and re-establishing sensitive coastal sage scrub vegetation as habitat to protect the California gnatcatcher. The conditions included the acquisition of off-site mitigation acreage for direct impacts to coastal sage scrub. The CIP results in estimated impacts to 180.9 acres of coastal sage scrub. The USFWS required an off-site mitigation ratio of 1:1 (one acre of off-site preservation for each acre of habitat disturbed) based on the implementation of a successful on-site revegetation program following construction. Off-site mitigation ratios can be as high as 3:1 without on-site restoration. The Authority purchased and now manages the Crestridge Habitat Mitigation Area located east of the city of El Cajon, San Diego County to meet the off-site mitigation requirement. A total of 261 acres were purchased for $2,300,000 on January 31, 1994.

Coastal Sage Scrub

Coastal sage scrub is a low brush-dominated vegetation type in coastal areas of central and southern California, and northern Baja California, Mexico. In San Diego County, this vegetation type is comprised mostly of drought deciduous shrubs from 3 to 6 feet in height, dominated by the shrub species: coastal sagebrush (*Artemisia californica*), flat-topped buckwheat (*Eriogonum fasciculatum*), black sage (*Salvia mellifera*), white sage (*Salvia apiana*), deer weed (*Lotus scoparius*), San Diego sunflower (*Viguiera laciniata*), bush sunflower (*Encelia californica*) and monkey flower (*Mimulus aurantiacus*). Frequently scattered throughout coastal sage scrub are larger more rigid evergreen shrubs growing to heights over 10 feet including: laurel leaf sumac (*Malosma laurina*), lemonade berry (*Rhus integrifolia*) and toyon (*Heteromeles arbutifolia*).

Total shrub cover of coastal sage scrub seldom reaches 100 percent. Native and non-native annual plant cover frequently comprises from 20 to 50 percent of total plant cover. Bare ground or rock is present beneath or between the shrub canopy in many instances. Native annuals, such as California poppy (*Eschscholzia californica*), lupine (*Lupinus* sp.), owl's clover (*Castilleja* sp.), gold fields (*Lasthenia chrysostoma*), tarplant (*Hemizonia fasciculata*) and phacelia (*Phacelia* sp.) are common components during the spring, especially after fires or other ground disturbances. Non-native annuals including wild mustard *(Brassica* sp.), wild brome and oats (*Bromus* sp. and *Avena* sp.), and filaree (*Erodium* sp.) are also commonly found in coastal sage scrub.

On-site Habitat Restoration

Prior to 1989, the principal revegetation for erosion control consisted of seeding with non-native cover crop plants, such as rye grass (*Lolium* sp.), barley (*Hordeum* sp.) and vetch (*Vicia* sp.). No other special revegetation provisions were made during construction, nor was any post-construction monitoring or follow-up restoration performed, save for an occasional repair due to slope erosion. With no further involvement, the coastal sage scrub re-established to approximate pre-construction conditions in an 8 to 12-year period.

To shorten the period for successful on-site restoration, more recent pipeline construction projects include new design, construction, and post-construction management techniques. These measures are designed to promote coastal sage scrub re-establishment in a five to eight years. Construction and post-construction techniques to achieve this include the development of standard and project-specific construction specifications for revegetation to ensure compliance by the contractor, and post-construction warranty requirements. These techniques are currently part of the Authority's mitigation monitoring program.

Based upon observations of past and current pipeline revegetation results, the Authority continues to modify construction techniques and revegetation methods to enhance coastal sage scrub re-establishment. Some of the variables normally considered when designing the revegetation measures on current Authority projects are shown in Table 1.

Activity	Variables
Potential for Erosion	Steepness of Slope, Soil Type, Length of Slope. Protection of Improvements/Developments
Topsoil Salvaging	Soil Depth, Storage Location, Duration of Storage
Visual Impacts	Viewshed and Adjacent Uses, Erosion Potential
Quality of Vegetation Type	Degree of Previous Disturbance, Native Plant Cover and Composition
Seed Mixes	Seed Availability, Cost, Time of Application
Irrigation Requirements	Water Availability, Irrigation Feasibility/Costs
Weed Control	Invasive Weed Potential, Maintenance Costs

Table 1. Revegetation Design Variables

Revegetation Methods

The three key elements of the Authority's on-site revegetation program include topsoil salvage, seeding, and post-construction weed control. Each of these elements are defined in the Authority's General Conditions and Standard Specifications (May, 1996). The standard specifications provide direction for clearing and grubbing, earthwork, and general revegetation procedures. Project specifications are tailored to meet the particular site requirements and expectations for each project, including a customized seed mix and quantities. Successful revegetation has been enhanced by employing procedures at the very beginning of construction, including the salvage and redistribution of topsoil. As with the rest of the mitigation monitoring program, the revegetation specifications are updated periodically to include new information or techniques that enhance revegetation success and reduce costs.

Topsoil Salvage

Topsoil salvage has been found to be one of the best methods for increasing revegetation success. Besides retaining the native seeds, topsoil also contains valuable microorganisms that aid in the establishment of the desired native shrubs.

Topsoil salvage is the first task performed by the contractor following mobilization of construction equipment on site, and precedes all on-site excavation work. Topsoil salvage includes mulching the standing vegetation into the topsoil by track rolling, or crushing with construction equipment. Mulching vegetation into the topsoil not only adds organic matter and native seed to the soil, but reduces the need to export brush trimmings to the landfill.

After mulching the existing vegetation into the topsoil, the upper 3 to 8 inches of soil is scraped from the surface and stockpiled in linear rows along the pipeline easement. Procedures and equipment used for stockpiling vary. Often a steel-tracked bulldozer scrapes the topsoil into windrows where an excavator lifts and places the topsoil to the edge of the construction limits (Figure 1). Linear stockpiling, or windrowing, is necessary to ensure that topsoil is replaced in the same area that it was removed, and to prevent mixing vegetative types. The windrows are marked with pipeline station limits, adding assurance that the soil will be replaced in the proper location.

Figure 1. Topsoil Stockpiling Operation

Normally, topsoil windrows have been limited to a height of 4 to 6 feet because of concerns associated with compaction and biological degradation due to anaerobic conditions. However, due to limited construction easement width on the Mission Trails Flow Regulatory Structure project, the topsoil was stockpiled to a depth of 10 feet, and subjected to monitoring for compaction and biological health. Tests revealed that after one year of stockpiling, there was no deleterious increase in relative compaction. Also, soil fungi (mycorrhizae) which are considered valuable in the restoration of native vegetation, retained substantial viability at all depths. The lack of compaction with increased depth and the retained mycorrhizae viability indicate topsoil can be stockpiled to depths up to 10 feet without significant degradation of physical and biological characteristics. These results are now being considered in the design of future projects allowing topsoil stockpiles to greater depth. This would reduce total construction impact, and avoid transporting topsoil to and from the project site.

Soon after the construction is complete, salvaged topsoil is carefully spread over all disturbed areas. The topsoil may need to be decompacted after spreading because the heavy equipment compresses the soil during spreading operations. Bulldozers with ripping teeth are typically used to decompact the topsoil and subsoil to a depth of 12 inches.

Topsoil salvage often has an immediate additional benefit to mitigate visual impacts of recently completed construction. Native topsoil is usually darker that the materials excavated from deeper in the trench. By spreading the topsoil over the finished excavation, the visual starkness after construction is greatly reduced.

Supplemental Seeding

The second key element to revegetation success is application of supplemental seed following topsoil placement. Even though salvaged topsoil contains some native plant seed, it is necessary to add supplemental seed to ensure coastal sage scrub establishment. Supplemental seeding is recommended immediately after topsoil replacement to take advantage of any rainfall that occurs. Otherwise, non-native weed seed germinates with the first substantial rainfall and will have an establishment advantage over the desired native plants.

Coastal sage scrub seed mixes are designed to include the dominant shrubs found on site, native annual wildflowers, and cover crop annuals. The seed mix is based on pre-construction vegetation surveys. Fortunately, many of the dominant shrub types germinate readily from applied seed. Native annual wildflower seed is added to provide visual enhancement. Since most of the coastal sage scrub shrub seedlings are slow growing, cover crop seed is added to provide erosion control until the shrubs become established. Cover crop plants are annual plants that grow rapidly after rainfall, binding the soil with fine fibrous roots. Non-native cover crop plants, such as plantago (*Plantago insularis*) are used because this species tends not to compete with the native shrub seedlings or persist on site after the first two years.

Native plant seed should be collected from as near to the construction site as possible. Local seed collection is preferred because local plants are usually better adapted to local soil and climate characteristics resulting in better performance. Local seed collection may need to be done one-to-two seasons in advance of seeding due to the annual variation in natural seed production. Native seed from distant collection sites is discouraged to reduce the chance of introducing poorly adapted or incorrect plant species.

A typical coastal sage scrub seed mix is shown in Table 2. It is important to specify pounds per acre, purity and germination of seed to ensure good quality seed is used. With these three seed factors expressed, adjustments in the number of pounds can be made to compensate for low purity or low germination seed lots. The plant type and pounds per acre is also adjusted for each project based upon, preconstruction vegetation, season of application, past experience with seed mixes, and seed cost.

Species/Common Name	% Purity/ % Germination	Lbs per Acre
Artemisia californica/coastal sagebrush	15/50	2.5
Encelia californica/bush sunflower	40/60	1.0
Eriogonum fasciculatum/flat-top buckwheat	10/65	5.0
Hemizonia fasciculata/tarplant	10/25	1.0
Lasthenia chrysostoma/goldfields	50/60	1.0
Lotus scoparius/deerweed	90/60	4.0
Lupinus succulentus/arroyo lupine	98/85	2.0
Mimulus aurantiacus/monkeyflower	5/70	1.0
Phacelia ramosissima/phacelia	95/85	1.5
Plantago insularis/plantago	98/75	30.0
Salvia mellifera/black sage	70/50	1.5

Table 2. Sample Coastal Sage Scrub Seed Material Specification

Once collected, cleaned, and tested for purity and germination, the seed is applied to the prepared topsoil. The most commonly used method is hydroseeding. The hydroseed mixture is sprayed onto the disturbed areas from a large specialized tank truck (Figure 2). The hydroseeding mixture includes the specified seed, wood fiber for erosion control and a tackifier to hold it on the slope. Fertilizer is not added to the seed mix in most cases, since undesirable weeds will benefit more from added fertilizer than the desired native plants.

Straw is applied to all slopes greater than 2:1 in steepness at a rate of 2 tons per acre, usually by a pneumatic straw blower. Straw is one of the most effective and least expensive temporary erosion control materials available. Rice straw is typically used to reduced weed seed content compared to barley or wheat straw. Although the primary function of straw is erosion control, straw appears to reduce the drying of the top layer of soil containing seed, increasing germination success.

Figure 2. Hydroseeding Operation

Weed Control

The third key element of revegetation is weed control. The desired native shrub seedlings grow slowly and are easily suppressed by rapidly growing non-native weeds. While fast growing non-native weed species provide early erosion control benefits, weed control is needed to provide a balance between erosion control and the need to establish slower growing native shrubs.

Weeds are categorized into two groups based upon the potential to compete with native plants and the ability to persist over time. The first category is annual weeds, such as wild mustard and radish *(Raphinus* sp.), non-native brome grass and wild oats, and filaree. Annual weeds in most situations diminish the second year after construction and do not significantly compete with native shrubs. The second category of weeds are the perennial invasives, including fennel (*Foeniculum vulgare*), tree tobacco (*Nicotiana glauca*), pampas grass (*Cortaderia* sp.), and artichoke thistle (*Cynara cardunculus*). These perennial weeds are long lived and often increase in abundance over time unless controlled.

Weed control is typically performed for a two-year period following construction. Perennial weeds are removed while annual weeds are generally left in place. After two years, most of the native shrubs are established and perennial weeds are less likely to take hold. Early weed control measures immediately after construction will result in reduced total long-term weed control effort and cost.

Research Opportunities

Practical opportunities to experiment and refine revegetation techniques often can be easily incorporated into construction projects. The main objectives for these research opportunities are to increase revegetation effectiveness and limit or reduce revegetation costs.

Seeding Trials

Seeding test plot trials were added as part of the revegetation program for the Mission Trails Flow Regulatory Structure project. This trial was designed to assess the potential for improving native shrub establishment from seed by comparing hydroseeding with hand-applied seeding and mechanical seeding using a seed drill. Preliminary results indicate that each of the three seeding methods are equally effective. These results will provide more flexibility and opportunities for cost savings on projects, since hydroseeding is typically a more expensive method of seeding.

Hand-applied seed covered with straw may be most cost cost-effective on small seeding projects due to low equipment and mobilization costs. Drill seeding can be practical on larger projects where mobilization costs are prorated over the total area seeded. Drill seeding may not be feasible or effective under certain site conditions, such as steep slopes or rocky soil conditions. Hydroseeding typically is more expensive due to equipment used, fiber mulch and other additives, but is more versatile under the difficult site conditions mentioned above.

A newer seeding device, termed an "imprinter" is now being used by other agencies and private developers. The imprinter places wedge shaped depression into the soil along with seed and mycorrhizal inoculum. The ability of the imprinter to place mycorrhizal inoculum in close contact with the seed is an advantage over other seeding methods. Imprinting equipment may have similar site condition limitations as drill seeding, especially in rocky or compacted soil conditions. Imprinting equipment is developing rapidly and this technology needs to be monitored for feasibility on large scale construction projects under varied site conditions.

Mycorrhizal inoculation is not necessary where topsoil has been properly salvaged and redistributed over the construction site. Native plants were substantially inoculated with mycorrhizae two years after topsoil redistributed on the Mission Trails Flow Regulatory Structure project.

Container-grown Plants

Although the Authority has been successful in establishing many of the coastal sage scrub species from seed, some of the evergreen shrubs have proven to be difficult. Laurel leaf sumac, lemonade berry, and toyon are routinely included in the hydroseed mix, but with little or no establishment success. A study of various practical methods to increase success of establishing these evergreen shrubs by seed is scheduled for fall 1998. Results will be used to determine if there are feasible alternatives to establishing shrubs from container-grown plants with irrigation.

The Authority is investigating the use of slow-release water-holding gels to increase the success rate of evergreen shrubs planted from containers. Packaged gels are placed adjacent to installed container-grown plants and slowly release water for a period of up to 90 days. These gels and other slow water release methods may provide a more practical alternative to installing irrigation systems or hand watering.

Conclusion

Thorough planning and proper implementation are essential to a successful revegetation project. Incorporating detailed revegetation specifications into the construction documents will allow for enforceable construction compliance. By focusing on improving revegetation methods through careful monitoring, opportunities to maximize success and reduce costs often arise.

It has been helpful to describe the revegetation process with three key elements: topsoil salvage, seeding and weed control. This approach takes into consideration the need to plan for the activities and materials necessary to accomplish these revegetation tasks. Postponing revegetation planning until the end of construction results in missed opportunities to increase revegetation success and could increase revegetation costs.

The benefits of a successful revegetation project go beyond basic mitigation measure compliance. Positive community and resource agency relations can lead to faster project permitting and reduced mitigation requirements. Today, the public expects a community-friendly approach on public construction projects that takes the interests of the environment and community into consideration.

References

U. S. Fish and Wildlife Service (July 19, 1993), Biological Opinion...San Diego County Water Authority Capital Improvement Program (1-6-93-F-28).

San Diego County Water Authority (May, 1996), General Conditions and Standard Specifications.

BIDDING PROJECTS WITH ALTERNATIVE PIPELINE MATERIALS

Karen Larson Henry, P.E., MASCE[1]
Dave Zoumaras, P.E.[2]
Timothy Pflum, P.E.[3]
Mark Giandoni, P.E., MASCE[4]

Abstract

This paper shares some of the City of San Diego's recent experience on pipeline projects that included contract documents allowing alternative pipe materials. These projects include the Bonita Pipeline, East Mission Gorge Trunk Sewer Relocation, the Middletown Trunk Sewer, and the North Mission Valley Interceptor Sewer.

Bonita Pipeline

The Bonita Pipeline replacement project was bid and constructed in two phases. The first phase consisted of 1,828 meters (6,000 feet) of 1,200 mm (48-inch) diameter water transmission main located within streets. The transmission main was intended to replace the deteriorated 700 mm (28-inch) diameter existing pipeline built prior to 1920. The first phase of the project was advertised for bids in June 1993. The bidding lasted several months as the result of amendments to the contract documents. The City of San Diego followed the current American Water Works Association (AWWA) standards and the City's own design standards for the design of all pipe materials for the project. The three pipe materials included were:

> **Concrete Cylinder Pipe (CCP):** AWWA C303, field welded, cement mortar lined and coated.

---

[1] Senior Civil Engineer, City of San Diego, 1010 Second Ave., San Diego, CA, 92101
[2] Senior Civil Engineer, City of San Diego, 9485 Aero Dr., San Diego, CA 92123
[3] Senior Project Manager, Corrao-Brady Group, 3838 Camino Del Rio North, San Diego, CA 92108
[4] Senior Project Engineer, Hirsch & Company, 3550 Camino Del Rio North, San Diego, CA 92108

Welded Steel Pipe (WSP): AWWA C200, field welded, cement mortar lined & coated or cement mortar lined and tape coated.

Ductile Iron Pipe (DIP): AWWA C151, mechanically restrained, cement mortar lined and epoxy coated.

The contract documents limited the contractor to bidding only one alternative pipe material.

For CCP and WSP, the City of San Diego's standard coating thickness is 32 mm (1-1/4 inches) over the rod wrapping or steel cylinder versus 20 mm (3/4-inch) per AWWA standards. Also, the City's standard lining thickness is 20 mm (3/4-inch) versus 15 mm (1/2-inch) per AWWA. The extra coating and lining thickness is desired for the purpose of corrosion protection only. Accordingly, the contract documents stated that only the AWWA standard coating and lining thickness could be used in the loading/deflection calculations. For DIP, the cement mortar lining thickness specified was doubled from the AWWA standard of 4 mm (1/8 inch) to 8 mm (1/4 inch). As dictated by the City's Corrosion Engineer, a liquid epoxy thickness of 24 mils per AWWA C210 was the specified coating and the use of poly-bags with pipe installation was specifically prohibited. The Ductile Iron Pipe Research Association (DIPRA) questioned these requirements in writing. However, no changes were made to the specifications thereby rendering DIP less competitive.

For DIP and WSP, deflection was limited to 2 percent of the internal diameter. Deflection of CCP was limited to the internal diameter squared divided by 4000 because it was treated as a rigid pipe. For pipe design to limit deflection, the Modified Iowa Formula was specified. A 30 percent allowance for transients was required or 30 percent of the working pressure. The design pressure was 689-1103 kilopascals (100 to 160 pounds per square inch). Since the pipeline was to be constructed in phases, portions of installed pipe would not be placed in service for up to a year. Therefore, a deflection lag factor of 1.5 was specified to limit long term deflection and possible cracking of the cement mortar coating.

The depth of cover ranged from 1.67 meters (5.5 feet) to 3.8 meters (12.5 feet). Backfill was specified as conventional aggregate bedding in all cases except CCP. For CCP, the allowable deflections were exceeded with conventional bedding for depths of cover greater than 2.74 meters (9 feet). Soil-cement bedding was specified in these areas to increase the pipe-soil resistance (E') so that deflection was held within allowable limits. This requirement was computed using the minimum wall thickness for internal pressure and hoop stress and the published material from the pipe manufacturer for class 150 and 200.

During bidding, one pipe manufacturer in Southern California made it known right off the bat that they desired to bid their C303 pipe because it had the greatest potential profit for them. As one might expect, a number of contract requirements were questioned in areas where it was more difficult for them to compete. They made numerous phone calls, sent letters, cited published papers, generated calculations, and met with the Project Manager's superiors in order to influence the City to change the specifications.

The manufacturer compared the CCP deflection criterion ($D^2/4000$) to that for DIP. Since CCP is a far more rigid than DIP, the deflection criterion for CCP is accordingly more restrictive than the criterion for DIP. Calculations provided by the manufacturer using values for the moment of inertia (I) were significantly higher than those published in their design manual. These higher values were the result of including the additional cement mortar coating thickness (in conflict with the contract documents). Another requirement questioned was the structural, soil-cement bedding for CCP. Comparisons were made by the CCP manufacturer to the DIP criteria, specifically, the allowable loads for DIP being less than for CCP with no soil-cement bedding requirement. The City's position was that DIP is designed as a flexible pipe, therefore it is not suitable to calculate an allowable load (for comparison with CCP). The manufacturer argued that the additional coating increased the pipe stiffness enough to be within the allowable deflection limit specified. However, by their own admission, they were concerned about supplying the CCP with thicker coating because they were not sure that shrinkage cracking could be controlled. In essence, the manufacturer wanted the City to rely on the additional coating thickness for structural/ flexural stiffness on the one hand, while questioning the integrity of the same coating on the other.

Another proposal for CCP was to include supplemental 14 gage wire reinforcing to control shrinkage cracking in the extra coating thickness and use the full coating thickness as part of the load bearing wall in lieu of soil-cement backfill. The City's Corrosion Engineer was concerned that the wire would act as a shield when the cathodic protection system was applied. The City initially accepted this proposal but required the manufacturer to electrically bond the supplemental wire reinforcement to each bell and spigot with a minimum of four continuous longitudinal bonding "straps". The bonding strap requirement was later rescinded when it became apparent that CCP could not be manufactured with the straps without significant modifications to the manufacturing process thereby increasing the likelihood of defects. After much discussion, specifying a thicker minimum wall thickness eliminated the soil cement requirement.

Relative to WSP, there was concern voiced by one manufacturer that the shrinkage control wire (for the coating) would not be located within the middle one third of the coating. This in turn would result in shrinkage cracks in the outer coating. The City was not concerned though since the additional coating was not counted on to provide structural/ flexural stiffness. In fact, the additional lining and coating was not to be used for determining the moment of inertia or other pipe properties.

Another manufacturer requested that the City allow the use of high yield strength steel for WSP. The option to use higher strength steel would potentially allow the manufacturer to be more competitive. After researching this proposal, the request was granted and the specifications were modified to allow "up to Grade 42". However, to avoid problems with brittle mortar coatings, flexible tape coatings were specified for WSP fabricated with Grade 36-42 steel. In addition, a minimum pipe wall thickness of 6.35 mm (0.25 inch) was specified for handling purposes.

Subsequent to this change, a competing pipe manufacturer contacted the City. They claimed that cement mortar lining experiences detrimental cracking when used with high yield strength steel citing "plant observed test data". Although the issue had been discussed at the AWWA C205 (cement mortar lining & coating) committee, test data to support the claim had not been presented. When the City requested the test data, an ASCE paper entitled "Observations On Mortar Lining of Steel Pipelines" was provided by the manufacturer. It was represented as a document that demonstrated the problems of cement mortar lining of welded steel pipe (i.e. detrimental offset cracking). The paper was, however, specific to large diameter penstocks subject to transients induced by improper operation and not found to be useful for the Bonita Pipeline application. The plant observed test data was never provided.

The bids were opened on November 2, 1993. The contract was awarded to the low bidder, T. C. Construction of Santee, California based upon a total bid amount of $1,569,220. California Steel Pressure Pipe Company of Riverside, California supplied WSP. The unit pipe cost bid for 1,200 mm (48-inch) water main (including excavation, bedding, backfill, and street restoration) was $630.97 per meter ($192.32 per linear foot). The project began construction in February 1994, and was completed in March 1995.

The second phase of the Bonita Pipeline was designed with 4,877 meters (16,000 feet) of 600 mm (24-inch) diameter transmission main. It was bid in June 1994, with the same three pipe materials. The contract documents contained basically the same specifications for the pipe materials. Very few inquiries during bidding resulted. It is assumed that this was due to the modification and refinement of the specifications during the bidding of the first phase. The contract was awarded to low bidder, E. J. Meyer Company of Highland, California based upon a total bid amount of $3,081,923. WSP was supplied by Ameron Concrete & Steel Pipe Systems Southern Division of Rancho Cucamonga, California. The unit pipe cost bid for 600 mm (24-inch) water main (including excavation, bedding, backfill, and street restoration) was $370.73 per meter ($113 per linear foot). The project began construction in September 1994, and was completed in October 1995.

East Mission Gorge Trunk Sewer Relocation

In early 1992 the California Department of Transportation (Caltrans) began coordinating with the City of San Diego and other public agencies for the relocation of Mission Gorge Road in Santee, California. The City of San Diego owned the East Mission Gorge Trunk Sewer (EMGTS), a 2,438 meters (8,000 foot) long, 1,050 mm (42-inch) diameter reinforced concrete pipeline which carried wastewater from the communities of Santee, El Cajon, Lakeside, and other unincorporated areas of east San Diego County. The EMGTS was connected at the east end of the project to the 1,050 mm (42-inch) County of San Diego owned Lakeside Interceptor Sewer and the City of El Cajon's El Cajon Interceptor Sewer through a diversion structure.

Caltrans began coordinating several years before construction of the SR-52 extension having learned from previous projects that coordination between the State and local city and utility agencies was a slow, difficult process. The City of San Diego was placed in a lead role for coordination of the trunk sewer and interceptor sewer relocations due to contract agreements the City had with the County of San Diego and the City of El Cajon to transport wastewater through the EMGTS.

The City of El Cajon insisted that their 550 meters (1,800 feet) of 1,050 mm (42-inch) vitrified clay pipe sewer be replaced in kind with no loss of capacity.

The three pipe material allowed in the contract documents were:

Vitrified Clay Pipe Extra Strength (VCES) – ASTM C700

Polyvinyl Chloride Closed Profile Pipe (PVCCP) – ASTM F794

Plastic Lined Reinforced Concrete Pipe (PLRCP) – ASTM C76

Only VCES and PLRCP were approved by the City of San Diego for use on large diameter sewers. The PVCCP was considered for this project due to the successful completion of the City's Encanto Trunk Sewer project. The contract specifications limited deflection to 5% of the minimum ID shown in ASTM F794. The contractor was required to perform a deflection test on the first 100 to 150 meters (300 to 400 feet) of pipe after it has been backfilled to grade in order to meet the requirements of the specifications. In addition, the PVCCP pipe was to be retested by the contractor for the maximum allowable deflection between 11 and 12 months after installation. Mechanical re-rounding of failed pipe was not allowed.

The contract was awarded to R. E. Hazard Construction Co. of San Diego, California. PVCCP pipe was supplied by Lamson Vylon Pipe of Cleveland, Ohio. The unit pipe cost for 1,050 mm (42-inch) Sewer was $384.45 per meter ($117.18 per linear foot), and $419.72 per meter ($127.93 per linear foot) for 1,200 mm (48-

inch) PVCCP. The utility relocation project began construction in June 1996, and was completed in March 1997.

Middletown Trunk Sewer

The Middletown Trunk Sewer was constructed in 1993 and involved the microtunnel installation of over 460 meters (1,500 feet) of 450 mm (18-inch) diameter sewer. The City of San Diego prepared contract documents that allowed either open trench construction or microtunneling. Since the City was inexperienced with microtunneling, the pipe specifications were based upon recommendations from a pipeline design consultant. The two pipe materials allowed in the contract documents for microtunneling were:

Vitrified Clay Pipe (VCP): Preliminary ASTM standard C 1208-01 (1991).

Fiberglass Reinforced Plastic Mortar (FRPM) Pipe: ASTM 3262 and AWWA C950.

For VCP, the maximum allowable jacking force was 72.5 metric tons (80 tons) with a factor of safety, with respect to the allowable pipe wall thrust, of 6.8. The calculations for the potential microtunnel jacking loads were overly conservative and resulted in a minimum allowable pipe wall thickness of (0.43 inches). This minimum pipe wall thickness was 0.25mm (0.01 inches) greater than the industry standard pipe wall thickness of 10.67mm (0.42 inches). Therefore, a higher strength classification of pipe was required to meet specifications.

For FRPM, the jacking load factor of safety was 6. The maximum jacking force allowed for the FRPM was 27.2 metric tons (30 tons). In addition to the overly conservative estimates for jacking loads, the specifications for the pipe material strength requirements and pipe wall thickness requirements were combined. The combination of pipe material and physical requirements rendered FRPM less competitive.

The issues encountered during bidding involved the pipe design and the choice of factors of safety. Pipe manufacturers, both for VCP and FRPM, challenged the requirements on a number of occasions. They wrote letters and visited the Project Manager and his superiors to influence the City to change the specifications. They contended that the safety factors used for this project exceeded the standards (safety factors of 2.5 to 3.5) used by the industry. Based on the advice of the pipe consultant the City did not change the pipe specifications.

The contract was awarded to Cass Construction of El Cajon, California based upon a total bid amount of $1,135,182. Mission Clay Products supplied the No-Dig vitrified clay microtunnel pipe. The unit price bid for microtunneling was $1,574 per meter ($480 per linear foot). The actual jacking forces encountered ranged from 1 to 50 metric tons (1 to 55 tons) with an average between 13 and 18 metric tons (15 to 20 tons). (Swallow, 1994)

North Mission Valley Interceptor Sewer

The City of San Diego's North Mission Valley Interceptor Sewer (NMVIS) was constructed in 1997 and involved the construction of over 3,650 meters (12,000-feet) of 1,950 mm (78-inch) diameter interceptor sewer and over 2,100 meters (7,000 feet) of 300 mm (12-inch) to 900 mm (36-inch) diameter collector sewer pipelines. The project was divided into three phases. The first phase was the pre-purchase of the 1,950 mm (78-inch) PLRCP. The second phase involved the installation of a majority of the large diameter pipeline. And, the third phase involved the construction of the balance of 1,950 mm (78-inch) diameter pipe, the required the collector sewers and the connections to the existing collection system. The total construction costs was approximately $20,000,000.

As part of the design studies for the North Mission Valley Interceptor Sewer, the City of San Diego's design consultant performed an alternative pipeline materials study. The designer reviewed numerous pipeline materials including, high density polyethylene (HDPE) pipes (both profile wall and smooth wall), fiberglass reinforced plastic mortar (FRPM), profile wall PVC pipe materials, steel pipe, reinforced concrete pipe, and pre-stressed concrete cylinder pipe. As a result of the studies completed, it was determined that only two (2) pipe material would be suitable for use in the NMVIS project, these were.

Plastic Lined Reinforced Concrete Pipe (PLRCP): ASTM C76 and C655

Fiberglass Reinforced Plastic Mortar Pipe (FRPM): ASTM 3262 and AWWA C950.

The selection process compared the pipe materials ability to meet the challenging installation conditions that were expected in the NMVIS project. These conditions included changing ground water levels, installation depths of over 12.2 meters (40-feet), embankment fills and low strength soils. The FRPM pipe (HOBAS) and the PLRCP were the only pipe materials capable of meeting the physical and structural requirements expected on this difficult project.

The major challenges involved with the bidding of the 1,950 mm (78-inches) pipe were associated with the protests of the concrete pipe manufacturers. The two southern California concrete pipe manufacturers questioned the FRPM pipe design and went directly to the City of San Diego's project administrators to protest. To address the PLRCP pipe manufacturer's concerns the designer held numerous meetings with the pipe manufacturers and provided detailed design information. Once the pipe manufacturers reviewed the design analysis and the process was explained to them they agreed with the City and the design consultant and no changes were made to the contract documents. A major factor that helped in the defense of the pipe material design was the fact that during the pipe design process the pipe strength calculations, for both concrete and FRPM pipe, had been submitted for an independent peer review. By completing a rational design and the peer review of the design, the NMVIS engineering consultant was able to provide a cost efficient alternative pipe material design.

The estimated cost savings to the City of San Diego for the purchase of the 1,950 mm (78-inch) pipe was over $1,000,000 (20% of the pipe materials estimated cost) for the $5,000,000 pipe purchase. The cost savings were based on estimates using pipe costs from other large diameter sewer projects and the concrete pipe suppliers' preliminary cost estimates. (Batta, 1998)

For the microtunneling portion of the project the contract documents allowed both FRPM pipe and VCP. These pipe materials are the typical materials used for installing microtunneled pipelines. However, the FRPM had not been used in San Diego up to that time. The estimated cost for the clay pipe was approximately $390.00 per meter ($120.00 per lineal foot). After the bids were opened, the contractor that submitted the low bid stated that the price he had been quoted (approximately $295.00 per meter, ($90.00 per lineal foot) for the clay microtunnel pipe was the lowest he had ever received on a project.

The Phase 1 portion of the NMVIS project involved the pre-purchase of the 1,950 mm (78-inch) PLRCP, was awarded to Ameron Concrete & Steel Pipe Systems, Rancho Cucamonga, California. Phase 2 involved the installation of the 1,950 mm (78-inch) PLRCP and was constructed by DLS Constructors, San Marcos, California. . The Phase 3 involved microtunneling of over 1,524 meters (5,000 feet) of 600 mm (24 inch) pipelines was awarded to BRH-Garver, San Diego. The pipe used for the Phase 3 microtunnel construction was VCP No-Dig pipe manufactured by Mission Clay Products.

Conclusion

Additional efforts are required to design a project when bidding with alternative pipe materials. However, by including alternative materials the pipe suppliers and contractors are challenged to bid competitively to get the job. When a project includes alternative pipe materials and is designed properly, the owner will benefit by a lower overall project cost.

REFERENCES

Batta, Harris, and Giandoni, *Appropriate Trenchless Technologies...*, 1998

Swallow, W.J. III, P.E., *Microtunneling Saves San Diego Sewer Project*, 1996

Horizontal Directionally Drilled River Crossing Meets Regional Water Needs

Daniel J. Nichols, PE.[1]
Thomas J. Meinhart, P.E.[2]
Steven R. Kramer, P.E.[3]

1.0 Background

Water District No. 2 of St. Charles County, Missouri is located in one of the fastest growing counties in the United States. Once a small rural water district, District No. 2 is now a major suburban water supplier. A 30-year planning study, completed in 1996, estimates that the average demand in 2015 will be 25 cfs (16 mgd), or three times the current demand. The average demand for the year 2025 is estimated at 37 cfs (24 mgd).

The Water District currently provides its 50,000 customers water from two sources — deep rock wells and treated surface water from a 50-year-old, County-owned treatment plant. Concerns about the future quantity and quality of water for this rapidly growing district led Water District officials to consider alternative sources of supply.

2.0 Alternative Sources of Supply

Previous studies identified five potential additional sources of supply to meet the district's growing need for reliable high quality water. These included:

1. Drill more deep wells
2. Construct a new water treatment plant
3. Purchase water from an investor-owned utility located adjacent to the district in St. Charles County

---

[1] Project Principal, Sverdrup Civil, Inc., 13723 Riverport Dr., St. Louis, MO 63043
[2] Project Manager, Sverdrup Civil, Inc., 13723 Riverport Dr., St. Louis, MO 63043
[3] Project Consultant, Jason Consultants International, 2000 Massachusetts Ave., NW. Washington, D.C. 20036

4. Purchase water from an investor-owned utility located across the Missouri River in St. Louis County
5. Purchase water from the City of St. Louis Water Division which operates the Howard Bend Water Treatment Plant across the Missouri River in St. Louis County.

The water from the District's existing wells is hard, and contains high concentrations of iron and manganese. The need for softer water would require capital expenditures for softening equipment and chemicals. The District is not staffed for the increase in operation and maintenance associated with these systems. For these reasons, drilling more deep wells was eliminated as an option for increased water supply.

Building a new surface water treatment plant, estimated at $40 to $50 million, was too costly an option for the District's current customer base. Future consideration may be given to constructing a new plant once revenues have reached an appropriate level for such an expenditure.

Purchasing water from the two nearby investor-owned utilities also proved to be more costly than initially expected. The City of St. Louis was able to offer a much lower wholesale rate than the other two utilities. The design team of Sverdrup and the Water District evaluated the facilities required to deliver water to the District's system from the City of St. Louis Howard Bend plant. This option was recommended to and accepted by the District's Board of Directors.

3.0 Water System Improvements

The design team determined that conveying finished water from the City of St. Louis Howard Bend Water Treatment Plant to the District's distribution system would require approximately 20,700 m (68,000 ft) of transmission main, a high service pumping station and ground storage reservoir, and an estimated 1036-m (3400-ft) crossing of the Missouri River. Four alternative transmission main alignments were evaluated including three options for crossing the river. The three options included:

- hanging the transmission main on the Daniel Boone Bridge which crosses the river between St. Louis and St. Charles Counties
- a traditional dredge and cover construction using specially designed ball and socket pipe
- a horizontal directionally drilled crossing

Based on factors such as potential for freezing, susceptibility for damage, and structural impact, the option of hanging the transmission main on the bridge was eliminated from further consideration. Of the three crossing options considered, the dredge and cover method posed the most severe environmental impact during construction. Coupled with the expected difficulty in constructing the crossing in deep, swift moving water with significant barge traffic, the dredge and cover method was also eliminated in favor of a horizontal directionally drilled crossing.

4.0 Design Challenges

The selected transmission main route, approximately 12,800 m (42,000 ft) in St. Louis County and another 7900 m (26,000 ft) in St. Charles County, requires major creek, highway, and flood protection levee crossings. Satisfying numerous review agencies' technical requirements set the challenge for the transmission main design. However, these regulatory issues paled in comparison to the hydraulic, geological and equipment issues presented in the design of one of the largest horizontal directional drilling projects in the U.S.

The design team evaluated the hydraulics of both the City's and the District's systems, as well as the projected water demand for a 30-year planning period. It was determined that a 1067-mm (42-in) diameter river crossing pipeline would be required to convey the projected peak flow of 51 cfs (33 mgd) with a minimum residual pressure of 138 KPa (20 psi). A 40-m (130-ft) elevation change between the two sides of the river, and a limestone bluff on the river's edge on the St. Charles County side mandated a thorough and detailed analysis of the challenging site conditions, and a complete review of the horizontal directional drilling equipment and technology available in the United States.

Following a detailed review of the area's topography, and identification of a number of construction constraints, the team began to gather the information needed to establish the optimal alignment. Using aerial photographs, the team studied alternative crossing alignments. It was apparent from the presence of the limestone bluff that bedrock would be encountered in a portion of the proposed crossing. This would become an important consideration in establishing the crossing's vertical alignment. The amount and quality of rock encountered would also greatly impact the construction cost, which had to be estimated during design.

4.1 Underground Investigation

To determine the nature and condition of the subsurface materials along the river crossing alignment, the design team worked with a geotechnical contractor to drill a total of 10 test borings at 90 to 120 m (300 to 400 ft) intervals. Five of the borings were taken in the river from a small barge. Because of barge traffic, U. S. Coast Guard requirements prevented borings from being taken in the main river channel. Borings over water were located with a GPS instrument and ground surface elevations at these locations were estimated from a gauge on the Daniel Boone bridge piers. All land borings were located and elevations determined by a surveyor.

Standard Penetration Resistance Tests (ASTM D1586) were conducted in each test boring using standard split-spoon samplers and a safety drive hammer. In addition, 76- mm (3-in) O.D. undisturbed, thin-walled (Shelby Tube) samples were obtained of representative fine grained soil strata and sealed for laboratory testing. The geologists took core samples (NX size) of the underlying bedrock for later examination and classification. Boring logs were prepared to show visual descriptions of the various subsurface strata encountered. The boring logs, all field sampling and test data, and photographs of the rock cores were assembled into a Geotechnical Investigation Report. The design team reviewed the report and made it available to horizontal directional drilling contractors during the bidding phase of the project.

An important aspect of the Geotechnical Investigation Report was the presentation of the bedrock properties determined by the Earth Mechanics Institute (EMI) of the Colorado School of Mines in Golden, Colorado. EMI determined the general bedrock shear strength properties with unconfined compression (Qu) tests. Point Load tests were also conducted to obtain further shear strength properties of the bedrock. Gradation analyses were performed on representative samples of the alluvium to confirm visual classifications. EMI also conducted Density, Uniaxial Compressive Strength, Brazilian Splitting Tensile, Punch Penetration and Cerchar Abrasivity Index Tests on the bedrock samples to estimate the tool wear and production rates for drilling through the rock strata encountered in the test borings. Microscopic analysis of the two formations encountered were also conducted to provide the design team and potential drilling contractors with as complete a subsurface investigation as possible.

4.2 Directional Drilling Design

Using aerial photographs, river soundings and test boring logs, the design team identified the project's challenging geological conditions. From the test borings, the team determined that the crossing would transition from alluvial material on the St. Louis County side to bedrock approximately midway under the Missouri River. This would present a challenge to the drilling contractor, but a review of various vertical alignments showed that this difficult transition was unavoidable. The 40-m (130-ft) elevation change between the two sides of the river and the limestone bluffs on the St. Charles County side required a sharp entrance/exit angle.

It was also a challenge to select and design an optimal drillpath for installing the pipe. A gradual radius of curvature was needed to avoid overstressing the 1067-mm (42-in) diameter steel pipeline. It was preferable to keep the drilling operation in the alluvial materials as long as possible to avoid the additional expense of drilling in the limestone formation. The high bank on the St. Charles County side of the river required a sharp entrance/exit angle to reach the preferred depth below the Missouri River. The minimum depth of 6.4 m (21ft) was necessary to minimize the likelihood of hydraulic fracture created by the drilling fluids.

Each of these factors required a careful study to create a feasible drillpath for installing the pipe. Using horizontal directional drilling (HDD) design software, the design team was able to evaluate multiple options. The team analyzed the impact of drilling operations and pipe installation to find a balance between constructibility and pipeline loads.

The final designed alignment will require a 1040-m (3413-ft) boring, 1370 to 1524 mm (54 to 60 in) in diameter, drilled for the 1067-mm (42-in) diameter product pipe, and entering at either 11° on the St. Louis County side or 12° on the St. Charles County side.

The specifications allow the contractor to choose from which side of the river the pilot hole is drilled. Arguments could be made for either side. Some contractors prefer to drill up hill against gravity, while others prefer to take advantage of the forces of gravity and drill downhill. The St. Louis side provided a larger space for assembling the pipe, but required the contractor to pull the pipe over a 40-m (130-ft) elevation change.

The HDD design software calculated the maximum installation tension and maximum installation stress. The expected values for stress and tension for the entire length of the bore are shown in Figures 1 and 2. The maximum tensile stress is projected to be 110 MPa (15,911 psi), which is below the tensile stress capacity of 290 MPa (42,000 psi). The maximum tension was computed to be 1.1 MN (244,692 lbs.), which is below the tension capacity of 16.3 MN (3,657,648 lbs). These calculations assume that the pipe is made neutrally buoyant during the installation process. Based on these calculations, the proposed installation should not produce any harmful loading or stress on the product pipe.

The software calculations are for an idealized situation. They do not account for small hole deviations that may occur during the drilling process. Therefore, normal practice is to allow a safety factor in the design based upon the calculated loads. Also, contractors may select a larger rig to provide capacity beyond the ideal calculated case.

5.0 Construction Challenges

At the time of publication, construction of the river crossing was just beginning. The design team anticipates that some of the more complex issues will include:

- transition from alluvial soil to hard rock while maintaining vertical and horizontal alignment

- reaming the pilot hole, in as many as six passes, to a diameter of 1370 mm (54 in)

- pulling 1036 m (3400 ft) of 1067-mm (42-in) steel pipeline against the 40-m (130-ft) elevation difference

For an updated version of this paper, including a detailed discussion of the construction phase activities, please contact the authors.

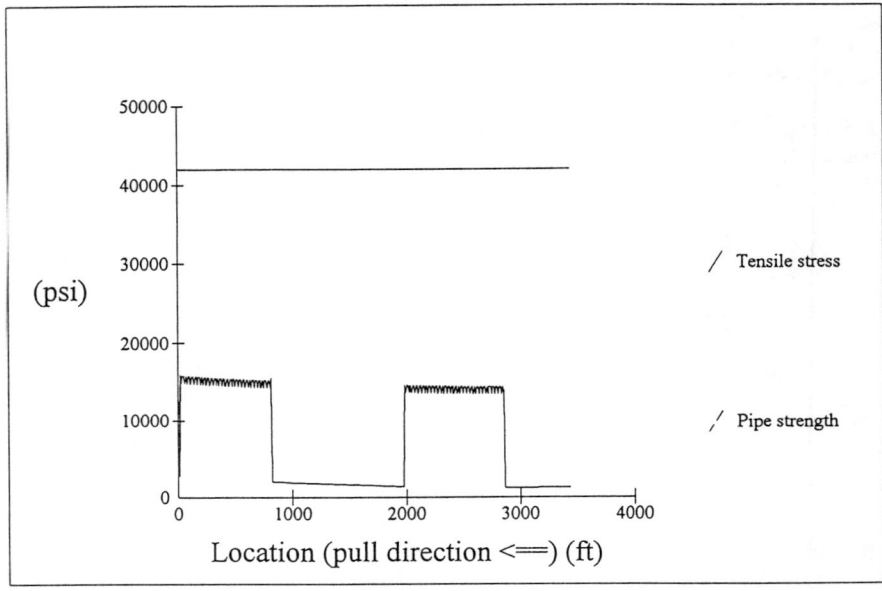

Figure 1. Projected installation tensile stress as a function of distance for the Missouri River Crossing Project

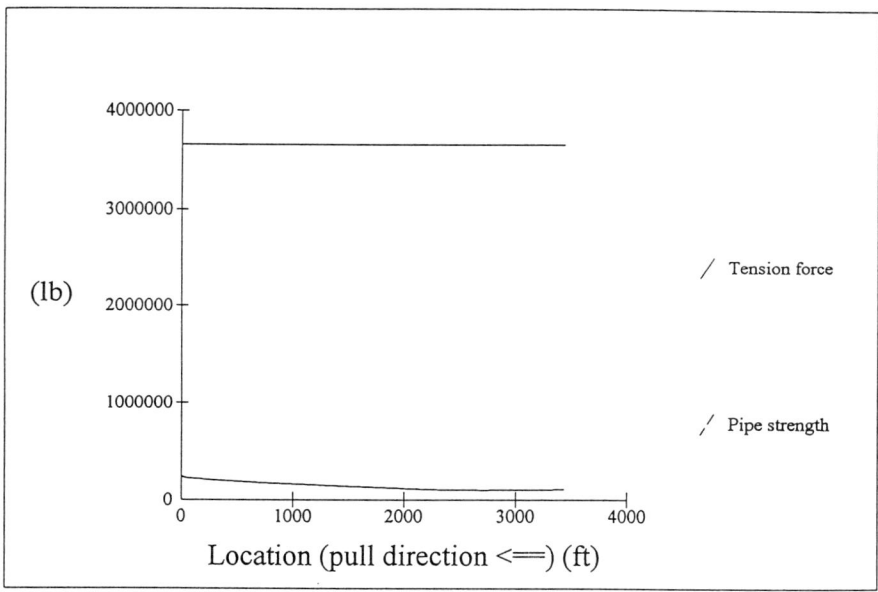

Figure 2. Projected installation tension as a function of distance for the Missouri River Crossing Project

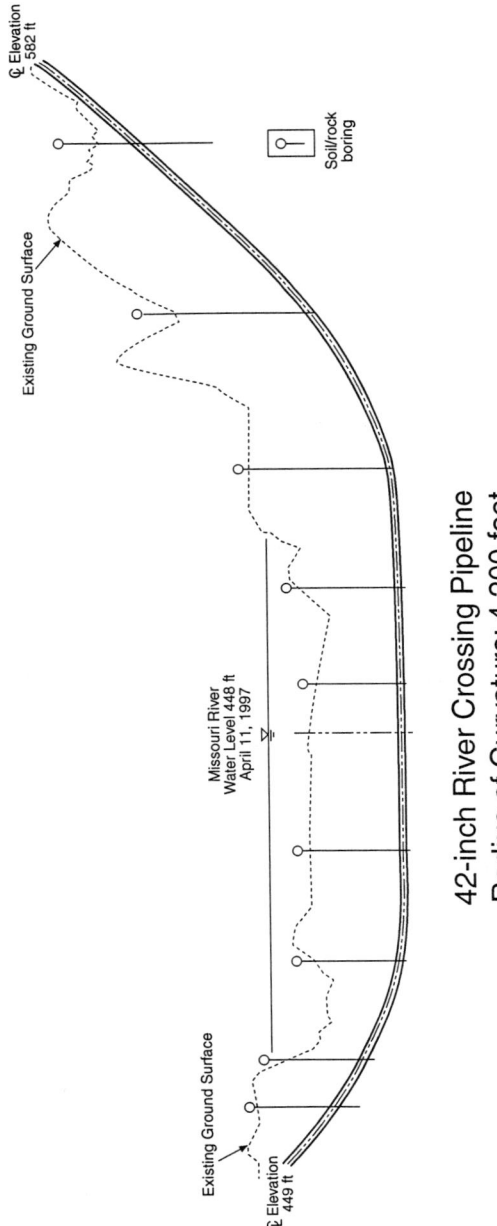

PIPELINES IN THE CONSTRUCTED ENVIRONMENT 171

HDD rig setup showing ancillary equipment

Drilling pilot hole

Electromagnetic Conductivity Survey of Pipeline Alignments for Soil Corrosivity Assessment and Corrosion Control Design for Large Diameter Water Pipelines

G.E.C. Bell[1], R.A. Pannell[2] and J.J. Galleher, Jr.[3]

Abstract

A cost effective methodology has been developed using electromagnetic conductivity survey (ECS) data along with other field and laboratory corrosivity data to design and assess the alignment corrosivity for water pipelines. The methodology results in higher data quality, improved understanding variations of alignment corrosivity and aides in material selection and design for corrosion control. Two case studies are presented to illustrate the methodology and its benefits.

Introduction and Background

New and existing pipeline alignments can require corrosivity investigations as a basis of design, replacement or repair. Traditionally, corrosivity assessments have been based on laboratory tests on soil samples from geotechnical investigations and field resistivity tests using the Wenner Four-Pin Method (WFP) per ASTM G57 [1]. Soil sampling and field resistivity test spacings vary from project to project, but generally are spaced between from 0.3 to 1.5 kilometers along any given alignment. Soil conditions and chemistry can change with in a few hundred feet. This is particularly true when alignments follow or traverse areas of periodic wet/dry conditions, which tend to concentrate soluble species (chloride, sulfate, etc.).

In the interest of improving corrosivity assessments for new and existing pipeline projects, a methodology was developed using ECS along with the

---

[1] President, M.J. Schiff & Associates, Inc., 1291 N Indian Hill Blvd., Claremont, CA 91711
[2] Senior Corrosion Technologist, M.J. Schiff & Associates, Inc., 1291 N Indian Hill Blvd., Claremont, CA 91711
[3] Assistant Engineer, San Diego County Water Authority, 610 W. Fifth Avenue, Escondido, CA 92025

traditional tools for alignment corrosivity assessments (i.e. soil sampling and ASTM G-57).

Electromagnetic Conductivity Surveys

ECS has traditionally been used by geophysicists to characterize subsurface conditions that manifest as changes in ground conductivity [2]. For example, geophysicist have used ECS for determination of bedrock depth, location of clay deposits, mapping of groundwater and salinity, locating geothermal areas, locating buried metal objects (pipelines, drums, etc.) and mapping ice in areas of permafrost.

A major factor in determining soil corrosivity is electrical resistivity [3]. Soil conductivity is the mathematical reciprocal of resistivity. The electrical resistivity of a soil is an intensive property of the soil and a measure of its resistance to the flow of electrical current. In general, corrosion of buried metal is an electrochemical process in which the amount of metal loss due to corrosion is directly proportional to the flow of electrical current (DC) from the metal into the soil. Corrosion currents, following Ohm's Law, are inversely proportional to soil resistivity. Lower electrical resistivities (higher conductivity) results from higher moisture and chemical contents and can indicate increased soil corrosivity. The corrosion engineer is primarily interested in areas along an alignment of low resistivity, conversely high conductivity. ECS is a tool that can be used by the corrosion engineer to rapidly identify areas of low resistivity along the alignment.

Operation of ECS is based on magnetic induction at low frequencies to measure earth conductivity in situ. The ECS equipment reports the average soil conductivity as a function of the distance between transmitting and receiving antennae and the antennae orientation. Commercially available ECS units can provide average soil conductivity information from between 2 to 15 meters below grade.

One drawback to ECS is that metallic structures such as fences, cars, overhead power and other pipelines will interfere with the instrumentation. Offsetting a short distance (10 to 20 m) makes it possible to minimize and mitigate such interferences.

Once collected, ECS data can be compared and correlated with field and laboratory resistivity tests per ASTM G-57. In addition, geotechnical boring logs may be used to correlate ECS data. One of the best uses of ECS is to identify additional locations for further soil corrosivity sampling and detailed resistivity investigations. Results from soil sampling and testing can be used for materials selection and design.

Once areas of low resistivity (high conductivity) are identified and, if materials of construction are such that cathodic protection is required, then ECS data can be used to locate the anode beds and design the cathodic protection system. This method has been shown to be useful when designing both shallow and deep anode beds [4].

Case Study #1: Corrosivity Assessment for an Existing Pipeline

An alignment contained parallel reinforced concrete pipelines approximately 34 miles long. Construction of the pipelines was monolithic, i.e. the pipelines were cast-in-place and have a steel cylinder with several inches of cover over the steel cylinder and at least 2-inches of cover over reinforcing steel in the concrete over the cylinder. Little or no corrosivity data were available for these pipelines. An assessment was conducted which included both ECS and WFP. The objective of the assessment was to provide a baseline of soil corrosivity data and identify areas for additional future investigation.

The ECS was conducted along the alignment using a Geonics Limited EM-31 unit. The coil spacing between transmitting and receiving coils are fixed such that average conductivity was measured to approximately 6.5 meters deep. Conductivity measurements were made such that the average interval was 3.3 meters, resulting in virtually a 100% sampling of average soil conductivity over the alignment.

Conductivity data were logged using the Geonics DL-720 data logger. The conductivity data were downloaded into a spreadsheet and converted to resistivity values in ohm-centimeters (ohm-cm) by dividing conductivity in milliSiemens/meter into 100,000. The conversion facilitates comparison of electromagnetic data with field and laboratory resistivity data. Further, soil resistivity is a more commonly used indication of soil corrosivity.

One person carried the ECS equipment and walked along an offset to the actual alignment. This particular survey was offset over the entire alignment to avoid interference from the pipeline under investigation as well as other aboveground structures that could interfere with the conductivity measurement. A second field person conducted WFP tests approximately every mile and shuttled the ECS between waypoint on the alignment.

The calculated resistivity values were plotted graphically versus pipe stationing. The y-axis scale of the plots was set at 0 to 10,000 ohm-cm, since values above 10,000 are not of interest from a corrosion standpoint. A sample plot of the data is shown in Figure 1 and is typical of data from the EMS method. It should be noted that not all overhead or underground metallic structures could be avoided. They show on the plots as spikes of very high or low resistivity value, depending on their orientation with respect to the transmitter and receiver.

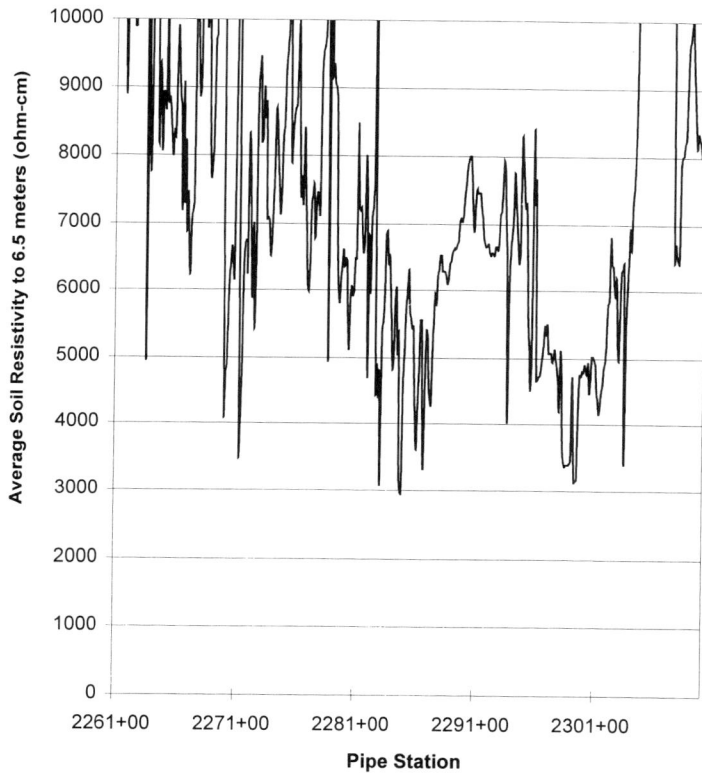

Figure 1: Sample of ECS Data

A comparison of the results from ECS and WFP methods is shown in Figure 2. Correlation between the two methods was good. No laboratory data were available for comparison from this assessment.

As a result of the assessment with ECS data, eleven areas of low resistivity were identified for further investigation

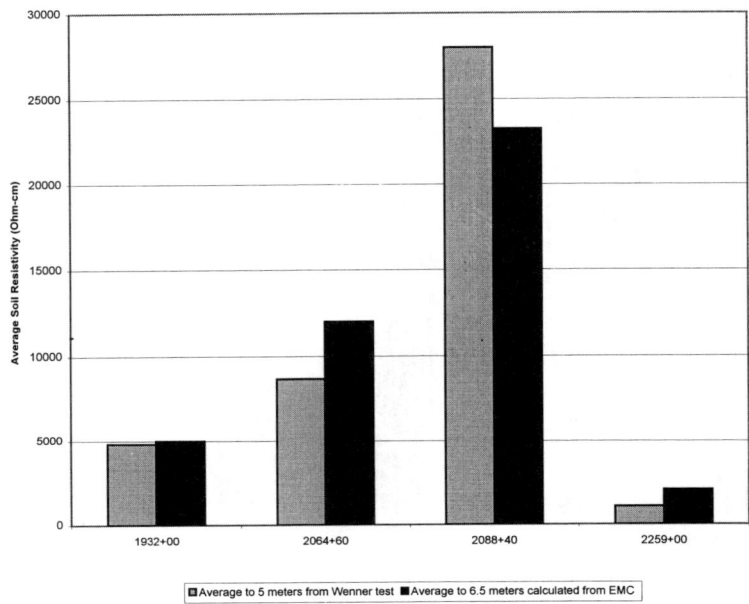

Figure 2: Comparison of ECS and WFP Average Resistivities

Case Study #2: Corrosivity Assessment for a New Pipeline

Construction of a new 3.65 meter diameter pipeline was planned. Pipeline materials under consideration were steel with a cement-mortar coating or steel pipe with a dielectric coating and cathodic protection. Total length of the pipeline was approximately sixteen miles.

As part of the soil corrosivity study, an ECS was conducted along the alignment using a Geonics Limited EM-34 unit. The EM-34 survey equipment is similar in operation to the EM-31 equipment used in the first case study with the exception that two people separated by a 10-meter cable carried the transmitter and receiver coils. With this arrangement, the EM-34 allowed simultaneous investigation of soil conductivity to 7 and 15 meters. This was necessary due to the anticipated depth of the large diameter pipe.

Average conductivity measurements to 7 and 15 meter depths were made at 3 to 35 meter intervals along the alignment using the horizontal and vertical dipoles. Larger intervals were used in continuous areas of similar electrical properties. Smaller intervals were used to define areas of changing electrical properties.

These data were logged using the Geonics DL-516 data logger. The conductivity data were downloaded into a spreadsheet and converted to resistivity values in ohm-centimeters (ohm-cm) by dividing conductivity in milliSiemens/meter into 100,000. The calculated resistivity values (average values to 7 and 15 meters deep) were plotted graphically versus stationing of the pipeline. A representative sample of these data are shown in Figure 3.

The data in Fig. 3 clearly shows the variation of soil resistivity with depth and the value of using ECS with two orientations for pipes with large soil overburens.

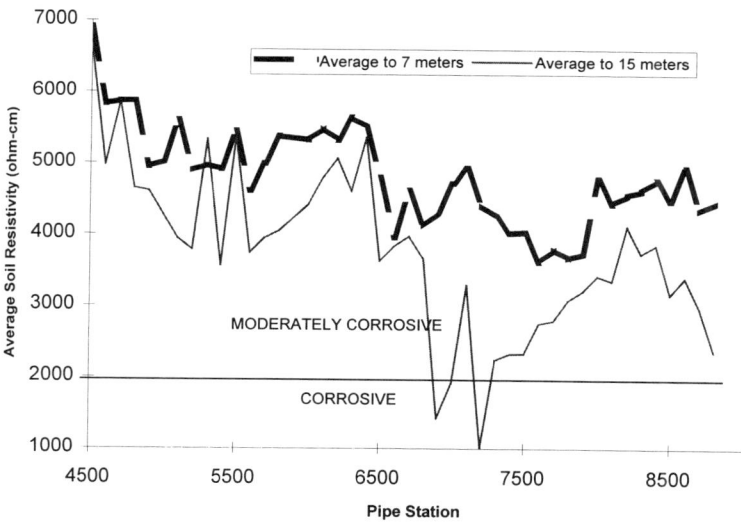

Figure 3: EM-34 Data from 3.65 meter diameter pipeline Alignment

In this case, the corrosivity of the soils at pipe depth were indicated by the ECS to be substantially more corrosive than the shallower soils. These results were confirmed by laboratory testing of soil samples from geotechnical borings. The increased corrosivity resulted in the decision by the designer/owner to install additional corrosion control in the vicinity of the change in corrosivity of the alignment.

Figure 4 shows the effects of pipeline crossings on ECS data. As with other in situ electrical property tests such as the WFP method, buried metal structure can and do interfere with the results of the method. Fortunately with the ECS, data are collected with such density that the effected areas a clearly visible in the data set and can be accounted for in the analysis and design.

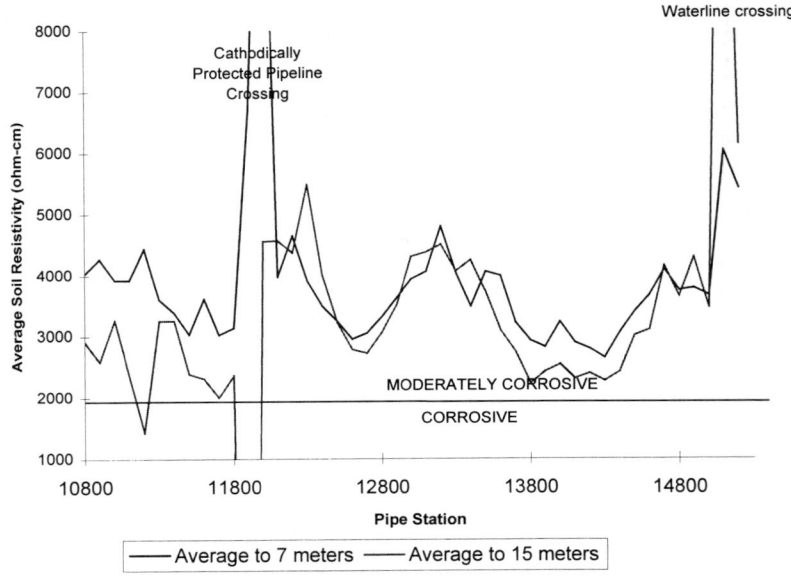

Figure 4: Effects of Buried Metal Pipelines on EM-34 Data

Summary and Conclusions:

A cost effective methodology has been developed using Electromagnetic Conductivity Surveys (ECS) data to design and assess the alignment corrosivity for water pipelines. The methodology results in higher data quality, improved understanding variations of alignment corrosivity and aides in material selection and design for corrosion control.

The following conclusions can be drawn from the case studies and information presented in this paper:

1. The ECS methodology was applicable for both new and existing alignments. The ECS methodology is particularly powerful when combined with traditional methods for alignment corrosivity analysis (i.e. soil sampling and Wenner Four Pin method).

2. The ECS provides a nearly continuous sampling of electrical conductivity along an alignment which provides the pipeline engineer with a more complete view of alignment corrosivity.
3. ECS data were in good agreement with WFP data.
4. ECS suffers from the same interferences due to buried metal structures as does the WFP method. However, the increased sampling frequency of the ECS improves interpretation and data analysis.

References:

1. "Standard Method for Field Measurement of Soil Resistivity Using the Wenner Four Electrode Method," *Amercian Society of Testing and Materials, Section 3.02, G-57,* Philadelphia, PA., 1993.
2. J.D. McNiell, "Electrical Conductivity of Soil and Rocks,"Geonics Ltd Techincal Note TN-5, Mississauga, Ontario, Canada, 1980.
3. Melvin J. Schiff, "What is Corrosive Soil?," Western States Corrosion Seminar, 1992.
4. F. W. Hewes, "Prediction of Shallow and Deep Groundbed Resistance Using Electro-Magnetic Conductivity Measurement Techniques," CORROSION/87, NACE International, Houston, TX, 1987.

POLYETHYLENE ENCASEMENT OF BURIED CONDUIT

George L. Ash[1]

Abstract

Typically, buried conduit consists of products manufactured of reinforced concrete, steel or iron pipe, polyethylene, or a combination of these products. The longevity of the buried system depends on several design and installation features, including resistance to corrosion. There are numerous causes of corrosion, some of which are oxygen replenishment, presence of sulfides, pH of surrounding soil environment, likelihood of stray direct electrical current, and the influence of microbiologically induced corrosion (MIC). These factors, either singularly or in combination, indicate the need for increased attention to adequate corrosion protection of the buried system.

Introduction

The selection of required conduit materials requires that proper attention be given to strength factors based on expectations of internal pressure, earth loads, and soil environment. The environment surrounding the installation may dictate that additional protection be provided to insure optimum system longevity against corrosive elements. Often this is determined by performing a survey to determine what conditions exist at several locations along the installation site. Once the conditions are defined the appropriate selection of protective materials can be determined. Encasing the conduit in polyethylene is an accepted practice that can provide long term passive protection against many aggressive elements, including stray direct electrical current.

When polyethylene encasement is the selected method of protection for the installation, it is important to select material that either meets established industry standards, or adequately define specific requirements to suppliers.

---

[1]Marketing Analyst, Fulton Enterprises, Inc., 108 Walter Davis Drive, Birmingham, AL 35209

Also, it is equally important that the selected material be properly installed. Proper installation requires that the conduit is clean, with no muck or debris that can become entrapped inside the encasement material. Also, any tears or damage must be repaired prior to back filling. Lastly, backfill should be accomplished in a manner that will not damage the installed polyethylene film.

Polyethylene Encasement Materials

Polyethylene film is generally classified as Low Density Polyethylene (LDPE) and High Density Polyethylene (HDPE). Low Density Polyethylene was developed nearly fifty years ago using ethylene gas to form ethylene molecules into a long, tough, polymer chain. This thermoplastic polymer is resistant to most chemicals and acids. It has good moisture resistance and excellent electrical resistance. Virgin LDPE resins provide very tough film characteristics, high dielectric strength, and low cost. It's limitations are that it performs poorly in the presence of Ultra Violet light and is therefore, subject to weathering, however, it's weathering characteristics can be improved by introducing carbon black as an Ultra Violet inhibitor. Also, LDPE does not perform well in resisting permeation by gases and vapors.

In the 1950's blown film technology allowed processes to be developed which could efficiently produce high quality polyethylene film. In this process, pelletized polyethylene is forced through extrusion dies under high heat and pressure to form a bubble of polyethylene. Inside the bubble, air is introduced to maintain an expanded tube which rises inside its production tower to a height of approximately sixty feet. At the top of the production tower the material is sufficiently cooled to allow it to feed into rollers which then carry the finished film through processes that result in rolls of finished polyethylene tube. While providing high efficiency and low manufacturing cost the process produces material that has higher physical strength characteristics in the direction of travel through the extrusion heads (Machine Direction) than it does in the transverse direction. Therefore the extruded material is more easily torn when force is applied in the transverse direction. Visual examination will reveal small die lines which easily identify the direction of travel. It is important to understand this relationship when specifying minimum tensile and yield strengths in blown film.

High Density Polyethylene (HDPE) was introduced in 1957. The processes of producing the two types of material are fundamentally different providing characteristics in HDPE that are not available in LDPE. High Density Polyethylene gives better chemical resistance than the low-density materials and is more resistant to permeation. Both of these enhanced properties are important to good protection of buried conduit. While exhibiting the enhanced resistance to chemical and permeation it is still important to know that it is also produced as a blown film and has similar strength characteristics in relation to direction of film travel in the manufacturing process.

Cross Laminated High Density Polyethylene is the most recently introduced material for encasement of buried conduit. While retaining the chemical and permeation resistance of HDPE, the Cross-Lamination process provides a toughness to withstand field conditions not found in other forms of polyethylene. The Cross-Laminated High Density Polyethylene film process begins with HDPE manufactured using the blown film process. Slitting the film tube, to form a sheet of film, which is then slit on the bias and stretched to reduce yield and increase tensile properties, further processes the material. The film is then laminated in a manner that provides the maximum tensile strength of one layer in the longitudinal direction and the other layer in the transverse direction. This process provides all the properties of HDPE and provides the highest tear and puncture resistance of the currently available film systems. Because of these processes it is practical to reduce the film thickness and resultant weight, enhance physical characteristics, and maintain required dielectric properties. See comparison between LDPE and CL/HDPE in Table 1.

In the past five to ten years there have been requirements for color-coded polyethylene encasement to help identify various types of water systems. While it is practical to color-code LDPE, provided that the required quantity is large enough to warrant a production run of material, this option is not available for Cross Laminated HDPE. The reason is that the color constituents have to be introduced at the blown film stage of production. Since the Cross-Laminated HDPE must be stretch annealed, any color that may have been in the blown film will be reduced to a near white color in the finished product. For these reasons, most LDPE films are furnished as black (using carbon black as an ultraviolet inhibitor) and colors (stabilized for ultraviolet conditions), and Cross Laminated HDPE films are furnished as white with ultraviolet stabilizers. It is important to note that even when polyethylene film has been manufactured with ultraviolet inhibitors it is not practical to expose it to sunlight for several weeks without some reduction in physical properties. Therefore, always protect the film from direct sunlight either by storage inside or by requiring that a separate covering material be supplied around the roll for protection from ultraviolet exposure until ready to install the material.

Specifications for Polyethylene Encasement Material

The ANSI/AWWA C105/A21.5 Standard [1] has been in existence since 1972. This standard followed research by the Cast Iron Pipe Research Association (CIPRA, currently DIPRA, Ductile Iron Pipe Research Association). The standard shows that polyethylene encasement studies began as early as 1951, which led to field installations and subsequent studies in 1958. This standard is the basis of many specifications relating to polyethylene encasement of buried conduit. Although this standard was specifically written for ductile iron pipe applications, it could be a valuable reference when designing other buried installations. When using an established standard, such as this one, or others, it is important to note that it is the users responsibility to determine if the suggested criteria is applicable to their particular requirement.

When specifying Polyethylene Encasement materials for corrosion protection it is important to emphasize the need for defining the required raw material, which in the case of ANSI/AWWA C105/A21.5-93 [1], the standard calls for virgin material meeting ASTM D1248-89 requirements. One reason is that much of the corrosion protection information available today comes from tests which began thirty to forty years ago in various test burial sites around the nation using the specified LDPE material. Even though the standard only required a minimum dielectric strength of 800 V/mil (31.5 V/μm) thickness, the specified material provided dielectric protection in excess of 2000 V/mil (64.4 V/μm). Depending upon the type of metal, direct current of one ampere discharging into the soil can remove approximately 20 pounds of metal per year. While in actual cases most current measurements are only in thousands of an ampere, the need for adequate protection from stray direct current is important.[2]

Another reason for defining the required raw material is that minimum tensile strengths are generally enhanced when using the virgin raw material. Tensile Strength at Break, using ASTM Test Method D-882, for 8 mil virgin LDPE which meets ASTM D-1248-89, is in excess of 2,500 psi (17.25 MPa) which is approximately twice the minimum shown in the C-105 Standard. The importance of proper tensile strength is the reduced likelihood that damage can occur during installation and back filling, that can reduce the effectiveness of the encasement installation.

In most instances you will find that suppliers of this material have established their product line using film thickness and tube diameters based on the ANSI/AWWA C105/A21.5-93 Standard. Therefore, if your design is for conduit up to approximately 64" in diameter there is an established range of sizes that will likely fit your installation. The existence of these standard products should not limit the engineer in selecting material to meet required film thickness or tube diameter.

How to insure delivery of required design properties

If polyethylene protection is selected for an installation, regardless of the reason or combination of reasons, it is important to determine that the design properties are represented in the material. One of the properties that should be checked is the film thickness. Since blown film techniques require that the film be produced as a bubble of film that rises from the extrusion equipment via a column of air, it is possible that one side of the bubble may be thicker or thinner than the other side. Therefore, it is good practice to check the thickness of the film at various points around the circumference rather than at one location only. Also, specify that certification documents specifically reference the properties that you requested in your design.

Polyethylene encasement film can be supplied to the job site and at the time of delivery meet every aspect of the design criteria and be certified as such. However, it is important that it be handled in storage for these properties to be applied in the installation. As tough as it is, polyethylene film is highly susceptible to ultraviolet damage, cuts, chafing, and abrasion. Verify that proper handling techniques are in place at the job site to insure that all of the value of the material is available at the time of installation.

Installation Requirements

Proper installation is no less important than requiring the proper encasement material. Since polyethylene encasement was first used in gray iron and ductile iron installations, most of the installation procedures are derived from those sources. Generally, ANSI/AWWA C105 Method A_1 is used and is practical on pipe with diameters up to approximately 64". Practical manufacture of blown film with a layflat dimension of above 121" is not available in the industry. Therefore, for pipe diameters above 64" the ANSI/AWWA C105 Method C_1 is more practical since it provides for the conduit to be wrapped in sheet material rather than encased in tubular material. As pipe diameters get larger it becomes more difficult to use methods which require that the conduit be encased before it is placed into the trench. When the pipe sizes are so large that it is not practical to wrap them and place them with the encasement material undamaged into the trench it is possible to use an open trench method. Here the polyethylene sheet is placed into the trench with the edges lying outside the trench, or folded inside the trench, to allow them to be pulled around the pipe after the pipe joint has been made.

Regardless of method, it is important that the installation process provide clean conduit with no dirt, mud, or other debris that may become encased inside the polyethylene enclosure. During the installation process, insure that any damage to the polyethylene film is repaired using polyethylene tape or patched using pieces of polyethylene encasement and polyethylene tape. Also, during installation, protect the film from damage by providing padding of wire rope slings, or avoid the use of wire rope slings and use wide nylon web slings to place the wrapped pipe into the trench. Finally, when installing back fill, prevent damage to the polyethylene film by removing rocks, boulders, limbs or sticks, or any material that would potentially damage the film.

In some installations cold weather or moisture can prevent polyethylene tape from adequately adhering to the polyethylene encasement. The installation can be made by using plastic tie straps, twine, or other materials that keep the polyethylene encasement snug around the pipe. Plastic tie strips are available that cinch tightly without buckles or clips that can speed up installation time in many instances.

Conclusions

Polyethylene encasement of buried conduit is a proven, effective, practice to reduce corrosion in buried conduit. The technology of polyethylene film has advanced from early low density polyethylene material to today's cross laminated high density films which provide higher tensile and dielectric strengths, and are less permeable than the low-density films.

Once the determination is made to use a polyethylene encasement system it is important to require certification that the various film requirements are in the finished product and to test to determine that the proper film thickness is provided throughout the film.

It is important to recognize the limitations of polyethylene products, especially thin films. Polyethylene does not perform well in the presence of ultraviolet light and must be protected. The film is also susceptible to damage from cuts and abrasion and should be examined prior to installation.

Effective installations require that the proper film be installed without damage to the film and that the conduit be free of dirt and debris that could become entrapped and lead to corrosion damage. Refer to established industry standards for installation or request installation guidelines from the film supplier.

Table 1
Physical Properties Of Virgin 8 mil Low Density Polyethylene Film Compared To Virgin 4 mil Cross Laminated High Density Polyethylene Film

Property	8 mil LDPE	4 mil C/L HDPE
Tensile Strength at Break ASTM D-882 (psi)	2745	8829
% Elongation at Break		
Machine Direction	475	368
Transverse Direction	573	203
Dielectric Strength V/Mil	2270	3420
Dart Impact ASTM D-1709.B gms	466	1229

Test by independent testing laboratory. Actual film thickness is 8.35 mils for LDPE and 3.7 mils for CL/HDPE.

References

1. ANSI/AWWA C-105/A21.5-93, American Water Works Association, 666 West Quincy Avenue, Denver, CO 80235

2. Richard W. Bonds, Causes, Investigation, and Mitigation of Stray Current Corrosion on Ductile Iron Pipe, Corrosion 91, NACE Annual Conference, Paper No. 516, P.O. Box 218340, Houston, Texas, 77218 (March, 1991)

George Ash received his undergraduate degree in Business Administration from the University of Montevallo, Alabama, in 1977. He has approximately fifteen years experience in purchasing polyethylene encasement products for the water works industry.

Cathodic Protection of Cast Iron Pipe

Nicholas J. Irias[1]
Marilyn L. Miller[2]

Abstract

Presents a pilot scale installation of cathodic protection on cast iron pipe. Pipe location, excavation and electrical bonding technologies are discussed. System measurements from the test installations are summarized, and an economic analysis is provided.

Introduction

The East Bay Municipal Utility District distribution system contains approximately 6000 km of water pipelines, of which 2500 km are constructed of cast iron. Over 75% of the pipe leaks experienced by the District occur on cast iron pipe, and the leaks on cast iron pipe are expected to increase unless remedial action is taken. The District has responded to this issue by launching a pilot program to determine whether cathodic protection (CP) systems can be retrofitted to cast iron piping, as a cost-effective means of extending the service life of the pipe. This report will describe the first phase of the pilot test and summarize test results.

Examples of CP installations on cast iron pipe in Canada

Corrosion engineers in Canada have installed CP on cast iron and ductile iron pipe networks in Canada, with excellent results. Leak rates on iron pipe in some municipalities were reportedly reduced by an order of magnitude following the installation of CP systems on that pipe.

It is often assumed that CP is not suited to cast iron pipe due to the pipe's inherent lack of electrical continuity. Even leaded pipe joints have high electrical resistance at the joints, and often lack continuity altogether. Many of the cast iron CP installations in

---

[1] Senior Civil Engineer, East Bay Municipal Utility District, 375 11th Street, Oakland, CA 94607
[2] Director of Engineering and Construction, East Bay Municipal Utility District

Canada overcome this problem by taking advantage of electrical grounding and bonding practices in that country. Typically, metallic water service laterals are not insulated from either the main or the meter, and are "tied" to the electric service neutral at the service panel. An electrical bond to a water main can therefore be established at any point where the electrical neutral is accessible. Because virtually every segment of iron pipe has at least one service lateral, the lack of continuity at pipe joints ceases to be a problem.

Considerations for CP installations in the United States

Installation of cast iron CP systems in the United States is complicated by plumbing practices that include the installation of insulators on service laterals, either at the water main or the water meter. Depending upon the insulation method used at each service, an electrical bond to the water main requires either a connection to a service lateral at the meter box or an excavation to allow direct bonding to the main. If the pipeline is to be made electrically continuous, a bond must be made to each pipe segment. Alternatively, bonding jumpers can be installed between the 5.5 m (18 ft) pipe segments.

Test CP System Description

The test included the installation of CP at three sites, on a total of 212 m (700 ft) of 10 cm (4") cast iron piping. The typical installation is shown in Figure 1.

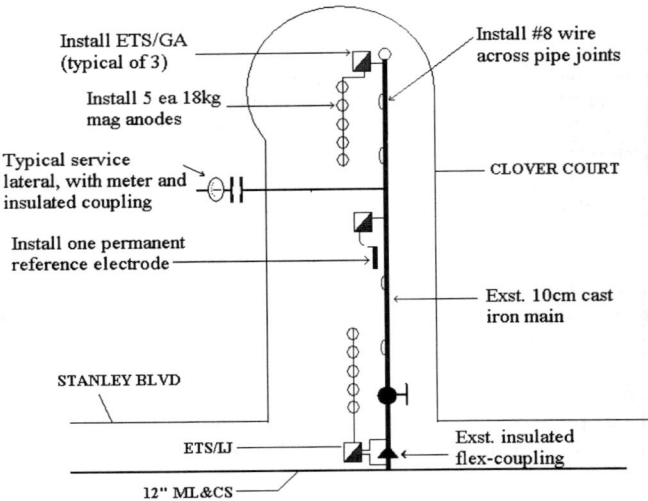

Figure 1. Typical Test CP Installation

The test sites were selected on residential cul-de-sacs, in an effort to minimize traffic impacts. The pipeline to be protected at each site began at the blow off valve at the end of the cul-de-sac, and continued over 100 meters to an insulating joint. Prior to the CP installation, all insulators at water service meters were tested and corrected as necessary, to avoid shorting the CP system to the electrical neutral. The CP system at each site consisted of five 18 kg (40 lbs) anodes installed at each end of the main, and a test station with reference electrode at the mid point of the main. All pipe joints were made electrically continuous with thermite-brazed bonding jumpers.

Design Considerations

The electrical resistance at joints in unbonded cast iron pipe varies substantially, making the direct measurement of CP system current requirements impossible prior to the installation of bonding jumpers. If a portable test rectifier were to be connected to the pipeline to measure the required current, the extent of the pipe surface area involved would be unclear. The tester would know with certainty that at least one pipe segment had been energized, but would not know whether adjacent pipe segments were energized or partially energized, due to the possibility of high electrical resistance across pipe joints.

To design the pilot test CP systems, the long-term current density of the structure was assumed to be 21 mA/m^2 (2 mA/ft^2). The current density and the surface area of the pipe were then used to calculate the current requirement of each system. To be conservative, total anode mass was calculated such that the anodes could produce twice the design current for a period of 20 years.

Each test installation was designed with test stations at the end points as well as at the mid-point of the main, to verify that complete protection had been achieved at pipe sections most remote from the anodes. The number of test stations installed for production CP installations would likely be reduced in order to minimize cost. Electrical continuity of the pipeline, established via bonding jumpers, allows routine monitoring of the system to be accomplished with a single test station.

Accurate Location of Pipe Joints

Installing bonding jumpers via small diameter excavations requires a means to accurately locate cast iron pipe joints. District records include as-built drawings of older pipeline installations, but these drawings rarely indicate the lengths of each pipe segment. Similarly, records of leak repairs do not indicate precisely where short lengths of non-metallic pipe may have been installed. The scarcity of good records makes radio locators indispensable when retrofitting CP to cast iron pipe.

At each site, the centerline of the pipe could be located by observing the location of valve pots on the main. A high frequency radio locator was also able to determine the centerline of the pipe. Capacitive coupling allows the high frequency signal to span joints that have poor conductivity. The high frequency signal can even span insulating joints, although with some attenuation of the signal.

In many cases, the distances of pipe joints from a valve are multiples of the length of one pipe segment (5.5 m). However, odd pipe segment lengths can make joint location more difficult. The high frequency locator is unable to locate joints and other minor discontinuities, because the signal is not appreciably attenuated when it spans a very short gap.

For the pilot test, a low frequency radio locator was used to verify the location of joints. The low frequency signal has very low capacitive coupling at electrical discontinuities; thus an attenuation or loss in the detected signal indicates a discontinuity. The low frequency locator was able to pinpoint most joints, and even located an electrical discontinuity beneath a stainless steel repair clamp. Only a few leaded pipe joints could not be located, likely due to a high degree of electrical continuity at the joints.

Excavation Methods

The selection of excavation tool affects the impact on customers, traffic control complications, and paving repair expenses. Two excavation methods were tested during the course of the CP system installations. At the first two sites, a contractor used vacuum excavation to install anodes and bonding jumpers. At the third site, District crews made anode and bonding jumper excavations with a conventional backhoe. In each case, a minimum 40 cm x 40 cm opening was first cut in the asphalt using pneumatic tools.

Each bonding jumper required an excavation approximately 30cm x 60cm at pipe joints, to the top of pipe, less than one meter deep in all cases. Each anode required an excavation 30cm x 30cm to a depth of 2.6 meters.

The vacuum excavator was a Miller Vac-Hoe operated by the contractor, Miller Pipeline. The excavator consists of a truck-mounted vacuum for soil collection, and a compressed air jet that displaces soil. A unique benefit of the vacuum excavator is that it is virtually impossible to damage buried cables and pipes, since the unit does not use cutters to remove soil.

Most of the vacuum excavations were made very quickly and with a moderate noise level. Exceptionally wet soils, hard clays and large rocks encountered at some locations clogged the vacuum, slowing the progress and producing extremely high noise levels. The average time required for each excavation, including time to install CP components, was 75 minutes.

The average time required for each backhoe excavation, including time to install CP components, was 45 minutes. When compared to the vacuum excavator, the time savings are considerable, however the excavations tend to be substantially larger. This requires a truck on site for trench spoils, and results in greater pavement repair costs.

Electrical Connections to Pipe

All electrical connections to pipe were accomplished using thermite brazing. In no case did this operation cause pipe failure due to thermal stresses.

While the brazing works well for shallow pipelines, this technique becomes more difficult as the depth to the top of pipe increases. Modified electric stud welders have been used with success in Canada, to bond to pipes as deep as 2.4 meters (8 ft), and the District will be investigating that technology in the future.

CP System Measurements

At each site, the soil resistivity and native potential of the pipe were measured prior to CP system installation. Immediately following installation, the pipe-to-soil potential was measured at all test stations, and the anode current measured at the galvanic anode test stations. These measurements were repeated after 6 days.

The observed pipe potentials indicate that effective cathodic protection has been achieved. Two months after installing the systems, the minimum pipe-to-soil "on" potential was 979 mV. Average current density has declined from approximately 47 mA/m^2 (5 mA/ft^2) to approximately 32 mA/m^2 (3 mA/ft^2).

The observed currents remain somewhat higher than the design current of 21 mA/m^2, but it is likely that the pipe is not yet fully polarized. When a CP system is activated, a galvanic cell is created, with hydrogen being evolved at the cathode (the pipe surface). In time, an insulating barrier is formed and the pipe is said to be polarized. The current required to produce a given pipe potential is much lower for a polarized pipe.

CP system measurements are available in the appendix.

Economic Analysis

An economic evaluation was performed, to compare the alternative of CP installation on cast iron pipe to replacement of the pipe. For this analysis, a number of assumptions have been made:

- The installation of CP on cast iron pipe will extend the life of the pipe by 20 years, rather than indefinitely extending the life of the pipe.
- The cost of the CP installation was $148 per meter ($45 per foot) for the cost of the test installation. This cost includes all design, construction and inspection costs.
- The cost of pipeline replacement is $620 per meter ($189 per foot), based on the District's annual appropriation for pipeline replacement and annual replacement rates. This cost includes all planning, design, permitting, construction and water quality testing costs. This cost is based on the aggregate of all pipeline replacements, most of which involve 15.2 cm (6") and 20.3 cm (8") dielectric coated steel pipe.
- The time value of money is taken as 6%/yr, based upon the District's current bond rate.

The analysis considers the case where CP is installed on a cast iron pipeline, allowing the replacement of that pipeline to be deferred. A simple present worth analysis follows:

Year	Task	Cost		Present Worth
0	Install CP	$148/m		$148/m
20	Replace Pipe	$620/m		$193/m
			Total:	$341/m

The cost of the CP alternative in present day dollars, is $341/m of pipe, as opposed to $620/m to immediately replace the pipe - a savings of approximately 50%. It is likely that the cost of installing large-scale CP systems would be significantly lower than the cost of the test installations resulting in additional savings.

Future Testing

In an effort to further reduce the cost of CP installations on cast iron piping, the District will be exploring the use of small impressed current systems with shallow anodes. The potential benefits of these systems are a reduction in the cost of anode materials, and a reduction in the size and number of anode excavations. In addition, impressed current systems offer increased control of over pipe potential, since the output of a rectifier can be easily adjusted.

Cathodic Protection of Cast Iron Pipe - CPS System Data

Clover Court: 50.9 meters of 10cm cast iron pipe
Surface area is approximately 24.4 sq-meters, soil resistivity=808 ohm-cm

Location	"Native" (mV)	Initial "On" (mV)	Initial Current (mA)	2-Month Instant Off" (mV)	2-Month Current (mA)	Density (mA/sq-m)
BO Test Station	588	927	1528	1123	770	
Center Test Sta.	553	650		889		
IJ Test Sta.	572	1205	1432	1076	700	
			Total 2960		Total 1470	60.2

Bonita Court: 141.7 meters of 10cm cast iron pipe
Surface area is approximately 68.0 sq-meters, soil resistivity=1375 ohm-cm

Location	"Native" (mV)	Initial "On" (mV)	Initial Current (mA)	2-Month Instant Off" (mV)	2-Month Current (mA)	Density (mA/sq-m)
BO Test Station	411	798	602	1202	742	
Center Test Sta.	433	525		998		
IJ Test Sta.	463	681	780	1225	712	
			Total 1382		Total 1454	21.4

Plumas Ave: 58.5 meters of 10cm cast iron pipe
Surface area is approximately 28.0 sq-meters, soil resistivity=3550 ohm-cm

Location	"Native" (mV)	Initial "On" (mV)	Initial Current (mA)	2-Month Instant Off" (mV)	2-Month Current (mA)	Density (mA/sq-m)
BO Test Station	546	850	202	N/A	N/A	
Center Test Sta.	543	650		N/A		
IJ Test Sta.	549	866	167	N/A	N/A	
			Total 369		Total N/A	N/A

Note: Only initial data available for Plumas Ave installation, due to later installation of CP system at that site.

Conclusions

Installation of CP on cast iron piping is a cost-effective means of extending the service life of the pipe. These installations can also provide increased flexibility to distribution system managers. The cost, installation time and impact to customers associated with CP installations are minimal when compared to pipeline replacement projects.

Installing CP on cast iron piping may not always be the best solution. Several points to consider in selecting candidate pipelines are:

- Is the pipe hydraulically adequate, or will a larger diameter pipe be required in the next several years?
- Have failures occurred on the pipeline due to mechanical action such as earthquake fault movement, a slide, or heavy traffic loads?
- Is the pipe responsible for water quality problems? Much cast iron pipe is unlined and unlined pipe is sometimes a factor in water quality problems.
- Was the pipe originally installed with deficient seals? Some cast iron pipe joints in the EBMUD distribution system were packed with sulfur-based sealants that become brittle with age, eventually causing leaks at joints.

Candidate pipelines for CP must therefore be selected with due consideration to these factors. A properly targeted CP installation program will augment a pipeline replacement program by allowing the replacement program to target pipelines that are most likely to cause impacts to customers.

Value Engineering Savings on Pipeline Project

Paul W. Johnson, P.E., Antonio J. Perez, P.E.,
Burt K. Yu, P.E., and Carlos de Leon, P.E., M. ASCE[1]

"Value Engineering is a process consisting of the systematic application of analytical, creative, and evaluative techniques on a multi-disciplined basis to achieve the desired function in a design or process while maximizing value."

Abstract

It is the Metropolitan Water District's policy to perform Value Engineering (VE) studies on all projects with an estimated cost greater than $2 million. The estimated cost of the San Bernardino Pipeline segment (Contracts No. 3 and No. 4) of the Inland Feeder Program is approximately $75 million.

This paper describes the project, the VE study, and the VE recommendations that were incorporated into the design. The resulting changes brought about alignment, environmental, and pipeline improvements, and produced a savings of $5.1 million. This represents a return on investment (ROI) of 63.8:1; i.e., for every dollar spent on the VE study, $63.8 dollars was saved. The most extensive design modifications accepted were revision of the Abbey Way alignment to minimize utility relocation, and the use of riprap in lieu of added depth at creek crossings. Two additional VE recommendations were at first rejected by the design teams and only later, after further consideration, accepted. These changes resulted in savings of an additional $400,000, for total savings of $5.5 million and an ROI of 68.8:1.

Introduction

Metropolitan is a public agency established by the Metropolitan Water District Act of 1928 for the purpose of developing, storing, and distributing water for domestic

---

[1] Senior Engineer, Quality and Value Engineering; Senior Engineer, Civil Engineering; Engineer, Civil Engineering; Associate Engineer, Quality and Value Engineering, Metropolitan Water District of Southern California, 350 South Grand. Los Angeles, CA 90071

and municipal purposes throughout Southern California. The agency's 13,310 square kilometer (5,139 square mile) service area reaches 113 kilometers (70.2 miles) inland from the coast, and extends along the Southern California coastal plain from the city of Oxnard south to the Mexican border. Metropolitan imports water from two sources: the Colorado River via the Colorado River Aqueduct, and Northern California via the State Water Project and its California Aqueduct. This water is delivered to 27 member agencies through a regional network of canals, pipelines, reservoirs, treatment plants, and pumping plants.

In 1993 Metropolitan's Board of Directors ratified an Environmental Impact Report which authorized the design and construction of the Inland Feeder Project. The Inland Feeder is a system of approximately 43 miles of tunnels and pipelines that traverses San Bernardino County and Riverside County to supply Metropolitan's Eastside Reservoir. It will connect the State Water Project at Devil Canyon with the 9-mile reach of Eastside pipeline currently under construction in the city of Hemet, California.

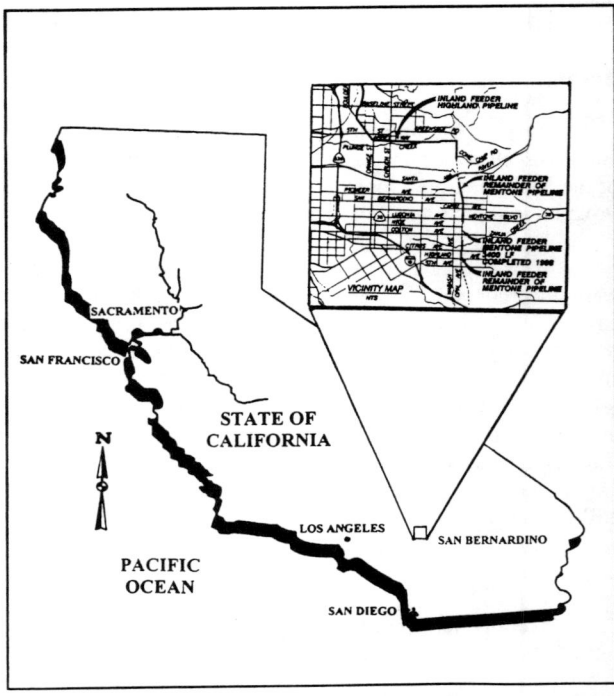

The increased capacity provided by the Inland Feeder will enhance Metropolitan's ability to deliver larger quantities of State Water Project water into its system. This will benefit Southern California by reducing its dependence on Colorado River water, improving water quality by mixing State Water Project water and Colorado River water, and providing additional capacity for growth in Southern California.

The San Bernardino Pipeline segment consists of 15.3 kilometers (9.5 miles) of 3.66 meter (12-foot) inside diameter welded steel pipe. It commences at the East Frontage Road of State Highway 330 in the city of San Bernardino, crosses the south branch of the San Andreas Fault, proceeds southerly through San Bernardino County's

City Creek, continues southerly through the city of Highland's Boulder Avenue and Abbey Way, tunnels under the Santa Ana River Wash, traverses southerly along Opal Avenue in the community of Mentone, and terminates at the North Portal of the Riverside Badlands Tunnel in Mentone. Portions of the project will be constructed and operated within street rights-of-way.

Design criteria for the San Bernardino Pipeline therefore address utility conflicts, right-of-way constraints, environmental and permitting concerns, constructability, and cost.

The VE Study

The VE study used the five-step VE process adopted by the Society of Value Engineering. The first of the VE phases is the Information Phase. During this phase the team gains an understanding of the project through a review of all available data. This was done by taking the VE team on a site visit on the first day of the study, followed by a formal presentation by the design engineers. This phase concluded with an intensive analysis of the project's function and identification of focus areas. The project was subdivided into 9 functions: alignment, operations and maintenance, hydraulics, constructability, pipeline, civil and site engineering, environmental, traffic, and program.

The second day of the study began with the Creative Phase and concluded with the Judgment Phase. The objective of the Creative Phase was for the VE team to generate, through a brainstorming session, numerous alternatives that could accomplish the primary functions of the project. This phase asks the question, "Is there a better way to perform the function?" The greater the number of solutions, the greater the chance of developing an excellent design recommendation. During this study, the VE team generated 103 ideas for design modifications that could potentially reduce the cost of the project.

The Judgment Phase evaluated and ranked the alternatives generated during the Creative Phase. The design engineers were invited to join the VE team for the Judgment Phase to ensure that the VE team focused its efforts on the ideas that would have the greatest likelihood of acceptance and cost benefit to the project. Involving the design engineers in the Evaluation Phase harmonized the working relationship between the VE and design teams and resulted in the design team supporting the VE team's efforts.

During the Development Phase the VE team spent approximately two and one-half days in developing 43 written recommendations based on the ideas that were approved in Evaluation Phase. These recommendations were formally presented to the design team on the last day of the study. The design team approved 22 of the VE recommendations.

Design Modifications

An important function of VE is to provide the design team with ideas that can be implemented with further development. An example of this is the team's proposal to provide an open cut trench in lieu of tunneling at State Hwy 38 and Mentone Blvd. Although the design team rejected it during the judgment phase, concerns about safety during tunneling operations led the team to reconsider the option to open cut a trench and detour traffic. Soil conditions at the proposed tunnel location, consisting of sands with cobbles and boulders, could create a "chimney" effect, with boulders being loosened during the tunneling operation. The boulders' movement could then cause the sandy soil to collapse and the roadway to settle. The design team elected to develop the open-cut trench design modification. By doing so an additional $200,000 was saved.

After studying the traffic patterns at this location, the VE team determined that it would be prudent to detour traffic. The design engineers prepared a traffic detour plan, and worked with CalTrans to obtain an encroachment permit for the open-cut construction of the pipeline crossing Mentone Blvd.

VE proposal EN 4.0 involved open cutting the Zanja Creek. Zanja Creek in the community of Mentone is part of San Bernardino County's flood control facility. The original design specified tunneling under Zanja Creek. During the VE study, the design team was against this proposal because of the delay to the project while the EIR was being modified. However, a staff member of Metropolitan's Environmental Section informed the team that revising the EIR would not delay the project. After the VE study, the design team determined that with proper environmental measures and permitting with local resource agencies, it would be more prudent to approach with the open-cut option. This idea had initially been rejected by MWD, but with further development of the design team, the idea was accepted and saved another $200,000.

VE proposals CN 1.0, CN 1.1, and CN 2.0 involved the installation of pipeline fronting Redlands High School. The newly constructed school, sited along the alignment of the Mentone Segment of the San Bernardino Pipeline, was scheduled to open in September 1997. The VE team provided the design team with three proposals for avoiding construction conflicts with the high school:

- Complete installation of the one-half mile of pipeline fronting the school prior to the school's opening, to avoid potential safety problems and traffic conflicts.
- Contract for 3400 lineal feet of prefabricated pipeline.
- Contract out the installation.

This VE design modification (CN 1.0) was accepted by the design team during the study. The accepted savings was $60,000; however, more importantly, safety and traffic conflicts with Redlands High School were avoided. Construction was successfully completed with no adverse impacts to the traffic flow of the new high school. The high school demonstrated its appreciation of Metropolitan's efforts by inviting the Assistant General Manager to the school's Grand Opening Ceremony.

Summary

The total implemented savings from the VE study was $5.5 million. This amount includes $400,000 in savings from the two VE design modifications that had initially been rejected during the study, for a revised ROI of 68.75:1. In addition to saving money, the potential for negative social and political impacts was minimized.

SUMMARY OF OPTIONS ORIGINALLY DEVELOPED (1 of 4)

Item	Potential Construction Cost Impact	Savings Approved by MWD	Resolution
Alignment			
AL 1.0 Revise Abbey Way alignment to minimize utility relocation	$208,000	$ 49,000	B
AL 2.0 Revise pipe alignment at EVWD pump site (Sta 121)	197,300	143,600	B
AL 3.0 Revise alignment at Highland to eliminate tunnel	(12,000)	0	C
Alignment Subtotal	**$393,300**	**$192,600**	
Civil/Site			
CS 1.0 Use riprap in lieu of added depth for creek crossing	$904,400	$904,400	B
CS 2.0 Regrade Mentone to solve drainage problem	(18,400)	0	C
CS 3.0 Alternative draining solutions at MTA property	DS	0	C
CS 4.0 Raise school site storm drain	DS	0	C
CS 5.0 Provide drainage solution at Richmond Tech	DS	0	C
Civil/Site Subtotal	**$886,000**	**$904,000**	

Legend: A = accepted as developed   D = warrants further study (30 day max.)
         B = accepted with modification   DS = "design suggestion," ideas not
         C = rejected   fully explored in this study

SUMMARY OF OPTIONS ORIGINALLY DEVELOPED (2 of 4)

Item	Potential Construction Cost Impact	Savings Approved by MWD	Resolution
Constructability			
CN 1.0 Advance purchase pipe material w/change order (Redlands H.S.)	DS	$ 60,000	A
CN 1.1 Use MWD to construct segment fronting high school	$ 37,100	0	C
CN 2.0 Provide contractor incentive for early completion of high school section	DS	0	C
CN 3.0 Prepare traffic plans during design	DS	0	A
CN 4.0 Use sheeted trench in lieu of tunnel at Zanja Creek	236,700	0	B
CN 5.0 Review alternatives for street crossing - high traffic	DS	0	A
Constructability Subtotal	**$237,800**	**$60,000**	
Environment			
EN 1.0 Perform hazmat check of MTA property	DS	$0	C
EN 2.0 Locate blow-off outside sensitive areas	DS	0	C
EN 3.0 Maintain easements	DS	0	A
EN 4.0 Open cut Zanja Creek crossing	DS	0	C
EN 5.0 Review schedule impact for permitting	DS	0	A
EN 6.0 Provide unified approach to C.O.E.	DS	0	C
EN 7.0 Procure C.O.E. permit	DS	0	C
EN 8.0 Assess cost of avoiding permits	DS	0	B
Environmental Subtotal	**$0**	**$0**	

SUMMARY OF OPTIONS ORIGINALLY DEVELOPED (3 of 4)

Item	Potential Construction Cost Impact	Savings Approved by MWD	Resolution
Operations/Maintenance			
OM 1.0 Use coated manway flanges in lieu of concrete caps	DS	$0	B
OM 2.0 Provide patrol access and maintenance areas	DS	0	A
OM 3.0 Add chlorine injection point(s) in pipe design	DS	0	D
Operations/Maintenance Subtotal	**$0**	**$0**	
Pipe/Pipeline			
PL 1.0 Modify standard trench details	$7,657,000	$3,191,600	D
PL 2.0 Reduce minimum pipe cover to 6'0"	1,539,600	0	C
PL 3.0 Reduce depth of scour cover from Sta 140 to Sta 180	751,400	751,400	B
PL 4.0 Modify design slope ratios to allowable	1,555,480	0	C
PL 5.0 Conduct slope stability checks at critical areas	DS	0	A
PL 6.0 Narrow width of new right-of-way acquisition to 40'	763,400	0	C
PL 7.0 Review phased epoxy pipe lining	DS	0	A
PL 8.0 Base steel pipe plate thickness design on surge pressure	DS	0	C
PL 9.0 Negotiate processing of excavated materials	DS	0	A
PL 10.0 Provide a park area on MTA property	DS	0	A
PL 10.1 Transfer ownership of MTA property	DS	0	A
PL 11.0 Review need for mitigation of height restrictions at Redlands Airport	DS	0	A
PL 12.0 Use Bailey sleeve valve in lieu of plug valve at blow-off	DS	0	C
PL 13.0 Provide friction slider system for fault crossing	DS	0	C
Pipe/Pipeline Subtotal	**$12,266,280**	**$3,943,000**	

SUMMARY OF OPTIONS ORIGINALLY DEVELOPED (4 of 4)

Item	Potential Construction Cost Impact	Savings Approved by MWD	Resolution
Program			
PR 1.0 Conduct peer review of Becker blow count	DS	$0	A
PR 2.0 Install seismic detection system	DS	0	A
PR 3.0 Install ground movement instrumentation as appropriate	DS	0	A
PR 4.0 Include VE incentive clause in contract	DS	0	C
Program Subtotal	**$0**	**$0**	

Total Study Potential Savings $13,819,380

Total Implemented Savings $5,100.000

Return on Investment   63.8:1

CAD Design Expedites Environmentally Friendly Project For CCWA

Dan Masnada[1]
Ken Ferguson[2]
Lyndel Melton[3]

The Central Coast Water Authority (CCWA) working in conjunction with the State of California Department of Water Resources (DWR) designed and recently completed construction of the 145-mile Coastal Branch of the State Water Project and Mission Hills and Santa Ynez Aqueduct Extensions. This system is designed to provide treated State Project Water to serve portions of San Luis Obispo and Santa Barbara Counties. Construction of the $575 million project was completed in 1997.

BACKGROUND

The primary purpose of the project is to provide supplemental water for San Luis Obispo and Santa Barbara Counties to address the issues of groundwater overdraft, water supply quality and new demand based on development foreseen in community general plans. Until late last year, Santa Barbara and San Luis Obispo Counties relied on a combination of local groundwater and surface water supplies. Also, water reclamation projects scattered throughout the counties provide limited quantities of water for irrigation purposes.

Another purpose of the project is to provide a "plumbing" system that would facilitate the future exchange or purchase of water from various sources throughout the State. The project serves as a connection between San Luis Obispo and Santa Barbara Counties and the State Water Project which delivers water to two-thirds of California's population. The State Water Project includes over 650 miles of transmission system (canals and pipelines) and 8 million acre-feet of storage.

---

[1] Manager, Central Coast Water Authority, 255 Industrial Way, Buellton, CA 93427
[2] Vice President, Montgomery Watson, 3014 Charleston Blvd, Las Vegas, NV 89102 [3] Principal, Raines, Melton and Carella, Inc. 3451 Golden Gate Way, Lafayette, CA 94549

Annual deliveries through this system have reached 4 million acre-feet and serve over 20 million people throughout the state.

In addition to the water supplies which are available directly through the State Water Project, there are also water supplies that can be "wheeled" through the system. Various "wheeling" agreements which have been considered in the past include an arrangement by which water could be transferred from Santa Barbara County's allotment to the Las Vegas Valley Water District through an agreement involving CCWA, DWR, Metropolitan Water District of Southern California and finally the Las Vegas Valley Water District. Although this agreement was not implemented it illustrates a particular water "wheeling" possibility.

CCWA PROJECT

The overall project is composed of the Coastal Branch of the State Water Project plus the Mission Hills and Santa Ynez Aqueduct Extensions. Essentially the DWR was responsible for the Coastal Branch design and construction while the CCWA was responsible for the design and construction of the Mission Hills and Santa Ynez Aqueduct Extensions and the Polonio Pass Water Treatment Plant.

The Coastal Branch portion of the project originally consisted of 102 miles of pipeline, 4 pumping stations, 5 tank sites, and an energy recovery facility. Water is delivered from the California Aqueduct near Kettlemen City to the Polonio Pass Water Treatment Plant east of the City of Paso Robles and then continues to Tank 5 on Vandenberg Air Force Base (VAFB). Also included were the Chorro Valley and Lopez Lake turnouts to serve San Luis Obispo County and the Guadalupe, Santa Maria and Southern California Water turnouts to serve northern Santa Barbara County.

The Mission Hills and Santa Ynez Aqueduct Extensions (designated Schedules A, B, & C) picks up the water at Tank 5 and delivers the water to downstream users, with a termination at Lake Cachuma. These facilities include 42 miles of pipeline, a pumping station, 2.5 million gallon tank and turnouts for service to VAFB, Buellton, Solvang, and the Santa Ynez River Water Conservation District, Improvement District No.1.

In February of 1995 CCWA entered into an agreement with DWR to design and construct the final 28 miles of the Coastal Branch portion of the project (designated Reaches 5B and 6). In order to keep the project on schedule, Reaches 5B and 6 had to be designed in 80 days. Although a preliminary design had been completed for these reaches, some substantial modifications to the preliminary design were incorporated into the final design including the elimination of the energy recovery station, two storage tank sites and a pumping station. This was a significant challenge which will be addressed later.

The construction costs of the Coastal Branch facilities was $461 million excluding the cost for the Polonio Pass Water Treatment Plant. The construction cost of the Polonio Pass Water Treatment Plant plus the Mission Hills and Santa Ynez Extension facilities was $114 million.

ENVIRONMENTAL ISSUES

The project alignment through San Luis Obispo and Santa Barbara Counties is a highly sensitive environmental area . Environmental issues encountered in this project included many biological issues and cultural resources.

Biological Issues.

Biological issues involved sensitive plants, animals and habitats.

Sensitive Animals of concern include:

California Tiger Salamander
Red Legged Frog
Southwestern Pond Turtle
Cooper's Hawk
Southwestern Willow Flycatcher
Unarmored Three Spined Stickleback

Some of these animals and fish are present only in one area along the alignment while the presence of others are more widespread. For example, the California Tiger Salamander for example is only present in a certain area known as the Campbell Vernal Pools and nowhere else. The same is true for the Unarmored Three Spined Stickleback which is only present in San Antonio Creek on VAFB. On the other hand, the Southwestern Willow Flycatcher is potentially present anywhere that willows grow.

Sensitive Plants include:

Seaside Birds Beak
Black Flowered Figwort
Purisima Manzanita
Hoover's Bentgrass

All of these plants are found in a particular habitat called Burton Mesa Chaparral. This habitat is found along a significant portion of Schedules A and B.

Sensitive Habitats such as Burton Mesa Chaparral and other habitats are present along the project alignment including:

Riparian habitats along creeks and rivers
Campbell Vernal Pools
Oak trees
Native grasses

Common riparian habitat include willow trees which provides habitat for the Southwestern Willow Flycatcher previously mentioned. The vernal pools provide a habitat for the California Tiger Salamander, the Red Legged Frog and the Southwestern Pond Turtle. Oak trees and native grasses are present along the alignment in many locations.

Cultural Resources

Cultural resources are present in many locations in San Luis Obispo and Santa Barbara Counties. In general these sites are classified as either Chumash Indian sites or Mission sites. In one location an old disposal area (dump) was classified as historical due to the age of the artifacts found. The Mission sites are the La Purisima Mission near Lompoc and the Santa Ynez Mission in Solvang. The Chumash Indian sites are widespread, which is consistent with the fact that they were nomadic and would set up camp at one site and then move to another site several years later. Logically, many of the sites occupied high ground or were near the streams and rivers.

Addressing Environmental Issues

Environmental issues were addressed on a case-by-case basis. To avoid impacts to Burton Mesa Chaparral, a number of different strategies were used. Avoidance was the chief strategy whenever possible without a significant cost impact. Where the costs of this strategy were too great, the width of the construction corridor was narrowed and damage to the chaparral was minimized.

The crossing of San Antonio Creek was another example of an environmentally sensitive area which required an innovative solution. The primary concern was the Unarmored Three Spined Stickleback, a fish which lives in the creek. Since this fish is highly sensitive to turbid water, a micro-tunnel crossing was constructed to avoid construction activities in the creek. The alignment of San Antonio Creek closely parallels an earthquake fault line. If the pipeline fails due to a seismic event, chloraminated water, which is toxic to fish, could be discharged into the creek. To minimize the impact of such an occurrence, vibration activated valves were installed on each side of the creek. These valves close during significant shaking. Following any such event, the valves can only be opened manually to ensure the opening process does not result in spilling chloraminated water.

DESIGN PROCESS (CAD vs Drafting Board)

CCWA's original design effort included the Mission Hills and Santa Ynez Aqueduct Extensions and involved a mostly rural, highly sensitive environmental setting. Flexibility to change alignments during design was a major concern. The environmental setting is very complex and there was a concern that environmental issues and concerns would be revealed on an ongoing basis during the design process. Also, there are many landowners along the alignment and property acquisition was a major issue. Consequently flexibility to re-align the project to minimize these impacts was also a concern.

The project was designed in three stages:

Conceptual Design - the design criteria was developed, general alignment corridors were confirmed, hazards and other major constraints were researched.

Preliminary Design - a 50% level design with plan and profile drawings was developed and additional alignment issues were defined.

Final Design - the final construction drawings and contract documents were developed.

To enhance the clarity and understanding of the final construction drawings, it was desirable to avoid many equation stations. Using traditional drafting board techniques, many equations stations were anticipated due to the inevitable alignment adjustments to avoid environmental issues and property acquisition issues.

The use of a CAD design approach was deemed to best respond to project concerns. Using Intergraph In-Roads software the whole Aqueduct Extension alignment of 42 miles was designed as a unit and separated into individual sheets when the design was complete, thus minimizing station equation. This approach also presented a definite design efficiency because the Preliminary Design drawings were usable in the Final Design phase with appropriate modifications and added detail, resulting in significant time and effort saving. Although 160 preliminary design drawings were produced for the Preliminary Design and many of these drawings required modifications and adjustments, the basic model was suitable for the final design process. As the Preliminary Design progressed and production of alignment and plan and profile sheets commenced, it was possible to produce the 160 drawings in a period of just six weeks. Although these drawings were advanced to only the 50 percent design level, this exercise was valuable experience for the later design of the 28- mile portion of the Coastal Branch.

As the design developed and the complexity of the issues were defined, CAD offered the project another great advantage. The base model for the pipeline plan was used to present the environmental constraints and requirements. The plan book consisted of environmental plans on the left facing page and the traditional plan and profile sheet on the right facing page, both drawn to the same scale and containing many "landmarks" connecting the engineering design with the environmental requirements. This approach provided the full range of information necessary to ensure the project was constructed in full compliance with both the engineering/operational requirements and the environmental/permitting requirements. This approach proved to be highly effective during the construction process.

DESIGN OF COASTAL BRANCH (Reaches 5B and 6)

As indicated previously, CCWA assumed responsibility for the design of the final 28 miles of the Coastal Branch (Reaches 5B and 6) in February of 1995. Reaches 5B and 6 included 28 miles of pipeline with pressures up to 400 psi, a pressure reducing station, four major turnouts, and two 2.5 million gallon tanks (Tank 5).

Critical to the assignment of responsibility to CCWA was the requirement that the design be completed in 80 days. Completion within this time frame was required to meet the overall project schedule of completing the majority of construction in the summer of 1996. Due to sensitive in-stream resources, stream crossings had to be completed prior to October 15. At the time this assignment was assumed, the environmental work had been completed and a pipeline alignment had been selected. There were, however, several areas along the alignment where additional environmental work was required prior to receipt of permits for construction. This work included cultural resource surveys, stream crossing surveys, stream crossing analyses and other related assessments.

Upon closer inspection of this portion of the project, CCWA determined that improvements in the preliminary design could be made which would reduce the environmental impacts and which could make the project more acceptable to the environmental community, the affected resource agencies, and the local landowners. Effecting these changes in the Preliminary Design had to be made in a manner that incorporated the issues and concerns of a wide range of disciplines.

It was decided to build on the experience gained in the design of Schedules A, B, and C, which had proven that CAD could greatly expedite the design process. However, to be successful on this project, another level of efficiency had to be attained. CCWA facilitated this effort by providing the environment in which timely decisions could be made by making information available from each area of the project. CCWA formed an on-site project team that consisted of CCWA staff and the design, construction management, environmental, cultural resource and right-of-way acquisition specialists. The on-site project team was housed in a trailer adjacent to CCWA's office. The design engineering team and a full complement of CAD facilities were included in the project trailer.

The CAD system was used to generate drawings and design concepts which could be used by all members of the project team to understand the issues in each area of the project and to develop alternatives which were consistent with the overall project objectives. Often, project team members used the CAD drawings to conduct field reviews of project issues and potential solutions.

The same approach was used to present environmental requirements and constraints as the Mission Hills and Santa Ynez Aqueduct Extensions. The compliance requirements were printed on the left facing page of the plan book, allowing the field teams to review both the engineering design and environmental compliance requirements on adjoining drawings. This facilitated the review process during design and greatly facilitated the ability of inspection crews to ensure contract compliance during construction.

Having all project team disciplines present at one location facilitated daily interaction and decision making, which was key in finishing this assignment in a timely manner. Weekly meetings, in which critical issues were identified, discussed and decisions were reached, were used to supplement the daily interaction. If a decision could not be reached, a schedule was established for making a decision and the information necessary for making the decision was identified.

In this manner, the project team was fully integrated and focused on a common objective: to complete the design in a manner that preserved the environmental, cost, schedule and quality goals of the project. By working together, this project team concept achieved all four project objectives for Reaches 5B and 6. The project design engineering was achieved at less than 6 percent of the construction cost and the construction cost was nearly 40 percent under the program level estimate.

CONCLUSIONS

The CAD design techniques pioneered on the 42-mile Mission Hills and Santa Ynez Aqueduct Extensions and refined during the design of the 28-mile Reaches 5B and 6 design were essential in achieving the project goals.

Utilization of the CAD design model as the presentation basis for environmental and permitting requirements and printing those requirement on adjacent facing pages with the design drawing was a very effective and convenient approach to facilitate design review, inspection, and construction contractor understanding.

Innovative organization and aggressive management coupled with the use of proven CAD tools allowed CCWA to achieve the goals of producing a cost effective and environmentally acceptable design of 28 miles of high pressure pipeline in a environmentally sensitive area in just 80 days.

PUBLIC OUTREACH
AND CONSTRUCTION STRATEGIES
For The
Mission Trails Pipeline and Flow Regulatory Structure

Zachary Ahinga[1] and Janice Collins[2]

Abstract

The public outreach challenges presented by construction of the Mission Trails Pipeline and Flow Regulatory Structure within the City of San Diego's pristine and mountainous Mission Trails Regional Park, as well as the challenges of accessing the project through the residential communities of Tierrasanta and San Carlos, were recently met by the San Diego County Water Authority. This paper discusses the public outreach program and construction strategies implemented with the communities and park officials to generate and maintain community support and resolve arising community issues.

Introduction

The San Diego County Water Authority (Authority) imports water and supplies it on a wholesale basis through its 23 retail member agencies to approximately 2.7 million people in San Diego County. The Authority in November 1996 completed the $30 million Mission Trails Pipeline (Pipeline) and Flow Regulatory Structure project. The 2.6 kilometer (1.6 mile), 2438 mm (96 inch) diameter Mission Trails Pipeline, including a 366-meter (1200-foot) long tunnel under the San Diego River and the 68.1 million liter (18-million gallon) underground Flow Regulatory Structure posed some of the most significant community relations challenges of any Authority pipeline construction project. This project was a vital final link in a new 48 kilometer (30-mile) water pipeline system for San Diego County, a region dependent on the Authority to import 90

---

[1] Zachary Ahinga, Senior Civil Engineer, San Diego County Water Authority, 3211 Fifth Avenue, San Diego, CA 92103, (619) 682-4100, Fax (619) 692-9356.

[2] Janice Collins, Public Affairs Supervisor, San Diego County Water Authority, 3211 Fifth Avenue, San Diego, CA 92103, (619) 682-4100, Fax (619) 683-3956.

percent of the water used locally. The project was constructed within Mission Trails Regional Park -- a pristine natural park and home to endangered bird species. The access route to the project site went through Tierrasanta, a politically active San Diego community located in the central eastern portion of the City of San Diego (see Figure 1 Location Map) that had just endured the construction of a highway at its northern border and another major Authority pipeline project. Undertaking this large water facility construction project through a populated area effected residents' daily routines; traffic safety and noise become major concerns that required an effective public outreach strategy to gain support for the project.

PUBLIC OUTREACH PROGRAM

The outreach program consisted of three phases: Research, Planning, and Execution. Each phase was fine-tuned throughout the project.

RESEARCH: The export of dirt and material deliveries for the project would require that up to 300 heavy vehicle trips be made along the access route, a street lined with schools housing 2,500 students. Adopting the philosophy that the community would know best how to meet its needs, the interviews with community leaders emphasized open-ended questions soliciting opinions and suggestions on issues.

Primary research included a review of the project's Environmental Impact Report to identify likely issues, discussions with project engineers and environmental specialists regarding potential issues, and touring the access routes and construction site to determine the neighborhoods most likely to be affected. Presentations were given to community and park advisory groups to explain the project and gather feedback. Separate meetings were conducted with Principals of five schools and two daycare centers located on or near construction access routes to determine specific safety issues. Group meetings were conducted with Principals, school district officials and police department representatives regarding school safety issues, and one-on-one meetings were conducted with community leaders. Secondary research included reviewing previous construction projects' community relations programs and studying other agencies' construction community relations programs.

Research identified three major issues:
- Safety of students crossing streets along the construction access route
- A strong Not In My Back Yard attitude in the community; and
- Middle school students were known to play "chicken" with traffic, imitating a stunt seen in a recent movie by sitting down in traffic lanes

PLANNING: The focus of the community outreach strategies were to ensure safety of schoolchildren, minimize the community's inconvenience, ensure the project's completion on time and within budget, strengthen existing relationships and seeking input with key publics, and inform and educate impacted audiences and advisory organizations about the importance of the project. The target audience or **Publics** were defined as schools, community leaders, local media, residents directly affected and residents indirectly affected. Strategies aimed not just at students and their parents, and community leaders, but at the construction truck drivers themselves -- advising them of the specific traffic habits of each school and encouraging extra caution.

Strategies developed for <u>schools</u> were to provide police department certified crossing guards to assist students across access routes (schools had no crossing guards before); inform truck drivers of school-safety issues and traffic habits of students at each school; provide Principals with concise written updates on traffic/construction for inclusion in school newsletters; hold safety/project information assemblies for students; provide a project information line number to which parents could be referred; and maintain direct two-way communication with Principals.

Strategies for <u>community leaders</u> were to provide tours of the project under construction; attend meetings on a regular basis with the area advisory and planning groups including the Tierrasanta Community Council (TCC), Mission Trails Regional Park Citizens Advisory Committee (CAC) and the Task Force Committee (TF) which included members of the the city council; provide written updates on traffic and construction; provide a project information line phone number to which parents and constituents could be referred; and maintain two-way communication.

Strategies for <u>local media</u> included touring the construction site; providing press releases as the project progressed; providing a project information line phone number for readers to call; maintaining direct two-way communication; and providing written updates on traffic/construction as well as any written materials sent to the community.

Strategies for <u>directly affected residents</u> included holding neighborhood coffees to present information, providing written updates and presentations on traffic/construction to homeowner associations, and establishing a process for answering questions or resolving concerns using the project information line.

Strategies for <u>indirectly affected residents</u> included distribution of project newsletter and use of media to relay project information.

The **key message** to be delivered through all strategies was: <u>the Mission Trails Pipeline and Flow Regulatory Structure was a fundamental part of the Authority's efforts to increase this region's water reliability and that it would be constructed in the safest and least disruptive manner possible.</u>

EXECUTION: The Authority implemented strategies to disseminate information and provide opportunities for community feedback. A community <u>contact list</u> was updated regularly that included school Principals, community groups, interested residents, businesses, elected officials, media, individuals who called the project for information, and others impacted by construction. A 24-hour <u>Project Information "hotline"</u> provided all publics with a way to express concerns, get information or ask questions even if it was not normal business hours and provided a valuable source of information on what project-related issues were creating community relations challenges. Approximately 1,500 Project Information Line cards were distributed.

Project Newsletters were developed to inform the entire affected community about the project and its importance and benefits to the community, as well as to issue safety reminders and other timely information. The newsletters were designed to look good but not expensive and to promote the public perception that funds were appropriately spent.

Project Updates were created to keep community leaders, Principals and others requesting them apprised of construction progress, traffic impacts and safety issues. Distributed twice monthly, updates were designed as one-page, two-sided, single color publications with a bullet-point format. This format was requested by school Principals so they could easily pull short, bullet-point items for their school newsletter that were sent home to parents. Approximately 5,200 updates were distributed.

Special Project Advisories were used to apprise residents of project construction changes that could affect them, safety alerts, or an explanation of work under way. Some of these changes included earlier work hours, weekend work, access route changes, and notification of the start of tunnel blasting. Advisories were created in a simple, one-page format on colored paper so they would be noticed. They were distributed either through direct mail or were delivered door to door as door hangers wrapped in a rubber band and placed on doorknobs. More than 21,000 advisories were distributed.

PUBLIC-SPECIFIC STRATEGIES

Special construction strategies were implemented to resolve design and arising construction issues. These strategies helped the project to proceed on schedule and within budget, at the same time responding to community and park concerns.

Construction Access Routes: To ensure the safety of the public along the construction vehicle access route (see Figure 2 Construction Access Route), construction contracts limited heavy construction traffic through Tierrasanta to the use of designated construction vehicle routes between the hours of 8:30 a.m. to 4:00 p.m., thus avoiding school start and end times, and commute times. Street crossing guards were hired for three affected schools during construction to safely assist students across the street during peak construction vehicle periods for the project's entire two year construction period. Traffic safety-oriented presentations were tailored to different grade levels and delivered at school assemblies to approximately 2,500 students. Articles were published prior to construction in the community newspaper discussing the construction access route and construction schedule and 30- and 7-day notices were delivered to residents within a 91 meter (300-foot) radius of the access points into the park.

During construction, teacher presentation materials were distributed (adapted for different grade levels) about the project and safety. A flashing speed indicator warning signal was placed along the construction route during periods of high construction traffic to remind motorists to reduce their speed. Tours were conducted of the project under construction for planning group members and the area's San Diego City Council member and staff, as well as for other community leaders. Meetings were held with the ownership of dirt export and material deliver firms to educate them about safety issues and community concerns, and fliers regarding safety were issued to all project drivers.

Efforts to "be a good neighbor" to the community included several good-will gestures. These included repairing and repaving pre-existing damage on a section of Clairemont Mesa Boulevard, the primary access route. As a good-will gesture to DePortola Middle School, which fronts Clairemont Mesa Boulevard, the Authority encouraged its contractors to work with the school Principal who wanted to have the school's dirt parking lot paved. The collaborative effort resulted in the parking lot being complete at an advantageous cost below market prices.

Access Bridge Meets Community and Park Needs: During the design of the Flow Regulatory Structure project, the Authority worked closely with the CAC, TF Committees, and the TCC to determine the best approach for crossing a deep canyon located at the beginning of the intended access route into the park. A year-round access road was required to complete the project on schedule. A 91.4 meter (300-foot) long temporary steel truss bridge with a 266,880 newton (30-ton) load capacity to accommodate construction vehicles was constructed, thereby avoiding harm to the canyon's sensitive habitat (see Figure 3). It was intended to be removed upon project completion. Following construction, community and park officials indicated their desire to establish a new park entrance at the end of Clairemont Mesa Boulevard and suggested the steel bridge be retained as a permanent pedestrian access bridge into the park. Perhaps the most significant gesture made to the community (following close coordination with the City of San Diego, CAC and TF committees, and TCC) was the transfer of ownership of the bridge to the City of San Diego in return for granting essential permanent access easements to the Authority for construction and maintenance of its facilities within the park. To prepare the area as a new entrance, the Authority graded a staging area for park events, and now the bridge and staging area are the centerpiece of a newly developed park entrance. Additionally, the Authority worked with the contractor so that an area graded for construction staging at the south end of the park near Mission Gorge Road was turned over to the park to provide improved access and parking for park visitors and to serve as additional parking for the Mission Trails Regional Park Visitors Center.

Modified Work Hours and Saturday Work: The Authority's original agreement with the communities and park officials was to conduct construction of the project on a regularly scheduled 8-hour day between the hours of 7:00 a.m. to 5:00 p.m., Monday through Friday with no work on the weekends. However, the wet winters of 1994 and 1995 caused construction delays due to inaccessible site conditions. To minimize the extra time required to complete the Pipeline and Flow Regulatory Structure, the Authority worked with the contractors to change the sequence of construction to take advantage of the longer daylight periods during the spring and summer months. The Authority was successful in obtaining approval for the schedule changes because the proposals addressed community and park concerns. Pipeline welding operations were extended until 7:00 p.m. and backfill operations were allowed on Saturdays from 8:30 a.m. to 5:00 p.m. The pipeline right-of-way was well isolated from the community, so conducting night welding and backfill work on Saturdays would have no effect on the community. Park patrons would not be impacted because all work would occur within a fenced area, confined to the Authority right-of-way, and monitored by Authority staff.

Dirt export and concrete delivery for the Flow Regulatory Structure was allowed to start an hour and a-half earlier at 7:00 a.m. along the construction vehicle route and within the park. This would not pose a safety hazard because the majority of students were on summer vacation; and the Authority assured the community that if the change caused problems the dirt export and concrete delivery schedule would be changed back to 8:30 a.m. Saturday work on the Flow Regulatory Structure was allowed including placement of reinforcing steel, construction of forms for concrete placement, concrete curing, and form removal. Adjacent residences would not be affected because the site is isolated, and no deliveries were allowed on the streets and within the park. The result of these schedule changes is that a construction change order was issued to the contractor for a time extension and no additional costs to the Authority.

Environmental Mitigation Achieved: <u>Aesthetic Design</u> --The Flow Regulatory Structure's function is to control water pressure surges in the Mission Trails Pipeline. It was built as an aesthetically pleasing alternative to constructing tall vents, as some existing vents extend 6 meters (20-feet) above the ground surface. Residents and park patrons were pleased that no more vents would be constructed. Working closely with the CAC and TF committees a 149-square meter (1,600-square foot) above ground access building was designed to include a stone-veneer exterior that fits in well with the natural surroundings (see Figure 4).

<u>Revegetation</u> -- To ensure successful re-establishment of the native coastal sage scrub vegetation within the park, an innovative approach was used that required collection and storage of seed-bearing topsoil for distribution during the revegetation operations and the application of a native coastal sage scrub seed mix. To protect the revegetated areas, the Authority worked closely with the park Ranger's staff to install 610 meters (2000 feet) of visually pleasing split-rail fencing to protect revegetated areas from unauthorized access during the vegetation re-establishment period.

<u>Riparian Habitat and Historic Artifact Protected</u> -- Planning studies indicated that open-cut trench construction was the most cost-effective construction method to cross the San Diego River area. However, the Authority decided to tunnel beneath the river and the historic Mission Flume to preserve the sensitive habitat of the least Bell's vireo and preserve the remnants of the flume, which was built in 1810 by the Kumeyaay Indian tribe as part of an extensive irrigation system used by Franciscan missionaries at the Mission San Diego de Alcala (see Figure 5).

Modified Tunnel Blasting Schedule: The 3.4-meter (11-foot) diameter, 366-meter (1,200-foot) long tunnel crossing under the San Diego River area was blasted through solid rock at a depth of 24.4 meters (80-feet). The tunnel surfaced at the south end in a 10.7-meter (35-foot) deep portal located within the park and adjacent to Mission Gorge Road and the San Carlos community. The construction contract specifications for the tunnel allowed tunnel blasting, Monday through Friday, from 7:00 a.m. to 7:00 p.m. and provided that blasting would be allowed on a 24-hour basis, Monday through Saturday, if the contractor obtained a noise variance from the City of San Diego to conduct tunnel blasting from 7:00 p.m. to 7:00 a.m. The contractor met all contract provisions and the Authority permitted the commencement of blasting at night and early in the morning.

Immediately, the San Carlos community issued complaints that the blasting created excessive vibrations and noise. Although the Authority's seismic ground vibration and noise monitoring programs indicated that all blasting met project specifications, the Authority responded by quickly scheduling a meeting to address the community's concerns. The Authority put together a team of staff members from several departments including public affairs, engineering, and risk management, who were lead by the Authority's Assistant General Manager and attended a meeting with the San Carlos Area Council and residents. The Authority provided information on the tunneling operations, received community complaints and suggestions, and discussed alternatives to resolve the complaints.

As a result, the Authority working with the tunneling contractor and the San Carlos Area Council, agreed that night blasting would only be conducted up to 8:30 p.m. on a daily basis. The contractor would be allowed to continue working through the night on removal of blasting debris from the tunnel and setting up for the next blast at 7:00 a.m. the following morning. This agreement resulted in the Authority issuing a negotiated minimal change order to the tunneling contractor for delays due to the schedule modification, amounting to 6 percent of the construction cost for the Mission Trails Pipeline project.

SUMMARY EVALUATION

The effectiveness of the Authority's public outreach program was evaluated by the project team monthly throughout the project in order to fine-tune the program. The number of people added to the community contact list; the number and type of inquiries received on the project information line, and the time taken to resolve concerns were noted to determine effectiveness. Ongoing oral and written solicitation of feedback from the community and its leaders on existing outreach efforts or refinements being considered helped redirect or heighten Authority efforts as necessary. Evaluations were constantly sought through one-on-one meetings, phone conversations and letters with community and school leaders; through formal planning group meetings, neighborhood and homeowners' associations meetings; and through Project Information Line calls. This approach with the community resulted in a project the Authority and community alike deemed a success in the final analysis.

Typically, audiences of a community relations campaign are very vocal when they do not like something a government organization is doing, but do not take the time to tell that organization if it is doing something well. The Authority received many calls on its Project Information Line from residents pleased with Authority efforts to minimize inconvenience and maximize safety. The area's trustee on the San Diego Unified School District Board of Trustees was also impressed with the results of the public outreach effort. On behalf of the Board of Trustees the Authority was recognized at a TCC community service awards ceremony for its efforts toward maintaining the safety of students.

The Authority met its community outreach goals on this project. There were no accidents involving schoolchildren and construction traffic. The project was completed on time and within budget. Residents were kept informed and their concerns were responded to immediately. Community ill-will was turned into appreciation for the Authority's efforts to minimize any inconvenience and maximize safety. The outreach efforts transformed the community's impression of the Authority from a faceless, uncaring government agency to concerned, trusted individuals.

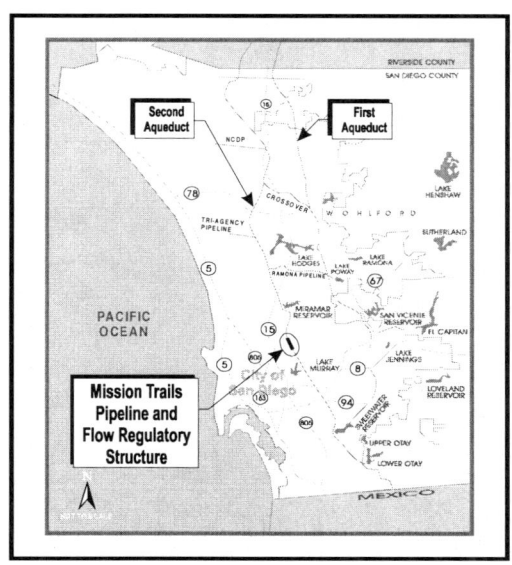

Figure 1
LOCATION MAP

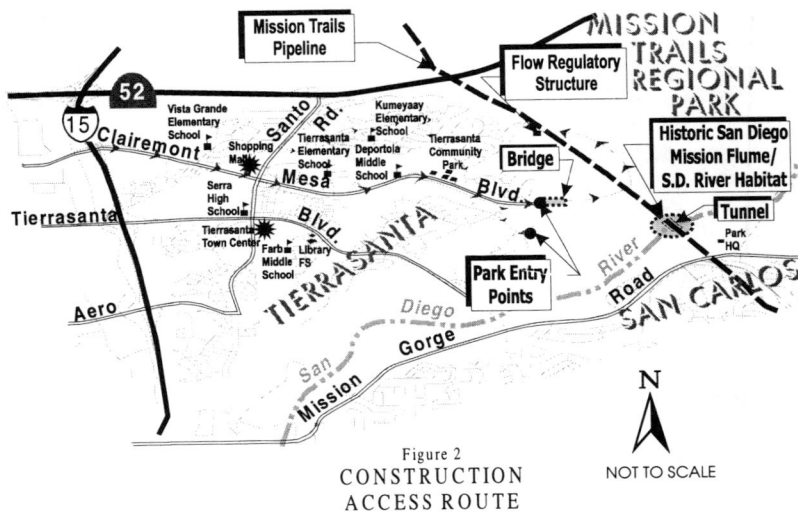

Figure 2
CONSTRUCTION ACCESS ROUTE

Figure 3
ACCESS BRIDGE

Figure 4
FLOW REGULATORY
STRUCTURE ACCESS
BUILDING

Figure 5
SAN DIEGO RIVER
SENSITIVE HABITAT

Author Biography

Zachary Ahinga

Zachary Ahinga is a senior civil engineer for San Diego County Water Authority, 3211 Fifth Avenue, San Diego, CA 92103. He managed design and construction of the Mission Trails Pipeline and Flow Regulatory Structure Project. Mr. Ahinga has worked for 16 years in engineering and construction for public utility and land development projects. He formerly worked for the San Diego Gas & Electric Company and a major southern California land development firm. He has a B.S. degree in civil engineering from San Diego State University. As an engineer, he enjoys challenging projects that better the quality of life.

Janice Collins

Janice Collins is a Public Affairs Supervisor for San Diego County Water Authority. She managed the public outreach campaign for the Mission Trails Pipeline and Flow Regulatory Structure project. Ms. Collins has worked in public relations and journalism for 14 years. She has a bachelor's degree in journalism from San Diego State University.

Rancho Penasquitos Pipeline Public Affairs Support

By James R. Melton, APR[1]

Abstract

Constructing a large diameter water supply pipeline through established suburban and rural neighborhoods is one thing, but gaining and maintaining public acceptance is quite another. This paper addresses various public affairs efforts the San Diego County Water Authority implemented to communicate with residents living adjacent to an existing right of way in which the 16-kilometer (10-mile), welded steel, 2.7-meter (108-inch) Rancho Penasquitos Pipeline was constructed during 1997 and 1998.

The Bottom Line

"We haven't gotten a lot of complaints from the approximately 6,000 residents and businesses impacted by this construction, and that is probably the best measurement of success for our public affairs efforts. It's not that people in the community enjoy inconvenience or hearing the roar of heavy equipment as we build our pipeline. We do have significant challenges dealing with residents, businesses, environmental groups and community leaders. But because people are kept informed of what's happening with the project, they seem to accept it--or at least have demonstrated they can live with it for a while."

These comments were made by Gary P. Stine, P.E., Principal Civil Engineer for the San Diego County Water Authority and Project Manager for the Rancho Penasquitos Pipeline Project. His supervisor, Assistant Director of Engineering Michael T. Stift, P.E., agrees.

"Before construction began, we put together a multi-disciplinary team of people from engineering, public affairs, right of way and environmental departments to outline our community relations needs," Stift said. "We looked at similar public affairs efforts we had implemented in recent years and considered the various groups we would impact with this new project. We concluded that the more we communicated with people during other projects, the more community acceptance

---

[1]Public Information Manager, San Diego County Water Authority, 610 W. 5th Ave., Escondido, CA 92025. Accredited Member, Public Relations Society of America.

we received. This wasn't a new idea, but it is one that construction managers sometimes don't pay enough attention to. Then they get in trouble with the community and wonder why people complain to city hall.

"Anyway," Stift continues, "we adopted an aggressive public affairs stance. Our in-house public affairs staff retained a public affairs consultant to help design and implement a community information plan. Staff and consultant work as a team, using different techniques to communicate with different groups. They stay on top of issues, make sure inquiries and complaints are handled quickly and authoritatively, and let us engineers focus on what we should be doing. The results have been very satisfying."

About the Project

To understand the character of the public affairs plan, one needs to understand the nature of the pipeline project. The San Diego County Water Authority is a wholesale water distribution agency which supplies water to about 97 percent of the 2.7 million population of San Diego County, which is in the extreme southwest corner of the United States, bordered by Mexico and the Pacific Ocean. The Authority, organized in 1944, imports raw and treated water from other agencies and in turn distributes it at wholesale rates to its 23 retail member agencies, which in turn distribute it to their customers. Water is delivered through four existing lines: Pipelines 1 and 2 are buried in the right of way known as the First San Diego Aqueduct, and Pipelines 3 and 4 are located several miles to the west in the Second San Diego Aqueduct (figure 1).

In the late 1980s, the Authority Board of Directors adopted a Capital Improvement Project (CIP) plan. The plan identified these objectives:

- Significantly increase the Authority's regional pipeline capacity, to meet both present as well as future demands, particularly during times of peak usage.
- Eliminate bottlenecks in the then-existing regional pipeline system.
- Increase reliability where water delivery was dependent on a single pipeline or source.
- Increase operational flexibility in order to facilitate pipeline maintenance.

Since the first CIP construction began in 1988, the Authority has installed eight regional pipelines that range in size from the 0.9-meter (36-inch) Ramona Pipeline to the 2.7-meter (108-inch) La Mesa-Lemon Grove Pipeline. This paper addresses the ninth pipeline, known as Pipeline 5 Extension, Phase 2 or the Rancho Penasquitos Pipeline, which is currently being constructed. The designation 5E-II will be used to identify the project for the remainder of this paper.

PIPELINES IN THE CONSTRUCTED ENVIRONMENT 223

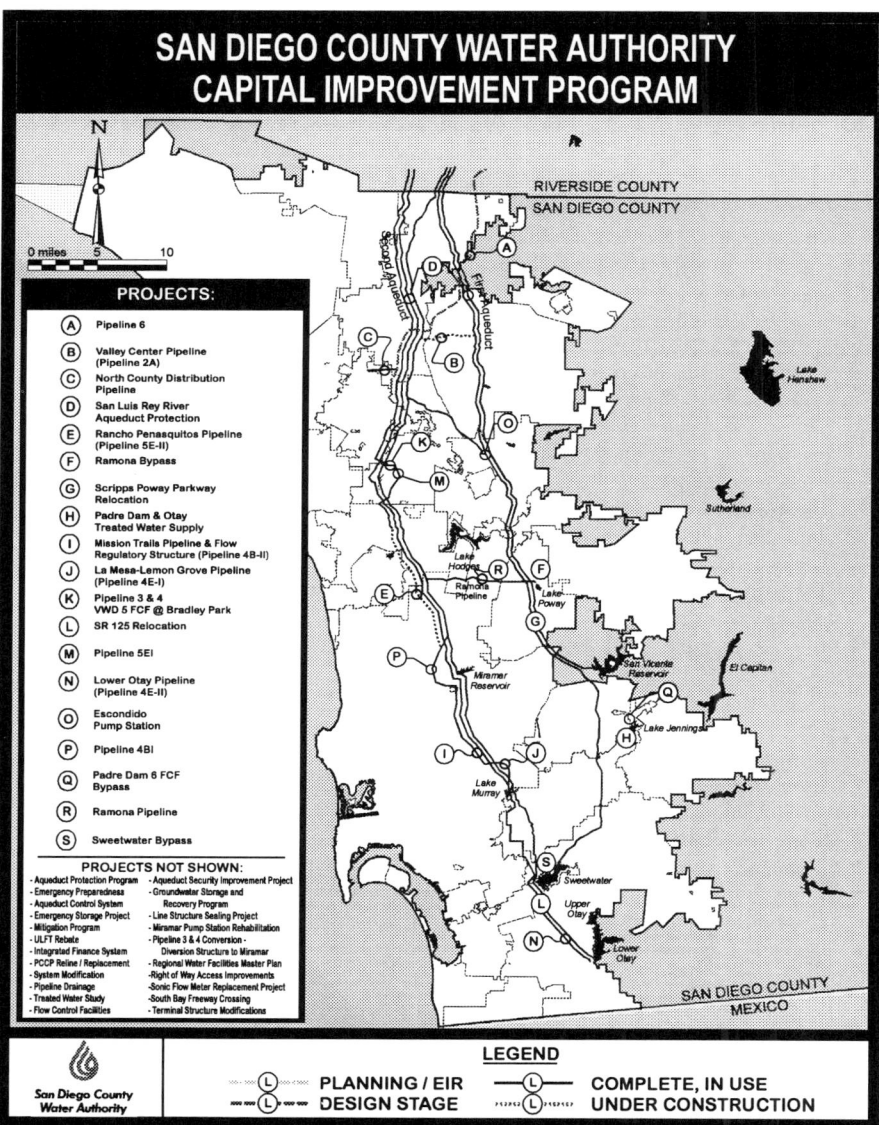

Figure 1

The 5E-II pipeline (figure 2) connects with the 5E-I project known as the San Marcos Pipeline. These lines comprise a new raw water pipeline that will improve the Authority's regional raw water distribution system. Upon Pipeline 5's completion, another large pipeline now used to deliver raw water will be converted to a treated water pipeline, thus increasing the amount of filtered water the Authority can deliver. While water in Pipeline 5 will normally flow by gravity from north to south, the pipe will be strong enough to withstand pumping that could send water in the opposite direction. This flexibility was designed into the line as part of the Authority's plans to build an Emergency Storage Project that calls for building a storage facility at the southern, low end of the otherwise gravity-flow distribution system.

The 5E-II project is about 16 kilometers (10 miles) in length and includes 1,478 pieces of 2.7-meter (108-inch) diameter, steel pipe primarily 12.2 meters (40 feet) in length. The pipeline is being constructed on the west side of the Authority's Second Aqueduct in a 42.7-meter-wide (140-foot) easement. It stretches from the top of a steep mountaintop grade (figure 3), where it connects with 5E-I, then moves south through plant nurseries and world-class horse-training facilities, tunnels under a major regional road, through the San Dieguito River, through miles of undeveloped cattle-grazing land, and finally through several miles of the highly urbanized Rancho Penasquitos community (figure 4) before it tunnels under Penasquitos Creek and another major road. Another significant challenge along the way was a large, operating pre-school sitting directly in the path of the pipeline.

Constructing Tunnels Under Roads, a Pre-School and Creek

Early in the pre-design process, the Authority became aware that its project was going to have considerable impact on the environment, on motorists, and even on hundreds of pre-school children. Not all of the issues were communications-oriented, but all could negatively affect project progress if not handled carefully and with sensitivity. For example, past agency experience had shown that the failure to properly manage an environmental issue and communicate with persons interested in such issues could bring unnecessary and potentially disastrous consequences to an entire project.

Del Dios Highway

One of the few major east-west routes in San Diego County is a curvy, two-lane road known as Del Dios Highway. This road connects coastal communities with inland population centers and is very heavily used by commuters who typically exceed a posted 24.6 meters/sec (55 mph) limit. Commuters don't like delays. Knowing fully well that open-trench construction could cause long backups, the Authority early in its planning opted to tunnel under the highway and install the pipeline through the tunnel..

PIPELINES IN THE CONSTRUCTED ENVIRONMENT 225

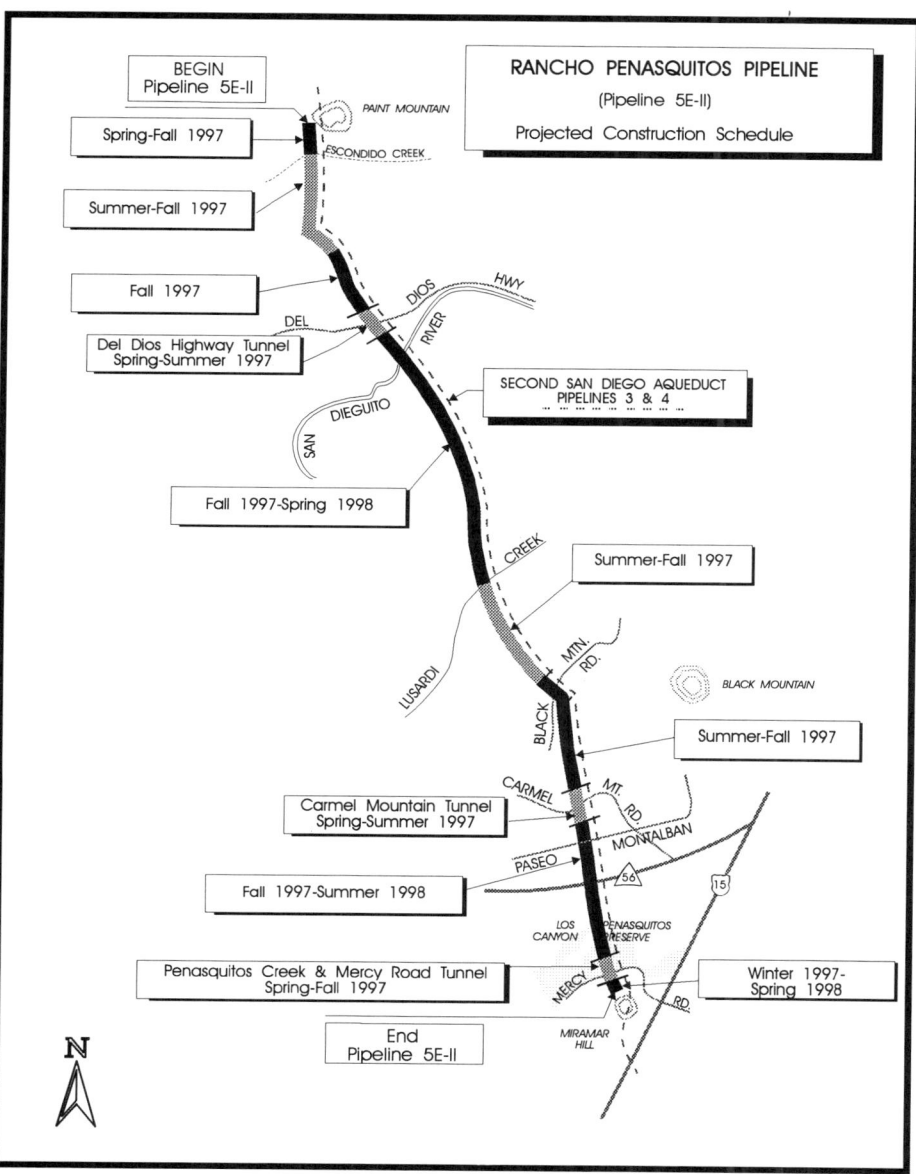

Figure 2

Even with a tunnel, though, construction was hampered by a steep grade and fractured rock on the north side of the road. K-rails had to be erected on both sides of the road, speeds were reduced to 17.9 meters/sec (40 mph), and traffic control flaggers had to be utilized for several months. The tunnel was completed early in 1998. It is 3.7 meters (12 feet) in diameter and 27 meters (90 feet) long.

Carmel Mountain Road and Pre-School

North of Penasquitos Creek, in the heart of suburban Rancho Penasquitos, is the Carmel Mountain Pre-School. The Authority's right of way passes on the west side of the school property in an easement. While most of the easement contains a driveway, there are active play areas within it. As project planning began, it was obvious that the only choices available to the Authority were to open-trench through the property or tunnel under it. (A decision to tunnel under a major adjacent artery, Carmel Mountain Road, had already been made). Trenching under the pre-school was the apparent least expensive alternative, but it also carried with it serious safety issues and negative impacts. Authority public affairs, engineering and right of way staff met on numerous occasions with the school operator, parents and city officials and ultimately arrived at a decision to tunnel under the pre-school as well as the road. This phase of the project was completed in late 1997. The 3.7-meter (12-foot) diameter tunnel is 137 meters (450 feet) long. Construction ultimately had no impact on the pre-school's operation, other than temporarily eliminating a few staff parking spaces. Children went about their normal activity, oblivious to the huge tunnel project going on almost under their feet.

Los Penasquitos Canyon Preserve

One of the area's most popular stream-oriented recreation areas is the Los Penasquitos Canyon Preserve, where Penasquitos Creek winds between suburban neighborhoods on its journey to the Pacific Ocean. An active City of San Diego-sponsored community committee meets regularly to discuss the preserve. When the Authority began meeting with the committee it became sensitized to the tremendous ecological and recreation value of the creek environment. This process resulted in the Authority making a decision to minimize damage to the creek habitat by tunneling under the area as well as under adjacent Mercy Road, a major artery. This decision was made well in advance of construction, but the Authority continued to make presentations to the committee even as construction continued. What could have been a matter of confrontation, litigation and project delay instead turned into communication, acceptance and a completed tunnel. The 3.7-meter (12-foot) diameter tunnel is 187 meters (615 feet) long.

Figure 3: The first 12.7-meter (40-foot) section of the 2.7-meter (108-inch), welded steel Rancho Penasquitos Pipeline is placed in a trench cut through Escondido Creek in 1997.

Figure 4: A major portion of the 5E-II project required work in the highly urbanized Rancho Penasquitos community. Here a section of pipe is placed near Oviedo Street in the Authority's 42.7-meter (140-foot-wide) Second Aqueduct easement.

The Bus Stops Here

Immediately adjacent to the Del Dios Tunnel is a wide spot along Del Dios Highway that for years has been used as a pick-up location and turn-around for school buses. The tunnel under the highway eliminated the bus stop and triggered a major challenge. When Authority public affairs and engineering staffs learned that the bus stop was being eliminated, the neighborhood homeowners association was contacted. A meeting was held at the site with school transportation officials, parents and neighborhood leaders, and a solution was agreed upon--or so everyone thought.

After a couple of weeks, school bus drivers determined they could not safely operate in the new location. Another meeting was held with the same parties, and another alternative was selected: the Authority and its contractor would erect a steel plate bridge over a short, narrow wooden bridge that crosses a wooden water-delivery flume. The plates would provide a safe passage for the school bus, which could then pick kids up in the neighborhood rather than up on the busy highway. This solution lasted throughout the project and earned the Authority not only respect and appreciation from the neighborhood, but a large box of chocolates presented by the president of the homeowners association! The public affairs issue here was that fast action, cooperation and communication solved what could have been a major conflict.

Constructing in Homeowners' Back Yards

Several years before construction began, the Authority right of way staff began working with homeowners whose homes were adjacent to the Second Aqueduct and 5E-II alignment. More correctly, the right of way is an easement, and nearly a hundred of these property owners owned to the center of the easement from one or both sides. Pipelines 3 and 4, built in the 1960s and 70s before the homes existed, require weekly inspection of the easement, so a patrol road is located along the entire alignment. Property owners are restricted to limited use of their property within the easement, meaning that perpendicular fences cannot be erected that inhibit patrol vehicles; swimming pools, permanent structures, large trees and other objects are prohibited.

In the months before construction began in 1997, right of way staff obtained agreements and understanding with the homeowners. One important issue was the removal/relocation of fencing and agreements concerning their re-establishment after construction. Not all homeowners were happy with the discussions, and some presented their case to the Authority's Board of Directors. The project public affairs team got involved as project construction neared. A community meeting was set. A major percentage of these homeowners whose property included the pipeline easement attended the meeting, packing the meeting room. Authority staff made a short presentation and spent two hours answering questions. Staff from the local city councilman's office attended and actively participated. When the meeting concluded, a whole different attitude prevailed. While no one was pleased about the prospects of construction in their backyards, they did express appreciation for the meeting and other communication. Their main request: construct as quickly as possible and keep them informed.

Communications Techniques

Meetings

In the months that followed the first community meeting, additional neighborhood meetings were held. Since the project stretched over 16 kilometers (10 miles), meetings were held in convenient neighborhood locations in the evening to address particular areas, not the whole project. Interestingly, very few residents attended these additional sessions. Those that did attend said they felt that residents were well enough informed so they didn't need to attend. As construction progressed, numerous meetings were held with individual property owners to discuss their issues. At other times, staff attended meetings of established groups such as homeowners associations.

Newsletters, Updates, Door-Hangers

In between meetings, detailed newsletters and simple project updates were mailed to affected residents, government officials community leaders, Authority Directors and staff and media to keep them up to speed with project progress and activity. Project photographs, drawings showing the size of the pipeline and construction equipment and other illustrations were used. As land-clearing became imminent, door-hanger notices were placed reminding people to remove fencing and other items from the easement.

Media

All local media, especially the community newspaper editor, were kept well informed of project progress and taken on several tours of project construction. Press kits including a fact sheet, detailed project information and information about the Water Authority were given to all media. Several articles were published.

Government Liaison

Knowing how important it is to keep local elected officials appraised of activity that affects their constituents, the Authority maintained a close relationship with them and their community staff aides. Aides were frequently taken on project tours and even invited to sit in on in-house construction meetings. They were also sent all newsletters, updates and other communications, as well as tipped off that newspaper articles would be appearing. One councilman's aide was particularly helpful to the Authority as he participated in on-going communications regarding placement of the pipeline between a high school and middle school. Before construction began, the Authority agreed to construct the pipeline in that area during summer months while students were gone. Unfortunately, the project reach wasn't completed until after school had resumed. But, because of close relationships established with school officials and the councilman's aide, work continued without much discussion.

Answerline

Probably the single most important communications technique was the establishment of the 24-hour Answerline. Residents and businesses were given the toll-free 1-800 number at the onset of construction, and it was included in every newsletter, update, door-hanger and newspaper article. When someone called the number, they were asked to record their question or comment and leave their name, address and phone number. Messages were retrieved by the public affairs consultant every morning except weekends and holidays. The consultant filled out an official Authority Resident Inquiry Form and faxed it to the Authority's public affairs staff and a senior engineering technician assigned to the 5E-II project. A few calls were handled by the consultant, and public affairs staff got involved in several difficult situations, but it was the engineering technician who handled nearly all inquiries. This person's responsibility was to return the resident's call the day or the day after it was received and to resolve issues. As a member of the construction team, he had detailed project knowledge and was able to handle inquiries authoritatively and effectively. His participation in the public affairs effort was invaluable.

Most Answerline calls were regarding noise, working hours, dust, traffic and project construction timelines. All inquiries were discussed at weekly meetings between the Authority and its contractor.

Summary

Rancho Penasquitos Pipeline was constructed during 1997 and 1998 in a 42.7-meter-wide (140-foot) existing easement that stretched about 16 kilometers (10 miles) through rural and suburban environments. To deal with many public affairs issues, the San Diego County Water Authority established a comprehensive communications program. This included community meetings, personal contact with residents, community leaders and business owners, newsletters, updates, door-hangers, governmental liaison, media relations and a 24-hour toll-free Answerline. While the project did not proceed without challenges, the communications effort was successful and most people accepted the project construction or were at least willing to put up with it for a while.

Biographical Sketch
James R. Melton, APR

Jim Melton is public information manager for the San Diego County Water Authority. Prior to joining the Water Authority in 1988 he served for 17 years as public information officer for the Santa Clara Valley Water District in San Jose.

Jim earned his BA in journalism from San Jose State University and his Masters in Public Administration from the University of San Francisco. He was a combat photographer in Vietnam and says he still takes pictures very quickly.

He is an accredited member of the Public Relations Society of America, where he was president of the San Francisco Peninsula Chapter. He recently served for one year as the voluntary chairman of public relations for the Poway Center for Performing Arts.

Jim handles the Water Authority's Capital Improvement Project public affairs program in addition to his other duties. He plans to retire from public service at the end of this year.

City of San Diego Mid-City Pipeline Project
Integrating a Public Outreach Program into the
Planning and Alignment Selection Process

Marc R. Weinberger[1]

Abstract

Obtaining public acceptance and support for construction of a major pipeline through a developed urban community can present more difficult challenges than the actual engineering elements of the project. Today, with the increasing public desire for a say about projects which impact their communities, obtaining public "buy in" can be one of the most challenging aspects of urban pipeline design. The days of "decide, design and defend" are gone, replaced by a process of community involvement in the design process.

The City of San Diego's Mid-City pipeline is an excellent example of bringing the community of this highly urbanized area in to active involvement with the engineers to achieve common goals. Achieving public support requires a commitment by both the designers and project owners to spend the time and resources to integrate the public into the decision making processes.

Introduction

The 1.2 meter (48-inch) diameter Mid-City Water Transmission Pipeline will be the first major transmission main project in the central (Mid-City) area of the City of San Diego in over 40 years. The $24 million 8 km-long (5-mile-long) pipeline will provide a back-up to the 40-year-old Trojan Pipeline that now serves the extensively developed Mid-City Area.

The area served by the Mid-City Pipeline Project is located entirely within the City of San Diego. This service area includes a number of communities within the City, including Rolando, Darnell, City Heights, Eastern Area, and the College Area.

---

[1]Vice President and Engineering Director, John Powell & Associates, Inc., 175 Calle Magdalena, Encinitas, CA 92024

This service area is a mature, developed, urban community. Land uses within this area include primarily residential and commercial development. Commercial development is concentrated largely along the El Cajon Boulevard corridor, which is a major transportation corridor through the area. Residential developments surround this major thoroughfare. Located within this area are facilities typical to urban or suburban residential and commercial areas, including schools, churches, colleges and other support facilities.

During the initial planning phases of the project, three major alignment corridors were identified. This includes the El Cajon Boulevard alignment, which is largely in commercial areas; the Orange Avenue alignment, which is predominantly in residential developments; and the University Avenue alignment, which is in a mixed development area consisting of both residential and commercial developments.

During initial studies, the City of San Diego Water Utilities Department identified the Orange Avenue alignment as being a preferred alignment for the location of the Mid-City Pipeline. This alignment was selected because of the desire to avoid El Cajon Boulevard, which is a heavily traveled arterial highway that contains extensive commercial development. The City believed that by locating the pipeline in predominantly residential streets, that adverse construction impacts to both traffic and adjacent businesses would be minimized.

Community Relations/Public Outreach Program

The City Water Utilities Department recognized that a successful public relations outreach program was key to the successful implementation of the Mid-City Pipeline Project. In order to successfully implement this program, a team of professionals would be assembled to plan and execute the outreach program. This team included the City of San Diego Water Utilities Department, and its civil engineering and public affairs consultants. Working as a unit, this team successfully conceived, planned and implemented a successful public outreach program.

As discussed earlier, land uses within the mid-city area are diverse. The mid-city area is also comprised of a number of distinct communities. There is also great cultural and ethnic diversity in the communities along the pipeline corridor. As a result, personal contact would be one of the keys to a successful community relations program. One of the objectives of the public relations program was to make direct personal contact and nurture and maintain positive relationships with those individuals who reside, own businesses, work and attend school or church in those area which will be affected by the project. The community relations outreach program included the following elements:

1. Stakeholder Analysis. People who will be affected by the outcome of the project, or have a "stake" in the project, are termed stakeholders. One of the first steps in the public outreach program was to identify

key decision makers, community leaders, school administrators, church officials, and business owners who are representative of the predominant interests in the affected communities. Once these stakeholders were identified, meetings were scheduled with them to explain the project, determine what their issues are, and assess their possible support or opposition to the project.

2. <u>Community Map Development</u>. A community map, or comprehensive list of people to contact within the community, was prepared. This list consisted of community and business leaders, government officials, and neighborhood civic and planning group members. In addition, a list of all schools affected by the project, and their schedules, was prepared. This list was continually updated with the names and information of individuals who attended the meetings or otherwise showed interest in the project.

3. <u>Community Leader Briefings</u>. Periodic meetings were held with community leaders, either individually or in small groups, to brief them on the project progress and keep the project team appraised of any new issues. Initial meetings were held with these community leaders at the beginning of the project so that they could disseminate accurate information to their constituents.

4. <u>Project Design Advisory Group</u>. Key community representatives were contacted and asked to form a Project Design Advisory Group. This group became the key element in the public outreach program, and was instrumental in gaining public acceptance of the proposed project. The group met five times to review the project plans, discuss design issues and offer input to enhance the design, suggest mitigation measures, and assure that community values are appropriately factored into the project design and mitigation plans. The group provided the necessary feedback from the community in identifying esthetic issues, traffic concerns, and other mitigation measures.

5. <u>Information Dissemination</u>. Methods to disseminate information to the public were also prepared. This included displays that could be shown at community meetings, as well as fact sheets that could be disseminated to the public at large.

Government Relations

It is also important that elected officials and other related government agencies understand the purpose and scope of the project. Facilitating an information exchange with these groups insured that accurate and consistent information about the project was disseminated. An open dialog also kept the project team aware of any issues or concerns that would potentially pose a threat to the success of the project. The government relations program included the following:

1. Government Leader Briefings. Meetings were scheduled with the San Diego City Council members and their representatives to determine the community needs and concerns, and to provide project information. These meetings were instrumental in assuring that elected officials were knowledgeable about the project progress and were aware of any significant community issues that were brought up during the public outreach process. These briefings helped assure that support of the project was obtained from the elected officials.

2. Inter-Agency Meetings. Meetings with other local and state agencies during the planning and design phase of the project helped insure that there are open lines of communication and project delays can be minimized. The agencies that were affected by the design of the Mid-City Pipeline Project included Caltrans, which was developing the nearby Interstate 15 corridor which is crossed by the Mid-City Pipeline; the Metropolitan Transit Development Board, which operates public transportation in the area; San Diego Unified School District, police and fire departments. These meetings made certain that the Mid-City Pipeline Project was coordinated with other possible projects in the area, as well as address the concerns of agencies operating within the service area. For example, extensive interaction with Caltrans was required to coordinate construction scheduling and pipeline tie-in locations.

Mid-City Pipeline Review Committee

As discussed earlier, the Mid-City Pipeline Review Committee was a key element in the public outreach program. It became the heart of this program, and the success of this committee largely contributed to the public acceptance of the Mid-City Pipeline Project. The Mid-City Pipeline Review Committee (MCPRC) consisted of interested parties from the mid-city area planning, school, church, business and residential groups. This committee consisted of members who came from local community councils, council persons offices, business improvement associations, planning committees, local schools and churches.

The purpose of the MCPRC was to help city staff and consultants evaluate potential alignments for the Mid-City Pipeline. This group was to help the City seek an alignment which is feasible from an engineering, construction and cost point of view, and identify feasible mitigation opportunities to lessen the disruption to the mid-city community.

Objective

The MCPRC assisted the City of San Diego Water Utilities Department in examining all aspects of the Mid-City Pipeline Project that may be of interest to mid-city residents, business owners and other interested parties. City staff and its consultants used this input to develop the final preferred alignment for the Mid-City Pipeline. The objectives of the MCPRC were to:

1. Identify potential community impacts, including noise, access limitations, traffic disruptions, construction times, construction scheduling, etc.

2. Identify mitigation opportunities.

3. Evaluate alignment selection criteria and weightings used by the City.

4. Discuss the cost considerations of one proposed pipeline alignment versus another.

5. Identify other potential public concerns.

Role of Committee Members

At the time the committee was formed, the engineering and preliminary design on several alignment options was proceeding. However, the final preferred alignment had not been determined. The MCPRC helped city staff consider all relevant community points of view before making a final determination on the alignment of the Mid-City Pipeline. As such, the role of the members was to:

1. Become familiar with the pipeline alignment considerations and limitations.

2. Investigate those issues that would be associated with the various pipeline alignment alternatives.

3. Insure that community values and concerns are reflected in the final alignment decision-making process.

Facilitator

A committee facilitator, as well as city staff and consultants, were included in the MCPRC. The facilitator served to chair and direct the committee meetings. However, the facilitator was neutral with respect to all of the issues related to selection of alignment and to any preferred pipeline alignment alternative.

Work Product

A summary report was prepared on behalf of the committee. The report included the following work products:

1. The scope and content of the committee's discussions.
2. Findings and conclusions on the issues considered.
3. Recommendations for mitigations for the project.
4. Individual opinions and observations that may not have been reflected in the main body of the report.
5. A matrix of weighted criteria evaluating each alignment alternative.
6. A suggested preferred alignment based upon the results of the evaluation matrix.

It should be noted that the deliberations and findings of the committee were advisory only. The City has the final decision regarding which is the preferred alignment. However, the results of the committee and its recommendations were considered very seriously in selection of the final preferred alignment.

The Review Committee Process

A total of 12 people were assembled from the community to form the Mid-City Pipeline Review Committee. This committee was supported by the City Water Utilities staff and their civil engineering and public affairs consultants.

This committee met five times over a four month period to research, review and evaluate the factors used to select a pipeline alignment.

The first meeting was spent presenting a general overview of the project. This included discussing the planning criteria that go into a pipeline project including; the need for the project, sources of water supply for the project, and major issues to be considered in selecting a pipeline alignment.

The second meeting was spent discussing how a major pipeline project is designed and constructed. During this meeting many of the impacts that the pipeline may have on the community during construction were revealed. This could include impacts such as noise, traffic, dust, conflict with other construction projects, and impacts to special facilities such as schools and churches. During this meeting while impacts were disclosed, potential mitigations to the impact on the community were also being discussed.

During this meeting the three potential alignments for the pipeline that were being considered were presented to the committee. These alignments included the Orange Avenue alignment, the El Cajon Boulevard alignment, and the University Avenue alignment. Initial advantages and disadvantages of each alignment was discussed.

The third meeting consisted of a field trip and a tour of each of the three alternatives. This occurred on a Saturday morning and a bus was rented for the day. All members of the committee were given the opportunity to tour each of the three alignments. This tour had many positive benefits. It gave committee members an opportunity to interact on more informal basis. It allowed for a more free interchange of ideas between the committee members and the City staff and the consultants. It was also an excellent opportunity to see first hand the alignments and some of the design and construction issues that had to be addressed in each of the alignments. It also gave the group an opportunity to see the types of facilities that may be impacted by the construction of the pipeline project. Everyone left this tour with a much better understanding of each others needs and concerns as well as a much better understanding of the project alignment and the issues related to each of these alignments.

Meeting number four was a very important meeting and became the heart of the review committee process. During this meeting two important documents were prepared. The first was a selection of pipeline alignment evaluation criteria. The second was a discussion of potential mitigation measures that would be used during the construction of the pipeline project.

After much discussion, the committee selected the following pipeline alignment selection criteria:

- Minimize commuter traffic impacts (this included impacts to through traffic in the area)
- Minimize local community impacts (this included such items as noise, safety, access to private property, impacts to local traffic, parking, economic impacts and conflicting construction activities)
- Capital costs
- Environmental impacts
- Hazardous substances impacts (this included contaminated soils that might be encountered when excavating close to old or abandoned gas stations, dry cleaning facilities or auto repair shops)
- System reliability (this includes the reliability of the existing water transmission and distribution system)
- System practicality (this will include such items as utility conflicts of constructability issues, proximity to connections to the existing distribution system and geotechnical engineering issues)

This list of selection criteria was incorporated into evaluation matrix used to rate each of the various alignment alternatives and select a recommended alignment.

The second major product of this meeting was the initiation of list of mitigation issues for the pipeline alignment. There were literally dozens of suggestions for ways to mitigate the disruption caused by the construction of a pipeline. Some major mitigation suggestions offered by the committee included the following:

- Constructing the pipeline at night in commercial areas,
- Constructing pipeline during daylights hours (between 8:30 a.m. and 3:30 p.m. in residential areas),
- Resurfacing of City streets to mitigate the damage to street surfaces caused by pipeline excavation and trench patches,
- Modifying construction schedules to work near schools when they are out of session,
- Combining construction of the Mid-City pipeline with other underground utility projects, and
- Implementing stringent safety plans when operating near schools or other areas of high pedestrian traffic

Meeting five was the final meeting for the committee and culminated in the preparation and review of an evaluation matrix and the selection of a recommended pipeline alignment. A draft alignment evaluation was prepared and distributed at the meeting. This evaluation matrix included the Orange Avenue alignment, the University Avenue alignment, the El Cajon Boulevard alignment, and an alignment which included constructing in El Cajon Boulevard at night.

The alignment evaluation matrix was prepared generally in the following manner. A initial matrix was prepared by the civil engineering consultant. The basic evaluation criteria of commuter traffic impacts, local community impacts, capital costs, environmental impacts, hazardous substances, system reliability and system practicality were set. Then with the help of the committee, the significance for weights for each criterion was assigned by the committee team. In addition, each committee member was asked to assign weights to the criteria based his or her individual perspectives. The higher the weight percentage the more important the criterion was to that individual.

The consultant and the City then established and assigned ratings for the criteria based on data collected for each alignment option. The rating scale ranged from 1-10 with 1 representing the lowest impact and 10 representing the highest impact. The ratings were multiplied by the weights and then totaled to achieve an overall score for each alignment option. A lower score represents the least overall impact of the criteria on the project or more feasible project.

It was evident from the initial evaluation summary that Orange Avenue and El Cajon Boulevard alignments received scores within several tenths of a percentage point from one another. The University Avenue alignment option score was significantly higher relative to the others and was therefore determined to be the least feasible alternative.

The matrix also considered the El Cajon Boulevard alignment option with construction activity to be scheduled only at night. Night construction would be less disruptive to traffic, business access and parking and have less of an economic impact then work done by day.

Recommendations

The City reviewed all of the documentation prepared by it's civil engineering consultant and the work done by the Mid-City Pipeline review committee. Based on this work and the results of the evaluation matrix, the City selected the El Cajon Boulevard alignment as the most feasible alternative. Further, the City decided that the El Cajon Boulevard alignment would be constructed at night.

Conclusion

Everyone involved in the project, including City staff, its consultants and the committee members felt that this was a very valuable and worthwhile process. The goals of the City Water Use Utilities Department and the goals of the community were both met. The pipeline alignment which achieved the City's goals of the distribution of water at a reasonable cost and providing a reliable system. At the same time, the community goals of achieving a reliable water supply while at the same time minimizing construction related impacts were also achieved. More importantly, the community felt that it was "heard" and was an integral part of the decision-making process. The result is a project that has the support of the City water utilities, the political leaders and the community at large.

Diagnosis of Buried Sewers: Tools and Methods, the French Experience

Youssef Georges Diab[1]

Abstract

The choice of an analytical method must be accomplished using a multitude of criteria that take into account the constraints of the pipeline facility, the objectives of the analysis and budget constraints.

Therefore, clients and engineers must, first make sure that the selected method (which could be conducted by readily available staff or outside specialists) is technically capable of achieving the desired results and fits within available budget constraints.

Introduction

The size and the length of buried pipes as well as the worksite conditions seem to be 'limitless'. Recent projects incorporate the progress in construction techniques, materials and the knowledge of design and construction companies.

Pipeline safety issues are increasingly important, especially in major metropolitan areas. Unfortunately, like any human achievement, these structures have aged and have modified over time to accommodate repairs and connections. The result is pipeline systems with varying conditions and possible failures that are difficult to predict.

They need to be monitored and incorporated into a preventative maintenance program. The major cities in France include Paris and it's suburbs, Lyon, Bordeaux and Nice. Faced with the management of such pipeline systems, the cities have created specialized 'Gerontology' departments in charge of these 'aging' facilities in order to reduce (or preferably avoid) serious failures with the associated disruption to the

---

[1] Youssef Georges Diab, Associate Professor, LGCU, Marne la Valleé University, Paris, 2 Rue Albert Enstein-Cite Descartes, 77420 Champs Sur Marne, France

surrounding community and discharge of raw wastewater. The objective is to increase the life span of aging pipelines and protect the local environment. The means selected and carried out by these agencies is the collection of data on the condition of the pipeline systems.

Analytical Factors

Several factors influence the deterioration of underground facilities and especially buried pipes:

- Age
- Material quality
- Construction method, surrounding ground and environment
- Characteristics of fluids carried

The relative importance of these factors varies significantly. A 100 year old sewer in quarry stoned masonry does not perform like a recent pipe in concrete formwork totally watertight. Establishing analytical techniques requires a good knowledge of the pipeline deterioration modes, a careful evaluation of available data on the pipelines operational history and testing techniques that are capable of providing additional data.

Analytical Goals

The work with "Gerontology Departments" within each city are focused on achieving the following goals:

Pipeline Evaluations: Provide assessments of the pipeline system conditions to determine the likelihood of a failure within an unacceptable period of time.

Operation and maintenance: Optimize operation and maintenance functions in order to obtain the data needed to determine the condition of the pipeline facilities economically.

Identify Major Repairs: Develop an accurate diagnostic technique that provides staff with information enabling each agency to set priorities and schedule repairs within budget cycles. Such scheduling is focused on setting high priority for repairs intended improve to safety conditions and avoid serious failures.

Safety Measures: An abnormality or the acceleration of a deterioration process detected while the analysis is being made may require immediate safety measures to ensure safety and avoid even worse disorders while waiting for an accurate intervention scheme.

Keeping of the functional performances of the pipeline and its safety level with regard to third parties, such as surrounding private property and other utilities.

Main Causes of the Deterioration of Surrounding Soil and the Pipe Wall

The study of sewer problems has shown a wide range of causes explaining the deterioration of the pipe system, the surrounding soil and the interface. Frequently several problem areas may be encountered. The challenge is to determine the most prominent failure mode associated with a pipeline system. The following is a list of potential problem areas:

1. Deterioration of the surrounding soil as a result of modified groundwater conditions
   - Washed out fine elements and particles
   - Reduction of mechanical properties of some soft soils when the water content increases
   - Swelling due to hydration (as encountered in some clays, chrysotiles, anhydrite and other similar conditions)
   - Shrinkage of clayish materials (and swelling if moisture returns)
   - Dissolution (gypsum and plaster stone (plaster of Paris), etc.)
   - Fluctuation of the level of water tables likely to causes settlements or heaves
   - Chemical and bacterial pollution of the water table following leaks from a sewer main

2. Deterioration of the soil from other causes
   - Creep under high stress
   - Redistribution of strains in the decompressed zone during construction especially with trenchless construction techniques
   - Voids around the tunnel leading to an erosion of the soil in the vicinity of the tunnel
   - General instability of the surrounding terrain (proximity of a slope, pipeline crossing a landslide, etc.)

3. Deterioration of the ground caused by external causes
   - Construction undertaken nearby (modification of the stress field by embankments or excavations, water table lowering, water inflow, etc.)
   - Consequences of the vibrations induced by rail or road traffic
   - Presence of neighboring underground works
   - Particular aspects of the structure (particular links, water supply pipes inside the pipe)
   - Chemical pollution of the soil

4. Deterioration of the pipe wall
   - Irrelevance of the initial design to the specific conditions of the site or due to further works, especially inappropriate repairs
   - Modification of the utilization conditions as a result of an increase of the flow through, or filling of a pipe initially with free flow, which might increase the stress in the wall
   - Worsening of construction defects (poor quality of materials, improper position of quarry stones, etc.)
   - Desegregated mortars due to aging hydraulic binders (lime, cement)
   - Fatigue due to repeated or alternated loads or strains (cracking)
   - Local fractures caused by excessive stresses (exfoliation or burst quarry stones)
   - Chemical reaction of the wall components (cements and sulfates for example) aggravated by the presence of water containing minerals that promote swelling (ettringites for example), thereby leading to alteration and strength loss
   - Punching of the terrain through the foundation of footings
   - Gel effect in cold cities (Diab, 1993)
   - Effects of temperature and hygrometry variations
   - Acceleration or aggravation of aging process of binders due to water
   - Action of bacteria on stones, concrete and binders
   - Rotting and /or swelling of supporting wood left near the pipeline facility during construction

Consequences - Failure Modes

From these differing conditions and factors, five types of failure modes have been found:

1. Instabilities more or less localized (partial collapse)
2. Deterioration of the wall (flaking, desegregation, cracking, disjoining, etc.)
3. Water infiltration in high groundwater areas or exfiltration into the surrounding soils
4. Reduction of the sewer performance
5. Damage risks incurred by third parties (personnel and equipment) in the vicinity of the sewer

Information and Data Required for the Analysis

- Unchanging data are essentially linked to the construction and environment of the facility and then to important events that have occurred on the pipeline
- An evaluation of the structures of the facility: all elements that enable to measure and visualize the modifications of the properties of the pipe and

the surrounding soil and, above all, to show the acceleration of the degradation
- Geological and geotechnical environment understanding: deterioration of the physical and mechanical properties of the surrounding soil
- Hydrogeology review: evaluation of flows and movements, water quality, condition of drainage and groundwater movement

Stages to Establish the Analysis

The approach followed by the "Gerontology Department" staffs generally includes the following tasks:

Establish the analysis file: All relevant data is collected and inserted in a project file. The information gathered normally includes dates, geology and hydrogeology, mechanical properties of the file, construction method (trench, embankment, tunneling.), materials, incidents, and repair work orders that might be found in the archives developed during the construction and maintenance of the facility. This file is updated upon every event, in particular visits and inspections.

Identification sheets: Depending on the case, it could be useful to sum up shortly in a sheet the essential elements that must be available to the operator for the routine servicing and maintenance of the facility.

Visiting and inspection of the pipe : In the framework of the routine operation which frequencies and modalities, in particular are defined for each type of network by instructions, recommendations or orders. At the end of this stage, a written inspection report specifies the conditions for continuation of operating.

Reinforcing monitoring: On stretches of the facility performing unfavorably (as determined by visits or inspections) or incurring great risks, it is designed to:
- increase the specific monitoring to ensure safety
- establish or confirm the diagnosis
- help define repair works and their range

Implementing facility diagnostics: these could include destructive or non-destructive methods, depending on the nature of the pipeline deterioration.

Diagnostic factors summary: The analysis must specify the cause of disorders and the incurred risks as regard to:
- The utilization of the sewer
- The change in the rate of deterioration
- The presence of factors contributing to the continuation of the observed deterioration

Problems Encountered with This Method

When we tried to apply this method to Paris (Diab, 1998), one major difficulty was encountered: How and where to begin? It is not easy to answer especially when there is 2,100 km (1,300 miles) of pipelines and we don't have enough information regarding the condition of these sewers. To find a solution, we added a step to the presented methodology. This step involves analyzing documents, information and files to define areas with a high-risk level. This multiple criteria analysis provides the review team with a method of defining priorities (lines from where visual inspection has to begin).

The selected analysis is based on identifying risks using a systematic process. The risks of concern for the Paris project are defined in table (1). On this table are presented the four categories of risk; geotechnical, hydraulic, endogenous and impact risks. For each risk, criteria were defined (five for the geotechnical factor, three for the hydraulic factor, two for the endogenous factor and four for the environmental factor). Then parameters allow one to evaluate each criteria: some examples are given on the third column of Table 1. For example, to illustrate the criteria 'presence of swelling clay' in the surrounding soil in the geotechnical risk (G), the note (a) is used. The criteria 'state of the sewer' in the endogenous risk (E) is noted (e), etc.

To quantify a given risk, this methodology should be followed:
- Attribution of a grade for each parameter: This grade is given in function of available information essentially in the database of the city.
- Combining the parameter notes for each criteria, which permit one to obtain the criteria note.
- Combining the notes of criteria characterizing the same category to obtain the note of the considered risk (G for the geotechnical risk, H for the hydraulic risk, E for the endogenous and I for the impact).

The Global note is obtained by following these three steps:
- Combination of notes H and E to obtain a note C for the sewer itself
- Combination of notes C and G to obtain a technical note T for all aspects
- Combination of notes T and I to obtain the global note N

RISK	CRITERIA	Note	PARAMETERS (examples)
1) related to the soil (geotechnical risks) NOTE G	1.1 movement of the soil particles 1.2 settlement 1.3 dissolution of plaster stone (plaster of Paris) 1.4 voids 1.5 swelling clay	f t d v a	- position of the sewer in comparison with the soil layer - importance of water circulation - thickness of the plaster stone layer - live load importance
2) flow risk (hydraulic risk) NOTE H	2.1 action mechanical action of the flow 2.2 physical and chemical action of the flow 2.3 hydraulic loads	m p c	- pipeline slope - sand transportation - invert shape - presence chemical products - load pipeline related to the flooding of the Seine
3) linked to the pipe (endogenous risks) NOTE E	3.1 geometry of the sewer 3.2 state of the sewer	g e	- shape of the pipe - thickness and nature of materials - number, lengths and cracks distribution
4) linked to the environment and the functioning of the sewer (impact risks) NOTE I	4.1 consequences on the functioning of the sewer 4.2 consequences on the functioning of other sewers 4.3 repercussion on the roads 4.4 consequences on the surrounding buildings and on social costs buildings	r x w z	- importance of the sewer (primary, secondary, main sewer..) - nature and importance of other networks - distance between the sewer and the road - importance of the road - presence de historical buildings

Table 1. Risk Notes

This methodology is showed on Figure 1. The risk level increases with the increasing of the note value:

1. No risk
2. Low risk
3. Medium risk
4. High risk
5. Declared risk (reserved to areas showing already disorders).

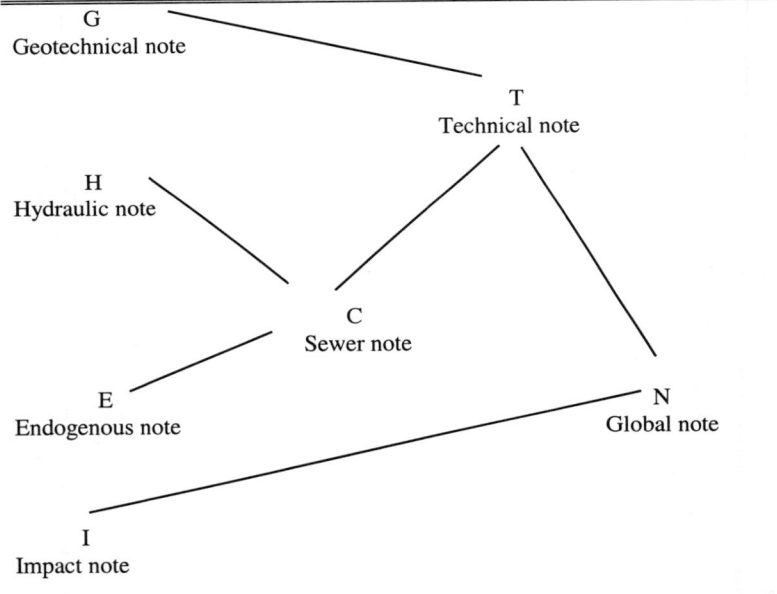

Figure 1. Calculation of the Global Note

Conclusions

This method is used to begin a pragmatic analysis on the Paris sewer network, which includes 2,100 km (1,300 miles) of sewer pipelines. The first results (on approximately 175-km (108 miles)) have allowed the team to determine a first estimation on sewers with a high risk of failure. They represent around 15 % of the entire system and it is early too fully evaluate the reliability of this approach.

As searchers we are looking to apply this approach to other cities in Europe and North America. We think that the methodology can be applied in other metropolitan areas. The risk criteria may require modification to accommodate specific factors unique to these communities.

Appendix - References

Diab Y. ' Reinforced and concrete pressure pipe diagnosis ' in pipeline infrastructure II, ASCE Conference, San Antonio (TX) August 1993. pp. 355-369.

Diab Y. ' La gestion du patrimoine enterré : approche multicritère '. Ecole Thématique CNRS Sol Urbain. Garchy, France, 1998, 4 p.

IMPACT ECHO TESTING OF IN-SITU PRECAST CONCRETE CYLINDER PIPE
Dennis A. Sack; Larry D. Olson, P.E. [1]

Abstract

This paper presents recent research and field work performed in the area of Nondestructive Testing (NDT) of Precast Concrete Cylinder Pipe (PCCP). The NDT was performed using the Impact Echo (IE) and the Spectral Analysis of Surface Waves (SASW) methods from the inside of PCCP sections to determine the integrity of the outer grout layer and bonding of the concrete to the embedded steel cylinder. Delamination of the outer grout layer or debonding of the steel cylinder from the concrete has been associated with corrosion and failure of the prestressing wires in this type of pipe. Testing of PCCP has been performed with the IE method using both a 4-channel scanning system as well as a hand-held single point IE device. The SASW testing of PCCP has been performed using discrete transducers, and examined both the material velocity characteristics of the pipe for use in IE measurements, the integrity of the pipe concrete and grout, and the material characteristics outside of the pipe.

The paper includes a brief description of the NDT methods and equipment used to perform the testing, as well as sample data for each method. The data was collected on in-situ PCCP sections under real-world conditions. This paper also presents a brief review of previous research conducted on NDT of in-situ PCCP.

Introduction

The use of concrete pipe has been shown to be an effective solution to the problem of transporting large quantities of fresh water and sewage over long distances. Concrete pipelines are generally reliable and of relatively low maintenance in most applications. There have been reports, however, of failures of this type of pipe under certain conditions, leading to the need to evaluate the condition of the pipe in-situ. Various technologies have been proposed and tried for testing PCCP in-situ, including a version of Acoustic Emission (AE) to listen for wire breaks, visual inspection and sounding from the inside, and NDT methods such as these presented here. Each of the techniques has been shown to have benefits and drawbacks, including cost, reliability,

---

[1] Olson Engineering, Inc. 5191 Ward Road, Suite 1, Wheat Ridge, CO 80033
303-423-1212

ease of use, and the information obtainable. In this paper, the use of the acoustic based NDT techniques of IE and SASW for evaluating the current condition and integrity of the pipe wall is presented.

Analysis of PCCP failures and previous research conducted by the Bureau of Reclamation have shown that many of the failures are due to corrosion of the prestressing wires and subsequent breakage, leading to loss of pressure capacity. The research has also shown that where the wires have corroded, there is also usually a failure (delamination) of the protective grout layer which is over the prestressing wires. The IE method is a nondestructive way to measure the effective thickness of the concrete pipe wall without requiring access to the outside of the pipe. This is important because excavation around a pipe can be expensive and time consuming.

Background and Previous Research

A previous paper by the authors presented the results of initial research into the use of the IE method in testing PCCP, as well as the use of a prototype IE scanner system in high-speed testing of this type of pipe. This previous research by the authors for the Bureau of Reclamation showed that the IE method is effective in measuring the wall thickness of sound, undamaged PCCP, even with an embedded steel cylinder and prestressing wires (Sack and Olson, 1994). The earlier work showed that where corroding wires have caused delamination of the grout layer, this damage is detectable with the IE method as long as the grout layer thickness is at least 4%-5% or more of the total wall thickness. This limitation is due to the resolution of the IE method in detecting small changes in wall thickness in practical applications.

The IE scanner system prototype developed as part of the previous research project used the IE method with scanning technology to measure the wall thickness of the pipe along given test lines on a near-continuous basis. The IE measured wall thickness could then be compared to the design wall thickness. Losses in wall thickness of 1.9 to 2.5 cm (3/4 to 1 inch) would normally be associated with the delamination of the outer mortar layer due to wire corrosion. The initial research into the scanner system used a buried section of damaged pipe for testing, followed by a field test in an in-situ pipeline. The scanner system worked relatively well for this pipeline, resulting in plots of wall thickness versus location along a number of lines in the pipe. The scanner did have problems, however, due to mud on the walls and the prototypical nature of the hardware.

Current Usage of the IE Method

Recent testing in PCCP sections under field conditions has shown that a practical application of the IE method for many pipeline evaluation scenarios is with an easily applied and used hand-held IE device. The hand held IE testing system has the advantage of being easy to bring into the pipe, and easy to transport within the pipe to areas of concern. The basic limitation of the hand held device is that only one point at a time can be tested, but data can be collected at rates of 4-5 seconds per point or more, depending on access conditions and the desired grid. Recent use of the pipe scanner prototype system has shown that rapid testing is possible, but that changes to the transducers may be required if testing is to be done under the wide range of surface conditions which can be encountered in real-world situations. A brief description of the IE method and the hand-held IE device is included in the next section.

IE Test Method Description

The IE test method and its use in testing concrete pipe have been described in detail in previous publications (Sack and Olson, 1994; Olson et al, 1992; Sansalone and Carino, 1986). The IE method is performed on a point-by-point basis by hitting the test surface at a given location with a small (90 gm (0.2 lb)) instrumented impulse hammer or impactor and recording the reflected wave energy with a displacement or accelerometer receiver mounted to or pressed against the test surface adjacent to the impact location. A simplified diagram of the method as applied to PCCP is shown in Fig. 1.

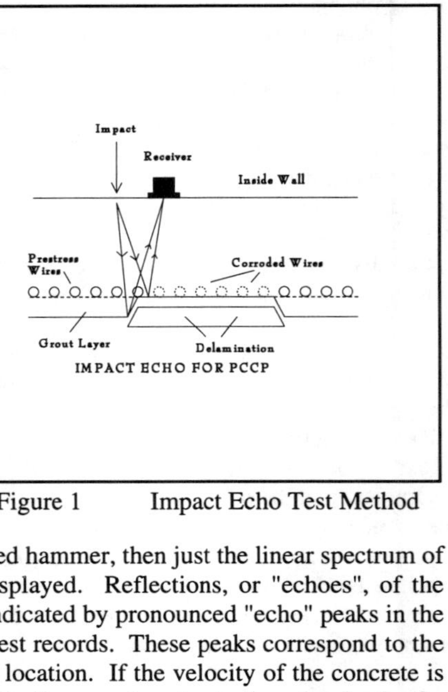

Reflections from sound areas of the pipe cover a longer path and thus take a longer time to reflect to the receiver. Over an area with an outer grout delamination, the signals cover a shorter path and thus reflect quicker to the receiver. Since the reflections are more easily identified in the frequency domain, the time domain test data of the impulse hammer (if measured) and receiver are processed by the data acquisition PC for frequency domain analyses. For data collected with the impulse hammer and accelerometer, a transfer function (system output/input) is then computed between the hammer (input) and receiver (output) as a function of frequency. If an impactor is used instead of an instrumented hammer, then just the linear spectrum of the receiver signal is computed and displayed. Reflections, or "echoes", of the compression wave energy are typically indicated by pronounced "echo" peaks in the transfer function or frequency spectrum test records. These peaks correspond to the effective thickness of the pipe at the test location. If the velocity of the concrete is known or can be measured (as is generally the case for pipes), then the depth of a reflector can be calculated from the echo peak frequency. This will be the design thickness at sound locations, or about 1.9 to 2.5 cm (3/4 to 1 inch) less than the design thickness at locations with a grout delamination.

Figure 1   Impact Echo Test Method

Single Point IE Test System

The single point IE testing system consists of a hand-held test head connected by a long (15-30m) cable to a battery powered data acquisition computer. The test head incorporates an electrically driven solenoid impactor and a piezoelectric displacement transducer which is used as a receiver for the received echoes. In operation, the test head is manually pressed against the wall of the pipe, and the solenoid is triggered. Triggering can be accomplished from either the data collection computer of from a switch on the test head. The received echoes are amplified and filtered at the data collection computer, and stored for analysis. Generally, two to three tests are done at each location the verify repeatability and assure high-quality data. The entire testing process takes about 3-5 seconds once the test head is pressed against the wall. Most of the testing time is taken up in moving between points and establishing the test grid.

Pipe Scanning System

The IE pipe scanning system has also been used in recent tests to evaluate its performance. The pipe scanning system was described in detail in the report on the previous research (Sack and Olson, 1994). Briefly, the pipe scanning system consists of 4 scanning heads which are spaced at 90 degrees around the interior of the pipe circumference, and are pressed against the walls pneumatically. The scanning heads use rolling transducer assemblies to receive the signals from solenoid impactors mounted adjacent to the receivers. As the scanner assembly is pulled down a pipe, data is collected sequentially from each scanner head, resulting in a linear test spacing of about 6 cm (2.5 inches) per test. The data is collected into a large data file and saved. Analysis of the data results in plots of distance versus effective pipe wall thickness for each of the scanner heads, which reveals any areas of delamination or other defects. Testing rates in clean pipe have been found to be about 6 m (20 feet)/ 5 minutes for the prototype system, not including initial set-up and take-down time.

Recent Pipe Test Results

The most recent field testing done on in-place PCCP were conducted by the author's firm on a pipeline in Texas for an American Water Works Association Research Foundation (AWWARF) research project by Texas Research Institute (TRI). This pipe is about 2 m (6 feet) I.D. and has a wall cross-section of about 13 cm (5.25 inches). Of this cross section, the outer 2-3 cm (0.75-1.25 inches) is the grout layer. The core has an embedded steel cylinder about 6 cm (2.5 in) from the inside wall face. Visual inspection of the inside face of this pipeline showed a generally clean concrete surface, but with small visible pock marks and aggregate. Scratching the surface of the concrete showed the surface material to be unusually soft.

Initial testing of a section of this pipe was done with the 4 channel pipe scanner. The results of this testing showed that, while some usable data was obtained, the soft surface conditions damped out the higher frequencies from the impactor required to test the pipe. The high receiver gains used to compensate for the damped impact force resulted in unacceptable levels of rolling noise riding on the received signals. The excessive noise resulted in signals which could not be analyzed by the automatic scanning software, and instead required manual interpretation.

Additional testing of the pipe was done with the previously described single point IE system at selected locations. The results of this testing showed that the single point system worked well, even on the softer concrete, due to the high-sensitivity transducer used and the lack of rolling noise (which the scanner showed) during testing. A typical result from a sound location in the pipe is presented in Fig. 2. This figure presents the receiver time domain raw data in the upper trace, and the frequency domain spectrum in the lower trace. As seen in the figure, very little information can be directly determined from looking at the time domain trace. The frequency spectrum, however, shows a single clear echo peak at 12,800 Hz, which corresponds to a thickness of 14 cm (5.6 in) using a compression wave velocity of 3,660 meters per second (12,000 feet per second, fps) in the impact echo equation:

$$T = V/(2 * Fp) \qquad (1)$$

Where: T = Thickness in meters
V = Velocity in meters/second
Fp = Frequency peak in Hz

This thickness echo corresponds to the expected total cross section of the pipe wall. This indicates that the tested point is sound with no delaminations.

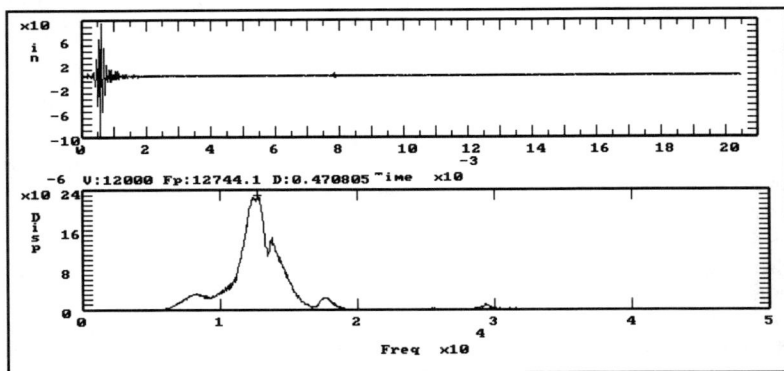

Figure 2   Sample Impact Echo Result for Sound Conditions

A typical test result from a location with a suspected delamination is shown in Fig. 3. Again, this figure presents the receiver time domain raw data in the upper trace, and the frequency domain spectrum in the lower trace. As seen in this figure, the frequency spectrum shows two distinct frequency peaks, one at about 18,000 Hz, and the other at 29,500 Hz. No frequency peak is seen in the 12,500 to 14,500 Hz range expected for sound locations. Using the impact echo equation, the lower frequency peak corresponds to a thickness of 10 cm (4.0 in), and the higher peak corresponds to a thickness of 6 cm (2.5 in). The thickness of the pipe wall without the grout is about

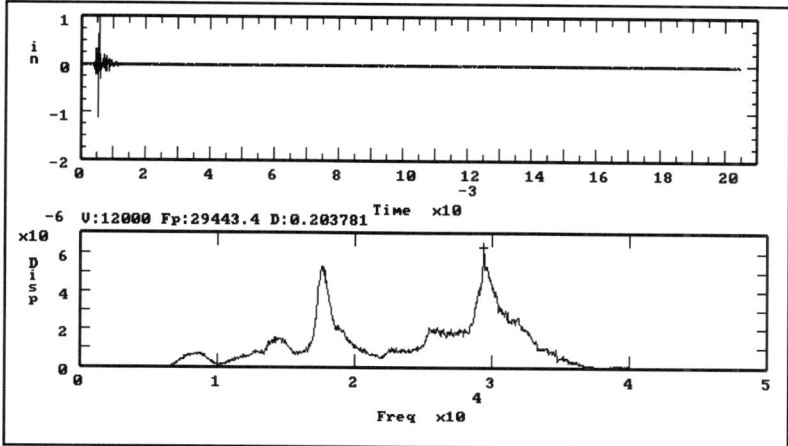

Figure 3　　Sample Impact Echo Test Record from Delaminated Conditions

10 cm, and the distance from the inner face to the embedded cylinder is about 6 cm. Thus, this record indicates that the grout layer is delaminated, and the concrete is beginning to debond from the embedded cylinder. Debonding can occur when the steel wires have failed and there is no longer prestress on the concrete.

The overall results of the single point testing are best presented in the format of Fig. 4. This is a plot of pipe wall thickness (in inches) and signal strength versus position (in feet) within the pipe. This type of plot can be easily made by the automated data scanning and plotting software for the scanning system as long as the data was collected along a line of points at equal intervals. Note that the data shows three distinct thickness ranges throughout the plot. The thicker points all correspond to echoes from the outer grout layer, and indicate sound conditions. The intermediate thicknesses shown correspond to echoes from the prestress wire level and indicate full or partial delamination of the grout from the wires and core. The final set of thicknesses seen are from echoes at the cylinder depth, indicating debonding of the steel cylinder from the concrete (likely from loss of prestress wires). Other thickness echoes are generally not seen, indicating that failures of these pipes occur in relatively predictable locations within the wall.

All of the single point IE testing for the most recent PCCP testing project was performed in areas where previous Acoustic Emissions (AE) monitoring indicated likely wire breaks, and thus was intended to find defects. Based on these results, one likely effective application of the various NDT methods would be the use of AE for long-term monitoring of pipes in service to listen for wire breaks, followed by intensive IE testing of suspect locations after the pipeline is dewatered for maintenance. This would allow confirmation and localization of the suspect pipe

256   PIPELINES IN THE CONSTRUCTED ENVIRONMENT

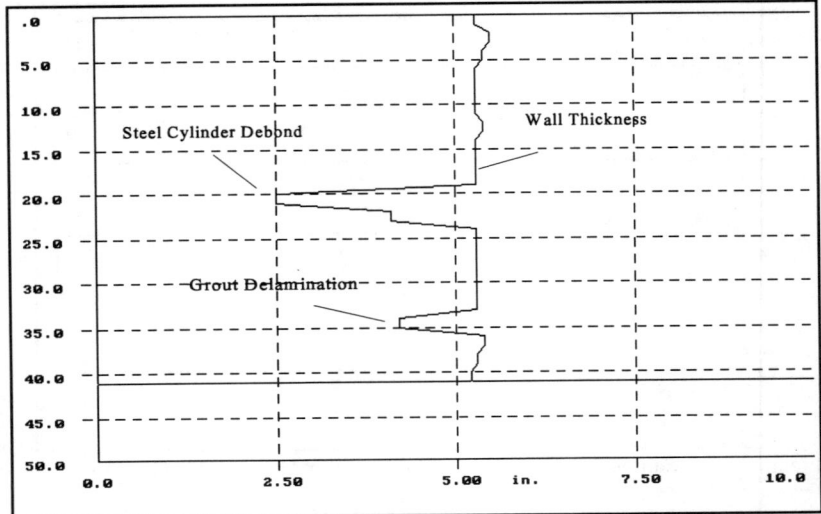

Figure 4   Sample Impact Echo Pipe Test Result - Wall Thickness vs. Location

section for repair. For this type of NDT program, the single point IE device would be the most applicable. If a baseline scan of unmonitored pipes is required, IE scanning could be used to locate areas which have already failed (and thus may not produce more wire breaks during AE testing).

Spectral Analysis of Surface Waves Test Method and Results

The Spectral Analysis of Surface Waves (SASW) test method was used at several locations on the pipe to measure concrete velocity for use in IE thickness calculations, as well as for investigating concrete conditions and soil conditions outside of the pipe.

The SASW method is based upon measuring surface waves propagating in layered elastic media and is pictured in Fig. 5. The ratio of surface wave velocity to shear wave velocity varies slightly with Poisson's ratio, but can be assumed to be equal to 0.90 with an error of less than five percent for most materials, including concrete. Measurement of the surface wave velocity with the SASW method similarly allows calculation of compression wave velocity for IE analysis. Surface wave velocity also equals 0.56 of the compression wave velocity for concrete (Poisson's ratio = 0.2). Knowledge of the seismic wave velocities and mass density of the material layers allows calculation of shear and Young's moduli for low strain amplitudes.

Surface wave (Rayleigh; R-wave) velocity varies with frequency in a layered system with velocity contrasts, and this frequency dependence of velocity is termed dispersion. A plot of surface wave velocity versus wavelength is a dispersion curve.

The SASW field tests consisted of impacting the test surface to generate surface wave energy at various frequencies that were transmitted through the material. Two accelerometer receivers were evenly spaced on the surface in line with the impact point to monitor the passage of the surface wave energy as illustrated in Fig. 5.

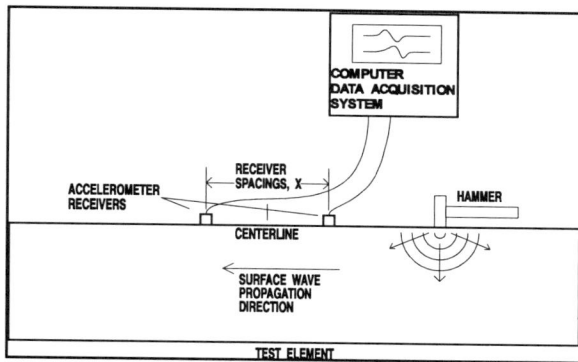

Figure 5   SASW Test Method Diagram

A PC data acquisition system digitizes the analog receiver outputs and records the signals for spectral (frequency) analyses to determine the phase information of the cross power spectrum between the two receivers for each frequency. The dispersion curve is developed by knowing the phase (ϕ) at a given frequency (f) and then calculating the travel time (t) between receivers of that frequency/wavelength by:

$$t = \phi / 360 * f \tag{2}$$

Surface wave velocity (Vr) is obtained by dividing the receiver spacing (X) by the travel time at a frequency:

$$Vr = X / t \tag{3}$$

The wavelength (Lr) is related to the phase velocity and frequency by:

$$Lr = Vr / f \tag{4}$$

By repeating the above procedure for any given frequency, the surface wave velocity corresponding to the given wavelength is evaluated, and the dispersion curve is determined.

A typical dispersion curve from an SASW test on a sound location of a PCCP section is shown in Fig. 6. This is a dispersion curve, or a plot of surface wave velocity versus wavelength. Note how the velocity is constant at a typical concrete surface wave velocity value (2,700 mps) until a wavelength corresponding to material outside of the pipe wall is reached. After this, there is a distinct drop to the apparent lower surface wave velocity (2,100 mps) of the material outside of the pipe wall. This result indicates that the pipe wall is sound with no apparent delaminations, and also that the material outside of the pipe wall is relatively dense and stiff compared to loose soils. Collection of data at longer wavelengths and forward modeling would be required to fully evaluate the characteristics of the material outside of the pipe.

Conclusions

The results of the ongoing research into the use of the IE method in testing PCCP lead to several conclusions. First, the IE method has been shown to be very effective in measuring pipe wall thickness and locating delaminations and cylinder debonding associated with wire breaks and strength loss. Scanning with the IE method does work, but requires more sensitive receiver transducers to work well in pipe where the inside surface has degraded. More sensitive transducers have been developed, and have been used in the single point IE test device, but have yet to be assembled into a rolling scanner head. The use of the single point IE device in recent tests has shown that this device is most effective when spot checking for quality, or when investigating areas where concern already exists. The use of AE monitoring during pipe operation can give an indication (by listening for wire breaks) of what areas to test intensively with IE once the pipe is dewatered. Finally, the SASW method, which has been used extensively in testing other types of concrete members, can be applied to the evaluation of PCCP conditions, although it is generally a slower method both during data collection and data analysis. The SASW method does, however, give information as to the condition of the material outside of the pipe wall (void, soft, stiff, etc.), which can be very valuable in certain circumstances.

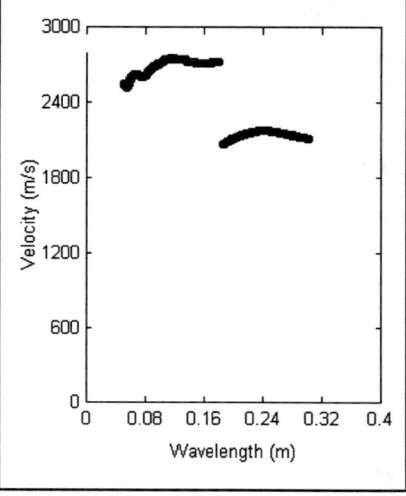

Figure 6 SASW Sample Result

References

Olson, L., Sack, D. and Phelps, G., "Sonic NDE of Bridges and Other Concrete Structures", National Science Foundation Conference on Nondestructive Evaluation of Civil Structures, University of Colorado at Boulder, 1992

Sack, D., and Olson L., "In-Situ Nondestructive Testing of Buried Precast Concrete Pipe", American Society of Civil Engineers Materials Engineering Conference, 1994

Sansalone, M., and Carino, N.J., "Impact-Echo: A Method for Flaw Detection in Concrete Using Transient Stress Waves," National Bureau of Standards Report NBSIR 86-3452, Gaithersburg, Maryland, September, 1986.

Inspection, The Key To Successful Rehabilitation

Paul J. Williams, PE.[1]

Abstract

Inspection of construction activities is most often considered a function to assure the owner of compliance with the contract or to protect the A/E from deviation in specifications and design. These are elements of inspection, but for inspections to be most effective there are other important and valuable consideration.

All to often inspection activities are minimized for the expressed purpose of lowering project costs. This is extremely short sighted and the reverse is more often true.

For inspection to be most effective, it must relate to the highest standards for the means, methods and materials of construction. The very best specifications are clearly defined, complete, concise and applicable to their subject. All to often, specifications are not truly workable at the construction level.

Inspections effect and benefit the owner, engineer and contractor. Yes, they do benefit the contractor. The owner's interests and investments can best be assured by proper inspection. The A/E's design and standards are monitored and documented. The contractor's benefits can be both direct and indirect. Direct benefits of good inspections to the contractor include:

- Documentation of contract compliance as the work progresses
- Verification of quantities for production and payment purposes

---

[1] Sr. Vice-President, ETS Liner, Inc., 4948 N. Orange Blossom Trail, Orlando, Florida 32810

- Verification of equipment and facilities usage for warrantee purposes
- Other

Owners have suffered the greatest losses over the years by lack of proper and in many cases no inspections. Today's emphasis on rehabilitation of pipe line systems has underscored the lack of workmanship when many of the current systems were installed. The nature of the faults being found routinely during rehab activities points directly to poor workmanship as the major cause of the faults. Typical problems include:

- Improper bedding
- Improper jointing of pipe lines
- Improper sealing of pipe lines at manholes
- Improper backfilling of pipe lines and manholes
- Improper installation to line and grade
- Installing pipe lines damaged before placement

The current sewer rehabilitation efforts must include an element of quality inspection to assure success. Review of the rehab efforts in the 1970's underscore that the major cause of failures was poor workmanship. At that time in our industry specifications were not all that great, but if followed would have produced a much better product. Proper inspection would have helped prevent the historically mass failures of that era.

Inspection involves experienced and trained personnel performing various tests and observation, and preparing detailed documentation of project events and activities. All three players in the industry benefit if inspection is properly performed and the efforts accurately documented. Owners, A/Es and contractors each view the inspection activity from different perspectives but because of the positive benefits received by each they should support the highest quality inspections for their projects.

Introduction

This paper is an organized collection of facts and opinions most likely known by most readers. The focus of the text will be on gravity pipe line rehabilitation but the message could easily apply to all elements of our infrastructure problems. We note that system failures found today relate in order of priority to poor workmanship in the original construction. Most could have been prevented by proper inspection of the work.

The purpose of this paper will be to draw attention to the critical need for proper inspections of our work and thereby prevent the repeat of our past failures.

Narrative

Inspection of construction activities is a vital step in the progress of a project from its conception, to construction completion, and finally its use by the owner. Rehabilitation projects develop through many steps before construction begins that utilize a substantial portion of the project funds to:

- Define the problem.
- Determine and specify the best solution
- Design and publish the plans for the solution
- Seek the funds which pay for the construction of the solution

The goal of all rehab projects is to provide the working solutions to the identified problems, not just report it, design it, etc. Almost everyone in our industry should agree that inspection is important but many times it is not emphasized.

Sewer system rehabilitation efforts in the 1970's were marginally successful at best. Current sewer rehab is addressing the repair of older sewers, sewers that were previously addressed in the 1970's era and sewers that have been constructed within the last 20 years. Through the current rehab efforts we are finding some interesting facts:
- Older sewers require typical and predictable repairs
- Sewers previously addressed in the 1970's are often in the worst condition of all three types
- Sewer systems installed in the last 20 years in many communities represent an unexpectedly high percentage their rehabilitation needs. Why?

We have reviewed elements of the older sewer systems including their means, methods and materials of construction and our industry has responded with improved materials and methods of placement. We attempted to correct many of our sewer system problems with the help of EPA funding in the 1970's and continue to promote the use of better materials, better means and better methods of new system construction. But even the newly constructed systems continue to fail; sometimes in as little as 5 years of their construction. Again, why?

The answer is two-fold. Until recently the engineering community had only limited opportunity to learn the proper application of means and methods to solve rehab problems. Literature provided by manufacturers and contractors was not always accurate and schools have just recently begun to teach rehab technology on a major scale. Many specifications used today are the result of trial and error. A review of several specifications related to construction and rehabilitation of pipe lines from a constructability and operations perspective resulted in the following comments:

- The materials are adequately specified

- Job site materials handling and other conditions are addressed but not fully detailed
- In-place performance is usually not clearly defined
- Testing to assure performance is sometimes required but not always related to the desired performance
- Tests are not always practical to perform as specified
- Acceptance criteria is often vague

The following are examples of an inadequate and a good working specification for manhole repair. For the sake of space, only performance and testing are addressed in these excerpts from typical specifications. Take note of which specification that, if followed, would assure the better quality of finished product.

Specification Example No. 1

Sanitary Sewer Manhole Rehabilitation

Control

1.1 Section Includes

Sanitary sewer manhole rehabilitation including:

- Rehabilitation and leak-proofing of manholes by lining with spray applied or centrifugally cast lightweight structural reinforced concrete, or spray applied urethane or epoxy resin systems.

- The repair and sealing of the manhole base, invert, walls, corbel/cone, and chimney of brick, block or precast manholes, including the removal of any unsound material.

- The inspection and testing of the various types of work to insure compliance

3.6 Manhole Rehabilitation Acceptance

A. After the manhole rehabilitation work has been completed, the manhole shall be visually inspected during high groundwater by the Contractor in the presence of the Owner's Representative and the work shall be accepted if found satisfactory to the Owner's Representative. The finished surface shall be free of blisters, "runs" or "sags" or other indications of uneven lining thickness. No evidence of visible leaks shall be allowed. In addition, at the Owner's request, the Contractor may be required within two years to visually inspect the manholes that were rehabilitated. Any work that has become defective within the two year period shall be redone by the Contractor at no additional expense to the Owner.

Specification Example No. 2

3.04 Clean-Up and Testing

A. Manholes shall be physically inspected and vacuum or hydrostatically tested to assure compliance with the Contract Specifications and the desired workmanship of the finished reconstruction has been finished.

B. Finished manholes shall be physically inspected and shall not exhibit runs, sags, blisters, voids, cracks, and free of leakage.

C. Manhole Vacuum Test

- All manholes shall be physically inspected, and all visible defects repaired before inspection.
- All manholes shall be subjected to a vacuum test of a minimum of ten (10) inches of mercury (Hg) prior to acceptance by the Owner. The test shall be considered acceptable if the vacuum remains at nine (9) inches of Hg or higher after the following times:

Manhole I.D. (inches)	48	60	72	84	96	120
Test Time for up to 8 Feet in Depth (Seconds)	60	70	80	90	100	100
Additional test Time for each 4 Foot added Depth (Seconds)	10	15	20	30	40	60

- Manholes failing the test shall be repaired by the Contractor and inspected by the Owners' representative and then retested. Should the manhole fail the test for a total of three (3) times, the manhole shall be repaired by pressure grouting the exterior and/or removing the previous applied repairs and beginning the reconstruction process again.
- The Contractor shall furnish all necessary equipment and personnel to conduct the tests in the presence of the Owner.
- Costs for initial testings shall be included within the incidental to the Contract Unit Price for manhole rehabilitation.
- Repairing, retesting, and/or pressure grouting of manholes failing tests/inspections shall be at the sole cost and responsibility of the Contractor, and shall be pursued in a timely manner to prevent disruption to the Project and/or sewer services.

- Manholes moved, displaced and/or damaged in any way during the finishing operations subsequent to successful testing shall be rested for acceptance as specified, at the sole cost of the Contractor

Both specification examples refer to quality rehabilitation of manholes but example one really does not give the inspector or contractor the proper basis for testing or determining the final quality of the contractor's efforts. Example two specifically details each step in the testing and the basis for establishing the acceptance criteria.

The second reason for construction failures was observed to be workmanship. As noted in the previous example specifications must be detailed and provide clear procedures for verification of quality control. The best specifications must be followed and/or enforced for the desired finished product to be achieved. In all three categories of sewers (old, new, previously rehabed) the failures are typical and similar. The most common ones are observed to be the result of::

- Improper bedding
- Improper jointing of pipe lines
- Improper sealing of pipe lines at manholes
- Improper backfilling of pipe lines and manholes
- Improper installation to line and grade
- Installing pipe lines damaged before placement

These faults are found routinely during current rehab activities and are the direct result of poor workmanship. Poor workmanship is preventable by the application of proper inspections, testing and documentation.

With the overwhelming evidence to support the need to improve workmanship in the construction element of our industry, why do we continue to have the problem? All to often it is perceived cost. After all the difficult work has been completed to plan and design the project and funds committed to its construction, we often limit the quality control efforts.

One of the most common misconceptions we utilize to lower cost is to employ younger engineers or technicians and retired engineers or construction workers. Both groups have the opportunity to provide excellent inspection services but we often handicap their potential. The younger group generally has the education necessary for the task and the retired group certainly has experience. Unfortunately we too often handicap their efforts by not providing the proper education and training to apply their skills effectively. There are other groups of inspection personnel that are similarly handicapped but all generally are employed at the lower end of the pay scale.

To place the labor cost of inspection into proper perspective, review of the following table provides an interesting comparison.

Examples of Cost of Good Inspection - VS - Construction and Total Payment Costs					
Project	Const Cost	Total Proj. Cost	Insp Cost	% Of Const.Cost	% Of Total Cost
1	800,000	913,400	50,400	6.3	5.2
2	1,200,000	1,351,500	69,300	4.8	4.3
3	430,000	496,500	31,000	6.2	4.7

These projects represent owners that chose to take the classical approach to project development and also to pay for quality engineering and financial assistance. The projects were well planned and specified and considerable time was given to selection of means, methods and materials. Note the percentage of the project costs assigned to inspection/testing. These projects were well inspected with excellent documentation of all activities on the project including testing and approvals of individual line segments, manholes, etc. From another perspective look at the 5-6 percent cost for inspection as the insurance to provide for the quality of the finished project.

Evaluation of rehab results from the 1970's and observations of newly constructed systems indicate failures after only 10 to 20 percent of their design life. Even failures at 50 percent of the design life represents a 100 percent increase or doubling of construction cost for a system to survive its design life. Compare the results of both approaches:

- Do it right the first time at an expense of 5-6 percent of the project cost
- Save a small percentage of the initial cost and potentially double the construction cost within the design life

Documentation of all of the activities associated with the project is critical. Inspections are necessary to assure the owner's interests and investments have been met and that the engineer's design and standards were delivered. Documenting these important elements is necessary for both contractual and quality control concerns. Contractors can also benefit by inspections and their proper documentation, including

- Documentation of contract compliance as the work progresses
- Verification of quantities for production and payment purposes
- Verification of the owner's usage of equipment and facilities for warrantee purposes
- Verification of testing and acceptance to minimize warrantee claims

A good friend defines "experience" as "what you get when you are looking for something else". I can only hope that your "experience" with inspections and the attempt to improve workmanship as our projects are constructed is better than mine has been.

- Engineers: Refine your specifications to be construction friendly and press for the best insurance (testing and inspection) to assure their compliance
- Owners: Don't sell yourself short by buying less expensive and/or an inadequate insurance (testing and inspection) to gamble that your project is delivering what you are paying for and deserve
- Contractors: Support realistic testing and inspections of your work. The best way to make a profit in the long run is to "do it right the first time". Once your work is completed, tested, approved and documented you have minimized your chances for warrantee recalls and its related costs

The facts and opinions presented herein are not new and, I don't believe anyone would really disagree, at least in concept. However, we as an industry all too often fail in our implementation of this major element of our work. I can only hope that the presentation of this matter will encourage your efforts to expand and improve the quality control of future projects, and the mistakes of our past will not be repeated.

ENVIRONMENTAL DESIGN FACTORS FOR CONCRETE PRESSURE PIPE

Wayne R. Brunzell, P.E.[1]
Member, A.S.C.E.

Abstract

As with any engineered and manufactured product, Concrete Pressure Pipe must be properly applied. Conditions which affect pipeline life need to be recognized in the investigation, planning and design process. The responsibility for these activities should be shared by the owner and the engineer. For all types of pipe, structural, operational and environmental characteristics of the project should be researched, understood and clearly stated. Environmental design conditions addressed in this paper include: extraordinary live loads, stray current interference, long term atmospheric exposure, connection to pipe of other materials, adverse foundation conditions, unusual installation methods and hostile soil environment. Soil environments which demand special design attention include chlorides, sulfates, acid, and wet-dry cycles. Design to actual burial conditions ensures that the fully intended life of the pipeline will be realized at the least cost.

The Owner-Engineer Design Team

The success of a pipeline project depends upon accurate prior knowledge of the intended service and environment into which the pipe will be placed. Information required to properly design a pipeline must come from a knowledgeable owner and an experienced engineer working cooperatively to enhance the project's success. With today's concrete pressure pipe technology, the manufacturer can design and build pipe that can stand up in hostile environments. The engineer must integrate concrete pipe quality and characteristics into its pipeline design to insure serviceability and longevity.

---

[1]Principal, Brunzell Associates, Ltd., 9933 Lawler Ave., Skokie, IL 60077

Cooperation and sharing of information by the owner and engineer is necessary to maximize the utility and value of the pipeline. Many water systems, which incorporate concrete pressure pipe have existed for a relatively long time. In these systems, owners have accumulated much service and environmental information surrounding the concrete pressure pipe, which comprises a significant portion of their system. Hopefully, much important data has been retained in pipeline project files and will serve as a valuable foundation of information for future projects. When a new concrete pressure pipeline project is first conceived, an owner should make all relevant historical information available to the engineering planning and design team.

The goals of the owner and consulting engineer separately should be the goals of the project team collectively. Project planning and engineering objectives should focus on operations, reliability, serviceability, longevity, cost and value. The engineer must function as the gate keeper of all information coming to the project development process. The consultant, working with the owner, must also assure that all operational and environmental concerns have been fully investigated for the specific project.

A strategy for assuring a successful concrete pressure pipeline project must include careful analysis and correct decisions in each step of the process. The major project facets include investigation, planning, design, construction and operations, plus geotechnical field investigation with applicable laboratory analysis of soil conditions.

The investigation phase must assemble all pertinent background information from the owner. If the engineer has worked in this particular system before, he should have valuable historical information to contribute also. He will also bring the wisdom gained from other projects with many of the same issues. The issues addressed in the planning process should include but not be necessarily limited to the following:

- Selecting Design Loads (external & internal)
- Stray Current Interference
- Long Term Atmospheric Exposure
- Connection to Pipe of Other Materials
- Adverse Foundation Conditions
- Unusual Installation Methods
- Hostile Soil Environment

These issues are addressed in more detail in the following discussions.

The planning process brings together the skills of the project team. In addition to the well known and necessary planning issues of terminal points, capacity, size, pressure rating, material, route, etc. that make a pipeline perform to expectations, careful planning for environmental issues make a pipeline reach and exceed its design life. Accounting for environmental factors through design, construction and the operations phases ensure that the pipeline will continue to provide reliable service throughout its expected life.

Selecting Design Loads

Intelligent long range planning of the overall water system will help anticipate internal pressures in the subject pipeline which might be greater in the distant future but still within the design life of the concrete pressure pipe. If and when the pipeline is called upon to operate at higher pressures or be subject to greater surges, it must have that capability without costly structural changes or awkward operational strategies required by a "weak link". Realizing that a properly designed pipeline should continue to serve for many years, it is necessary to look and plan as far into the future as possible.

Likewise, planning and designing for future increased external loads will avoid costly encasements or pipeline relocations. Predicting future development trends and events is admittedly difficult. But planning for the design life of the pipeline will help it serve effectively for that period of time.

Stray Current Interference

Stray direct current from electric railway or subway systems and impressed-current cathodic protection systems may be accepted and later discharged by buried pipelines with metallic components. Pinholes and other flaws in an organically coated pipeline result in coating delamination and corrosion pitting at these stray current discharge points. However, if stray current discharges from steel rods or bars encased in concrete or mortar, the discharges will be spread over larger surface areas and will consume excess alkalinity before the steel begins corroding. It is this built-in mitigating factor which allows concrete cylinder pipelines to tolerate stray current interference without the onset of corrosion for a longer period of time than organically coated steel pipelines.

The detrimental effects of stray current interference can be eliminated or mitigated by following one or more of the following strategies:

- Reduce or eliminate the source of the stray current

- Supplement the pipes concrete with an additional dielectric coating to increase the pipes electrical resistance

- Design a system of electrical continuity bonding as part of the installation process. Determine if stray current is being accepted or discharged. If so, design a system to safely discharge the stray current.

Long Term Atmospheric Exposure

In the course of a complete pipeline design, it may be preferable and beneficial to locate all or some part of the system above ground. Accordingly, the above-ground portion of the pipeline is subject to a different environment than a buried system. The factors to which the above-ground line is exposed include wetting and drying cycles, potentially large temperature fluctuations, as well as the possibility of freezing and thawing cycles, which over an extended period of time can adversely affect the protective properties of the concrete or mortar coating.

An above-ground pipeline is also exposed to atmospheric carbonation, that over time can reach the reinforcing wires or bars and allow the steel to corrode. In such installations, depending on the nature of the resident atmosphere, a light colored barrier coating over the pipe exterior may be used. The barrier should be inspected on a regular basis and renewed as necessary to maintain and preserve the integrity of the underlying mortar coating and reinforcing steel.

Connection to Pipelines of Other Materials

The electrical potential of steel reinforced concrete pressure pipe is 300 to 500 mV less negative than a bare iron or steel pipe or one with organic coating, when the pipes of dissimilar materials are connected together. The concrete pipe can be protected by the sacrificial corrosion of the ferrous pipe. In most cases the problem does not reach serious proportions. However, to completely eliminate the problem, the connecting joint between the two materials should be of an insulated type or the ferrous pipe should be encased in concrete or mortar to equalize the potentials of both pipe materials.

Adverse Foundation Conditions

Design of concrete pressure pipelines for dead and live loads and a variety of trench, tunnel and embankment conditions is well understood and applied by system owners and consulting engineers. However, there may be unstable foundation conditions for a particular route which, if not properly addressed, may damage the pipeline or render it unusable.

Saturated soils which can be found in many areas along coastlines or in swamps are often unstable or nearly liquid. Special foundations such as pile-supported piers should be considered if it is determined that more conventional methods are not appropriate. For most pipelines, the composite weight of concrete and water ranges from 80 to 95 lb./ cu.ft. This loading is usually less than the in situ material. If a trench can be cut in this unstable material and a pipe and backfill of about the same weight installed, the pipe will "float in equilibrium. In this environment larger pipelines in particular may be subject to buoyancy when empty and should be carefully studied for this possibility.

Pipe which must be installed in an unstable bog or swamp can be supported on a specially constructed trench bottom which will spread the load rather than concentrating it along the invert. A spread bedding of shells or granular material 1 1/2 to 2 pipe diameters wide have been successfully used under these conditions, as have timber mats underlying the bedding material. A geotextile envelope encasing the seashells or granular material can help spread the pipe load over the unstable trench bottom.

Some of the methods mentioned above may also be necessary for support of construction equipment or to maintain an open ditch. "Floating" installations are appropriate for soils that will remain in equilibrium. If it is expected that the soil will shift or not remain static, other pipe installation procedures should be considered. Loads and moments applied to a pipeline by shifting soils can be extreme to the point of pipeline damage. In such cased, elevating the pipe on a bridge or piers or otherwise rerouting the pipeline around the shifting soil may be the most economical installation.

Unusual Installation Methods

If certain features of the pipeline result in installation procedures which are not "normal", special design of the concrete pressure pipe should be considered. The necessary installation procedures may require that the pipe be loaded in a way for which it is not normally designed. These abnormal installation conditions can include point loads, bending beam action, special stresses on joints, and abrasion of the protective coating. The issues of construction loading should be carefully considered so as to prevent damage to the pipeline even before it is placed in normal operating service.

Hostile Soil Environments

Low soil resistivities are an indication that high chloride or sulfate concentrations may be present. Since soil resistivity is more easily measured than soil chemistry, it is recommended that a soil resistivity survey be performed along the pipeline's rights-of-way to locate soils that are potentially aggressive. As a guideline, soils of resistivity less than 1,500 ohm-cm should be tested for aggressive salts.

High-Chloride Environments

The necessary components to initiate corrosion include a sufficiently high concentration of chloride ions, in combination with oxygen at the steel surface. These factors will disrupt the passivation effect of the steel embedded in concrete or mortar. The replenishment of oxygen at the steel surface will allow corrosion to continue and flourish.

Pipeline routing that requires concrete cylinder pipe to be buried in soils with significant chloride concentrations should be carefully evaluated to assure long term serviceability. When concrete pipe is installed in soil zones that do not become totally dry, the mortar coating remains moist, significantly restricting the amount of oxygen that can reach the steel reinforcement. However, when the soil's water soluble chlorine content exceeds 1,000 ppm, specific design, manufacturing and construction steps are strongly recommended. In addition to the mortar coating, a moisture barrier should be used to restrict the movement of chloride ion laden water. Also, silica fume amounting to 8 to 10 percent of the cement weight in the mortar coating mix should be included or electrical continuity bonding of the pipe should be added to permit monitoring of the pipeline for corrosion.

In arid regions in particular, concrete cylinder pipe in soils may be subject to cyclical wetting and substantial drying. This environment can cause the chloride ion content to become concentrated in the mortar coating and subject the steel to the chloride ion. Oxygen will also be pumped in and out of the surface in the wetting and drying cycles. If these circumstances exist, water soluble chloride concentrations exceeding 150 ppm can present a danger to the integrity of the pipe. The exterior surfaces of the pipeline should be protected with a barrier coating or electrical continuity bonding should be provided to permit monitoring for corrosion.

Oxygen transfer is the key element of corrosion. Pipelines submerged in seawater with chloride concentrations of 20,000 ppm do not experience damage to their embedded steel reinforcement due to the extremely low rate of oxygen diffusion through the saturated mortar coating. Consequently, additional protection over the mortar coating is not required for continuously submerged concrete cylinder pipe. Exposed steel joint rings should have an applied barrier coating unless they are encased in portland cement mortar or grout.

High Sulfate Environments

The term alkali is applied to soils with high concentrations of sodium, magnesium or calcium sulfates. In partially buried concrete structures, high sulfate soils can attack using the mechanisms of capillary action and surface evaporation to build up high sulfate concentrations in the mortar or concrete. Sulfate build up is normally not a problem in pipelines which are completely buried. Resistance to sulfate attack is further enhanced by high cement contents which are typical of the mortar coating of concrete cylinder pipe, usually in excess of 10 sacks per cubic yard.

Low permeability of the placed concrete or mortar significantly reduces sulfate attack so that even high tricalcium aluminate cements can be used to make concrete or mortar with good sulfate resistance.

For totally buried concrete pipelines installed in soils with 2,000 ppm or less soluble sulfates or for concrete pipe continuously submerged in seawater, standard Type II portland cement containing a maximum of 8 percent tricalcium aluminate can be used successfully. For installations containing more than 2,000 ppm water soluble sulfates, portland cement with 5 percent or less tricalcium aluminate or silica fume in the range of 8 to 10 percent of cement weight should be considered. Another approach is to use a barrier coating on the concrete or mortar rather than special cements or additives.

Severe Acid Environments

It is rare that the exterior of concrete pipe is attacked by naturally occurring acid soils. More likely, an acid problem is caused by mine wastes, acid spills or industrial discharges of certain mineral acids such as sulfuric acid exhibiting low pH and high total acidity values.

Soils with low permeability allow the acid to be neutralized by the alkalinity of the mortar or concrete with very slow acid replacement. Therefore, in fine grained silts or clays supplemental protection against acid conditions is not normally necessary. Additional measures to protect the concrete or mortar exterior of the pipe should be considered in granular acid soils.

Guidelines have been developed for evaluating the aggressive nature of acid soils and the need for supplemental protection of concrete pipe. A pH below 5 in granular soils should trigger a test for total acidity. If the total acidity exceeds 25 milliequivalents/100 gm of dry soil, one of the following protective methods should be used. The pipe zone should be backfilled with limestone or consolidated clay material. The protective ability of the mortar coating may be enhanced by incorporating 8 to 10 percent silica fume in the cement.

Concrete pipe should be installed in an acid resistant membrane or enveloped in nonaggressive consolidated clay where the soil pH level is below 4.

Summary

The preceding discussion has attempted to identify major issues which should be addressed in the planning, design, manufacture and installation of concrete pressure pipe, leading to successful, long term operation. The owner and engineer should consider the following parameters and recommendations in the development of their pipeline project:

- External and Internal Design Loads -- Consider the complete spectrum of present and future loading possibilities.

- Stray Current Interference -- Reduce, eliminate, or design for safe current discharge.

- Long Term Atmospheric Exposure -- Provide a supplemental barrier coating.

- Connection to Pipelines of Other Materials -- Use an insulated joint or encase the ferrous pipe in concrete to equalize the potentials.

- Adverse Foundation Conditions -- Study soil conditions and design for long term stability to avoid damaging concentrated stresses.

- Unusual Installation Methods -- Carefully analyze and design for possible excessive loads experienced during construction.

- Hostile Soil Environment -- Test soils for high levels of chlorides, sulfates, and acidity.

Although each pipeline project will not be faced with all of the factors possible in the burial environment, it is well to use these factors as a check list. The ultimate success of the pipeline is dependent upon good information and good decisions throughout the process. The owner and consulting engineer should identify and design for all specialized facets of the burial environment and clearly transmit the design requirements through the drawings and specifications. Because of their depth of specific experience in designing and manufacturing concrete pressure pipe for various burial applications, the manufacturer's engineering staff is a valuable resource to the owner-consultant team.

References

"CONCRETE PIPE HANDBOOK", American Concrete Pipe Association, 1988, Vienna, Virginia

"CONCRETE PRESSURE PIPE", American Water Works Association Manual of Water Supply Practices - M9, 1995, Denver, Colorado

"GUIDELINES FOR THE PROTECTION OF CONCRETE CYLINDER PIPE", American Concrete Pressure Pipe Association, Reston, Virginia

PIPELINE CONSTRUCTION FOR DESIGNERS

Deon T. Fowles, P.E.[1]

Abstract

There are a number of materials promoted for use in underground pipeline service. Each material is unique in it's own ability to resist loads and function as intended. The designer must consider the strength of the manufactured product and relate that to the construction operation. Designing pipes as a composite of pipe material and earth support dictates that an adequate safety factor be applied due to the many unknowns of future earth and traffic loads and the quality of construction.

This paper focuses on construction practices of pipelines, in particular ductile iron pipe, and what a designer can do to improve specifications to help ensure a satisfactory project.

Introduction

In the process of designing a pipeline system, the Design Engineer considers the technical aspects of hydraulics, sizing, alignment, existing utilities, connections, etc., and the contract documents which govern the project. Less frequently considered are the effects of construction methods and operation. Ease of building a system is a factor in construction costs and time to completion. Use of strong, properly designed materials can reduce the amount of time and effort in construction, thus achieving a reduction of cost. For instance, the right selection of pipe material and wall thickness can reduce the amount of trench bedding and

---

[1]Senior Regional Engineer and NACE Certified Corrosion Specialist, Ductile Iron Pipe Research Association, 10920 East Powers Court, Englewood, Colorado 80111.

compaction required. Consideration of this, along with adequately specifying the proper materials for incidental items, such as bolts, connections and fittings, should provide a quality installation.

Standards

A number of recognized standards are available for guidance in design and construction. Reference to such standards is suggested and widely accepted. However, these standards are intended to set a minimum level of performance and the designer should satisfy himself as to their adequacy. The designer should also become aware of the design factors and recommended procedures in a particular standard, remembering that strong pipe sections require less construction effort. The completed design and the project specifications are the contract documents that ultimately will govern.

The following are areas where design considerations should be carefully reviewed and factors chosen that will coordinate with anticipated construction effort.

Trenching

Ground conditions through which the trench will travel should be identified. Solid rock should be over-excavated to provide for a cushion of select material on which to lay the pipe. Trench width should be sufficient to accommodate the pipe and allow for working room, usually the diameter of the pipe plus two feet. Narrower trenches can be used as long as proper installation can be achieved. The use of a pipe with high ring strength does not rely as much on the trench support as much and would be a better choice for a narrow trench where placement and compaction of backfill are more difficult.

Where unstable sub-grade conditions are encountered, overexcavation and placement of coarse material is usually used to help control the unstable area. This material should be placed below the bedding zone required for various pipe materials. Ductile iron pipe requires less material for a bedding zone and, thus, reduces the amount of extra bedding and trenching. Excessive ground water or sand will require special considerations.

Trench Backfill Materials

Pipelines placed underground can be designed with different types of material as bedding and backfill. The design of either rigid or flexible pipes uses factors produced by the trench conditions in selection of pipe thickness. Most pipe materials for water and wastewater pipelines, including ductile iron pipe, are flexible pipe materials. Only flexible pipes will be considered in this discussion.

Material for backfill can vary. For instance, a gradation of fines to coarse gravel with rocks in larger diameters can be accommodated in backfill material by ductile iron pipe. Less coarse material with smaller size stones in the backfill would be required for weaker pipe materials. Cost savings can be achieved by using the backfill material which is most widely available. In most instances, native excavated material can be used for backfill around ductile iron pipe.

Trench Conditions

A round conduit in soil is acted upon by the weight of the backfill material and the internal pressure. In the design of ductile iron pipe, these two conditions, internal pressure and external load, are considered separately.

The well-constructed trench with compacted backfill will help maintain the flexible pipe in a round condition. The ring strength developed by the pipe will also help maintain a round condition. Generally, an acceptable amount of ring deflection is allowed in design. Therefore, backfill support and material strength combine to maintain an acceptable ring deflection. The Iowa Deflection Formula, modified by Watkins, for determining the trench effects on the design thickness of a flexible pipe is given in Equation 1.

$$\Delta X = D_L \left[\frac{KWr^3}{EI + 0.061E'r^3} \right]$$

Equation 1

where:

ΔX	=	Deflection (in.)
D_L	=	Deflection lag factor
K	=	Deflection coefficient (dependent on bedding angle)
E	=	Modulus of elasticity (psi)
E'	=	Modulus of soil reaction (psi) (dependent on sidefill soil support)
I	=	Moment of inertia of pipe wall (in.^{3})
r	=	Mean radius of pipe (in.)
W	=	Earth load (lbs/in.)

A similar formula is used to determine invert stress in a pipe due to bending. The modulus of soil reaction, E' (sidefill soil support), and bedding coefficients, in addition to other factors, are used in the equations. The E' value and the bedding coefficient cannot be measured by testing a soil sample, but are empirically determined.

If a pipe has thin wall sections and/or is made of a weak pipe material, then greater compaction and larger bedding angles are required in the trench to maintain an acceptable deflection and bending stress. Examples of suggested E' values are listed below:

	E' (psi)
Steel pipe	200 to 3,000
PVC pipe	200 to 3,000
Ductile iron pipe	150 to 700

In general, the higher the E' value, the more the pipe's reliance on sidefill soil for support and the greater the compaction requirements for backfill soil material in the pipe zone. However, limits to the amount of support given must be considered. For instance, certain types of soil will compact more easily than others. Other factors affecting the degree of compaction attained are the methods and diligence actually applied by the contractor in constructing the trench.

The designer should carefully review recommended trench values as given in standards for different pipe materials and compare those to what may be reasonable to achieve. The degree of on-site inspection that is going to be required must also be considered. The same formulas are used for all flexible pipe materials so trench factors should be considered equally.

Ductile iron pipe has a great amount of ring strength in the thicknesses produced. Therefore, it has less dependence on trench compaction of the backfill, and can rely on assigned trench conditions that are easily obtained in the field. Project specifications should take advantage of the strength of pipe materials. Where ductile iron pipe is used, lesser trench conditions can be specified. Even with minimum pressure class ductile iron pipe in most typical installations, high compaction is not required.

Care must be exercised in backfilling if bonded coatings or loose polyethylene encasement is used on the pipe. Once the backfill is in place, compaction equipment must be used carefully to avoid damage to the coating.

Handling

All pipes should be handled correctly. Damage to pipe coatings and linings can result from impact due to dropping pipe sections on other pipe, rocks or hard ground.

Damage can also occur to pipe linings and coatings due to flexing the pipe as it is handled. Thin pipe walls may be adequate for trench and pressure loads, but may not provide support for the pipe against flexure due to handling conditions. The designer should call for adequate pipe wall thicknesses to resist excessive flexure.

Storage of ductile iron pipe in utility yards or on-site is permissible for long periods of time since sunlight does not affect it. Other materials may require storage in shaded conditions.

Permeation

The pipeline route should be carefully studied for evidence of ground pollutants comprised of low molecular weight petroleum products or organic solvents or their vapors. Research has documented that pipe materials such as polyethylene, polybutylene, polyvinyl chloride and asbestos cement are subject to permeation through the pipe walls. Taste and odor problems can occur within months of service and damage to the pipe may occur. Internal pressure does not alleviate this phenomenon.

Gasket materials used at joints may also allow permeation. However, the available area exposed to the contaminates is much smaller when considering the exposed gasket area versus the total pipe area. Gasket materials are available that resist permeation. The right choice of gaskets in the project specifications will avoid a problem.

Bolts

Any bolted system for underground service should use the best choice of material for strength and corrosion resistance of the bolts and nuts. For ductile iron pipe, a choice is found in the ANSI/AWWA C111/A21.51 Standard which allows either cast iron bolts with a minimum of 0.50% copper or high strength, low alloy steel. It is recommended that high strength, low alloy steel be specified. Tests have shown that these bolts are cathodic to iron and, consequently, more resistant to corrosion. For bolts used on systems with other pipe materials, corrosion protection is recommended in some form.

Line Drawings

Some pipe materials cannot be field cut and consequently each pipe length is designed to be placed in a certain location and sequence of construction. Thus, a line drawing fills this need.

Line drawings are not required for ductile iron pipe due to the flexibility of its joint and the fact that ductile iron pipe can be cut to length in the field. Line drawings specify and require markings on each piece of pipe identifying it as to its required location in the project. This is to ensure that valves, fittings, hydrants and curves occur where desired. Since the gaskets for ductile iron pipe push-on and mechanical joints are placed in the bell of the pipe, the spigot end can be cut and fit together in the field to meet changing requirements such as avoiding existing utilities or locating hydrants, fittings, etc., where they belong.

Consequently, ductile iron pipe sections can be used randomly by the contractor. The only exception to this is when a ductile iron pipe length has a feature welded on at the factory for a particular purpose, such as a welded bead for a restrained joint. This situation is easily accommodated by cutting an adjacent length of pipe to the restrained section to the proper length so the restrained pipes are placed where intended.

Joints

The ability of an underground pipeline to flex at the joints is a big asset. Flexible joints facilitate construction and help reduce bending stresses that may tend to build in the pipe wall. Ductile iron pipe uses push-on joints for connecting pipe sections and often uses mechanical joints at fittings, valves, etc. The flexibility available at these joints allows the pipe to deflect, thus avoiding stress that could have built up by soil movements over time.

Joint deflection also allows the pipe alignment to be changed gradually without the use of fittings. Special fittings are designed and needed for pipes other than ductile iron to make slight alignment changes. If an existing utility line is found out of expected position or was not anticipated, then ductile iron pipe can be realigned by using the joint deflection. Other materials could require a special fitting be fabricated with loss of time and extra cost.

Project Records

Inspection of a project during construction is generally required to some extent. All inspectors should be instructed to keep good field notes, listing the things that are done correctly as well as the items that have some deficiencies.

These notes are most important if an argument arises as to proper methods used and if the end result is not as desired. Document all important conversations, instructions given and correspondence.

Miscellaneous

A designer's specification should provide enough information to identify the performance required from any item used and the material strength standards accepted for that item. Hopefully, this would eliminate the use of sub-standard materials. In some instances, fittings and other accessories of foreign manufacture supplied to a job site have been sub-standard, even though they are represented to be the same quality as intended in the specifications. Failures have resulted not long after installation.

It is recommended that job specifications require that all items used on a project be made in America unless the designer has previously studied and accepted particular items supplied from other sources. Markings on all items should be required to identify their manufacturer's name, place of origin and strength of materials or class of service. In the case of small items such as bolts, connectors, etc., warranties could be called for to insure that proper items are supplied.

Summary

A federal court has ruled in a Virginia case that engineers clearly have the authority to write their specifications in what they believe to be in the best interest of their client, and that a city does not have to accept the "low price" if, in its opinion, other factors beyond price come into play.

Many factors go into the considerations for the proper design and achieving a satisfactory service life for a pipeline project. The technical aspects of design should reflect a thoughtful review of the construction process. Items allowed for use in the project should be covered adequately in the job specifications so that the intended materials and manufacturers are obtained. The designer should question recommendations made for design parameters and study the options. The ability to obtain a good, final project should balance all of these aspects.

References

ANSI/AWWA C150/A21.50, "American National Standard for the Thickness Design of Ductile Iron Pipe," AWWA, 1991.

AWWA M-11, "Steel Pipe - A Guide for Design and Installation, 3rd Edition," AWWA, 1989.

"Uni-Bell Handbook of PVC Pipe Design and Construction," Uni-Bell PVC Pipe Association, 1991.

RETHINKING THE APPROACH TO ENGINEERING PIPELINES
Andrew E. Romer, PE[1]
Member, ASCE

Abstract

Economic pressures on pipeline owners have led to a growing attempt to reduce project costs by minimizing the up-front design engineering costs. The consequence of this is the bidding of professional engineering services with subsequent inadequate design fees for many pipeline projects. Some consulting engineering firms have attempted to compensate for inadequate fees by specifying that the detailed engineering of the pipe is the responsibility of the contractor and the pipe manufacturer. Some AWWA pipe standards seem to encourage this approach. This paper challenges the assumption that reducing design engineering costs reduces overall project costs.

Introduction

Consider a publicly bid project where the pipe design is specified to be the responsibility of the contractor. The contractor's risk assumptions, and thus bid price, must not only include the project unknowns, but the unknown of what interpretation of the plans by the pipe manufacturer the engineer will accept during the shop drawing review. Contract documents showing complete pipe design however, level the playing field for the bidders, who can equitably compete on the basis of manufacturing and construction efficiencies, rather than by cutting design corners. Another advantage is the potential for significant reductions in construction schedule as a result of eliminating the time necessary for the manufacturer to design the pipe and to get that design approved by the consulting engineer of record.

Responsibility for Civil Engineering Projects

It has become common to specify that pipe and other manufacturers provide *design* calculations sealed by a professional engineer registered to practice in the state wherein the project is to be constructed. Statutory definitions of responsibility for civil engineering plans and specifications effectively limit the engineer of record from

---

[1] Senior Engineer, Boyle Engineering Corporation, 1501 Quail Street, Newport Beach, California 92658-9020. Telephone 949-476-3546, Facsimile 949-721-7482.

shifting *design* responsibility by specification to either the contractor or the manufacturer. California's Professional Engineer's Act, for example requires:

> "*All civil engineering plans, specifications, reports or documents shall be prepared by a registered civil engineer or by a subordinate employee under his direction, and shall be signed by him to indicate his responsibility for them.*"[2]

It is clear from this assignment of responsibility that that responsibility extends to pipeline materials and products selected and specified by the design engineer. That assignment of responsibility is based upon the assumption that that person exercised "*...independent control and direction, by the use of initiative, skill, and independent judgment, of the investigation or design of professional engineering work...*"[3] How can an engineer employed by a manufacturer or fabricator thus be held responsible for the work?

Plans, specifications, and the calculations substantiating the engineering judgment reflected in them are the responsibility of the consulting engineer. The Professional Engineers Act limits the ability of the design engineer to "pass the buck" to the manufacturer: "*This chapter, except for those provisions which apply to civil engineers and civil engineering, shall not be applicable to the performance of engineering work by a manufacturing, ... or other industrial corporation or by employees of such corporation, provided such work is in connection with or incidental to the products, systems, or services of such corporation or its affiliates.*"[4] In other words, the engineer employed by a pipe manufacturer is not exempt from the act if he practices civil engineering. If he designs the pipeline he is practicing civil engineering. But the act also allows him to refuse to sign and seal calculations, which are merely substantiating compliance with the materials and products specified by the design engineer.

Reallocation of Risk

The consequences of selection of professional engineering services on the basis of price are similar to the consequences of selection of the lowest bid price for construction contracts. When faced with the need to have as low a price as necessary to get the design project, there is a high likelihood of inadequate design fees. Like a general contractor, the low-bid engineer will only do the minimum amount of work that can be done to minimize his own liability in order to complete the engineering documents. Design decisions will then be dictated not necessarily by "*...the laws, phenomena and forces of nature.*"[5] Instead, by shifting design decisions to a

---

[2] The Professional Engineers Act, Chapter 7, Division 3, of the California Business and Professions Code, Section 6735.
[3] ibid, Section 6703.
[4] ibid, Section 6747.
[5] ibid, Section 6730.

contractor who was selected on the basis of price, the design decisions will be made on the basis of price. The natural reaction to the risk inherent with that approach is to try to place all the design risk on the contractor.

Shifting design engineering costs by specification to the manufacturer increases the risk of liability for the results of that task and cost shift. The engineers who take this approach attempt to limit design liability by writing criteria defining (performance) specifications. Those specifications rely to a great extent upon contractor's and manufacturer's certifications. Inclusion of "Pontius Pilate" clauses in project specifications is a further attempt to reallocate risk inherent in design to the contractor. An example of this hand washing is:

> "All materials employed shall be suitable for the intended application; materials not specifically called for shall be high-grade, standard commercial quality, free from all defects and imperfections that might affect the product for the purpose for which is intended."[6]

As noted above, the design of the pipe is the responsibility of the engineer of record. This includes the responsibility for determining the suitability of all materials for the intended application. Those materials should be what are specified. Similarly nothing should be utilized that is not specified, unless it has been reviewed by the engineer and accepted as a change to the contract by the owner. Certainly "high-grade" is subjective; and "commercial quality" is the lowest quality product commonly available! What is a defect or imperfection that *will* affect a pipe for the purpose for which it is intended that cannot be delimited in a specification?

This type of specification is unfair to the contractors and manufacturers. Without resorting to litigation, it places them at the mercy of the design engineer. The design engineer can hold the project hostage until a component is furnished to his interpretation of such subjective language. Even when the contractor's and manufacturer's specification interpretations are accepted by the engineer, what really determines the quality of the constructed product? It is apparent that the manufacturer's efforts to minimize his overall cost determine the level of quality, instead of the design engineer of record.

Reallocation of Costs

Someone must design the pipeline; regardless when in the project it is done. As noted previously, some engineers have attempted to compensate for inadequate fees by specifying that the design engineering of the pipe is the responsibility of the contractor and the pipe manufacturer. On the surface, this is relatively simple task, because the AWWA pipe standards most often specified for waterworks are structured to allow this. For example:

---

[6] An actual specification!

"*If the manufacturer is required to determine the wall thickness, the manufacturer's calculations of wall thickness shall be submitted to and accepted by the purchaser.*"[7]

When the standards are referenced in project specifications the *manufacturing* of the pipe should be adequate for most applications. Specifying that the manufacturer does it, however, does not necessarily reduce the cost of the design of the pipe. It takes just as many man-hours for the engineer of record to properly design a pipeline before the bid as it does by the pipe manufacturer after the bid. Consider a pipeline that is designed by the manufacturer: The design must be submitted, and checked by the engineer of record. Any disagreement with the manufacturer submitted design must then be communicated to the pipeline designer, who must either redesign it in accordance with the comments of the engineer of record, or respond in defense of his design. The most likely scenario is the latter, for his design represents the assumptions made by the manufacturer when preparing his bid.

The likelihood of design changes which will result in increased costs to the manufacturer are slim unless accompanied by an approved change order. This always will result in increased total project costs for the owner. This will also likely result in delays from re-ordering materials, re-manufacturing time sensitive components, and a longer production schedule. Where then are the savings from reducing design effort before issuing contract documents for bid?

The Level Playing Field

The consulting engineer who represents the owner's interest, and is most familiar with the project, is in the strongest position to engineer the most appropriate and economical pipe design. When the consulting engineer has adequate time and fee to properly engineer the pipeline, there is no incentive to cut any design corners that might otherwise result in increased risk or cost to the owner, as described above. Some of these include:

- resolution of unbalanced forces through length of restrained joints,
- crossing of existing utilities,
- backfill materials and compaction,
- details of outlets and bulkheads,
- pipe joints,
- linings and coatings,
- and the pipe cylinder.

Describing the design complexity of a pipeline project is beyond the scope of this paper. The time spent throughout the design process by the engineer is significantly

---

[7] AWWA Standard for Steel Water Pipe-6-In. (150mm) and Larger": Section 4.4

greater than that made available to the contractors for assembling their bids. The time is utilized to determine the many things that the pipe manufacturer is unaware, especially at the time of bid. These include hydraulic transients, geotechnical and constructability considerations, groundwater and soil corrosivity, and cathodic protection. At the time of bid each contractor should be most effectively devoting time to assembling his price on the basis of the same well defined design.

Describing the design complexity of a pipeline project is beyond the scope of pipeline standards also. For example, AWWA Standard C301-98 states *"Purchasers are advised that, while this standard presents information on materials and procedures for manufacture of pipe, it does not contain all of theengineering information needed to prepare a complete specification for a particular pipeline installation."*

What are the benefits of showing a completely engineered pipeline in the contract documents? First and foremost, each contractor and manufacturer can assemble his price on the basis of manufacturing and construction economies. Each pipeline contractor has unique approach to pipeline construction as does each pipe manufacturer have unique capabilities. When bid prices are based upon a well-defined set of contract documents, no short-cuts regarding the design need be assumed to be the successful low bidder. After all, what the engineer will accept has been completely defined. There is nothing to guess about.

The experience of the author's employer in over two decades of showing complete design for pipeline and other civil engineering projects has been good. Historical price difference between the two lowest bidders has been less than one-half percent. Change orders have been two percent or less, with most of the change orders due to items (such as paving quantities) unrelated to the design. In doing the research for this paper the author attempted to find data to substantiate that completely engineered pipeline projects have lower change orders than those merely specified. Of course no owner has actually bid the same project both ways, and few have the experience of bidding similar scope projects both ways. The author found the best apparent correlation with two public agencies that have recently constructed reclaimed (recycled) water pipeline networks. Both have employed many consulting engineering firms with different approaches to engineering pipelines. However, the data did not differentiate owner-initiated change orders uniformly. (Recycled water pipeline construction seems to encourage users to contract for water, which leads to construction of more service laterals issued as change orders.)

A more defensible claim to the benefit of completely engineering the pipeline prior to bid is the potential for reduction of the construction schedule when necessary. Or looking at it another way, the time available to the contractor to actually install the pipeline can be maximized. **Figure 1** illustrates this with an actual pipeline project. That project, a 60-inch diameter pipeline about 3.5 miles in length, was bid with completely engineered welded steel pipe and ductile iron pipe alternatives for contractor selection. The welded steel pipe option was selected by the low-bid contractor, and is illustrated. The owner's schedule required that the pipeline be in service by September 25, 1998.

Construction Schedule Savings

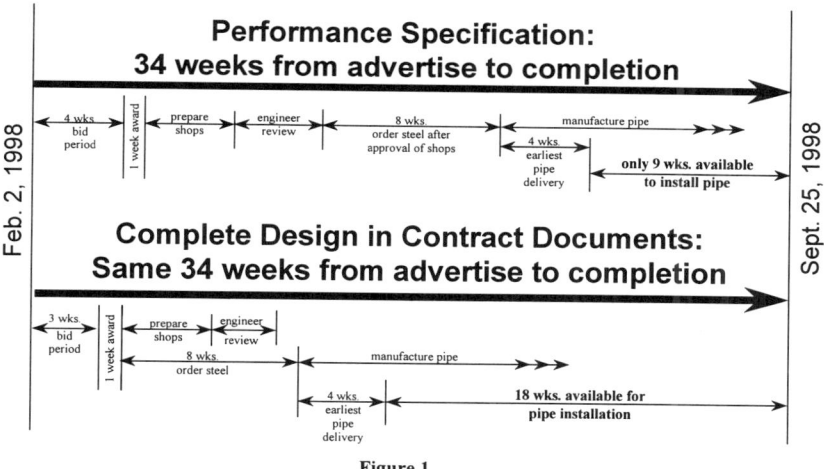

Figure 1

If the pipeline had been performance-specified, the top time-line would be anticipated. That anticipated allowing four weeks for the contractors to assemble their bids, and one week for the owner to examine the bids, the board to approve the contract, and preliminary notice awarded. It was assumed that at least four weeks would be required to prepare the design by the pipe manufacturer, and another four weeks for engineer reviews, and concurrent preparation of and approval of shop drawings. Welded steel pipe manufacturers typically place steel orders after receipt of owner-approved shop drawings. Assuming eight weeks for earliest delivery of steel (about average), and four weeks for manufacturing early delivery of pipe, that would leave only 9 weeks to install the pipe.

This was obviously unacceptable. The owner recognized the schedule benefits of showing a complete pipeline design in the contract documents. For this approach, the bottom time-line was anticipated. Notice that only three weeks were allowed for the contractors to assemble their bids, because there was no time required for additional engineering. Because the steel was completely engineered, it could be (and was) ordered immediately upon preliminary notice of award. This allowed production of detailed shop drawings to proceed immediately, for the risk of receiving approval for contractor's design is eliminated. Engineer's review is now reduced in time (and to double checking geometry of the pipeline layout) and it is out of the critical path. The result is at least double the available time for installation of the pipe. And note that the construction schedule was not reduced in favor of any additional time necessary to engineer the pipeline in advance of the bid. The design contract was awarded in September of 1997, which allowed less than four months to design the pipeline!

Conclusion

Reduction of pipeline project costs by minimizing the up-front engineering costs lead to the bidding of professional engineering services with subsequent inadequate design fees for many pipeline projects. Compensation for inadequate fees by specifying that the detailed engineering of the pipe is the responsibility of the contractor and the pipe manufacturer has been shown to be inconsistent with civil engineering registration rules. Contract documents showing complete pipe design however, level the playing field for the bidders, who can equitably compete on the basis of manufacturing and construction efficiencies, rather than by cutting design corners. Another advantage is the potential for significant reductions in total project schedule as a result of eliminating the time necessary for the manufacturer to design the pipe and to get that design approved by the consulting engineer of record.

PIPELINE PLANNING AND DESIGN
CITY OF SAN DIEGO WATER DEPARTMENT CIP

Frank X. Collins[1] and Steven W. Wallace[2]

Abstract

This paper outlines the Program Management approach used to control and manage the pipeline planning, design, and construction activities associated with the City of San Diego Water Department's $773 million Capital Improvements Program (CIP). This paper also provides information on the cost estimating technics used to develop budgetary figures on the CIP Pipeline Projects to be implemented between 1998-2006.

Introduction

The City of San Diego Water Department has embarked on a $773 million Capital Improvements Program (CIP) to ensure it can continue to reliably deliver safe, high quality water to its customers well into the next century. Prior to establishing the CIP, the City developed a detailed Strategic Plan for Water Supply. The Strategic Plan, which included extensive public involvement through a Public Advisory Group (GAP), outlines the City's current and future water needs, future water conservation measures, water reclamation, and local water resources. The Strategic Plan also addresses the City's vital need to repair, replace, upgrade and expand its water storage, treatment, and delivery systems.

The Water Department is the lead agency with overall responsibility for implementing this Capital Improvements Program. Funding will be provided through revenue-bonds which will be repaid by an increase in water rates. This increase was approved by the San Diego City Council in August 1997. Once the facilities and improvements are completed, the Water Department will own, operate and maintain the facilities.

---

[1] Parsons Engineering Science,Inc., 9404 Genesee Ave., Suite 540, La Jolla, CA, Phone No. (619) 453-9650

[2] Water Department, 600 B Street, City of San Diego, CA, Phone No. (619) 533-4102

CIP Program Organization

A CIP Program Management Team has been established in the Water Department to implement this program. The program is being headed by an executive management position within the Water Department. Parsons Infrastructure & Technology is providing technical and administrative support to the Water Department Program Manager.

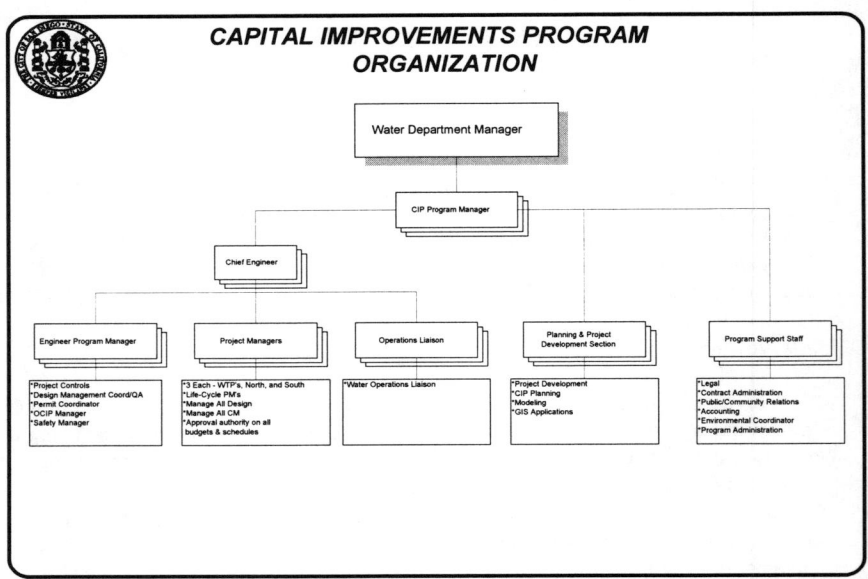

Figure 1 - Capital Improvement Program Organization

The overall management structure for implementing the CIP along with the roles and responsibilities of the individual components are shown in Figure 1.

The CIP is scheduled to take place over the next eight years (1998-2006) and will upgrade, expand or develop new Water Department facilities in the following areas:

- Pipelines and Water Mains($280 million)
- Treatment Plants($260 million)
- Dams and Reservoirs($10 million)
- Potable Water Storage Facilities($70 million)

- Pump Plants($65 million)
- Reclaimed Water Distribution Pipelines($10 million)
- Other System Improvements($75 million)

The Planning and Project Development Section is responsible for both the management and execution of all planning. This starts with the system master planning which progresses through project definition and development, to an approximate 10% level of design. Once a project is approved for inclusion in the CIP, the project manager assumes a life-cycle/direct management responsibility.

The design execution responsibility will be assigned to either a design consultant under contract with the City of San Diego or an in-house (City of San Diego engineering Department) design team. The determination of the selected option is a decision of the Water Department Manager. Performance standards, budgets and schedules will be developed for each project by the CIP management team. Design reviews will be conducted per the current City Standard requirement at 30%, 75%, 90%, and 100% designs.

Figure 2 presents graphically the group or individual who has direct management responsibility for a given phase of the project, as well as who has the execution responsibility.

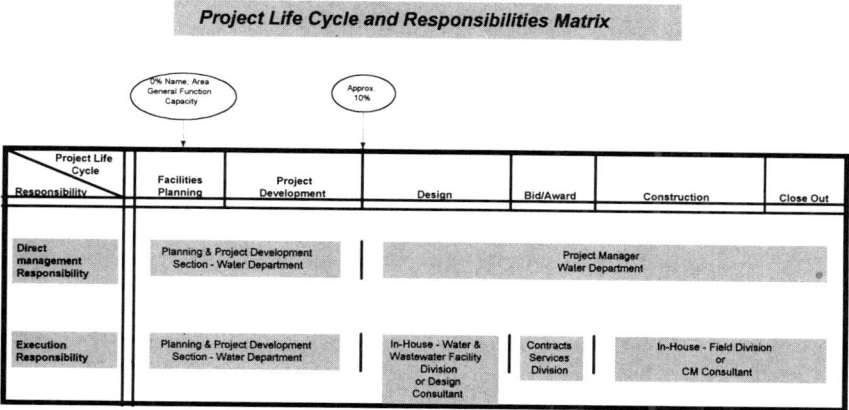

Figure 2 - Project Life Cycle and Responsibility Matrix

CIP Pipeline Projects

The CIP Pipeline Projects ($280 million) are a major portion of the CIP Program and are being managed by two Program Management Teams, the Northern Section CIP Management Team and the Southern Section CIP Management Team.

The major CIP Pipeline Project names, estimated costs, and design/construction schedules are shown in Figure 3 below.

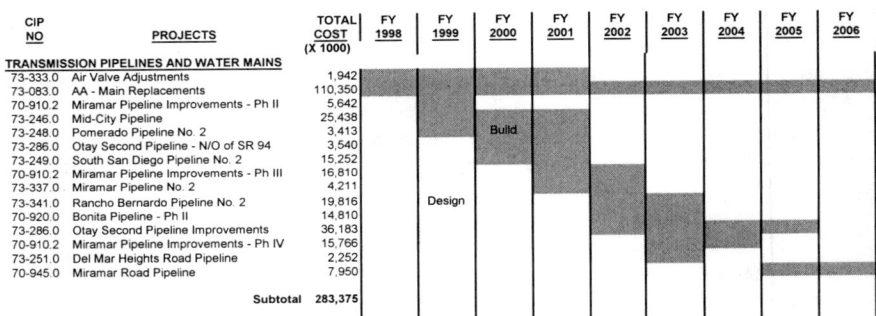

Figure 3 - Transmission Pipeline and Water Main Project CIP List

A major portion of the CIP pipeline/transmission system expenditures will be CIP 73-083.0, "Water Main Replacements," which will include the replacement of approximately 100 miles of aging cast iron water mains in the older sections of the City. Some of the cast iron water mains slated for replacement were installed at the turn of the century and have significantly exceeded their intended service life of 50 years.

Replacement of these pipelines (which have be plagued by recent breaks resulting in service disruptions, public inconvenience, and property damage) will increase service reliability, reduce maintenance cost, and bring this component of the distribution system up to current standards for water supply infrastructure.

CIP Pipeline Cost Estimating

Cost Estimating Guideline was developed and applied to each pipeline project in the CIP to produce a total program pipeline cost estimate and cash flow forecast. The purpose of this guideline was to develop a consistent cost estimating methodology for programming project budgets on projects with little or no detailed engineering completed.

The cost estimating models developed were based on historical contract bid cost information from the City of San Diego's Engineering and Capital Projects Department, cost information from similar agencies within the Western United States, and Means Heavy Construction Cost data.

The small diameter pipeline unit-cost equation was generated using construction cost data from the City of San Diego Water Department (SDWD) construction projects data base with bid dates ranging from 1988 to 1997. All costs were adjusted to an ENR-CCI for Mid 1997 (6580).

The unit cost equation for a large diameter pipeline was based on a cost estimating model developed for the Southern Nevada Water Authority. This model was verified with SDWD construction cost data were possible. All cost data was restricted to construction projects utilizing pipeline materials considered applicable to the SDWD program.

These pipeline materials include:

- PVC where appropriate;
- Mortar Lined and Coal Tar Coated Steel;
- Mortar Lined and Tape Coated Steel; and
- Mortar Lined and Coated Steel

Small pipeline diameters ranged from 6 to 20 inches with average trench depths reaching 7 feet. Pipe pressure class ranged from 150 to 200 psi, with the majority of the data points in the 150 psi range. The unit cost equation for small diameter pipelines is as follows:

$$\$/LF = 17.21 * D^{0.60}$$

where D is the nominal pipe diameter of pipe (inches), and $/LF is the unit cost of pipe represented as a cost (dollars) per lineal foot of piping. The unit cost equation includes a 15 percent allowance for those variables which cannot be quantified for the project at the time of developing the estimate.

Large diameter pipeline diameters ranged from 24 to 108 inches with average trench depths reaching 16 feet. Pipe pressure class ranged from 150 to 450 psi, with the majority of the data points in the 200 psi range. The unit cost equation for large diameter pipelines is as follows:

$$\$/LF = 0.60 * D^{1.67}$$

where D is the nominal pipe diameter (inches), and $/LF is the unit cost of pipe represented as a cost (dollars) per lineal foot of piping. The unit cost equation includes a

10 percent allowance for those variables which cannot be quantified for the project at the time of developing the estimate.

Pipeline installation costs are affected by complex factors such as soil conditions, traffic control, existing utilities, trenching and backfill requirements, paving requirements, and environmental requirements. All of these factors were considered to the extent possible when preparing cost estimates for a pipeline project.

The above pipeline cost equations are exclusive of main line valves and other major appurtenances.

CIP Pipeline Valve Cost Estimating

A cost equation was developed for main line valves using SDWD construction cost data. The valve sizes ranged from 2 to 48 inches. Vendor costs were used to develop valve costs for valves larger than 48 inches. The valve cost equation is based on the assumption that valves will be placed every 1200 or 1600 lineal feet per the SDWD Water and Sewer Design Guide.

The unit cost equation for valves sizes ranging from 2 to 60 inches is as follows:

$$\$/LF = 0.26 e^{0.09D}$$

The unit cost equation for valves size larger than 60 inches is as follows:

$$\$/LF = 1.0 * D$$

where D is the nominal valve size (inches), and \$/LF is the unit cost of valves represented as a cost of valves (dollars) per lineal foot of piping.

CIP Traffic Control Cost Estimating

Traffic control also has some significant cost impacts on the overall construction cost. In some cases the project area will have high traffic impact and will require construction to be done at night, which will increase the cost of construction.

From existing bid data, construction cost estimates, and discussions with contractors the following traffic costs were developed:

1. Traffic control for light traffic is 0.5% of the construction cost.
2. Traffic control for heavy traffic is 1.0% of the construction cost.
3. Allowance for night work is 10% of the construction cost.

Pipeline Jack and Boring Cost Estimating

The same level of historical information was not available for jacking and boring as for pipelines, therefore, a different approach has been used to develop an estimating equation for jack and bore construction. Historical costs from a contractor were averaged to a cost of (dollars) per diameter inch per lineal foot, and compared to Means Heavy construction costs. The construction costs for these operations were developed for a range of diameters (20 to 72 inches) and are anticipated to be approximately:

$9.25/ diameter inch/LF for jacking and boring

These costs have been adjusted to include an allowance of 30 percent for those variables which cannot be quantified for this project at this time.

Conclusion

The pipeline project cost estimates developed using this approach have been used to establish the pipeline projects budgetary baseline for this Program and have been used as the basis for establishing the required pipeline projects funding through revenue bonds.

During the Project Development Phase these budgets will be reevaluated based on the results of detailed alignment studies, preliminary environmental studies, preliminary utilities reviews and any other preliminary design studies conducted. During this preliminary planning phase (10% design work) a revised cost estimate for construction and a revised project budget will be prepared. This updated project budget will be baselined. All forecast and actual activity costs will be compared against this revised baseline budget, as will subsequent changes in the Engineer's estimate of construction during the detailed design phase. This will allow the program to measure it's performance against an established baseline and compare it's performance against the original budgetary cost estimates developed using the technics outlined in this paper.

BALANCING ENVIRONMENT AND RELIABILITY:
EAST MISSION GORGE TRUNK SEWER REHABILITATION PROJECT CASE STUDY

Michael E. Conner, P.E.,M. ASCE and Marnel Hale, P.E., M. ASCE
Engineering & Capital Projects Department
City of San Diego, California

Stephen L. Deering, P.E., M. ASCE and D. Michael Metts, P.E., M. ASCE
Dudek & Associates, Inc.

Abstract

The East Mission Gorge Trunk Sewer was first put into service in 1963. The pipeline alignment follows the San Diego River through Mission Trail Regional Park, to the City of San Diego limits, traversing a distance of approximately eight miles and ranging in diameter from 36 to 42 inches. The rehabilitated pipeline ranges in slope between 0.23 and 6.03 percent, contains approximately 10,000 linear feet of curve ranging between 150 to 2,000 foot in radius, and a total of 106 manholes. The rehabilitation project comprised four distinct operations including pipeline cleaning, pipeline rehabilitation, manhole rehabilitation, and pipeline protection. Design operations required pipeline access for implementation, as well as successful negotiation of numerous other challenges. Two rehabilitation techniques were determined to have specific advantages, including Cured-in-Place Pipe and Spiral-Wound Pipe alternatives. Each of these alternatives provided demonstrated ability to cover long lengths with minimal access at relatively low cost, and could be accomplished without the need to excavate. The lessons learned during the development and completion of the EMGTS project are applicable to all large diameter rehabilitation projects in environmentally sensitive areas. Construction of the EMGTS project is an ideal example of insitu repair of existing infrastructure, while managing conflicting objectives of environmental, community, and historical groups. The City's successful implementation of this $11 million rehabilitation project is a tribute to its dedication to community involvement and environmental preservation, as well as customer service and system integrity.

Introduction

Throughout history, communal growth has occurred in and around natural freshwater streams, rivers, and lakes. These areas provide settlers with essential drinking water, transportation and sanitary uses, as well as natural habitat for propagation of a variety of animal and plant life. However, the traditional function of these natural waterways has been displaced by man-made waterways. Communities are provided potable water through a vast network of water supply pipelines. Wastewater is transported through an equally large network of wastewater collection pipelines. The role of natural stream courses has been transformed from a necessity of survival to a haven for environmental preservation and recreational activity. As a result, the engineering and construction watch words for the 1990's have proven to be "Environmental Impact Avoidance and Mitigation."

The City of San Diego, recognizing this need to protect local natural habitat from increasing development, created Mission Trails Regional Park (MTRP). MTRP, the largest urban park in the United States, is located in the eastern portion of the City, covering a total area of approximately 5,700 acres. The park is one of the area's most prolific wildlife and habitat preservation areas, providing breeding habitat to a number of federally and locally endangered species including the least Bell's vireo and the California Gnatcatcher. MTRP has a history tracing back to the first Native Americans in the region, as witnessed by numerous archeological artifacts throughout the park. Additional cultural resources include the Old Mission Dam and Flume, constructed to convey water to the historic Mission San Diego de Alcala located approximately 10 miles west of the park. Simply stated, MTRP is one of the most treasured resources of the City of San Diego . . . and a sewer runs through it.

The East Mission Gorge Trunk Sewer (EMGTS) was first put into service in 1963, and has transported wastewater from users east of San Diego to the Metropolitan (Metro) Wastewater System since that time. The EMGTS remains the only gravity conveyance system which links the surrounding East County communities to the Metro system. The pipeline alignment follows the San Diego River through MTRP, to the City of San Diego limits, traversing a distance of approximately eight miles and ranging in diameter from 36 to 42 inches. The pipeline's original design conveyance capacity of approximately 30 million gallons per day (mgd) has been reduced over time, approximately 15 percent, by pipeline deterioration and debris deposition. Rehabilitation of the lower 37,000 linear feet is the focus of the EMGTS project. This portion of the pipeline ranges in slope between 0.23 and 6.03 percent, contains approximately 10,000 linear feet of curve ranging between 150 to 2,000 foot in radius, and a total of 106 manholes.

The EMGTS project area includes the San Diego River channel throughout most of its length. The pipeline alignment also traverses varying terrain, including urban development areas, a U.S. Navy golf course, a light industrial park, a local rock quarry, and the MTRP. The pipeline alignment is located primarily within the river's floodplain and adjacent riparian and coastal sage scrub habitats. The varying terrain has made pipeline maintenance access difficult. Limited maintenance access and increasing wastewater volumes contributed to the City's concern over the EMGTS's condition, considering the potential impact that a pipeline breach could have on MTRP and the San Diego River.

Project Development

The City initiated the EMGTS Rehabilitation Project to address concerns regarding the condition of the aging EMGTS. Initial considerations which led to the development of the EMGTS project included:

- The newly constructed parallel EMGTS pump station and forcemain system provided a window of opportunity for dewatering of the EMGTS. Given that the EMGTS is located within a gorge and in sensitive habitat, paralleling the trunk sewer for additional conveyance capacity was not an option. Therefore, a pump station and forcemain was constructed to convey peak wastewater flows in excess of available EMGTS capacity. The pump station and forcemain, being constructed to ultimate capacity, had capacity to fully divert upstream EMGTS wastewater flows for the first few years. This allowed a small window of time in which dewatering could be conducted to facilitate rehabilitation of the EMGTS.

- Infiltration and inflow (I/I) for the EMGTS was a concern. Over the years, the unlined reinforced concrete pipe used in the construction of the EMGTS has experienced corrosion throughout its length. Flow monitoring and visual inspection indicated that I/I was present and required attention.

- External damage to the EMGTS was discovered. Several portions of the EMGTS were found to be exposed by long-term scour of the riverbed, leaving sections of pipeline visible at grade with the possibility of significant damage caused by falling rocks and debris. Furthermore, many of the manholes along the pipeline's length had been vandalized, providing additional sources of inflow, rocks and other debris.

- The EMGTS has relatively few sewer or lateral connections. Development in the vicinity of the EMGTS alignment was limited due to the presence of MTRP and other large land area uses. This fact made rehabilitation a favorable alternative, as the majority of the flow diversion operations could be accomplished at the upstream end of the project.

The early phases of the EMGTS project development focused on definition of the true need for rehabilitation of the EMGTS. The City's initial project objectives included definition of the pipeline's condition, and the determination of the need, feasibility, capacity, schedule and cost of the project, considering various rehabilitation methods. The results of the initial project efforts were documented in a Basis of Design Report, containing flow monitoring data, physical inspection reports, video inspection summaries, rehabilitation technology reviews, pipeline defect information and Environmental Constraints/Biological Resources Mapping. Pipeline inspection techniques used included a walking inventory of the pipeline alignment, smoke testing, visual and photographic inspection of manholes, and internal video inspection of the pipeline. The overwhelming conclusion of the preliminary investigations was that limited future access and the need for a highly reliable system mandated that the EMGTS receive needed repair and rehabilitation.

Project Challenges

Having definitively established the need for the EMGTS Rehabilitation Project, the next step was the development of construction plans and specifications for the project. The rehabilitation project comprised four distinct operations including pipeline cleaning, pipeline rehabilitation, manhole rehabilitation, and pipeline protection. Design considerations addressed pipeline access for implementation, as well as successful negotiation of numerous other challenges. The following discussions provide an overview of the more critical project challenges and how they were accommodated.

Pipeline Alignment Access

As discussed previously, the location and alignment of the EMGTS created extreme challenges with regard to all aspects of the project. Simply finding the pipeline was difficult in the dense brush and rocky terrain of Mission Gorge. Project access was to be a determining factor in the success of the project. Access to the pipeline alignment would include grading, filling and cutting operations, mechanical brush clearing, and construction of bridge crossings for the river in selected areas of the project.

The first step in developing access for the project was to map biological and cultural resources throughout the project area. In this manner, alternative access plans could be developed which minimized impact to the protected resources. The developed access alternatives were then presented to numerous jurisdictional agencies and committees to gain approval of a preferred access plan.

Because much of the work took place in MTRP, acceptance of the access plan rested with those individuals involved with the park and its care. Activities in the park are overseen by three separate political bodies. The MTRP Citizens Advisory Committee is the lead organization that oversees policies and activities in the park. The group meets regularly and responds to the needs of the public, as well as the City's Park and Recreation Department. Acceptance of the proposed access plan required several sessions with the committee to gain their approval.

The MRTP Task Force is a City Council subcommittee, chaired by Councilwoman Judy McCarty. The Task Force includes City Council members from San Diego an the City of Santee, as well as two San Diego County Supervisors. In addition to setting policy for the park, the Task Force views the park as a resource from a regional perspective. The Task Force required regular updates of the project prior to giving approval of the access plan.

The MTRP Foundation is the money generating political body for the park, which is very active in lobbying for grant funds and local tax moneys to fund park improvements. This organization meets regularly and required certain additional park improvements to be incorporated into the EMGTS project to compensate for damages cause by project access. To facilitate project access approval, a project access plan was developed that maximized use of existing park trails, minimized environmental impact and mitigation, and minimized overall access requirements.

It was determined that all project access, other than that obtained from existing project trails, would be considered temporary access and would be fully revegetated to its original state after use. This decision impacted design of the project and selection of the rehabilitation techniques. Pipeline rehabilitation techniques which provided extended lengths without intermediate access were considered more favorable over typical "manhole to manhole" rehabilitation alternatives. Furthermore, the location of access corridors were strategically selected to minimize the length of corridor needed to provide required access for rehabilitation equipment. In this manner, the long-term cost of planting, monitoring and maintaining revegetation areas was significantly reduced.

Flow Diversion

Dewatering of the EMGTS was a key consideration throughout the project. Initially, it was determined that the upstream EMGTS pump station and forcemain system could divert approximately 95 percent of the tributary wastewater flow. However, consideration had to be given to possible loss of the pump station and/or forcemain during project operations. Furthermore, the upstream portions of the EMGTS, between the EMGTS Rehabilitation Project and the pump station, were being relocated under a state highway project. The coordination of flow diversion activities required a significant amount of effort.

Considering the large flow volume, the pump station and forcemain system was allowed in the project documents as a means of bypassing flow for the Contractor. The Contractor was required to develop and provide to the Construction Management team a comprehensive Flow Diversion Plan, outlining the dates and methods to be used in dewatering the EMGTS. In this way, the City could effectively coordinate and control the demands on the pump station and forcemain. Also, coordination efforts with the upstream construction activities were facilitated.

Having diverted the majority of the flow at the upstream end of the pipeline, the Contractor was free to focus on the diversion of tributary flow from approximately 11 downstream connections. These connections totaled to a flow rate of approximately 1.0 mgd. As with the upstream diversions, the Contractor was required to provide a complete plan for review and approval. Submitted plans included both gravity and pumped flow diversion operations, dependent on the location and accessibility constraints.

Environmental Constraints

The environmental constraints proved to be by far the most confining of the challenges during the design of the project. An environmental constraints analysis was performed during the preliminary design phase of the project to assist in the design team's understanding and avoidance of environmentally sensitive areas. Unfortunately, the vast majority of the project area was protected in some manner or the other. Therefore, the challenge became how to minimize impacts and protect the City from public relations issues and excessive construction claims.

The first step was the development of an Environmental Impact Report (EIR) for the project. It was determined that a Mitigated Negative Declaration could suffice for the project. However, should the public feel that the City had not fully investigated some portion of the project, the City could be forced into the development of a full EIR at a later date. Therefore, it was determined that a full EIR would be developed for the project, thereby eliminating the possibility of delays prior to or during construction. A comprehensive EIR was prepared to establish limitations on construction activities and to define other environmental restrictions. Design and construction activities were coordinated with other City departments, including the Park & Recreation and Metro Wastewater Departments. Various local, state and federal resources agencies were also consulted to define all aspects of the project' constraints. Permitting was required from various agencies, including the Army Corps of Engineers, California Department of Fish and Game, U.S. Federal Wildlife Service, and the City's Development Services Department.

One of the more restrictive constraints was related to project scheduling. Construction schedules were restricted to avoid disturbance of nesting seasons within project areas, which occur between March 15 and September 1 of each year. In this one constraint, the construction period was reduced to approximately one-half of the year. Further complicating the project, construction times included the winter months which are traditionally the wetter months of the year. And to make the project even more challenging, the project was scheduled to be conducted during a time when the weather services were predicting the current "El Nino" weather pattern. The contract documents needed to protect the City from excessive construction claims with construction occurring in the river channel during one of the wettest winter seasons in the region in recent history.

Cost

As with all capital improvement projects, cost is always an issue of concern. Because the EMGTS project was not a standard rehabilitation project, cost estimate development was complicated. Almost every aspect of the project led directly to increased cost for the project. The challenge was to effectively write the project specifications to allow Contractors to bid the project, without leaving the City open to numerous and expensive construction add-ons. Every aspect of the project had to be fully evaluated to make certain that the desired construction operations could be completed. Some of the more expensive issues to consider were the project access corridors, flow diversion issues, and environmental protection issues.

Rehabilitation Alternatives

Having defined the project challenges and several means of addressing them, the project focus was turned to identification of available rehabilitation alternatives. Numerous alternatives were considered for the project, each with its own advantages and disadvantages. The goal of the evaluation was to select several rehabilitation alternatives which would provide for a competitive bidding process while addressing the project challenges at the lowest project cost. The following is a list of evaluated rehabilitation techniques and there associated rehabilitation limitations:

OVERALL REHABILITATION ALTERNATIVES SUMMARY			
Rehabilitation Method	Diameter Range (in)	Typical Length (ft)	Applicable?
Cured-in-Place Pipe			
Inverted/Winched in Place	4 to 108	2,000	Yes
Spray-on-Lining	3 to 180	500	Yes
Sliplining			
Segmental	4 to 158	1,000	No

OVERALL REHABILITATION ALTERNATIVES SUMMARY			
Continuous	4 to 63	1,000	Yes
Spiral Wound	4 to 100	1,000	Yes
In-Line Replacement			
Pipe Displacement	4 to 24	750	No
Pipe Removal	36	300	No
Deformed and Reshaped			
Deformed Pipe	4 to 24	1,500	No
Drawdown	3 to 24	1,000	No
Roll Down	3 to 24	1,000	No
Point Source Repair	Any	n/a	Yes
Manhole Rehabilitation	Any	n/a	Yes

It was determined that a sufficient number of Contractors existed for each of the applicable alternatives to assure the City of a competitive bid process. Therefore, the selection of rehabilitation techniques for the project revolved around how well each alternative accommodate the environmental constraints of the project. Considering these limitations, two rehabilitation techniques were determined to have specific advantages, including Cured-in-Place Pipe and Spiral-Wound Pipe alternatives. Each of these alternatives provided demonstrated ability to cover long lengths with minimal access requirements, relatively low cost, and could be accomplished without the need to excavate along the pipeline alignment. For these reasons, the project specifications were confined to these two alternative rehabilitation techniques.

Construction Documentation

Having determined the appropriate rehabilitation techniques and having addressed the various project issues, development of the project documents became the next step in the process. The initial challenge was to accurately provide Contractors with a reasonable representation of the alignment and profile of the existing pipeline. "As-built" drawings were not available for the EMGTS. The original design drawings and historic photos of the pipeline were available for project use. Surveying teams were used, in conjunction with the preliminary pipeline investigation teams, to identify manhole locations and invert elevations. This information was combined with the original pipeline design drawings to clearly identify the approximate length, size, location, slope and alignment of the existing EMGTS.

Plan and profile sheets (40-scale) were developed from a project aerial survey of the project area for the purpose of both documenting the existing alignment and for general project use. A set of 200-scale maps were developed for use in identifying environmentally sensitive areas, and for delineating project access alignments. Comprehensive details were provided for access corridor construction and alignment, thereby reducing the likelihood of inadvertent collateral damage to protected habitat along the pipeline alignment. Specific discussions on how access corridors were to be constructed and marked were included in the project specifications to avoid the possibility of construction claims for inadvertent environmental damage.

The EIR included the development of a Mitigation and Monitoring Plan. This plan defined the limits of acceptable environmental disruption, the locations of that disruption, and the restoration necessary for mitigation of that disruption. Specific language was included in the specifications to notify the bidding Contractors of the environmental limitations of the project and to inform the Contractors that collateral damage outside the allowed areas, without approval, was not allowed. Also, that collateral damage would be the responsibility of the Contractor, including all multiple ratio mitigation imposed by environmental resource agencies. Inclusion of this clause in the project specifications was for both the protection of the City and the Contractor. By clearly stating the consequences of variance from the project plans, each party could confidently move forward with the project.

Environmental permitting is a long process, with numerous reviews and public evaluation periods. It was crucial that all necessary permitting be planned well in advance to avoid delay of the project. With the EMGTS project, the environmentally sensitive nature of the project only increased the importance of the permitting process. Preliminary meetings were conducted with all permitting agencies to inform them of the project and to anticipate resulting permitting requirements. Creative solutions, such a giant cane (Arundo Donax) removal throughout the project area was incorporated into the project to provide the most benefit to environmentally sensitive areas, while limiting cost and project schedule. The resulting environmental permits were included in the project specifications as appendices, and enforced throughout the project.

Bidding and Construction

So, was all the planning and evaluation worth the effort? It certainly was! Bidding, award and construction of the project was significantly enhanced by the planning performed in the early stages of the project. The City benefitted greatly from the bidding process and subsequent implementation of the project documents. However, no amount of planning can completely anticipate the ingenuity of today's Contractors. The following discussions identify several issues which were critical to the success of the project.

Bidding Wars

The first challenge to the project specifications came only one day into the advertisement period. The Cured-in-Place Pipe (CIPP) manufacturers immediately began a lobbying campaign with regard to the thermosetting resins and chemical resistance test which would be used for the project. The specifications allowed both polyester and vinylester resins. The two principal manufacturers of CIPP rehabilitation processes bidding the project included Insituform Technologies, Inc., and InLiner USA, Inc. Each of these manufacturers claimed that the resin used by the other was inferior or unnecessary. Furthermore, the some proposed resins did not meet the local southern California "Greenbook" chemical resistance testing requirements, where the others did. The entire issue became a discussion of the

Greenbook requirements versus the ASTM F1216 standard for CIPP installations, each of which could be reasonably applied to the project. The result was a flood of memos and letters from both manufacturers trying to eliminate the competitor's process. This frenzy of activity was anticipated and expected, as it happens on almost every project where these two manufacturers compete.

The normal size of a large CIPP rehabilitation project is in the $1 to $5 million range. The EMGTS project was an $11 million project. This fact fanned the fire of competition. The manufacturer elevated into the political arena. This led to several meetings and numerous discussions. In the end, the project moved forward without significant delay. The lesson learned . . . clearly establish the type resin and testing methods to be used for the project, and stick with them. Each manufacturer has the ability to use any materials necessary to meet the specification. The competition between the various manufacturers assures the owner of competitive bids, resulting in lower overall project costs.

In the case of the EMGTS project, we also included the Spiral-Wound Pipe (SWP) manufacturers in the project specifications. The primary manufacturer interest came from Danby of North America. The SWP process is typically lower in unit cost than the CIPP under normal circumstances. However, considering the environmentally sensitive nature of the project, it was projected that the unit cost of the SWP process would approach that of the CIPP process. The SWP manufacturer's reaction to the project specifications was associated with the chemical resistance testing. The SWP manufacturers objected to changes in the original specifications as a result of lobbying by the CIPP manufacturers. Of course, this was because the original specification language was more favorable to the SWP process. In the end, the SWP manufacturers did not make much attempt to change the project specifications.

The bidding period ended with only two addenda issued. The first was issued to allow both the Greenbook and ASTM F1216 chemical resistance tests to be used. The second merely extended the bid opening by a week. In the end, the bidding documents held up! And the bid results were surprising. The engineer's estimate for the project was $11.6 million. Danby of North America was the first bid opened at a price of $12.6 million. InLiner USA did not bid the project, and Insituform Technologies bid $10.4 million. The bidding strategy was successful, resulting in a $1 to $2 million saving for the City on the project.

Environmental Protection

The contract documents were successful in both protecting the environment and the City from excess environmental impact. The project experienced only minor instances of increased environmental impact. The Contractor required some additional impact area in several project access locations, and less in others. The construction management team successfully negotiated these increased amounts of impact. The ability to add impact area was anticipated in the development of the project documents and the environmental documentation. For this reason, negotiations during the permitting process included establishment of a total environmental impact for the project. This allowed the Contractor to increase impact in one area and offset that added impact with less impact at other locations. Overall, the environmental impact of the project was as anticipated in the development of the project.

Project Schedule

Of all the challenges facing the project, the schedule was the issue over which we had the least control. The weather was a variable which we could not write specifications to control. Therefore, we focused on defining the conditions under which the project might be impacted. The specifications were written to allow the Contractor sufficient time for delays caused by rain or inability to work, but protected the City from excessive extension claims. The result was a partnering approach between the Contractor and the Owner, which worked well throughout the project.

This is not to say that the weather did not delay the project. Rain lead to numerous delays on the project, both for the rehabilitation contractor and the revegetation contractor. Access roads were flooded and washed out several times, necessitating additional time for construction. Also, during the revegetation period for the access corridors, the rain led to the loss of an entire revegetation area. However, these issues were anticipated and the project specifications protected the City and the Contractor.

The flow diversion issues did result in some difficulties. The Contractor devised a method of diverting wastewater to the upstream pump station by reversing flow in the gravity sewer. The process involved plugging the sewer and backing the wastewater up through the gravity system. This process was evaluated and determined to work satisfactorily. However, complications resulting from the upstream state highway project to interference between the two projects. Through careful coordination, the bypass issues were resolved and work was completed successfully.

The final result was that project construction, originally scheduled to span a two year period (two winter seasons), will be requiring a portion of an additional winter season to complete. The weather resulting from the recent winter storms was instrumental in the need for this added time to the project.

Summary / Conclusions

The lessons learned during the development and completion of the EMGTS project are applicable to all large diameter rehabilitation projects in environmentally sensitive areas. Attention to details, such as project location, flow diversion, and environmental constraints, during the planning and design phase of a project will save money during construction. Early consultation with regulatory / resource agencies and the public provide insight into how impacts can be avoided or minimized. And project specifications can be used to protect the Owner and Engineer, while also minimizing risk for the Contractor.

The City of San Diego needed to improve the reliability and capacity of a key component of its wastewater collection and conveyance system. They accomplished that task, while protecting and improving the environment. Construction of the EMGTS project is an ideal example of the insitu repair of existing infrastructure, while managing conflicting objectives of environmental, community, and historical groups. The City's successful implementation of this $11 million rehabilitation project is a tribute to its dedication to community involvement and environmental preservation, as well as customer service and system integrity.

Construction of the East Mission Gorge Trunk Sewer Rehabilitation Project, San Diego California

Denis M. O'Malley[1], Peter J. Barden[2], Michael E. Conner[3], Marnell L. Hale[4]

Abstract

Construct the East Mission Gorge Trunk Sewer Rehabilitation Project in accordance with stringent environmental constraints, during the fall and winter months only, in a river gorge, within a federally protected habitat inside the Mission Trails Regional Park, as well as under the watchful eyes of environmentalists, the Regional Park Rangers, the US Fish and Wildlife Service (and other federal, state, and local agencies) — all while keeping the trunk sewer in service. This was the challenge faced by the City of San Diego, Brown and Caldwell (the Construction Manager retained by the City), and Insituform Technologies, Inc. (the contractor)

Challenges that faced the Construction Management team and the contractor for this project included:

- Extensive coordination with multiple parties: federal, state, and local agencies; City Departments; community organizations; and the public.
- Trenchless rehabilitation of the existing line and manholes while maintaining the pipeline in service to convey wastewater flows
- Work in a sensitive, protected habitat that contains nesting habitat for *Vireo Bellii* (least Bell's Vireo) and *Polioptila Californica* (California Gnatcatcher), both protected bird species; the largest known field of *Ambrosia Pumilla,* a protected plant species; as well as Native American and Spanish Mission artifacts.

---

[1] Construction Manager, Brown and Caldwell, 9040 Friars Road, Suite 220, San Diego, CA 92108
[2] Senior Inspector, Brown and Caldwell, 9040 Friars Road, Suite 220, San Diego, CA 92108
[3] Project Manager, City of San Diego, Water Utilities Department, Engineering Division, 600 B Street, San Diego, CA 92101-4502
[4] Resident Engineer, City of San Diego, Engineering & Capital Projects Department, 9485 Aero Drive, San Diego, CA 92123

- Mitigation of environmental impacts in the protected area, including removal of *Arundo Donax* (the Giant Reed, an exotic species), and revegetation of selected areas within the protected area
- Challenging working conditions, including restriction of work to the period from September through March, work in the river gorge, lack of vehicular access to approximately two-thirds of the existing pipeline, and the necessity for installing both permanent and temporary river crossings during periods of potentially high river flows.
- Incomplete record documents for the original construction of the pipeline in 1963, necessitating field investigation and coordination with the design engineer and City staff prior to construction activities.

Introduction

In 1963, The City of San Diego constructed the East Mission Gorge Trunk Sewer (EMGTS). The purpose of this trunk sewer was to convey wastewater from the Cities of Santee, El Cajon, and Alpine, as well as from several adjacent sanitary districts, to the San Diego Metropolitan Wastewater System. The total length of the line is approximately 13 kilometers (8 miles).

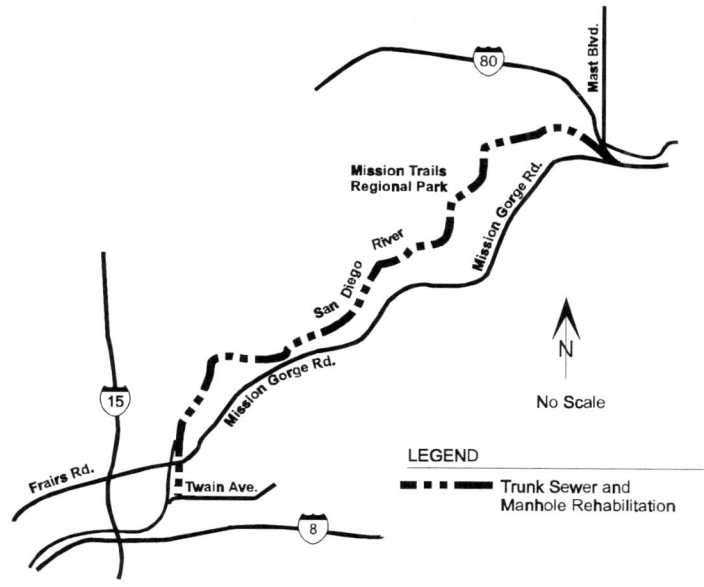

Figure 1. Project Location, San Diego, California

In 1994, deterioration of the line and deposition of debris reduced the EMGTS capacity, and the City determined that rehabilitation was necessary. To facilitate rehabilitation, the City constructed a parallel force main to provide relief during the rehabilitation project.

Original Construction

The EMGTS was constructed in 1963 using 920- to 1080-mm (36- to 42-inch) Reinforced Concrete Pipe (RCP) and Vitrified Clay Pipe (VCP) and for most of its length followed the bed of the San Diego River, crossing it 10 times in 8 kilometers (5 miles). The contractor used large earth moving equipment for clearing, grading and excavation. The project required extensive blasting for virtually the entire length of the project. Excavated rock was spoiled along or adjacent to the pipeline trace. (The spoil eventually created challenges for access and habitat restoration for the current rehabilitation project.) Habitat restoration efforts appeared to have been minimal on the original project. However, the native riparian habitat gradually reasserted itself over the following thirty years or so. Eventually, the project area became a significant breeding ground for many native animal species, including several protected and endangered species

The Need for Rehabilitation

By 1994, deterioration of the line and deposition of debris had reduced the capacity of the line from its original computed full flow capacity of $1.3 M^3/sec$ (29 mgd) to an estimated $1.1 M^3/sec$ (25 mgd). The City determined that urgent rehabilitation of the line was necessary. The City retained Dudek & Associates, Inc., in San Diego, to design the rehabilitation project and to prepare the Environmental Impact Report (EIR) for the project.

Construction of Parallel Force Main to Facilitate Rehabilitation

In anticipation of rehabilitation of the EMGTS, and considering work commissioned by CalTrans in the area upstream from this project, the City constructed a parallel force main along existing roadways and clear of the river gorge. The purpose of this parallel force main, relative to this project, was to facilitate rehabilitation by providing relief for the deteriorating line and by providing an alternative conveyance system during construction and rehabilitation.

Basis of Design

The basis of design for this project has been discussed in a separate, contemporaneous technical paper (*Metts et al*). The project objectives were:

1. Extend the service life of the EMGTS by rehabilitation.
2. Maintain ultimate sewage conveyance capacity.
3. Protect environmentally sensitive areas along the EMGTS alignment.

Because of environmental restrictions coupled with physical lack of access, the EMGTS had to be rehabilitated with virtually no excavation. The only excavation permitted was immediately adjacent to manholes, to repair and/or replace broken manhole cone and barrel sections, or for temporary removal of manhole cones for access.

Slip lining was not acceptable because the depth and size of the line would have required insertion pits far in excess of the limitations on allowable excavation. Moreover, the reduction in diameter would have been unacceptable for peak flows. Similarly, conditions militated against the use of Plastic Reinforced Mortar Pipe. The internal diameter of the line, 1100 mm. (42 inches) maximum, precluded methods that required extensive work to be performed from inside the pipe (e.g., Mastic Bonded PVC liner).

These constraints effectively reduced the rehabilitation options to two: Cured in Place Pipe (CIPP), and Spiral Wound Polyvinyl Chloride (SWPVC). The bid documents were formulated to allow uniform and competitive bidding by contractors using either of the two methods, or some combination of each.

Development of the Construction Management Team

Prior to completion of the design, the City retained Brown and Caldwell to be the construction manager (CM) for the EMGTS project. Prior to bidding, the CM assembled a constructability review team to review the contract documents and provided comments and recommendations to the City and the design engineer. After consideration of the comments and recommendations, the City and design engineer revised the documents accordingly.

Once construction commenced, the CM team comprised the construction manager and lead inspector (from Brown and Caldwell); the City's project manager and a resident engineer from the City; and a project biologist, a project archaeologist, and a noise mitigation specialist (from Affinis, Inc., a subconsultant to Brown and Caldwell). The City provided additional inspectors when necessary to ensure thorough inspection when the contractor used multiple shifts during the rehabilitation activities.

Environmental Mitigation and Monitoring Requirements

The Environmental Impact Report (EIR) for the project addressed the concerns of the federal, state, and local agencies having jurisdiction. The EIR contained the Mitigation and Monitoring Program (MMP), which identified the following mitigation and remedial measures:

1. Work within Mission Gorge was restricted to the period September 1 to March 15 to preclude any impact on or interference with protected and endangered species during their breeding season.

2. Special provisions were prescribed for protection of native flora along the project pipeline trace. These provisions included minimizing the length and width of temporary access corridors needed for construction.

3. Mitigation measures were prescribed for the project, including removal of invasive exotic species from the project area, particularly *Arundo Donax* (the Giant Reed), plus replanting, temporary irrigation, and maintenance of replanted vegetation of native species.

Results of Bidding

The City received two bids, one for each of the two methods. Insituform Technologies, Inc., using the CIPP or "inversion lining" method was the successful low bidder at $10.2 million, or approximately $2 million less than bid using the SWPVC method. Examination of each of the bids showed that access for construction was a substantial constraint for each of the two methods. In three sections of the existing line ranging in length from 930 meters up to 2,300 meters (3,000 to 7,500 feet) each, for a total of 4,700 meters (15,400 feet), there had been no vehicular access since the completion of construction in 1963.

Coordination with City Departments and the Public:

To facilitate public relations, the City's project manager met with the various local planning groups early in the design of the project. Because much of the construction and rehabilitation work took place in Mission Trails Regional Park (MTRP), the public relations effort during construction concentrated on the people involved with the park. Three sensitive political bodies oversee activities in the park: the MTRP Citizen's Advisory Committee, the MTRP Task Force, and the Mission Trails Foundation

The *MTRP Citizen's Advisory Committee* is the lead organization that oversees policies and activities within the MTRP. The group meets regularly and responds to the needs of the public as well as the City's Parks and Recreation Department. The rehabilitation project required ongoing communication among the CM team, the City's project manager, the MTRP park rangers, and the contractor to coordinate construction activities with the various public events in the park.

Part of the public relations effort required periodic project updates before the MTRP Citizen's Advisory Committee at the MTRP Visitor's and Interpretive Center. Because the Visitor's Center is the focal point for park users, a large display board was constructed and newsletters were made available to inform the public of the ongoing project.

The *MTRP Task Force* is a City Council subcommittee that includes City Council members from San Diego and Santee, as well as two San Diego County Supervisors. In addition to setting policies for the park, the Task Force views the park as a resource from a regional perspective. The Task Force required regular updates of the project prior to and during construction.

The *Mission Trails Foundation* is the money-generating political body that is instrumental in obtaining grant funds and local taxes to fund improvements for the MTRP. A decomposed granite walkway was incorporated into the project on the part of the pipeline alignment near the historic Old Mission Dam. The City's project manager and the CM team coordinated with members of the Mission Trails Foundation to ensure timely completion of the walkway, which was a prerequisite to subsequent improvements within the Park.

Father Junipero Serra Trail serves the MTRP, a gated park road that begins at the MTRP Visitor's Center and terminates near the Old Mission Dam Historic Site. The two-lane road is divided by an asphalt berm. One half is dedicated to bicycle and pedestrian traffic, while the other half is dedicated to one-way vehicular traffic. Construction crews used the road for access to several manholes and some of the inversion points. This required ongoing coordination with the public and the park rangers. The gates to the road are open from sunrise to sunset. However, the Park is open to the public 24 hours per day. Construction activities (principally "inversion points") near the Old Mission Dam drew many curious onlookers, which required heightened vigilance by the contractor and the CM team to ensure public safety. The CM team coordinated periodic closures to the road to aid in safe deliveries of equipment and materials. Vigorous enforcement of the traffic control plans resulted in minimal complaints and preservation of the public safety. Coordination with the park rangers was key to disseminating information to the public and promoting their safety.

Mitigation of Environmental Impacts

One of the primary objectives of the project was to protect the environmentally sensitive habitats along the EMGTS alignment and the San Diego River. Native vegetation within the MTRP includes riparian or stream side wetland habitats along the San Diego River, coastal sage scrub on the steep slopes, coast live oak woodlands, and mixed chaparral. The vegetation is dense and of high biological quality. Sensitive animal species living within the riparian habitats include the federally listed endangered least Bell's Vireo. The coastal sage scrub is known to support the

Figure 2. Contractor's Crew Locating Manholes Prior to Inversion

California Gnatcatcher.

The project's primary challenge was the limited access to the existing pipeline and manholes due to its location within the San Diego River channel (Figure 2). The construction documents specified that construction activities minimize impacts to sensitive habitat. Access corridors were restricted to a width of 3.7 meters (12 feet) through native habitat, with an allowable 7.6-meter (25-foot) wide area around manholes. A mitigation ratio of 3:1 for impacts upon riparian habitat provided the incentive for the contractor to perform the work with minimal disruption to the natural surroundings. Impacts upon coastal sage scrub were mitigated by revegetation at a 2:1 ratio. (The CIPP method employed by the contractor was appropriate because extensive lengths of the EMGTS could be rehabilitated with relatively minimal access and staging requirements.)

The EIR had been prepared for the project in consultation with local, state, and federal resource agencies. One form of mitigation to which all parties agreed was the removal from the riparian habitat of *Arundo Donax* (Giant Cane), an exotic (not native) plant species that has invaded native riparian habitat throughout the country. *Arundo* is a tall, feathery, bamboo-like plant that grows very dense and spreads rapidly, choking out more desirable habitats.

Permits from the U.S. Fish & Wildlife Service and Army Corps of Engineers required the removal of all large stands of *Arundo* within a 90-meter by 30-meter (300-foot by 100-foot) area around each manhole. To facilitate the removal of other non-native plant species within MTRP, temporary access corridors were directed through areas of non-native plant growth. MTRP has an ongoing *Arundo* eradication program. The mitigation plans for the EMGTS project were coordinated with the MTRP park rangers. The objective was for the contractor to remove the large patches of *Arundo*, and then for the volunteers to remove small remaining patches. A riparian forest that was being taken over by *Arundo* was rescued from certain domination. The net result was a dramatically upgraded habitat that will support important biological species.

Biological and Archeological Monitoring

Working under the direction of the CM, the project biologist conducted: field monitoring of manholes and access points/corridors, including route selection or providing input to route selection; staking or flagging of habitats or populations; monitoring of contractor's compliance with conditions of the EIR and other agencies requirements; field checking in response to concerns raised by the CM, the City, or one of the permitting agencies; and consultation with the park rangers. The project biologist conducted these activities on an almost-daily basis. Because of the valuable cultural resources evident in the project area (the Old Mission Dam and the flume that conveyed water to the settlements), and the potential for uncovering cultural resources during construction, a project archaeologist was also assigned to the CM team on an as-needed basis. No artifacts were uncovered during the project.

Pipeline Rehabilitation

Rehabilitation by the CIPP inversion method can be accomplished in either of two ways. In each, the felt liner must first be impregnated with the thermally reactive resin. In a "factory wet out" inversion, the felt liner is impregnated with resin at the contractor's shop and transported to the inversion site packed in ice. A crane is required to facilitate removing the liner from the transport trailer for placement in the pipeline. This is commonly called a "crane inversion." This is generally quicker and easier, but there is a severe practical limit on the weight of "wet bag" that can be transported on the highway.

The alternative is an "over-the-hole wet out" in which a dry liner is brought to the site and impregnated with resin as it is inserted. Because of access limitations, the contractor wished to maximize the length of individual inversions. Therefore, for the longer, less accessible reaches, the contractor was obliged to use the over-the-hole method. This required setting up what is essentially a miniature factory at each site, using conveyor belts, bulk resin storage tanks, a mobile laboratory, and other ancillary equipment. The felt liner and resin were brought to the site. Because of the thermal reactivity of the resin, these inversions were around-the-clock operations that presented both the contractor and the CM team challenges for quality control: ambient temperature variations from –2 degrees C to 28 degrees C (28 F to 85 F); night monitoring; and spill prevention (fluids from boilers, generators, hydraulic power units, resin mixing and injection equipment). The maximum length for a single insertion of CIPP is approximately 760 meters (2500 feet) for the size of the line to be rehabilitated on this project. Consequently, temporary access corridors and river crossings were constructed in at least three locations.

Figure 3. Over-the-Hole Inversion in Temporary Access Corridor

Bypass Pumping

Each inversion necessitated bypass pumping: generally some combination of flow diversion to the parallel force main plus temporary bypass pumping lines installed by the contractor. Prior to installation of the temporary bypass pumping, the project biologist walked the proposed bypass line alignment with the contractor to assess and flag that alignment to ensure the least impact upon the area. Following the alignment determination, the contractor installed the bypass line and conducted a clean-water bypass-pumping test. All leaks, if any, were repaired and the pumping test repeated until the bypass line was free from leaks. Following successful completion of the clean water test, the contractor bypassed raw sewage during the installation of the CIPP.

Testing and Inspection

Following an inversion, the newly installed CIPP was hydrostatically tested for leakage. Internal TV and physical inspection of the finished product was conducted to ensure conformance with the contract documents. Physical inspection included holiday testing (17,000 to 20,000 volts) to ensure the finished product was free from holidays and pinholes. Structural integrity was confirmed using coupons and plate samples tested by the City's testing laboratory.

Manhole Rehabilitation

Since the original construction in 1963, cleaning and maintenance technologies had improved substantially. Consequently, the City and the design engineer determined that a number of the original manholes (approximately 24 of a total of 107 manholes) was no longer necessary. The contractor installed the CIPP through the manholes that were to be abandoned, and then filled the manholes above the new CIPP.

Locating some of the manholes presented a substantial challenge - many of them had been grown over or buried since the original construction (Figure 2). While the coordinates shown on the original record drawings proved to be generally accurate, re-establishment of the manhole locations required painstaking work by the City's survey parties. The high walls of the Mission Gorge virtually prevented the use of Global Positioning Systems (GPS). Consequently, survey controls had to be brought in, often through thick brush and dense riparian forest.

Rehabilitation of manholes was subject to the same environmental and access constraints as the main line; approximately 25 percent of the manholes had no vehicular access. For those that were accessible by vehicle, the contractor used a spray applied urethane liner. For the inaccessible manholes, the contractor used a trowel-applied epoxy mortar that could be transported by workers into the manhole sites. In both cases, the finished lining was visually inspected and subjected to a holiday testing (17,000- 20,000 volts) to detect pinholes and holidays.

In the upper reaches of the EMGTS project, several manholes were in the middle of a field of *Ambrosia Pumilla*, and endangered plant species. In fact, this field is the largest known field of *Ambrosia Pumilla* in North America. Access to these manholes was restricted to foot traffic. The contractor's workers, under the direction of the project biologist, generally achieved access by means of temporary plywood walkways to avoid damage to the plants.

Installation of new frames and covers for rehabilitated manholes and concrete capping of manholes selected for abandonment presented different challenges, not the least of which was the difficulty in walking to the manholes. Frames and covers weighed about 280 kg (600 lbs), while concrete for capping weighed about 360 kg (800 lbs). The contractor's workers carried new frames and covers on stretchers into the inaccessible manholes that were rehabilitated. The contractor proposed, and the CM team accepted, the use of a two-part urethane foam in lieu of concrete for filling abandoned manholes, and incorporation of the existing frames and covers in the abandoned manholes.

Inspection Challenges

The CM team faced many inspection challenges. The lack of contiguous access to the work, and the frequent preclusion of vehicular access due to environmental constraints, necessitated that inspectors walk into the work areas, often over a mile. The excavation spoil from the original construction, the dense foliage, and the constantly changing width of the river challenged each CM team member's pioneering, rock-climbing, and wading skills – and several longed for the simian tail of which evolution had deprived them.

Coupled with the physical obstacles were the environmental challenges from rattlesnakes, insects, wide variations in temperature, and solar exposure. In addition, to the usual tools (hard hat and orange vest), the CM team wore gaiters, sunscreen, and insect repellent. Mountain lion sitings were reported, however, no encounters occurred. The contractor and the CM team have had no lost-time accidents.

Phases of the Work

At the time of the initial notice to proceed, the City had planned two phases to complete the EMGTS rehabilitation project. The first phase commenced in November, 1996, and completed in April, 1997. The second phase commenced in September, 1997, and was planned to complete in March, 1998.

The second phase had progressed well into December 1997, and concerns about the climatological impact from El Niño seemed overly pessimistic. While other parts of California, as well as Oregon and Washington, had been affected, the San Diego area had been largely unaffected. About the time local climatologists were talking of a weakening, El Niño proved them wrong. By the end of January, 1998, rainfall in the San Diego area was over 175 percent of normal, and the expectations were that the series of short duration, high intensity storms would continue into April. Flow in the San Diego River was high and substantial areas within the EMGTS project were severely inundated. Most of the areas in the project in which worked remained to be done were inaccessible.

The CM team, the City, and the Contractor discussed the alternatives available for the project. Ultimately, the uncertainty of the weather for the foreseeable future led to the pragmatic agreement to resume work in a third phase beginning in September, 1998.

Evaluation of 70-Year-Old Non-Reinforced Concrete Sewers

James Biery, P.E. [1]
Alison Ratliff [2]
Sylvia Hall, P.E. [3]

Abstract

The City of South Gate, California undertook an evaluation of their 70-year-old, small-diameter, non-reinforced concrete sanitary sewers in 1997. The City inspected approximately 121,920 meters (400,000 linear feet) of concrete pipe primarily using cleaning and closed-circuit television (CCTV) inspection methods. In order to correlate the image viewed on the CCTV tape to the actual physical condition of the pipe, sections of pipe were excavated and replaced. The removed sections were sent to a laboratory for structural, physical, and chemical evaluation. These sections were selected to represent poor, moderate, and good sewer conditions observed on the CCTV image. In addition, wastewater samples were taken from the sewer immediately upstream of the test sections. Soil conditions were also tested during excavation. Physical testing included identification of wall cracks, thickness dimensions, compressive strength, absorption and specific gravity of the concrete, and the external crushing strength of each pipe using a three-edge bearing load test. Wastewater samples were tested for temperature, pH, dissolved oxygen, biochemical oxygen demand, total sulfides, and dissolved sulfides. Soil conditions included moisture content and degree of compaction. This information was used to evaluate the relative priority of all sewers scheduled for rehabilitation according to the predicted condition based on the samples tested. The results of this analysis and the correlation to the images shown on the CCTV tapes are discussed in this paper.

Introduction

The City of South Gate (hereinafter referred to as the "City"), California, maintains more than 186 kilometers (116 miles) of sanitary sewer pipeline ranging in size from 200 to 60 millimeters (8 to 24 inches) in diameter, including about 2,350 manholes. The City experienced a number of sewer collapses that created sink holes

---

[1] Public Works Director, City of South Gate, 8650 California Avenue, South Gate, California 90280-3075, ASCE Member Grade
[2] Senior Engineer, CH2M HILL, 3 Hutton Centre Drive, Suite 200, Santa Ana, CA 92707, ASCE Member Grade
[3] Director, Engineering Development Center, Ameron International, 8627 S. Atlantic Avenue, South Gate, CA 90280

at the surface. Over 90 percent of the City's system was built before 1940 with about 75 percent of the system constructed using non-reinforced cement-mortar concrete. Industry experience with this pipe material indicates an expected service life of about 30 years. This means that over 75 percent of the sanitary sewers are currently serving beyond the normal life expectancy.

In 1992 through 1995, the City rehabilitated several concrete sewers using cured-in-place pipe (CIPP). The projects were successful and proved that rehabilitation of the sanitary sewers could be constructed using trenchless replacement methods. The City began a multi-year program to systematically inspect and rehabilitate the sewer system. Previous closed-circuit television inspection tapes were reviewed to identify sections of pipe with the most severe deterioration for immediate replacement to avoid costly collapses. A total of 2,286 meters (7,500 linear feet) of pipe were replaced in 1995.

Current Project

In 1996, the City contracted with CH2M HILL and Insituform Technologies, Inc. to clean, inspect, and evaluate approximately 127,437 meters (418,100 linear feet) of concrete sewers (excluding any that were previously inspected). The concrete consists of portland cement and fine aggregate indicating that the material is actually a "mortar" rather than "concrete", which contains coarse aggregate. As part of this project, a number of sections were selected for structural, physical, and chemical evaluation so that a correlation could be made between the image on the inspection tape and the estimated remaining service life of the pipe.

During excavation to remove sections of pipe for structural evaluation, ground conditions were observed and moisture content and density of the surrounding soil were determined. This was done to correlate trench conditions to pipe condition. Additionally, wastewater samples were obtained immediately upstream of removed pipe sections for testing and analyzed in an independent laboratory.

The original hypothesis was that the sewer system was deteriorating due to the corrosive internal environment. Based on the CCTV images, the condition of the pipe sections was rated as good, moderate or poor, except for those that were scheduled for immediate repair. There was a need to determine if the pipe rated as moderate to good condition could be placed on a low priority. The pipe sections selected for testing are numbered in Table 1 and indicate the condition of the pipe based on CCTV inspection images, shown in Photographs 1 to 3.

**Table 1
Condition of Pipe Sections Selected for Evaluation**

Pipe No.	Condition
1	Good
2	Good
3	Good
4	Moderate
5	Poor

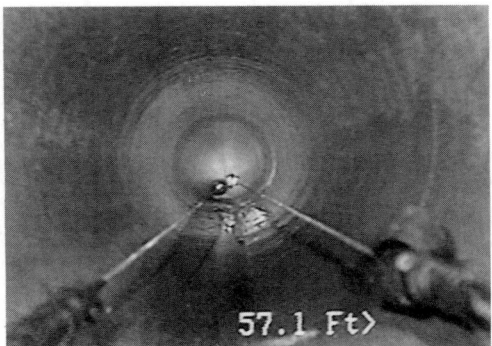

Photograph 1
CCTV image showing minor corrosion on the interior wall.

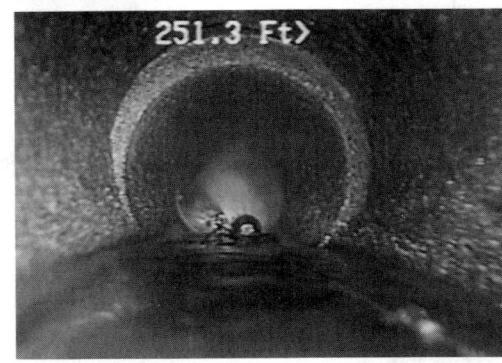

Photograph 2
CCTV image showing moderate corrosion on the interior wall.

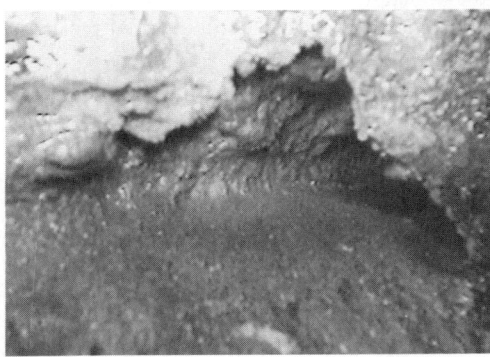

Photograph 3
CCTV image showing severe corrosion on the interior wall.

Structural, Physical, and Chemical Evaluation of the Pipe Sections

Five pipe sections that represented a cross-section of pipe in good to poor condition were selected for testing. The test sections were 1.22-meter (4-foot)-long sections of 200-millimeter (8-inch)-diameter pipe and were installed in 1927. They were visually examined for cracks and deterioration. Pipes 1 to 4 were delivered for testing intact and Pipe 5 was delivered in two pieces. The structural, physical, and chemical evaluation of the pipe sections was performed by Ameron International. Pipes 2 and 5 are shown in Photographs 4 and 5.

Photograph 4
Exterior surface of Pipe 2 after 70 years of service in the sanitary sewer system. A longitudinal crack along the entire length of the pipe section is shown.

Photograph 5
Pipe 5 was delivered in two pieces since it fell apart when removed from the trench.

Sections that were 0.3 meters (1 foot) long of Pipes 1, 3, and 4 were tested using a three-edge bearing test. Due to the longitudinal cracks that extended the entire length of Pipe 2, the bell end was removed leaving a 571.5-millimeter (22.5-inch)-long section with the spigot intact for the three-edge bearing test.

The external load crushing strengths of each 0.3-meter (1-foot) sections of Pipes 1, 3, and 4 and the 0.6-meter (2-foot) section of Pipe 2 was determined in accordance with ASTM C497-96 and are presented in Table 2. Some sections contained full-length and half-length longitudinal cracks prior to the loading test. Except for one specimen, crushing strengths were greater than the minimum 2,231 kilograms/meter (1,500 pounds/linear foot) required for an ASTM C14, Class I pipe. Crushing strength for the sections with no existing cracks was greater than 3,720 kilograms/meter (2,500 pounds/linear foot) required for a Class III pipe. The class or specification of the pipe sections was not known.

Because the exterior of the pipe is ribbed, the loading was uneven. An attempt to redistribute the load for the bearing test reduced the crushing strength by 44 percent.

Table 2
INSIDE DIAMETER, WALL THICKNESS, AND EXTERNAL CRUSHING STRENGTH OF PIPE SPECIMENS AND ABSORPTION, SPECIFIC GRAVITY, CUBE COMPRESSIVE STRENGTH, pH AND CARBONATION OF CONCRETE

Pipe-Specimen No.	1-1		1-2		2-1		3-1		4-1		5-1/2		4-2		5-1	
Depth of Cover, feet	7-1/2				6-3/4		6								5	
Inside Diameter (in) From: Top to Bottom - A to E; 45° to 225° - B to F; Springline - C to G; 135° to 315° - D to H	Spigot Side 8.10 8.02 8.04 8.07 8.16	Bell Side 8.10 8.01 7.97 7.94 8.00	Spigot Side 8.04 8.04 8.07 8.18	Bell Side 8.10 8.16 8.08 8.01		Bell Side 8.06 8.00 8.05 8.10	Spigot Side 8.14 8.07 8.03 8.14		Spigot Side 8.10 8.04 7.98 8.10	Bell Side 8.10 8.02 7.97 8.04			Spigot Side 8.17 8.07 8.02 8.15	Bell Side 8.10 8.08 8.09 8.00	Not determined; specimen was received broken in half longitudinally	
Average	8.09	7.98	8.10	8.09	8.05		8.10	8.07	8.05	8.01			8.10	8.07		
Standard Deviation	0.06	0.03	0.06	0.06	0.04		0.05	0.04	0.06	0.03			0.07	0.05		
Wall Thickness (in) At: Top - A; 45° - B; Springline - C; 135° - D; Bottom - E; 225° - F; Springline - G; 315° - H	Spigot Side 1.03 0.90 0.86 0.90 0.90 0.92 0.94 0.96	Bell Side 0.99 1.02 1.00 1.10 1.06 0.97 0.95	Spigot Side 1.07 1.00 0.91 0.95 0.95 0.92 0.96	Bell Side 0.94 0.92 0.89 0.97 0.87 0.84 0.93		Bell Side 0.97 0.89 0.90 0.93 0.99 0.96 1.00	Spigot Side 0.89 0.85 0.87 0.88 0.89 0.97 0.90	Bell Side 0.90 0.90 0.85 0.94 0.94 0.85 0.86 0.88	Spigot Side 0.92 0.94 0.96 0.95 0.93 0.93 1.01 0.94	Bell Side 0.97 1.02 1.08 1.11 1.07 0.94 0.93			Spigot Side 0.91 0.93 0.91 0.91 0.89 0.93 0.92	Bell Side 0.88 0.95 0.99 0.92 0.93 0.95 0.92	Spigot Side 0.88 0.88 0.99 0.94 0.95 1.00 0.93 0.94	
Average	0.93	1.02	0.96	0.91	0.94		0.90	0.90	0.95	1.01			0.91	0.93	0.94	
Standard Deviation	0.05	0.05	0.05	0.04	0.04		0.04	0.05	0.03	0.07			0.01	0.04		
External Crushing Strength, lbs/in.ft (No. & length of cracks)	1537 (1 full & 1 half-length cracks)		863* (2 half-length cracks)		1968 (2 full-length & 2 short cracks)		1978 (No cracks)		2598 (No cracks)				2604 (No cracks)		Not determined; specimens was received broken in half	
Absorption (%)	Upper 5.10 5.69	Lower 6.71 6.19	Upper 4.83 4.84	Lower 6.20 5.86	Upper 5.03 5.00	Lower 7.15 7.66	Upper 6.21 6.09	Lower 7.68 7.30	Upper 5.16 5.71	Lower 5.39 5.26			Upper 5.51 5.14	Lower 5.98 6.25	Upper 12.1** 11.7**	Lower 11.2** 11.0**
Average	5.40	6.45	4.84	6.03	5.01	7.41	6.15	7.49	5.44	5.32			5.32	6.11	11.9	11.1
Standard Deviation	0.42	0.37	0.007	0.24	0.02	0.36	0.08	0.27	0.39	0.09			0.26	0.19	0.28	0.14
Specific Gravity (Saturated Surface Dry, SSD)	Upper 2.30 2.30	Lower 2.29 2.29	Upper 2.32 2.32	Lower 2.28 2.28	Upper 2.28 2.29	Lower 2.28 2.28	Upper 2.32 2.33	Lower 2.30 2.31	Upper 2.29 2.29	Lower 2.30 2.30			Upper 2.28 2.28	Lower 2.28 2.28	Upper 2.27** 2.28**	Lower 2.26** 2.26**
Average/Standard Deviation	2.30 / 0.006		2.30 / 0.02		2.28 / 0.005		2.31 / 0.01		2.30 / 0.006				2.28 / 0.00		2.27 / 0.01**	
Cube Compressive Strength (psi)	Upper 13,000 11,100 11,800	Lower 9,390 10,200 10,800	Upper 11,900 12,900 13,200	Lower 12,700 12,200 12,300	Upper 13,200 12,100 13,000	Lower 12,900 12,300 10,800	Upper 12,800 12,500 14,000	Lower 12,900 12,400 12,300	Upper 14,100 13,500 13,000	Lower 14,400 12,500 13,800			Upper 12,300 14,500 12,300	Lower 12,600 13,900 11,700	Upper 10,900 9,890 10,400	Lower 11,400 8,720 8,420
Average	12,300	10,100	12,700	12,400	12,700	12,000	13,100	12,600	13,500	13,600			13,000	12,700	10,200	9,500
Standard Deviation	1,500	700	600	200	600	1,100	800	300	500	990			1,300	1,100	300	1,600
Depth of Carbonation or Leaching of Pipe Interior Based on Color Change	1/8" at top; 1/4" at bottom				1/16" at top; 1/4" at bottom		1/8" at top; 3/16" at bottom		No carbonation						>1/4" exterior & interior at top & bottom	
pH at Interior Surface***	5 to 6				5 to 6		5 to 6		5 to 6				5 to 6		5 to 6	

* Sand bags were used to equalize the load; ** Cracks were present in the absorption/specific gravity specimens; *** pH of the distilled water ranged from 5 to 6.

Pipes 1 through 4 exhibited a slight discoloration and abrasion of the bottom interior surface indicating a typical flow pattern and level. This abrasion did not appear to affect the structural integrity of the pipe. Except for the cracking, the pipe sections appeared to be in good condition.

The inside diameter and wall thickness at 45-degree increments around the pipe circumference at each end of the 1-foot-long sections of Pipes 1, 3, and 4, the cut end of Pipe 2, and the broken end of Pipe 5 are presented in Table 2. Wall thickness ranged from 21.3 to 28.2 millimeters (0.84 to 1.11 inches). This variance was more a result of the rib profile of the existing pipe. Based on the ASTM C14-95 specification for non-reinforced concrete pipe, a wall thickness of 22.2 millimeters (7/8 inches) places this pipe in a Class II category.

The absorption and specific gravity of two specimens from the upper and lower portions of each specimen after loading to failure were determined in accordance with ASTM C497-96, Method A, are also presented in Table 2. The compressive strength of approximately 19.1- to 22.6-millimeter (0.75- to 0.89-inch) cubes were determined using a procedure in AWWA C301. The absorption, specific gravity, and cube compressive strength of the concrete removed from the upper and lower locations of the pipe sections after the bearing test indicated that Pipes 1 to 4 are substantially below the maximum 9 percent allowed in ASTM C14-95 for adsorption. Adsorption of the concrete in Pipe 5 was greater than 9 percent. However, cracks in this section increased the surface area exposed during testing and invalidated the values. The specific gravity of Pipes 1 to 4 ranged from 2.28 to 2.33. A slightly lower range of 2.26 to 2.28 was determined for Pipe 5, which is consistent with the existence of cracks. The cube compressive strengths of Pipes 1 through 4 ranged from 64,742 kilo-Pascal (9,390 psi) to 99,974 kilo-Pascal (14,500 psi) and are considered excellent values. The cube compressive strengths of Pipe 5 were lower and ranged from 58,054 kilo-Pascal (8,420 psi) to 78,372 kilo-Pascal (11,400 psi), which are still considered good values. The lower results are consistent with the cracks found in Pipe 5.

The pH of the interior surface at the top and bottom of each pipe were determined by placing a few milliliters of distilled water on the pipe surface, waiting for approximately 10 to 15 seconds, and then placing a pH strip against the wet surface for 5 to 10 seconds. The color change to the strip was compared with a color charge that converted the color to a pH value. The degree of carbonation, leaching, and/or acid attack through the pipe wall was determined by applying a phenolphthalein solution to a freshly broken piece of concrete and measuring the wall thickness that did not turn pink. Concrete, due to the portland cement, has a typical pH of greater than 12.5. When a phenolphthalein solution contacts a surface with a pH greater than approximately 9, a pink to purplish color results. In concrete, the pH and, hence, the hydroxide ions can be reduced by carbonation in which carbon dioxide or bicarbonate ions can react with the hydroxide ions and calcium in the concrete to form calcium carbonate. The pH can also be reduced by neutralizing the hydroxide ions with the acids that can occur due to bacterial action in a sewer system. In addition, the pH can be reduced by simply leaching hydroxide ions from the concrete.

The depth of carbonation or leaching of the pipe wall and the pH of the interior surface are given in Table 2 and shown in Photograph 6. Insignificant carbonation or leaching occurred to the interior or exterior surface of Pipe 4. Substantial carbonation or leaching occurred to the interior and exterior of Pipe 5. The amount of carbonation or leaching that occurred to the interior of Pipes 1 through 3 is consistent with the age of the pipe, with only slight amounts occurring to the exterior. No signs of acid attack due to bacterial action were found. Insignificant amounts of carbonation or leaching had occurred to the exterior surface of all pipes.

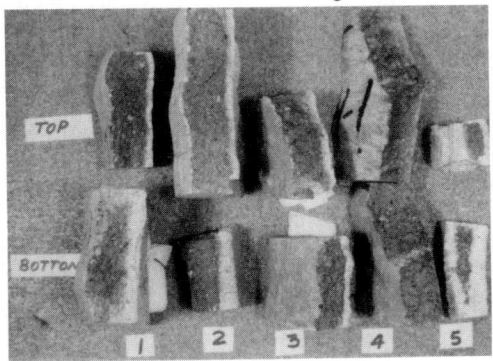

Photo 6
Carbonation or leaching that occurred through the concrete wall at the crown, and invert of the five pipe sections. The interior wall of the pipe is to the right. The dark (pink to purplish color in the test) area indicates that the pH is greater than 9.

Wastewater Sampling

Wastewater samples were taken at three of the pipe section locations in the month of June, 1997 at low flow periods when samples are expected to be the most concentrated. Results were essentially the same for all three locations. Ambient hydrogen sulfide was not detected at all three sites, oxygen was at atmospheric conditions, and the ambient and flow temperatures ranged between 25 to 34 degrees Celsius. The pH of the wastewater ranged between 8.0 and 8.2. Biological oxygen demand (BOD) values ranged between 107 and 154 mg/l. Total sulfides and dissolved sulfides were not at detectable limits in the stream. Dissolved oxygen (DO) ranged 5.6 to 6.8 mg/l for the three samples.

Sulfate, organic matter, and sulfate-reducing bacteria are present in virtually all wastewater. If sulfate is limited, sulfide generation is proportional to the nutrient concentration in the wastewater, as measured by the BOD[1]. BOD values appear to be low from the laboratory results.

The critical DO in wastewater below which sulfate reduction can occur is 0.1 to 1.0 mg/l. Sulfide is generally not produced until most of the DO and nitrates in the wastewater are depleted. With DO at 5.6 to 6.8 mg/l in the samples taken, sulfate reduction was not likely to occur.

The pH of the wastewater influences the relative proportion of dissolved hydrogen sulfide (H_2S) and hydrogen sulfide ion (HS^-). At a pH of 7 and a temperature of 25 degrees Celsius, the dissolved sulfides exist as approximately 50 percent hydrogen sulfide ion and 50 percent dissolved hydrogen sulfide[2]. The pH of the wastewater also influences the biological activity of the sulfate-reducing bacteria, which are most active at a pH of 5.5 to 9.0. At pH of 8.0 to 8.2, HS⁻ represents about 80 percent of the dissolved sulfides and a small concentration of sulfide ions ($S^=$) is present[5]. Dissolved sulfides were not detected.

Trench Conditions

Soils tested at the five test sites are uniform consisting of silty sands and poorly graded sands with relative compaction between 75 and 84 percent of ASTM D1557, Modified Proctor, except for Site 5 where the pipe was in poor condition and the surrounding soils exhibited lower soil density and much lower moisture content. Moisture contents of the first four locations were slightly below optimum moisture content ranging from 8 percent at 1.2 meters (4 feet) below the surface at Site 4 to 12.6 percent at 1.8 meters (6 feet) below the surface at Site 1. Soil values were fairly consistent vertically through the trenches. Soil around the pipe were observed during the removal of the pipe and found to be stable except for Pipe 5, which was the only pipe with bedding material comprised of 25.4-millimeter (1-inch) maximum rock.

Summary of Conditions at Test Sections

The five 200-millimeter (8-inch)-diameter pipe sections were in poor to excellent structural condition after approximately 70 years of service. Pipe 4 was in excellent condition with no cracks present, no carbonation or leaching, and with a crushing strength greater than required for an ASTM C14, Class III pipe. Pipe 5 was in poor condition due to the collapsed state when removed from the ground, contained many cracks, and a high level of carbonation or leaching. The absorption and compressive strength of the concrete were considered very good. The results for Pipes 1 through 3 were similar to those values for Pipe 4.

Acid attack at the crown of the pipe does not appear to have occurred since the depth at which the phenolphthalein solution turned pink was less than the depth at the pipe invert. This is consistent with the non-detected level of hydrogen sulfide in the wastewater sample. The amount of carbonation or leaching that occurred at the pipe invert is consistent with longer contact time of sewage with the wetted perimeter. The level of carbonation or leaching in the interior of Pipes 1, 2, and 3 is consistent with the age of the pipe. No carbonation or leaching has occurred to Pipe 4, which may indicate that this pipe has not been exposed to much flow during the 70 years of service, or that it is a newer pipe. Considerable carbonation or leaching occurred to the exterior and interior wall of Pipe 5 indicating that this pipe may be older than the other pipe sections or was consistently exposed to more sewage than the other pipe sections.

Cause of Failure

The anticipated problem with the pipeline prior to testing was that acid attack to the top and springline of the pipe from bacterial action was causing failures in the system. Based on the evaluation described, acid attack was not the primary cause of failure in the sections tested.

Instead, cracks were found near or at the springline of the pipe indicating that external loading caused the cracking. The streets of South Gate have experienced a much greater traffic loading over the years that were probably not anticipated during the late 1920s when the pipe was manufactured and installed. The existence of larger and heavier trucks and equipment during the 70 years of service increased the external loading on the pipe leading to cracking and eventual collapse of the non-reinforced concrete pipeline.

Also, once cracks formed and started to open, then fine soils from outside the pipe were allowed to enter through the cracks. This undermined the external support for the pipe, thus contributing further to failure of the pipe at some locations.

Calculation of dead and live loads on the trench using the Marston Equation and H-20 wheel loadings result in loads that could account for failure of an aging pipe. The depth to pipe ranged from 1.2 to 2.1 meter (4 to 7 feet). The total load due to dead and live loads ranges from about 1,207 to 1,570 kilograms per meter (811 to 1,055 pounds per foot) of pipe for a single axle live load, and from about 1,442 to 1,652 kilograms per meter (969 to 1,100 pounds per foot) of pipe for a double-axle live load. The external crushing strength of Pipe 1, Specimen 2, was 1,284 kilograms per meter (863 pounds per foot). Other specimens ranged from 2,287 to 3,866 kilograms per meter (1,537 to 2,598 pounds per foot). As the pipe deteriorates, its becomes more susceptible to cracking from superimposed loads above the pipe. When the soil becomes saturated during significant rain events, another 74.4 kilograms per meter (50 pounds per foot) or more of load is added.

Recommendations

Based on the conclusion that loading conditions are the primary cause of failures in the concrete pipe, outlining a repair program includes consideration of this finding. Phasing of the repair and rehabilitation program can be extended because of the confidence that pipe sections not in imminent danger of collapse are suitable for rehabilitation by trenchless methods, or reinspection at a later date. Without this information, pipe sections that appear moderately corroded on the CCTV image, would have been scheduled for replacement in the early phasing of the project, but can now be rehabilitated and phased over a longer period of time. Pipe sections that have the appearance of minor or no corrosion, can be rescheduled for inspection in 5 to 10 years. Emergency repairs have been scheduled for all pipe sections where the existence of holes in the pipe or voids outside the pipe may cause the pipe to collapse in the near future.

By undergoing this inspection program, supplemented by the special testing, the concern for the structural integrity of the system has been considerably reduced. As shown in Table 3 below, about 81 percent (combination of Phases 4 and 5) of the this 70-year-old system, is expected to perform for another 10 to 20 years at current loading levels. Phase 1 represents the emergency repairs that need to be undertaken. Phases 2 and 3, or 18 percent of the system, can be rehabilitated using trenchless technologies. In Phase 4, it is expected that about the same percent of the footage scheduled for reinspection will require replacement or rehabilitation, based on the current results (21 percent of the system from Phases 1, 2, and 3).

Table 3
Schedule for Repair, Rehabilitation, and Reinspection of the Sewer System

Phase	Year	Description	Percent of Program
1	1998-1999	Emergency Repair of Sections in Danger of Eminent Collapse	3
2	1999-2000	Rehabilitation of sections with severe corrosion, numerous open joint cracks and longitudinal cracking	5
3	2000-2002	Rehabilitation of sections with moderate corrosion, numerous cracked joints, some longitudinal cracking	13
4	2003-2007	Reinspection of the sections with minor corrosion, some joint cracking Replacement or rehabilitation of approximately 21 percent of the concrete sewer system (based on the combined percent of Phases 1, 2, and 3)	32
5	2005-2007	Reinspection of those sections that exhibited no corrosion and only minor defects	49

The City is currently evaluating funding options to proceed with the replacement and rehabilitation program for the non-reinforced concrete pipe in the sanitary sewer system.

References

[1] EPA, *Design Manual, Odor and Corrosion Control in Sanitary Sewerage Systems and Treatment Plants*, October 1985.
[2] Pomeroy, Richard D. and Parkhurst, John D., *The Forecasting of Sulfide Buildup Rates in Sewers*, Progress in Water Technology, Volume 9 Pergamon Press, 1977

Water Line Splitting In Gainesville, Florida

John S. Gifford, P.E.
Larry J. Ruffin

Abstract

This paper describes Gainesville Regional Utilities demonstration project that explored the use of traditional open trenching methods versus the trenchless method of pipe splitting for the rehabilitation of 152 meters (500 feet) of existing 50 mm (2 inch) Schedule 40 galvanized steel pipe in service as a potable water line. The galvanized steel pipe was replaced by using the pipe splitting method and pulling new 50 mm outside diameter polyethylene pipe into place. The hydraulics of the existing galvanized steel pipe and new polyethylene pipe are compared. Economic comparisons of the traditional method versus the trenchless method of pipe splitting are presented. Construction advantages and other considerations associated with the trenchless method demonstrated are also discussed.

Introduction

In Gainesville, Florida, the combination of a small town personality, big city opportunities, and the nationally respected University of Florida, results in a highly desirable environment for both living and learning. Perhaps it is the long-standing relationship between the City of Gainesville, Gainesville Regional Utilities (GRU), and the University of Florida that contributes to the open-minded consideration of emerging new construction technologies.

The project described within this paper was initiated when GRU responded to reports of reduced water pressure and water taste concerns from homeowners served by a water line extending along a dead-end street located within the Gainesville City Limits. The existing potable water line of interest was 152 meters of 50 mm Schedule 40 galvanized steel pipe

---

Senior Utility Engineer, Gainesville Regional Utilities, P.O. Box 147117, Gainesville, FL 32614-7117, Non Current ASCE Society Member.
Principal, L.J. Ruffin and Associates, 7928 Sebago Court, Orlando, FL 32835, Non Current ASCE Society Member.

with four existing residential service connections. Each connection consisted of a 19 mm (3/4 inch) diameter service line with a 16mm X 19 mm (5/8 inch X 3/4 inch) water meter assembly. A fifth home being constructed on the last available lot at the cul-de-sac had also requested GRU water service.

The existing residences along the street have well manicured lawns, including decorative landscaping. Each home also has a concrete driveway and sidewalk that extends over the existing 50 mm galvanized steel water line. One of GRU's project goals was to minimize water service disruption to the homeowners coincident with mitigating construction activity disturbance to their existing lawns, landscaping, driveways, and sidewalks. As with most municipal projects another necessary consideration involved completing the work while maintaining a reasonable construction cost.

As part of investigating the reported reduced water pressure, GRU was able to isolate a portion of the galvanized water line and remove a 0.9 meter (3 feet) section for inspection. Inspection revealed that the line was significantly deteriorated and contained appreciable scale build up. It was estimated that the original inside area of the 50 mm Schedule 40 galvanized steel pipe had been reduced by approximately fifty-five percent. A fifty-five percent reduction in pipe inside area approximately equals a 35 mm (1.39 inch) inside diameter pipe. For comparison of hydraulic conditions within new Schedule 40 galvanized steel pipe and the existing water line of interest consider the following information:

New 50 mm (2 inch) Schedule 40 Galvanized Steel Pipe

Selected Flow Rate = 189 liters per minute (50 gallons per minute)
Assumed Hazen-Williams "C" Factor = 120
Pipe Inside Diameter = 52.5 mm (2.067 inch)
Pipe Length Considered = 30.5 meters (100 feet)

Calculated Head Loss = 1.83 meters (6 feet)
Calculated Flow Velocity = 1.46 meters per second (4.8 feet per second)

Existing 50 mm (2 inch) Schedule 40 Galvanized Steel Pipe

Selected Flow Rate = 189 liters per minute (50 gallons per minute)
Assumed Hazen-Williams "C" Factor = 100
Pipe Inside Diameter = 35.3 mm (1.39 inches)
Pipe Length Considered = 30.5 meters (100 feet)

Calculated Head Loss = 17 meters (56 feet)
Calculated Flow Velocity = 3.2 meters per second (10.4 feet per second)

Primarily, the above example illustrates that at the selected flow rate, the head loss in the existing water line is greater than 9 times the head loss calculated for a new 50 mm (2 inch) galvanized steel water line. This head loss relationship holds true over the range of flow rates that would be anticipated for this water line. Therefore the validity of the reduced water pressure reports associated with the existing water line was confirmed through both visual observation and hydraulic calculations. It was readily apparent that the existing galvanized steel water line was in need of rehabilitation.

GRU's traditional method to rehabilitate a water line of this size would be pipe replacement using open trenching. Due to constraints outlined above a trenchless rehabilitation approach was considered for this project. GRU investigated the economics (see the cost comparison below) of a "partnering" approach with L. J. Ruffin and Associates using a trenchless technology to rehabilitate the existing 50 mm galvanized steel water line. The selected trenchless technology was replacement of the existing galvanized steel pipe with blue color-coded polyethylene pipe using the pipe splitting method. For comparison with new Schedule 40 galvanized steel pipe consider the hydraulic characteristics of the polyethylene pipe as shown below:

New 50 mm (2 inch) Polyethylene Pipe

Selected Flow Rate = 189 liters per minute (50 gallons per minute)
Assumed Hazen-Williams "C" Factor = 135
Pipe Inside Diameter = 48 mm (1.89 inch)
Pipe Length Considered = 30.5 meters (100 feet)

Calculated Head Loss = 2.3 meters (7.5 feet)
Calculated Flow Velocity = 1.7 meters per second (5.7 feet per second)

As calculated above, the head loss characteristics of the polyethylene pipe are essentially equal to that calculated above for new 50 mm Schedule 40 galvanized steel pipe. While new 50 mm polyethylene pipe has a smaller inside diameter compared to new Schedule 40 galvanized steel pipe, the higher "C" factor associated with the polyethylene pipe results in similar calculated head loss as compared to the galvanized steel pipe. It is anticipated that the polyethylene material may also be less susceptible to a long term "C" factor reduction when compared to galvanized steel pipe. The above calculated flow velocity in the new 50 mm polyethylene pipe exceeds that for new 50 mm galvanized steel pipe, but the velocity is still within reasonable engineering design criteria.

As referenced above, construction cost is an important aspect of municipal projects. This is particularly applicable for unplanned or emergency projects that may utilize capital funds from a non-specific project budget category such as "rehabilitation and replacement of water lines". The following cost comparison for traditional versus trenchless construction methods was performed for this project.

Estimated Cost For Traditional Water Line Rehabilitation ($10,500.00)

Project Assumptions:

GRU Water Construction Crew Size = 2 crews with 3 persons per crew
Estimated Construction Time = 40 hours (4 days at 10 hours per day)
Required Equipment = 1 backhoe, 1 flatbed truck, and 2 service trucks
Pipe Materials = 152 meters of galvanized steel pipe and required fittings
Estimated "Clean" Backfill Material Required = 23 cubic meters (30 cubic yards)
Concrete Replacement Required = 79 square meters (850 square feet)
St. Augustine Sod Required = 8 pallets

Using actual GRU crew, equipment, and material costs, along with recent typical costs for the remaining items above results in an estimated cost of approximately $10,500.00 to complete the work using a traditional construction approach.

Estimated Cost For Trenchless Water Line Rehabilitation ($8,500.00)

Project Assumptions:

GRU Water Construction Crew Size = 1 crew with 3 persons
Estimated Construction Time = 15 hours
Required Equipment = 1 backhoe, 1 flatbed truck, and 1 service truck
St. Augustine Sod Required = 2 pallets
Pipe Splitting and Pulling Equipment = Complete Supply by L. J. Ruffin and Associates
Pipe Materials = Complete Supply by L. J. Ruffin and Associates

GRU's crew, equipment, and material costs, are estimated to be approximately $2,000.00 to perform the required activities. A lump sum price of $7,000.00 for providing the required trenchless equipment and labor was received from L. J. Ruffin and Associates. The lump sum amount included the cost for the polyethylene pipe pulling equipment, the bullet to split the existing galvanized steel pipe, the polyethylene pipe and fittings, and travel expenses. The estimated total cost of the trenchless approach for rehabilitating the water line is $9,000.00.

Therefore, for this particular project the construction costs favored the trenchless technology when compared to GRU's traditional water line rehabilitation approach. The trenchless technology method resulted in a unit cost of approximately $18.04/foot of pipe compared to a unit cost of approximately $20.84/foot of pipe for the traditional method. The trenchless method offered greater than a ten percent construction cost reduction for this project.

Additional Considerations Related To Traditional Versus Trenchless Construction Methods

As stated above, the construction costs favored the trenchless technology when compared to GRU's traditional water line rehabilitation approach. Identified construction advantages of the trenchless method are summarized below. In addition, construction considerations related to utilities that do not possess trenchless equipment are summarized below:

Identified Construction Advantages Of The Trenchless Method

Reduced construction time by greater than fifty percent
Reduced construction crew requirements by fifty percent
Minimized vehicular and pedestrian traffic disruption
Minimized disturbance to homeowner's property
Minimized downtime of the water line and customer services
Minimized post-construction "dress-up" time and cost

Other Construction Considerations Related To The Trenchless Method

Trenchless construction equipment required
Trenchless construction company assistance required

Conclusions

Specialists within the trenchless construction industry have observed that a method such as pipe splitting can result in up to a fifteen percent reduction in construction cost after introducing the method to a given market. In the example described within this paper, this new trenchless technology was introduced in the field on the day of the project to the GRU construction crews. Even though each crew member had to obtain training "on the fly" from L. J. Ruffin and Associates in the implementation of the pipe splitting method, the trenchless approach was still more cost effective. The actual construction time for this project was recorded at just ten hours (including the training). This trenchless technology demonstration project generated considerable interest by attracting representatives from five surrounding municipalities and one local engineering firm. A greater than normal construction time for a project of this size was required in order to allow for inspection by the visitors and explanation of procedures being used. Two additional demonstration projects in Tampa and Miami using the pipe splitting method are being designed as a direct result of the GRU project.

This particular project demonstrated the advantages of a trenchless method when compared with traditional construction methods. Considering the identified construction advantages discussed above for the trenchless method, pipe splitting may also prove to be the preferred method even if the costs exceeded traditional methods. For this project the trenchless method reduced the required construction crew size, reduced construction time by more than fifty percent, reduced the amount of equipment required, and reduced post-construction "dress-up" activities required.

The trenchless method described herein provided a realizable economic advantage to GRU and created a "win-win" situation for both the customers GRU serves and to the utility. When construction was completed and the water line was placed back into service water pressure was dramatically improved and the water quality issues have been eliminated. By every measure the trenchless method using pipe splitting was a resounding success for this project. The possibility for utilizing trenchless construction methods for future GRU projects appears very realistic.

PRINCIPAL AUTHOR BIOGRAPHY

For the past 2 years, John S. Gifford, P.E., has been employed with GRU. He is a Senior Utility Engineer in GRU's Water/Wastewater (W/WW) Engineering Department and manages the W/WW Engineering Design group. He is responsible for directing the design and construction phases of GRU's Water/Wastewater capital improvement program related to pipelines, lift stations, and treatment plant project.

Mr. Gifford received a BSCE degree (structural emphasis) in 1984 and was employed with Baskerville-Donovan Engineers, Inc., in Pensacola, FL for approximately 1-1/2 years. In 1988 Mr. Gifford received a MSCE (environmental engineering emphasis) from Tennessee Technological University. Mr. Gifford subsequently joined CH2M HILL and became a State of Florida licensed professional engineer in1991. While employed for 8 years with CH2M HILL, Mr. Gifford was involved in the design of 14 Water Treatment Plant expansion projects and more than 9 municipal/industrial Wastewater Treatment Plant upgrade projects.

WASTEWATER FORCE MAINS PROBLEMS AND A SOLUTION

BY: ROGER M CIMBORA, SR.

ABSTRACT

Wastewater force mains are rarely considered cost effective candidates for internal rehabilitation. However, when they are cleaned and restored to their maximum flow capacity the results are often dramatic as pump run times and their corresponding energy consumption are significantly reduced.

As consequential decreases in operating costs can be expected from the rehabilitation of a force main this can also provide a very advantageous "pay back" period. Thoroughly and properly cleaning a force main can produce other benefits including the purging of harmful gas accumulations from the system, allowing applied chemical treatments to be effective, diminishing the effect of unexpected high septic loads at the treatment plant and relieving the operators and management of these systems of the ringing red light "sweats"!

The cleaning of a force main with the poly pig method is the procedure of choice by those experienced in wastewater piping rehabilitation. For one, and perhaps most importantly, the system is never off line or out of service while it is being cleaned. For most systems the wastewater it normally transports is used for propelling the cleaning poly pig through the system. Depending on line conditions this flow may be augmented particularly for force mains whose planned high volume use never materialized and whose piping has now evolved into an elongated settling basin.

INTRODUCTION

As engineers, designers and managers of wastewater collection and treatment facilities are becoming increasingly aware, the efficiency and effectiveness of the entire system is clearly dependent upon all of its components functioning as they are required and expected to do. Traditionally, wastewater systems can be divided into three segments, i.e., collection,

General Manager
Professional Piping Services, Inc.
Wesley Chapel, FL

delivery and treatment/disposal. Conventional collection piping design includes manholes strategically located for the necessary access into these systems for cleaning and rehabilitation purposes. Modern treatment plants in response to logistical, mechanical and regulatory requirements are now designed to readily comply with rehabilitative efforts on an as needed basis. However, the usual connecting link between the collection system and the treatment facility, the force mains, are typically given little or no regard for cleaning or rehabilitation. This may occur because of all the effort expended and difficulties encountered to simply find and design a route for the force main. In many applications it may have to cross under railroad tracks, canals, D.O.T right of ways, highways, or through parking lots, public and private easements, environmentally sensitive areas, or climb mountains, just to connect the lift station to its disposal point. Or it may be a reflection of the prevailing philosophy that the concept of attending to flow related problems in a force main starts with a shudder by the operators and managers of these systems.

THE PROBLEMS

In the past this reluctance to clean a force main may have been understandable. The apparent mechanical and logistical differences between gravity and force main systems would appear to present problems for force main cleaning that could make the costs exceed the benefits. Problems include a lack of strategically placed access points, a constantly flooded and pressurized system with little or no downtime available for taking it out of service, routing through and under non accessible areas, and the concerns that one misstep could result in a massive "Honey Wagon" bill or worse, create expensive causes for regulatory agencies to develop writers cramp!. These perceived difficulties may have led to the consensus of opinion that when the cleaning of a force main is considered, it might be better to, "live with it, than mess with it."

In common with the flow difficulties usually associated with the operation and maintenance of collection or gravity systems, force mains are also subject to a wider variety of unique problems. These include the following determinations which can cause a force main to malfunction or become "dirty"!

Cause– Typical wastewater flow has physical properties which will allow fluid borne material to adhere, cling or deposit on the interior pipe walls

Effect– Reduction in system laminar flow characteristics and eventual creation of serious and consequential impediments to full volume flow

Cause– Force main is not cleaned as part of the completion procedure to prepare a system to go "on line". Potential for construction oriented debris, bedding material, sand, etc., is now locked into the system.

Effect– Inline materials find a "home" in system low spots, valving and size change areas. Create flow impeding dams and serious flow restrictions.

Cause– The systems wet well or holding tank has a volume capacity, substantially less than that of its connecting force main. This factor creates an incremental discharge flow pattern in the force main resulting in no sustained scouring or purging discharge flow

Effect– Solids precipitate out of the flow when pump cycles to off position. Dropped solids are difficult to restore to suspension in intermittent flow. This factor contributes to the build up of flow impeding materials inside the pipe.

Cause– The diameter of the force main is predicated on high or future high volume capacity being necessary, which doesn't materialize. System never experiences full bore flow or scouring velocity.

Effect– With limited flow transported through an oversized main it is slowly converted into an elongated settling basin.

Cause— Treatment chemicals of varying functions are entered into a force main where they create unanticipated or unknown compounds inside the system.

Effect— Further puzzling flow reactions are created and can produce more impediments to laminar flow characteristics.

Cause— Unauthorized dumping of unapproved materials into the force main

Effect— See above

Cause— Force main changes diameter without a corresponding increase, at this site, in velocity and volume for the flow in the system.

Effect— Material in suspension tends to fall out of the flow, creating a damming effect within the system.

Cause— Force main does not sustain full volume flow for its entire length and thereby never can completely purge itself. This inability provides the opportunities for the development of air and other gas pockets.

Effect— Collected air will impede flow and other gases, notably, hydrogen sulphide, will evolve into pipe penetrating acids.

Cause— Maintenance of air release valves, vacuum breakers and other such appurtenances is not done properly or not at all.

Effect— See previous comment

Cause— Bio-films develop on interior pipe walls in layers, film or deposits, caused by lack of scouring velocity and purging ability in the system.

Effect— Rampant odor producing bacterial growths contribute to the development of poor flow movement in the piping.

Cause— Smaller force mains connecting to one larger force main as in a common manifold upstream of a treatment plant headworks quickly lose their flow motivation to keep solids in suspension.

Effect— Concentrated solids fall out of the carrying flow and begin to clog and collect at these sites and can cause backups in the system.

Cause— System profile allows gravity flow in that portion of the force main leading to its discharge point, air then periodically enters the piping.

Effect— Top and sides of piping develop carbonized and oxidized scale deposits which threaten pipes integrity and flow capacity.

Cause— System is dormant or little used for extended periods of time.

Effect— Inline material congeals, pipe attacking acids have a field day.

What all of these familiar and frequent conditions produce are force mains that can be best described as "DIRTY". A dirty force main may then be characterized by some or all of the following.

* Severe restriction of its designed flow capacity.
* "C" Factor drops below tolerable values.
* Major increases in pump(s) "run time"
* Corresponding significant increases in energy consumption, pump maintenance and replacement
* Erratic delivery of flow and septic load to treatment facility
* Loss or absence of designed system expansion capacity
* Increases in pump discharge head pressure
* Extended wet well evacuation time
* Operational staff develops a shuddering aversion to ringing red lights
* Potential threats to systems integrity and operational capacity materialize
* High level or standby pumps have as much run time as the primary pumps
* Upgrading the size and horsepower of the lift station pumps results in minimal system improvement

That wastewater force mains can become dirty and continue to be operated in this mode when this interior condition is discovered or acknowledged is fairly indisputable, particularly to those who daily work and cope with these systems. Most managers, once the wincing has stopped, usually consider a very short list of options of how to deal with a malfunctioning force main, i.e., change out the pumps, replace the pipe or live with it. Some systems certainly can benefit from a pump upgrade and general rehab of their lift stations. Replacing the pipe in some systems can mean eliminating the monotony of road patches and their covered pipe repairs and at long last having valving and other inline appurtenances that are operational. However, if an accurate analysis of the entire system indicates that the primary problem is the loss of carrying capacity in the force main, (an element of the analysis equation that can be overlooked or ignored), then replacing the pipe may mean throwing away perfectly serviceable conduits whose only sin is that they are dirty!

The inline conditions that can cause a force main to provide limited service and perform as a dirty system can be corrected by cleaning and restoring the system to its maximum flow capacity. The procedure to do this must be in compliance with the following criteria.

* The system must remain in or be available for service at all time during its cleaning
* The wastewater flow that the system normally transports must be used in the cleaning process to prevent backing up of fluids in the system
* The force main must be cleaned in its entirety, from lift station to discharge point, as one operation. This will eliminate potential environmental and public health concerns that can occur if the system has to be opened and entered in short sections.
* Cleaning through a multi dimensional system with inline valving and connecting force mains should pose no difficulties for the cleaning process
* The cleaning procedure must be able to effect the removal of all the adhering, accumulative and deposited material in the system without subjecting it to stresses and pressures that exceed its designed operating range.

* The means and methods used for initially cleaning the system should be easily duplicated by the utility staff. This will allow the force main to be maintained on an as needed basis or at its peak efficiency as a matter of routine maintenance.

A SOLUTION

One procedure that meets all of the necessary criteria for successfully cleaning a force main is the poly pig method of cleaning a piping system. Poly pigs are pipe cleaning tools specifically designed to effect the removal of solids, fluids, gases and adhering and accumulative deposits from a piping system. In liquid applications, they are independently propelled and motivated by the fluids normally transported through the piping. This process means that a poly pig passage through a system effects the dewatering of the entire system while the propelling flow is refilling the system. Therefore to push one poly pig through a piping system will require the volume of flow that the piping contains.

Properly selected and applied poly pigs will readily navigate through piping and their fittings, valving and diameter and material composition changes. At an inline velocity of 0.914 (M/S2), (3 FPS), a poly pig can enter and exit 8.05 KM, (5 miles) of piping in approximately two and half hours. These valuable features make poly pigs excellent pipe cleaning tools and for force mains, the unparalleled means and method of choice.

Cleaning a force main using the poly pig procedure requires only one line size entry port into the system. This is usually located between the lift station pumps and the downstream valve that prevents dewatering of the system. It can take the form of a pipe fitting prudently previously installed for rehabilitation purposes, or the emergency clean out connection for small diameter system's or if necessary, modifications of the existing piping configuration. This entry port is then used for temporarily connecting a poly pig launching assembly to the system. The launcher is used to insert into the force main, by auxiliary supplied pressure, the various poly pigs to be used for the cleaning, for the mounting of the pressure gauges and other devices for inline monitoring purposes and for potentially supplementing the flow into the system.

In virtually all applications the termination of the force main cleaning is where the force main ends, i.e., manhole, bar screen or wet well. This factor is most important for it eliminates any valid concerns about having to cope with potential problems that the disposal of fluids and solids might otherwise present.

A typical force main cleaning procedure may well fit into the following scenario. The operators of this system have detected various clues that it isn't working as well as it should or as it use to do. They have noticed that the wet well takes longer to evacuate. When observing the flow at the discharge manhole they now see a half pipe flow where it use to be full bore flow coming into it. The occasional opening of the vent or blow off valves results in long sustained venting of the system or they may have gotten a call from an observant person in the accounts payable department who is questioning why the electrical consumption for the lift stations pumps is getting higher every month.

Preparations would include;

a) A site and system inspection and evaluation. Getting out the as built drawings, (also known as the "as imagined" drawings for some systems), and checking the system.
b) Locating, identifying, marking and very importantly, operating the inline or connecting valves to confirm that they exist and will function properly when needed.
c) Notifying and conferring with the treatment plant staff so that changes in flow volume or the septic load won't catch them by surprise.
d) Setting up the mechanical requirements, i.e., finding or creating an entry port, making arrangements if necessary for capturing or collecting solids at the cleaning termination point, and finding an auxiliary fluid supply to supplement the systems flow if its normal operation will not provide the volume of flow needed for cleaning on a sustained basis. For example, 1.6 KM, (1 mile) of 150 MM (6") force main can hold over 29,148 litres, (7,700 gallons US). The pigging process will use this volume while propelling the pig through the system in thirty minutes. Instead of waiting for the collection system to provide an additional 29,148 litres, for the next

cleaning run, it is usually more productive to find and add this flow on an as needed basis. Of course, the first priority is to always use the system supplied fluids.

e) Determining the types, sizes and numbers of the poly pigs to be used. As with any first time exposure to a new procedure it is beneficial to use the knowledge and experience of others.

f) Additional preparations could include, making system related repairs, replacing a bad or suspect valve found during the site and system inspection, setting the work schedule for the staff, arranging for equipment to be on site, etc.

With the prepatory work properly done and completed, system cleaning can now begin.

The first poly pig is loaded into the launcher and inserted into the system. The system is now activated by turning on the pumps. The poly pig and the flow pushing it begin their journey to the retrieval site.

For the first cleaning pass through the system it is always advisable to use a low density and flexible poly pig to "prove" the system. This pig will confirm by its passage through the piping, velocity, pressure ranges and contents of the piping. It can then also confirm what the external inspection and evaluation had predetermined would be the internal condition and contents of the piping system. It also demonstrates that the cleaning pigs can traverse the system from point "A" the launch site, to point "B" the retrieval site, which is not always a given!

Subsequent runs through the system may be used to;

a) Further remove the deposited, adhering or accumulative material in it that has exceeded initial expectations, (not an unusual occurrence).

b) To abrade or scrape the entire periphery of the pipes interior.

c) To effectively remove everything that is in the pipe that shouldn't be in it until the conduit has been restored to its desired pristine interior condition.

Poly pigs are not plows, they don't push the solids out of the piping in one mass. What they are very capable of doing is putting solids into suspension and incorporating them into the flow. Therefore it is the hydraulic capacity of the system that effects the removal of what's in the piping and it is the system's flow capacity that carries it through and out of the system.

As the cleaning of the force main progresses, certain measurable effects can be noted.

a) Empirical evidence. Visually inspecting the discharging flow during the initial stages of cleaning a force main will usually detect vast amounts of sand, bedding materials, hunks and chunks of grease, solids of many descriptions, "surprising souvenirs" and enormous quantities of black, viscous and very septic fluid. All of which is common as the contents of a force main and all of which, once removed from it, will enhance its operational capability. As the cleaning, properly applied, progresses, the volume of solids and septic material removed should be significantly reduced and this decrease discernable to the eye. In most applications combining the reduced operating head pressure with the measurable increase in velocity and the observation, "That nothing more is coming out", usually proves sufficient to pragmatically judge a system to be clean.

b) Of course flow measurements are advisable to be done prior to and upon completion of the cleaning process for comparative purposes. There's nothing like a good before and after "C" factor to make everyone happy!

c) Another method for measuring flow and effiency in a force main is to compare the volume of fluid pumped and the electrical energy expended to do this. This can be done by a simple equation where volume of gallons pumped divided by the kilowatt hours of energy used equals gallons pumped per kilowatt hour. This method can create a historical record for comparison purposes, indicate the current system effectiveness and establish a base line for restoration and rehabilitation purposes. It can also imply the amount of money that can be saved if a system is restored to its maximum flow capacity and this data correlated for establishing the "pay back" period anticipated.

Force mains can be restored to or maintained at their maximum flow effiency by the simple process of cleaning them. This procedure will relieve the threats to their functioning of collected inline gases and of flow impeding deposits and material accumulations. It will sustain the costs of their service at designed levels and most importantly, provide the means to maintain, at will, full control of the systems operating characteristics. A dirty force main no longer has to be tolerated or "lived with!"

ABOUT THE AUTHOR:

Roger M Cimbora is the General Manager of Professional Piping Services, Inc., Wesley Chapel, Florida, a specialty company that has successfully cleaned over nine million feet of previously malfunctioning piping.

Mr. Cimbora has been employed in varying capacities in the piping industry for over forty years from pipe fitter to pipe pigger.

Mr. Cimbora is the author of many published articles on piping system rehabilitation including, contributing to The ASCE Publication - "Water Supply System Rehabilitation". When someone says pig to him its never Porky or Petunia, but always "POLY".

EVALUATING A PROPOSED RIGHT-OF-WAY FOR PCCP

Summary:

Soil resistivity measurements coupled with field tests including soil chemistry moisture content and soil composition are advisable to evaluate a proposed right-of-way for prestressed concrete cylinder pipe (PCCP) in arid or semi-arid areas such as the southwestern United States. Other field tests such as pH, total acidity and Redox can also be helpful in special cases. Cyclical wetting and drying and chlorides pose the most serious threat to rapid corrosion of PCCP. Avoidance of rights-of-way where cyclical wetting and drying can occur is desirable. A field examination and failure analysis of a large PCCP project is related that shows the interrelationships of mortar quality, soil chlorides and soil moisture, particle size and distribution characteristics which contributed to PCCP failures that occurred within four years of pipeline burial. The relationship of mortar thickness and quality variables such as absorption, specific gravity, pH, and total alkalinity is related to soil chlorides, the concentration of chlorides in the mortar, and the degree of corrosion noted on the prestressed wire in this project.

Soil Resistivity:

The most common method employed to determine if a right-of-way (ROW) is potentially corrosive to prestressed concrete cylinder pipe (PCCP) is a soil resistivity survey. This is always a good first step. As shown in Table I, a soil resistivity survey is often classified into resistivity categories of "severely corrosive", "corrosive", "moderately corrosive", and "mildly corrosive" soils. Sometimes overlooked is that these terms were borrowed from soil classifications pertaining to dielectrically-coated or bare ferrous pipe material such as steel or ductile iron. Without additional information, these resistivity ranges are not necessarily indicative of a propensity for corrosion on cementitious-coated pipe including PCCP. However, the soil resistivity range is a useful first step in determining what additional tests may be required the achieve the desired life expectancy of PCCP in the right-of-way under investigation.

Soil Resistivity (ohm-centimeters)			Soil Corrosivity
0	to	1000	Severely corrosive
1,000	to	2000	Corrosive
2,000	to	10,000	Moderately corrosive
Over		10,000	Mildly corrosive

Table 1
Soil Resistivity vs. Corrosivity (Steel)

A soil resistivity survey in a semi-arid area like Southern California should be conducted after a substantial rain.

Soil Moisture:

The moisture content of all soil samples should be measured, both in the as-received and saturated condition. The importance of this will be expanded on in the section on How to Prevent Cyclical Wetting and Drying.

Chlorides and Sulfates:

Chlorides are usually responsible for the most serious corrosion problems noted on PCCP. An analysis for chlorides should be conducted on any saturated soil sample with a resistivity less than 3,000 ohm cm. Sulfates are usually not a threat to totally buried PCCP, but wind can uncover a PCCP line in arid areas. Any soil with a resistivity of less than 3,000 ohm cm should be checked for sulfates.

Simplified Soil Analysis:

A small soil sample can be placed in a clear plastic container and agitated with distilled water. The relative percentages of clay-silt, sand, gravel, and clear or turbid water are measured after settlement. The significance of the ratios of sand to clay-silt will be discussed in the How to Prevent Cyclical Wetting and Drying section.

pH and Total Acidity:

Soils with either low pH or high in total acidity, especially where a fluctuating water table exists, can cause rapid carbonation of mortar coatings used on PCCP. Normally, it would be expedient to check for pH in any initial soil survey. If a pH less than 5.5 is found, then analyze for total acidity. Soils with high total acidity invariably contain organic acids. Total acidity is expressed as milliequivalents per

100 gm of soil. A soil with a very high total acidity may exceed 20 milliequivalents per 100 gm of soil.

Oxidation Reduction Potentials (Redox):

Oxidation reduction potentials may indicate a hostile environment for PCCP when the mortar coating has been physically damaged or chemically altered by carbonation. Redox potentials lower than about 200 millivolts indicate a reducing environment. A reducing environment is capable of supporting sulfate reducing bacteria (SRB). Hydrogen sulfide is a by-product of SRB metabolism. Hydrogen sulfide (H_2S) is a "poison" which, if it can reach the prestressed wire, will inhibit the formation of hydrogen gas (H_2), that might otherwise be harmlessly released. Poisons like H_2S sustain the formation of monomolecular hydrogen (H) which can result from a corrosion reaction or from cathodic protection.

Monomolecular hydrogen has a sufficiently small diameter that permits it to diffuse into the interstices of the steel lattice in prestressed wires. Hall[1] cites circumstances under which this monomolecular hydrogen may also be harmless by diffusing out of the prestress wire. Clift[2] has shown that prestress wire detrimentally strain-aged, e.g., from excessive drawing temperature can become hydrogen embrittled if the monomolecular hydrogen (H) combines to form molecular hydrogen (H_2) within a strain-aged lattice.

At least six PCCP failures have occurred from embrittled wire where excessively negative potentials from cathodic protection pre-existed. In at least three of these cases hydrogen sulfide was also known to be present. Therefore, H_2S could also have been a contributing factor to the embrittlement failure at imperfections in the mortar coating or where the mortar coating had been degraded by physical damage or carbonation.

There appears to be near unanimity on certain hydrogen-related failures on prestressed wire. For example, there appears to be general agreement that extremely high strength wire like the Type IV wire formerly used by Interpace and excessive drawing temperatures can promote hydrogen embrittlement in prestress wire. Since the mid-1980's, these practices no longer are permitted by Industry standards. However, a divergence of opinion remains as to what additional standards or guidelines are required to prevent hydrogen embrittlement in varying environments, and also whether the acceptable potential levels of cathodic protection should be altered for special environments or on prestress wire manufactured prior to the time when the standards and guidelines for prestress wire were improved.

AC and DC Stray Current:

To date, our examination of PCCP with induced AC currents, has not revealed any AC-induced corrosion. An examination of a large diameter PCCP pipe under a high voltage AC transmission line indicated maximum induced AC voltages of between two and three volts. Removal of the mortar over the prestressed wires at the points of highest-induced AC voltage revealed no corrosion of the wires. Several dielectric-coated steel pipeline failures from AC corrosion have been reported in Europe, particularly in Germany[3]. Dielectric-coated pipelines also experience high-induced voltages from overhead AC power lines. In extreme cases, over two hundred volts have been reported. While high AC voltages are primarily an electrocution hazard, rather than a corrosion problem, any source of induced-AC should be noted in the survey, both as a potential source for corrosion, and to avoid electrocution potentials to construction and maintenance personnel, as well as the public.

Cyclical Wetting and Drying:

Cyclical wetting and drying is the most commonly encountered mechanism responsible for rapid corrosion attack on PCCP. The author[4] first reported on this mechanism on PCCP in 1985. Cyclical wetting and drying is most prevalent in arid or semi-arid climates such as those that exist in the southwestern part of the United States. Cyclical wetting and drying degrades the mortar coating by carbonation and attacks the prestress wires both by concentrating the chlorides and introducing atmospheric oxygen into the mortar capillary system. According to Broomfield's remarks during the NACE Symposium Corrosion 98, San Diego, California, entitled "Corrosion and Corrosion Control of Reinforced Concrete Structures", the resulting attack forms a "front" that can initiate corrosion on steel embedments in mortar/concrete. The effect of carbonation reduces the pH of the mortar on PCCP. As the mortar pH is reduced, chlorides are released that have been chemically bound by hydrated cement constituents such as tricalcium aluminate (commonly referred to as C_3A). This combined effect of lowered pH and "released" chloride and atmospheric oxygen is the "front" that rapidly penetrates into the mortar capillary system of PCCP. According to Sagues[5], a relative humidity of 60% is ideal for rapid carbonation of mortar/concrete. At significantly higher humidities, moisture will block further intrusion of carbon dioxide. At significantly lower humidities, there is insufficient moisture for the carbonation action to proceed. Sagues[5] reports that carbonation-induced corrosion tends to develop later and proceeds at slower rates than chloride-induced corrosion.

Table 2 shows how the corrosion threshold on steel embedments in mortar is related to both the pH and chloride concentration in the presence of atmospheric oxygen, according to Hausmann[6]. Note that when only dissolved oxygen is a factor, a much higher chloride concentration is required for corrosion initiation.

pH	Chloride (Mg/Kg)
13.2	8,900
12.5	700
11.6	71

NOTE: Hausmann[6] also showed that with only dissolved oxygen present, the corrosion threshold at pH of 11.6 is 890 Mg/Kg chloride.

Table 2
Corrosion Threshold pH vs. Chloride Concentration in the Presence of Atmospheric Oxygen (Hausmann[6])

In our experience, the net result of the "front" created by cycles of wetting and drying, can increase the chloride concentration in the mortar capillary system by a factor of at least five fold. Accordingly, a soil with only 140 milligrams per kilogram (Mg/Kg) of chloride can result in a concentration to 700 Mg/Kg chloride in the mortar capillary adjacent to a prestress wire at a pH of 12.5. As can be seen from Table 2, this chloride/pH combination will result in corrosion initiation on the prestress wire when atmospheric oxygen is present. Once corrosion has initiated, there is a further stimulus to corrosion caused by the acidic corrosion products of iron, e.g., ferrous and ferric chloride which are a corrosive environment in and of themselves. Ferrous and ferric chlorides have a pH of four and two respectively which can neutralize the remaining alkalinity in the mortar. Loss of alkalinity will then permit initiation of corrosion on adjacent helices of prestress wire.

Fluctuating water tables and tidal conditions can also promote cyclical wetting and drying, and thereby accelerate the ingress of chlorides into the capillary system of PCCP via the "front" mechanism. Cyclical wetting and drying on PCCP was first identified by the author[4]. However, "ponding", which incorporates the same wetting and drying mechanism, has been used by investigators for many years as an accelerated method of evaluating corrosion of steel reinforcements in concrete[7].

How to Prevent Cyclical Wetting and Drying:

Cyclical wetting and drying is obviously eliminated if the environment is either constantly wet; as in a PCCP outfall in seawater; or is always dry, conceivably in some desert. However, selection of an optimum right-of-way in acceptable soils or selection of favorable bedding and burial materials, can also reduce or eliminate cyclical wetting and drying. A constantly moist clay environment will eliminate cyclical wetting and drying. Unfortunately, moist clay often will not support the required foundation loads of some PCCP projects.

Determination of permeable and impermeable soils like sand and clay can be identified by geotechnical consultants with soil boring logs and subsequent sieve analysis. Our simplified test for classifying soils was previously described. This simple test has two advantages over a sieve analysis. It is far less expensive. Also, it can be performed in the field with cores from a hand or power auger at pipeline depth. Uniform soils as for example all clay or all sand (in the absence of a fluctuating water table) eliminate the possibility of corrosion from cyclical wetting and drying. The most intense corrosion on prestress wires from cyclical wetting and drying generally occurs when the sand to clay-silt ratio is greater than about 2:1. Corrosion of prestress wires occurs under the clay layer and often initiates under the clay immediately adjacent to the layer of sand-clay interface.

A Case History:

Soil stratification into dense layers of clay in one layer, and permeable sand in an adjacent layer, in the presence of chlorides and cyclical wetting and drying, can lead to rapid failure of PCCP. An interesting example of this occurred on a large PCCP aqueduct project in Mexico. This project consisted of a prestressed concrete pipe which originated from a pumping station west of Mexicali, elevation sea level. Through a series of elevation changes, Colorado River water is pumped by six pumping stations up over a 4000' plateau with gravity flow back down to the city of Tijuana, Mexico. A corrosion failure on this 60 inch PCCP occurred near the eastern end at Km 4+209 in 1980 during hydrostatic testing of the line and prior to placing the line into service. The corrosion failure occurred about four years after the pipe had been installed. Our investigation determined that one major cause for the failure was the importation of bedding material for the pipeline which contained over 12,000 Mg/Kg of chloride mined from an ancient seabed. The corrosion failure that occurred resulted in a vacuum being formed approximately 100 feet upstream, which caused an implosion of another prestressed pipe section. Our examination of the prestressed wires at the second failure indicated that it was not corrosion-related. The subsequent vacuum shock wave responsible for the second PCCP failure also effectively destroyed several hundred meters of 54" diameter steel pipe another 100 meters upstream from the corrosion failure. The vacuum collapse was opined and documented by a hydraulics expert from the Bureau of Reclamation.

Parameters of Cyclical Wetting and Drying:

After identifying the causes of the failure at Km 4+209, we set about to locate additional areas of likely corrosion on the prestressed aqueduct. The PCCP was not bonded at the bell and spigot joints and, hence, not electrically continuous. This precluded a meaningful close interval pipe-to-soil potential survey. In the early 1980's, cell-to-cell surveys were not well developed, so initially soil resistivity was used to locate additional corroded PCCP sections. The resistivity surveys were performed by the Wenner Four Pin method by layers to pipe invert depth. Four 100 meter sections, with the lowest soil resistivity, were then excavated for a detailed visual and tactile examination of the PCCP. Extensive corrosion of the prestressed wire was noted on numerous pipe joints in three of the four 100 meter excavations. "Sounding" of the mortar for 270° of the pipe circumference was conducted. Mortar was then removed at "drummy" areas. In most cases, removal of drummy mortar revealed incipient to advanced corrosion on the prestress wires. In a few cases, the #6 prestress wire had already failed from corrosion. Soil samples were analyzed. High concentrations of both chlorides and sulfates were noted on these three 100 meter excavations. Typically, the soil layers near the crown of the pipe were predominately dry sand, whereas, moist sand and clay were encountered at deeper layers, generally below the springline. The clay layers retained the moisture which a consulting geologist indicated could raise water by capillary action for a distance of up to 40 feet. The moisture at the clay-sand interface was found to be only a few degrees warmer during the day, permitting a drying action. During the drying cycle, atmospheric oxygen (21% concentration) and carbon dioxide (0.03% concentration) entered the mortar capillary system. Moisture containing dissolved chlorides was then reintroduced from the clay layer to the clay-sand interface during the slightly lower temperatures at night. This explained the heavy prestress wire corrosion which occurred due to cyclical wetting and drying on three of the four 100 meter sections.

Of equal interest was the fourth 100 meter excavation where no corrosion of the prestress wire occurred. Surprisingly, this excavation had the lowest soil resistivity and correspondingly the highest concentration of chlorides and sulfates. Damp to wet clay was noted on the side wall of this excavation. The moist clay layer extended from at least two feet above the pipe crown to below the pipe invert. Since this clay layer continuously held the moisture, there was no drying cycle within the pipe burial zone during the day. Without cycles of drying, there was no corrosion of prestress wire on this 100 meter segment of PCCP.

Locating Underground Corrosion on PCCP:

This observation is valuable because it means that, while soil resistivity can sometimes be used to locate corrosion as it was in locating corrosion of prestress wires at three of the four 100 meter excavations, more accurate corrosion detection and quantification occurs when soil resistivity is used in conjunction with pipe-to-soil potential or cell-to-cell surveys. Currently, we combine these two measurements (soil resistivity and pipe-to-soil potential) to calculate an Apparent Corrosion Intensity Profile^R (ACIP^R). Using ACIP^R greatly improves the reliability of locating corrosion on PCCP.

Degradation of Mortar by Cyclical Wetting and Drying:

Tables 3 - 5 show the correlation between a number of mortar variables measured on the Mexican aqueduct including: compressive strength, thickness, density, absorption, pH, and total alkalinity versus chlorides and degree of corrosion on the prestress wires. The mortar samples were sectioned to provide an interior sample next to the prestress wire, and an exterior sample next to the soil environment. Chloride concentrations were analyzed on mortar samples, both interior and exterior. It should be noted that on all mortar samples, where heavy corrosion of the prestress wire was detected, the chloride concentration at the wire side of the mortar was substantially greater than on the soil side of the mortar. This observation is typical of advanced corrosion of steel embedments in mortar/concrete. The explanation is that the chloride anion (Cl^-) is electrochemically attracted to the anodic (corroding) area. Based on the manufacturing data available, we estimate that the original design mortar mix would have resulted in an alkalinity at the time of manufacture of about 28% $CaCO_3$ and a pH of about 12.6. Based on this pH and total alkalinity at the time of manufacture, Table 3 shows that the mortar samples located near the corrosion failure at Km 4+209 have been severely carbonated. Also note that the pH and total alkalinity data indicate a lowering of pH from the soil side to the wire side, with a corresponding loss of mortar alkalinity.

Sample No.	pH	Total Alkalinity % $CaCO_3$	Chlorides Mg/Kg	Mortar Location On Pipe	Wire Corrosion
			NEAR BREAK @ Km 4+209		
2I *	11.1	4.6	3910	Pipe Crown	Heavy
2E **	10.9	4.4	2350		
1I	11.1	7.0	3410	Springline	Heavy
1E	11.1	2.7	1957		
5I	11.9	6.5	3260	Below Springline	Heavy
5E	11.7	3.7	1180	(SEVERAL WIRES SEVERED)	
			@ Km 4+109		
3I	11.6	6.1	231	Pipe Crown	None
3E	11.5	3.2	340		
4I	11.7	7.0	162	Springline	Slight
4E	11.6	2.7	370		
			@ Km 0+000		
6I	11.9	16.2	649	Unknown	Moderate
6E	11.5	3.9	650	(PIPE REMOVED PRIOR TO INSPECTION)	

* I = INTERIOR MORTAR SAMPLE NEXT TO WIRE ** E = EXTERIOR MORTAR SAMPLE NEXT TO SOIL

**Table 3
Comparison of Wire Corrosion to Mortar pH, Total Alkalinity and Chlorides at Three Locations**

PIPELINES IN THE CONSTRUCTED ENVIRONMENT 353

Sample No.	Absorption %	Specific Gravity	Chlorides Mg/Kg	Mortar Location On Pipe	Wire Corrosion
			NEAR BREAK @ Km 4+209		
2I*	11.31	2.288	3910	Pipe Crown	Heavy
2E**			2350		
1I	11.47	2.283	3410	Springline	Heavy
1E			1957		
5I	13.65	2.238	3260	Below Springline	Heavy
5E			1180		(SEVERAL WIRES SEVERED)
			@ Km 4+109		
3I	13.28	2.231	231	Pipe Crown	None
3E			340		
4I	13.82	2.220	162	Springline	Slight
4E			370		
			@ Km 0+000		
6I			649	Unknown	Moderate
6E			650	(PIPE REMOVED PRIOR TO INSPECTION)	

* I = INTERIOR MORTAR SAMPLE NEXT TO WIRE ** E = EXTERIOR MORTAR SAMPLE NEXT TO SOIL

Table 4
Comparison of Wire Corrosion to Absorption, Specific Gravity and Chlorides at Three Locations

Sample No.	Compressive Strength MPa	Mortar Cover mm	Chlorides Mg/Kg	Mortar Location of Pipes	Wire Corrosion
			NEAR BREAK @ Km 4+209		
2I *	55.6	18.5	3910	Pipe Crown	Heavy
2E **	57.1		2350		
1I	51.5	15.5	3410	Springline	Heavy
1E	57.5		1957		
5I	59.8	14.7	3260	Below Springline	Heavy
5E	55.1		1180		(SEVERAL WIRES SEVERED)
			@ Km 4+109		
3I	51.0	14.2	231	Pipe Crown	None
3E	51.2		340		
4I	47.6	19.1	162	Springline	Slight
4E	45.8		370		
			@ Km 0+000		
6I		12.5	649	Unknown	Moderate
6E			650	(PIPE REMOVED PRIOR TO INSPECTION)	

* I = INTERIOR MORTAR SAMPLE NEXT TO WIRE ** E = EXTERIOR MORTAR SAMPLE NEXT TO SOIL

Table 5
Comparison of Wire Corrosion to Compressive Strength, Mortar Cover and Chlorides at Three Locations

The above observations support the "front" associated with cyclical wetting and drying previously described. Also note the inverse correlation between the mortar absorption and specific gravity with the highest absorption corresponding to the mortar with the lowest specific gravity (see Table 4). A minimum absorption value is sometimes specified by the owner as a quality control standard on PCCP. For the relatively short period of exposure (about 4 years), the mortar thickness (cover) was not as important a variable in preventing wire corrosion as might be expected as long as a minimum of about 14mm of mortar cover was maintained (see Table 5). There was no apparent correlation between mortar compressive strength and prevention of wire corrosion (see Table 5).

In Conclusion:

To avoid the worst environments for PCCP, parameters in a right-of-way survey should include soil resistivity, soil moisture, chlorides, and a soil classification analysis. A totally moist (or totally dry) environment is benign to PCCP. The worst environment most commonly encountered for PCCP is where many cycles of wetting and drying occur in soils where chlorides exceed 140 Mg/Kg, concentrate in the mortar capillary system, and initiate corrosion of the prestress wire. Carbonation caused by many cycles of wetting and drying can also initiate corrosion on prestress wire even without chlorides, but the rate of corrosion is much less in the absence of chlorides. Typically, such cyclical wet and dry environments are found in semi-arid areas, for example, in the southwestern United States. If the right-of-way is subject to numerous cycles of wetting and drying, then supplemental protection will be required on PCCP.

ACKNOWLEDGMENT

My appreciation is expressed to my long-time associate, Jerry Ott, who not only performed much of the field and laboratory work on the Mexican project, but also whose computer skills made this paper possible.

REFERENCES

1. Hall, Sylvia C.; *Cathodic Protection Criteria for Prestressed Concrete Pipe*, Page 637, Corrosion 98, San Diego, California.

2. Clift, James S., PCCP - *A Perspective on Performance*, 1991 ACE Proceedings Copyright©, AWWA.

3. Gummow, R. A.; et al, *AC Corrosion - A New Challenge to Pipeline Integrity*, Paper 566, Corrosion 98, San Diego, California.

4. Benedict, R. L.; *Corrosion Protective Properties of Concrete Cylinder Pipe;* AWWA Distribution System Symposium Proceedings 20196, Seattle, Washington, September 1985.

5. Sagues, A.A.; *Carbonation Induced Corrosion of Blended Cement Concrete Mix Designs for Highway Structures;* Paper 635, Corrosion 98, San Diego, California, March 1998.

6. Hausmann, D. A.; *"Steel Corrosion in Concrete, How Does It Occur?",* Materials Performance, November 1967.

7. Clear, K. S., FHWA-RD-76-70, April, 1976.

CORROSION CONTROL OF PRESTRESSED CONCRETE CYLINDER PIPE

Sylvia C. Hall, PE[1]

Abstract

Recommendations for corrosion control of prestressed concrete cylinder pipe (PCCP) lines are given to ensure that the passivating (corrosion-inhibiting) film is maintained, that the pipeline can be monitored, and that, if necessary, cathodic protection can be applied. Standard recommendations for corrosion control of PCCP are discussed first, followed by special recommendations for PCCP installed in unusually corrosive environments. These environments include acidic, high sulfate-containing, or high chloride-containing soils and waters, stray current interference, and connections to other pipelines. Other less commonly encountered environments include subaqueous installations, long-term atmospheric exposure, and transition from soil or water to air.

Introduction

PCCP is a rigid and durable pressure pipe designed to take optimum advantage of the compressive strength and corrosion-inhibiting property of portland cement concrete and the tensile strength of steel. Most PCCP produced in the United States and Canada is designed and manufactured in accordance with American Water Works Association Standard C301 and C304 (AWWA 1992). It includes a steel cylinder, a rigid concrete core, circumferentially wrapped prestressing wire, and a protective mortar coating. The components of an installed PCCP are identified in Figure 1.

PCCP was first installed in the United States in 1942 (AWWA C301 1992). From 1942 to 1995, 30,700 km (19,100 miles) of PCCP have been installed in North America (Clift 1991; Prosser 1996), principally for the transmission of water for municipal, industrial, and agricultural uses.

---

[1]Director, Ameron International Corporation, Engineering Development Center, 8627 South Atlantic Avenue, South Gate, CA 90280

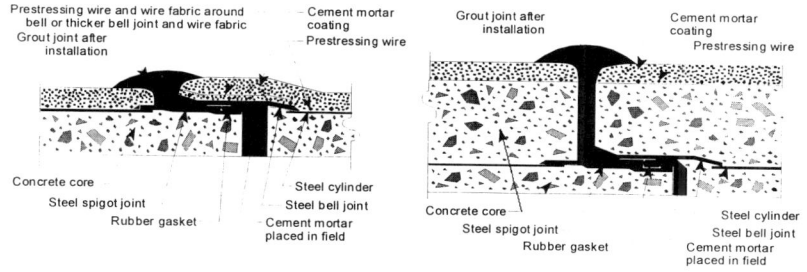

Figure 1. Components of installed PCCP

PCCP is remarkably resistant to physical damage and corrosion under a wide range of environmental conditions. This performance is due in large measure to the effects of circumferential prestressing which places the concrete core in compression. This compression induced in the core makes it possible to design PCCP to withstand the combined effects of internal pressure and external load without exceeding the tensile strength of the concrete core. Experience and extensive testing have shown that this design approach also ensures that the protective cement mortar coating will be visibly crack-free under operating conditions. In addition, corrosion of steel elements in the pipe is prevented by a passivating iron oxide film which quickly forms and is maintained in the highly alkaline concrete core and cement-mortar coating (Scott, 1965; Hausmann, 1964).

Unusual environmental conditions can be recognized and effectively dealt with before damage occurs. These conditions include sulfate or acid attack, high chloride-containing soils and waters, stray current electrolysis, and connections to other pipelines. Other less commonly encountered conditions include subaqueous installations, long-term atmospheric exposure, and transition from soil or water to air.

The recommendations that follow ensure that the passivating film on all steel elements is maintained, that the pipeline can be monitored, and that, if necessary, cathodic protection can be applied. Standard recommendations for corrosion control of PCCP are discussed first, followed by special recommendations for PCCP installed in unusually corrosive environments. The recommendations are summarized in Table 1.

Recommended Corrosion Control Provisions

The following standard corrosion control provisions are recommended as part of the manufacture and installation of all PCCP buried pipelines:
- ♦ Provide a steel shorting strap under the prestressing wire
- ♦ Apply cement slurry to the core during both prestressing and mortar coating operations

TABLE 1
RECOMMENDATIONS FOR PROTECTION OF PCCP
IN CORROSIVE ENVIRONMENTS

Condition	Criteria	Recommendations
Normal exposure	Non-corrosive environments	1, 2, 3 and 4
High sulfate soils/waters	Soils containing more than 0.2% $SO_4^=$ or waters containing more than 2000 ppm $SO_4^=$	1, 2, 3, 4 and 5
Acidic soils or waters	Soil pH less than 5.0	1, 2, 3, 4 and 6
High chloride soils/waters	Chloride content in soil or water greater than 350 ppm. As a guideline, soils with resistivities less than 1500 ohm·cm should be analyzed for chloride content.	1, 2, 3, 4 and 6
Stray-current interference	Prolonged discharge of direct current picked up from cathodic protection systems or other DC systems	1, 2, 3, 4 and 6 or 1, 2, 3, 4, and 7
Corrosive water	Water with a pH less than 5.5 conveyed or containing chemicals corrosive to concrete	1, 2, 3, 4 and 8
Subaqueous installations	Continuous immersion	1, 2, 3 and 4 or 1, 2, 4, 7 and 9
Atmospheric exposure	Continuous atmospheric exposure for more than 5 years	1*, 2, 10 and 11
Transition from buried to atmospheric exposure		1, 2, 3, 4 and 12
Connections to organically-coated steel pipelines	Buried pipelines	13

*Shorting strap and electrical continuity not required.

Recommendations:
1 = Provide a steel shorting strap under the prestressing wire; apply cement slurry to the core at time of prestressing and mortar coating; apply a 3/4-inch-thick (19 mm) mortar coating over the prestressing wire; make all steel components in the pipe electrically continuous.
2 = Fill interior joint recesses with cement mortar.
3 = Fill exterior joint recesses with cement mortar grout confined in polyethylene-foam-lined grout bands.
4 = Make all pipeline joints electrically continuous with low-resistance bonds; provide test stations to monitor pipe potentials and current flow.
5 = Use portland cement containing not more than 5 percent tricalcium aluminate for all concrete and cement mortar components.
6 = Seal the pipe exterior with a high-build barrier seal coating.
7 = Apply cathodic protection.
8 = Line the pipe with polyvinyl chloride sheet.
9 = Coat steel joint rings with a high-build barrier seal coating applied over an epoxy polyamide primer.
10 = Coat steel joint rings with a high-solids epoxy coating applied over an inorganic zinc primer.
11 = Seal the pipe exterior with an acrylic latex coating; recoat as necessary to maintain coating integrity.
12 = Seal the pipe exterior with a high-build barrier seal coating from 0.9 m (3 ft) below to 0.3 m (1 ft) above the ground surface.
13 = Electrically insulate the connecting pipelines.

- Apply a 19 mm (3/4-inch) thick mortar coating over the prestressing wire
- Electrically bond the joints of installed pipe
- Fill interior joint recesses with cement mortar
- Fill exterior joint recesses with cement mortar grout confined in polyethylene-foam-lined grout bands.

The first four provisions are incorporated during manufacture. The shorting strap is a thin steel band which electrically contacts each wrap of prestressing wire. Its purpose is to reduce the voltage drop in the long prestressing wire if cathodic protection is later applied. Making all steel elements in each pipe electrically continuous is a requirement for both corrosion monitoring and cathodic protection. It is accomplished by welding the prestress anchor assemblies to the steel joint rings during pipe manufacture.

Cement slurry having the consistency of thick cream is applied at time of prestressing and again at time of mortar coating to ensure complete encasement of the prestressing wire in a highly alkaline environment.

Bonding the steel joint rings makes the entire pipeline electrically continuous for monitoring and, if necessary, for cathodic protection. Suggested bonding details are shown in Figure 2 (AWWA 1995).

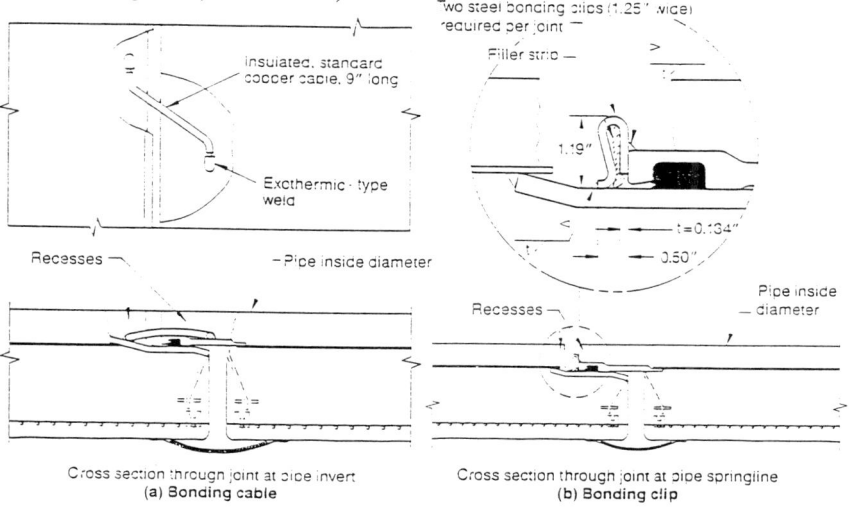

Figure 2. Suggested joint bonding details for PCCP.

Filling the interior and exterior joint recesses with cement mortar prevents corrosion of the joint rings. Cracks which may later occur in the interior joint mortar heal through deposition of calcium carbonate after the pipe is filled with water, a process often referred to as autogenous healing (Dhir et al 1973; Wagner 1974). Cracks in the exterior joint mortar generally heal by the same process, but additional protection against groundwater infiltration is provided by the polyethylene-foam-lined grout band shown in Figure 3.

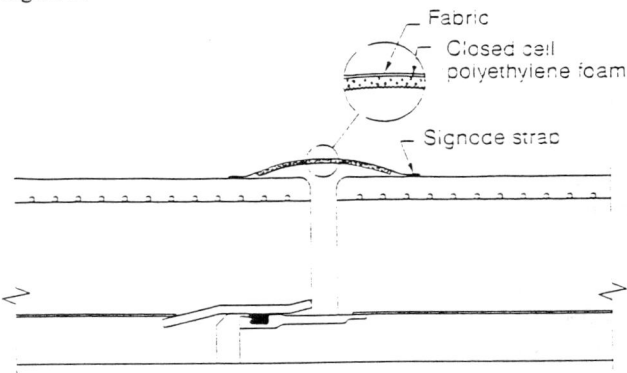

Figure 3. Joint grout band recommended for use on PCCP.

Unusual Environmental Conditions

Unusual environmental conditions can be recognized and effectively dealt with before damage occurs. These conditions include soils and waters which contain high sulfate or chloride contents or are acidic, stray current interference, and connections to other pipelines. Other less commonly encountered conditions include subaqueous installations, long-term atmospheric exposure, and transition from soil or water to air. Criteria and recommendations for these unusual conditions are given in Table 1 and are discussed in greater detail in the sections which follow.

High Sulfate-Containing Soils and Waters

Experience has shown that most soils and waters have little chemical effect on the concrete and cement mortar components of buried pipelines. However, soils or waters with high sulfate content are potentially damaging to the portland cement matrix. To increase resistance to sulfate attack, the use of a portland cement containing not more than 5 percent tricalcium aluminate is recommended for pipe installed in soil containing more than 0.2 percent water-soluble sulfate ion or in waters containing more than 2000 ppm sulfate ion (USBR 1975). While these recommendations have been applied for many years to all types of concrete structures, practically all known incidents of sulfate attack have involved the buildup of exceedingly high sulfate concentrations on partially buried or partially submerged structures. Buried or subaqueous concrete pipelines have proved to be unaffected by this type of attack. For example, sulfate attack does not occur on concrete pipelines submerged in the ocean even though seawater contains approximately 2700 ppm sulfate ion.

Acidic Soils and Waters

When exposed to soils of pH less than 5.0 (Am. Concrete Pipe Assoc. 1980), all exterior pipe surfaces should be sealed with a high-build barrier coating to prevent acid attack.

If water or liquid conveyed by the pipeline has a pH less than 5.5 or contains chemicals corrosive to concrete, the pipe interior should be protected with a polyvinyl chloride sheet having T-shaped longitudinal fins on one side which mechanically lock the sheet to the interior surface of the concrete core. After the pipe is installed, the liner sheets are connected at each joint by heat welding.

High Chloride-Containing Soils and Waters

The protective mechanism of portland cement will prevent corrosion of steel in concrete or cement mortar as long as:

- A highly alkaline environment (pH greater than 11.5) is maintained at steel surfaces, and
- Both chlorides (Cl^-) and oxygen (O_2) are prevented from reaching steel surfaces in quantities sufficient to initiate corrosion cells (Scott 1965; Hausmann 1964, 1967).

In most environments, the above conditions are satisfied for the life of the pipeline, and consequently, corrosion caused by chloride ions seldom occurs. Supplemental protection against corrosion in high chloride soils and waters can be provided by use of a high-build barrier coating applied over the mortar coating (Hall at al 1996). This provision is recommended in soils containing more than 350 ppm chloride ions (Cl^-). A more stringent limit of 150 ppm may be appropriate in soils subject to cyclical wetting and drying, an occurrence which accelerates the buildup of chlorides in the mortar coating.

High-chloride soils have low electrical resistivity. Since soil resistivity can be measured rapidly in the field, it is recommended that a soil resistivity survey be performed to identify soils of potentially high-chloride content along the pipeline right-of-way. As a guideline, soils of resistivity less than 1500 ohm cm should be analyzed for chloride content.

Stray-Current Interference

Buried metallic pipelines may collect and discharge stray currents which originate from nearby cathodically protected pipelines, electric railways, and other sources of direct current. The discharge of such currents from organically coated steel pipelines occurs at pinholes or other flaws in the coating and results in steel pitting at these locations. The effects of current discharge are quite different for steel encased in concrete or cement mortar. In this case, current discharge is opposed by polarization effects, occurs over large surface areas, and initially consumes alkalinity at steel surfaces rather than the steel itself (Hausmann 1964; Scott 1965; Hall et al 1996).

Current discharge from PCCP can be tolerated for a long time before the protective alkalinity is consumed and steel corrosion occurs (Hall et al 1996). Nevertheless, it is recommended that current discharge be avoided or controlled when recognized. Coating the pipe exterior with a high-build coal tar epoxy greatly reduces both current pickup and discharge and is recommended on new pipelines when interference is anticipated. The length of pipeline to be coated should be determined by a pipeline corrosion engineer. PCCP pipelines crossing cathodically protected steel pipelines should be coated for some distance on both sides of the crossing. Distances of 60 to 120 m (200 to 400 feet) are typically specified. Damage to an existing PCCP pipeline due to an interfering cathodic protection system can, in some cases, be prevented by relocating or adjusting the interfering system. Alternatively, cathodic protection can be applied to the PCCP line.

Connections to Other Pipelines

The potential of steel in concrete normally lies between -50 mV and -250 mV (CSE) which is approximately 300 mV more noble than the potential of bare steel. Consequently, if an uninsulated connection is made between a PCCP line and an organically coated steel pipeline, the steel pipeline may corrode sacrificially, protecting the PCCP line. The problem can be avoided by insulating the connection between the two dissimilar piping materials or by mortar coating the steel pipeline so that the steel in both pipelines is at the same potential.

Subaqueous Installations

In subaqueous pipelines, the preferred method of joint protection is cement mortar inside and out. Where this method of protection is impractical, it may be possible to mortar interior joint recesses and cathodically protect exterior steel joint ring surfaces. To reduce the current required for cathodic protection, it is recommended that the joint rings be coated with a high-build coating applied over an epoxy polyamide primer.

For low-pressure pipelines, typical of many seawater cooling systems, prestressed concrete pipe can be produced without an embedded steel cylinder and with an all-concrete joint which does not require supplemental protection.

Atmospheric Exposure

Application of an acrylic latex seal coat to the pipe exterior is recommended for PCCP continuously exposed to the atmosphere for more than 5 years. In this case, the function of the seal coat is to reduce drying shrinkage in the mortar coating and to prevent loss of coating alkalinity due to reaction between calcium hydroxide in the mortar and carbon dioxide in the atmosphere, a process known as carbonation. Recoating may be necessary every 10 to 15 years to maintain coating effectiveness.

Transition from Soil to Air

PCCP lines extending from below ground to above ground should be protected with an exterior seal coat in the transition zone to prevent the accumulation of soil chemicals by wicking action in mortar coating. Application of a high-build barrier coating is recommended from 0.9 m (3 feet) below to 0.3 m (1 foot) above the ground surface.

Monitoring

Buried pipelines should be monitored periodically to determine pipeline potentials, current flow, and effectiveness of insulated connections. Measurements of pipeline potentials is especially significant because of the unique potential of steel in portland cement environments. As noted earlier, the potential of steel in concrete or cement mortar normally lies between -50 and -250 mV versus a copper-copper sulfate reference electrode (CSE). Pipeline potentials in this range usually indicate that exterior steel surfaces in the pipeline are passivated and protected from corrosion. However, a sudden shift in potential from the apparent baseline of the survey, although still within the -50 to -250 mV range, can also indicate that corrosion is occurring (Hall 1994). Potentials more positive than approximately +50 mV (CSE) are usually indicative of direct current discharge from pipe to soil.

Potentials more negative than approximately -300 mV (CSE) indicate either that steel corrosion is occurring somewhere on the pipeline or that the pipeline is collecting current from a cathodic protection system or some other current source. The occurrence of cathodic interference can be verified by determining the magnitude and direction of current flow along the pipeline. Identification of corrosion activity may require visual inspection of the pipeline at selected locations, supplemented by pipe potential measurements made directly against the mortar coating.

To utilize monitoring procedures effectively, pipe joints in the installed pipeline should be electrically bonded and connections for test leads should be provided at convenient intervals, such as every 300 m (1000 feet) along the pipeline (Carlson 1995; Hall 1994).

Cathodic Protection

Cathodic protection of PCCP pipelines is seldom required and should not be applied indiscriminately. However, the following circumstances may justify the use of cathodic protection:

- *Stray current electrolysis.* Cathodic protection is required only at sites of current discharge and only at levels sufficient to counter the discharge.
- *Damaged pipe.* The option of repairing a damaged or defective pipeline should be carefully considered; cathodic protection, if used, is required only at damaged areas.
- *Bare or organically coated steel.* Corrosion of uninsulated steel appurtenances or pipelines can be prevented by local application of cathodic protection.

Cathodic protection of limited areas is often obtained most economically with galvanic anodes. Impressed current cathodic protection may be suitable if more extensive coverage is required.

There are no universally accepted criteria for cathodic protection of steel in concrete. A commonly used criterion for cathodic protection of bare or organically coated steel is a steel potential of -850 mV (CSE) (RP0169-96 1996). This criterion can be applied to PCCP lines where bare steel is exposed or is no longer passivated in an alkaline environment. Two other criteria for cathodic protection are a minimum negative (cathodic) polarization shift or decay of 100 mV (RP0169-96 1996; RP0290-90 1990). These criteria have been used successfully in the protection of concrete pressure pipelines (Benedict 1989; Hall & Mathew 1995; Hall 1998). The maximum interrupted-current potential should not exceed -1000 mV (CSE) on PCCP to avoid evolution of hydrogen (Scott 1965; Hausmann 1964) and possible embrittlement of the prestressing wire (Hall & Mathew 1995; Hall et al 1996; Hall 1998).

Current density requirements of operating cathodic protection systems of unsealed PCCP lines have been shown to range from 100 to 1000 microampere/m^2 (10 to 100 microampere/ft^2) to achieve the 100 mV minimum shift and -1000 mV (CSE) maximum criteria (Hall 1998). Current density requirements on PCCP lines sealed with a barrier coating have been shown to be approximately 3 to 4 times less than that of unsealed lines (Hall et al 1994; Hall 1998).

Planning

Proper planning early in the design of a PCCP pipeline can prevent corrosion problems. Planning should include a survey to determine soil resistivity, pH, and chloride and sulfate contents along the right-of-way. Locations of parallel and crossing pipelines, cathodic protection and other DC sources, and connections to other pipelines should also be determined. A qualified corrosion engineer should be retained to interpret the information and to recommend corrosion control provisions.

References

Benedict, R. L. (1989). "Corrosion Protection of Concrete Cylinder Pipe." *CORROSION/89*, paper no. 368, NACE Int'l, Houston, TX.

Carlson, E. J., Stringfellow, R. G., and Hall, S. C. (1995). "Finite Element Modeling of Ground Level Potential Measurements of Galvanic Cells on Concrete Pipe." *Techniques to Assess the Corrosion Activity of Steel Reinforced Concrete Structures, ASTM STP 1276*," ASTM, Philadelphia, PA.

Clift, J. S. (1991). "PCCP - A Perspective on Performance." *1991 ACE Proceedings*, Denver, CO.

Concrete Manual. (1975). U. S. Dept of the Interior, USBR, eighth ed., 11.

Concrete Pipe Handbook. (1980). American Concrete Pipe Assoc., Vienna, Virginia, 6-13.

Concrete Pressure Pipe - AWWA M9. (1995). Am. Water Works Assoc., Denver, CO.

Dhir, R. K., Sangha, C. M., and Munday, J. G. L. (1973). "Strength and Deformation Properties of Autogenously Healed Mortars." *ACI Journal*, March, 231.

Hall, S. C. (1994). "Analysis of Monitoring Techniques for Prestressed Concrete Cylinder Pipe." *CORROSION/94*, paper no.510, NACE Int'l, Houston, TX.

Hall, S. C. (1998). "Cathodic Protection Criteria of Prestressed Concrete Pipe - An Update." *CORROSION/98*, paper no. 637, NACE Int'l, Houston, TX.

Hall, S. C., Carlson, E. J., and Stringfellow, R. G. (1994). "Cathodic Protection Studies on Coal Tar Epoxy-Coated Concrete Pressure Pipe." *Materials Performance* 33 (10), 29.

Hall, S. C., and Mathew, I. (1995). "Cathodic Protection Requirements of Prestressed Concrete Cylinder Pipe." *Second International Conference - Advances in Underground Pipeline Engineering*, ASCE, New York, 168.

Hall, S. C., Mathew, I., and Sheng, Q. (1996). "Prestressed Concrete Pipe Corrosion Research: A Summary of a Decade of Activities." *CORROSION/96*, paper no. 330, NACE Int'l, Houston, TX.

Hausmann, D. A. (1964). "Electrochemical Behavior of Steel in Concrete." *J. American Concrete Institute, Proceedings*, 61 (2), 171.

Hausmann, D. A.(1967)."Steel Corrosion in Concrete."*Materials Protection*, 6 (11), 19.

Prosser, D. P. (1996). "Research, Product Improvement, New AWWA Standards, Compliance Certification & Performance Contribute to the Increased Usage of Prestressed Concrete Cylinder Pipe." *1996 ACE Proceedings*, Denver, CO.

"RP0169-96 - Control of External Corrosion on Underground or Submerged Metallic Piping Systems." (1996). NACE Int'l, Houston, TX.

"RP0290-90 - Cathodic Protection of Reinforcing Steel in Atmospherically Exposed Concrete Structures." (1990). NACE Int'l, Houston, TX.

Scott, G. N. (1965). "Corrosion Protection Properties of Portland Cement." *J. American Water Works Assoc.*, 57 (8), 1038.

"Standard for Prestressed Concrete Pressure Pipe, Steel-Cylinder Type, for Water and Other Liquids, C301-92." (1992). Am. Water Works Assoc., Denver, Colorado, vi.

"Standard for Design of Prestressed Concrete Cylinder Pipe - C304-92." (1992). Am. Water Works Assoc., Denver, CO.

Wagner, E. F. (1974). "Autogenous Healing of Cracks in Cement-Mortar Linings of Gray-Iron and Ductile-Iron Water Pipe." *Journal AWWA*, June 1974, 358.

EXTENDING THE LIFE OF PRESTRESSED CONCRETE CYLINDER PIPE WITH PULSE CATHODIC PROTECTION

Ted Doniguian, PE; Harry Kipps, PE; John Barnes, CET,CCT
Robert Bein, William Frost & Associates
27555 Ynez Road, Suite 400, Temecula, CA 92591

Introduction

Large diameter Prestressed Concrete Cylinder Pipelines (PCCP) are not usually prone to failure (1). Failures may occur, however if the concrete protecting the steel reinforcing members, or the metal cylinder, has been compromised. Once the protective concrete begins deteriorating, the steel components encased within the concrete, become susceptible to corrosion and embrittlement failure. Because of their complex construction PCCP pipelines are very difficult to protect with conventional cathodic protection. Pulse Cathodic Protection has been used since 1969 to successfully extend the life of many miles of PCCP pipelines.

Corrosion and Embrittlement

Pipelines buried in soils are exposed to groundwater. Dissolved oxygen molecules and/or hydrogen ions present in the groundwater will absorb onto the surfaces of the steel members of the pipelines. As depicted in Figure 1., iron within the steel member will corrode when acceptors are found for the two iron electrons. The absorbed oxygen molecules and/or hydrogen ions readily accept the iron electrons, thus allowing the iron to corrode. If not arrested, corrosion of the prestressing wires and/or the cylinder, will lead to the eventual failure of the pipeline (2).

The corrosion process also produces ample quantities of hydrogen atoms. Most of the hydrogen atoms combine to form molecules and dissolve into the groundwater. Some hydrogen atoms produced by the corrosion process, however, will dissolve into the steel members. If the prestressing wire is susceptible to hydrogen embrittlement, the pipeline may fail simultaneously due to corrosion and embrittlement (3).

Conventional Cathodic Protection

With Conventional Cathodic Protection (conventional CP), steady state direct electrical current (DC) is caused to flow continuously onto the outer surfaces of the structure. Typically, the negative lead from a DC power supply (called a rectifier) is connected to the corroding structure, and the positive lead is connected to an anode bed. The anode bed usually consists of one or more rods, (called anodes), fabricated from expendable materials such as graphite or silicon iron. The anodes are buried in deep wells, or shallow holes, or trenches, or are submerged, in the vicinity of the structure to be protected.

The DC current reduces the oxygen molecules and/or hydrogen ions continuously as they become absorbed on the steel surfaces. Once reduced, they are not available to accept the iron electrons, thus the iron does not corrode.

Conventional CP Drawbacks

There are significant drawbacks to the use of conventional CP, however. Electric current flows in the soil continuously, in a relatively large three dimensional envelope. The current envelope surrounds the anode bed and the structure being protected. If there are other metallic structures located within the current envelope, that are not electrically bonded to the pipeline being protected, the foreign structures will very likely incur metal loss and be damaged. If the prestressing wire of the PCCP pipeline is not electrically connected to the cylinder, the prestressing wire will act much like a foreign structure and be damaged as well.

Also, during earth movement, or if the mortar coating at the pipeline joints deteriorates, the bonds, that connect the pipe joints together electrically, may become disconnected or may exhibit high electrical resistance. Bonding failures, on joints protected with conventional CP, are susceptible to metal loss at the joints. Another drawback to conventional CP is that, the "throw", or distance along the pipeline, that can be protected, is limited by the magnitude of the current reaching the more distant corroding surfaces.

Pulse Cathodic Protection

With the development of the Pulse Cathodic Protection (Pulse CP) (4, 5, 6, 7) the conventional CP drawbacks have been significantly overcome. The Pulse CP rectifier applies very short duration, but very high voltage DC pulses between the anode bed and the buried structure. The very high voltage pulses result in very high peak currents, which instantaneously reduce the oxygen molecule and/or the hydrogen ions absorbed on the metal surfaces. The very short "on" pulses are separated by relatively long "off" intervals. The pulse voltage is applied typically for less than 10% of the total time. Thus, the continuous conventional CP current envelope and the resulting interference on foreign structures and across pipeline joints is greatly reduced.

Since "throw" is a function of current magnitude, the very high pulsed currents result in a more complete and greatly improved "throw" throughout, and down the length, of a complex structure. Also, during the short "on time" of the high voltage pulse, a significant voltage develops across the structure-to-soil interface. Immediately after the high voltage pulse is turned off, the interface voltage discharges, resulting in the redistribution of protective currents to points of lower voltage on the structure (7). Thus, the greatly improved high current "throw" is further enhanced by the more uniform distribution of the protective currents during the "off" periods when no voltage is applied.

Application

Several large diameter transmission mains located in Orange County, California are depicted in Figure 2. The Tri-Cities Transmission Main is a major water source for Southern Orange County. It extends approximately 22 miles from central Orange County to the San Onofre Nuclear Power Generating Station, located in San Diego County. Approximately nine miles of the Northern reach is constructed of 60-inch diameter PCCP joints.

Originally constructed in 1961, the northern reach experienced two corrosion failures in 1964. Three conventional cathodic protection stations installed in 1967 exhibited only limited protection. The conventional stations were subsequently replaced with three Pulse CP stations in 1972. Eighteen additional Pulse CP stations have since been installed over the years. No corrosion failures have occurred to date since the pipeline has been protected with Pulse Cathodic Protection.

The Laguna Canyon Transmission Main, also depicted on Figure 2, is a 42-inch diameter Concrete Cylinder Pipe (CCP) pipeline constructed in 1972. Reach One was constructed alongside, and parallels the Tri-Cities PCCP pipeline, for a distance of approximately six miles. The two pipelines are currently bonded together electrically at fourteen locations, and protected jointly with five Pulse CP stations. Reach two of the Laguna pipeline, which follows a southwesterly direction and terminates in the City of Laguna Beach, is protected with three additional Pulse CP stations.

The Pipe

The Tri-Cities PCCP and the Laguna CCP joints are depicted schematically in Figure 3. Both of these types of pipeline joints have been manufactured in the United States since 1942 and primarily used for transmission main applications requiring large diameter pipe (8), (9).

The CCP joint consists of a steel cylinder with a hot-rolled steel bar wrapped spirally directly onto the cylinder, using a moderate tension in the bar. The bar wrap is welded to the steel cylinder at each end. The cylinder and the bar wrap are then coated with cement. The bar wrap is also in physical contact with the cylinder throughout its length, thus providing for electrical continuity between the bar wrap and the cylinder.

The PCCP joint consists of a steel cylinder embedded in concrete. After the concrete was cured, the pipe was spirally wrapped with a hard-drawn Class I wire, with a yield strength of 105K-psi, using a stress of 75% of the minimum specified tensile strength. The wrapped wire was then coated with an outer layer of cement. The wire is anchored in the outer concrete core at each end of the joint. This method of anchoring the wire does not provide for electrical continuity between the prestressing wire and the cylinder.

It is suspected that the lack of electrical continuity between the prestressing wire and the cylinder has contributed to the recent failures of similar PCCP systems that were subjected to conventional CP systems. In order to mitigate this problem, more recently manufactured PCCP joints include longitudinal shorting bars that are electrically connected to the cylinder at both ends, and to the wrapped wire at regular intervals.

The ever present interference and limited throw problems associated with conventional CP systems make it very difficult to protect CCP and PCCP pipelines. Metal loss, at marginally conducting or electrically open joint bonds, is a continuing and ever present problem with jointed pipelines. Although a majority of the multiple steel members may be receiving adequate current to prevent corrosion, some steel members may be receiving too little current, whereas other members may be overprotected.

If the steel members protected with conventional CP are of prestressed construction and are susceptible to embrittlement, it is very possible for some parts of the pipeline to be experiencing deterioration due to corrosion, and other parts of the same pipeline to be failing simultaneously due to embrittlement.

The greatly reduced possibility of unbonded interference, and the greater, and more uniform distribution of currents provided by the Pulse CP systems, is essential to protecting the complex design of the CCP and PCCP jointed pipelines.

Data

Pipe-To-Soil potentials recorded over the years for the five mile portion of Reach One of the Tri-Cities PCCP pipeline are depicted on Figure 4. The five mile portion is protected by four Pulse CP rectifiers located at stations TC-03★1, TC-06★2, TC-

10★3, and TC-12★4. The potential readings indicate that the pipeline continues to be more than adequately protected by the four Pulse CP systems.

Pipe Relocation and Inspection

In order to accommodate the construction of a hospital, approximately one mile long sections of the parallel portions of the Tri-Cities and Laguna pipelines were excavated for realignment in 1994. Samples from the sections are depicted in these photographs. The realignment provided a fortuitus opportunity to carefully examine the effects of Pulse Cathodic Protection on large diameter CCP and PCCP water transmission mains. Random sections of the removed joints of both pipelines were selected for inspection and subjected to several tests. Sample coupons of the pipe wall were obtained by core sawing. The retrieved coupons exposed a cross section of pipe wall allowing access to the concrete matrix, the steel cylinder, and the spirally wrapped reinforcing wire.

PCCP Sample Coupon

After having been protected with Pulse CP for periods exceeding two and three decades respectively, all of the excavated parts of the CCP Laguna and PCCP Tri-Cities pipelines, including the prestressed steel members and cylinder samples, were found to be in excellent condition.

In order to plan for the eventual replacement of the Tri-Cities pipeline, the pipeline was de-watered and inspected internally in March 1997. After 35 years of service, the prestressed pipeline was determined to be serviceable for many years to come. The anticipated date for the replacement of this pipeline has been extended indefinitely.

CCP Reinforcing Wire and Cylinder

CORROSION REACTIONS

$$Fe^{o} \dashrightarrow Fe^{++} + 2e \qquad HOH \underset{\dashleftarrow}{\dashrightarrow} H^{+} + OH^{-} \qquad 2H^{+} + 2e \dashrightarrow H_2$$

$$2HOH + O_2 + 4e \dashrightarrow 4OH^{-}$$

CONVENTIONAL CATHODIC PROTECTION

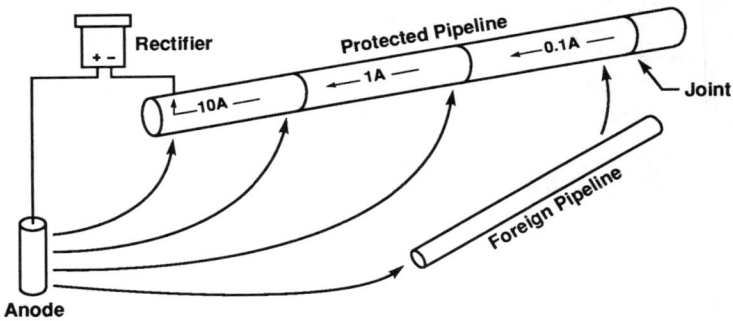

PULSE CATHODIC PROTECTION

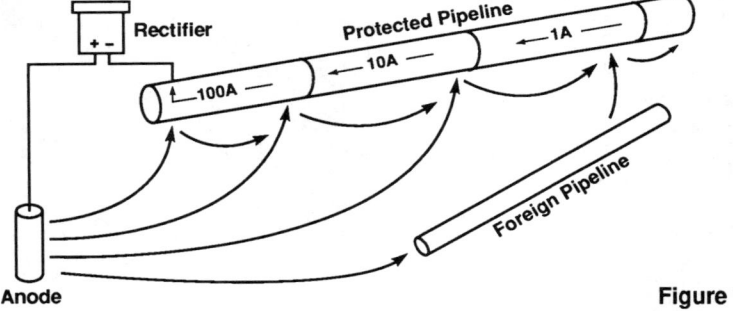

Figure 1

PIPELINES IN THE CONSTRUCTED ENVIRONMENT

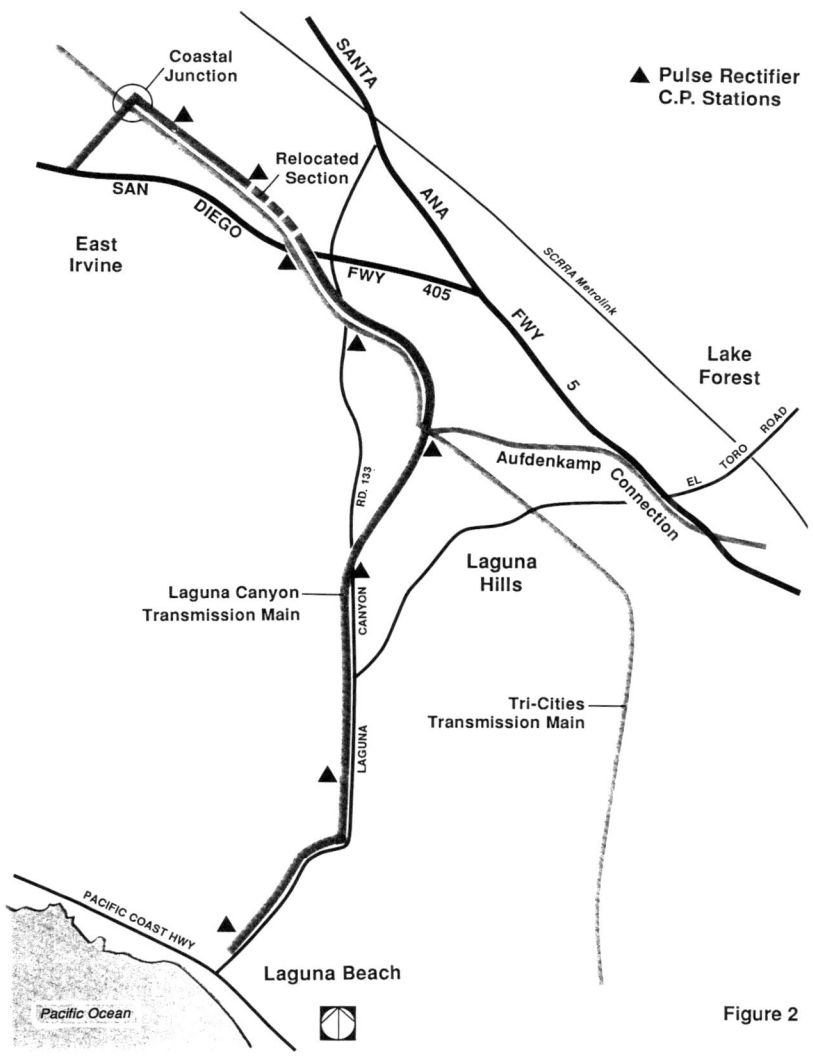

Figure 2

Concrete Pressure Pipe

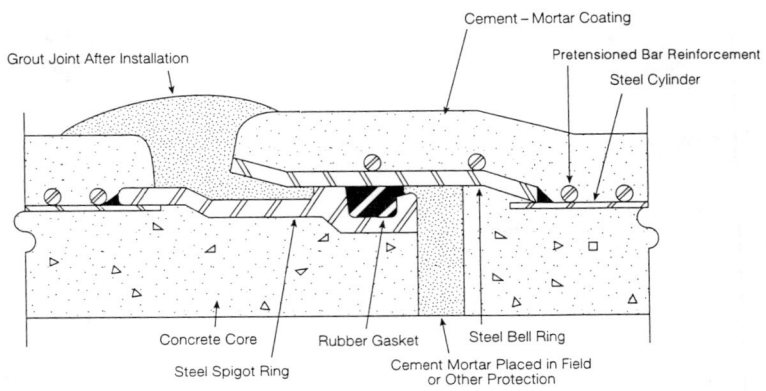

Concrete Cylinder Pipe

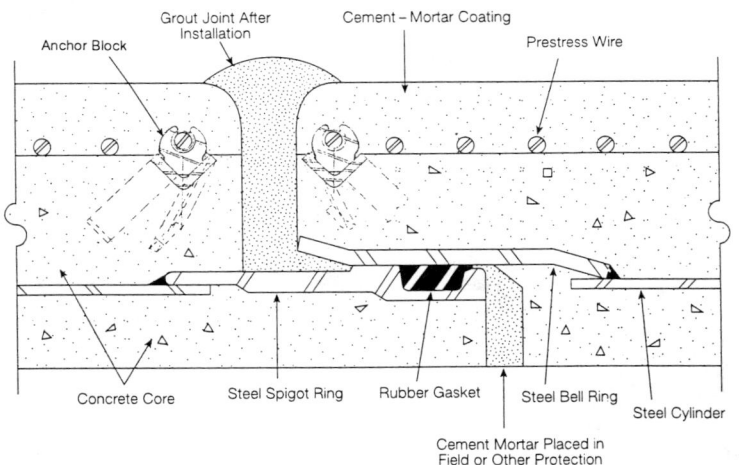

Prestressed Concrete Cylinder Pipe

Figure 3

Tri-Cities Water Transmission Main - Reach one
Test Station Readings (CU/CUS04 to Cylinder)

STATION	SURVEY	1975	22SEP79	10FEB92	10JAN94	18JAN95
TC-01	72+28	295	420	460	519	730
TC-02	72+69	420	420	440	692	MISSING
TC-03*1	99+69	1400	1050	1533	1480	2179
TC-04	128+95	460	860	1004	1000	1116
TC-05	153+80	975	970	MISSING	MISSING	MISSING
TC-06*2	155+70	1450	1280	1558	1390	1496
TC-07	157+10	220	1090	1040	980	771
TC-08	170+45	120	410	810	MISSING	737
TC-09	184+66	130	360	1225	1000	1058
TC-10*3	191+90	320	355	1355	1350	2504
TC-11	221+50	320	365	660	850	1006
TC-12*4	246+75	1450	1240	1440	1455	1605
TC-13	249+00	805	865	1080	MISSING	800

Figure 4

Bibliography

1. Scott, G. N., "Corrosion Protection Properties of Portland Cement Concrete", J. Am. Water Works Assoc., Vol. 52, No. 8(1965), pp. 1038-1052.

2. Hausmann, D. A., "Steel Corrosion in Concrete", Materials Protection, Vol. 6, No. 11 (1967), pp. 19-23.

3. Assefpour-Dezfuly, M., "Hydrogen Atom Ingress in Ferrous Metals", Australian Corrosion Association, Inc. Conference 24, Rotorua, NZ., 1984.

4. Doniguian, T. M., "Pulse Rectifier Improves Cathodic Protection", Oil and Gas Journal, 26 July 1982, pp. 221-229.

5. Doniguian, T. M., "Pulsed Cathodic Protection Apparatus and Method", U.S. Patent No. 3,612,898, 12 October 1971.

6. Kipps, H. J., "Cathodic Protection Systems", U.S. Patent No. 3,692,650, 19 September 1972.

7. Doniguian, T. M., "Pulse Cathodic Protection System", U.S. Patent No. 5,324,405, 28 June 1994.

8. Ameron, Concrete Cylinder Pipe, Design Manual 303, Ameron, Monterey Park, Calif., 1st Edition, 1988.

9. Ameron, Concrete Cylinder Pipe, Design Manual 301, Ameron, Monterey Park, Calif., 2nd Edition, 1987.

EXTERNAL CORROSION CONTROL OF WATER MAINS TO MAXIMIZE OPERATING LIFE

Bryan M. Bradish, P.E.[1]
Michael J. Szeliga, P.E.[2]

Abstract

Water pipelines in the Eastern part of The United States have been installed with the notion that they would last 100 years, with little concern for the environment that they were installed in and limited concern for the pipe material. Many of the older systems in the East have pipelines that prove this notion. However, they also have pipelines that have had to be replaced in much less than 100 years which disproves the notion .

At this time, many pipelines have reached the end of their useful life and it has generated interest in the industry. According to AWWA, water distribution systems are being replaced on an average rate of only one half of one percent per year. Case studies of breaks reported by a 1989 ASCE Task Committee study, show that pipeline break rates are 5 to 25 per 100 miles of mains per year. The AWWA Research Foundation has awarded two contracts recently to help the industry to address this problem.

Newport News Waterworks has attempted to keep abreast of the industry's best methods for identifying necessary pipeline replacements. They have also tried to

---

[1]Chief Engineer, Newport News Waterworks, 2600 Washington Avenue, 5th Floor, Newport News, VA 23607

[2]Chief Engineer, Russell Corrosion Consultants, Inc., P. O. Box 197, Simpsonville, MD 21150

learn from their past experience of pipeline performance and failures.

This paper covers the methods that Newport News Waterworks has implemented corrosion protection to address issues associated with existing and with new pipelines. The installation of all new pipelines addresses the concerns of useful life, the need to minimize the cost of these pipelines and the need to minimize customer interruption. Each new pipe is designed and constructed in the Newport News Waterworks system with a commitment to the careful consideration of corrosion control principles to enhance the expected life of the pipeline. The investigation of existing corrosion related pipeline failures provides a means of reducing future failures and provides information necessary to schedule pipeline replacements.

Recognizing that there is no single solution to the challenge of insuring pipeline integrity, Newport News Waterworks has categorized their water mains in order to address on a more individual basis to cost effectively maximize their life.

In each case, Newport News Waterworks utilizes state-of-the-art technology in order to make informed, technically sound and cost effective decisions in establishing reliable pipeline systems. This paper covers the specific techniques that Newport News Waterworks utilizes in assessing the condition of existing pipelines and the need for corrosion control for new pipelines. The general costs associated with their program along with data with regard to the dollars saved versus dollars spent on corrosion control are also presented.

Introduction

Water pipelines in the Eastern part of this country have been installed with the notion that they would last 100 years, with little concern for the environment they were installed in and limited concern for the pipe material. The older systems in the East have pipelines that prove this notion. However, they also have pipelines that have had to be replaced in much less than 100 years which disproves the notion.

At this time, many pipelines have reached the end of their supposed useful life and it has generated interest in the industry. According to AWWA, water distribution systems are being replaced at an average rate of one percent per year. Case studies of water system breaks shows that pipeline break rates are between 5 to 25 per 100 miles of mains per year. In addition, the AWWA Research Foundation has recently awarded two contracts to help the industry address these problems.

- RFP461-MAIN BREAK, PREDICTION, PREVENTION AND CONTROL

- RFP459-DEVELOP DECISION CRITERIA TO PRIORITIZE REPLACEMENT AND REHABILITATION OF MAINS AND APPURTENANCES

Newport News Waterworks has attempted to keep abreast of the industry's best methods for identifying necessary pipeline replacements. They have also tried to learn from their past experience of pipeline performance and failures. The installation of all new pipelines addresses the concerns of useful life, the need to minimize the cost of these pipelines and the need to minimize customer interruption. It appears that prevention of exterior corrosion is a primary element that was missed in the past installation of pipelines; therefore, Newport News Waterworks has developed a corrosion control program to address exterior corrosion that divides their pipelines into two categories. Category 1 includes transmission pipelines typically 24 inch and larger. These pipelines could affect thousands of customers. These have been installed with a proactive engineered corrosion protection system. Category 2 includes distribution pipelines typically 16 inch and smaller that have passive corrosion protection or no corrosion protection. A failure in these pipelines could affect hundreds of customers. In Category 1, corrosion control engineering studies and designs are made a part of new pipeline designs to assure that proper corrosion control techniques are applied. Field inspections during construction and final acceptance testing are performed by the project's corrosion engineer to assure that the corrosion control systems are properly installed. In addition, the corrosion control systems are routinely tested to verify proper operation. Since 1980, Category 2 pipelines have had a passive corrosion protection system applied (polywrap) as needed, based on field soil testing. This category also includes many pipelines that do not have any corrosion control, and these are being evaluated whenever they fail. The Category 2 failure evaluation is presently being implemented and it includes; type of corrosion, soils testing, and review of remedial measures to reduce future failures. The location of the failure and the soil data will be entered into a GIS mapping system to identify the location of corrosion related failures. Also, the location of where polywrap has been installed will be entered in the GIS. This information will be evaluated to assist in the scheduling of replacement of existing pipe and to identify what type of corrosion protection would be best by location in the system. The corrosion control program is being developed as an integrated effort with input collected from all levels of the water department's staff. This program has been under development for over 15 years and it is viewed that it is a part of the water system's routine construction and maintenance.

System Background

The Newport News Department of Public Utilities (Waterworks) is among the top one hundred water utilities in the United States. It provides potable water over a 250 square mile area to a population of more than 350,000 people living and working on the Virginia Peninsula. The department delivers over fifty million gallons of water per day to its retail customers. Annually, about 1,600 new connections are added. The pipelines in the system range in size from 2 inch to 54 inch with a total length of over 1,650 miles. The pipe materials include cast iron, ductile iron, concrete and steel. Service connections are copper and galvanized steel. Newport News Waterworks' system is similar to any underground pipe system with more than 1,000 miles of pipe. Some of the major pipeline challenges are the scheduling of pipe replacement and the design and installation of new pipe to maximize service life. These challenges are being addressed in the Newport News System with the use of an external corrosion control program.

Category 1 - New Pipelines

In 1990, a corrosion control program was initiated on all new transmission pipelines. Some of the existing transmission pipes in the system are over 60 years old and the condition of these pipes are unknown. They are assumed to be good because they are not failing. Without excavation of the entire pipeline or internal inspection, there are few methods to determine the condition of these pipes to predict their need for replacement. There is proven corrosion engineering technology available that can identify areas of highest potential for corrosion damage. These areas are identified by field testing corrosion potentials of the pipe and the corrosivity of the soil. It has become clear, there is a critical need to forecast the condition of a transmission pipe in the Newport News System, because each of these pipes affects service to a large percentage of the customers. Waiting for failure of these pipelines that cost over $250 per foot to replace is not reasonable. It takes a minimum of three years to budget, design and construct a replacement for these pipelines; therefore, the Category 1 pipelines are the more aggressive in proactive corrosion control. The Category 1 program objectives are to provide a cost effective method to extend the life of pipes, the ability to forecast condition of pipes, and the potential to improve customer service by avoiding pipeline failures. The program's objectives are being met to-date, based on our assessment, but actual data will not be available to prove this for many years. It is conservatively estimated that the life of the new pipelines should be doubled at an estimated increase in construction cost of less than 20%.

A transmission pipeline failure has the potential to affect more than 50% of the customers. Anytime a transmission pipe failure is avoided, customer service is improved. In addition, recent transmission failures across the country have indicated that the cost of a single failure could pay for hundreds of feet of replacement if the potential failure could be identified before the pipe breaks.

Category 1 - Exterior Corrosion Evaluation

All new transmission pipelines have an exterior corrosion evaluation performed as part of the design process. The evaluation of the site soil corrosivity is conducted through in-situ testing as well as laboratory analyses of soil samples obtained along the proposed pipeline route. The soil corrosivity tests include resistivity measurements, pH measurements, chloride concentrations, and sulfate concentrations. These four soil characteristics are considered the absolute minimum data required to evaluate soil corrosivity. In some areas, and for some pipe materials, additional site testing is also performed that may include testing of the ground waters. Where ground waters are tested, the testing includes the four major parameters listed above along with aggressive CO_2 content and hardness. Stray current testing is also routinely conducted along the pipeline route to evaluate the potential for stray DC current corrosion from operating impressed current cathodic protection systems or other sources of DC current.

Category 1 - Exterior Corrosion Control Design

Site specific corrosion control and/or monitoring systems are designed for each pipeline. The level of corrosion control varies with the aggressiveness of the site soil and ground water as well as the type of pipe material to be installed. Steel piping, ductile iron piping and prestressed concrete piping all have different levels of susceptibility to corrosion. The corrosion control design considers the susceptibility of each pipe material in selecting the corrosion control methods to be installed on a particular pipeline. The possible corrosion control measures that may be implemented include, but are not limited to the following: electrical isolation, electrical continuity, external bonded coatings, electrical sectionalization, test facilities, stray current control devices and cathodic protection. The selection of the appropriate corrosion control system is made only after a full review of the site soil and ground water conditions is completed with regards to the proposed pipeline material. Consideration is also given to the desired operating life of the pipeline along with the pipeline's initial cost and the cost of future repairs if the pipeline was to fail due to corrosion. The analysis of costs associated with the failure also considers the level of disruption to local water supplies due to a shutdown of the pipeline.

The detailed corrosion control designs minimize corrosion on the pipelines and maximize their expected operating life. Additionally, the corrosion control designs provide a basis to make alternate pipeline materials' service lives comparable. The design criteria for corrosion of all pipe materials is the same. The replacement elements of the corrosion control system for each pipe material are based on a 50 year life. The inspection and preventive maintenance of the systems will be similar for different materials.

Category 1 - Site Inspection

The installation of the corrosion control system is inspected by the corrosion control design engineer. The corrosion control engineer is a licensed engineer that is certified by the National Association of Corrosion Engineers and has applied experience in pipelines. The engineer routinely has to provide site training to the contractor's personnel and explains the purpose of the corrosion control measures being installed. This greatly enhances the quality of workmanship by the contractor because they understand the purpose of the corrosion control. It normally takes only one site meeting for the contractor to understand the importance of the corrosion control system and to no longer consider it to be a "nuisance" item that slows pipe installation.

In general, the corrosion control measures should be inspected by individuals trained in corrosion control. The individuals may be corrosion control engineers or inspectors who have been trained in the particulars of the corrosion control devices being installed on the project pipeline. Periodic site inspections by the corrosion control design engineer are recommended to assure that the corrosion control measures are properly installed and that the site inspectors are familiar with proper installation procedures.

Category 1 - Acceptance Testing

Part of the pipeline contract specification includes a performance requirement of the corrosion control system. However, the contractor is not typically responsible for performing the final acceptance testing. This testing is the responsibility of the corrosion control design engineer who is qualified to verify that the system has been installed properly, and is operating properly. This test is similar to the pressure test of the pipeline that verifies the quality of craftsmanship by the contractor. The contractor is required to correct any and all deficiencies in the corrosion control system that are discovered as part of the final acceptance testing.

The corrosion control design engineer submits the results of the final acceptance testing to the owner along with specific recommendations for the periodic maintenance of the pipeline's corrosion control and/or monitoring system.

Category 1 - Maintenance of Pipelines with Corrosion Control Systems

The proper maintenance of each of these new pipelines' corrosion control systems is required to meet the objective of the corrosion control for the transmission pipe.

The level and frequency of maintenance applied to a corrosion control system varies with the type of corrosion control system installed, the corrosivity of the site soil and ground water, the type of pipe material and the amount of third party construction adjacent to the pipe. Pipelines with corrosion prevention systems (bonded coatings and galvanic cathodic protection) should be evaluated on a yearly basis. The testing includes pipe-to-soil potentials at all test stations, measurement of anode current outputs and verification of electrical isolation where appropriate. Pipelines with corrosion monitoring systems (bonded joints and permanent test facilities) are evaluated on a two to five year cycle depending upon the corrosivity of the site soil and ground water and the potential for stray current damage. Testing is performed on a two year basis in more corrosive areas. Testing is performed on a five year basis in less corrosive areas. The testing includes a close interval potential survey than can document areas where active corrosion is occurring. The data is compared to previous surveys to determine the need for remedial measures prior to a corrosion failure of the pipeline. Some pipelines are provided with stray current control measures to protect them from the damaging effects of another utility company's impressed current cathodic protection systems. In these cases, the pipeline is provided with bonded joints, in-line electrical sectionalization and galvanic anodes for stray current drainage. These stray current control systems are routinely tested on a yearly basis. Additional testing should be performed when construction is performed near these pipelines and damage is suspected. These tests will help verify third party damage so the damaging party can pay for the repair to the corrosion protection system as is done when a pipeline is broken by a third party.

Category 2

The majority of the existing distribution pipelines in the Newport News system were installed without an active corrosion control systems. These pipelines are typically cast or ductile iron and they have provided good service for the most part. It was a corrosion investigation of failures that began the Waterworks' corrosion program. In the early 1980's, three sections of ductile iron pipe had to be replaced because of corrosion failures. Two of the sections were 12 inch pipes each over 1,000 ft. in length. The third section was a 4 inch pipe over 500 ft. in length. These pipelines were only ten years old and they were in different areas of the system miles apart. The replacement study for these pipelines included testing of soils to identify the scope of the replacement. The same soil test performed to identify the area of these replacements in 1980 has been used by the Waterworks' pipeline inspection personnel since that time to identify where polyethylene wrap will be used as corrosion protection for ductile iron pipe. This corrosion control method does not provide data on the condition of the pipe, but it does not require any routine maintenance and it extends the life of pipe. Defining the limits to replace these three pipelines was the beginning of the corrosion program for the Waterworks and the development of Category 1 was the second step of the program. The third step in the program includes enhancements to Category 2, which includes the following:

- Provide education to repair personnel to help accurately identify corrosion.
- Map number of breaks, corrosion failures and soil characteristics.
- Define areas of greatest corrosion and prescribe standard methods for corrosion protection and materials to be used.
- Identify new methods of repair that have potential to reduce future failures.

This step of Category 2 is currently being implemented and it will assist in better identifying sections of pipelines that need replacement. It should assure that the known corrosion problem areas in the system include a design for corrosion protection in future pipe replacements that will help the next generation with system maintenance. Education of personnel to apply corrosion protection consistently at all levels of pipeline maintenance is the goal of this step.

Internal Corrosion

No known pipe failures in the Newport News Waterworks system are the result of internal corrosion. The only internal pipeline corrosion problem that has been identified in the system is water quality. The quality problems with internal corrosion have been successfully addressed for the majority of the system with cement mortar lining. The majority of new pipelines are lined with cement mortar. Some of the new steel pipelines are lined with epoxy. Other actions taken to address internal corrosion include: a material standard change for 2 inch distribution pipeline has been made to replace galvanized steel with Schedule 80 plastic; and copper is used for meter connections. This has been done to primarily address the discolored water concerns so that this component of the system maintenance is manageable.

There are sections of unlined cast iron pipe that will need to be scheduled for replacement, cleaning and lining or just cleaning. The advances that have been made to enhance water quality with chemical treatment has reduced internal corrosion. This may make cleaning of older pipelines without lining an acceptable maintenance process so long as the exterior of the pipe is sound.

Summary

Presently, each new pipe is designed and constructed in the Newport News Waterworks system with a commitment to the careful consideration of corrosion control principles to enhance the expected life of the pipeline. This includes different degrees of protection and maintenance for different size pipes in specific locations. We are applying practical corrosion protection in the repair of pipelines and a display of the specific areas of corrosive soils in the system.

New transmission pipe receives the greatest degree of corrosion engineering and corrosion preventive maintenance. A predesign corrosion study evaluates site soil, site ground water and stray current interference potential so that proper corrosion control methods are implemented as part of the project designs. The corrosion design is specific for each pipeline material. Corrosion control engineering provides a basis to assure competitive bidding on an equivalent life for different pipeline materials.

Proper corrosion control will assure the increase in service life of these new transmission pipes. In addition, the testing of corrosion control systems will provide the Waterworks with a method to investigate and quantify the condition of a pipeline without excavation. These two factors are the key reasons that more aggressive corrosion control has been included as part of new transmission pipes because it provides a method to avoid breaks and provides a method to quantify the pipe's condition. This is important on the pipes that will affect the most customers when they fail.

The investigation of existing corrosion related pipeline failures provides a means of reducing future failures and provides information necessary to schedule identify sections of pipeline replacements. This information will also assist in providing a replacement pipeline that will provide longer service. The mapping of failures along with the soil corrosivity data provides a valuable tool in long term planning of system maintenance. When this concept is completely realized, more repairs can be planned and reaction maintenance will be reduced.

While there is a cost associated with the different degrees of corrosion control, the major benefit of these systems is the extension of the life and the ability to monitor and determine the condition of the pipeline in the future. The greatest degree of cathodic protection can allow a pipe to last forever. The proper use of any corrosion protection provides additional service life to a pipeline and this needs to be evaluated to determine what degree of protection is best for each pipe.

The costs associated with corrosion control will vary somewhat from pipeline to pipeline depending upon the exact corrosion control measures required. However, in general terms based on bid data, the highest degree of corrosion control for a new transmission main will usually be 3% to 7% of the pipeline cost. Annual maintenance costs will typically be less than $0.05 per foot and has the potential to more than double the operating life of the pipeline. When this cost is compared to the average replacement cost for a new major transmission main (more than $250 a linear foot), and the cost of a large diameter pipeline break of $50,000, the financial benefit of the corrosion control program is readily evident.

The minimum degree of corrosion protection on distribution pipe is the price of equipment and education of personnel. This investment can result in doubling the life of water pipes. With no investment in corrosion protection, the pipe may last 10 years as experienced by Newport News Waterworks. There are unknowns involved in corrosion protection because its actual value or cost cannot be calculated until the next generation. These authors believe that the savings of not using corrosion protection can be calculated today, but the related cost in potential pipe failure cannot be calculated. It is clear that technology has provided an opportunity to do more to make water pipes last longer. This warrants an investment in corrosion protection of the water industry's pipelines for the future.

Reference:

Cullinance, M.J., Jr., James Goodrich and Ian Goulter, "Water Distribution System Evaluation," in Reliability Analysis of Water Distribution Systems, Larry Mays, Ed., Report of the ASCE Task Committee on Risk and Reliability Analysis of Water Distribution Systems, ASCE, 1989

Internal Inspection of 20-inch Natural Gas Pipeline

Barbara B. Ostrander, P.E.[1]

Abstract

In June of 1996, Sacramento Municipal Utility District (SMUD) commissioned its fifty mile, 20-inch diameter natural gas pipeline. At a nominal operating pressure of 700 psig, the pipeline serves three natural gas-fired cogeneration plants, which generate up to a maximum of 375 MW for SMUD.

In December of 1997, SMUD contracted with BJ Pipeline Services (BJ) to provide a high accuracy internal inspection of the in-service pipeline. Information gathered during this inspection provides base-line data for comparison with future inspections. The inspection data becomes the basis for pipeline integrity analyses and also confirms or corrects plan and profile as-builts.

Introduction

Rather than focusing on the technology and engineering principles employed by various internal inspection devices, this paper will describe the field procedures utilized to prepare for and complete the internal inspection of the SMUD pipeline. It is hoped that this description will be useful to those preparing to perform an internal pipeline inspection on their own facility. The references at the end of this paper provide more detailed background, which would be useful in preparing to manage a smart pigging operation.

Background

A "pig" is any device that is inserted into a pipeline to travel freely, driven by product flow. A "smart pig" is any pig which electronically records and reports information about a pipeline. The chief objectives a pipeline operator may

---

[1] Principal Mechanical Engineer, Sacramento Municipal Utility District, P.O. Box 15830, MS-B355, Sacramento, CA, 95852-1830

have for conducting an internal inspection are:

- metal loss survey
- geometry services
- leak detection and location
- bend measurement
- crack detection
- photographic/video inspection
- line cover and spanning
- plan and profile monitoring and mapping

Smart pigs available today meet these objectives by employing one or more of the following technologies:

- magnetic flux leakage
- ultrasonics
- eddy currents
- electro-mechanical tools
- inertial navigation systems
- gyroscopes
- odometers
- accelerometers
- sonars
- photographic/video recorders

Selection of the employed technology will be determined by the information required during the survey. Smart pigging can provide information concerning internal diameter; wall thickness changes; welds; location, size and shape of anomalies such as dents, wrinkles or flaws; pipeline curvature; ovality; metal loss; position; and attitude.

Once the information is collected, many types of analysis are possible:

- Fitness for purpose
- Numerical analyses
- Deformation
- Fracture mechanics
- Modeling
- Failure mode
- Experimentation
- Repair prioritization
- Code conformance

The chief objective of the internal inspection of the SMUD pipeline was plan and profile monitoring and mapping. SMUD selected BJ Pipeline Services (formerly NOWSCO of Calgary, Alberta, Canada) to conduct the smart pigging with their patented GEOPIG. The GEOPIG utilizes an inertial navigation system, along with ultrasonic and sonar calipers to collect data.

Information Required Before Inspection

Before the pigging contractor can prepare for the project, the following information must be provided:
- **Company details**: Name, address, contact name and phone
- **Pipeline details**: Pipeline size, locations for launching and receiving traps, total trap-to-trap length of line, wall thicknesses in the line, welded or seamless, pipe specifications

- **Pipe fitting details**: Minimum bend radius, nominal bore bend, mitre bends, valve types, minimum bore of valves, valve locations, check valves, barred tees, branches
- **Trap details**: Dimensions, orientation, concentric or eccentric reducers, valving arrangement, pressure equalization line, availability of lifting facilities and power
- **Product details**: Product type, composition, operating temperatures, pressures, range of flow rates during pigging run, levels of wax, CO_2, sand, condensate, saltwater, H_2S content
- **Pipeline history, records**: Age of line, availability of maps, details of last internal inspection, expected internal corrosion, pipeline cleaning procedures used, internal coatings

Prepare for the Pipeline Inspection

All precautions must be taken to ensure that the pig does not become damaged or lodged in the pipeline and have to be cut out. There are also logistical details to be planned. Pipeline clearance procedures must be prepared to ensure that the required activities are completed safely.

Minimum bore diameter and pipeline cleanliness are the most important factors affecting the resulting condition and effectiveness of the smart pig. A gauging pig run should be completed to verify there will be no obstructions to damage the smart pig. For a pipeline that has been in service for a long time, it is worthwhile to conduct cleaning operations before putting the smart pig into the line. A damaged smart pig becomes the property of the pipeline owner. With a typical cost of about $ 250,000, this unplanned expense could easily overrun a project budget.

The SMUD pipeline was designed and constructed to permit smart pigging. Pig launchers and receivers are permanent facilities, mainline valves are full-opening ball valves and pipeline bends have a minimum radius of three times the diameter. After construction, the pipeline was dewatered, cleaned out with brush pigs, and then packed with dehydrated air until it was placed into service, about six months later.

For optimal data collection, flow conditions during the internal inspection should be maintained at a steady state. The pigging contractor can give a range of flow rates that will yield acceptable results. If not maintained at a uniform speed, the pig may slow down, until it is propelled at a high velocity by the pressure build-up behind it. Data collected during this high velocity excursion may not be reliable.

Before scheduling personnel and equipment for the inspection, it is helpful to estimate how long it will take the smart pig to travel to the receiver, once it is launched. The time required for the inspection can be determined from the pig's estimated velocity, once the line pressure and expected flow rate are known. The equations below can be used prior to the pigging, to get a feel for how long the pig will be in the pipeline. It is also worthwhile to use these equations on the day of the pigging to update the estimates.

Keeping units consistent,

$$V = \frac{\pi d^2 L R}{4}$$

$$v = \frac{1.02 Q L}{V}$$

$$t = \frac{L}{v} = \frac{V}{1.02 Q}$$

Where
- V = Volume inside pipe, at pipeline pressure
- d = Inside diameter of pipe
- L = Length of pipeline pig will travel
- R = Ratio of absolute line pressure to absolute atmospheric pressure
- v = Pig Velocity
- Q = Flow rate (standard conditions) expected during the pig run
- t = Time for pig to travel through the section of pipe
- 1.02 = Factor to compensate for estimated 2% bypass around the pig

SMUD controlled pipeline flow rate by appropriately dispatching the downstream cogeneration power plants. A planning meeting was held with the Power System Operations group in advance of the pigging, so that the energy supply portfolio reflected the pipeline needs.

BJ also required the use of a climate-controlled shop near the pipeline, so that instrument calibrations, data downloads, and other related activities could be conducted. SMUD provided the maintenance shop at one of the cogeneration power plants for BJ's use (Figure 1). The shop was adjacent to one of the receiver traps, and proved to be a very convenient location.

Figure 1. Maintenance Shop for Calibration and Data Downloading (GEOPIG in Foreground)

Conduct the Internal Inspection

The SMUD pipeline is designed with two piggable sections. Line 700 A, about forty miles long, starts at the launcher in Winters, CA, and ends at the receiver in Sacramento, CA. Line 700 B, all in Sacramento, starts at the second launcher and ends about ten miles away at the second receiver. (Figure 2.)

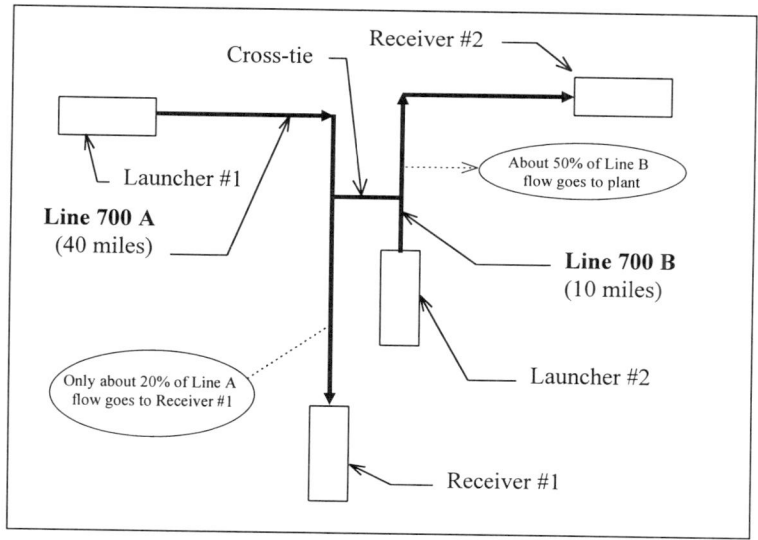

Figure 2. SMUD Pipeline Schematic Diagram

The SMUD pipeline has mainline valves with automatic line rupture control. Before the pig was expected to reach a valve site, the automatic feature was disabled. After the pig was verified to have passed the valve, it was restored to normal.

For each section of pipeline, a gauging pig run preceded the GEOPIG run. The main reason for doing so was to ensure that the GEOPIG would pass safely through the pipeline. The gauging pig consisted of three circular plates, mounted on a shaft, with urethane cups on the leading and trailing ends. One plate was 90% of the smallest internal diameter. The other two plates were sized to pass through bends with a minimum radius of three and five times the diameter. The GEOPIG was not inserted into a section of pipeline until the gauging pig was inspected after its run.

The plate to test for bends of five times the diameter emerged from the pipeline bent. This was expected, and did not preclude the use of the GEOPIG. The other two plates were undamaged. If either of these two had emerged from the pipeline bent, BJ would not put the GEOPIG into the pipeline, because it would very likely become damaged and/or lodged in the pipeline. The general condition of the gauging plates also gives an indication of pipeline cleanliness. The only debris found on the gauging pig was valve lubricant. The gauging pig runs for both sections proved that the pipeline was clean and could accommodate the GEOPIG without degrading data quality (Figures 3 and 4).

Figure 3. Gauging Pig -- Before   Figure 4. Gauging Pig -- After

The following table summarizes the sequence of events during the pigging:

	ACTIVITY DESCRIPTION
Day 1	Equipment arrives at shop. BJ performs setup and calibration procedures
Day 2	Gauging pig launched into Line B (10 miles long), and inspected after removing from receiver
Day 3	➢ Gauging pig launched into Line A (40 miles long) ➢ GEOPIG launched into Line B ➢ GEOPIG removed from Line B (average speed: 2 mph) ➢ Data downloaded from GEOPIG
Day 4	Gauging pig removed from Line A and inspected.
Day 5	➢ GEOPIG launched into Line A ➢ GEOPIG removed from Line A (average speed: 4 mph) ➢ Data downloaded from GEOPIG ➢ Field report delivered to SMUD

On Day 2, when the gauging pig passed through a mainline valve location, there was a momentary, nominal pressure drop indicated on our SCADA system. This is a typical signature for a passing pig, and can be used to update the timing estimates. We were expecting to observe the same signature as the gauging pig passed through the valves in Line A on Day 3. However, there were no fluctuations to be observed. The gauging pig did not arrive at the receiver when it was expected.

We later discovered that the pig became lodged in the tee at the cross-tie between Lines A and B. At this location, about eighty percent of the flow is diverted, leaving little flow for the pig to negotiate the last mile to the receiver. Increasing the load at the receiver end of the pipeline was insufficient to dislodge the pig. A combination of valve manipulation and blowdowns was required to produce sufficient flow to move the pig into the receiver.

On Day 5, we dispatched personnel to each mainline valve in advance of the pig's expected arrival there, to positively verify the pig's progress through the line. As the pig travels along the pipe, there is a distinctive "click-click" noise, indicating the pig's leading and trailing cups are passing across pipeline welds. Each mainline valve is buried, but blowdown piping extends above ground. Pig travel could be heard by putting one's ear to the pipe wall. In addition, BJ verified pig passage by picking up the pig's transmitter signals with their portable receiver. Valving and blowdown procedures were again necessary to move the GEOPIG into the receiver downstream of the cross-tie.

Figure 5. GEOPIG into Launcher

Figure 6. GEOPIG out of Receiver

Review the Data

Before returning to Calgary, BJ provided SMUD with a field report, describing the preliminary check of the data. It was determined that both GEOPIG runs delivered high quality data. This first check of the data indicated that there were no dents in the pipeline six percent or greater than its outer diameter. If there had been any significant dents in the pipeline, this early data report would give the pipeline operator the ability to investigate and take corrective action immediately.

Final data processing incorporates data from a Global Positioning System (GPS) survey taken at several points along the pipeline. Once incorporated into the data file, the pipeline location will be referenced to the California Grid coordinate system, complete with a three-dimensional position description. This correlated data can then become part of a Geographical Information System (GIS).

The final data for the internal inspection is recorded on a CD-ROM disk. As part of the inspection service, BJ provided SMUD with their proprietary

software used to interpret the pipeline data. Most internal inspection companies will provide their own software interpret the collected data. The PC-based software provided by BJ allows the pipeline operator to:
- Locate deformations based on user-input parameters
- View the deformations in three dimensions
- Generate reports to aid in prioritizing investigations
- Locate all pipeline welds
- View plan and profile views of the pipeline centerline
- Evaluate pipeline strain data
- Compare multiple runs on the same pipeline, to check for pipeline movement and strain changes
- Verify the accuracy of as-built drawings

Conclusion

Prior to the inspection technology currently available, the alternatives for assessing pipeline fitness were typically (1) excavate and replace, or (2) take a system out of service for hydrotesting. The data gathered by this internal inspection is a valuable base line for future comparisons. Data from another pig run can be compared to this baseline data, to determine if the pipeline's integrity has been compromised.

Every pipeline operator is responsible for continuous diligence in assuring the integrity of the pipeline. Smart pigging data can be one of the most cost effective and efficient means of evaluating a pipeline's condition. With minimal interference to pipeline service, internal inspections can provide accurate data for analyses, as opposed to making assumptions and applying unreasonably large factors of safety. In fact, smart pigging is playing a prominent role in the Department of Transportation's demonstration program for Pipeline Risk Management (DOT - Federal Register, 10/10/97). In this program, selected pipeline companies will be employing new methods in pipeline management to reduce overall risk to public safety. Smart pigging is one of the methods that will be implemented and evaluated for its impact on pipeline safety.

References
1. Cataford, Guy & Joerissen, Dan. (1995). "Pipeline mapping using an inertial inspection tool." First National Oil & Gas Technical Conference, Calgary, March 27-28.
2. Cordell, Jim & Vanzant, Hershel. (1996). All About Pigging. Gloucestershire, UK: On-Stream Systems, Ltd.
3. Cordell, J. L. (1991, July). In-line inspection: What technology is best? Pipe Line Industry, 47-56.
4. Cox, Michael, Garrigus, Andrew, Walker, William, and Wade, Ron. (1995). "Pipeline monitoring and remedial action from inertial geometry surveys in buried pipelines." Pipeline Pigging Conference, Houston, February 13-16.
5. Czyz, Jaroslaw A., Fraccaroli, Constantino and Sergeant, Alan P. (1996). Measuring pipeline movement in geotechnically unstable areas using an inertial geometry pipeline inspection pig.
6. Department of Transportation, Research and Special Programs Administration. (1997). Pipeline Safety: Remaining candidates for the pipeline risk management demonstration program. Federal Register Notice. October 10, 1997. Vol. 62, No. 197, 53052.
7. Gas Research Institute. (1997, August). R&D is helping to make smart pigs even smarter. Pipeline & Gas Journal, 65-68.
8. Macpherson, Charles. (1997, November). Smart pigging project benefits from detailed contingency plan. Pipe Line & Gas Industry, 71-74.
9. Mitchell, Jesse L. (1996, June). Smart pigs getting smarter to meet operator demands. Pipeline & Gas Industry, 37-43.
10. Palmer, A. & Jee, T. (1990, April). Why pig a pipeline? Pipeline Industry, 50-54.
11. PSE&G uses Geopig to inspect Crown Central gas pipeline. (1997, April). Pipeline & Gas Journal, 54-58.
12. Schaefer, Ed. (1992, July). Considerations for intelligent inspection from start to finish. Pipe Line Industry, 33-36.
13. Shamblin, Terry. (1996, June). Columbia Gas steps up annual MFL line inspection program. Pipeline & Gas Industry, 23-27.
14. Tiratsoo, J.N.H. (Ed.) (1989). Pipeline Pigging Technology. Exeter, UK: Short Run Press.
15. Wade, R.L. & Adams, J.R. (1995). "An integrated approach for pipeline fitness for purpose determination using corrosion and geometry pig inspection systems." Pipe Tech, Thailand, September 20-21.

The Use of Submersible Remotely Operated Vehicles for the Inspection of Water-Filled Pipelines and Tunnels

Ronald E. Heffron, P.E., M. ASCE[1]

Abstract

Advanced underwater robotic technology is providing an alternative to the dewatering process for the inspection, evaluation and repair of deep and long water-conveyance tunnel/pipeline systems. The traditional process of dewatering has many drawbacks in the eyes of owners of these facilities, including the extended disruption of operations. In addition, inspections performed under full-pressure conditions yield data which is much more representative of the normal operating water pressure conditions.

At Virginia Power's Bath County Pumped Storage Station, the largest station of its kind in the world, this robotic technology was pioneered through the development of a purpose-built Remotely Operated Vehicle (ROV) to gain access to three large tunnels. Unmanned ROVs have been used by the offshore oilfield community and the military since their inception in the early 1970s, and have seen significant improvements over the years. The HYDROVER System is the first ROV ever built specifically for inspecting long tunnels.

The HYDROVER System was successful in performing a record tunnel excursion of 6100 ft at a depth of over a mile. One important element in this success was the development of a 7000-ft neutrally buoyant umbilical. The umbilical provides power and telemetry to the ROV and transfers data back to the surface control station. By making the umbilical neutrally buoyant, frictional drag of the umbilical on the concrete surfaces is minimized. This translates into lower power requirements of the ROV and permits the physical size of the ROV to be reduced.

The use of fiber optic technology and other enhancements now permit similar excursions as long as 35,000 ft or more. The real key to the success of

---

[1] Regional Manager, Han-Padron Associates, LLP, 100 Oceangate, Suite 650, Long Beach, CA 90802, Tel 562-690-6032, Fax 562-590-6042, E-Mail Rheffron@ix.netcom.com

this technology however, lies not simply is getting to the end of the tunnel or pipe, but rather on the techniques used and the experience of the engineers operating the system and performing the inspection work. Specific inspection techniques, equipment and procedures have been borrowed from related industries and adapted to the pipeline/tunnel environment, to facilitate a systematic approach to the inspections.

The HYDROVER System incorporates an array of video, stereoscopic and photogrammetric cameras arranged to complement the unique requirements of tunnel inspections. An on-board dye release system has proven to be very useful to the engineers in detecting and measuring the movement of water through cracks in the tunnel or pipe. In addition, sophisticated sonar techniques have facilitated efficient inspections where visual means have fallen short. Other enhancements, such as cathodic protection potential measurements, ultrasonic steel thickness measurements and laser line scale imagery can also improve the quality of the inspection results.

The HYDROVER System has conducted the 6100 ft excursion to the end of the Bath County tunnel system over 50 times to date. This very successful application of this advanced technology, along with refinements in inspection techniques and methodologies, have proven the ROV to be a safe, practical and very economical alternative for the inspection of water-conveyance tunnels.

Introduction

The inspection of water conveyance pipelines and tunnels has been a problem for owners of these facilities worldwide. The pipelines and tunnels are generally too long and/or deep for divers and dewatering can be very expensive and disruptive. Such was the problem faced by the Bath County Pumped Storage Station in rural western Virginia.

The Bath County Pumped Storage Station, with a combined generating capacity of 2100 megawatts, is the largest station of its kind in the world. Owned by Virginia Power and their partner, Allegheny Power System, the station was placed into operation in 1985.

Since the station had been operation, the condition of the three tunnels had been assumed to be satisfactory based on performance, though the actual condition was largely unknown. When it was determined that the tunnels should be inspected, it became readily apparent that the use of an ROV would provide the safest and most efficient means of inspection.

Inspection Alternatives

Alternative means of conducting the inspection which were considered included saturation divers, manned submersibles, and dewatering the tunnels. As indicated in Table 1, the diving and manned submersible options were eliminated due to safety concerns.

TABLE 1
COMPARISON OF INSPECTION ALTERNATIVES

METHOD	ADVANTAGES	DISADVANTAGES
Commercial Saturation Divers	Allows U/W Inspection	Unsafe
	Allows firsthand observation	Limited by Diver's inspection qualifications
Manned Submersibles	Allows U/W inspection	Unsafe compared to other methods
	Allows first-hand observation	Limited productive time due to batteries and air supply
		Limited by pilot's inspection qualifications
Remotely Operated Vehicle	Allows U/W Inspection	Requires development of new equipment
	Allows Professional Engineers direct control over inspection	
	Safe and economical	
Dewatering	Allows first-hand observation	Requires additional time for draining and refilling
	Allows Professional Engineers direct access to tunnels for inspection	Requires setting up lighting and scaffolding for access to crown
		Possible damage to liner from high external pressure
		Possible loss of crack caulking and silt build-up in cracks due to groundwater inflow
		Not representative of operating conditions due to relaxed pressure

Dewatering the tunnels posed concerns as well, including the disruption of operations while the tunnels are drained, inspected, and then refilled. Dewatering must be done at a slow rate so as to balance internal and external pressures. High external pressure forcing inward on the dewatered tunnel could cause damage to the tunnel liner. In addition, dewatering could cause groundwater inflow into the tunnels which could dislodge crack caulking or wash out silt built-up in the rock joint, both of which help control water egress. The extensive down time required for the dewatering process makes this option unattractive economically since replacing the lost generating capacity is very costly.

Dewatering the tunnels also relaxes the outward pressure on the tunnel liner, allowing cracks, if present, to close. The resulting inspection would not be representative of operating water pressures.

The use of an ROV offers the advantages of a safe and economical inspection with the least amount of disruption to operations. The ROV also offers the advantage of being able to view and plot water egress from the tunnels under normal operating water pressures.

The Technology

The goal in developing the system for the inspection of the Bath County tunnels was to overcome the deficiencies associated with the use of ROVs normally deployed in offshore or military applications. These deficiencies include:

- Offshore work typically involves deploying downward to great depths, with only limited horizontal excursions. Pipeline/tunnel deployments are just the opposite.
- Deploying over long horizontal distances requires an umbilical which is neutrally buoyant, thus limiting drag on the pipeline/tunnel liner.
- Most ROVs are built with forward-looking camera systems. The inspection of pipelines/tunnels requires the ability to view surfaces which are overhead, beneath and to the sides of the ROV. Turning or maneuvering the vehicle to view at odd angles is a time-consuming proposition.

For these reasons, the owners of the Bath County tunnels decided to develop an ROV system custom-designed for the inspection of long pipelines/tunnels. No such system had ever been built before. The result of this development effort is the 'HYDROVER' ROV shown in Figure 1.

Fig. 1 - The HYDROVER ROV

The potential for limited visibility water conditions and the requirement for close-up detailed inspection of concrete crack and steel corrosion characteristics dictated that the inspection camera be no more than twelve inches, at maximum, from any portion of the tunnel wall. This was accomplished by mounting the primary video inspection camera on a swing arm that rotates 360 degrees around a horizontal axis through the longitudinal center of the vehicle. The unique arrangement of this color video camera facilitates the close-up inspection of the crown, invert, and side walls of the tunnel while maintaining the vehicle in a heading parallel with the tunnel direction. This may seem to be a minor factor, but by eliminating the need to turn the vehicle 90 degrees in each direction to face the walls, the speed of the inspection as well as the accuracy of ROV positioning is greatly increased.

In addition to the rotating camera, two other video cameras are included. The bow-mounted color camera points forward and has an internal pan, tilt, and rotate mechanism for the lens. This camera is used for navigation and for inspecting vertical surfaces. The third camera is black and white and is mounted aft. It is used primarily for tracking the position of the umbilical behind the vehicle.

A stereo photo camera is used for photographing cracks, spalls, or other features in the tunnels. The resulting slide pairs are mounted to produce a 3-dimensional effect for depth perception.

Among HYDROVER's other unique capabiltiites, a dye release system is used to release a small quantity of highly concentrated red dye near a crack to observe potential water egress through the tunnel liner. A mechanical arm, or manipulator, is used for accurately positioning the dye port. The manipulator, equipped with a stiff bristle brush, is also used for cleaning surfaces to allow closer inspection. This was especially useful in evaluating corrosion in the steel lined penstocks.

Of particular importance to navigating and positioning in the tunnels are the dual head sonar system and the gyrocompass. The sonar proved to be so valuable that, without it, the documentation of observed cracks and features could not have been performed to an acceptable degree of accuracy.

The most significant difference between the HYDROVER System and conventional ROVs is the 7000 ft umbilical, shown in Fig. 2, which is neutrally buoyant over the entire length. Manufactured in the U.K., this is the longest neutrally buoyant umbilical known to exist. At a diameter of 1.75 inches, the umbilical is jacketed with a pressure resistant material of low specific gravity to counter balance the weight of the

Fig. 2 – The HYDROVER System Deployment with 7000 ft Umbilical

conductors such that the umbilical neither floats nor sinks under pressure. Neutral buoyancy was critical in order to keep the umbilical from dragging against the concrete liner of the tunnel.

Frictional resistance due to dragging was kept to a minimum with a neutral umbilical. Low frictional resistance thus reduced the forward thrust requirements and hence the physical size of the vehicle.

The horizontal position of the ROV within the tunnel is very accurately determined by measuring the amount of umbilical deployed. Slight forward thrust on the vehicle is used to maintain a tight umbilical, eliminating slack which could lead to measurement errors. An electronic counter on the winch transmits the layout to the pilot's heads-up video display. A back-up mechanical counter is used for redundancy. In addition, the umbilical is marked in 5 ft increments to provide further assurance of positioning accuracy.

The Inspection Techniques

Operation of the vehicle is provided from a control station, shown in Fig. 3, which contains the control console, sonar processor, power distribution unit, annotation keyboard, and the video tape recorders and monitors. The pilot controls the vehicle functions while the navigator/documentor keeps track of vehicle position, maps the inspection results, operates the video equipment and narrates the video.

Fig. 3 – Engineers Document the Tunnel's Condition at the Control Station

The inspection results are documented on a data acquisition map which provides a continuous presentation of the tunnel layout. Crack widths and other feature dimensions were determined from the monitors which were equipped with graduated engineering scales, corrected for the effects of camera position and underwater magnification.

Unlike inspections conducted above water where the entire structure is typically visible, underwater inspections must be accomplished with a limited field of vision. If the goal of the inspection is a "General Condition Assessment," then it is impractical to attempt to inspect 100 percent of the structure. Therefore, the inspection should focus on statistically representative sampling of sections of the pipeline/tunnel. In areas where problems are suspected or anomalies are observed, then more detailed inspection coverage, up to 100% visual examination, may be warranted. A typical inspection program is shown in Table 2.

TABLE 2
TYPICAL INSPECTION PROCEDURES FOR THE GENERAL CONDITION ASSESSMENT OF PIPELINES/TUNNELS

1. Perform 360 degree visual circumferential profiles of pipeline/tunnel lining at nominal 50 foot intervals utilizing rotating color camera oriented directly at the inspected surface.
   - stand-off distance of ½ to 2 feet
   - evaluate lining/coating integrity

2. Transit between circumferential profile stations along pipeline/tunnel invert or along longitudinal crack if present
   - follow each visible crack to points of origin/termination
   - visually determine extent of debris or siltation along pipeline/tunnel inverts

3. Perform 360 degree visual circumferential profile of intake headers, bifurcations, manifolds, concrete/steel liner interface areas, surge tanks/chambers, and areas requiring special attention at 20, 10, or 5 foot intervals as appropriate
   - stand-off distance of ½ to 2 feet

4. Perform visual inspection of intake gates, gate guides, trash racks, bifurcation/manifold nose piers, and valves
   - determine structural condition
   - evaluate coating integrity
   - evaluate base metal for scoring, pitting, corrosion, etc

5. Document all observed defects, including cracks, spalls, holes, etc. on the continuous scrolling data acquisition map of the tunnel layout
   - continuously observe and evaluate potential crack/spall patterns
   - note evidence of movement and/or offsets in observed cracks
   - document each entry on map with station or elevation, date, umbilical payout, video tape number and elapsed time
   - label each crack or other feature with dimensional parameters using inspection monitors calibrated for camera position and underwater magnification

6. Test cracks as appropriate for water egress using dye injected near crack opening
   - quantify water egress on an empirical relative scale

7. Photograph critical features using stereo photogrammetry techniques for evaluation of extent and depth of feature

8. Maintain accuracy of vehicle positioning to within +/- 3 feet vertical and +/- 5 feet horizontal

9. Index video tapes and stereo photos for cross reference with data acquisition maps

10. Annotate the video record with specific station/elevation and inspection notes for engineering documentation/retrieval

11. Prepare comprehensive bound inspection report
    - include methods, documentation, inspection results, conclusions, and recommendations
    - include record of data acquisition inspection maps
    - include stereo and mono photographs

12. Prepare edited video inspection report tapes
    - Provide concise visual record of highlights of inspection results
    - supplements the written report

Conclusions

The success of the Bath County project has confirmed the effectiveness of the use of ROVs for pipeline/tunnel inspection work. The reader is cautioned however that all ROVs are not alike. Attempting to perform this type of work with an ROV system that has not been designed or specially adapted for working in the pipelines/tunnels will likely result in disappointment.

Ultimately, the results of the inspection will be dependent upon the skill and knowledge of the inspectors. Consistent with an international trend among Departments of Transportation, military branches and utilities with underwater structures, the owners of the Bath County Station elected to require that registered Professional Engineers perform the underwater inspection work. This trend reflects the increased emphasis on the importance of civil works inspections in light of dam, pipeline, bridge and powerhouse collapses in recent years.

Requiring that engineers conduct inspection work of this nature significantly enhances the value of the work since engineers trained in structural mechanics and materials properties have the ability to quickly recognize load paths, failure modes and the structural significance of what they observe. They can then document and communicate the results of their work using the technical vocabulary and terminology that client/owners understand, eliminating errors resulting from miscommunication.

SEISMICALLY UPGRADING THE MOKELUMNE AQUEDUCTS

David L. Pratt[1], M. ASCE
Christopher F. Dodge[2], M. ASCE
Frederick N. Brovold[3], M. ASCE
and
Howard O. Wilson[4], M. ASCE

ABSTRACT

The East Bay Municipal Utility District (EBMUD), which provides potable water for approximately 1.2 million people in San Francisco's East Bay, is currently constructing seismic upgrades for the portion of the Mokelumne No. 3 Aqueduct that crosses the Sacramento-San Joaquin Delta. This section of the Aqueduct includes 3 major river crossings, 14.4 km (9 miles) of elevated pipeline, and 8 km (5 miles) of buried pipeline. Studies of the anticipated post-earthquake water demands would exceed the terminal storage reservoir capacities. The overall purpose of the work is to improve the seismic performance of the Aqueduct No. 3 during a strong earthquake, and to provide a system that meets EBMUD's performance criteria.

This paper will describe the initial studies, development of alternatives, design, and environmental permitting of the project.

INTRODUCTION

The project area includes the portion of the Mokelumne Aqueduct that crosses the Sacramento-San Joaquin Delta. The area is approximately 24 km (15 miles) long and extends from the outskirts of Stockton at the east end to EBMUD's storage yard at Bixler at the west end. Figure 1 shows the limits of the study area and the portion of the aqueduct crossing the

---

[1] Manager of Design Division, East Bay Municipal Utility District (EBMUD), 375 Eleventh Street, Oakland CA 94607-4240
[2] Associate Civil Engineer, EBMUD
[3] Branch Manager, GEI Consultants, Inc., 565 Commercial Street, 3rd floor, San Francisco, CA 94111
[4] Vice President, CH2M Hill, 1111 Broadway, Suite 1200, Oakland CA 94607-4046

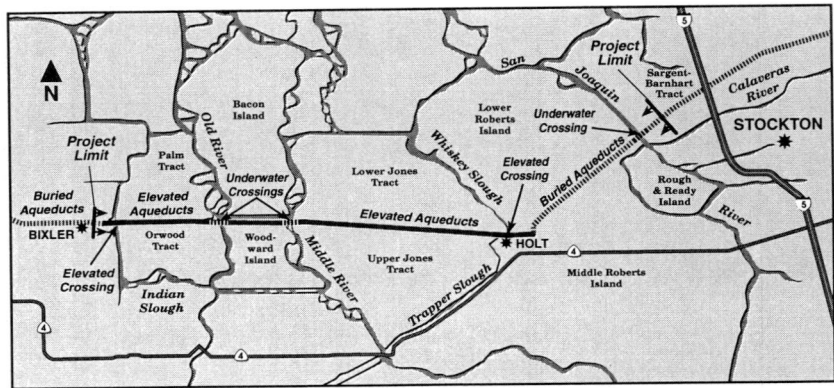

Figure 1. Aqueduct Route Across the Delta

Delta islands and tracts. EBMUD's main source of water supply is Pardee Reservoir which is located on the Mokelumne River in the Sierra Nevada foothills. Water is conveyed by three aqueducts from Pardee Reservoir, across the San Joaquin Valley, to the terminal storage reservoirs located in the East Bay.

Aqueduct No. 3, which was built in 1963, is a 2.2 meter (87-inch)-diameter pipeline. It is the primary water transport facility in the Mokelumne Aqueduct System, and was the focus of this study. Aqueduct No. 3 is the third of three parallel water supply pipelines carrying water to the East Bay. This pipeline typically conveys 4.6 cubic m/sec (100 MGD), approximately one half of EBMUD's demand. Aqueduct No. 1 is typically 1.5 meters (61 inches) in diameter, built in 1929, and Aqueduct No. 2 is about 1.7 meters (67 inches) in diameter, built in 1949.

The preliminary design study indicated a high probability that one of the main aqueduct's performance criteria, an aqueduct outage of less than six months, could not be achieved if a 500-year return period, or greater, earthquake occurred unless the aqueducts were upgraded. Widespread structural damage to the buried and elevated pipe sections, as well as flooding caused by levee breaks are predicted. Estimated costs for replacing the heavily damaged aqueduct with a new post-earthquake-constructed aqueduct range from $210 to $300 million. Post-earthquake construction costs in flooded conditions are estimated to cost two to three times the construction cost in an un-flooded condition. The aqueduct outage time required to construct a replacement aqueduct is estimated to be 1-1/2 to 3 years. An outage of this duration would greatly exceed the capacity of EBMUD's system of reservoirs on the west side of the San Joaquin River Valley.

Another important conclusion of the study is that the condition of the pipe joints and levees at the main river crossings seriously threatens the aqueducts. Pipe joint failures have occurred in the past and could potentially have caused a breach of the levees if the failures had not been promptly repaired. One of these joint failures occurred on Aqueduct No. 3 in January 1992 on the Upper Jones Tract side of the Middle River crossing and washed away portions of the land side of the levee. A similar pipe joint failure occurred at the levee on Woodward Island in March 1997. The effects of these breaks were detected in time to shut down the aqueduct before more significant damage occurred.

The three major alternatives for seismically upgrading the Delta crossing section of Aqueduct No. 3 are:

- No upgrading with a future cost of $210 to $300 million to replace the aqueduct after post-earthquake damages plus unacceptable outage times,

- Upgrading and retrofitting the elevated pipe, buried pipe and river crossing sections at an estimated total project cost of $38 million,

- Replacing Aqueduct No. 3 with a new pipeline at an estimated cost of $140 million.

Other seismic upgrading alternatives evaluated included tunneling the three river crossings and replacing the buried pipe. Estimated project costs are $30 million for three new tunneled crossings compared to $19 million for the selected river crossing alternative. Estimated total project costs are $24 million for replacing the buried pipe compared to $8 million for retrofitting the buried pipe joints.

Since aqueduct seismic upgrading alternatives are available at lower cost to meet the project serviceability criteria, the additional costs for tunneling the three river crossings and replacing the buried pipe are not considered necessary for a design based on a 500-year seismic event. Both of these alternatives, however, would increase the level of aqueduct security.

SEISMIC ASSESSMENT

Potential seismic sources in the study area include several right-lateral strike-slip faults in the Bay Area, such as the San Andreas, Hayward, and Calaveras faults, and a blind thrust fault, which is postulated to lie under the west end of the study area near Bixler. The blind thrust fault, called the Coast Range-Central Valley (CRCV) fault, is broadly accepted by seismologists who have studied the area. The CRCV fault was found to have a dominant impact on the seismic hazard at the site because of its proximity.

The potential seismic sources discussed above were used to estimate the probability that an event with certain characteristics will occur in a year. These characteristics were then matched against those of a single earthquake, which provided design parameters for structural dynamic analyses.

Various earthquake return periods were studied to represent reasonable lower, intermediate and upper bound earthquake events. The purpose of studying multiple events was to permit a probabalistic approach to the subsequent seismic analysis and design process. The seismic shaking is greater at the longer return periods; therefore, the longer the return period used for the aqueduct upgrading, the more secure the aqueduct will be against seismic damage. However, the cost of designing to satisfy the longer return periods will be greater, and the necessity to do so may be questionable. EBMUD's performance criteria for the aqueduct do not include continuous service (an outage of six months is acceptable); therefore, increasing the design return period beyond 500 years was considered not to be warranted on that basis.

For the 500-year return period earthquake, the model predicts a peak ground acceleration of 0.37 g or greater at the west end of the site near Bixler. This level of shaking could be represented by a Magnitude 6.7 earthquake on the CRCV fault, rupturing 10 km (6 miles) from Bixler.

Liquefaction Hazards

The soils within the upper 15 meters (50 ft) of ground surface generally consist of levee fills, Holocene peat-organic soils, Holocene alluvium, and Pleistocene alluvium. Holocene refers to soils deposited within the last 11,000 years. Sediments from this period are mostly fine-grained and have nearly always been under water. The deposits are relatively soft and compressible. Most recently, in terms of geologic time, organic sediments (peat) have developed from dense, decaying tule and reedy plants. Soils deposited during the earlier Pleistocene epoch are both fine and coarse grained; some have been desiccated (dried) during low sea-level (glacial) periods, and generally are over-consolidated and stiff.

Liquefaction analyses indicates that an earthquake with a peak ground acceleration of 0.20 g will cause most of the sand and silty sand deposits in the Holocene alluvium to liquefy. The dense sands and silty sands in the Pleistocene alluvium are not expected to liquefy, even due to ground shaking from the larger design-level earthquake. There are extensive Holocene sand deposits at and in the immediate vicinity of the rivers and sloughs, especially near the Old and Middle Rivers. The geotechnical data do not indicate extensive Holocene sand deposits away from the waterways.

Liquefaction at the site vicinity could cause levee failures and localized settlement of the buried pipe. While liquefied soil can exert large buoyant forces because the soil behaves as a dense fluid, liquefaction is not expected to cause the buried pipe to be buoyed up out of the ground. Analysis indicates that the existing overburden soils will prevent the pipe from floating.

Inundation Hazards

The study analyzed the static and pseudo-static stability of the six levees that comprise the three river crossings. Each levee was analyzed for the stability of both the land side and waterside slope. In general, the static stability of all six levees is relatively poor, with safety factors typically about 1.2.

Pseudo-static analyses indicate that all of the levees will deform under pseudo-static forces of 0.2 g or less. Deformations are expected to be larger at the levees that have only marginal static stability. Significant damage to the pipeline and the levee are possible since displacements on the order of 30 cm (1 ft) would threaten the structural integrity of the pipeline.

Previous levee breaches in the Delta have resulted in 90 to 180 meters (300 to 600 ft) of levee being eroded away, and the development of large scour holes which are often 60 m

(200 ft) wide and long, and 12 to 18 meters (40 to 60 ft) below island ground elevations. These breaches could cost $8 million to $12 million to repair, and require large quantities of dumped rock, and up to one year to fill. The time to repair a levee breach results from the combination of placing a large volume of rock and soil fill, and island dewatering by pumping. The islands are expected to be flooded after a 100-year or greater earthquake event.

SEISMIC UPGRADES DESIGNED

Seismic upgrades have been developed for the river crossings and buried pipe section. EBMUD implemented interim upgrades to the elevated reach of pipeline in 1993. The design of the long-term upgrades for the elevated pipeline have been deferred until the year 2000.

River Crossings

As shown by Figure 2, the selected levee strengthening includes:

- Lengthening the existing sheet pile wall along the longitudinal axis of the levee.

- Constructing a shoring and bracing system so that the land side portion of the pipe can be removed and a new pile supported concrete pipe foundation installed.

- Replacing the pipe and backfilling around and above the pipe.

- Flattening the levee land side slope and adding a toe berm for about 75 meters (250 ft) along each side of the pipe.

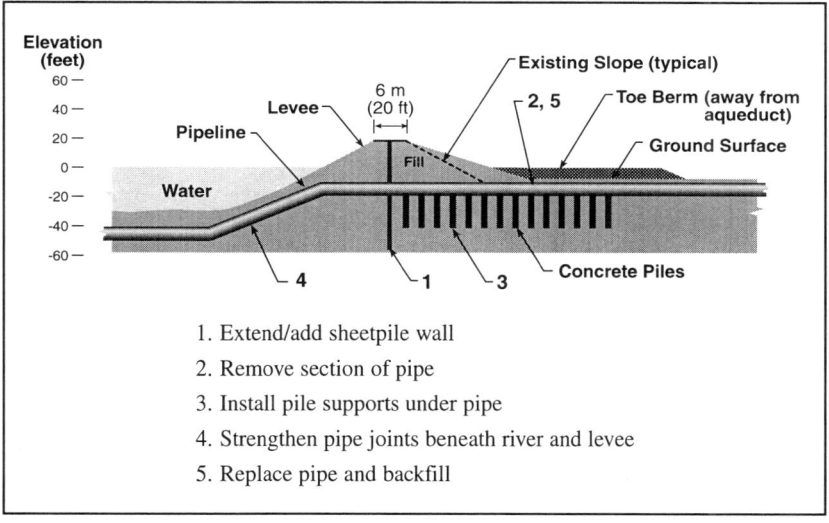

Figure 2. Secure River Crossing – Section View

- Strengthening the existing pipe joints below the river and adjacent buried pipeline by welding an internal saw-tooth patterned inside butt strap across approximately 250 existing bell and spigot joints.

The function of the sheet pile wall is to provide an effective barrier to waterside slope movements or failures. It is intended to act as a levee protection barrier that can undergo large movements and plastic deformations in the event of waterside slope failures and prevent the slope failures from breaching the levee at the aqueduct location. It is expected that the levee waterside slope will be reconstructed in the event of slope movement. The pipe foundation will be improved on the land side of the levee to provide vertical support for the pipe and to reduce the potential for long-term total and differential settlements caused by either compression of the Holocene peat-organic soil or by liquefaction of the Holocene alluvium. The seismic upgrading for the river crossing includes adding an internal welded butt strap to increase the strength of the joints similar to that described for the buried pipe joints below. The stability of the land side levee slopes will be significantly improved by constructing a levee toe berm at the levee back slope.

Buried Pipe

Static analyses were performed to evaluate the stresses in the pipeline system under normal working loads (gravity, thermal and water pressure loads). These analyses indicated that the working stresses approach, but do not exceed, current allowable code limits. The single fillet weld at the bell-and-spigot joints along the buried pipe have little reserve strength to resist additional stresses from seismic loads. The working stresses at the welds may be even higher than calculated because of unknown localized curvature in the pipe caused by settlement, leaving even less reserve strength to resist seismic loads.

The selected upgrading alternative for the buried pipe includes:

- Welding an internal straight patterned butt strap across approximately 700 existing bell and spigot joints, as shown in Figure 3.

- Providing increased access to the pipeline with new concrete vaults and manholes.

The section of buried pipe selected for seismic upgrading begins west of the San Joaquin River and extends to Holt where the pipe becomes elevated. This section was selected because depths of flooding, and post-earthquake repair costs and difficulty would be greater

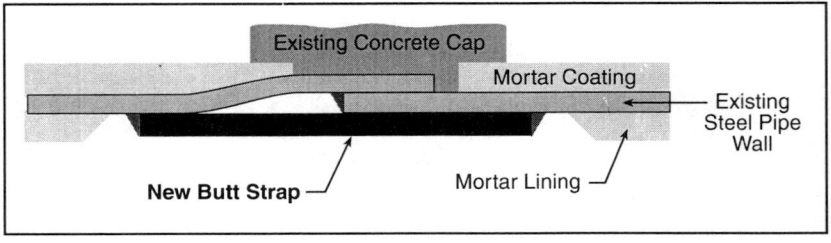

Figure 3. Reinforce Buried Pipe Joints

than for the buried pipe section eastward of the San Joaquin River. The existing bell and spigot pipe joints are inherently weak because of weld eccentricities and the small axial length of the weld. These joints currently exhibit a joint efficiency of approximately 30%. The proposed butt strap will improve the joint strength to approximately 70% of the pipe wall strength.

The new concrete vaults will be 4.8 meter (16 feet) square, with a removable pipe spool. These vaults will be the primary material delivery points and will be spaced approximately 1.6 km (1 mile) apart. Intermediate 76 cm (30 inch) manholes, at 210 meter (700 feet) spacing will also be installed for ventilation and weld lead purposes.

Environmental Documentation

The project identified the appropriate level of environmental documentation and permitting required for its construction. This consisted of early scoping meetings and site visits with the federal, state, county and regional regulatory officers. The information obtained during these early efforts were important factors in the selection of the upgrade alternatives. The selected levee strengthening alternative focuses its work on the land side of the levees. If the work were primarily on the water side, more environmentally damaging impacts would occur, resulting in more stringent permit conditions and schedules of work not congruent with EBMUD's needs.

A public outreach program was also initiated. These efforts helped in the process of obtaining the construction easements and the purchase of the land needed to construct the levee toe berms.

EBMUD's Board of Directors approved the project-specific Mitigated Negative Declaration for the project in May 1996. This environmental document included proposed mitigations to locate, monitor and protect certain biological resources adjacent to the work sites.

Construction Phase

The contract documents were advertised for bid consideration in September 1997. The bids were opened in October 1997. Notice to Proceed was given to the contractor in December 1997. The construction work completed to date includes installation of the levee toe berms, steel sheet piles, and concrete access vaults.

The shutdown of Aqueduct No. 3, the pipeline foundation, and internal butt strap work is scheduled to begin in October 1998. This work is planned for the low water demand period, since Aqueduct No. 3 typically conveys 50 percent of EBMUD's water supply. The project requires that the pipeline be returned to service by March 15, 1999 to meet EBMUD's water storage requirements for fire suppression. The planned 5-1/2 month shutdown will also require additional pumping on the other pipelines to meet customer demand. All construction activities will be completed by July 1999.

CONCLUSIONS

The project studies indicated that there is a high probability that one of the main aqueduct performance criteria, an outage of less than six months, could not be achieved if a Magnitude 6.7 or larger earthquake occurred unless Aqueduct No.3 is upgraded. Widespread structural damage to both the elevated pipe and buried pipe sections, as well as flooding caused by levee breaks are predicted. Another important conclusion of the study is that the existing condition of the pipe joints and levees at the main river crossings seriously threatens the aqueducts.

Estimated costs for replacing the heavily damaged aqueduct with a new post-earthquake constructed aqueduct range from $210 to $300 million. The aqueduct outage time required to construct a replacement aqueduct is estimated to be 1-1/2 to 3 years. The lowest cost and selected alternative provides an adequate level of aqueduct security for a 500-year seismic event. The upgrades are currently under construction and include:

- River Crossing Upgrading - strengthen the levees and retrofit the pipe joints at three major river crossings.

- Retrofitting Buried Pipe Joints - retrofit the pipe joints and provide new internal access points along an 8 km (5 mile) reach of pipe.

REFERENCES

GEI Consultants, Inc./Roger Foott Associates Division, 1995, "Mokelumne Aqueduct Seismic Upgrade Project", Preliminary Design Phase - Final Report, prepared for EBMUD.

Michael Brandman Associates/EDAW, Inc., 1996, "Initial Study and Mitigated Negative Declaration for the Mokelumne Aqueduct Seismic Upgrade Project", Environmental Documentation Phase, prepared for EBMUD.

Specification 1732 - Mokelumne No.3 Aqueduct Seismic Upgrade Project, 1997, prepared for EBMUD by the Engineer-of-Record, CH2M- Hill, Inc.

SEISMIC DESIGN OF PUERTO RICO'S NORTH COAST SUPERAQUEDUCT

Mehdi S. Zarghamee[1], F. ASCE, Rajesh S. Rao[1], M. ASCE.
Michael A. Yako[2], M. ASCE, Edward M. Motley[3], M. ASCE
Felix Garcia[4], and Anibal Camacho[5]

Abstract

The North Coast Superaqueduct is a 72-inch diameter prestressed concrete pipeline that runs along the north coast of Puerto Rico over the rugged karst terrain. The Superaqueduct is constructed mostly in the right-of-way of a highway running at various locations along the top, side, or toe of the highway embankment. The paper discusses the seismic hazard along the Superaqueduct. For the design seismic event, the effect of vertical acceleration on the design loads on the pipe, and the effect of the ground motion on a straight section of pipeline with restrained and unrestrained joints, at a bend with harnessed joints, at rock-to-soil transitions, and at the connections of the pipeline to large structures are discussed. The stability and lateral ground displacements of the highway embankment slopes and their impacts on the pipeline in the design seismic event, and the conditions under which changes in the alignment are needed to reduce the risk of failure of the pipeline are discussed.

Introduction

During the 1990's San Juan, Puerto Rico experienced critical water supply shortages. The Puerto Rico Aqueduct and Sewer Authority (PRASA), as a part of their approach to solve this critical water supply shortage, decided to develop and use a supplemental water supply from the existing reservoirs in the Arecibo River basin, some 50 miles west of San Juan. This project is known as the North Coast Superaqueduct Project. Thames-Dick Superaqueduct Partners were selected to design, construct, and operate the North Coast Superaqueduct Project. Quiõnes, Diez and Silva Associates (QDSA) of San Juan were responsible for the design of the raw water supply system and the treated water pipeline and storage system. Chiang, Patel and Yerby, Inc. of Dallas, Texas was a subconsultant to QDSA, providing design support services. Simpson Gumpertz & Heger Inc. of Arlington, Massachusetts was a specialty subconsultant to QDSA, providing seismic and prestressed concrete cylinder pipe (PCCP) design expertise.

---

[1] Principal and Senior Engineer, respectively, Simpson Gumpertz & Heger Inc., Arlington, MA
[2] Project Manager, GEI Consultants, Inc., Winchester, MA
[3] Senior Vice President, Chiang, Patel & Yerby, Inc., Dallas, TX
[4] Managing Partner, Quiõnes, Diez, and Silva Asociados, San Juan, PR
[5] Assistant to the Executive Director, Puerto Rico Aqueduct and Sewer Authority, San Juan, PR

The North Coast Superaqueduct Project (NCS) is comprised of a river intake located on the Arecibo River; a 300 mg bank-side storage reservoir; a 110 mgd raw water pump station; a 10 mile, 72-inch raw water pipeline; a 100 mgd water treatment plant; a 40 mile, 72-inch treated water pipeline; and two 10 mg treated water storage reservoirs. PCCP was the selected pipe for both the raw water and treated water pipelines, because of its availability on the island from a local manufacturer, the familiarity of local pipe contractors with the installation of PCCP, and its excellent performance history on the island.

The geology of the northern part of Puerto Rico consists of Aymamón and Aguada limestone foundations, overlain in different parts of the island by blanket deposits, and alluvial and river terrace deposits. Outcrops of Aymamón weathers to steep-sided conical hills and magotes. Outcrops of Aguada contain closely spaced sinkholes. Along its alignment, the pipeline passes, for the most part, along the embankment of Highway PR-22 and encounters transitions in soil types, descends steep slopes, ascends the sides of magotes, and passes through flat alluvial plains. The geotechnical data along the pipeline shows that the soil varies and consists mainly of limestone, blanket deposits, river terrace deposits and alluvium, and fill. The blanket deposit is clayey, ferruginous, fine to medium sand, typically 0 to 30 m thick, typically found between ridges of limestone outcroppings. River terrace deposits and alluvium are thick masses (sometimes more than 100 m) of clay, sand, gravel and boulder deposits in the alluvial plains of the rivers. Typically, the soil in the fill area and blanket deposits are quite similar, both are silty sand (ML), medium to dense.

Puerto Rico is the easternmost island of the Greater Antilles, and is the focus of frequent seismic activities. To the north, the Puerto Rico Trench is a major tectonic feature associated with the highly oblique subduction of the North American Plate under the Caribbean Plate. The maximum horizontal ground acceleration of soft rock for the design seismic event, with 10% probability of exceedance in 50 years, is predicted in a report by Ricardo Dobry Consultants (1996) to be 0.29 g, based on an earthquake in the Puerto Rico Trench of magnitude $M = 7.5$ at a focal distance of 50 km using an attenuation formula corresponding to the worldwide mean plus one standard deviation. Dobry's maximum ground acceleration and a design response spectrum for ground motion at the bedrock level were used for all points along the Superaqueduct.

Ground Motion

During an earthquake, seismic waves propagating through the soil produce strains in the soil which induce axial and bending stresses in the buried pipeline. Seismic waves are composed of body waves (compression and shear waves) and surface waves (Love and Rayleigh waves). Love waves are generally not important in terms of calculating pipeline responses. The longitudinal strain ϵ_a and the curvature κ in the ground due to wave propagation effect are given by:

$$\epsilon_a = \frac{v_p}{C} \quad (1)$$

$$\kappa = \frac{a}{C^2} \quad (2)$$

where v_p = maximum particle velocity, including soil amplification effect, a = peak ground acceleration and C = apparent wave velocity. Bending strains due to the curvature in the ground are much lower than the axial strains. The value of the apparent wave velocity is much higher than the measured body and shear wave velocities in the surface soils. This is because the seismic body waves traveling from the source are expected to undergo refraction at the soil-rock interface, and the refracted waves would travel in a more vertical direction as they approach the ground surface. This gives rise to an apparent body wave velocity in the horizontal plane which is much larger than the actual body wave velocity. From the seismic refraction data gathered by Suelos Inc. at the proposed treated water storage tank site, the actual compression wave velocity was measured to be 1,000 m/s (3,280 fps) for the blanket deposit and 1,800 m/s (5,900 fps) for the limestone rock. From the measured compressive wave velocity, the shear wave velocity can be computed at 525 m/s (1,722 fps) for blanket deposit and 945 m/s (3,100 fps) for the limestone. Using the methodology provided by O'Rourke et al. (1982) and using the measured shear wave velocities we estimated the apparent wave velocity for blanket deposits to be 5,840 fps. Several documents provide guidelines for selecting the design apparent wave velocity and recommend values for C (ASCE (1986)). Based on these recommendations and measurement of wave velocities, we selected a conservative value of C = 3000 ft/sec for the apparent wave velocity.

The earthquake motion from the bedrock level can be amplified as it travels through the soil layers above the bedrock. The extent of soil amplification depends on the depth of soil, soil properties, and intensity of ground motion. For pipe laid over blanket deposit layers, the peak ground acceleration was calculated at the top of the soil using a ground motion amplification factor of 1.1 as recommended by 1994 NEHRP Recommended Provisions FEMA (1995) for soft rock. For pipe laid on limestone and blanket deposits, the maximum particle velocity was calculated using a ratio of 110 cm/sec/g as recommended by Seed and Idriss (1982). The NEHRP soil amplification factors and the ratio of 110 cm/sec/g can be overly conservative for soft soils such as alluvium deposits and fill. Hence, for pipe laid on alluvium deposits and fill, a time-history analysis was performed on the soil profile by applying an input ground acceleration at the bedrock (limestone) layer and calculating the ground acceleration time history at the top of the soil profile. The soil properties and the depth of the soil above the bedrock for constructing the soil profile for the alluvium deposits and fill were obtained using the soil boring data. The time history analysis was performed using the computer program SHAKE. For a given soil profile, the program SHAKE performs a vertical shear-wave propagation analysis and determines the acceleration time history at the top of each soil layer. The acceleration time history at the bed rock level was generated using a ground motion simulation program SIMQKE. The simulated ground motion time history has a peak rock acceleration of 0.29 g and matches the velocity response spectrum for 5% damping.

In the vertical direction, the intensity of the ground motion is typically between one-half and two-thirds of the intensity of the horizontal ground motion for the range of periods between 0 and 6 seconds (Newmark and Hall 1982). A ratio of vertical-to-horizontal acceleration of 2/3 was used in our analysis. For calculating the peak ground strain, we combined the maximum particle velocity in the horizontal and vertical direction. We calculated the maximum ground strains calculated for different soil profiles to be 410 microstrains (με) for limestone, 460 με for blanket deposits, 460 με for fill and 500 με for river terrace and alluvium.

Behavior of Pipeline with Restrained Joints

When a seismic wave travels along the pipeline, the longitudinal strain and curvature of the soil induces strain in the pipe through friction forces between the pipe and the soil. When sufficient length of pipe has joint restraints that prevent relative motion of the adjoining pipe segments, and sufficient friction can develop over a quarter of wavelength, the entire longitudinal strain and curvature change in the soil will be transferred to the pipe producing longitudinal stress for which the pipe has to be designed. Full restraint is expected to result from welded joints, and partial restraint from mechanically harnessed joints. Mechanically restrained joints, such as the snap-ring and the C-clamp harness joints, used in the Superaqueduct, do not provide full restraint against the relative motion of the adjoining pipe segments. Tests performed by Price Brothers show that harnessed joints after grouting can allow a motion of 1/16 in. to 1/8 in. Therefore, away from the thrust restraint points, each joint can accommodate 1/16 in. of relative motion of the adjoining pipe segments. Such a relative motion at a joint will relieve the strains in the pipe and reduce the strains and the stresses in a 20 ft long pipe segment by 260 microstrains.

During a compression cycle of the wave propagation, compressive stresses are induced in the pipe wall. The maximum compressive stresses of up to 2,012 psi in the concrete core and 14.6 ksi in the steel cylinder are expected for pipe segments with welded joints and up to 960 psi in concrete and 7 ksi in steel for pipe segments with mechanically harnessed joints. The allowable compressive stress is 2,925 psi in concrete (0.65 f'_c for transient loading as per AWWA C-304), and 26.3 ksi in steel (0.6 F_y increased by 33% for seismic loading as per AISC-ASD Specification).

When tensile strain is induced in the pipe wall during the tension cycle of the wave propagation, the concrete core is subjected to tensile softening. With increasing tensile strain, the average tensile stress in concrete initially increases up to a strain of about 120 microstrain and then decreases gradually as concrete develops microcracking in its fracture process zones (Zarghamee and Fok 1990). At a crack in the concrete core or at the weld line joining the steel cylinder to the spigot ring, the concrete core is not able to transfer any stress across the crack, and the stress resultant in the softened concrete has to be carried by the steel cylinder. For a longitudinal concrete stress equal to the tensile strength of concrete of about 470 psi, the tensile stress in the steel cylinder at the weld line to the spigot ring can exceed yield strength of the steel cylinder. However, the post-yield deformation resulting from the yielding of the steel cylinder should not exceed 1/16 in. For a proper weld between the steel cylinder and the spigot ring, this post yield deformation is not expected to cause rupture of the steel cylinder.

Behavior of Pipeline with Unrestrained Joints

When the bell-and-spigot joints between the pipe segments are not mechanically harnessed, the strain imparted to the pipe from the soil is accommodated by the relative motion between the adjacent pipe segments. During the tension cycle of the seismic wave, the adjacent pipe segments would be pulled apart resulting in opening of the unrestrained bell-and-spigot joints. The joint opening (δ_{joint}) due to axial strain ϵ_a is calculated by:

$$\delta_{joint} = \epsilon_a L_{pipe\ segment} \tag{3}$$

The effect of ground curvature is negligibly small. If there is excessive opening of the joints, the joints could disengage and leak during a seismic event. We calculated that the maximum joint opening in 20 ft pipe segments during a design seismic event would be 0.12 in. The standard bell-and-spigot joint geometry allows for a hypothetical motion of 2 in. (Fig. 1). Allowing 1-1/4 in. for installation and alignment adjustment, there is an allowance of 3/4 in. for the seismic motion before the gasket pressure is reduced by the flare in the bell ring. For a minimum safety factor of 2, the maximum allowable limit for seismic motion is 0.375 in.

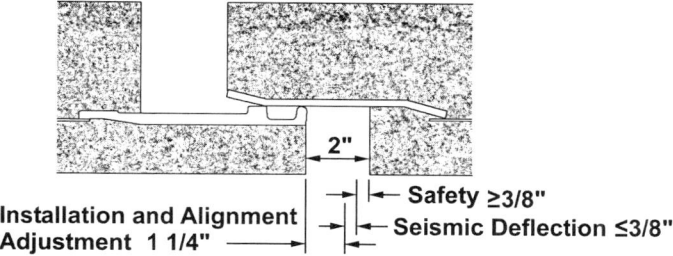

Figure 1. Deformation Capability of Bell-and-Spigot Joint

When a stretch of the pipeline has restrained joints, the unrestrained joints at either end of the stretch would undergo joint openings larger than 0.12 in. When the calculated joint opening exceeded the maximum allowable limit, bell-and-spigot joints with extra deep joint rings were used. These joints allow a joint opening of 1-3/8 in. with a minimum factor of safety of 2. It must be noted that during the compression cycle, the mortar in the joint would prevent axial motion of the joint and the response of the pipe segments would be similar to those with mechanically harnessed joints.

Behavior of Pipeline at Bends

The pipeline has several sharp bends. To resist the thrust resulting from the internal pressure at the bends, the joints of the pipeline are mechanically restrained for a specified length on either side of the bends. When a seismic wave travels along the length of the pipe, the axial force induced in a leg of the bend would cause bending moment in the pipeline near the bends and cause large longitudinal stresses in the pipe walls. A finite element models (FEM) was developed to model the complex soil-structure interaction effect when a seismic wave propagates along the pipeline.

To illustrate the FEM of a bend, a compound bend consisting of three vertical bends was selected for analysis. The FEM of the pipe bend is shown in

Fig. 2. The pipe itself is modeled with linear elastic beam elements. In addition to the restrained length of the pipe, the FEM model included three 20 ft long pipe segments with bell-and-spigot joints on either side of the restrained length of the pipe. For mechanically restrained joints, the joint model provides stiff load paths for transfer of axial force, bending moment, and shear force across the joint. For the bell-and-spigot joints, the model provides for shear transfer between the adjoining pipe segments, but allows their relative rotational motion and axial motion in tension. The soil springs model soil resistance to motion of the pipe perpendicular to its length. The longitudinal frictional resistance of the soil surrounding the pipe was modeled by frictional gap elements placed parallel to the pipe which were connected to the centerline of beam elements by rigid bars. The maximum load that the frictional gap elements carry is defined by the maximum frictional force between the pipe and the soil. A coefficient of friction of $\mu = 1.0$ was used to represent the combined effect of cohesion and friction between the clayey soil and the pipe walls.

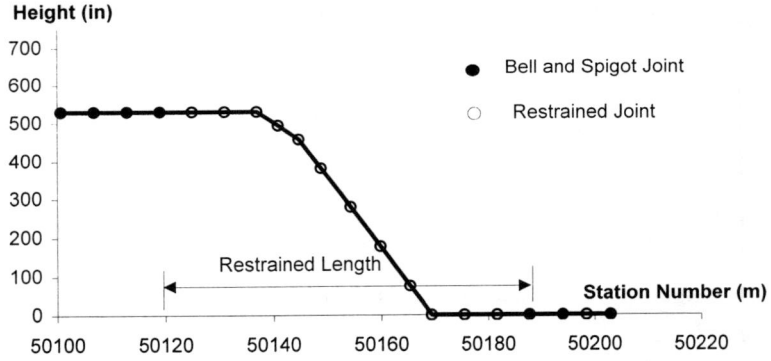

Figure 2. Schematic View of Finite Element Model of Vertical Pipe Bend

In order to determine the stiffness of the soil for vertical pipe motion, the nonlinear soil-structure interaction analysis program CANDE was used. Using a detailed finite-element model of the pipe and the surrounding soil, vertical loads were incrementally applied to the pipe at the springline in both upward and downward directions to calculate the stiffness of the soil. The stiffness of the soil resisting pipe displacement was calculated and used to model soil springs in the bend model.

The seismic loading on the pipe due to wave propagation effect was simulated by imposing a longitudinally linearly varying ground displacement along the length of the pipe, corresponding to a constant ground strain of 460 microstrains expected in the design seismic event. (The actual seismic wave is in the form of a sinusoidal ground displacement; however, since the wavelength is large, the assumption of linearly increasing ground displacement is conservative.) A tension cycle and a compression cycle of the wave propagating along the pipeline were considered. In addition to the seismic loading, the pipeline model is subjected to pressure-induced thrusts at the bends and gravity loads which are needed to develop the pipe-soil friction.

The results of our analysis show that the transient joint opening at the end of the harnessed section of pipe during the design seismic event is 0.41 in. and 0.43 in. at the two ends of the pipe. Therefore, bell and spigot joints with extra deep joint rings were required at the ends of the harnessed lengths at this bend. The seismic effect will increase the thrust in the pipe wall by 6,952 lb/in. (strain of 312 microstrains), which is in addition to the static thrust of 1,253 lb/in. resulting from the pressure-induced thrust. The combined effect of seismically induced strain and pressure induced thrust would result in circumferential microcracking and softening of concrete core and yielding of the cylinder near the spigot ring. The calculated post-yield deformation of the steel cylinder at the spigot ring is 0.023 in. This level of deformation is not large enough to cause fracture of the steel cylinder.

Connection to Large Structures

The connections of pipeline to large monolithic structures, such as tank foundations, are subjected to large relative motions. In a seismic event, the tank slab moves as a rigid disk while the supporting soil deforms with the seismic wave. The influent and effluent piping, being embedded in the soil, will move with the soil. The relative motion between the tank and soil at the connection of the influent, the effluent and the overflow pipes were calculated to be in the range of 1.25 to 1.5 in. occurring simultaneously along and normal to the axis of the pipe. The results of calculations showed that the first joint must accommodate about 1.5 in. and the second joint, 0.5 in. of opening. A special connection detail consisting of pipe joints that can accommodate 3.6 in. of opening without failure was used.

To allow for seismic compressive strain between the tank slab and the pipes, the joints were set with 1.5 in. of initial opening. Consideration was also given to the length of the first pipe segment so that motion normal to the axis of pipe does not produce excessive joint opening.

Soil Transitions

The pipeline was laid in different soil types including blanket deposit, Aymamón limestone, and river terrace deposits and alluvium with transitions from blanket deposit to rock, fill to rock or blanket deposit, and river terrace deposits and alluvium to rock or blanket deposit. To illustrate the seismic analysis for such transitions, a rock-to-fill transition is selected. The subsurface profile of the transition consists of a limestone outcropping sloping at 1H:1V and the depth of fill above the limestone away from the transition is 30 ft (Fig. 3). Thus, one pipe is entirely supported by limestone, and the adjacent pipe segments are supported on fill up to 30 ft in depth. When the limestone underlying a fill is subjected to a seismic ground motion, the pipe segments placed in fill will respond differently than the pipe segment on limestone. This results in relative motions that have to be accommodated by the joints near the transition. The motion at the top of the fill was calculated by applying the simulated bedrock acceleration time history at the top of the limestone layer (Point A), using the soil dynamic analysis program SHAKE. The acceleration time history was then integrated to obtain the displacement at the top of the fill (Point C) and the limestone outcropping (Point B). The maximum relative displacement between the top of the fill and the limestone is 0.25 in. for 30 ft fill height. It was conservatively assumed that this relative motion occurs at a single joint in the fill-to-rock transition area. Superimposing the effect of wave motion and the fill-to-rock transition effect, for shear waves causing particle motion along the pipe axis, the maximum relative motion of the adjoining 20 ft pipe segments is expected to be 0.36 in. The joint opening due to shear waves causing lateral pipe motion can be similarly calculated. Based on this calculation, when depth of fill is 30 ft, bell-and-spigot joints with extra deep joint rings were used for all pipe segments that are installed partly on fill and partly on limestone and for the joint on the adjoining pipe segment installed entirely on fill.

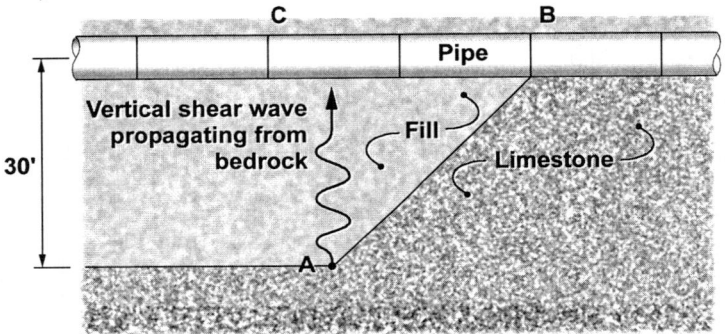

Figure 3. **Fill-to-Rock Transition**

Vertical Acceleration Loading on Pipeline

The design earthquake has a vertical acceleration of 0.21 g for pipe laid in blanket deposit and 0.28 g for pipe laid in 30 ft of fill. That is, the gravity effects, i.e., pipe and fluid weights and earth load, will increase by 21% to 28% in the design seismic event. The additional gravity effect is considered as a transient live load. The pipe design was evaluated accounting for the additional seismic load resulting from the effect of vertical acceleration on earth load, pipe weight, and fluid weight. The results of our evaluation showed that the pipe designs developed for static load combinations were adequate for the additional live load resulting from the vertical acceleration of the design seismic event.

Slope Stability and Lateral Displacement

The Superaqueduct runs at various locations along the top or toe of Highway PR-22 embankment or ascends or descends steep slopes at major crossings. Seismically-induced slope instability and lateral ground displacement could cause pipeline failure and possibly highway embankment failure in major seismic events.

The study of seismically-induced instability and lateral ground displacement was performed in two phases. In the first phase, preliminary analyses of slope stability and seismically-induced displacement of the embankment were performed using conservative values for shear strengths of the embankment fill and the natural soils. In the second phase, additional field explorations and laboratory testing were performed at potential failure sites to determine the shear strengths of the embankment fill and the underlaying natural soils. Final analyses of slope stability and seismically-induced lateral displacement were performed using the measured soil properties.

The field exploration included additional borings and observation wells at the top and toe of the highway embankment slopes. Boring depths ranged from 15 to 45 ft. Split spoon samples were obtained at intervals of 2 to 5 ft and undisturbed Shelby tube samples of the natural soil deposits were collected. Water levels were measured in the observation wells.

The laboratory testing included unconsolidated undrained triaxial compression tests performed on undisturbed tube samples of the natural soil deposits. The peak undrained shear strength determined for each test represents either the maximum shear stress mobilized during the test or the value at 15 percent strain, whichever occurred first. Shear stresses were calculated taking into account the change in average areas of the specimen during the test. The peak undrained shear strength of the natural soil deposits (blanket soils) varied form 1.6 to 1.9 kips per square foot (ksf) for samples collected at the toe of the embankment and from 4.2 to 4.3 ksf for samples collected from beneath the embankment. The samples collected from beneath the embankment exhibited higher strengths due to consolidation under the weight of the embankment.

Anisotrophically consolidated undrained triaxial compression tests were performed on recompacted samples of the embankment fill. Samples of the embankment fill were collected from the auger cuttings from one boring between depths of 15 to 20 ft and compacted to a target density and a moisture content. The recompacted samples were saturated under back pressure in the triaxial cells. The samples were then consolidated isotropically to the estimated horizontal effective stresses and anisotrophically to the estimated vertical effective stresses at the selected locations on the failure plane. After consolidation was complete, the samples were sheared undrained. Changes in pore pressure during shearing were recorded. The samples exhibited a peak undrained shear strength of 1.28 ksf and a steady-state strength of approximately 1 ksf, which was used in the stability analyses.

The stability analysis was performed as follows:

1. Developed a soil profile for each site based on the plan and profile drawings and the subsoil explorations performed.

2. Estimated depth to ground water based on the water content data reported on the boring logs.

3. Performed pseudo static slope stability analyses to estimate the yield acceleration, i.e., the average horizontal acceleration on the sliding mass resulting in a factor of safety of 1.0) for the highway embankment.

4. Estimated the permanent displacements based on the ratio of the yield acceleration to the design peak ground surface acceleration of 0.32 g and the laboratory measured shear strengths of the natural soil deposit and of the embankment fill. Displacements were estimated using published data from Makdisi and Seed (1978) and Hynes-Griffin and Franklin (1984).

The results of these analyses (Fig. 4) produced major and minor changes in the pipeline alignment to avoid problem areas.

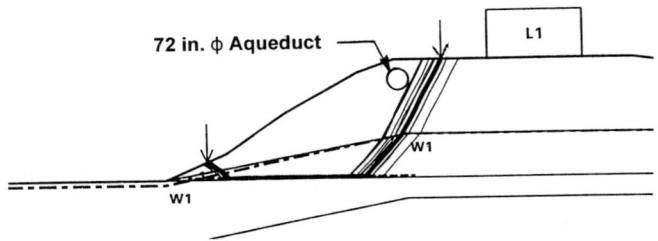

Figure 4. Critical Planes of Failure in Pseudo Static Slope Stability Analysis

Conclusions

The comprehensive approach for evaluation and control of seismic risk for the north Coast Superaqueduct is presented. The seismic safety of the Superaqueduct is especially of concern because of the frequent seismic activity at the site, the special features of the geology and topography of the site, and the risk of its failure and the consequential damages to other infrastructures of the island. Although the results are site-specific, the methodology can be applied to seismic risk management of other pipelines.

References

ASCE (1986), *Seismic Analysis Safety-Related Nuclear Structures and Commentary on Standard for Seismic Analysis of Safety Related Nuclear Structures*, ASCE Standard 4-86, American Society of Civil Engineers, September.

Hynes-Griffin, M.E., and Franklin, A.G. (1984), "Rationalizing the Seismic Coefficient Method," Miscellaneous Paper GL-84-13, U.S. Army Corps of Engineers, Vicksburg, MI, July.

Makdisi, F.I., and Seed, H.B. (1978), "Simplified Procedure for Estimating Dam and Embankment Earthquake-Induced Deformations," *ASCE Geotechnical Journal*, Vol. 104, No. GT7, July, pp. 849-867.

O'Rourke, M.J., Bloom, and Dobry, R. (1982), "Apparent Propagation Velocity of Body Waves," *Earthquake Engineering and Structural Dynamics*, Vol. 10, pp. 283-294.

Seed, H.B., and Idriss, I.M. (1982), "Ground Motions and Soil Liquefaction During Earthquakes," Monograph Series, Earthquake Engineering Research Institute, Berkeley, CA.

Zarghamee, M.S., and Fok, K.L. (1990), "Analysis of Prestressed Concrete Pipe under Combined Loads," *Journal of Structural Engineering*, Vol. 116, No. 7, pp. 2022-2039.

Case Study: Crossing of Existing PCCP Aqueducts A and B
Southern Nevada Water Authority
Las Vegas, Nevada

Philip K. Ryan[1]   R. Ted Davis[2]
Member   Member

Abstract

The Southern Nevada Water Authority (SNWA) is the principal wholesaler of water to purveyors in the Las Vegas Valley. Approximately 85 percent of the potable water supplied by SNWA is currently conveyed from the Alfred Merritt Smith Water Treatment Facility at Lake Mead to the Las Vegas Valley via Aqueducts A and B. These aqueducts are two parallel large diameter pipelines (2,438 mm [96-inch]) constructed of prestressed concrete cylinder pipe (PCCP). A third large diameter (2,286 and 2.438 mm [90 and 96-inch]) water supply aqueduct is being constructed to increase the capacity of SNWA's system. The alignment selected for this new aqueduct requires crossing the existing PCCP aqueducts at two locations.

The existing aqueducts were designed for about 1.5 to 3 meters (5 to 10 feet) of earth cover at both crossing locations. Also, neither aqueduct can be taken out of service for more than a few days and that outage may only take place during the winter low demand period. The new pipe (full) would weigh more than 4,470 Kg/lineal meter (3,000 lb/lineal foot) and could cause an undesirable loading condition if placed directly over the existing aqueducts. A crossing configuration that does not pose a significant structural or construction related risk to the performance of the existing aqueducts is required.

As part of design activities, several alternatives were considered for crossing over or under the existing aqueducts. They included:

- Open Cut Undercrossing
- Tunneled Undercrossing in In-Situ Soils
- Tunneled Undercrossing in Chemically Hardened Soils
- Structurally Supported Overcrossing

---

[1] Civil Engineer, CH2M HILL, 2030 E. Flamingo Road, Suite A, Las Vegas, NV, 89119
[2] Civil Engineer, CH2M HILL, 1525 Airpark Drive, Redding, CA, 96002
[1,2] (Both assigned to Montgomery Watson/CH2M HILL Joint Venture for SNWA Project)

The structurally supported overcrossing was selected because of its lower cost and lower risk characteristics relative to other alternatives.

The structurally supported overcrossing consists of a structural support, earth cover fill material, and a welded steel pipe overcrossing. The structural support includes a corrugated arch with a reinforced concrete load carrying cap straddling each existing aqueduct. The earth cover includes mounded and compacted native soils filled to a configuration that provides visual and thermal cover over the entire crossing. The welded steel pipe overcrossing includes appropriate air release and blowoff appurtenances. The design concept included provisions for excavating the trench for the new pipe into the compacted soil cover and using a controlled low strength material (CLSM) for pipe zone and bedding.

Design development information and sketches depicting the structural support and the Aqueduct C overcrossing are presented.

Background

Conveyance of treated water through Aqueducts A and B of the Southern Nevada Water System (SNWS) accounted for about 85 percent of the water supply for the Las Vegas Valley in 1997. As part of a regional water supply expansion program, a third water supply pipeline, Aqueduct C, is being added to the system. Aqueducts A and B are parallel 2,438 mm (96-inch) diameter prestressed concrete cylinder pipes (PCCP-embedded cylinder type). The pipelines have single rubber gasket "Carnegie" type bell and spigot joints. Aqueduct A was constructed in the late 1960s and Aqueduct B was constructed on an immediately parallel alignment (generally about 15 m [50 feet] away) in the mid 1970s. Both aqueducts have been inspected recently at several locations and found to be in excellent condition. In fact, inspections of sections of the aqueducts removed at nearby tie-in locations conducted as part of related construction projects showed the pipe to be in "like new" condition.

The new Aqueduct C system includes a lower reach, Aqueduct C1, and an upper reach, Aqueduct C2. Aqueduct C1 is a 2,286 mm (90-inch) diameter welded steel pipe with mortar lining and a polyethylene tape coating with a mortar rock shield. Aqueduct C2 is a 2,438 mm (96-inch) diameter welded steel pipe, lined and coated the same as Aqueduct C1. Two new pumping stations are included along Aqueduct C; one at the beginning of the aqueduct at the existing Alfred Merritt Smith Water Treatment Facility (AMSWTF), and one at about the system midpoint that separates Aqueducts C1 and C2.

An extensive route evaluation was conducted to select the best route for Aqueduct C from the AMSWTF to the east portal of the existing River Mountains Tunnels. The route selected for Aqueducts C1 and C2 required crossing existing Aqueducts A and B at two locations; one each at the upstream end of both Aqueducts C1 and C2.

Crossing Alternatives

The important water supply function of the existing aqueducts dictated that any crossing by Aqueduct C must be conducted using a method that reduced the risk of structural damage both during and after construction to an insignificant level. Also, since these aqueducts supply the majority of water to the Las Vegas Valley, operational parameters dictated that any single aqueduct could not be taken out of service for more than about 60-hours at a time. One aqueduct must also remain in service at all times. Due to the high demand for water in the Las Vegas Valley, it was also necessary that all service interruptions take place during the lowest demand portion of the year; typically December and January.

A variety of alternatives to accomplish the crossing were initially considered. The following four alternatives were selected for a comparative evaluation:

- Open Cut Undercrossing—This alternative included supporting the existing aqueducts in an open cut excavation while Aqueduct C is installed underneath.
- Tunneled Undercrossing in In-Situ Soils—This alternative included utilizing conventional tunneling techniques in the native geology to install Aqueduct C beneath the existing aqueducts while they remained in place and undisturbed.
- Tunneled Undercrossing in Chemically Hardened Soils—This alternative included utilizing conventional tunneling techniques in chemically hardened soils to install Aqueduct C beneath the existing aqueducts while they remained in place and undisturbed.
- Structurally Supported Overcrossing—This alternative included constructing Aqueduct C over the top of the existing aqueducts and using a structural support to prevent the additional load from being carried by the existing pipes.

Description of Alternatives

Open Cut Undercrossing. For this alternative, open cut excavations and conventional pipe laying would be employed to install Aqueduct C underneath the existing aqueducts. Installing the new pipeline beneath the existing aqueducts using an open cut trench would require a carefully excavated trench at least 3.4 to 3.7 meters (11 to 12 feet) wide. This width assumes that vertical trench walls could be maintained with shoring or bracing. Assuming the trench could be made stable beneath the existing pipes, of equal concern is supporting the existing 2,438 mm (96-inch) diameter aqueducts across the open trench. The existing aqueducts were constructed using 7.3-meter (24-foot) long sections, and therefore it is likely that a joint would be exposed in the excavations. Due to the importance of continued service in the existing aqueducts, a temporary structural support system would be required for the existing aqueducts during construction. Also, high strength backfill, such as concrete, would be required between the new and existing pipes to help eliminate settlement after construction.

<u>Tunneled Undercrossing in In-Situ Soils</u>. For this alternative, conventional tunneling techniques would be used to install Aqueduct C underneath the existing Aqueducts. Tunneling would be conducted such that disturbance to the ground surface and the existing aqueducts is avoided. Tunneling methods for pipelines the size of Aqueduct C generally include microtunneling, pipe jacking, and utility tunneling.

Microtunneling is performed using a remotely controlled tunnel boring machine (TBM). Based on the mode of operation, microtunneling can be subdivided into two groups; slurry method or auger method.

In the slurry method, soil is cut at the face of the TBM and removed by hydraulic action. A slurry is circulated from the face of the machine, out to a settling pond on the ground surface, and back into the face of the TBM. In addition to removing the cuttings, the pressurized slurry acts to stabilize soil at the face of the cutter head. This method can be helpful in unstable soil conditions that would be prone to cave-in using other tunneling methods. The face of the articulating TBM is controlled and steered with a system of hydraulic rams.

As in the slurry method, a cutter head is also used with the auger method. The spoils, however, are transferred to the installation pit with an auger. A clam shell or excavator at the installation pit is then used to transfer the spoils to the surface. The auger system and boring head can be sealed and air pressure applied to help balance hydrostatic groundwater conditions thereby providing some stability of the excavation. This method, however, is not as effective in stabilizing the excavation as the slurry method in granular soils that run freely with or without groundwater flow.

Pipe jacking typically involves workers working within and at the head of the casing pipe and either removing the material by hand or with mechanical equipment such as a boring head or small hydraulic backhoe. Spoils are removed by carts. As the excavation proceeds, the casing pipe is jacked through the excavation with hydraulic rams.

Utility tunneling is similar to pipe jacking except tunnel liner plate is normally installed as the excavation proceeds instead of jacking a casing pipe through the excavation.

It was assumed that all of these tunneling methods would be allowed by the contract documents for this alternative. For all of these methods, a casing pipe or tunnel liner would be installed under the existing aqueducts providing a conduit for installation of the new pipeline, or carrier pipe. A pit would be excavated at one end of the crossing for operation of the tunneling equipment and installation of the casing and carrier pipe. After installation of the carrier pipe, the annular space between the casing and the carrier pipe would be filled with sand or grout. The approximate length of the pipe installed by tunneling would be about 43 lineal meters (140 lineal feet) for each crossing. This length allows a single tunneling operation to cross beneath both existing aqueducts.

Tunneled Undercrossing in Chemically Hardened Soils. For this alternative, conventional tunneling techniques similar to those described for the previous alternative would be used to install Aqueduct C underneath the existing Aqueducts. However, the existing site soils would be hardened using chemical grouts prior to tunneling. Chemical hardening was included to provide a homogenous soil column for tunneling, to stabilize the soil column against cave-in, and to support the pipelines against shock or movement. Other details regarding this alternative would be the same as those described for the previous alternative.

Structurally Supported Overcrossing. This alternative includes installing Aqueduct C over the top of the existing aqueducts. Figure 1 shows a conceptual layout of this alternative.

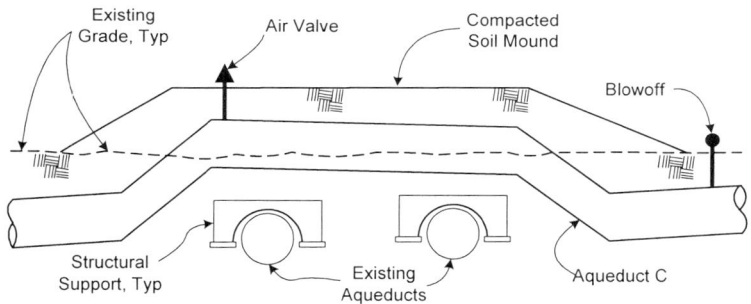

Figure 1. Structurally Supported Overcrossing Alternative

The Structurally Supported Overcrossing includes the following main elements:

- Excavate to springline of the existing pipes and construct structural supports over the existing aqueducts to carry the load of the new pipe and increased earth loads. These supports would be about 37 to 43 meters (120 to 140 feet) long and extend along the existing aqueducts the full length of the area to be modified above them. The supports would be constructed using a reinforced concrete cap over an arch culvert support.
- Mound sufficient compacted soil over the top of the existing aqueducts (over the structural supports) to provide bedding and backfill for the new pipe as well as providing earth cover for thermal, visual, and structural protection.
- Install the new pipeline through the compacted soil mound. The new pipe configuration would include two bends of approximately 25- to 45-degree deflection on each side of the existing aqueducts to bring the Aqueduct C pipeline up, over, and back down while crossing the existing aqueducts. Also included would be a combination air/release valve to vent accumulated air at the high point and a minor blowoff at the low point to allow for draining of the pipeline.

Geotechnical Setting

Geotechnical explorations at the crossing locations indicate that native soils are a mixture of sands and gravels with the potential for larger rocks and cobbles. For the C1 crossing, the sands and gravels overlay weathered conglomerate of the Muddy Creek Formation, a cemented alluvium. Therefore, layers of cemented material are likely to occur in the soil column. During excavation of test pits, refusal occurred at depths from 8 to 12 feet substantiating the existence of these hard layers.

Near the C2 crossing, the subsurface profile consists of fill and alluvium underlain by weathered and altered volcanic bedrock. During excavation of test pits, backhoe refusal occurred between 15 and 20 feet deep. Test trenches also confirmed these depths to bedrock.

Non-Cost Evaluation and Screening of Alternatives

<u>Open Cut Undercrossing</u>. This alternative would require trenching to depths of about 4.6 meters (15 feet) below the existing aqueducts for a total depth of about 9.1 meters (30 feet). Excavations for the Aqueduct C1 crossing would likely encounter zones of cemented soils. Excavations for the Aqueduct C2 crossing would likely encounter rock beneath the existing pipes. Experience in the area indicates that rock excavation techniques are likely to be required for both the cemented zones and the rock zones. Since the existing aqueducts are important water supply facilities for the Las Vegas Valley, blasting close to them is considered an unacceptable risk. Therefore, rock excavation would need to be conducted using mechanical methods or hydraulic splitting. Hydraulic splitting methods were not expected to be particularly effective at this site and where expected to be slow and costly. Mechanical rock excavation was considered the likely method that would be used.

Also, trench support systems and a structural support system for the existing PCCP pipes would need to be installed as excavations progressed. Each temporary support system for the existing pipes would need to be capable of supporting a 2,286 mm (96-inch) PCCP full of water. A support system design to support the pipe from above would need to span the width of the excavations and be compatible with conducting those excavations. Critical criteria would include not imposing additional shear loads on adjacent pipe joints. Also, only minimal deflections could be tolerated for either the pipe or the support system. Excess deflection could result in failure of the rubber gasket pipe joints.

Overall, the proposed excavations and support systems needed for an open cut operation were considered difficult to successfully accomplish. This difficulty would increase the potential for loss of pipe support and possible pipe failure during construction. Since other alternatives were available, the open cut undercrossing was dropped from further consideration due to it's higher risk potential.

<u>Tunneled Undercrossing in In-Situ Soils</u>. Several site specific factors need to be considered when reviewing the appropriate tunneling alternative. For the Aqueduct C

crossings, it is important that the existing pipelines are not damaged by loss of pipe support or by jarring from shock waves or earth movement. Also, the expected underground conditions for the site will affect the choice of tunneling method.

TBM techniques work best in generally homogeneous soil columns. Due to the variability and layering of the native soils at the crossing sites (layers of rock, cemented gravels, and cobbles), non-homogeneous conditions at the boring head are expected. This complicates control of the boring head. If a rock boring head is used and gravels are encountered in a portion of the tunnel, the gravels tend to bind against the rock and prevent rotation of the head. Additionally, for the size of casing involved with this project (3,048 mm [120-inch] diameter), and the short tunneling runs, mobilization of a TBM can be very expensive and the equipment difficult to schedule. Normally, use of pipe jacking or utility tunneling methods would be more economical than mobilizing a TBM. However, due to mixed soil conditions, including gravels that could run into the excavation, the need to excavate rock and cemented material, and the need to maintain support for the existing pipe, these methods would probably be more risky than microtunneling. Since blasting is not recommended, and even small movements of the existing pipes are unacceptable, the apparent approach for tunneling alternatives is to lower the tunnel entirely into bedrock or a homogeneous cemented material.

It was concluded that implementation of this alternative would indeed require lowering the tunnel grade at both crossing locations. Another advantage to lowering the tunnel elevation is to provide additional soil column depth beneath the existing aqueducts for soil bridging in case minor caving occurred. At the Aqueduct C1 location, mixed face tunneling would probably still be required because the variable geotechnical conditions were observed in nearby excavations and borings to a considerable depth. Due to the variability of the soils, pipe jacking or utility tunneling would be the likely method employed. At the Aqueduct C2 location, it was expected that the tunnel could be lowered into a completely rock column. However, the limits on blasting would make pipe jacking or utility tunneling difficult and microtunneling was considered the most feasible method. The complexity of utilizing two different tunneling methods, and the need to install deep pipe fittings to adjust Aqueduct C to the lower tunneling grade were expected to make this alternative significantly more costly than others available. Therefore, the conventional tunneling alternative was dropped from further consideration.

<u>Tunneled Undercrossing in Chemically Hardened Soils.</u> This alternative includes stabilizing the soil surrounding the proposed tunnel area by chemical grouting. This feature would provide a homogenous soil column for tunneling, stabilize the soil column against cave-in, and support the pipelines against shock or movement. The hardened soil column would allow tunneling by TBM at a shallower grade beneath the existing aqueducts thereby reducing excavation and material costs. Since this alternative eliminates most of the significant risks and complex implementation challenges previously identified for tunneling operations, this alternative was further considered on a cost basis.

<u>Structurally Supported Overcrossing</u>. The principal issues related to this alternative involve the ability to support the additional loads imposed by the new pipe and compacted soil mound. Also, the welded steel pipe will require side support in the trench as it crosses over the existing aqueducts. These issues were addressed by including the structural support system over the existing aqueducts and requiring that the pipe trench be excavated in a previously placed compacted soil mound with a pipe bedding and zone constructed of a soil/cement slurry. Cover over the top of the pipe was minimized to 0.6 to 0.9 meters (2 to 3 feet) to provide thermal and visual cover only. The low depth of cover facilitated structural design of the pipe to resist buckling from less than ideal side support.

This alternative could also be constructed using conventional means and methods expected to be used for other aspects of the project. There is no chance for loss of soil support under existing Aqueducts A and B. The overcrossing is not effected by variable underground conditions, thereby reducing uncertainty regarding construction methods and hopefully reducing bid contingencies. Trench shoring and bracing would only be required if excavations could not sloped. However, suitable space is available at both sites to slope all excavations required for this alternative.

Since there are no significant unresolved risks or issue related to this alternative, it was further considered on a cost basis.

Cost Comparison

The Tunneled Undercrossing in Chemically Hardened Soils and the Structurally Supported Overcrossing were compared on a cost basis to help select a final alternative for implementation. Order-of-magnitude costs were developed using the configurations of the alternatives described above. The mobilization cost for the Structurally Supported Overcrossing was assumed to be 10 percent of the subtotal construction estimate. The mobilization cost for the Tunneled Undercrossing in Chemically Hardened Soils was assumed to be 20 percent of the subtotal construction estimate to account for the special tunneling equipment, customizing a TBM for the work, and additional equipment shipping requirements.

The cost of the Aqueduct C pipe was common to both methods and not included in the estimate. The cost of the four pipe bend fittings on the overcrossing were included since they were particular to that alternative.

The estimated comparative costs for the alternatives are:

- Tunneled Undercrossing in Chemically Hardened Soils $ 1,127,000
- Structurally Supported Overcrossing $ 707,000

Selected Alternative

The Structurally Supported Overcrossing alternative was selected for implementation because it had the lowest cost, had low risk of damage to the existing aqueducts, could be constructed with construction equipment and methods already intended for the work, and had less uncertainties. All of these factors were expected to contribute to lower bid contingencies and higher degree of safety against damage.

Design Development Information

Figures 2 and 3 (following page) show more detailed drawings for the Structurally Supported Overcrossing alternative. The following criteria were used to prepare the final design for the selected alternative.

Structural Supports

- Structural supports to extend along existing aqueducts for full length of soil mound above
- Footing set on undisturbed earth and located low enough to transfer the loading influence zone beneath the existing pipes
- Structural support includes a multi-plate arch and a reinforced concrete cap
- Design includes an air gap over existing pipe to allow some deflection of the arch and concrete cap under load without transferring loads to existing pipes
- Both the arch and concrete cap designed for full design load
- Design load equal to full height of cover plus an 36,364 Kg (80,000 lb) wheel load (note that soil cover weighs more than full Aqueduct C pipe)
- Special designed CLSM end caps used to prevent backfill material from entering air gap beneath support and potentially transferring loads to pipe.

Existing Aqueducts

- Sand blast to clean outer pipe surface, coat with coal tar, and cover with polyethylene sheeting to provide additional die-electric corrosion protection

Aqueduct C

- Backfill between top of structural supports and bottom of Aqueduct C with controlled low strength material (CLSM)
- Excavate trench in previously placed compacted soil mound for better support
- Use CLSM pipe bedding and zone
- Design for full vacuum at 70% relative compaction of pipe side support and 0.9 meters (3 feet) of earth cover or for full vacuum in exposed condition, whichever is worse
- Provide air release and vacuum protection with full redundancy
- Coordinate blowoff appurtenances for drainage; coordinate layout with remainder of Aqueduct C configuration

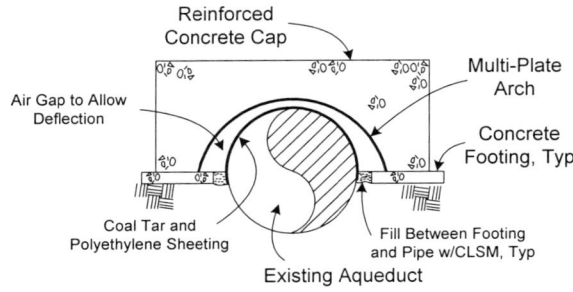

Figure 2. Structural Support Detail

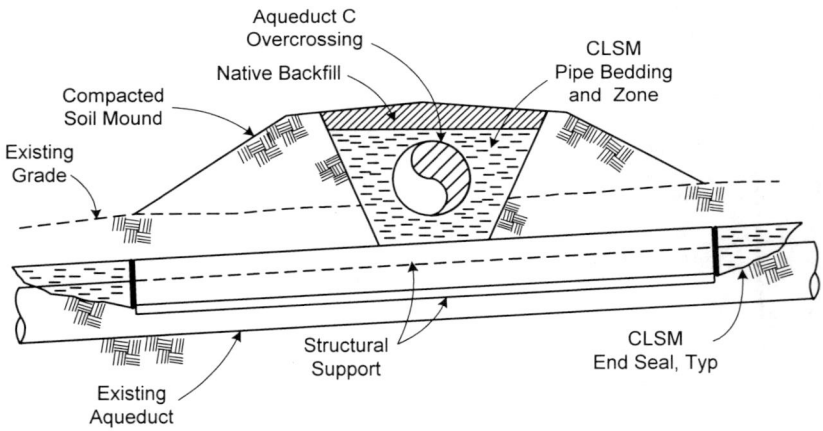

Figure 3. Structurally Supported Overcrossing Section

Pipe Joint Failure Caused By an Inadequately Specified Constructed Environment

Kenneth K. Kienow, P.E., G.E., MASCE[1]

Abstract

A number of 96-inch diameter AWWA C-302 concrete pressure pipe joints failed during installation. The failures were caused by an AWWA specification used in the wrong environment, lack of adequate pipe inspection, failure to understand the implications of the pipe specification joint tolerances, and failure to take into account pipe specification dimensional tolerances in the specified constructed pipeline environment. The pipe specified was AWWA C-302 Concrete Pressure Pipe, with "Carnegie" steel joint rings. Some of the contributing causes were improperly specified pipe bedding, a pipe standard specification which was inappropriate for the project conditions, failure to adequately inspect the pipe, and failure to store the pipe in the specified vertical position. The author identifies the causative factors and recommends steps which can be taken to avoid similar expensive pipe and bedding modifications, repairs, jobsite delays, and remedial measures which were required on this project.

Introduction

The specified pipe was 96-inch diameter, 16-inch thick wall, AWWA C-302 Concrete Pressure Pipe which was to serve as the outlet pipe under a large flood control dam, not yet constructed. The installation specification called for the pipe to be installed on precast concrete cradle bedding units. Kienow Associates, Inc. (KAI)

---

[1] President, Kienow Associates, Inc. PO Box 121110, Big Bear Lake, CA 92315   phone 909-866-8636
E-mail: kkienow@bigbear.net

was called to the site by the contractor, when he experienced difficulty joining the pipes. Pipe bells and spigots were breaking, massive forces were being applied to the pipe in attempts to make the joints, and when a joint could be made up, the joints leaked under the hydro joint test.

An on-site inspection was made, and the project plans and specifications were reviewed. The objective of our review was to determine the cause or causes of the difficulty in assembling the pipe joints, the cracked and leaking pipe joints, to observe a hydrostatic test, and to evaluate several sections of pipe which had been joined and then removed after failing the hydrostatic test.

In addition to the on-site review, the pipe manufacturer's shop submittal drawings, the pipe manufacturer's design calculations for the 96-inch pipe, the pipe portion of the project specifications, and the pipe manufacturer's submitted gasket calculations were reviewed.

The pipe barrel length and the distance from the bell and spigot rings to pipe outside diameter was measured on six pipe stored on site. The purpose of these measurements was to determine the variation in length of the pipe barrel and also the variation in distance from the outside pipe wall O.D. to the bell ring I.D. and spigot ring O.D. Since the pipe rested on already installed to grade precast concrete cradles, any variation in the joint ring to pipe outside diameter dimension will result in a similar variation in the alignment of the joint mating surfaces as the pipe is installed and resting on the cradles, and pulled home. The joint alignment is critical to laying the pipe without damaging the integrity of the pipe or joint.

The maximum differentials found in six pieces of randomly selected pipe were as follows:

Pipe barrel length outside: 3/4-inch difference
Pipe barrel length inside: 5/8-inch difference
Spigot ring to pipe O.D.:   1/2-inch difference

The project specifications required that the precast concrete cradles be set to grade within ± 0.01 foot, or ± 1/8-inch. The pipe and precast cradles are shown in Figure 1.

FIGURE 1
96 inch Concrete Pipe on Precast Cradles

One section of pipe joined and later removed was cracked at the bell end, with the crack originating at the bell ring, radiating diagonally across the bell face and longitudinally along the outside of the pipe for twelve to eighteen inches. There was no evidence of any impact which might have caused this type of cracking damage.

The cracked pipe bell end was photographed (Figure 2), as was other pipe with visible cracks on the joint ends. While some cracks appeared to be shrinkage cracks, others were ring or hoop tension cracks in the bell from attempting to force a too large spigot ring into the smaller bell. Some of the cracks were observed to bleed or leak water during the hydrostatic test, with some actually emitting a steady stream of water through the bell face when under hydrostatic pressure. The water apparently migrated along the joint ring, past the ring and out the joint face. In addition, some disbonded or unbonded areas were evident at the steel/concrete interface.

Additional Observations

1. The manufacturer's pipe strength calculations conformed to the project specifications.

2. The project specifications required that the pipe be stored in the vertical position in the pipe yard, and strutted with timbers if stored horizontally on the jobsite, in order to prevent "egging", or out-of round deflection of the pipe. According to the pipe manufacturer, the pipe was stored horizontally in the plant. The pipe was not strutted, as called for in the specifications for horizontal storage. Diameter measurements indicated that only minimal deflection had taken place, and most pipe were within the C-302 allowable out-of-round tolerance.

3. The random sample concrete pipe barrel length differentials were within the allowable tolerances of AWWA C-302. The joint ring to pipe O.D. (wall thickness) variation was also within the C-302 tolerance. The barrel length variation made it difficult and in some cases impossible to maintain a uniform 5/8-inch inside joint space as required by the project specifications.

**FIGURE 2
Bell Section Cracking**

4. The precast concrete cradle units were installed at or near the ± 1/8-inch grade requirement.

5. An out-of-round or ellipse condition was observed in some of the bell and spigot rings, partially due to ellipsing of the pipe. The degree of out-of-round was checked using a micrometer extension pole and calipers, made for this type of measurement. The pipe on the site were found to be within ASTM C-302 tolerances.

6. The large pipe size, 96 inch diameter, and 16 inch thick pipe wall resulted in a weight per eight foot joint length of pipe of about 25 tons. The concrete thickness at the spigot end was about 4 3/4-inches. The spigot (or bell) is susceptible to damage from any misalignment of the joint. The spigot (or bell) was not designed to, or capable of, carrying half the pipe weight of 12 1/2 tons. Yet, due to variations in the spigot ring to pipe O.D. dimensions, in some cases the pipe was "hanging" or suspended by the joint ring surfaces, rather than bearing on the concrete cradle. The potential damage from such misalignment varied from spigot or bell ring deformation, cracking or spalling of the concrete behind the joint ring, or failure of the bond between the concrete and the joint ring. Any one of these conditions may result in a joint leak, as water migrates along the concrete/steel interface, through the concrete, and out the bell face or other joint surface.

7. The variation of as much as 5/8-inch in the distance from the pipe O.D. to the spigot ring O.D. or bell ring I.D. is not consistent with the required joint alignment accuracy needed to successfully assemble the joint, since the pipe rests either on the cradle and the joint is misaligned, or the pipe "hangs" on the joint ring and does not rest firmly on the cradle. The result in either case is to put severe shear and bending strain on the joint rings which they are not designed to withstand.

Initial Remedy Attempts

The following jobsite procedures were instituted in an attempt to alleviate the pipe jointing and leaking problems:

1. Epoxy injection was used to repair the damaged bell

ends, as well as any joints which showed signs of lack of intimate contact and bond between the joint ring and the concrete. Such areas were observed visually in some cases and in other cases were identified by "sounding" the joint area gently with a hammer.

2. Pipe sections were rotated in the field before laying to match the spigot to O.D. dimension to the bell to O.D. dimension at the invert of the pipe last laid. Each subsequent spigot to O.D. dimension of the pipe last laid was then matched. Selecting individual pipe from among those stored on the jobsite, and rotating of the 25 ton pipe sections to match a spigot with a bell it would fit into proved to be very expensive and time consuming.

3. The grade of the top of the concrete cradles was lowered by one or two inches by removing concrete with a bush hammer, or by reinstalling the cradles. Lubricated 1/8-inch Masonite shim pads were placed between the cradles and the pipe to make up the difference in spigot OD to pipe OD dimensions so that the bell and spigot rings were made to align as closely as possible to the concentric position before bringing the joint home.

4. Care was taken to avoid tipping the pipe spigot up or down to gain entry to the bell, and the joint was not pulled home unless the spigot and bell ring were aligned and concentric. A pulling force of five tons maximum was estimated to be sufficient to overcome friction of the cradles and apply sufficient force to home the joint. Forces in excess of five tons would further disbond and distort the rings resulting in leaking joints.

5. Consideration was given to omitting the cradles completely, and laying the pipe on firm but yielding granular bedding material.

6. The contractor was asked to consider a pipe handling method other than a sling. One suggested possibility was to core one or two lift holes in the pipe at the top, and use a device inside the pipe which will spread the load adequately. The lift holes could be repaired with an epoxy mortar mix after laying.

More Jobsite Pipe Failures

The above changes were instituted, and attempts to join the pipe were again begun, with little success.

Specifically, the changes made included marking the pipe prior to laying and attempting to match the spigot ring to pipe O.D. dimensions to the pipe bell already laid. Due to the variation in this dimension (wall thickness variation) longitudinally along the pipe, the joint ring would not hold at a constant elevation as the pipe was pulled home. The force required to make the joint was such that the bolts were completely sheared off on the tugger used to attempt to pull the joints home. The tugger was capable of pulling with 60,000 pounds force, ten percent greater than the pipe section weight, but joint closure could still not be achieved, even at these extreme pressures. Two pipe sections were removed and the cradles bush hammered to allow 1/2 inch clearance between the pipe OD and the cradle, to allow for placement of shims between the cradle and the pipe (Figure 3).

Observation of the second pipe removed showed that the paint was scraped off the bell surface down to bare metal at approximately the 2, 4, 7, and 10 o'clock positions, indicating metal to metal joint ring contact at these points. The gasket was severely cut for the bottom 20 degrees of the joint, due to being pinched between the back of the spigot groove and the bell ring. The paint scrape locations were consistent with the "squaring" of the spigot observed in the joint micrometer measurements made earlier.

The principal problem up to this point appeared to be the AWWA C-302 dimensional requirements and associated tolerances, particularly the fact that there is no plus tolerance required on the design pipe wall thickness. The allowable variation in wall thickness was not consistent with the project design which required laying on concrete cradles preset to grade. In addition, there were deficiencies in the pipe steel joint ring/concrete bond and concrete porosity which contributed to the leaking problem.

After lowering the tops of the cradles, another attempt was made to join the pipe. As the joint closed, a very loud "pop" was heard, and the concrete cracked at the pipe joint. Pipe laying efforts were stopped at this point. Lowering the cradles and carefully matching the OD/joint ring dimension did not eliminate the difficulty in making the joint. The pipe manufacturer continued to insist that the pipe met the project and AWWA C-302

PIPELINES IN THE CONSTRUCTED ENVIRONMENT 441

FIGURE 3
Lowering Concrete Cradles With Bush Hammer

specifications and blamed the cracking and leaking problem on the contractor's jobsite procedures in laying the pipe.

Numerous additional micrometer measurements were made on the pipe joint rings on the jobsite. In many cases, it became apparent that the spigot O.D. dimension at some point on the circumference was actually larger than some of the measured bell dimensions, in one case by as much as 3/16 inch. For example, comparing diameter "A" on the bell of pipe #9 with diameter "C" on pipe #10 indicated that the spigot diameter was greater than the bell diameter by 0.163 inches, or 5/32 of an inch. Similarly, pipe #12 spigot diameter exceeded pipe #18 bell diameter by over 1/8 inch, and pipe #11 spigot diameter exceeded pipe #9 bell diameter by .044 inches. A bell joint ring confined by 10-3/4 inches of concrete behind it is not going to deform to accept a spigot 1/8 to 3/16 inches larger than the bell, without causing a concrete ring tension cracking failure at the joint. Neither is the spigot going to be able to deform with nearly five inches of concrete inside it.

The micrometer measurements also indicated a very small average difference when the spigot average diameter was compared to the bell average diameter. Average annular clearances measured were: pipe #11: .024 inches; pipe #12: .026 inches; pipe #13: .019 inches; and pipe #16: .015 inches.

Figure 4 (page 14) indicates typical joint interference, specifically for pipe sections #22 spigot and #27 bell. The radial measurements are exaggerated, but they are to scale.

The above factors, particularly the fact that random spigot outside diameters significantly exceeded some bell inside diameters, and the failure of the attempts to join the pipe with a tugger capable of exerting 60,000 pounds pull, led to the conclusion that the AWWA C-302 pipe, regardless of the fact that it met the specification "tolerances", could not be laid to form a watertight rubber gasket joint pipeline as required by the project specifications.

Jobsite Conference

Further meetings between the owner, design engineer,

contractor and pipe supplier concluded that the problems would be resolved as follows:

1. The pipe manufacturer would remove approximately thirty thousandths all around (total .060) of the O.D. register portion of the spigot ring, located at the back edge of the gasket groove.

2. The gasket diameter would be reduced to 3/4-inch from 25/32-inch, and the gasket volume decreased to 100%. The Shore hardness of the gasket would be reduced to the lower range of the allowable AWWA C-302 Shore hardness, providing a softer gasket.

3. All the pipe would be installed prior to hydro testing, rather than testing each joint as it is laid. Hydro testing would be conducted when laying was nearing completion to avoid the test pressure disjointing the pipe.

4. If any joints failed the hydro test, they would be repaired by dry-packing epoxy grout in the inside joint space with SIKA 212 non-shrink grout or equal, applied per manufacturer's directions. Epoxy grout was to extend to within 3/4-inch of the inside surface of the pipe. The balance of the inside annular joint space was to be filled with SIKAFLEX 2C polyurethane base elastomeric joint sealant or HORNFLEX-L polysulfide rubber sealant to flush with the pipe inside diameter. Concrete surfaces were to be primed in accordance with manufacturer's directions with primer suitable for submerged concrete surfaces (Sika 429 for Sikaflex, Hornflex 2c for Hornflex-L) prior to application of sealant.

5. All joints which passed the hydro test were to receive the specified cement mortar in the inside annular space, applied to within 3/4-inch of the pipe inside diameter. The balance of the inside annular space was to be filled with an elastomeric sealant as specified in #4 above to flush with the pipe inside diameter.

6. Rather than remove and reinstall (lower) all the precast cradles, the pipe invert grade was to be adjusted by raising the pipe invert elevation by 1/2-inch in a distance of 24 feet, or three pipe sections, and then carry the .04/100 grade per plans in order to allow clearance between the cradles and the pipe O.D. for shimming to align joint rings.

More Joint Failures

A few days after the above methods were agreed upon by all parties, the pipe manufacturer balked, and suggested that they would assemble pipe in their yard to demonstrate that the pipe was acceptable (capable of being joined) as furnished, and asserting again that the problem was due to the contractor's procedures. The test turned out to be a failure, as two 40 to 50 thousand pound capacity fork lifts and "come-alongs" could not pull the pipe joints closed, even with the pipe on wooden supports in the manufacturer's pipe yard.

The pipe manufacturer had not complied with the specification requirement which requires that dimensions and tolerances of the gasket contact surfaces shall be submitted for approval. The engineer, inspector, owner and contractor therefore had no way of inspecting the joint dimensions for conformance to the required joint tolerances.

Conclusions and Recommendations

1. The project plans called for fixed elevation precast concrete cradle support. The design was not compatible with the AWWA C-302 pipe specification which requires no plus tolerance on wall thickness, and allows unlimited plus variations in the wall thickness. The design should have called for and required a specific distance from the pipe O.D. to the joint ring mating surfaces in addition to the general C-302 requirements, or required shim space between the cradle and the pipe O.D. However, the pipe manufacturer presumably bid the pipe with full knowledge of the proposed installation method, and should have recognized the need to hold the O.D. to joint ring dimension to a very close tolerance, and either produced the pipe accordingly, or recommended a different specification.

2. Similarly, the pipe manufacturer, as an experienced producer of AWWA C-302 pipe, should have known that the double thick 16-inch pipe wall, with eleven inches of concrete outside the bell, and five inches of concrete inside of the spigot, was so stiff from a deformation standpoint that the C-302 "out-of-round" tolerance of 1/2 inch would result in a joint that could either not be

joined, or could be joined only with the application of such extreme forces that serious damage to the bell and spigot would result, with significant delays to the project schedule and significant extra costs incurred on the jobsite. Figure 4 below illustrates one of the typical joint interference problems found.

3. Failure of the pipe manufacturer to store the pipe vertically, or to strut the pipe if stored horizontally, exacerbated an already serious joint tolerance problem.

4. At the heart of the C-302 specification problem in this case is the fact that the diametrical dimensional control is stated on a circumferential basis, as opposed to a diametrical basis. Apparently underlying this approach is the assumption that the joints will deform diametrically so that the joint will make up. The assumption is that so long as the inner circumference of the bell ring is slightly greater than the outer circumference of the spigot ring, the joint will go together. In smaller diameter pipe, or for pipe with thinner walls, this is probably a valid assumption.

"B" wall concrete pipe has a ratio of wall thickness to diameter ratio of 1:12. The 96-inch pipe on this project had a wall to diameter ratio of 1:6, far too stiff a structure to deform any significant amount. As far as the joint configuration itself, a spigot ring with nearly five inches of concrete inside of it will not deform any significant amount, and neither will a bell ring with eleven inches of concrete outside of it. This was probably the most serious flaw in the project. AWWA C-302 was the wrong pipe specification for this project, given the thick wall required to meet the structural requirements.

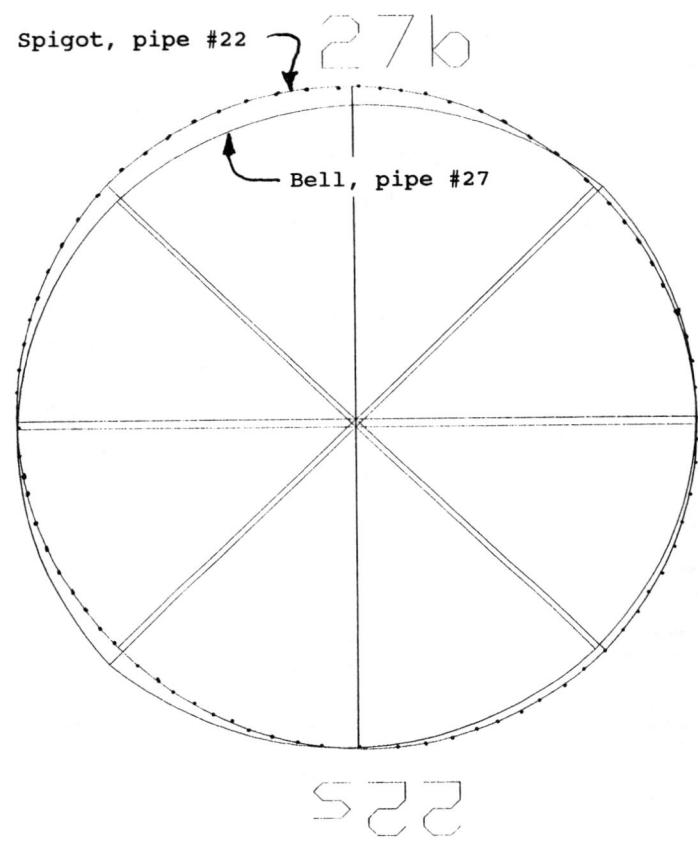

FIGURE 4
Typical Joint Interference
Radial scale magnified for illustration, joint interference of 1/8 to 3/16 inches.

APPENDIX A

Specifications Review
96" AWWA C-302 with Carnegie Joint

AWWA C-302 specification tolerance review and summary:

Section 3.1.3. Limits the allowable out of roundness of the pipe barrel by limiting the difference between maximum and minimum pipe internal diameters to one percent, i.e., 0.96 inches for 96" pipe.

Pipe ends must be square with the axis of the pipe within 3/8-inch. This will allow a difference in length of 3/4-inch on two opposite sides of the pipe.

Section 3.1.4. Limits diameter tolerance on the I.D. to minus 3/4-inch for pipe 84-inches and larger. There is no plus tolerance on pipe I.D.

Section 3.1.5. Establishes the tolerances on pipe wall thickness as "not less than the design thickness by more than 3/8-inch" for pipe larger than 72-inches. There is no plus tolerance on design wall thickness specified, therefore there is no tolerance required which controls the pipe O.D. to joint ring distance.

This makes it impossible for the design engineer to specify a cradle or pipe support system with fixed elevations for the pipe support. It also precludes laying pipe on a fixed surface such as a concrete slab or precast concrete cradles, since lining up the pipe O.D's on grade does not necessarily bring the joint rings into alignment.

Section 3.3.3.1. States that the inside bell ring circumference cannot exceed the outside spigot ring circumference by more than 1/4 inch. This implies a maximum joint clearance of 0.25 inches divided by pi, or .079 inches, or between 1/16 and 3/32 of an inch. This is inconsistent with the 1/2 percent difference in diameters allowed, or 1/2 inch maximum. It is obvious that a joint with a clearance of 1/16th inch maximum will not mate if it is out-of-round by 1/2 inch, unless the mating joint is out-of-round by the same amount on exactly the same axis, or if the bell or spigot ring is able to deform significantly. Section 3.3.3.1. also limits the joint ring out-of-round dimension or the joint ring diameter difference between the maximum and minimum ring diameter to one percent or not more than 1/2-inch.

Section 3.9.2. The hydrostatic test section states that no leakage is allowed in a 20 minute test period at 120 percent of the design pressure.

APPENDIX B

Field Data and Measurements

TABLE 1

Table of Differences, bell average diameter minus spigot average diameter, bell and spigot of same pipe, in inches:

Pipe #	Bell/Spigot Difference
9	.044
10	.060
11	.024
12	.026
13	.019
14	.057
15	.031
16	.015

TABLE 2

Measured Interferences/Clearances[1]

```
       Bell # 9    .405
     Spigot # 10   .568
  Interference: (-.163 inch)

       Bell # 18   .427
     Spigot # 12   .565
  Interference: (-.138 inch)

       Bell # 15   .455
     Spigot # 12   .565
  Interference   (-.110 inch)
```

[1] Note: Smallest bell diameter measured and largest spigot diameter, each pipe.

Observed Joint Interferences

Joint #17 spigot into #18 bell; 106.554 into 106.427; .127" interference. "B" into "D" on micrometer readings.

Joint #12 spigot into #13 bell; 106.565 into 106.460; .105" interference. "D" into "B".

Joint #16 spigot into #11 bell; .516 into .436; .080" interference. "A" into "A". Vertical interference.

Joint #13 average bell diameter is 106.503 and the average spigot diameter is 106.518, spigot exceeds bell by .015 inches.

NOTE: These interferences are measured at the bell entry slope beginning, and the bell is actually smaller at the gasket bearing surface than indicated above, due to the slope of the bell ring gasket bearing surface.

Manufacturer's Bell I.D.

The pipe manufacturer's submittal indicated the minimum bell ring inside diameter to be 106.530 inches. Many of the pipe measurements were significantly below this minimum. For example, the average diameter for the bell of pipe #13 was 106.503, #11 was 106.526, #9 was 106.507, #17 was 106.519, #18 was 106.487. The design spigot diameter and bell ring diameter, tolerances, and groove dimensions and tolerances are not known, since no joint gasket bearing surface dimensions and tolerances were submitted as required by the specifications.

APPENDIX "C: References

1. AWWA C-302-87 "AWWA Standard for Reinforced Concrete Pressure Pipe, Noncylinder Type, for Water and Other Liquids.

2. SSPWC "Standard Specification for Public Works Construction", Southern California APWA/AGC Joint Specification Committee, BNI Books, Building News Inc., Los Angeles, CA 1994.

UNIFIED DESIGN METHODOLOGY FOR MOST PIPELINE MATERIALS

Dr. Jey K. Jeyapalan, P.E. and Dr. Sri K. Rajah, P.E.

Pipeline Engineering Consultants
21 Meetinghouse Terrace, New Milford, CT, USA 06776
Phone: 1-860-354-7299
email: jkjeyapalan@snet.net

Abstract

The pipe materials used for either new pipe or pipe renovation are usually designed to meet some minimum consensus standard within ASTM, AWWA, AASHTO, or other in USA. Despite their claims of balanced representations among pipe manufacturers, city officials, and general interest members such as those from the consulting community, the standards historically have been documents developed with mostly vendor input and distributed mainly to protect the interests of the pipe manufacturers and their raw material suppliers. Consider for example the standards within AWWA for pressure pipe design for water transmission design and distribution. The designers and City officials have to look up C150 for Ductile, C200 for Welded Steel, C300, C301, C302, or C304 depending on the type of concrete pipe, C900 or C905 for PVC, C901 or C906 for HDPE, C950 for fiberglass, and to add to all this confusion we have M9 for design of concrete, M11 for Welded Steel, and M45 for Fiberglass. Many more such design manuals not part of the pipe standards are in the works with the philosophy that the more in print the more confusion they create that the City officials and the designers have to turn to leaving the design and development of the specifications to the pipe suppliers. How elegant this is for the pipe manufacturers in that they are designing and writing specifications they know they will have to meet when the projects are bid on behalf of the public. This paper presents details of what design steps are required in a unified engineering design standard based on sound engineering know-how to protect the major investment we make in the pipeline assets.

Introduction

The primary standard writing bodies in America are ASTM, AASHTO, AWWA, ASCE, and ASME. All these serve only as minimum standards, and additional requirements must be set for all pipe materials in order for the engineer to receive good dependable pipe supply and pipe-soil systems. In many cases other design tools need to be used to supplement the above standards. The availability of strong material, design, testing, quality control, and installation standards and engineering guidance also need to be considered as part of the selection criteria for all pipe materials.

Most standard writing activities only on the face of them appear to the user of pipe materials as if these standards are the works of well-trained engineers and scientists specializing in those pipe materials. To the contrary, despite the printed and highly publicized claims, these standards are usually the brainchild of most pipe vendors, their raw material suppliers, and others who stand to benefit from the increased sale of such materials. Among the standards, the ASTM standards on one hand are the most well written when it comes to details, while on the other, the standards with almost no major engineering requirements for the pipe vendors to meet. In a way, some of the ASTM standards are written with the philosophy that these are mere manner in which pipe suppliers of the same type of pipe material have agreed on how to conduct their business. Then there are somewhat stronger standards coming from AWWA on fiberglass pipe or prestressed concrete pipe where many important engineering requirements are laid down for the pipe supplier and the design engineer to rely on to come up with an acceptable design and QA/QC of the engineering pipe. Still most American Standards suffer from a major flaw in that each pipe material has its own standard driven by the pipe material rather than some widely accepted engineering concept. This is illustrated in Table 1 summarizing the AWWA standards for pressure pipe design for water transmission and distribution. This paper presents a critical comparison of these AWWA standards with a view of promoting a uniform non-conflicting design philosophy to be adopted by the city officials and engineers.

Design Philosophy

The structural design procedure for buried pipe involves establishment of design conditions, selection of pipe classes and corresponding pipe properties, selection of installation parameters, and performance of pertinent calculations to assure that the design requirements are satisfied. An outline of this general unified design philosophy proposed is presented in the following paragraphs.

Design Conditions

Design conditions are largely determined by required flow rate and flow velocity limitations, hydraulics, pipeline elevations and associated geology and topography, and available rights-of-way. The design conditions such as, Nominal pipe size, Working pressure (P_w), Surge pressure (P_s), Soil conditions for the pipe zone embedment and native material at pipe depth, Soil specific

TABLE 1: AWWA Standards and design manuals for different pipe materials

Pipe Material	Standard	Design
Ductile Iron	AWWA C151/A21.51	AWWA C150/ A21.50
Welded Steel	AWWA C200	AWWA M11
Concrete Pipe		
PCCP	AWWA C301	AWWA C304
RCP, bar wrapped, Steel Cylinder	AWWA C303	AWWA M9
RCP, Steel Cylinder	AWWA C300	AWWA M9
RCP, non-Cylinder	AWWA C302	AWWA M9
Fiberglass	AWWA C950	AWWA M45
PVC	AWWA C900 AWWA C905	AWWA M23 AWWA C905, Appendix A
HDPE	AWWA C901 AWWA C906	AWWA M46

weight (γ_s), Depth of cover (minimum and maximum), Vehicular traffic load (P), Internal vacuum pressure (P_v), and average and maximum service temperature have to be established before performing structural design calculations. Hydraulic head loss due to pipe friction should be calculated using a friction formula such as Manning's or Hazen-Williams'. Surge pressures should be calculated on the basis of the pipe hoop modulus and thickness-to-diameter ratio for given system design parameters. Excessive surge pressures should be identified in the design phase, and the causative condition should be eliminated or automatic surge-pressure relief provided or a higher pressure class should be selected.

Selection of Pipe-Class and Pipe Properties

Preliminary pipe-pressure-class selection can usually be made on the basis of working pressure, surge pressure and external loads established. In some cases, selection may involve preliminary evaluation of pipe properties for several classes.

Properties at the anticipated average and maximum service temperature for a given class of a specific pipe product should be obtained from the manufacturer or the manufacturer's literature. Values for ring stiffness, axial strength, and hoop tensile strength given in appropriate standards are minimum requirements only, and specific pipe products may have significantly higher values for these properties. Pipe properties necessary to perform design calculations include: 1) Nominal reinforced wall thickness, t and liner thickness, t_L; 2) Hoop tensile modulus of elasticity, E_H; 3) Hydrostatic design basis, HDB; 4) Ring flexural modulus of elasticity E; 5) Minimum pipe stiffness, $F/\Delta y$ (i.e., load on pipe unit length/predicted vertical pipe deflection); ; 6) Long-term, ring-bending strain, S_b; and 7) Poisson's ratios in the directions of applied hoop stress and applied longitudinal stress, v_{hl}, v_{lh}.

The primary installation parameters that must be selected according to the site conditions and planned installation are the type of backfill soil immediately around the pipe (pipe-zone backfill), degree of compaction and the native soil characteristics at the pipe elevation. Initial selection of these parameters may be controlled by prevailing standard specifications, the project soils boring report, manufacturers' recommendations, or past experience. A given combination of soil type and degree of compaction will largely determine the following values required for design calculations: 1) Bedding coefficient, K_x; 2) Modulus of soil reaction, E' (Tables 5, 6, and 7).; and 3) Deflection lag factor, D_l.

Design Procedure

With conditions, properties, and installation parameters established, satisfaction of the requirements is checked by design calculations. The calculations may be made using either stress or strain, depending on the basis used to establish a particular product performance limit. In general, the major design checks and the corresponding calculation steps to determine whether pipe meets the requirements are as follows:

1. Internal Pressure
    1.1. Calculate the maximum sustained pressure, P_c, from HDB
    1.2. Check working pressure P_w
    1.3. Check surge pressure P_s
2. Ring Bending
    2.1. Calculate allowable deflection from ring bending
3. Deflection
    3.1. Determine soil loads W_c and live loads W_L
    3.2. Calculate the composite modulus of soil reaction E'
    3.3. Check deflection prediction $\Delta y/D$
4. Combined Loading
    4.1. Check combined loading
5. Buckling
    5.1. Check buckling
6. Handling Strength and Stiffness

Of these important design concepts, not all are available in pipeline design standards even by AWWA for different materials. A summary of the design checks for AWWA standards corresponding to different pipe materials is given in Table 2.

Design Calculations

Its interesting to note in Table 2 that the AWWA standard for fiberglass pipe design includes all major design checks, although this is the least widely used pipe material. The remainder of this paper, therefore, will use only AWWA C-950 and M-45 wherever possible.

Internal Pressure

The pressure class, P_c, the maximum sustained pressure for which the pipe is designed is related to the long-term hydrostatic hoop strength, HDB, of the pipe as follows:

- For stress basis HDB

$$P_c \leq \left(\frac{HDB}{FS}\right)\left(\frac{2t}{D}\right) \quad (1)$$

- For strain basis HDB

$$P_c \leq \left(\frac{HDB}{FS}\right)\left(\frac{2E_H t}{D}\right) \quad (2)$$

Where, the minimum design factor, FS for fiberglass design is = 1.8.

Inside and outside diameters, D_i and D_o, for the pipe can be obtained from the design tables.

- For inside-diameter (ID) series pipe

$$D = D_i + 2t_L + t$$

- For outside-diameter (OD) series

$$D = D_o - t$$

The Hydrostatic Design Basis, HDB of pipe varies for different products, depending on the materials and composition used in the reinforced wall and in the liner. Further, the HDB may be defined in terms of reinforced wall hoop stress, average stress (apparent glass-fiber stress for fiberglass materials), or hoop strain on the inside surface. The relationship between HDB in terms of composite stress and reinforcement stress is:

$$HDB = \left(\frac{t_R}{t}\right)(HDB)_R \quad (3)$$

In the case of fiberglass materials, for example, the effective thickness of hoop reinforcement need to be adjusted for helix angle.

The HDB at ambient temperature must be established by tests in accordance with the appropriate standards for each pipe product by each manufacturer. The required practice is to define projected product-performance limits at 50 years. Performance limits at elevated temperature depend on the materials and type of pipe wall construction used. The manufacturer should be consulted for HDB values appropriate for elevated-temperature service.

Table 2: AWWA STANDARDS ON WATER PIPE

Design Check	Ductile iron	Steel	PVC	HDPE	Fiberglass
Internal Pressure	Yes	Yes	Yes	Yes	Yes
External Load Induced-Bending	Yes	No	No	No	Yes
External Load Induced-Deflection	Yes	Yes	Yes	Yes	Yes
Buckling	No	Uses C950	No	No	Yes
Handling Stiffness & Strength	No	Wrong	No	No	Yes
Combined Stress under External and Internal Loads	No	No	No	No	Yes

Design factors

Prudent design of pipe to withstand internal hydrostatic pressure requires consideration of two separate design factors. The first design factor is the ratio of short term ultimate hoop tensile strength, S_i to hoop tensile stress, S_h at pressure class, P_c. This factor assures that short-term peak pressure conditions do not exceed the short-term hydrostatic strength of the pipe. These values reflect a minimum design factor of 4.0 on initial hydrostatic strength. The second design factor is the ratio of HDB to hoop stress or strain, S_r at pressure class, P_c. This factor assures that stress or strain due to sustained working pressure does not exceed the long-term hoop strength of the pipe as defined by HDB. For fiberglass pipe design, this minimum design factor is 1.8.

Both design factors should be checked. Either design factor may govern pipe design, depending on long-term strength regression characteristics of the particular pipe product. Prudent engineering may dictate an increase or decrease in either design factor, depending on the certainty of the known service conditions.

Working pressure, P_w

The working pressure in the system shall not exceed the pressure class of the pipe, as follows:

$$P_w \leq P_c \tag{4}$$

Surge pressure, P_s

The maximum pressure in the system, due to working pressure plus surge pressure, shall not exceed 1.4 times the pressure class of the pipe.

$$P_w + P_s \leq 1.4 P_c \tag{5}$$

The above treatment of surge pressures reflects the characteristics of fiberglass pipes covered by AWWA M-45, and other materials could be designed to meet the same criterion. Field or factory hydrotesting at pressures up to $2P_c$ is acceptable and does not violate equations (4) and (5), for some pipe materials.

The surge-pressure calculations shall be performed using recognized and accepted theories. Calculated surge-pressure magnitudes are highly dependent on the hoop tensile elastic modulus and thickness-to-diameter $\left(t/D\right)$ ratio of the pipe. Because of this, the designer should generally expect higher calculated surge pressures for pipes with higher modulus or thicker wall or both.

The surge allowance, P_{sa}

The portion of the surge pressure that can be accommodated without changing pressure class is called surge allowance. The surge allowance is intended to provide for rapid transient pressure increases normally encountered in transmission systems. The surge pressure allowance is based on the increased strength of a pipe under rapid strain rates. Special consideration should be given to the design of systems subjected to rapid cyclic service. The manufacturer should be consulted for specific recommendations.

Ring Bending

The maximum allowable long-term vertical pipe deflection shall not result in a ring-bending strain (or stress) that exceeds the long-term ring-bending strain (or stress) capability of the pipe reduced by an appropriate design factor. Satisfaction of this requirement is assured by using one of the following formulas:

- For stress basis:

$$\sigma_b = D_f E\left(\frac{\Delta y_a}{D}\right)\left(\frac{t_t}{D}\right) \leq \frac{S_b E}{FS} \tag{6}$$

- For strain basis:

$$\varepsilon_b = D_f \left(\frac{\Delta y_a}{D}\right)\left(\frac{t_t}{D}\right) \leq \frac{S_b}{FS} \tag{7}$$

Where, t_t is the total thickness of the pipe (=t + t_L); D_f is shape factor; σ_b is maximum ring bending stress due to deflection; ε_b is maximum ring bending strain due to deflection; and Δy_a is maximum allowable long-term vertical pipe deflection. For fiberglass pipe design, the long-term ring-bending strain, S_b for the pipe is used and the design factor, FS is 1.5.

Shape factor, D_f

The shape factor relates pipe deflection to bending stress or strain and is a function of pipe stiffness, pipe-zone backfill material and compaction, haunching, native soil conditions, and level of deflection. Table 3 gives values for D_f assuming inconsistent haunching, deflections of at least 2 - 3 percent, and stable native soils or adjustments to trench width to offset poor conditions. Values given in Table 3 are for typical pipe-zone backfill materials. For other pipe-zone backfill materials, use the highest value for each pipe stiffness.

Table 3: SHAPE FACTORS, D_f

Pipe Stiffness		Pipe Zone Material Types	
		GW,GP, SW, SP, GM, GC, SM, SC	
psi	kPa	SPD < 85%	SPD = or > 85%
9	62	6	8
18	125	5	7
36	250	4	6
72	500	3	5

Long term ring-bending strain, S_b

The long-term ring-bending strain varies for different products, depending on materials and type of construction used in the pipe wall. Long-term ring-bending strain shall be determined from the appropriate standards.

Bending design factor

Prudent design of pipe to withstand bending requires consideration of two separate design factors. The first design consideration is comparison of initial deflection at failure to the maximum allowed installed deflection. The ring stiffness test subjects a pipe ring to deflections far exceeding those permitted in use. For this particular case, the test requirement demonstrates a design factor of at least 2.5 on initial bending strain. The second design factor is the ratio of long-term bending stress or strain to the bending stress or strain at the maximum allowable long-term deflection. For fiberglass pipe design, this minimum design factor is 1.5.

Deflection

Buried pipe should be installed in a manner that will ensure that external loads will not cause a long-term decrease in the vertical diameter of the pipe exceeding the maximum allowable deflection established using ring bending criteria or as required by the engineer or manufacturer ($\delta d / D$) whichever is less. This requirement may be stated as follows:

$$\frac{\Delta y}{D} \leq \frac{\delta d}{D} \leq \frac{\Delta y_a}{D} \tag{8}$$

Where, δd is maximum permitted long-term installed deflection and,

$$\frac{\Delta y}{D} = \frac{(D_l W_c + W_L) K_x}{(0.149 PS + 0.061 E')} \tag{9}$$

The parameters are defined as follows: deflection lag factor, D_l; bedding coefficient, K_x; vertical soil load on the pipe, W_c; live load on the pipe, W_L; pipe stiffness, PS; and modulus of soil reaction of the pipe-soil system, E'.

Design calculations requiring deflection as an input parameter shall show the predicted deflection $\Delta y / D$ as well as the maximum allowable deflection $\Delta y_a / D$ at which the allowable design stress or strain is not exceeded. The maximum allowable deflection shall be used in all design calculations.

Deflection prediction

When installed in the ground, all flexible pipe will undergo deflection, used here to mean a decrease in vertical diameter. The amount of deflection is a function of the soil load, live load, native soil characteristics at pipe elevation, pipe embedment material and density, trench width, haunching, and pipe stiffness. Many theories have been proposed to predict deflection levels; however, in actual field conditions, pipe deflections may vary from calculated values because the actual installation achieved may vary from the installation planned. These variations include the inherent variability of native ground conditions and variations in methods, materials, and equipment used to install a buried pipe.

Field personnel responsible for pipe installation must follow procedures designed to assure that the long-term pipe deflection is less than Δy_a as determined in ring bending criteria, or as required by the engineer or manufacturer, whichever is less. As presented above and as augmented by information provided in the following sections, Eq 9 serves as a guideline for estimating the expected level of short-term and long-term deflection that can be anticipated in the field. Eq 9 is a form of the Iowa formula, first published by Spangler in 1941. This equation is the best known and documented of a multitude of deflection-prediction equations that have been proposed. As presented here, the Iowa formula treats the major aspects of pipe-soil interaction with sufficient accuracy to produce reasonable estimates of load induced field deflection levels.

Not addressed by this method are pipe deflection due to self weight and initial ovalization due to pipe backfill embedment placement and compaction. These deflections are typically small for pipe stiffness above 9 to 18 *psi* [62 to 124 kPa] (depending on installation conditions). For pipe stiffness below these values, consideration of these items may be required to achieve an accurate deflection prediction.

Application of this method is based on the assumption that the design values used for bedding, backfill and compaction levels will be achieved with good practice and with appropriate equipment in the field. Experience has shown that deflection levels of any flexible conduit can be higher or lower than predicted by calculation if the design assumptions are not achieved.

Deflection lag factor, D_l

The deflection lag factor converts the immediate deflection of the pipe to the deflection of the pipe after many years. The primary cause of increasing pipe deflection with time is the increase in overburden load as soil "arching" is gradually lost. The vast majority of this phenomenon occurs during the first few months of burial up to a couple of years depending on the consolidation characteristics of the ground conditions and wet/dry cycles. These causes are generally of much less significance than increasing load and may not contribute to the deflection for pipes buried in relatively stiff native soils with dense granular pipe zone surrounds. For long-term deflection prediction, a D_l value of greater than 1.00 is appropriate.

Bedding coefficient, K_x

The bedding coefficient reflects the degree of support provided by the soil at the bottom of the pipe and over which the bottom reaction is distributed. Assuming an inconsistent haunch achievement (typical direct bury condition), a K_x value of 0.1 should be used.

Vertical soil load on the pipe, W_c

The vertical soil load on the pipe may be considered as the weight of the rectangular prism of soil directly above the pipe. The soil prism would have a height equal to the depth of earth cover and a width equal to the pipe outside diameter.

$$W_c = \frac{\gamma_s H}{144}$$

where, H is the burial depth to the top of the pipe and the units of W_c, γ_s, and H are *psi*, *pcf*, and *ft.* respectively.

$$W_c = \gamma_s H \qquad (10)$$

where, H is the burial depth to the top of the pipe and the units of W_c, γ_s, and H are *kPa*, kN/m^3, and *m.* respectively.

Live loads on the pipe, W_L

The calculation below assumes a four lane road with an AASHTO HS-20 truck (10-inch x 20-inch = 0.254 *m* x 0.508 *m* contact area; sloping at a rate of 0.875H;)

centered in each 12 ft (3.66 m) wide lane. The pipe may be perpendicular or parallel to the direction of truck travel, or any intermediate position. Other design truck loads can be specified as required by project needs and local practice.

- Compute L1, load width parallel to direction of travel.

  $L1 = 0.83 + 1.75H$ $L1 = 0.254 + 1.75H$ (11)

  where, $L1$ is in ft.   where, $L1$ is in m.

- Compute $L2$, load width perpendicular to direction of travel.

  $\Rightarrow 2' < H < 2.48'$ $\Rightarrow 0.610m < H < 0.756m$

  $L2 = 1.67 + 1.75H$ $L2 = 0.509 + 1.75H$ (12)

  $\Rightarrow H \geq 2.48'$ $\Rightarrow H \geq 0.756m$

  $L2 = \left(\dfrac{43.67 + 1.75H}{8}\right)$ $L2 = \left(\dfrac{13.31 + 1.75H}{8}\right)$ (13)

- Compute W_L:

  $W_L = \dfrac{PI_f}{144(L1)(L2)}$ $W_L = \dfrac{PI_f}{L1 \times L2}$ (14)

  Where: Where:

  P = 16,000 lbs. (HS-20 load) P = 71.17 kN. (HS-20 load)
  I_f = impact factor I_f = impact factor
     = 1.1 for 2 ft. < H < 3 ft.   = 1.1 for 0.61m.<H<0.91 m
     = 1.0 for H ≥ 3ft.   = 1.0 for H ≥ 0.91 m

This computation is independent of pipe diameter and the resulting live loads are tabulated in Table 4.

Pipe stiffness, PS

The pipe stiffness is the product of the ring flexural modulus of elasticity, E of the pipe-wall material, and the moment of inertia, I of a unit length of pipe divided by the quantity 0.149 times the mean radius, r, cubed (see Eq 15). The moment of inertia is equal to $t_t^3 / 12$, where t_t is the total wall thickness. When other than uniform wall construction is used, consult the manufacturer for the proper moment of inertia equation.

$$PS = \dfrac{EI}{0.149r^3} \quad (15)$$

The pipe stiffness can be determined by conducting parallel-plate loading tests in accordance with ASTM D2412, Standard Test Method for Determination of External Loading Characteristics of Plastic Pipe by Parallel-Plate Loading. During the parallel-plate loading test, deflection due to loads on the top and bottom of the pipe is measured, and pipe stiffness is calculated from the following equation:

$$PS = \frac{F}{\Delta y} \tag{16}$$

Table 4: HS-20 Live Loads

DEPTH		W_L		DEPTH		W_L	
(ft)	(m)	(psi)	(kPa)	(ft)	(m)	(psi)	(kPa)
2	0.601	6	41.37	10	3.048	0.8	5.52
2.5	0.763	3.9	26.89	12	3.658	0.6	4.14
3	0.914	3.3	22.75	16	4.877	0.5	3.45
3.5	1.067	2.6	17.93	20	6.096	0.4	2.76
4	1.219	2.2	15.17	27	8.230	0.2	1.38
6	1.829	1.5	10.34	40	12.19	0.1	0.69

Modulus of soil reaction, E'

The vertical loads on a flexible pipe cause a decrease in the vertical diameter and an increase in the horizontal diameter. The horizontal movement develops a passive soil resistance that acts to help support the pipe. The passive soil resistance varies depending on the soil type and the degree of compaction of the pipe-zone backfill material, the native soil characteristics, the cover depth and the trench width. To determine E' for a buried pipe, separate E' values for the native soil, E'_n and the pipe backfill surround, E'_b must be determined and then combined using Eq 17.

$$E' = S_c E'_b \tag{17}$$

The value of the soil support combining factor, S_c, is given in Table 5, as a function of E'_n / E'_b and B_d / D (B_d is the trench width at pipe springline). The values of modulus of soil reaction of the pipe zone-backfill embedment, E'_b, can be obtained from Table 6. And the modulus of soil reaction of the native soil at pipe elevation, E'_n are obtained from Table 7. For the following special circumstances, the E' value will be estimated as described: **Geotextiles**: When a geotextile pipe zone wrap is used, E'_n values for poor soils can be greater than shown in Table 7, if proper specifications are followed; **Solid sheeting:** When permanent solid sheeting designed to last the life of the pipeline is used in the pipe zone, E' shall be based solely on E'_b; **Cement stabilized sand:** When cement stabilized sand is used as the pipe zone surround, initial deflections shall be based on a sand installation and the long-term E'_b = 25,000 psi(170 Mpa); **For**

embankment installation: $E'_b = E'_n = E'$. *Note:* It is important for the pipe supplier to use the material property data that could be defended in a legal discovery instead of claiming that this procedure described herein is too technical for them and that special material properties do not apply to the pipe material they have marketed.

TABLE 5: VALUES FOR SOIL SUPPORT COMBINING FACTOR, S_C

E'_n/E'_b	B_d/D					
	1.5	2	2.5	3	4	5
0.1	0.15	0.30	0.60	0.80	0.90	1.00
0.2	0.30	0.45	0.70	0.85	0.92	1.00
0.4	0.50	0.60	0.80	0.90	0.95	1.00
0.6	0.70	0.80	0.90	0.95	1.00	1.00
0.8	0.85	0.90	0.95	0.98	1.00	1.00
1.0	1.00	1.00	1.00	1.00	1.00	1.00
1.5	1.30	1.15	1.10	1.05	1.00	1.00
2.0	1.50	1.30	1.15	1.10	1.05	1.00
3.0	1.75	1.45	1.30	1.20	1.08	1.00
≥ 5.0	2.00	1.60	1.40	1.25	1.10	1.00

Combined Loading

The maximum stress or stain resulting from the combined effects of internal pressure and deflection shall meet Eq's 18 and 19 or Eq's 20 and 21 as follows:

- For stress basis HDB and S_b:

$$\frac{\sigma_{pr}}{HDB} \leq \frac{1 - \left(\dfrac{\sigma_b r_c}{S_b E}\right)}{FS_{pr}} \qquad (18)$$

- For strain basis HDB and S_b:

$$\frac{\sigma_b r_c}{S_b E} \leq \frac{1 - \left(\dfrac{\sigma_{pr}}{HDB}\right)}{FS_b} \qquad (19)$$

$$\frac{\varepsilon_{pr}}{HDB} \leq \frac{1 - \left(\dfrac{\varepsilon_b r_c}{S_b}\right)}{FS_{pr}} \qquad (20)$$

Table 6: MODULUS OF SOIL REACTION, E'_b in psi(Mpa)

Type of Soil	Standard Proctor Relative Compaction Density (SPD)			
	85%	90%	95%	100%
CL, ML, CL-ML	500(3.4)	700(4.8)	1000(6.8)	1400(9.6)
SM, SC	600(4.1)	900(6.2)	1350(9.3)	2000(13.6)
SP, SW, GP, GW	700(4.8)	1000(6.8)	1500(10.2)	2300(15.3)

Notes: The above values apply when the soil cover is between 0 and 5 feet(1.5m). The designer may increase the above values by 100 psi(0.7 Mpa), when the SPD is 100% and by 25 psi(0.17Mpa), when the SPD is 85%, for every foot(0.3m) above the basic 5 feet(1.5m) of soil cover over the pipe crown for which the above values are given. It is important to recognize that E' for the soil-pipe system is also affected by the size of the pipe and pipe-soil stiffness ratio. If the pipe designed is either smaller than 4 feet(1.2m) in diameter or if the pipe-soil stiffness ratio computed using PSR= PS/.061E' is lower than 0.05, the above values of E'b need be lowered

$$\frac{\varepsilon_b r_c}{S_b} \leq \frac{1 - \left(\dfrac{\varepsilon_{pr}}{HDB}\right)}{FS_b} \tag{21}$$

Where:

FS_{pr} = pressure design factor, 1.8

FS_b = bending design factor, 1.5

σ_{pr} = working stress due to internal pressure, [psi; kPa]

= $\dfrac{P_w D}{2t}$

σ_b = bending stress due to maximum allowable deflection, [psi; kPa]

= $D_f E \left(\dfrac{\delta d}{D}\right)\left(\dfrac{t_t}{D}\right)$

r_c = rerounding coefficient, (dimension-less)

= $\left(1 - \dfrac{P_w}{435}\right); \cdots P_w \leq 435 psi$ = $\left(1 - \dfrac{P_w}{3000}\right); \cdots P_w \leq 3000 kPa$

ε_{pr} = working strain due to internal pressure, [in./in.; m/m]

$$= \frac{P_w D}{2tEH}$$

ε_b = bending strain due to maximum allowable deflection

$$= D_f \left(\frac{\delta d}{D}\right)\left(\frac{t_t}{D}\right)$$

Buckling

Buckling theory

Buried pipe is subjected to radial external loads composed of vertical loads and possibly the hydrostatic pressure of groundwater and internal vacuum, if the latter two are present. External radial pressure sufficient to buckle buried pipe is many times higher than the pressure causing buckling of the same pipe in a fluid environment, due to the restraining influence of the soil.

Buckling Calculations

The summation of appropriate external loads shall be equal to or less than the allowable buckling pressure. The allowable buckling pressure, q_a, is determined by the following equation:

$$q_a = \left(\frac{1}{FS}\right)\left[32 R_w B'E' \frac{EI}{D^3}\right]^{\frac{1}{2}} \qquad (22)$$

in which, the design factor, $FS = 2.5$; the water buoyancy factor, R_w and empirical coefficient of elastic support, B', are given by,

$$R_w = 1 - 0.33\left(\frac{h_w}{H}\right); 0 \le h_w \le H \qquad (23)$$

where, h_w is the height of the water surface above the top of the pipe.

$$B' = \frac{1}{1 + 4e^{-0.065H}} \qquad B' = \frac{1}{1 + 4e^{-0.2133H}} \qquad (24)$$

NOTE: Eq 22 is valid under the following conditions:

w/o internal vacuum: $2 ft. \le H \le 80 ft.$ w/o internal vacuum: $0.6m \le H \le 24m$
with internal vacuum: $4 ft. \le H \le 80 ft.$ with internal vacuum: $1.2m \le H \le 24m$

Where internal vacuum occurs with cover depths less than 4 ft(1.2 m) but not less than 2 ft(0.6 m), q_a in Eq 25 may be determined as the critical buckling pressure given by the Von Mises formula. The 2-4 ft of soil cover provides a safety factor in excess of the recommended 2.5 value. In the 2-ft to 4-ft depth range, live loads plus dead loads should be checked by Eq 26 to determine the governing required wall thickness. The manufacturer should be consulted for further recommendations in this depth range.

The Von Mises formula is:

$$q_a = \left(\frac{2Et_t}{D(n^2-1)(1+K^2)}\right) + \left[n^2 - 1 + \left(\frac{2n^2-1-v_{hl}}{1+K}\right)\right]\left(\frac{8EI}{D^3(1-v_{hl}v_{lh})}\right) \quad (25)$$

Where,

$$K = \left(\frac{2nL}{\pi D}\right)^2$$

The value of n, the number of lobes formed at buckling, must give the minimum value of q_a obtained by iterative solution.

Note: For solid-wall (non-ribbed) pipes, the distance between rigid ring stiffeners, L shall be the distance between joints, such as bells, couplings, flanges, and the like.

Satisfaction of the buckling requirement is assured for normal pipe installations by using the following equation:

$$\gamma_w h_w + R_w W_c + P_v \leq q_a \quad (26)$$

The specific weight of water (γ_w) is 0.0361 lb/in³ (= 9.81 kN/m³) and the internal vacuum pressure, P_v is atmospheric pressure less absolute pressure inside the pipe.

In some situations, consideration of live loads in addition to dead loads may be appropriate. However, simultaneous application of live-load and internal-vacuum transients need not normally be considered. If live loads are considered, the buckling requirement is assured by:

$$\gamma_w h_w + R_w W_c + W_L \leq q_a \quad (27)$$

Summary

The paper identifies the essential design steps and sample design calculations for concept oriented pipeline design standards. A careful comparison of even the most useful AWWA design standards for different pipe materials reveal that the present standards are material specific and most do not consist even the essential design concepts. The existence of too many standards for each pipe material and the inconsistency with which the design standards are printed make it very confusing. Efforts within standard writing bodies for a more consistent and technically sound set of pipe design standards are needed towards a better design standard. This paper presents a unified methodology where the same standard could be applied to the design of gravity and pressure pipelines made of different materials.

Table 7: values of Modulus of Soil Reaction, E'_n for the Native Soil at Pipe Zone Elevation

NATIVE IN SITU SOILS				E'_n	
GRANULAR		COHESIVE			
SPT Blows/ft.	Description	q_u (Tsf)	Description	(psi)	(MPa)
>0 - 1	v. v. loose	>0 - .125	v. v. soft	100	0.69
1 - 2	very	.125 - .25	very soft	300	2.07
2 - 4	loose	.25 - .50	soft	700	4.83
4 - 8	loose	.50 - 1.0	medium	1,500	10.34
8 - 15	slightly compact	1.0 - 2.0	stiff	3,000	20.68
15 - 30	compact	2.0 - 4.0	very stiff	5,000	34.47
30 - 50	dense	4.0 - 6.0	hard	10,000	68.94
> 50	very dense	> 6.0	very hard	20,000	137.89
ROCK				≥ 50,000	≥ 344.72

Acoustic Monitoring of Prestressed Concrete Cylinder Pipe

Mark Holley, A.Sc.T. - Marketing Manager, Pipelines
Doug Buchanan, P.E. - Operations Manager

Pure Technologies Inc.
1050, 340 12th Avenue S.W.
Calgary, Alberta T2R 1L5
Canada

Phone : 1-800-537-2806

Abstract

Corrosion of prestressing wire in Prestressed Concrete Cylinder Pipe (PCCP) and other structures is a widespread concern for owners and managers of these facilities. The general inaccessibility of the prestressing wire makes evaluation difficult, costly and often inconclusive. Random examination of prestressing wires by excavating or internal inspection of the pipe wall gives only a very localized knowledge of the prestressing wire condition, conventional investigations can be misleading, often resulting in an underestimate of the extent of corrosion, deterioration or wire failure. The operation of a recently developed acoustic system for monitoring PCCP to determine the time and location of pipe deterioration is discussed. Case studies where the system has been used to identify deterioration in PCCP are presented.

Introduction

Prestressed Concrete Cylinder Pipe (PCCP) consists of a steel cylinder lined with concrete or cement mortar, then helically wrapped with a wire and coated with a dense mortar. The prestressing wires provide the pipe with sufficient strength to withstand water pressures as high as 300 pounds per square inch.

The stress imposed on these wires during the manufacturing process is greater than 70 percent of ultimate breaking strength of the wire. Therefore, a strand of 1/4 inch Class III wire will possess more than 7,000 pounds of tension when wrapped on the core. A relatively small amount of corrosion will cause a wire to break, resulting in a sudden release of strand energy. (detected by the acoustic monitoring system.)

Research done in the late 1980's and early 1990's by the United States Department of The Interior, Bureau of Reclamation heralded the use of continuous acoustic monitoring to track the deterioration of prestressed concrete pipelines. Results from work done at the Aqua Fria pipeline in Arizona indicated that deliberate wire cuts generated large distinctive acoustic anomalies, which could be recorded by suitable equipment.

In 1993, Pure Technologies Inc. independently began to use continuous acoustic monitoring to track the failure of unbonded post-tensioning strands in concrete building and parking structures. The size and complexity of these structures required the development of specialized equipment and software to collect, manage, and analyze the large amounts of data flowing from these sites. These programs, techniques and equipment designs have been applied to the monitoring of prestressed concrete pipe.

The Corrosion Process

When deterioration of PCCP occurs, it usually originates from the exterior of the pipe. The corrosion occurs due to a breakdown in the protective properties of the cement mortar. In high pH concrete the embedded steel wire is protected from galvanic corrosion. However, when the protective mortar is chemically or physically compromised, the passivating quality of the high pH cement breaksdown. This loss of protection allows corrosion to initiate on the highly stressed wires. With very little corrosion of the wires, failure can result in the form of galvanic corrosion and/or stress corrosion cracking. If enough wires fail, the structural capacity of the PCCP is compromised and the pipe is in danger of a catastrophic rupture. When a prestressing wire breaks, it immediately redevelops bond in the surrounding mortar. Hundreds of wire breaks may occur before the pipe is weakened to the point of failure. The wire failures release a high level of energy that can be detected by the Acoustic Monitoring System. This early detection of areas on a pipeline that are exhibiting some deterioration provides the owner with time to proactively address the deteriorating section prior to a catastrophic failure. The proactive owner can now address the suspect pipe sections as a remedial maintenance item rather than incurring the large capital expenditure of a pipeline replacement project.

How Acoustic Monitoring Works - Single Stations vs. Long Arrays

Single Stations

Pure Technologies set out to develop an acoustic monitoring system that would provide accurate reliable information on the rate of deterioration in prestressed concrete cylinder pipe. Every effort was made to design a system whereby the owner of the line could maintain full operation of the pipeline during the installation and monitoring period.

Our initial system design was based on work done by the United States Department of the Interior, Bureau of Reclamation and included a dual hydrophone station. The original system concept required multiple stations inserted at regular intervals over the length of the pipeline being monitored. These locations required wet tap or access through existing valves.

An obvious advantage of this system is that single monitoring stations can be installed fairly easily if there is access to the outside of the pipe at regular intervals. Another advantage is the apparently low cost of isolated stations.

Location of wire break events is usually done by comparing the times of arrival of the acoustic wave as it encounters sensors on either side of the break. This requires that at least two sensors be within the detection range of the wire break. The range over which the Bureau was able to detect wire breaks in the Aqua Fria was sometimes thousands of feet. This was encouraging news as the number of stations per mile required to monitor a section of pipe might be small.

Until recently, disadvantages of these single stations included the difficulty of interconnecting sensors on the surface, protection of several different access points, damage caused by the hot taps, and flow noise caused by the hydrophone position normal to the flow. In addition, stations may not be suitable for smaller diameter pipelines due to attenuation of the acoustic signal over relatively short distance. Pure Technologies has found correlation between pipe diameter vs. sensor spacing. In fact, small diameter pipe may require sensor spacing so tight (approx. 100' on center) that it would be cost prohibitive for the owner to provide the required number of access location to the pipe to facilitate the survey. To overcome this problem, Pure Technologies has developed long hydrophone arrays that can provide a cost effective alternative for small diameter pipe.

Long Arrays

A hydrophone array consists of a long cable inserted into active sewer or water pipeline at an existing valve or wet tap location. The hydrophone array can be manufactured in various lengths depending on the project requirements. To date, we have installed arrays ranging in length from 1500' to over 6000' from one insertion point. Once in place, the hydrophones continuously monitor the pipe for acoustic events that exhibit properties characteristic of prestressed wire failures. These events are recorded by a data acquisition unit and compared to preset acoustic criteria. If the acoustic event recorded meets the established criteria, the event is uploaded to a remote site for further evaluation by a trained technician, using proprietary processing software.

When an event has been determined to have all the acoustic characteristics of a prestressed wire failure, the analytical software further evaluates the signal to allow for accurate location of the event origin. The speed of sound in water is known as well as the spacing between the hydrophones. By comparing the arrival time between two adjacent hydrophones, the signal processor is able to accurately determine the location of the wire break. Advantages of the hydrophone array over the single station are many. They include:

1. The array requires only one insertion point for up to 6000' of pipeline.
2. The cable can be manufactured in various lengths to facilitate specific project requirements.
3. The location and spacing of the hydrophones can be varied to provide adequate acoustic coverage in various pipe sizes.

Information provided by the system allows utilities to proactively plan and budget for scheduled maintenance. Through periodic maintenance of the PCCP pipe the life of the structure can be extended by years or decades offsetting large capital expenditures associated with the replacement of the pipeline.

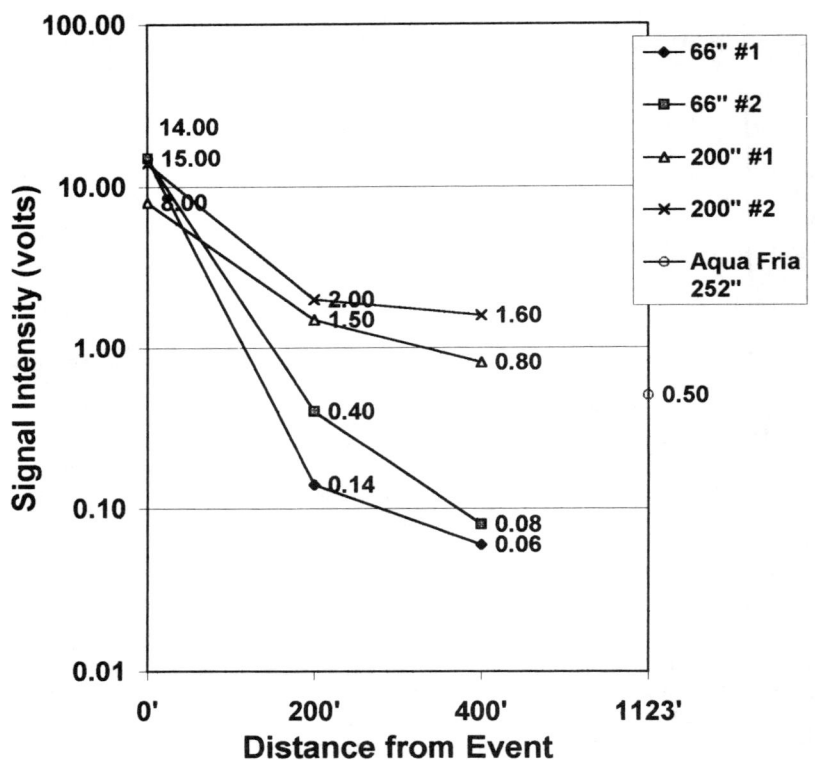

Signal Attenuation vs. Distance From Event

Generic Problems of Acoustic Technology

Providing power, secure enclosures, and data transmission links are basic requirements that must be available to any electronic-based data acquisition system. At remote sites, in difficult climates, within high vandalism areas, and in areas beneath roadways, providing these fundamentals can be a challenge.

The systems also require that the data be secure, duplicated, and be capable of rapid and efficient analysis. All data logged must be reduced to a reporting format useful to the operators of the pipeline. Potable water standards must be met with all procedures and materials which contact the water.

Identification of wire breaks as opposed to other events must be demonstrably correct. This identification becomes more and more difficult with increasing distance to the sensors as the dispersion of the frequencies generated by the event will distort and disguise its character.

Unattended operation of complex data-logging systems requires that the systems be interactive through the data link. System uptime must be known to effectively evaluate the statistical significance of the number of wire breaks in a known interval. System problems must be identified promptly to allow repair.

The rate of breakage in various sections requires that the sections be monitored for an interval sufficiently long to detect enough breaks to realistically assess the deterioration. This can vary from minutes to years, and requires that the stakeholder make some engineering judgment of the significance of the rate reported. Signal to noise ratios must be sufficiently high to permit consistent identification and analysis of significant data.

Generic Advantages of Acoustic Technology

The biggest advantage of the technology is the fact that it is possible to monitor every single wire break in a section of pipeline. This one hundred percent sample changes the confidence that the stakeholders may have in the data available to them, and hence in their ability to safely manage the pipeline. Future experience will indicate over what interval one must monitor a pipe section to gain a reasonable assessment of the rate of breakage. This will be determined by the temporal variability in the rate experienced over longer term monitoring of some sections.

Preliminary Results from Various Projects

Pure Technologies Inc. is presently involved with a number of commercial and research projects. The following is a generic description and summary of a number of project results:

Structure #1

This pipeline transfers treated municipal sewage effluent for approximately 36.9 miles. The pipeline is comprised of three spans of differing diameters. The first two spans are gravity fed and are comprised of 6.1 miles of 114" diameter pipe and 22.6 miles of 96" diameter pipe respectively. A pump station that delivers the water to the plant site feeds the remaining 8.2 miles of 66" diameter pipe.

Since the installation of our acoustic monitoring equipment, Pure Technologies Inc. has recorded events in over 15,500 feet of this pipe. Each section of monitored pipe is detailed on the table below.

Monitoring Period	Classification	Approx. Weeks	Pipe Diameter (inch)	Events
01/23/98- 02/16/98	1.Possible Wire Break	3	96	0
	2.Probable Wire Break			0
	3.Possible Delamination			1
02/28/98- 03/22/98	1.Possible Wire Break	3	96	1
	2.Probable Wire Break			5
	3.Possible Delamination			3
04/28/98- 05/14/98	1.Possible Wire Break	3	96	7
	2.Probable Wire Break			13
	3.Possible Delamination			16

During the monitoring period of 04/28/98-05/14/98, a number of wire breaks recorded in a short time interval prompted the owner to excavate and investigate three pipe sections. The sketch below shows the wire breaks recorded for each pipe section as well as, delaminated concrete coating found after the excavation. The owner has chosen to proactively wrap these pipe sections with tendons in an effort to deter the possible future rupture of this reach of pipe.

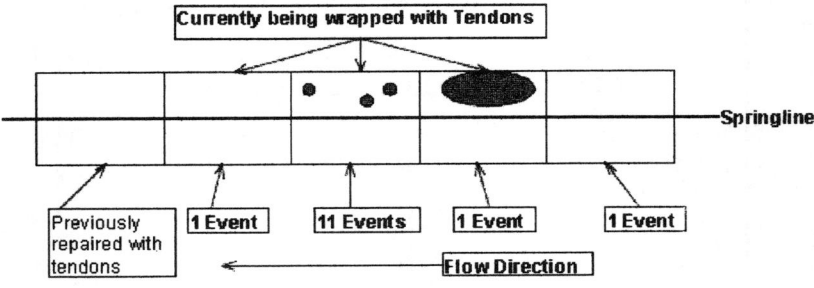

Summary
The owner of this structure will continue to rely on the information generated by the acoustic monitoring equipment to proactively address the condition of their prestressed pipe inventory.

Structure #2

This pipeline, located in Texas was remotely monitored for approximately seven weeks. The array was installed at a manhole station and approximately 4,700 feet of hydrophone array was deployed into this 72" diameter PCCP. The array was installed without interruption to service, through an existing 4" gate valve under a pressure of 165 p.s.i.. The flow conditions of the pipe are approximately 1.5 feet per second.

Pressure at the array insertion point was continuously monitored with a pressure transducer that produced a DC offset on one of the data acquisition channels. The owner provided power, phone and a temporary structure for the data acquisition system.

Summary

1. Acoustic events which appear to be related to failed prestressing wire can be detected, classified and located on the 72" diameter prestressed concrete cylinder pipe.
2. A span greater than 800 feet between hydrophones on this 72" diameter pipe, with a flow rate of approximately 1.5 ft/s, would have been too large for locating all events of interest.
3. Although perhaps operationally desirable, it does not appear that acoustic monitoring only during forced pressure changes will identify all areas of potential distress. However, this procedure does appear to increase the event count and therefore may provide pipeline condition assessment data at a faster rate than passive acoustic monitoring on its own.

Monitoring Period	Classification	Approx. Weeks	Pipe Diameter (inch)	Events
12/10/97-01/27/98	Probable Wire Break	7	72 "	3
	Large Delam Noise at Pressure Change			3
	Delam Noise at Pressure Change			27
	Pressure Anomaly			4

Conclusions

Continuous acoustic monitoring of PCCP structures offers reliable and confirmable data about the rate of breakage of prestressing wires in a structure. Repairs can be planned and budgeted based on observed rates of breakage.

Over the summer months of 1998 Pure Technologies Inc. will have the opportunity to monitor additional PCCP structures throughout North America. The data generated from these projects will provide important information about the longevity of these structures.

More data need to be gathered to test the significance of the current rate of breakage in forecasting and back casting beyond the monitoring interval.

Within a structure, there generally exist local areas, which exhibit higher breakage rates than surrounding areas. The acoustic detection of these breaks allows stakeholders to identify and repair localized areas of potential structural deficiency, and to identify and remedy the conditions which are causing the localized corrosion.

References

1. Holley, Mark, W. "Continuous Acoustic Nondestructive evaluation of Prestressed Concrete Structures", NACE Conference, Victoria, BC, December 1998.
2. Paulson, Peter, O., "Acoustic Monitoring of Prestressed Concrete Pipe", NACE Conference, Palm Springs, CA, November 1998.

Acknowledgments

Pure Technologies Inc. and the writer thank those organizations and individuals that have contributed to the advancement of the Science and the Technology of Continuous Acoustic Monitoring of Prestressed Concrete Pipe.

These include:

The Arizona Public Service
through Sarah Rittenhouse and Ed Barfoot

The Metropolitan Water District of Southern California
through Leonard Grasha, Gary Shipley, Subramanian Iyer, Danny Wang and Chris Beggs

The United States Department of the Interior, Bureau of Reclamation
through David Harris, Mike Peabody, and
Fred Travers

An Update on Acoustic Emission Testing of PCCP

by

Will Worthington, P.E. Member ASCE [1]

I.   Abstract

At the 1993 PCCP Pipe Users Forum in San Diego, the Bureau of Reclamation introduced the concept and development of Acoustic Emission Testing on Prestressed Concrete Cylinder Pipe (PCCP). Since 1993 a number of improvements have been incorporated into the technology as it is being applied commercially This paper will provide an update on the application of AET to PCCP pipelines.

II.   Background

At the 1993 meeting of PCCP Pipe Users Forum in San Diego, the Bureau of Reclamation provided further information on their continuing investigation into the premature failure of Prestressed Concrete Cylinder Pipe (PCCP) in the Central Arizona Project. They reported on improved, very effective methods of refurbishing PCCP that was found to be in a state of distress. However in the absence of a reliable method of detecting areas of deterioration, the painful decision had been made to replace portions of the world's largest diameter PCCP.

In their search for a means of determining the condition of PCCP, Reclamation had perceived that acoustic emission technology might hold promise. (Worthington) Early experiments had demonstrated that the prestressing wire in PCCP emits a sharp sound when it breaks. Reclamation had detected and recorded these sounds using underwater microphones, or hydrophones, at a distance of 1,524 meters (5000 feet). Armed with this information, Reclamation was in the process of developing and installing an acoustic emission test (AET) system on the PCCP inverted siphon at the Agua Fria River in Arizona.

1.   Will Worthington, President, Pipeline Technologies, Inc. 1425 North Hayden Road, Scottsdale, AZ 85257

Early interest in AET stemmed from operational aspects of this technology as well as its technical potential. It was realized that it would not be sufficient to focus attention on a few pieces of empty PCCP in a laboratory environment. The pipe owner's problems are different. They face a pipeline which is buried, which is usually difficult to take out of service, and which is up to a hundred miles long. So a system that might look good to a scientist may not fill the bill for a pipeline owner. The owner needs a pipeline inspection method which:

- is reliable
- is quick
- performs without taking the pipeline out of service
- performs without uncovering the pipeline
- will detect areas of distress early enough to permit refurbishment
- is cost-effective

Many technologies and methods which held promise of meeting several of these criteria but AET appeared to have the potential to meet all six.

III. Principles of Acoustic Emission Testing

Passive acoustic emission detection technology has been recently adapted to concrete pressure pipelines. This method of inspection is based on the acoustic emissions made by the prestressed reinforcing wire as it releases its energy.

Prestressed concrete pipe is reinforced by spirally wrapping high strength wire around a concrete cylinder. If the pipe is in a state of distress, the prestressing wire will be involved. When this occurs, the wire will break in a relatively brittle fashion, with an instant release of the tensile force up to 5,000 kg (11,000 pounds) in that strand of wire.

Much of this energy is in the form of sound energy which propagates through the pipe core and into the column of water within the pipe. The broken ends of the wire are immediately re-anchored in the protective mortar due to friction and the Poisson effect. If the deterioration continues, the protective mortar will be further compromised and the stored energy within the prestressing wire will be released in a series of discrete events. As more wires in the area of distress are involved, they too will break. The process of deterioration leading up to a corrosion-related failure takes several years to run its course, and it is a very noisy process.

Pipeline acoustic emission technology draws heavily from the field of anti-submarine warfare. In anti-submarine warfare and in pipeline testing, acoustic emissions of interest are detected by a series of hydrophones and screened for the known acoustic signatures that are emitted by an event of interest. The opening of a torpedo door is an event of interest to submariners. The release of energy by the prestressing steel is an event of interest to PCCP owners. These and most other events have unique acoustic properties which allow these events to be distinguished.

The precise identification of the arrival times of these signals at a series of hydrophones is used to locate the source of the events. Sound travels through water at a known and constant speed. That speed is approximately one mile per second, or about five times the velocity of sound in air at sea level. The time it takes for a sound to arrive at a hydrophone is directly related to the distance it travels. The greater the distance, the longer the time. Therefore, the physical location of a wire break can be determined by comparing the arrival times of that event at both hydrophones.

The figure below highlights the components of an acoustic emission detection system. The illustration simulates a wire break close to the left hydrophone. The sound from this event will be detected first at the left hydrophone and momentarily later at the right hydrophone. By comparing the difference in arrival times between the two hydrophones to millisecond accuracy, the location of the event can be determined.

IV. Early Experiments

In 1991, the U.S. Bureau of Reclamation performed the first experiments to establish the viability of using AET to identify and locate structural distress in PCCP. A single strand of prestressing wire was cut while a hydrophone one mile away monitored acoustic signals in the water in the pipeline. The sound of the breaking wire was distinctly identified. This opened the door to the adaption of AET to PCCP pipelines.

To avoid "reinventing the wheel", Reclamation actually worked with the U. S. Navy's Naval Research Laboratory (NRL) to assess the feasibility of the technology. NRL analyzed several sample acoustic signals and advised that AET is fully capable of doing what Reclamation was trying to do. NRL assisted in establishing parameters for a system to meet pipeline conditions.

Following further testing in 1991 and 1992, a system was developed and permanently installed on a 3.2 km (2-mile) inverted siphon in Arizona. That system consisted of 12 hydrophones inserted through valves added to the pipe at 305 meter intervals (1000-foot) for that purpose. All signals were transmitted real-time to a signal processor located on site. Signals were classified and localized for immediate use by the system operator.

Many important lessons were learned from this system. First, the quantity of data was surprisingly high. As a result, the condition of the pipeline began to come into focus within a few days. It did not require months of testing as had been anticipated. This is attributed to the relatively noisiness of the pipe deterioration process, and suggested that a permanently installed system may not be necessary. A portable system with an approximate test period of days or weeks was discussed. This, however, would require some method of replacing the buried cable in order to be practical. Another important lesson had to do with the sensor spacing interval. The hydrophones in this system were spaced at 305 meter (1000 foot) intervals. The analysis of signal detection indicated that this spacing was probably overly conservative, and that spacing greater than 305 meters (1000 feet) was justifiable based on the data gathered. (Reclamation)

V. Current Acoustic Emission Test Equipment -

Systems have been developed commercially which take advantage of these lessons learned. One acoustic emission detection system (See Figure) in commercial use new is the proprietary AH-1 Pipeline Test System. This system consists of:

- A series of two or more sensitive hydrophones are used to detect noise in the pipeline. These sensors are mounted on the end of a stainless steel shaft which is inserted into the pipeline through a series of seals and valves while the pipeline is in service at operating pressure. The hydrophones are usually installed in the pipeline through existing air valves after temporarily removing the air handling mechanism.

- Signals from these hydrophones are monitored by a small computer located close to the hydrophone. This battery-operated computer screens all acoustic activity against the acoustic signature of prestressing wire-related emissions. The computer records all signals matching the wire signal characteristics on data storage disks for later processing.

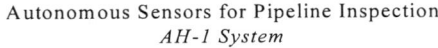

Autonomous Sensors for Pipeline Inspection
AH-1 System

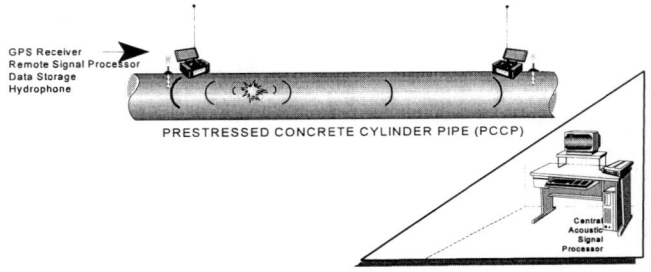

- A third component of the system, and the key to elimination of the cumbersome telemetry, is the global positioning system (GPS) antenna and processor which is incorporated into the acoustic system. This feature accomplishes two purposes. Primarily it serves as a very accurate clock. It determines the precise time of passage of the signal to an accuracy greater than a thousandth of a second. This precise time of passage is compared to the same information at adjacent hydrophones to determine the point of origin of the sound. Coincidentally it provides the location of the hydrophone in latitude and longitude so that there is no ambiguity as to where a signal was detected.

VI. AET Performance to Date -

We have been involved in the analysis of acoustic data from ten different pipelines. In eight cases, the testing has confirmed that the PCCP is in good shape. In these instances, the results have proven useful to the pipe owners even though no bad pipe has been detected. For example, one pipe owner was faced with a decision as to the advisability of replacing a 20-year-old, 8 km (5 mile) long pipeline due to a road widening project. The pipeline in question was manufactured by now-defunct Interpace at a time when quality was a problem. AET testing showed this particular line was in good condition and not actively deteriorating. Based on the results of this testing, the line was left in place at a savings of approximately $6 million. This decision was based in large part on the confidence that pipe owner gained from spot checks using AET for which he paid less than $50k.

In two instances, our conclusion was that the pipe was in an advanced state of deterioration. Excavation for visual inspection of the pipe surface was recommended. It is worth noting that in both instances, the pipe had been tested by a visual interior inspection (with sounding) by experienced personnel. There was no indication of distress whatsoever. In both instances, the pipe had been evaluated using over-the-line potential methods, again with no indication of distress. In both instances, our recommendation was accepted and the pipe was excavated for visual inspection.

The distress in the 252-inch pipe was detected based on less than 20 hours of data. (Reclamation) Distress in the 72-inch pipe was based on 160 hours of data collected by a pair of hydrophones spaced at 590 meters (1940 feet). (Marshall)

**Photo 1 - Reclamation's 6.4 meter (252-inch) PCCP Pipe
Distress Detected Acoustically**

**Photo 2 - 1.8 meter (72-inch) PCCP Pipe
Distress Detected Acoustically**

VII. The Future -

We feel the future is bright for PCCP pipe owners, insofar as condition assessment and rehabilitation is concerned. Five years ago at the last meeting of the PCCP Users' Group, there was no proven technology available to give reliable condition assessment of PCCP lines. AET is available commercially now. We expect continued advances in the AET systems. Testing and research will further increase the confidence we can place in the results. As we continue to gain experience, we will become more efficient in its use. Additional technologies will also become practicable. Developments will continue in active acoustics - impact echo and ultra-sound systems. Remote field eddy current technology shows much promise. In some instances these technologies will compliment AET to give additional information. In some instances there will be systems that compete with AET.

Pipe owners by nature are conservative. In our discussions with pipe owners regarding the use of this technology, several thoughts are frequently expressed. First, they would like to see a long list of satisfied customers before signing up to try this technology. That was a tough nut to crack, getting the first commercial customer. The list may not be long now, but there is a list and it is growing, and they are well-satisfied.

A second concern is that the price seems too high and if they wait a few years it will come down. The development of this technology is not inexpensive. There has no research or development money available in the US to defray the development costs - not federal, not AWWA, not ASCE, not ACPPA, not SBIR. These costs must be recovered. It should be noted that both AWWA Research Foundation and ACPPA have undertaken studies to evaluate various technologies, and this is a step in the right direction. Clearly if all pipe owners take the position of waiting until there is a long list of customers, and until others have absorbed the costs of technology development, it will not happen. The future for AET and other technologies will be bleak. I am sure all of us involved in the development of systems aimed at pipeline condition assessment will agree on this point: We need your support, and we need the opportunity to demonstrate that AET works now for owners of PCCP.

There is a compelling analogy between what we do with AET and what a cardiologist does with the EKG - the electrocardiogram. Both entail the use of sensors placed on a pipe to measure and analyze acoustic data. Both give a condition assessment in a relatively short test period. I give my prognosis for the future of AET:

Within ten years the use of AET in the PCCP pipelines will be as commonplace as the use of the EKG in health care today.

Demonstrations and refinements of this technology will develop the confidence and support of PCCP users in order for this to occur. AET will not be the only technology used in this role, but it will continue to evolve as a very important tool.

VIII. References

Marshall, David H. and Will Worthington, "Increasing the Reliability of Concrete Pressure Pipe," ASCE Pipeline Conference, Boston, June 1997

US Bureau of Reclamation; "Acoustic Monitoring of Prestressed Concrete Pipe at the Agua Fria River Siphon", December 1994

Worthington, Will; "Prestressed Concrete Pipe Inspection and Monitoring Methods", Proceedings of the Conference of Nondestructive Evaluation of Civil Structures and Materials, Boulder. CO; May 1992

Pipeline Rehabilitation & Repair

Tim Gwaltney, P.E.[1]

Abstract

As our cities continue to age, we are faced with the ongoing challenges of renovation and repair. One of the biggest problems facing our cities today is the deterioration of underground piping systems.

In the past, excavation was used almost exclusively to restore crumbling pipe. However, as cities became more congested, the disruption and expense associated with excavation became less acceptable. Because of this, other forms of pipeline reconstruction with less excavation were developed.

Today, there are many trenchless and semi-trenchless methods for reconstructing deteriorated pipelines. Two of the most used methods of true trenchless pipeline rehabilitation are cured-in-place pipe (CIPP) and fold & formed pipe (FFP). Two of the better known methods of pipeline rehabilitation that fall into the semi-trenchless category are sliplining and pipe bursting. With the development of these trenchless/semi-trenchless forms of pipeline reconstruction, renewal of piping systems can now be accomplished with only minimal surface and sub-surface disruption.

Each of the methods mentioned has its place in the reconstruction of deteriorated pipelines. However, there are circumstances when one rehabilitation method may be better suited for a particular project than another. This paper will focus on the applicability of each method to a particular project, materials, installation techniques and the various ASTM standards that cover each rehabilitation method.

Introduction

The first step of any sewer reconstruction program should include an inventory and evaluation of the existing storm and sanitary sewer systems. This involves physical inspection and evaluation of physical conditions of all pipelines and manholes. Once

---

[1] Project Engineer, Insituform Technologies, Inc.
702 Spirit 40 Park Drive, Chesterfield, MO 63005

the inventory and evaluation are complete, various reconstruction technologies are recommended for deteriorated portions of the system. There are several software packages available that can be used to organize and inventory data collected in the field. Some software packages even evaluate different trenchless reconstruction techniques.

The problems associated with pipeline rehabilitation are complex and interrelated, with no two projects being exactly alike. Due to the importance of rehabilitation and extending the life of our infrastructure, engineers must be educated as to the methods and limitations of sewer collection system reconstruction.

TRENCHLESS METHODS

CURED-IN-PLACE PIPE (CIPP)

Deteriorated sewers in the United States have been reconstructed by CIPP for over 20 years. CIPP methods installed by the inversion method are described in ASTM F1216- Standard Practice for Rehabilitation of Existing Pipelines and Conduits by the Inversion and Curing of a Resin-Impregnated Tube. This proven technology is able to reconstruct pipelines from 100 mm (4 inches) to 2,750 mm (108 inches) in diameter. CIPP uses a flexible non-woven felt tube coated on the outside with a tough elastomeric layer. This tube is manufactured to the diameter and length required by the pipeline to be reconstructed. The wall thickness is determined by standard buried flexible pipeline equations and can be increased by simply adding more layers of felt. This felt, which has a high content of voids, is vacuum impregnated with a liquid thermosetting resin. A wide variety of resins can be used to fit the application. Different types of resins include polyester, vinyl ester, and epoxy. For standard sanitary sewers, the resin of choice is generally an isophthalic polyester resin. Polyester resins tend to be the most cost-effective thermosetting resins and have excellent corrosion resistance to low pH environments. This is beneficial in the sewer environment where hydrogen sulfide gas can lead to the formation of sulfuric acid. The sulfuric acid generated in sanitary sewers has minimal corrosion effects on the isophthalic polyester resins used with CIPP. In more demanding industrial applications and/or in pressure pipes, vinyl ester or epoxy resins may be considered.

Since the installation of CIPP temporarily plugs the existing pipeline, the portion of pipeline to be reconstructed must be segregated from the collection system. In low flow conditions, this can be accomplished by plugging the sewer just upstream from the reach to be reconstructed. When the flows are higher and plugging the sewer is not feasible, over pumping is required. Once the existing pipeline is segregated, it is inspected internally and cleaned. Because CIPP is soft and pliable during the installation phase, any debris left in the pipeline will cause an irregularity in the finished CIPP. Therefore, it is important that the existing pipeline be thoroughly cleaned.

Prior to installation, the CIPP felt tube is vacuum impregnated with the liquid thermosetting resin. When the existing pipeline is segregated and cleaned, the resin impregnated tube CIPP is ready to be inserted into the pipeline. The most common installation technique for CIPP is hydraulic inversion. This method inverts or turns the tube inside out through energy provided by a column of water. Using water to invert CIPP provides a buoyant effect that virtually eliminates friction between the uninverted material and the inverted tube. Another benefit associated with the inversion process is sequential or progressive expansion. As the tube is inverted, it progressively moves through the pipe. Wherever the material is turned inside out is its final position. The Insitutube unrolls. Therefore, there is no pulling, tearing or abrasion between an inverted tube and the host pipe. In addition, standing water is pushed in front of the inverting face as it moves through the host pipe and is discharged into the terminating manhole. This prevents the trapping of water between the tube and the existing pipe which can occur if the inversion process is not utilized. Through the inversion process, resin impregnated felt is forced tightly against the existing pipe. This allows some resin to migrate into cracks and joints thus locking the CIPP into place. Also, utilization of the CIPP hydraulic inversion process allows reconstruction of irregular shaped sections and installations in excess of 650 m (2,000 feet).

Once the inversion process is completed, the water used to invert the tube is then heated. The heat initiates polymerization which converts the pliable tube into a hard CIPP. Flow capacity is typically increased because of the resulting smooth plastic pipe with no joints. In large diameter pipes, workers can enter the CIPP to do trim work and open side connections. In non-man entry pipes, the reinstatement of service laterals sealed by the installation process is accomplished by a remote controlled robotic cutter.

Force main reconstruction (pressure pipes) requires CIPP methods with end seal technology. End seals prevent tracking of effluent between the CIPP and the existing pipe. Force mains can be designed either as a liner to seal leaks and control internal corrosion, or as a free-standing structural pipe.

Fold and Formed Pipe (FFP)

The next category of trenchless reconstruction is the Fold and Formed Pipe technology (FFP). Several processes use a folded thermoplastic pipe that is pulled into place and then expanded to conform to the internal diameter of the existing pipe. This type of reconstruction can be thought of as an improved form of sliplining. Unlike sliplining, excavation is not required since installation can be accomplished through manholes, and lateral reinstatement is accomplished internally. The finished pipe has no joints, minimal annular space and typically will increase flow capacity in spite of a slight reduction in cross-sectional area. These FFP processes have less versatility than CIPP in terms of diameter range and installation length, and only slight offsets and bends can be negotiated. An advantage that fold and formed has over CIPP is speed of installation; there is no curing process like that for CIPP. Fold and formed processes are suited for circular pipe cross-sections with diameters from 150 mm (6") through 600 mm (24") with typical installation lengths of 90 to 180 linear meters (300 to 600 linear feet). Installations are generally limited to pipes with offset joints no greater than 12.5% of the inside diameter or lines with only gentle radius bends.

Today, polyvinyl chloride (PVC) is one of the most commonly installed sewer pipe materials in the world for new construction in small diameters. Fold and formed processes have been adapted to allow this material to be used in trenchless reconstruction for small diameter sewers. For design to site specific conditions, dimension ratios of 26 through 50 are available. In concept, the process is quite simple. A PVC pipe is extruded in a folded configuration. The folded pipe is flexible when hot but rigid at room temperatures. The pipe is placed on large spools during manufacture while it is hot and flexible. This is done for ease of handling, transportation to the job site, and to facilitate the installation process.

Much like CIPP, reconstruction with the folded PVC pipe requires that the pipeline segment in question be segregated from the collection system. Depending upon the flow through the existing pipe, this can be accomplished by either plugging at an upstream manhole or by over pumping. Since the PVC material is soft and pliable during the rerounding phase of installation, the new pipe will simply take the shape of the existing conduit. For this reason, it is important to thoroughly clean the existing pipe before reconstruction.

In one particular PVC process (PVC cell class 13223-B as per ASTM D 1784), spools of the folded pipe are transported to the job site where the material is reheated to make it flexible. While flexible, the material is pulled into the existing pipeline. It is again heated both internally and externally using hot water, steam or hot air. Once the appropriate temperature has been reached, the folded PVC pipe is pressurized internally to unfold and expand to conform to the existing pipe using a progressive rounding device to expel standing water and to ensure a tight fit. The new PVC pipe is held under pressure until it cools and regains its strength. This process is described in more detail in ASTM F 1504, Standard Specification for Folded PVC Pipe for Existing Sewer and Conduit Rehabilitation. Service lateral reconnections are made by cutting holes in the PVC pipe from within by using a robotic cutting device. When heated and expanded, the PVC pipe dimples at each service so that the connections can be located for internal cutting.

High Density Polyethylene (HDPE) is another material commonly used in the FFP process. HDPE products are usually extruded in a circular shape then deformed into the folded, or "U" shape configuration. The deformed pipe is pulled into the existing pipe then reformed to a circular shape using heat and pressure. This process is described in more detail in ASTM F 1606, Standard Practice for Rehabilitation of Existing Sewers and Conduits with Deformed Polyethylene Liner and ASTM F 1533, Standard Specification for Deformed Polyethylene Liner. Service laterals are reinstated internally using CCTV and robotic cutters.

There are different formulations of PVC and HDPE. PVC and HDPE products available for FFP and deformed and reformed rehabilitation projects have different cell classifications. As mentioned previously, PVC cell classifications are determined as per ASTM D 1784. HDPE cell classifications are determined as per ASTM D 3350. The different formulations of each material exibit different characteristics as far as physical properties, thermal coefficient of expansion and long-term structural capabilities. The specifications for a particular rehabilitation project should require each product manufacturer to include that product's cell classification in the submittal package.

SEMI-TRENCHLESS METHODS

SLIPLINING

Sliplining is one of the oldest forms of pipeline rehabilitation. Sliplining is a rehabilitation method by which continuous or segmented lengths of smaller diameter pipe are inserted into an existing deteriorated pipe. The annular gap between the slipline pipe and the existing pipe is grouted after the service laterals have been reinstated. This form of pipeline rehabilitation is considered semi-trenchless because excavation is required for an insertion pit and reinstatement of service laterals. Unlike CIPP and FFP processes, the sliplining process can be accomplished without bypass pumping in most cases. Advertised diameter ranges for sliplining products is 300 mm - 2500 mm (12" - 96").

Continuous slipliner insertion involves standard 12 m (40') sections of pipe joined together (usually Polyethylene by butt fusion) in one continuous length. Before the installation process, the entire length of required slipliner pipe is butt fused end-to-end and laid along the ground in-line with the existing pipe. An insertion pit is required for installation. The crown of the existing pipe is removed to the springline. After the existing pipe has been cleaned and prior to installation, a short length of pipe equal in diameter to the slipline pipe is pulled through the line. This ensures the slipline pipe will pass through the line. The entire length of slipline pipe is then inserted into the existing pipe in one continuous operation. The slipliner pipe is either pulled or pushed into position using a winch and cable system. After the lining system is in place, service laterals are reinstated externally and the annular space is grouted. Due to the reduction in cross-sectional area, flow capacity may be reduced. HDPE is the material most commonly used for continuous sliplining. ASTM F 585 Standard Practice for Insertion of Flexible Polyethylene Pipe Into Existing Sewers describes this process in more detail and discusses design considerations. Depending on the diameter of the slipliner and depth of the host pipe, the size of the insertion pit may be quite large. According to section 5.2 of ASTM F 585, the insertion pit should have an entry slope of 2½:1. The length of the level excavation should be at least 12 times the diameter of the polyethylene pipe being inserted.

Segmented slipliner insertion involves standard sections of pipe assembled at the bottom of an insertion pit (typically 2.5 m x 6.0 m; 8' x 20'). The top of the existing pipe is removed down to the springline. A short length of similar diameter pipe is then pulled through the line. This ensures that the slipline pipe will make it through the existing pipe. One section of slipline pipe is lowered into the pit and pushed into the existing pipe using a hydraulic ram or simple cable and pulley system. The next section of pipe is lowered into the pit. The two pipe sections are joined together, usually with a rubber gasket system, and pushed further into the existing pipe. This process is repeated until the project termination length has been reached. The service connections are reinstated externally, then the annular gap between the liner and existing pipe is bulkheaded at each manhole interface and grouted. Due to the reduction in cross-sectional area, flow capacity may be reduced. If the pipe is flowing less than half full, bypass pumping is usually not required. Common materials used in segmental sliplining are Reinforced Plastic Mortar, Centrifugally Cast Fiberglass Pipe (ASTM D 3262 - Standard Specification for "Fiberglass" Glass-Reinforced Thermosetting-Resin Sewer Pipe), PVC (minimum cell class designation as per ASTM D 1784), PVC profile wall (minimum cell class as per ASTM F 1803) and profile wall HDPE (minimum cell class designation as per ASTM F 894).

As stated earlier, there are different formulations of PVC and HDPE. PVC and HDPE products available for sliplining projects have different cell classifications. The different formulations of each material exibit different characteristics as far as physical properties, thermal coefficient of expansion and long-term structural capabilities. The specifications for a particular sliplining project should match equal products or match the design wall thickness to the differing structural properties of each product.

PIPE BURSTING

Pipe bursting is the only pipeline rehabilitation technology which allows size-for-size or up-size pipeline replacement. The name "pipe bursting" describes the process. The existing pipe (100 mm - 915 mm; 4 - 36" diameter) is burst and pushed aside into the surrounding soil by a head that is winched through the pipe. The replacement pipe (usually HDPE) is attached to the head. Therefore, bursting and replacement is simultaneous. The HDPE replacement pipe can be the same diameter or larger than the existing pipe and the entire project length is butt fused end-to-end and placed in-line with the pipe to be replaced. Service laterals are reestablished at excavation pits with external saddles. The extent of diameter upsizing is dependent on soil conditions, surrounding utilities and depth of the pipe. Similar to sliplining, this method requires excavation for an insertion pit and for service lateral reinstatement. Unlike sliplining, the flow in the line must be by-passed, or if conditions permit, the line may be plugged upstream.

Typical pipe bursting heads on the market are static, pneumatic and hydraulic. Static heads have no moving parts. The head is simply pulled through the old pipe by a heavy-duty winch. Pneumatic heads use pulsating air pressure to drive them forward and burst the old pipe. A small pulling device guides the head. Hydraulic heads expand as they are pulled through, bursting the existing pipe.

Pipe bursting requires almost no cleaning of the host pipe since the host pipe is going to be destroyed. As long as a cable can be strung from one access point to the other, the process can begin. However, excavations are required at the insertion end, each service lateral and usually at the termination end. Therefore, the time saved in cleaning is generally offset by the excavation time. Since this process offers upsizing capabilities, flow capacity can be increased dramatically. In relatively shallow pipelines (<1,5 m; <5 feet) soil heave may cause surface problems such as cracked pavement.

Summary

Technologies exist to reconstruct deteriorated pipelines with little or no excavation. Cured-in-Place Pipe, Fold and Formed Pipe, Sliplining and Pipe Bursting have been used successfully in pipeline rehabilitation projects for many years. There are many parameters that determine which process or processes is best suited for a particular project. These parameters include, but are not limited to, the following: host pipe diameter, host pipe material, condition of the host pipe, number and angle of bends, location of surrounding utilities, soil conditions, flow capacity increase/decrease requirements, cross-sectional shape, location and layout of the project and infiltration/inflow reduction requirements. Following is a chart summarizing the published capabilities and advantages/disadvantages of each process:

	CIPP	FFP	Sliplining	Pipe Bursting
Diameter Range	100 mm - 2,750 mm (4" - 108")	100 mm - 600 mm (4" - 24")	300 mm - 2500 mm (12"- 96")	100 mm - 900 mm (4"-36")
Host Pipe Condition	60% of original flow area required	<12.5% offsetts, <1" protruding taps	Must have sufficient clearance between host and slipliner	No collapsed sections
Installation Lengths	Over 610 m (2000')	180m-245m (~600'-800')	Over 305 m (1000')	305 m (~1000')
Bends	Multipe up to 90°	Multiple <22.5°	Slight Bends	Slight Bends
Excavation Pits	N/A	N/A	Required	Required
Lateral Reinstatement	Internal	Internal	External	External
Impact on Surrounding Utilities	No Impact	No Impact	No Impact	Potential Impact
Grouting	N/A	N/A	Required	N/A
Existing Cross-Sectional Area	Circular/ non-Circular	Circular	Circular or Non-circular	Circular
Impact on Flow Capacity	Equal to or Greater Than Existing	Equal to or Greater Than Existing	Equal to or Less Than Existing	Equal to or Greater Than Existing
Host Pipe Material	Any	Any	Any	Non-ductile materials*

* Consult manufacturers concerning ductile materials such as Steel and Ductile Iron

**Tim Gwaltney's Bio for
ASCE Conference
San Diego, CA
Aug. 23 - 26**

Tim Gwaltney is a Project Engineer for Insituform Technologies, Inc. based in St. Louis, MO. He joined the company in 1993 and has provided technical support to Insituform's contracting companies worldwide. These services include technical marketing, project reviews, sales training and design analyses. He is an active member of the ASCE.

Tim received his BS in Civil Engineering from the University of Memphis in 1992. He is a registered Professional Engineer in the states of TN and MO.

"Horizontal Directional Drilling with Ductile Iron Pipes"

Randall C. Conner[1]

Abstract

Designers and contractors plan and install most projects with ductile iron and other pipes with traditional trench or "cut and cover" construction. This, of course, generally involves successive joining and installation of pipe sections in final position in an open trench. However, even long sections of fully assembled restrained joint ductile iron pipes are also pull-installed as units in various special environments. Subaqueous crossings and also intake or outfall installations are examples of such ductile iron pipe installations.[2,3] In some of these applications, contractors pull assembled sections of ball joint or other restrained joint pipes, or combinations of ball joint and less flexibly joined restrained pipes, along the bottom or the bottom of a trench in the body of water. On other projects, they pull and float similar sections of pipe into position and then with uniform control sink them to the bottom, etc. Contractors also pull or push joined segments of ductile iron pipelines as carrier pipes up inside larger pipes. Such applications include conventional casing and carrier pipe (under highways, railroad tracks, etc.) installations, sliplining, and various other crossings.

Contractors now also apply the proven advantages of ductile iron pipelines and the strength, flexibility, and quality of many joining systems available with ductile iron to various types of newer trenchless technology and construction. Contractors around the world now push or jack assembled sections of variously joined ductile iron pipes directly through holes in the ground as in pipe jacking, microtunneling, special pipe bursting, etc. Some restrained joint ductile iron pipelines are now also being installed by horizontal directional drilling (HDD) methods where the pipeline is pulled rather than pushed. This paper briefly discusses case histories of ductile iron pipelines successfully installed by HDD in Avon, CT and St. Louis, MO. Also, this paper provides additional information and illustrations concerning an installation of a

---

[1] Member ASCE, Research Engineer, Technical Division, American Cast Iron Pipe Company(ACIPCO), PO Box 2727, Birmingham, AL 35202, USA

152 mm (6 inch) restrained joint ductile iron pipeline by horizontal directional drilling in Grand Rapids, MI.

In addition to discussions of several installations, this paper also discusses the subject of frictional resistance to the movement of ductile iron pipes relative to soil. American Cast Iron Pipe Company (ACIPCO) has conducted model tests with and without conventional bentonite-based drilling fluid as per API RP13B requirements and with and without loose polyethylene encasement as per ANSI/AWWA C105/A21.5.[4] This paper compares and contrasts the resulting values obtained in these model tests to the results of older research. The manual, "Thrust Restraint Design for Ductile Iron Pipe," published by the Ductile Iron Pipe Research Association (DIPRA) in various versions for more than 20 years,[5,6] forms the primary basis of these comparisons.

History and Basic Challenges of Pulling Pipe
In the traditional trenchless applications of ductile iron pipes including casing-carrier pipe installations and sliplining, contractors most often assemble the pipes by pushing the joints together and thereafter further pushing the assemblies into the casings or into the old pipelines. Ductile iron pipes and the traditional joints thereof possess substantial columnar strength and eminent suitability for these purposes. Now, new bell-less joints are available that maximize the strength and minimize the pushing profile of ductile iron pipes for even wider trenchless applications.[7,8]

The application of pulling ductile iron pipelines in horizontal directional drilling applications is a quite natural progression from a one time use of threaded joint ductile iron pipes for well casings[9] and horizontal road borings and also its ongoing use in casing-carrier pipe, sliplining, subaqueous crossing, intake, and outfall applications. Horizontal directional drilling is similar to underwater pulling installation of pipe without a diver or barge, in the sense that one cannot see what is going on around the pipe in pulling operations. These can understandably be somewhat "nerve-wracking" construction procedures for all involved.

ASCE Manual of Practice No.89 (MOP No.89) is a relatively new publication that describes horizontal directional drilling as well as other "crossing type" construction procedures.[10] MOP No. 89 and also current ASCE seminars go into HDD designs and construction procedures in significant detail; however, what follows are the basic unique aspects of this construction procedure:

1. The contractor surveys the pipeline profile, centerline, and entry and exit locations.
2. The contractor sets up required persons, equipment, etc., and he also excavates the entry and exit drilling mud pits.
3. The contractor drills a "pilot" hole of 102 mm (4 inch) - 305 mm (12 inch) diameter in "steerable fashion" from the drilling rig side to the pipe side. The contractor and engineer consider numerous factors in establishing the target drilling trajectory or path. It is obvious that the drill path must avoid obstacles. Also, the

drill path by design should penetrate available soil layers, types, and/or conditions supporting HDD. Additionally, it is normally best to follow a path of very gradual curvature or near straight alignment to minimize friction and also to stay within the allowable "bend radius" of the pipe. This minimizes the chance of getting the pipeline "hung up" in the soil or damaging the pipe.

4. The contractor enlarges the pilot hole by pulling back gradually larger reamers or reaming heads from the pipe side to the rig side. The contractor reams the hole either slightly larger or even up to several centimeters (several inches) larger than the largest pipe diameter involved, depending on his experience and the job. The largest diameter is of course the "bell outside diameter" in the case of ductile iron pipe.

5. The contractor attaches the carrier or product pipe, normally with some sort of swivel head, to the reamer head or drill pipe before pulling the largest and final reamer back through the hole.

6. The carrier or product pipeline is then pulled back to the rig side through the reamed HDD hole.

7. The Contractor hydrostatically tests and "ties in" the new HDD pipeline, adjusting the approach areas into alignment with the carrier pipeline as necessary.

The contractor basically wants to get a good, maintained open hole, effectively join pipe sections, and then pull-install the pipeline, all as smoothly and quickly as possible. This minimizes the chances of getting the line hung up and/or damaging the pipe. He then, of course, wants to get a good hydrostatic test on the pipe. In the case of steel and polyethylene pipes, the contractor most often pre-welds the pipeline section together in one or more relatively long segment(s) ahead of time. This is due to the significant time required to position and weld individual pipe sections. With long pre-welded pipeline sections, of course, it is necessary also to have substantial space available to pre-assemble the pipe above ground. The contractor places the assembled pipe section(s) in substantially straight alignment with the end of the drill path. Sometimes, this also requires setup and positioning of the pipe assemblies on rollers. The contractor normally accomplishes this ahead of time, of course before he attaches the end of the pipe to the final reamer and drill pipe for the pull.

In the case of ductile iron pipe, it is generally possible to assemble pipe in a mud pit. This requires significantly lesser space or right of way requirements on the "pipe side." With some modern near "automatic" assembling ductile iron pipe restrained joint systems, contractors can assemble the pipes very quickly. The contractor can do this efficiently as he retracts the drill pipe and progressively installs the pipe. Normally, this requires little more time than it takes to disassemble the normally relatively short drill pipe sections as they are pulled back and stored on the rack. The case studies later in this paper illustrate these points.

Assuming properly designed and prepared sites and drill paths, HDD pulling operations are in a sense also safer for the installation of ductile iron restrained joint pipe. Contractors cannot as readily overdeflect (or damage with applied bending moments) the joints in the HDD pulling operation. In the processes of uncontrolled pulling, floating, or sinking open-cut subaqueous crossings, contractors can and do at times unintentionally beam-load pipe and joints. The normally close-fitting HDD reamed hole effectively restricts lateral movement of the joints and pipes.

The restrained joints of ductile iron pipes applied to HDD are flexible (in effect hinges). Conforming to properly designed radii of curvature of ductile iron pipe drill paths and applying pulling load creates little or no additive tensile stress effect on the walls due to bending moment. With continuous welded pipes bent and pulled around a curve, bending and pulling loads combine to increase total stresses in the pipe wall. Bending loads thus can and do affect the tensile strength and the collapse resistance of continuously welded pipe segments as explained in references and experience[11,12] and discussed in the next paragraph. In other words, with ductile iron pipes and a properly designed drill path the primary design concern relative to pulling the pipe in HDD is basically just the pulling load, including any impact effects if applicable.

Another significant design provision in some HDD installations is inward buckling or collapse resistance of the pipes due to external hydrostatic pressures. These pressures can be significant and damaging to some pipes in some deep and/or long pulls. External pressures can be due simply to groundwater or drilling fluid head around the pipe, or also grouting pressures, if required in the application.[13] Vacuum conditions inside the line, if encountered in later service or maintenance, can also increase the external pressures. The buckling or collapse resistance of pipes is normally a function of the (long-term or short-term as applicable) modulus of elasticity of the pipe material, the pipe size, and the pipe wall thickness. Ductile iron pipe is generally the stiffest, and therefore most collapse resistant, of the available flexible pipes. The rather high short and long-term modulus of elasticity of 166,000 mPa (24×10^6 psi) and commonly employed, substantial wall thicknesses combine for beneficial results in the formulae.[14] For this reason, buckling design very rarely controls for ductile iron pipe.

Environmental Issues
HDD normally is less disruptive to existing land, structures, environment, etc. than open cut construction procedures. The choice of the HDD construction procedure in some cases possibly also minimizes problems in obtaining easements, right-of-ways and permits. HDD may likewise reduce the required widths and/or breadths of easements in many cases from those required for open cuts.

As previously mentioned, contractors often use "drilling mud" (normally a mixture of fresh water, bentonite (sodium montmorillinite) clay, and in some cases other materials such as polymers) in these operations to stabilize and lubricate the drilled hole. The base materials used to mix the mud are in some cases listed for direct contact with potable water as per ANSI/NSF Standard 60.[15] Suppliers generally pursue these listings due to applications in vertical drilling of water wells, etc. While many have thus argued that drilling mud is even by such standards relatively harmless, most states require that waste drilling mud be disposed of in an environmentally approved facility. In some states, however, contractors can dispose of the mud by dewatering and farming into the owner's right-of-way, if approved by the landowners. ASCE Manual No. 89 and other references contain more complete discussions of these issues.

Friction Theory and Laboratory Friction Tests
Several researchers have attempted on the basis of laboratory experimentation to predict the amount of skin friction between various types of construction materials and soil, including pipes and piles. All of this experimentation and the results obtained are valuable information and have limited application to actual pipe work in the field. While the tests and analyses as presented in this paper are not claimed to represent specific field pulling circumstances, it is hoped they are very helpful to increase understandings of behaviors encountered.

In conventional, simplistic frictional force calculations as per Newton and Coulomb, the frictional resistance is equal to a "coefficient of friction" between the surfaces multiplied by the normal force. Potyondy tested these principles using steel panels and various soils and published a research paper in 1961 that defined a similar cohesionless soil/construction material coefficient effectively as equal to the tangent of "δ" in non-cohesive soils.[16] δ was in turn defined as the product of the internal friction angle of the soil, "φ", and the ratio of pipe friction angle to soil friction angle, "f_ϕ". Potyondy went on to show that in certain other soils there was an additive component of cohesive and adhesive forces between the soil and pipe, referred to as "cohesion" (C). This additive resisting force was directly proportional to the contact area. In such cohesive soils, this increased the "frictional resistance" (F_s) to movement between the two engineering materials. The amount of additional frictional resistance was equal to the product of the soil cohesion as determined by triaxial shear tests ,"C_s", and the ratio of pipe cohesion to soil cohesion, "f_C", multiplied by the contact area (A_P). In other words, Potyondy thus defined the unit frictional resistance (F_s) for steel and other panels relative to various soils as follows:

$$F_s = W\tan\delta = W\tan(\phi f_\phi) \quad \text{for non-cohesive soils,}$$
and
$$F_s = A_P C + W\tan\delta = A_P f_C C_S + W\tan(\phi f_\phi) \quad \text{for cohesive soils,}$$
where W represents the normal force (a function of earthload and pipe weights)

For nearly a quarter century, designers have successfully employed the principles developed by Potyondy and later adopted to thrust restraint designs of ductile iron pipe systems by the Ductile Iron Pipe Research Association. Contemporary ductile iron pipe has a "peened" appearance surface texture (reflecting the peening operation on the metal molds used to cast the pipe) which is generally rougher than the mill surface of rolled steel. Generally also, the surface profile of this peening increases as the pipe size increases. As Potyondy's work involved relatively smooth steel panels, the calculation of required lengths of restrained joint pipe for thrust restraint design of ductile iron pipe systems based on values developed for generally smoother steel represents conservative design. Also, the aforementioned DIPRA thrust restraint design manual conservatively applies "saturated" cohesion and internal friction angle values of soils, and it also applies reductions to the value of f_ϕ when lesser trench types 1 and 2 per ANSI/AWWA C150/A21.50 are assumed for design purposes. Even further reductions are applied when loose polyethylene encasement is employed. In effect, the suggested design coefficient of friction thus varies from a low of 0.18 in a relatively unconsolidated silt (ML & ML-CL[17]) soil with polyethylene encasement to a high of *0.51* for non-polywrapped pipe in compacted, clean sand (SW & SP[17]).

The design objectives are of course different when determining the maximum load required to purposefully drag a pipeline over soil or through a hole in the soil,. That is why we performed these tests. We cut rectangular coupons from an ACIPCO 152 mm (6 inch) and a 1500 mm (60 inch) ductile iron pipe and also from a 203 mm (8 inch) HDPE pipe for comparison. These coupons were roughly 102 mm (4 inches) in (arc) width and 254 mm (10 inches) in (axial) length dimension. Two shaped beds of soil were constructed. One bed consisted a relatively clean, fine sand and one consisted a clay with some sand content. In the friction tests, we pressed these coupons into the surface of formed beds of soil and then horizontally pulled them across the surface of the soils with the known mass (by normal gravity) holding the coupons against the soil. We used a tensile digital load cell (a Nor mark "Weigh-In" 4.5 kg (10 lb) rated electronic digital scale) to measure the "breakout" static forces required to first move the panels (see Figure 1). These values, of course, were used to calculate a static "coefficient of friction" for various test conditions. A 2.3 kg (5 lb) surcharge weight was added to the middle of the 152 mm (6 inch) DIP and HDPE coupons to achieve readable pull values. The test conditions included very wet and rather dry soils and contact surfaces, with and without polyethylene wrap, and also with and without the presence of a bentonite-based drilling fluid film on the soil surface. We conducted four pull tests for every condition and averaged the results. I have summarized the results of these tests in Table 1.

Figure1.

Case Studies

Avon, CT Experience:
In 1995 Hemlock Directional Boring Company of Torrington, Connecticut, became the first company in the United States to successfully install ductile iron pipe by traffic intersection. The Water Company needed to install the water main without

Table 1. Laboratory Friction Tests of Pipe Coupons on Soil

Pipe Coupon From Pipe	Soil Type	Moisture Content	Drilling Mud	Poly wrap	Normal Force (LB)	Force/ Move Avg. (LB)	Static Coeff. Friction
6 Inch DIP	Clean Sand	Dry	No	No	7.56	4.75	0.63
60 Inch DIP	Clean Sand	Dry	No	No	4.44	7.10	1.60
8 Inch HDPE	Clean Sand	Dry	No	No	6.00	3.95	0.66
6 Inch DIP	Clean Sand	Wet	No	No	7.56	2.58	0.34
60 Inch DIP	Clean Sand	Wet	No	No	4.44	1.27	0.29
8 Inch HDPE	Clean Sand	Wet	No	No	6.00	1.33	0.22
6 Inch DIP	Clean Sand	Dry	No	Yes	7.56	0.75	0.10
60 Inch DIP	Clean Sand	Dry	No	Yes	4.44	0.58	0.13
6 Inch DIP	Clean Sand	Wet	Yes	No	7.56	2.28	0.3
60 Inch DIP	Clean Sand	Wet	Yes	No	4.44	1.3	0.29
8 Inch HDPE	Clean Sand	Wet	Yes	No	6.00	1.41	0.24
6 Inch DIP	Clean Sand	Wet	No	Yes	7.56	0.88	0.12
60 Inch DIP	Clean Sand	Wet	No	Yes	4.44	0.48	0.11
6 Inch DIP	Clean Sand	Wet	Yes	Yes	7.56	0.64	0.08
60 Inch DIP	Clean Sand	Wet	Yes	Yes	4.44	0.39	0.09
6 Inch DIP	Clay w/ Sand	Dry	No	No	7.56	7.52	0.99
60 Inch DIP	Clay w/ Sand	Dry	No	No	4.44	7.08	1.59
8 Inch HDPE	Clay w/ Sand	Dry	No	No	6.00	2.28	0.38
6 Inch DIP	Clay w/ Sand	Wet	No	No	7.56	4.73	0.63
60 Inch DIP	Clay w/ Sand	Wet	No	No	4.44	4.38	0.99
8 Inch HDPE	Clay w/ Sand	Wet	No	No	6.00	2.95	0.49
6 Inch DIP	Clay w/ Sand	Dry	No	Yes	7.56	3.81	0.5
60 Inch DIP	Clay w/ Sand	Dry	No	Yes	4.44	1.97	0.44
6 Inch DIP	Clay w/ Sand	Wet	Yes	No	7.56	1.5	0.2
60 Inch DIP	Clay w/ Sand	Wet	Yes	No	4.44	2.06	0.46
8 Inch HDPE	Clay w/ Sand	Wet	Yes	No	6.00	0.45	0.08
6 Inch DIP	Clay w/ Sand	Wet	No	Yes	7.56	0.39	0.05
60 Inch DIP	Clay w/ Sand	Wet	No	Yes	4.44	1.14	0.26
6 Inch DIP	Clay w/ Sand	Wet	Yes	Yes	7.56	0.39	0.05
60 Inch DIP	Clay w/ Sand	Wet	Yes	Yes	4.44	0.59	0.13

traffic disruption or decreasing the integrity of the road. The project was completed using an Ardco DBS 1000 Boreking drilling machine. This drill rig was accompanied by an Ardco model 15OO XHD Drill Fluid Recirculating mud unit allowing retrieval and recycling of all excess Wyo-Ben high yield bentonite drill fluid. The initial pilot hole was drilled through a gravel and cobble medium utilizing a 152 mm (6") diameter bit produced by Geological Boring Company. The pipe pullback was completed by enlarging the pilot hole through backreaming, utilizing a combination of a 356 mm (14") reamer, a 457 mm (18") reamer and a 356 mm (14") packer. Flexible restrained joint ductile iron pipe was used because of its inherent ability to simultaneously address axial joint restraint and flexibility.

The entire project proceeded on schedule with the first day being dedicated to the drilling of the pilot hole. With the special considerations of having straight segments of flexibly joined ductile iron piping, extensive planning and care were exercised in the bore. One long horizontal curve and three vertical curves were experienced in the pilot bore which made for a long first day, but fell within the flexibility parameters of the pipe. The backreaming and pulling performed on the second day proceeded without any problems. The ductile iron pipe was then subject to a pressure test to determine if the integrity of the joints was maintained throughout the pull. The pipe passed the test with no problems, finalizing the project as an unqualified success.

St. Louis, Mo. Experience
In 1996, approximately 305m (1,000 feet) in 91 m (300 feet) or more segments of polyethylene wrapped 152 mm (6 inch) ductile iron pipe was pulled in behind the 254 mm (10 inch) back ream of a directional bore[18]. The 102 microns (4-mil) thick cross-laminated polyethylene wrap per ANSI/AWWA C105/A21.5 was used for these installations. This wrap, which is very tear resistant, was held in place with special rubber-sheathed steel bands near the bell shoulders. These bands reportedly kept the wrap from bunching up in the hole entrance, and the wrap also reportedly remained intact as it emerged in the exit pit.[19] This project was installed using a Bor-Mor machine with the pipe pulled back through a bentonite slurry filled hole. The pipes were joined in a pipe entry pit approximately 8m (25 feet) long, and the pipe joints were effectively restrained from separation during the retraction process by gripping gaskets. The only problems reported in these operations were some "break outs" of the boring mud/slurry to the surface and one unplanned damage in the drilling operation to an existing main. The slurry break outs were due to the relatively shallow cover depth (less than 1.8 m or 6 feet), and also the presence of probe holes and disturbed soil from a prior main maintenance operation. St. Louis County Water reports positive benefits including significant improvements in the public relations aspects of their replacement program as a result of such construction. Also, they note they were able in trenchless fashion to install easily tapped, polyethylene wrapped ductile iron pipe, which they specify exclusively for their distribution system.[17]

Grand Rapids, MI Experience
Also in 1996, the City of Grand Rapids, MI commissioned another pioneering HDD installation of approximately 67 m (220 feet) of 152 mm (6 inch) ductile iron pipe watermain. The contractor, Miller Pipeline Services, drilled and reamed a hole and installed the pipe with a Vermeer Navigator 50x100 machine, also with the aid of drilling mud. The contractor installed the pipe within a trench box in an entry pit similar to as described in the St. Louis work, and also similar to this work the pipe was installed at less than 1.8 m (6 feet) of cover. Like the St. Louis work, some drilling mud emerged to the surface through some cracks in the asphalt paving, but in this case it was not enough to even "sweep up." In preparing for this project, ACIPCO fabricated a special pulling head for the contractor to attach to his drill pipe for installing the flexibly joined American Flex-Ring®[20] restrained joint pipe. This

pulling head was fabricated from a short ductile iron Flex-Ring bell piece with a steel pulling flange welded in the barrel end. This pulling head also served as a hydrostatic testing bulkhead once the line was installed (see Figure 2). Flex-Ring ductile iron pipe is a quick assembling, locking ring restrained joint capable of substantial deflection. By virtue of near continuous bearing between the bell and spigot members in straight alignment and in deflected positions, the Flex-Ring joint can withstand very substantial thrust loads.[21] The maximum standard working pressure ratings, bell outside diameters,

Figure 2. allowable pulling loads, maximum allowable deflections, and minimum allowable turning radii for curves employing Flex-Ring joint pipes of all sizes are shown in Table 2.

Table 2. Standard** Capabilities of American Flex-Ring Joint Pipes

Nominal Size mm(inches)	Bell OD cm (in.)	Allowable Pull Force kgs(kips)	Maximum Allowable Joint Deflection (degrees)	Minimum Allowable Radius of Curve m(ft)
102(4)	20(7.88)	3,800(8.3)	5.00	70(230)
152(6)	25(9.75)	6,000(13.1)	5.00	70(230)
203(8)	30(11.88)	10,200(22.5)	5.00	70(230)
254(10)	36(14.13)	15,400(33.9)	5.00	70(230)
305(12)	42(16.63)	21,800(47.9)	5.00	70(230)
356(14)	48(19.02)	29,200(64.3)	4.00	87(285)
406(16)	54(21.14)	37,800(83.2)	3.75	93(305)
457(18)	59(23.36)	47,700(105)	3.75	93(305)
508(20)	65(25.48)	58,200(128)	3.50	100(330)
610(24)	77(30.35)	83,200(183)	3.00	116(380)
762(30)	93(36.69)	91,400(201)	2.50	140(460)
914(36)	110(43.19)	131,000(288)	2.00	174(570)

**Contact ACIPCO for needs in excess of standard product capabilities.

The contractor was able to readily join the pipes in the insertion pit. In the early going the laborers had a minor problem due to the pipe bells resting directly on a "sliding board." This did not allow much room to tap the snap rings into the bells (he had never before assembled this type of joint). The contractor addressed this with caulking technique, and it did not develop into a significant problem. The first 8-10 pipes slid very easily and quickly up into the bore (see Figure 3), but as the last couple pipes were installed the machine pulling load drastically increased. Over a period of roughly 45 minutes, and in a valiant effort so that all might see the pulling head emerge through the exit pit wall, the contractor repeatedly pushed and pulled with all the hydraulic force the machine could muster. It was calculated from hydraulic pressure gauge readings on the machine that approximately 25,000 kg (55,000 pounds) of pulling force was repeatedly applied to the 152 mm (6 inch) pipe section. The line would not move. The contractor then moved his backhoe over near the exit pit and simply excavated the last few feet of trench needed to expose the pulling and testing head. In spite of the repeated tremendous pulling load application, the subsequent successful hydrostatic test of this line to 11 bars (160 psi) in this position resulted in no leakage. This assured proper performance of the line as required. After required connections to existing piping were completed, the line went to work for the City of Grand Rapids. Resulting Figure 4 illustrates an accomplished site objective of successful waterline installation with no significant damage to flora and fauna, yards, driveways, and streets.

Figure3. **Figure 4.**

Conclusions and Discussion

I conclude the following concerning the laboratory friction tests. I note that with normal moisture content soil, with all ductile iron pipe and HDPE pipe tested, and without drilling mud or polywrap lubrication, the friction coefficients are generally higher than assumed for thrust restraint design in even the best soils and trenches in the DIPRA Thrust Restraint Design Manual. This is true even though the laboratory soil/pipe contact times were consistently very short and the contact pressures low compared to many field circumstances. It is probable that with greater contact times and pressures, allowing the soil grains to mold to, adhere, and otherwise interlock with the pipe surfaces, that the coefficients of friction would be even higher in these conditions. It is suggested that contractors could readily run their own simple tests with specific project pipes and soils for better information. It is obvious also that commonly employed drilling mud and also polyethylene wrap, such as employed with ductile iron pipe in the St. Louis HDD case study, have expected lubricating effects.

The friction coefficients were obviously reduced by the drilling mud in the laboratory tests, most dramatically in the sand soil. With regard to field experience, I think it is possible and perhaps most common for pipes in HDD to slide in very easily. However, it is also possible for them to become hung up very solidly as occurred near the very end of the Grand Rapids pull. With regard to the ease of pulling some jobs, I think that if the reamed hole is reasonably clear and full or nearly full of essentially liquid drilling mud, this is quite understandable. Drilling mud has a density of approximately 990 kg/m^3 (65 lb/ft^3). The effective density in air and full of air for example, of 152 mm (6 inch) Pressure Class 350 ductile iron pipe with standard cementlining, is 1,050 kg/m^3 (69 lb/ft^3). According to Archimedes principle and with the drilled hole full of liquid mud around the pipe, the calculated unit normal force in a relatively straight section of pipe would thus be only 1.6 kg/m (one lb/ft)↓ of pipeline, assuming buoyant conditions and no caving or squeezing inward of the soil against the pipe as it is advanced. With such very small normal forces and the friction testing results contained herein, slight pulling loads in some cases should perhaps not be surprising. Similar calculations can be done for other sizes, pipe materials, and with or without assuming buoyant conditions, etc.

In examining the soil borings for the Grand Rapids site after the project was completed, I notice that the site sand soil in the pipe zone was significantly less stiff based on blow count near the exit pit where the line apparently got hung up than it was elsewhere. It is possible that this is a factor in the great difference encountered in moving the pipe from one end to the other. It would appear that thorough geotechnical investigations are critical to the planning for this construction as they are to so many others. It would also appear that the ability of the contractor to build and maintain a clear hole with proper delivery of drilling mud, etc. along the hole is likewise critical. These studies indicate that high friction, perhaps even more than predicted by the DIPRA thrust restraint manual, is possible if native soils are allowed to encroach, collapse, or squeeze in on the pipe surface (particularly if the surrounding soil encasement is effectively "non-liquid"). Future work and testing should provide for better documentation of actual pulling loads encountered in various soil conditions, and I'm sure also cost data for such work is also needed.

Bibliography and Endnotes
2 "Ductile Iron Pipe for Underwater Service", by Ben Helton, Proceedings of the ASCE Pipelines in Adverse Environments II Specialty Conference, San Diego, CA, 1983
3 "Ductile Iron Pipe Subaqueous Crossings", technical brochure published by the Ductile Iron Pipe Research Association, 1987
4 American National Standard for Polyethylene Encasements for Ductile-Iron Pipe Systems, ANSI/AWWA C105/A21.5-93, published by the American Waterworks Association, 1993
5 "Thrust Restraint of Buried Ductile Iron Pipe", by Randall C. Conner, Proceedings of the ASCE Pipeline Infrastructure Specialty Conference, Boston, MA, 1988

6 "Thrust Restraint Design for Ductile Iron Pipe", Fourth Edition, published by the Ductile Iron Pipe Research Association, 1997
7 "Ductile Iron Microtunneling Pipe, Non-Traditional Installation Applications", by Ralph R. Carpenter and Randall C. Conner, Proceedings of the ASCE Pipeline Crossings Specialty Conference, Burlington, VT, 1996
8 "Ductile Iron Pipe Subaqueous Crossings", Technical Brochure published by the Ductile Iron Pipe Research Association, 1987
9 American Pipe Manual, 17th edition, by American Cast Iron Pipe Company, 1994
10 ASCE Manuals and Reports on Engineering Practice No. 89, Pipeline Crossings, by American Society of Civil Engineers, 1996
11 "Investigation of Pipeline Buckle Failure in a Horizontally Directional Drilled Installation", by Hugh W. O'Donnell, PE, Proceedings of the ASCE Pipeline Crossings Specialty Conference, Burlington, VT, 1996
12 "Design of a 610-mm Water Pipeline Across Providence Harbor", by David E. Hairston, Pasquale DeLise, and William Skerpan, Jr., Proceedings of the ASCE Pipeline Crossings Specialty Conference, Burlington, VT, 1996
13 "Horizontal Directional Drilling for By-Pass Tunnels", by Michael J. Robinson, Tracy J. Lyman, Larry Erwin, and Steve Wolfman, Proceedings of the ASCE Pipeline Crossings Specialty Conference, Burlington, VT, 1996
14 "Critical Buckling Pressure for Ductile Iron Pipe", by Richard W. Bonds, technical paper published by the Ductile Iron Pipe Research Association, 1992
15 NSF Listing, Drinking Water Additives - Health Effects, published by NSF International, 1997
16 "Skin Friction Between Various Soils and Construction Materials", by V. G. Potyondy, Geotechnique, London, England, 1961
17 ASTM D2487-Classification of Soils for Engineering Purposes
18 "Case Studies in Trenchless Technologies Water Main", by Kent Alms, Proceedings of the ASCE Trenchless Technology Specialty Conference, Boston, MA, 1997
19 "Using Directional Boring to Install Ductile Iron Pipe", by Donovan Larson, paper presented at the AWWA Distribution System Symposium, Norfolk, VA, 1997
20 Flex-Ring® is a registered trademark of American Cast Iron Pipe Company.
21 "Approval Report - Ductile Iron Pipe - Flex-Ring Pipe with Fastite Gasket", Job I. D. #0Y2A0.AH, by Factory Mutual Research Corporation, pg. 3-5, 1995

Acknowledgments

I extend special thanks to the following individuals and organizations for special vision and information helpful in the development of this paper:
Mr. Ken Traub - Hemlock Directional Boring Company
Mr. Kent Alms and Mr. Donovan Larson - St. Louis County Water Department
Mr. Fred Bloom (now retired) and others- City of Grand Rapids Water Department
Etna Supply Company - Grand Rapids, MI
Mr. John Cianci - Miller Pipeline Services
Mr. Joe Dobry - Baroid Corporation

Municipal Infrastructure - Innovative Trenchless Replacement Method
Utilizing Bell-Less Ductile Iron Pipe, Case Studies

Al Tenbusch[1] and Ralph Carpenter, Member ASCE[2]

Abstract

Misaligned joints, cracked pipe sections, and either complete or partial line collapse are all conditions that exist in virtually all municipal wastewater systems. Utility owners are increasingly searching for alternatives to traditional open-cut replacement methods for their failing infrastructure. This paper will discuss two (2) innovative developments in the trenchless construction industry. The innovations include a new trenchless replacement method called the Tenbusch Insertion Method and two (2) new "bell-less" ductile iron pipes, referred to as MT and GS Push Pipe, that can be used in lieu of other pipe materials.

Introduction

Engineering News Record recently reported that a construction market survey, completed by *Underground Construction* magazine, points to increasing interest in trenchless technologies. "These (trenchless technologies) will account for 17% ($731 million) of total pipeline spending ($4.3 billion) in 1998 and 27% ($567 million) of what will be spent on rehabilitation ($2.1 billion)."[3] This astonishingly strong demand for trenchless rehabilitation technologies, fueled by the deterioration of municipal wastewater infrastructure, has brought a demand for new installation methodologies and, from many owners, a demand for pipe material alternatives to high density polyethylene (HDPE) pipe. This paper will discuss the development and field testing of one pipe material option that was developed in response to owners preferring the conservative design of ductile iron pipe as well as it's inherently superior pressure and external load capabilities. It will also discuss the Tenbusch Insertion Method that utilizes the capability of ductile iron pipe to withstand substantial axial compression in a direct jacking installation and several

---

[1] Al Tenbusch – Owner, Tenbusch Contracting Inc.
[2] Ralph Carpenter - Marketing Specialist, American Cast Iron Pipe Company, Birmingham, AL

operational advantages over systems that utilize fusion welded high density polyethylene.

Installation Methodology

A new and innovative trenchless replacement method is the Tenbusch Insertion Method (TIM™)[4].

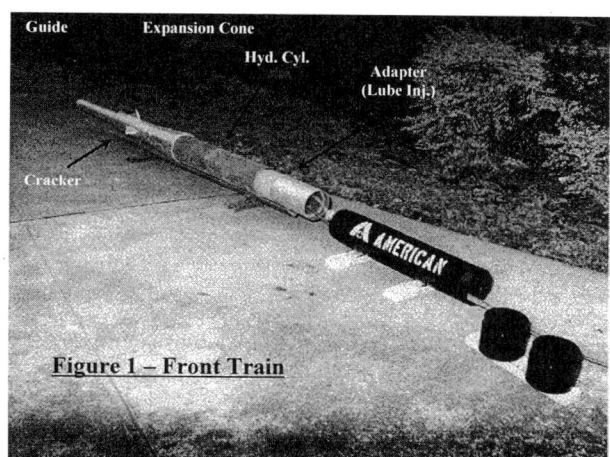

Figure 1 – Front Train

TIM was designed and developed in 1993 by one of the co-authors of this paper. The first patent for the TIM method was applied for in the summer of 1993.

The TIM system is uniquely different from other replacement methodologies, in that TIM pushes or jacks new pipe into the existing deteriorated or sub-standard pipe as it advances. This trenchless replacement method essentially uses the old pipeline as the guide for installing the same size or larger pipe without traditional open-cut trenching. Unlike other current trenchless methods, TIM utilizes the columnar strength of segmented "bell-less" ductile iron pipe to advance the lead "train" through the old pipe. The lead train (see Figure No. 1) consists of a heavy steel guide pipe [approximately 1 m (3 feet) long] that has the responsibility of maintaining the alignment within the center of the old pipe; the cracker which fractures the old pipe; and the expansion cone which radially expands the fractured line into the surrounding soil. This expansion cone is designed to create a minimal amount of annular space for the advancement of the new pipeline. The next piece in the lead train is a hydraulic cylinder capable of axial thrust force followed by an adapter piece that allows the end of the hydraulic cylinder to mate with the new pipe. The adapter piece is fitted with a lubricant injection port(s), whereby lubricant (polymer or bentonite) can be injected into the annular space. This ability of the TIM system makes it possible to efficiently replace existing pipes even in soft sticky clays or wet sands. Dual flexible hose sections that transport lubricant and hydraulic fluid to the front train are fed through each new pipe section. Each new hose section is connected to the previous sections and to the control panel with quick-connect couplings. Using the new pipe as a support column, the front hydraulic cylinder can advance the lead train into the old pipe independent of the advance of the new pipe column. The new pipe is jacked behind the lead train piece by piece by the jacking frame from a secured operating pit. The primary jacking

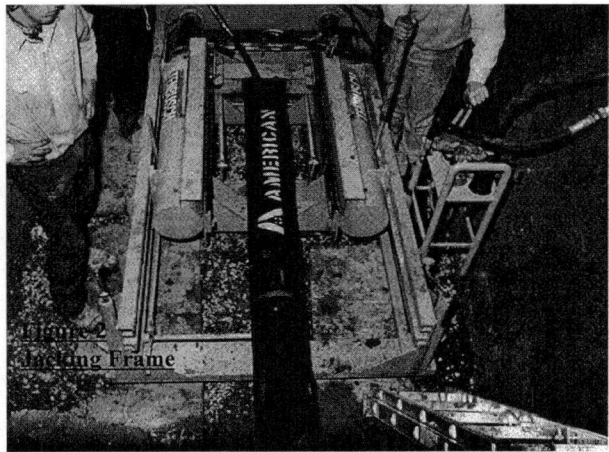

Figure 2
Jacking Frame

frame, affectionately called "TIM" (see Figure No. 2), applies the required thrust to advance the new pipe column and lead pipe train. Instrumentation and controls on the jacking frame allow the operator to "feel" his way through the existing pipe. The new pipe string and the front train (forward hydraulic cylinder) can be moved through the existing lines much like an "inch worm". Upon completion of the line replacement the lead train can be disassembled easily inside a typical 1.2 m (4 feet) diameter receiving manhole and the new pipe can be jacked into its final position.

Operating pit size is a function of the site conditions, and to a lesser extent local social and political impact, for any given installation. The TIM system offers operational options to the owner that can lessen the impact to customers and their neighborhoods or to production personnel when used within a manufacturing facility. Based on site conditions the operating pit can be as small as 2.1 meters (7 feet) square. For typical urban areas the operating pit size, when replacing deteriorated or up-sizing sub-standard municipal wastewater lines, is approximately 3 meters (9 feet) square. This size operating pit can accommodate joint lengths up to approximately 1.25 m to 1.5 m (4 feet to 5 feet). However, even when using longer pipe lengths the operating pit required is only about 2 m (6 feet) longer than the pipe length specified.

This system utilizes a concentrated staging area and with segmented pipe lengths minimizes the impact to the owners' customers. The entire operation (other than the pit itself) can be staged off of the back of a truck or trailer, i.e. the pipe can arrive at the last minute on a trailer and not be unloaded unless it is immediately used. This is in contrast to systems that use pipe materials requiring pre-assembly of joints, e.g. fusion welded high-density polyethylene or butt welded steel pipe, into continuous lengths. In many installations these continuous lengths may stretch several hundred meters, potentially blocking cross streets, business accesses, and customers' access to their residences.

With the Tenbusch Insertion Method the system allows the owner the flexibility of designing more than one line replacement from the same operating pit. These operating pits are generally proposed to be located at existing manholes or at critical lateral connections. TIM has the capability of replacing deteriorated or sub-standard existing pipe materials such as: vitrified clay, reinforced concrete, plastic, asbestos cement, corrugated metal, and cast iron, with same size or larger segmented "bell-less" ductile iron pipe.

"Bell-Less" Ductile Iron Pipe

In late 1992, the American Cast Iron Pipe Company (American) began developing a new and innovative "bell-less" ductile iron pipe for trenchless applications.[5] This joint, referred to as MT Push Pipe (MT denoting the pipe's capability as a microtunneling pipe, and Push Pipe suggesting the pipe's versatility to be installed by other trenchless construction methodologies), was designed to maintain a uniform outside diameter and efficiently transfer axial thrust loads across the joint. By mid-1994, American had completed two (2) phases of testing to verify the integrity of design for MT Push Pipe. The first phase was in-house testing and the second phase was field testing. This pipe is presently being marketed for both water and wastewater services in sizes 102 mm (4 inch) through 610 mm (24 inch) with pressure capabilities up to 2.4 MPa (350 psi). The MT Push pipe joint is fabricated by machining a counterbore in both ends of a special thickness class 55 ductile iron pipe. The connection between the two (2) counterbores on adjacent pipe sections is made with a specially machined internal, double-rubber-gasketed, coupling.

From the experience gained in the development and testing of MT Push Pipe, American' Research Engineers developed the next generation joint for gravity service applications beginning in March 1996. This joint (see Figure No. 3) is referred to as GS Push Pipe (GS denoting the gravity service capacity and Push Pipe providing a verbal illustration of the primary method for trenchless installation). GS Push Pipe is presently being marketed for water or wastewater gravity service up to 0.3 MPa (43 psi or 100 feet of head) in sizes 102 mm (4 inch) through 406 mm (16 inch).

**Figure 3
Cut-Away of GS Push Pipe**

GS Push Pipe consists of machining a counterbore in one (1) end of a special thickness class 56 ductile iron pipe, thus forming the bell of the joint. The other end of the pipe is also machined to form a mating spigot with double O-ring gasket grooves turned into the spigot surface.

Similar to the development process used for MT Push Pipe, GS Push Pipe underwent two (2) distinct phases of testing. Phase I testing was in-plant testing and phase II was field testing. The first phase involved a battery of three (3) procedures and/or tests, which included:

1. Finite Element Analysis (FEA)
2. Columnar Load Test
3. Hydrostatic Test of Joint Assembly

Finite element analysis was completed by modeling the bell (counterbored end) and spigot of a 203 mm (8 inch) GS Push Pipe independently. This was viewed as being the most conservative, as the deformation of the bell and spigot would be constrained by the assembly of one into the other. The FEA would assist in determining the areas of the bell and spigot with the highest stress levels. Three (3) models were analyzed so that American's Research Engineers could make a conservative correlation between the columnar load test results with an allowable axial thrust rating for the pipe.

Three (3) columnar load tests were performed on 1.47 m (4.83 feet) sections of 203 mm (8 inch) GS Push Pipe. Two (2) tests were completed using an assembled test configuration involving two (2) GS Push Pipe sections; one of these tests required that the pipe joints be deflected 9.5 mm (0.375 inch). The third test required an assembly configuration that used four (4) GS Push Pipe joint sections in straight alignment. These three (3) tests resulted in an average effective strength of 312.61 MPa (45,338 psi). Based on this average strength, and the evaluation of the FEA, American's Research Engineers determined an allowable axial thrust rating for GS Push Pipe (see Table No.1). The rated thrust loads have a minimum explicit safety factor of 2.0 to the minimum yield of ductile iron, which is 289.59 MPa (42,000 psi). Based on the columnar load testing results and actual effective cross-sectional areas of the joint, actual safety factors are no less then 2.25.

Table No. 1 - Allowable Thrust Load		
Pipe Size (mm (inch))	Rated Thrust (N)	Rated Thrust (Lb.)
102 (4)	266,880	60,000
152 (6)	551,552	124,000
203 (8)	782,848	176,000
254 (10)	1,031,936	232,000
305 (12)	1,307,712	294,000
356 (14)	1,610,176	362,000
406 (16)	1,885,952	424,000

The final test in phase I testing required that the assembled pipe joints be subjected to a hydrostatic test to qualify the pressure service rating. GS Push Pipe was tested in both straight and deflected alignment. In straight alignment the pipe sustained an internal hydrostatic pressure of 4.21 MPa (610 psi). However, the deflected condition for GS Push pipe proved to be the controlling parameter. With a conservative safety factor of 2.0, the pipe qualified for a rating of 0.3 MPa (43 psi or 100 feet of head).

With phase I testing completed, and based on the experience and performance of MT Push Pipe, American released GS Push Pipe for field testing, phase II. The first installation of GS Push Pipe is described in Case 2.

Case I: Birmingham, Alabama – American's Plant Site

Case I involved an existing gravity sewer line that was approximately 76 m (250 feet) of 8 inch vitrified clay pipe and served several office buildings at the American Cast Iron Pipe Company's North Birmingham, Alabama, plant. Typical of similarly aged sewer lines the existing pipe had been damaged by tree roots and ground settlement. Approximately two (2) months prior to the line replacement, which took place during a routine holiday plant shut-down in January 1996, the existing line experienced a total line collapse and blockage in a small section of the line. American' Construction Department personnel were able to replace approximately 3 m (9 feet) of deteriorated pipe by traditional open-cut trenching. After a detailed evaluation of the line that included an attempted internal video inspection

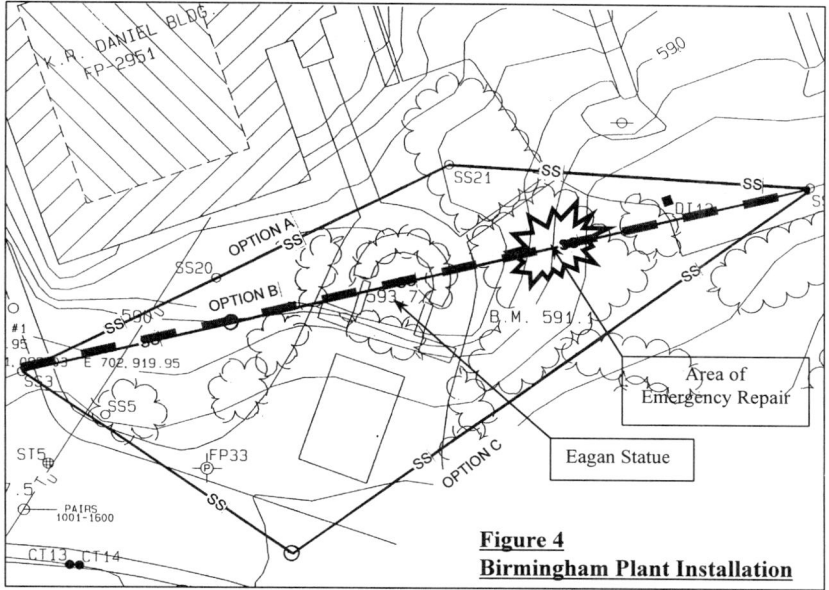

**Figure 4
Birmingham Plant Installation**

(unsuccessful due to severe deterioration) it was determined that the line would have to be replaced.

The layout of the line (see Figure No. 4, heavy dashed line) shows the alignment of the existing sewer service line, which happen to run directly under a memorial statue of American's founder, John J. Eagan. American's Engineering Department personnel were tasked with evaluating multiple options (see Figure No. 4, Option A, B, and C) for replacing the 3.05 m (10 feet) deep sewer. Existing ground conditions indicated mostly wet-sticky clay material with medium to high plasticity. Backfill at the point repair, made two (2) months earlier, was sandy soil to approximately 0.5 m (1.5 feet) above the top of the replacement pipe.

Option A was for a scenario involving jacking and boring a steel casing (open cut in this area was prohibitive because of numerous 50+ year old shade trees) and installing a new 254 mm (10 inch) ductile iron sewer line inside the casing. Option A was estimated at $105,000 and would have required by-pass pumping for approximately four (4) days.

Option B was a trenchless replacement option that would utilize the Tenbusch Insertion Method. This option would replace and upsize the existing 8 inch vitrified clay pipe sewer line with new 254 mm (10 inch) MT Push Pipe. It was estimated that the line could be replaced in approximately two (2) days and would have the least impact to this sensitive area of American's campus. The cost estimate for the Tenbusch method was $65,000.

Option C would have the new line re-routed into pavement areas servicing numerous buildings at the American' complex. It was estimated that this option would have taken approximately six (6) days to complete the line installation. By-pass pumping would have only been required during the actual tie-ins to the existing manholes at both ends. This option (Option C) was the most expensive, estimated to cost approximately $120,000.

As the cost estimates shown above indicate, the selection of Option B was the most cost effective alternative both initially and in retrospect with a final installed cost of $61,000. Using the Tenbusch Insertion Method, the project was completed and placed in service in less than two (2) working days. TIM accomplished the up-sizing (8 inch to 10 inch) in two pushes using 254 mm (10 inch) American MT Push Pipe, bell-less ductile iron pipe. Each push was made from a single, secured operating pit. The location of the operating pit was determined by the location of a critical service lateral.

The first push was approximately 15.25 m (only 50 feet) and was completed in less than 1.5 hours. The lead train was removed immediately outside of the receiving manhole. TIM was used to jack the new MT Push Pipe directly into the manhole prior to rotating TIM 180 degrees for the next push. Although prepared for polymer injection at the lead train, none was used due to the extremely low jacking loads encountered.

The second push was approximately 61 m (200 feet) and was the section that transitioned directly under the memorial statue and through the 3 m (9 feet) point repair. It was in this section that TIM utilized its unique ability to inject polymer to ease the jacking pressures. American' Construction Department personnel chose not to receive the front train in the receiving manhole and instead constructed a secured receiving pit immediately adjacent to the existing manhole (the manhole was only 1 m (3 feet) in diameter). This section was completed in approximately four hours.

As mentioned previously jacking loads were kept low by using a polymer mix as a lubricant. The lubricant was injected at the front train continuously as the new pipe advanced. With approximately 2/3 of the push complete, a delay of approximately 45 minutes occurred. With the existing clay soils there is tendency for the clay to adhere to the new pipe, typically, start-up jacking loads can be substantial as a result of this adhesion. This phenomenon was inhibited by the presence of the polymer lubricant. When the push was resumed only a very slight increase in initial jacking pressure was experienced. With the effective use of lubricant, the overall jacking loads were kept to a minimum.

Case 2: Fairfield, Jefferson County, Alabama – Residential Sewer Line Replacement

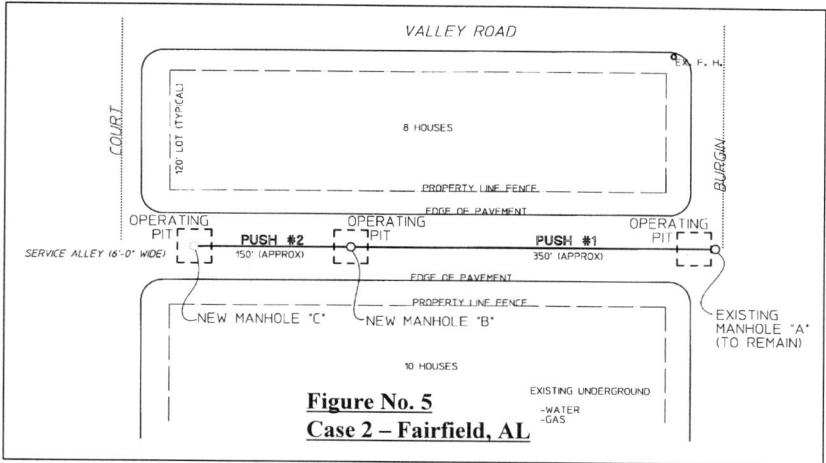

The existing line was approximately 137 m (450 feet) of 6 inch vitrified clay pipe located in the center of a narrow paved alley serving a residential area of Fairfield, Alabama, just West of downtown Birmingham. The existing pipe was in typical condition for lines installed during the early 1950's. The vitrified clay pipe had deteriorating mortar joints that allowed root intrusion and ground water infiltration.

Existing ground conditions indicated dry clay soils of low to medium plasticity. The narrow service alley in which the line was centered provided service laterals and was the access to 18 residential units and their garages. This line was of fairly shallow bury averaging approximately 1.25 m to 1.5 m (4 feet to 5 feet).

In July 1997, the Jefferson County Department of Environmental Services decided to up-size the deteriorated and sub-standard 6 inch vitrified clay line with new 203 mm (8 inch) American' "bell-less" ductile iron GS Push Pipe. The operating pit for the first push was located at the intersection of the service alley and Burgin Street, as shown in Figure No. 5, and immediately adjacent to the downstream manhole (MH "A"). This manhole would remain in-service throughout the installation.

TIM's first push was approximately 100 m (330 feet) and terminated upstream at an existing manhole location (MH "B"). Manhole "B" was replaced with a new precast manhole. This portion of the project was completed in approximately 5 hours. Because of the existing clay soil, the decision was made to continuously inject polymer lubricant. In retrospect, it was unnecessary, as jacking loads were relatively low for a push of this length.

The second and final push was made from a pit that was excavated next to the new manhole MH "B". This push was approximately 37 m (120 feet) and was completed in approximately 1.5 hours.

Prior to starting any installation of the GS Push Pipe the crew located each lateral by excavating down to each in the grassy area between the paved alley surface and the property line. After the existing 100 mm (4inch) laterals were exposed and severed, the crew set a 380 mm (15 inch) diameter plastic pipe [approximately 600 mm (24 inch) long] vertically into the excavated hole. These plastic pipes were used for temporary "catch basins" serviced by a 50 mm (2 inch) submersible electric pump in each of the plastic pipe. The crew then temporarily buried the pump and pipe in clean crushed rock. The pump hoses were all connected to a manifold that carried the flow on top of the ground to the downstream manhole. The lateral reconnections were then made one at a time during normal working hours. In spite of heavy rains that delayed some of the lateral reconnects, there were no complaints from any of the residents.

Conclusions:

This paper has discussed two (2) innovative developments in the trenchless construction industry, the Tenbusch Insertion Method and American's MT and GS Push Pipes. These innovations offer owners the ability to replace and or upsize deteriorated gravity sewer lines with new "bell-less" ductile iron pipe. The TIM system has proven it's ability to replace and or upsize existing lines with misaligned joints, cracked pipe sections, and either complete or partial line collapse while minimizing the impact to the general public and the environment by utilizing compact work areas. The TIM system utilizes the new "bell-less" ductile iron push pipe, providing owners with the confidence of a proven product and an alternative to plastic piping systems.

Acknowledgement

The authors of this paper would like to recognize the project site support provided by the following individuals:
Mr. Ron Wilson, P.E. – Jefferson County Environmental Services Department
Mr. Harry Chandler, P.E. – Jefferson County Environmental Services Department
Mr. Tommy Smith – Russo Corporation
Mr. Keith Brown – Russo Corporation
Mr. Mike Kitchens, P.E. – American Cast Iron Pipe Company

---

[3] ENR, Engineering New Record *"Infrastructure, Water-Sewer Pipeline Work is Growing With the Flow"*, January 19, 1998, page 11.
[4] Tenbusch Insertion Method, TIM™, US Patent Number 5482404, with other Patents pending in United States and abroad.
[5] Carpenter and Croxton, "The Development and Installation of Ductile Iron Microtunneling Pipe", presented at the Second International Conference on Advances in Underground Pipeline Engineering, PLD-ASCE, Bellvue, Washington, June 1995.

Geotechnical Investigations for
Tunneling & Pipe Jacking

Gregory L. Raines [1]

Abstract

Subsurface soil, rock, and groundwater conditions present one of the most important issues associated with construction of underground projects. The conditions can have significant cost implications in the initial stages, based on the pre-constructed understanding and more importantly during the construction phase, if significant differences occur. This is even more pronounced when constructing through the urban environments where there is little room to accommodate changes in the construction approach. The purpose of this paper is to continue dissemination of information on these practices. Major points which will be addressed include:

- Specific field investigation and laboratory testing procedures appropriate to small diameter (e.g. less than 10 feet) tunneling, microtunneling, and pipe jacking.
- Discussion of the geotechnical issues associated with tunneling, microtunneling, and pipe jacking which need to be evaluated during the design stage of the project to aid in successful construction.
- Recommendations for report preparation with emphasis towards preparing documents that are valuable to all parties including the owner, designer, and contractor.

Introduction

This paper addresses site investigation issues that are important for detailed design and construction of tunneling and pipe jacking projects. It also includes details on conducting site investigations, laboratory testing, and field testing for microtunneling and pipe jacking projects. Many of the elements contained in this paper were developed through the author's experience and from the draft ASCE guidelines for tunneling currently being developed. (ASCE, 1998)

---

[1] Senior Tunnel Engineer, Haley & Aldrich, 2918 Fifth Avenue, Suite 202, San Diego, CA 92107

Accurate understanding of geotechnical conditions is essential for:

- The Engineer to assess the feasibility of tunneling methods, to develop bid documents, and to prepare an accurate cost estimate and schedule.
- The Contractor to plan and schedule construction work, to select the proper equipment, and to develop a responsive construction bid.
- The Owner to get a cost-effective, operational project completed on time, within budget, and without significant unanticipated cost increases during construction.
- For All Parties to recognize the inherent risk and uncertainties of ground conditions.

Microtunneling, tunneling, and pipe jacking equipment can be selected, set-up, and operated to provide satisfactory results under a wide variety of ground conditions. However, a given combination of equipment, set-up, and operating practice cannot be expected to perform effectively in all possible ground conditions. It is improtant that predominate ground conditions and their range of variability be reliably established so the project can be designed, the Contract Documents prepared, and the appropriate equipment selected. Identifying problematic ground conditions early enables the designer to confirm selection of tunneling approach and prevents lost time pursuing an option that will not be successful. The potential range and likelihood of ground conditions that can adversely impact the performance of tunneling and pipe jacking equipment must be identified as early as possible during design.

The subsurface information required can be obtained, interpreted, and applied to tunneling and pipe jacking with the proper application of existing technology. As with all underground projects, experience and good judgment are critical ingredients throughout the site investigation development, implementation, and interpretation processes.

The key geotechnical factors to achieve these goals include (1) perform an adequate subsurface exploration program, and (2) consider anticipated geotechnical conditions in developing pipeline design so that construction can be performed with the least cost and risk.

This paper contains a discussing of field and laboratory methods to obtain pertinent information, geotechnical issues to address, and approaches to presenting the information and issues in a geotechnical report.

Field Investigations

Detailed site investigations should be used to obtain factual information about the distribution and engineering characteristics of soil, rock, and groundwater at a site to an accuracy which allows for economical design and reliable construction planning for the work at a budget/risk level which meet the client and owners needs. Several approaches and methods to include are shown on the following page:

- Information review
- Geologic mapping
- Aerial photography
- Test pits
- Trenches
- Borings
- Large diameter drilled holes
- Cone penetrometer testing
- Geophysical surveys
- Ground penetrating radar

Due to the high costs of a geotechnical investigation program, it is desirable to use a phased approach so as to get the better information and increase the value of the effort spent. Where budget and schedule limitations allow, a useful concept for conducting detailed site investigation is to do the work in phases whereby each phase of the work provides the information needed to plan the next phase of study for site-specific problems. The first phase simply includes a review of existing information and a field reconnaissance. The second phase may consist of a few test borings and an observation well or two, with soil/water samples obtained for laboratory testing purposes. Upon completion of this work, it is then possible to establish a working understanding of the site and to go back to the field with a considerable amount of additional information about soil properties and stratigraphy. Depending on the outcome of this work, the importance of specific conclusions relative to subsurface conditions, or possible project changes, it could be prudent to go back for a third phase of even more detailed investigations to resolve remaining uncertainties. Finally, the geotechnical conditions should be monitored during construction. The four phases include the following:

>Phase I - Literature review, aerial photographs, mapping.
>Phase II - Preliminary drilling and laboratory testing.
>Phase III - Detailed drilling and laboratory testing and geophysics.
>Phase IV - Monitor during construction.

During Phase I, initial site investigation work, which is typically conducted during the planning stage of a project, should include a site visit and a review of all readily available sources of additional information concerning the site, such as geologic and topographic maps, aerial photographs, soil survey reports, previous investigation information, construction case histories, professional papers about the area, historical documents concerning prior site usage, and discussion with local building officials and local contractors. In general, it is prudent to find out as much about the site as possible prior to planning and implementing subsurface investigations. After review of the available information, geologic mapping should be done to identify surface exposures (if pertinent), rock mass properties, and to measure strike and dip of rock discontinuities.

Phase II and III subsurface exploration methods include:

- Small diameter borings (hollow stem auger, rotary drilling, or rock coring)
- Large diameter borings in ground containing cobbles and boulders
- Test Pits and Trenches
- Cone Penetrometer Tests
- Geophysical Methods
- Horizontal Directional Drilling

Small diameter borings are the key element of any subsurface exploration program. They have a relatively low cost/benefit ratio when considering the potential for cost overruns during construction. The first decision prior to subsurface investigation is how many test borings to drill and to what depth. As a minimum, test borings should be taken at all jacking and receiving shaft locations and at a spacing of about 300 to 500 feet along the alignments. (Essex, 1993) The actual spacing will vary depending on site and geologic complexity. Railroad, road, and creek crossings should attempt to get at least three borings with one at each shaft and one in the middle. For railroads, the nearest accessible location adjacent to the tracks should be attained. Creeks or waterways may require barge mounted or all terrain type vehicles to access if reasonably possible.

It must be pointed out that the final decision on the location, number, and depth of borings should be based on the importance and complexity of the project and the nature and complexity of the site and geologic conditions. The more difficult and costly the proposed project and the more complex the ground conditions, the more extensive the subsurface investigation program should be. Similarly, the more uniform the site conditions or the more costly the borings due to access or depth restrictions, the wider the spacing may be. It is also desirable to keep the borings off the alignment so that they do not affect the tunnel during excavation. If the holes are drilled along the alignment then they can be grouted with a low strength grout.

The borings should extend to at least two pipe diameters below proposed pipe invert. Borings at shafts should extend approximately 1.5 times the depth of the shaft to accommodate shoring system and shaft foundation design, especially in soft soils.

Two improtant decisions that must be made relative to the test borings are the sampling interval and the need for field testing. The frequency and type of samples that must be obtained are a function of the project type and ground conditions. Samples should be typically taken at intervals no greater than 5 feet and at changes in strata. Continuous sampling from one diameter above the pipe to one diameter below the pipe is prudent in order to identify thin strata that could have significant impacts on tunneling and pipe jacking. In addition, it is prudent to perform nearly continuous sampling at shafts in stratified deposits to identify thin unstable units which may affect the performance of the shaft during excavation. Groundwater piezometers are useful at shaft locations which are below groundwater. Water level readings, recovery tests, or even long-term pump tests are useful in evaluating hydraulic parameters. They can also be used to measure gas content (e.g. lower explosive limit, hydrogen sulfide content, etc.) in the casing at the time of groundwater measurements.

Site investigation strategy for projects where rock is anticipated should focus on determining:

- Depth and extent of rock
- Rock type and weathering characteristics
- Rock properties including unconfined compressive and tensile strength, together with abrasiveness and hardness
- Rock discontinuities (bedding, joints, faults, fractures, etc.)

During the field investigation program disturbed samples may be taken for classification whereas undisturbed samples should be obtained for laboratory density, strength, and compressibility tests. Field tests such as Standard Penetration Test blow counts (as a minimum) should be obtained. In addition, in-situ vane testing, pressuremeter testing or in-situ pumping tests can provide valuable to develop in-situ parameters. Indirect strength tests such as pocket penetrometer and pocket vane shear tests may be performed on tube samples. Short term rising and falling head permeability tests may be done in well casings for a general assessment of soil permeability.

At some sites it may be prudent to supplement a small diameter test boring program with other forms of subsurface investigation such as cone penetrometer tests, large diameter drill holes, and test pits. Cone Penetrometer Test (CPT) may be added to the program to supplement the boring information particularly in soft variable soils where it gives a continuous record indicating soil types and properties at a very low cost. CPT's are limited by refusal on gravel beds, cobbles/boulders, and cemented materials.

Large diameter drill holes can provide for sampling gravels, cobbles, and boulders. They also provide direct evidence about geologic structure and possible ground behavior characteristics such as running, raveling, squeezing, or groundwater inflow in the vicinity of the large diameter boring. Test pits provide information similar to large diameter borings but are limited in depth to approximately 15 to 20 feet. General parameters for cobbles and boulders include:

- Maximum cobble/boulder size from large diameter borings and an understanding of the local and regional geology
- Percentage of gravel, cobbles, and boulders
- Unconfined strength of cobbles/boulders which can be obtained by obtained by testing core samples taken from the cobble/boulder.
- Potential for nested boulders and expected percentage to be encountered.

There are also some innovative and cost effective methods to investigate the ground by geophysical means. These investigations can be particularly effective in helping to interpolate subsurface conditions between the test borings, identifying the soil/rock interface or cobble/boulder units, and depth of groundwater. Some common tests and their applications are as follows:

- Seismic Refraction which is useful to identify soil/rock interfaces. The estimated soil/rock interface is accurate to about +/- 20 percent of estimated depth.
- Seismic Reflection which is useful for identifying distinct units such as gravel/cobble units or faults
- Ground Penetrating Radar which is useful for identifying individual obstructions which have distinctive signature properties embedded in soils which can transmit the radar signal.
- Down-hole Seismic Refraction to aid in correlating strata between boreholes or from the borehole to the ground surface.
- Electric/resistivity logs which are useful for characterizing materials in the immediate vicinity of the borehole.

It should be noted that geophysical results may be difficult to interpret under certain conditions. In the appropriate setting, such applications should be used if they can provide supplemental subsurface information at a cost which provides value relative to proposed design and construction requirements and costs. Geophysics should not be relied upon as the sole field investigation method; borings should always be included.

Horizontal directional drilling may be used to (1) probe ahead in advance of the tunneling; (2) install a high density polyethylene casing for geophysical testing in the zone of the tunnel; and (3) collect samples in rock materials. These applications can prove cost effective especially for crossings where surface access is limited or not allowed.

Another importance aspect to the success of a subsurface investigation program includes continuous on-site monitoring of the work by a trained professional representing the designer. Thorough observation of field activities must be provided so the results of the field work can be accurately and correctly interpreted.

Laboratory Testing

Detailed laboratory testing has to be performed to provide accurate soils information. Relying on visual descriptions alone is not sufficient for accurate evaluations because they may be inaccurate and specific test values are needed for detailed design and construction. Tests need to be run to not only characterize the materials for design but to select the proper construction means and methods. Table 1, on the following page, summarizes important tests to consider for cohesionless soils (predominately sands and gravels), cohesive soils (clays and silts), and rock materials (Klein, et. al., 1996):

Table 1
Summary of Soil and Rock Laboratory Tests

Test Parameter	Cohesionless Soils	Cohesive Soils	Hard Rock Material
Sieve Analysis	X	X	
Hydrometer Test		X	
Moisture Content	X	X	
Dry Density		X	X
Plasticity Index		X	
Unconfined Compressive Strength		X	X
Direct Shear Test	X		
Permeability	X		
Compressibility		X	
Swell		X	
Slaking		X	X
Tensile Strength			X
Abrasivity			X
Hardness			X
Modulus of Elasticity		X	X
Resilience			X
Poisson's Ratio			X
Point Load Test			X
Vane Shear		X	
Pocket Penetrometer		X	
Mineralogy		X	X

Laboratory testing and soil classifications should be done according to established procedures such as those developed by the American Society for Testing and Materials.

The test results are used to perform detailed design and evaluation of items such as pipe wall thickness, long-term pipe and manhole settlement, shoring systems, tunneling machine/shield make-up, pipe friction, slurry separation, etc. I few of the specific items are included in the next section.

Geotechnical Issues

Geotechnical issues to assess include the following:

- Soil/Rock types and distribution
- Soil/groundwater contamination
- Alignment Selection
- Settlement/heave damage potential
- Long-term settlement potential
- Ground improvement/stabilization
- Obstructions
- Soil material properties
- Groundwater inflows
- Tunnel Excavation Methods
- Ground behavior at tunnel heading
- Cutterhead selections
- Tunnel lining systems
- Jacking Pipe
- Pipe Friction
- Alignment and steering
- Muck transportation
- Slurry separation
- Pipe design parameters
- Ground loads on tunnel lining
- Shaft and portal construction
- Shaft shoring pressure diagrams
- Portal slope stability
- Shaft deflection
- Breaking in/out of shafts
- Thrust block resistance
- Shaft backfilling
- Instrumentation and monitoring

Due to space limitations all of these issues cannot be discussed in detail. A few of the key issues are discussed in more detail below:

Soil and Rock Conditions

Anticipated geologic conditions should be evaluated to make an assessment of the information within and between borings based on all the collected information. The accuracy of interpretations between borings is enhanced by closely spaced explorations, geophysics, and a thorough understanding of the local and regional geology. The geotechnical engineer is in the best position to make these assessments for both the designer, contractor, and construction manager to use on the project. The best available information should be used along with good judgement. All assessments which are made for the contractor should be done in accordance with the owner's acceptance of the associated liability. In general the following geologic elements should be discussed and presented in the geotechnical report:

- Evaluate geologic conditions and map on plan and profile sheets.
- Prepare geologic cross section along alignment.
- Develop geologic unit descriptions and engineering parameters.
- Evaluate potential for obstructions and wood (natural/man-made).
- Develop groundwater model.
- Evaluate geologic discontinuities.
- Evaluate potential for gaseous ground conditions.

The Tunnelman's Ground Classification system, first described by Terzaghi (1950) and slightly refined by Heuer (1974), may be used to describe ground conditions and their impact on conventionally constructed small diameter soft ground tunnels and is a useful tool for evaluating soft ground conditions as they relate to microtunneling, also. General ground conditions include raveling, running, flowing, squeezing, and swelling. These descriptions are applied to gain an understanding of how the ground behaves at the heading during excavation.

Groundwater

Groundwater conditions will have a significant influence on ground behavior and thus on jacking and receiving shaft design and construction, tunnel boring machine or shield selection and operation.

The groundwater level and piezometric levels in aquifers should be determined by installing observation wells and/or piezometers at selected shaft locations and at intermediate points along the alignment for longer drives. In addition, the hydraulic conductivity of water bearing strata should also be presented.

Hydraulic conductivity of water bearing strata can be estimated using grain size correlation's and short-term borehole permeability tests. For larger projects where highly permeable soils are anticipated along with dewatering for groundwater control during construction, long-term pumping tests may be warranted.

Potential for Obstructions

The likelihood of buried objects, their nature and relative sizes, should be established by the site investigation. Obstructions can generally be defined as objects or features that lie completely or partially within the cross-sectional area of the planned pipeline excavation and which prevent continued forward progress of the tunneling. They may typically include the following:

- Isolated Boulders
- Cobble/Boulder Units
- Bedrock
- Concretions
- Corestones
- Old foundations, shoring, tiebacks
- Utilities
- Wood & Organic Material

Typically when obstructions are encountered the face must be exposed to allow removal. Access to the face may be gained by use of an open face machine or shield, by sinking a rescue shaft, by sinking a shaft off-line and hand mining to the face, or by hand-mining from the reception shaft.

In addition to some of the field exploration techniques described earlier which can aid in identifying obstructions, other less direct techniques such as a thorough understanding of the local and regional geology, case histories, historical documents, old aerial photographs, etc. can provide valuable clues regarding the potential for obstructions. Although there are many approaches for locating potential obstructions, owners must be recognized that locating all of them along a given alignment is often not possible with practical budget and time constraints for site investigations. This situation can be addressed with owner approval by developing plans for dealing with obstructions, technically and contractually, before construction begins.

Man Made and Environmentally Sensitive Features

Existing site features that could be impacted by tunneling and pipe jacking operations (including shaft construction) should be identified during the site investigation. Existing utilities may be located from as-built records, although locations of abandoned utilities are often not known. "Potholes", which may be open cut or auger or vacuum extraction borings excavated for investigation purposes, should be used to confirm locations of utilities that are near or intersected by the planned pipeline. Historic buildings and environmentally sensitive areas usually require evaluation on a case by case basis to ensure protection.

Contaminated Groundwater or Soil

Determination of the potential for encountering contaminants and determination of the nature and extent of contaminants, if present, should be done during the site investigation. Sources or information regarding the potential for contamination include: health department records; fire, planning and building departments; and local or regional pollution control or water quality agencies. In areas where there are indication of the potential for contamination, chemical analyses of soil and groundwater samples to identify their nature and concentration should be undertaken. This information is necessary to determine the appropriate tunneling method, and spoil disposal procedures during construction.

Report Preparation

All subsurface data collected in connection with the site investigation, professional interpretations thereof, and design and construction considerations should be summarized in project reports. Depending on the size and complexity of the project, reports may consist of Geotechnical Investigation Reports, Geotechnical Data Reports, Geotechnical Design Memorandum, and Geotechnical Baseline Reports (ASCE, 1997) which are summarized as follows:

- Geotechnical Investigation Report is a standard report written primarily for designer and hopefully for the contractor
- Geotechnical Data Report is a statement of facts only for all to use
- Geotechnical Design Memorandum is a summary of basis for alignment, design, loads, etc.

- Geotechnical Baseline Report is a ground specification and is included as part of contract documents

Traditional Geotechnical Investigation Reports may be prepared during the design phase. It is generally recommended that these reports not only address the factors important to the design but they should also include discussions of important construction aspects of the project. It is important to know the contractual status of the document while preparing it, so that a consistent approach can be maintained between the geotechnical and other contract documents.

It may be advantageous to prepare geotechnical reports in two sections or volumes, depending on the size of the project. One volume, commonly referred to as the Geotechnical Data Report, presents only the factual data collected during the site investigations. The other section, commonly referred to as the Geotechnical Interpretive Report or Geotechnical Design Memorandum, presents the geological assessment of the office studies and field explorations, evaluation of soil and rock characteristics, interpreted subsurface profiles and recommendations for design and construction. This report should include a summary of the anticipated ground conditions and behavior, design criteria and construction considerations.

In recent years a fourth type of report, now commonly referred to as a Geotechnical Baseline Report has come into use on larger tunnel projects. The purpose of this report is to be the sole document that presents geotechnical interpretations and information upon which the contractor should and may rely. The Geotechnical Baseline Report is typically used in connection with dispute resolution during construction although this is not required. A scaled down version of this approach may prove to be useful on smaller tunneling and pipe jacking projects, especially where significant footage is expected to be installed.

Budget and Technical Limitations

The budget to perform the geotechnical investigations, evaluations, and report preparation should be sufficient to address the issues to an acceptable level. For a medium sized project (approximately $5,000,000) inclusion of all the elements discussed in this paper can typically run between 3 and 5 percent of the construction budget. For complicated small projects the percentage could go as high as 8 percent. For a larger less complicated project it could go as low as 1 percent. There is therefore no single set number which should be targeted. The budget and scope needs to be established with an understanding of the complexity of the project and the owner's desires for obtaining information useful to the contractor. The geotechnical report should summarize important assumptions and state the intent and limitations. It should be recognized, however, that even with the most extensive geotechnical investigations, analyses, and evaluations that there is always a potential for conditions to be different than expected. It is therefore prudent to monitor the geotechnical conditions, ground behavior, and performance during construction to provide a more complete understanding. Additional explorations may be also be useful in the early phases of construction to address any additional aspects not covered during the design phase.

Summary

Geotechnical investigations along with careful evaluation of the information and presentation in formats which are useful to the owner, designer, contractor, and construction manager are essential to successful completion of tunneling, microtunneling, and pipe jacking projects. This is even more important in the urban environment where the potential for man made features, soil and groundwater contamination, and third party impacts are more pronounced. Experienced practitioners can bring a wealth of knowledge from past projects to make future projects more successful for all the parties. Some lessons learned include the following (Essex, 1993):

- Provide enough geotechnical data for the designer to design and the contractor to construct the project.
- Anticipate the unanticipated based on past experiences.
- Target exploration budget at a level which will allow sufficient evaluations.
- Prepare concise useable geotechnical reports which meet the owners desire for risk assessment.
- A little more data can go a long way to avoid delays, cost overruns, and claims
- Monitor during construction as it is the final phase of exploration.

References

American Society for Civil Engineer, 1997, "Geotechnical Baseline Reports for Underground Construction, Guidelines and Practices," New York, NY.

American Society of Civil Engineers, 1998, "DRAFT Standard Construction Guidelines for Microtunneling".

Essex, R. J., 1993, "Subsurface Exploration Considerations for Microtunneling/Pipe Jacking Projects," Trenchless Techonology, An Advanced Technical Seminar: for Trenchless Pipeline Rehabilitation, Horizontal Directional Drilling, Microtunneling, pp. 275-288

Heur, R.E., 1974, "Important Ground Parameters in Soft Ground Tunneling," in Subsurface Exploration for Underground Excavation and Heavy Construction, ASCE, New York, NY, pp. 41-55.

Klein, S. J., Nagel, G. S., Raines, G. L., 1996, "Important Geotechnical Consideration for Microtunneling," No-Dig Engineering, July/August.

Terzaghi, K., 1950, "Geologic Aspects of Soft Ground Tunneling," Chapter 11 in Applied Sedimentation, ed. P. Trask, John Wiley and Sons, New York, NY.

ROLE OF SURGE AND POSSIBLE MITIGATION IN PCCP DESIGN

Dr. Sri K. Rajah, P.E. and Dr. Jey K. Jeyapalan, P.E
Pipeline Engineering Consultants
21 Meetinghouse Terrace, New Milford, CT, USA 06776
Phone 1-860-354-7299
email: jkjeyapalan@snet.net

Abstract

In a number of prestressed concrete cylinder pipe(PCCP) failures, the unfortunate events seem to have been precipitated by over-stressing of the pipeline due to surge events such as power failure, pump trip, and either closure or opening of the gates in the line faster than permissible. Historical accumulation of stresses due to one or more of these above operational problems also trigger the failure. In many cases, design engineers simply failed to do any form of surge calculations and failed to provide adequate surge protection measures as part of the overall design of the project. This paper provides a concise summary of sources of surge events, surge evaluation techniques, and surge mitigation measures for typical situations. Sizing and locating surge mitigation devices as well as the causes of secondary surge are also discussed in this paper.

Introduction

In a closed conduit, any sudden change in fluid velocity causes pressure waves to travel upstream and downstream from the source of disturbance. The pressure changes in the pipeline referred to as water hammer, surge, or transient pressure, is proportional to the wave speed and the velocity change. Of the resulting positive and negative surge pressures, positive surge pressures have to be accommodated in the PCCP design. Negative surge pressures do not affect the design of PCCP unlike in the case of the design of a flexible pipe as long as vapor pressure is not reached. Extensive vapor cavities may result under vapor pressure and the collapse of these cavities may result in extremely high positive surge pressures. A detailed treatment of the theory of surge is not within the scope of this paper and is left to the textbooks. Instead, we will concentrate on the sources of surge events, mitigation, and the associated issues for a successful surge alleviation system design.

Sources of Surge

As to the causes of surge, pump trip, operations of the valves and gates, pump startup, pump shutdown, and collapse of vapor cavities caused by negative pressures are some of the primary ones. Detailed surge analysis of the system should be conducted during the design stage for all potential operating conditions that could cause surge in the system. The nominal surge allowance the PCCP design, given in standards, provides usually is adequate for surge. However, additional allowance should be provided if a detailed analysis warrants it. Alternatively, an appropriate surge alleviation system could be designed using a detailed surge analysis. This requires a detailed understanding on the part of the design engineer of the different mitigation devices, ability to model such devices using an analysis software, and the capability to design the location, type, and size of such devices. Inappropriate design and use of mitigation devices can be another primary source of surge in the system. While these devices serve as surge preventive mechanisms in some designs, they become the primary source of surge causing failure in others. Air venting of the pipeline during filling; improper operation of pressure relief valves, operation of large-orifice air and vacuum valves are some of the often overlooked but critical causes of surge.

Surge Mitigation Devices

It is rather impractical and uneconomical to design a pipeline for the highest surge pressure accounting for all possible surge situations. However, it is within the designer's choice to select appropriate hydraulic components along the pipeline to minimize the surge effects. Hydraulic components placed on the pipeline primarily to control and to suppress the effects of surge are called surge mitigation devices. Conventional surge mitigation devices include 1) Surge tanks (open surge tanks, air vessels/closed surge tanks, and one-way surge tanks); 2) Pressure relief valves; 3) Surge anticipation valves; 4) bypass lines; and 5) Air release and Vacuum Valves. The mechanism with which each of these surge devices controls the surge is different and determines its intended use. A brief summary of the function of a few of these devices is provided in the following paragraphs.

The use of surge tanks prevents high pressures and cavitations by receiving liquid from and supplying liquid to the pipe, respectively. As the name indicates, open surge tanks are open to the atmosphere limiting its use to locations where either static pressure heads are small or tall tanks are feasible. Closed surge tanks, also known as air vessels, can be installed anywhere along the pipe. The effect of closed surge tanks on the resulting surge pressures primarily depends on the location, vessel size, entrance resistance, initial gas volume, and initial gas pressure. One-way surge tanks, also known as feed tanks, can be either an open or a closed surge tank equipped with a check valve allowing flow only into the pipe system. This can be installed anywhere on the pipe system and its function is to avoid cavitation by feeding fluid into the pipe system at low pressures. Recharging of these systems is required after each surge event.

The function of pressure relief valves is to eject liquid out to prevent excessive high-pressure surges. The valves are activated when the pressure at a specified location in the pipeline exceeds a preset value. The valve closes at another preset value of pressure at the specified point. The surge anticipation valve is a special pressure relief valve designed to open on a down-surge in anticipation of an up-surge to follow.

Pumps are often equipped with check valves to prevent flow reversal. Often, check valves are selected only with the primary purpose in mind, i.e., to prevent flow reversal. Most check valve designs require reversal of flow to activate the mechanism to close. This also means that the liquid is already in motion and the abrupt closure of the check valve results in additional transient pressures. Bypass lines are used to prevent this situation and are activated when the pump suction head exceeds the discharge head. They prevent high-pressure buildup on the pump suction side and prevent cavitation on the pump discharge side.

High points in a pipeline are susceptible to the buildup of air and vapor cavities and the use of air release and vacuum valves at these locations are used to control the surge. Air is taken into the pipe when the line pressure drops below atmospheric value and the air is expelled when the line pressure exceeds atmospheric. These valves are also known as air relief valves, air inlet valves, and dual-acting air valves. These valves, while helping reduce surge effects in pipeline can also serve as a primary source of surge. A discussion on the function of the valve, selection and sizing is given below.

Selection of air release valves has been specified traditionally by the nominal inlet diameter only. Considering the fact that the characteristics of an air release valves is dependent on several other factors including the internal configuration and the outlet diameter of the valve. The performance of air valves from different manufacturers with the same inlet diameters is significantly different. During low pressures in the pipeline it is primarily the magnitude of the atmospheric pressure that causes air to flow in and the reason for the importance of the value of inlet area. However, when the air flows out of the pipe the air pressure increases in order to expel the air. In order to protect the pipe, the outlet must be appropriately sized to limit the pressure increase to allowable limits. The recommended safe pressure across the large orifice without the risk of pipe wall collapse or damage to joint seals or other damages due to negative internal pressures depends on the pipeline material. Accurate selection of valves can only be made when the performance data for the particular valve is available. Reliance on manufacturers data for these performance, as most engineers do, can be quite dangerous, as they often publish these curves based on some theoretical calculations without performing any actual tests.

Surge Analysis

The selection of pipeline components, flow control devices as well as surge mitigation devices, have a major impact on the magnitude of the surge pressures on the pipeline. The components' function, mechanism, location, and sizing, all of which contribute to the resulting surge. Therefore, it is essential that none of these aspects and the resulting effects be overlooked in the representation of the individual hydraulic components in the surge analysis model. In most cases, however, the software is not equipped to accommodate a detailed representation of every hydraulic component. Besides, the design engineers almost always rely on the manufacturers for these data, which may have been obtained under totally different test conditions. Therefore, comprehensive analysis of surge phenomena using computer software is difficult due to the lack of proper performance data and the lack of boundary conditions in the analysis software.

Besides, the resulting flow due to surge is governed by non-linear partial differential equations requiring complex solution schemes. The air entrapped in the pipelines makes the analysis more complex resulting in design engineers resorting to simplified design procedures as the basis of the analysis software. This leads to improperly designed pipeline systems with, 1) possible over design of the pipe; 2) inadequate and improper surge protection; and 3) lower performance than the one pipeline was designed for.

Case History

The following case history illustrates the use of the above principles. The reservoir is located in a township, on the east coast. The pipeline is one of the main components in a system developing the river basin water resource. The primary function of the reservoir and the release pipeline is to augment the natural low flow of the river during the dry season. The release pipeline was originally designed to provide a maximum discharge of 400 mgd at a reservoir elevation of 286 feet and was constructed by a construction company in 1976. The release pipeline runs easterly from the north dam and terminates approximately 3.6 miles downstream at a discharge structure on south branch creek in another township. The release pipeline ruptured at *Station 174+74* on the morning of June 15, 1988.

Release Control Works, Release Pipeline, and Discharge Structure

At the base of the north dam of the reservoir, the 108 inch diameter PCCP pipeline is connected to the twin 72 inch diameter welded steel pipelines of the North Dam Release Structure. At Blackhouse station, the 108 inch pipeline is reduced to a 96 inch diameter pipeline, which has been bulkheaded to provide the connection point for the proposed future Confluence Force Main. The length of the 108 inch diameter PCCP release pipeline is approximately 18,400 feet. From the Blackhouse station, the release pipeline becomes an 84 inch diameter PCCP heading northward for approximately 1,100 feet and where it is connected to the discharge structure on the south branch Creek. The 84 inch pipeline is provided with a 72 inch Tee for a possible connection to the proposed future pipeline.

Release Control Works

The reservoir water enters the release pipeline by way of an outlet tower at the north dam which controls the flow into the pipeline by a series of 5 ft. x 6 ft. and 3 ft. x 5 ft. sluice gates, which are operated locally from the outlet tower operating floor. There are two 5 ft. x 6 ft. gates each at elevations 270 and 307 feet, and two 3 ft. x 5 ft. gates at elevation 357 ft. leading to a vertical outlet shaft. The various sluice gate elevations are used to select the proper temperature for the released flow augmentation waters. The sluice gates are manually operated or alternatively by a gasoline engine powered motor operator. From the vertical outlet tower shaft, the waters enter twin 72 inch diameter steel pipelines passing under north dam and ending at a valve chamber at the downstream toe of the dam where each of the two lines is controlled by a 72 inch diameter electric motor-operated butterfly valve. Downstream of the butterfly valves the 72 inch diameter pipes are interconnected by a 48 inch diameter tie line controlled by a 48 inch electric motor-operated butterfly valve. Downstream of these points, the 72 inch lines continue to a metering vault, where the easterly pipe is fitted with a 72 inch by 48 inch venturi meter. Further downstream, the two 72 inch diameter branches are combined into the single 108 inch diameter release pipeline.

Discharge Structure

The 84 inch line, approximately 1,160 feet long ends at a discharge structure where the pipeline divides into four branches each controlled by a motor-operated butterfly valve and an energy dissipating free discharge valve. Two of these branches are 48 inches in diameter and have 48 inch diameter butterfly valves and 24 inch diameter free discharge valves. The other two branches are each 36 inches in diameter, and are controlled by 36 inch diameter butterfly valves and 18 inch diameter free discharge valves. All of the valves are locally electric motor-operated. Each of the four discharge branches have magnetic flow meters located between the butterfly and free discharge valves. Flow is recorded locally on charts.

Hydraulic Conditions at the Time of Failure

On June 15, 1988, the reservoir pool was nearly full at an elevation of 383 feet. During the failure, the top two sluice gates in the outlet tower were in the fully open position. Hydraulic properties of the pipeline could be estimated from the design drawing of the release pipeline set of the contract drawings. It is observed that at the design outflow of 400 mgd from the reservoir 10.8 feet of hydraulic pressure head is lost between the reservoir surface and start of the 108 inch diameter pipe. With the design flow, 2.6 feet of hydraulic pressure head is lost in 1000 linear feet of 108 inch diameter pipe.

Hydraulic Transient Analyses

In this study, the computer program "SURGE" was used to analyze the hydraulic transients in the release pipeline. The entire release pipeline from the reservoir to the discharge structure was modeled in the analysis. The layouts of the pipeline, release structure, and discharge structure are obtained from the design drawings. The hydraulic properties of the pipelines and other hydraulic components are obtained from the design drawings and other investigative reports done prior to the authors being retained by the lawyers representing the contractor. From the design drawings it could be noted that the head lost in a 1000 ft linear length of 108 inch diameter pipeline under the design outflow of 400 mgd is 2.6 feet. This is equivalent to a friction factor "n" of 0.012 or a Hazen Williams "C" value of 125.29. Also from the design drawings the head losses in the release structure under the design flow of 400 mgd is noted as 10.8 feet. Friction losses in the twin 72-inch diameter pipeline, the main component of the release structure, is modeled with the Hazen Williams "C" value of 120.0 estimated from the design tables for welded steel pipelines. The remaining friction losses could be lumped as minor losses of the various components such as valves, metering vaults, etc.. in the release structure. The minor loss coefficients associated with these losses can be estimated by calibrating the pipeline under design flow of 400 mgd, for which the head loss in the release structure is given in the design drawings as 10.8 feet. The discharge structure can be modeled appropriately and the friction losses can be accounted with a Hazen Williams "C" value of 120.0 corresponding to welded steel pipelines. Friction losses across the Howel Bunger valves can be estimated with a coefficient of discharge (C_d) of 0.85.

Preliminary Surge Analyses

In the following set of surge analyses, all four Howell Bunger valves are open at the same percentage stroke initially to maintain the required amount of flow. The surge conditions are created by closing the valves simultaneously. According to the research, the closure time required to close all four valves simultaneously from fully open position is 166.6 seconds. The critical surge conditions under the given operating conditions are expected with a valve closure time of 12.06 seconds, which corresponds to the time taken for a reflected wave to travel from the discharge structure to the reservoir and back to the discharge structure. Considering the closure time used in prior set of analyses, 12.06 seconds closure time corresponds to 7.2 % of valve stroke. Five different valve openings, 1.0, 3.0, 5.0, 7.2, and 10.0 percent stroke were considered in the preliminary surge analyses with a reservoir elevation of 286.0 feet using the data shown in Table 1. Surge analysis was repeated for each valve opening with a closure time estimated linearly from the closure time from the fully open position. Also, for 7.2% valve stroke the surge analysis was repeated with 12.02 sec closure time. The maximum pressure head experienced at Station 147+74 from the surge analyses performed are summarized in the following table.

Table 1: Valve Closure Characteristics

% Stroke	Closure Time (sec)	Head at Station 147+74 (feet)	Flow (mgd)
1.0	1.67	195.2	21.55
3.0	5.0	607.4	63.88
5.0	8.33	735.8	104.07
7.2	12.06	826.8	145.33
10.0	16.66	600.1	189.75
5.0	12.06	714.5	104.07

From the design drawings it could be noted that the design maximum pressure at the failure location is 350 feet. With the allowance of 40 % for surge effects, the allowable pressure at the failure location is 490.0 feet. For the surge analyses performed on the release pipeline, the failed section was subjected to a maximum pressure of 826.8 feet. This pressure head is 96 % more than the allowable head for this section. For a flow of approximately 100 mgd the maximum pressure head at the failure location is 735.8 feet and 714.5 feet respectively for 8.33 sec and 12.06 sec closure times. These values are 70 % and 64 % more than the allowable head respectively. These results are in contrast to the findings and claims by other experts.

Summary

The paper details the role of surge and the surge mitigation devices in PCCP failures. Also, the paper identifies improper use of surge mitigation devices as one of the primary sources of surge and the potential of failure of PCCP pipelines. To illustrate the above discussion on surge control, the principles are applied to a specific case history. From the hydraulic transient analyses of the PCCP pipeline chosen for the case history, the following preliminary conclusions can be made:

- hydraulic transients caused by improper operating conditions of the pipeline could be the primary cause of failure.

- further work is needed in the area of surge analyses using better valve operating curves from the manufacturers and for other operating conditions not covered in this preliminary study.

LARGE DIAMETER REINFORCED CONCRETE PIPE BEDDING DESIGN AND INSTALLATION IN ADVERSE SOIL CONDITIONS

Mark Giandoni, P.E., MASCE[1]

ABSTRACT

This paper discusses and presents information obtained from studies completed on four large diameter gravity pipelines constructed in weak soils and high groundwater. Two of the reinforced concrete pipeline projects included pipe bedding designs that addressed the potential for poor installation conditions, while the other two appeared to be based only on the basic industry standard reinforced concrete pipe design methods.

The pipeline designs that had contract documents that addressed the potential for poor installation conditions had minimal defects in the constructed pipeline. The projects that used the basic industry standards and did not address the potential failure modes of the poor installation conditions had considerable problems. These problems included joint seal failures, joint cracks, excessive settlements, excessive leakage and loss of design flow capacity.

The adverse soil conditions that were encountered in all four of the pipeline alignments included high groundwater, weak soils, perched water tables, and low permeability soils that were difficult or impossible to dewater.

The primary focus of this paper will be to, 1) show the results of pipeline designs that do not address poor soil conditions, 2) show that industry standard design methods do not address poor soil conditions with respect to bedding and foundation design, and 3) to provide information that can be used to help address the potential problems associated with the design and installation of large diameter pipelines in poor soil conditions.

---

[1] Principal Engineer, Advanced Infrastructure Systems, El Cajon, CA (619) 447-5380

1.0 PIPELINE DESIGN

The design process for large diameter gravity pipeline projects generally follows a linear procedure typical to the following:

1) Alignment Selection
2) Design Studies
3) Pipeline Design

The pipeline project begins with an alignment selection, where several alternative alignments are analyzed and the most efficient alternative is selected.

With the alignment selected, additional engineering studies are completed to provide the detailed design data required for the final pipe design. The engineering studies and the data obtained will be used in the detailed design of the pipeline to determine whether any special design requirements need to be addressed. These engineering studies typically include but are not limited to geotechnical investigations and studies, geological studies, and groundwater investigations. The studies should identify and quantify specific design parameters that must be addressed in the pipeline design. The typical pipeline design parameters include, soil types, soil conditions, ground water, potential for seismic activity, faulting, and other special construction requirements. To complete a successful design the designer must address all design parameters and any potential failure modes that are associated with the adverse soil conditions.

This paper discusses and compares four large diameter reinforced concrete pipeline projects and shows the adverse results from final contract documents that do not address the potential failure modes due to adverse soil conditions. Specifically this paper will investigate the effects of adverse soil conditions of high ground water, weak soils and how difficulties in construction dewatering will impact the installation and performance of a large diameter pipeline.

Typical Concrete Pipe Design Method

The reinforced concrete pipe strength calculations are the typical design calculations promulgated by the concrete pipe manufacturers. The calculations are used by designers to determine the earth loads and the pipe wall strength requirements. During this standard design process a bedding design is assumed and then the appropriate bedding factor is used to calculate the pipe wall D-Load requirements.

The following are the typical "Trench Installation" calculations for concrete pipe design. First the trench earth loads (W_d) on the pipe are computed using the Marston Equation.

$$W_d = C_d w B_d^2$$

C_d is further described as:

$$C_d = \frac{1 - e^{-2K\mu'H/B_d}}{2K\mu'}$$

W_d = Trench Earthloads (plf)
C_d = Load coefficient for trench installation
w = Unit weight of backfill (pcf)
B_d = Width of trench at top of pipe (feet)
e = base for natural log
K = Conjugate ratio for backfill or fill material
u = Poison's Ratio
H = Height of Backfill or fill above top of pipe (feet)

With an earthload calculated the designer then selects the most cost efficient bedding and associated bedding factor. With the bedding factor determined the pipe D-load rating of the concrete pipe is calculated.

$$D\text{-}load = \frac{W_d + W_L}{B_f X D} XF.S.$$

D-load = Supporting Strength of Pipe
W_d = Earthload
W_L = Live Loads
B_f = Bedding Factor
D = Inside Diameter of Pipe
F.S. = Factor of Safety

The above pipeline design calculations provide the necessary information to design the concrete pipe wall but, they do not address the potential failure modes associated with the physical conditions of the installed pipe. The industry standard concrete pipe design methods do not provide design criteria for bedding or foundation design or how to address adverse soil conditions. The typical bedding "designs" do not address weak soils or the impacts that groundwater will have on the foundation or bedding of the pipe.

In a review of available pipeline design references the author found only minimal information on the design and impacts of high ground water or weak soils conditions with respect to bedding and foundation design. Most design guides infer that a "solid" foundation and bedding are important, but provide no method for quantifying the quality of the bedding or foundation. Only one design guide provided a method to estimate the quality of the bedding and that was related to a person walking on the bedding and estimating the quality of the bedding by the depth of the indentation (Howard, 1996).

2.0 PIPELINES STUDIED

Pipeline 1 – 1,350 mm (54") Trunk Sewer

Pipeline 1 consisted of over (4000-feet) of 1,350 mm (54-inch) plastic lined reinforced concrete pipe. The majority of the pipeline alignment was in sensitive riparian habitat and impacts to the area were to be minimized. This required a design that limited access and time of construction. The soils were made up of aluviums and marine deposits. The soil types ranged from fat clays, to sandy silts and gravely silts. The widely varying soil types should have been a sign that the design would need to address changing conditions.

The ground water level was at the surface for much of the alignment and ranged down to (13-feet) with the groundwater level always above the pipe invert. Perched water tables were encountered during construction. The dewatering specification required that the water table be lowered to a depth of (five feet) below the trench excavation. With the groundwater levels near the surface and excavations up to 6.8 meters (20') deep special consideration should have been given to the dewatering requirements.

The typical trench section (see Figure 1) provided for an ACPA type C bedding design, which is a shaped subgrade with no additional bedding or foundation material. A detail for overexcavation was provided that included a (12-inch) gravel bedding section (see Figure 2). Geotextile fabric was required around gravel or rock bedding material.

The contract documents for the trench, bedding and foundation design did not address the poor installation conditions that were known to exist at this site. The result of this "standard design" was settlement of up (16-inches). This caused pipe joints to pull apart and leak. The constructed pipeline had numerous broken pipe joints and a loss of over 30% of the pipeline hydraulic capacity.

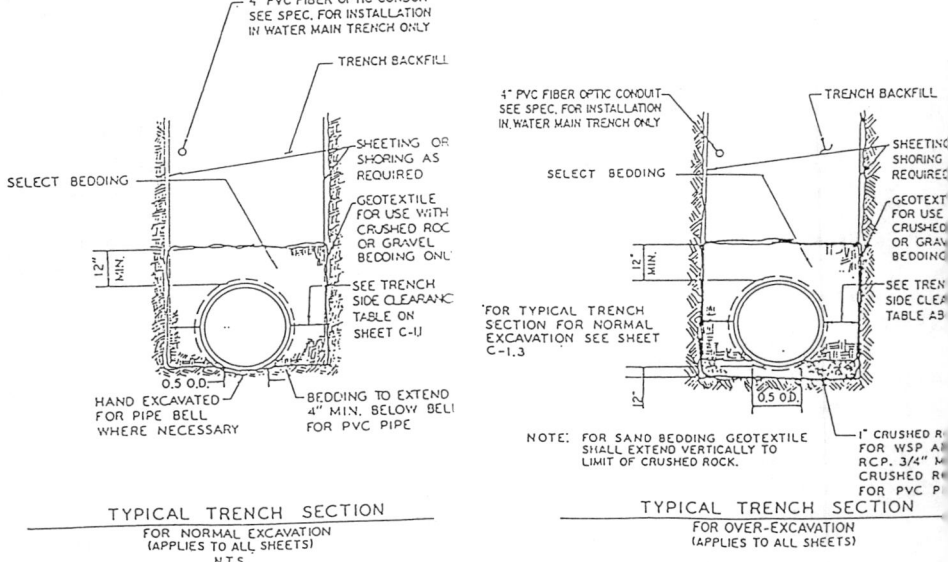

Figure 1 – Pipeline 1 Typical Trench Design

Figure 2 – Pipeline 1 Overexcavation Trench Design

Pipeline 2 – 1,950 mm (78") Interceptor Sewer (North Mission Valley Interceptor Sewer Phase 2)

Pipeline 2 involved the installation of over (10,000-feet) of 1,950 mm (78-inch) diameter plastic lined reinforced concrete pipe. The project was constructed in the San Diego River valley where the soils were made up of alluviums that included silts, silty sand, organics, and clays. The groundwater level was above the pipeline invert and trench dewatering was required along the entire alignment. Perched water tables were encountered in several areas of the project and the installed dewatering system was ineffective. To mitigate the inability to dewater the excavation the design required over-excavation and up to four feet gravel bedding. The foundation and bedding design addressed the potential for poor installation conditions and there were minimal defects in the constructed project. In the areas where weak soils with high clay content or organics were encountered the contractor was directed to over excavate and place an additional four-feet of rock foundation material. The foundation material was a one-inch crushed rock wrapped with geotextile fabric. On top of the foundation material (one to two foot) of bedding material was placed. The bedding material was mechanically compacted to minimize the potential for settlement of the gravel material.

The geotextile material provided a filter barrier that minimized fines migration. Overexcavation and the rock and gravel refill was effective in mitigating the high ground water levels when dewatering was not possible and week soils.

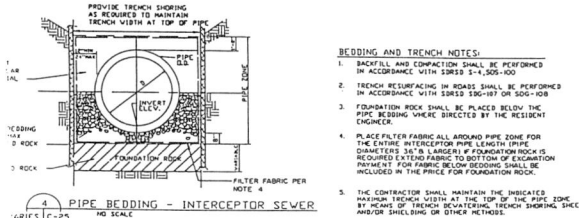

Figure 3 – Pipeline 2 Trench Design

Pipeline 3 – 2,700 mm (108") Interceptor Sewer

Pipeline 3 involved the construction of over 3,657 meters (12,000-feet) of 2,700 millimeter (108") plastic lined reinforced concrete pipe. The pipeline alignment crossed a river and passed through urban areas built on hydraulic fills and aluviums. The soils encountered included silty sands, silty clays and bay muds. The groundwater level was above the pipeline for most of the pipeline alignment. Dewatering for this project was accomplished with drilled wells and trench sumps. The dewatering was successful for most of the project however groundwater was apparent in the trench in some areas and overexcavation was necessary.

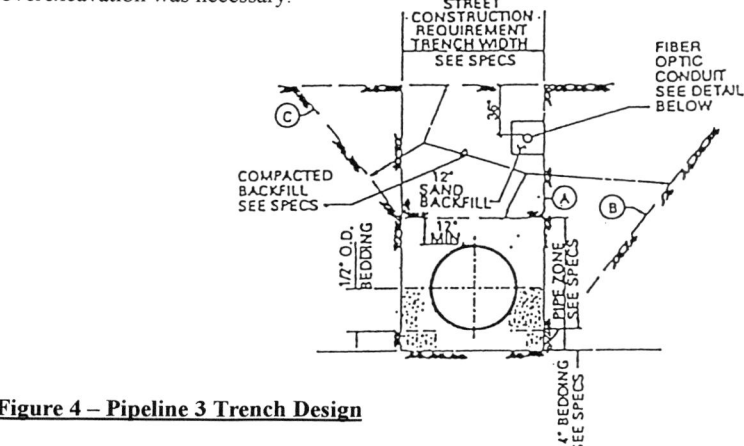

Figure 4 – Pipeline 3 Trench Design

The standard foundation and bedding for this pipeline required a minimum of twenty-four inches of select backfill. The contractor was allowed to use 20 millimeter (¾") gravel. The design did not require the use of geotextile fabrics. Settlement of the pipeline was observed after installation and backfill of the pipeline. The settlement of the pipeline caused leaking joints that were repaired with joint seals.

Pipeline 4 – 1,800 mm (72") Interceptor Sewer

Pipeline 4 required the installation of over 2,43 meters (9,000 feet) of 1,800 millimeter (72") plastic lined reinforced concrete pipe. This pipeline was also constructed in the San Diego River valley. The soils encountered included silts, running sands, silty clays and clays. The pipeline bedding and foundation design required a minimum of 200 millimeter (8-inches) of (¾") gravel and provided for overexcavation and rock refill to mitigate high groundwater and/or poor soil conditions. In one area where the trench had standing water and the contractor was not required to overexcavate the pipe settled nearly (3 inches). After this section was laid and the pipe settled the contractor was required to overexcavate whenever water rose above the trench bottom. Several sections of the pipeline alignment encountered perched water tables where the groundwater would flow into the trench above the trench bottom. In these situations the deep wells were ineffective and overexcavation mitigated groundwater in the trench and the potential for excessive settlement. Overexcavation was used on much of the alignment where soils with high clay contents were encountered. Geotextile fabric was used to completely wrap the gravel bedding to prevent fines migration into the bedding materials.

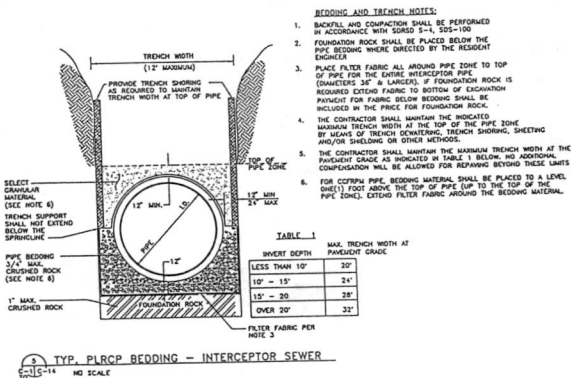

Figure 5 – Pipeline 4 Trench Design

The following Tables 3.1 to 3.4 provide summaries of the existing site conditions, construction requirements, and post construction pipeline conditions of the studied pipelines.

3.1 PIPELINE PRECONSTRUCTION CONDITIONS

	PIPELINE 1 1,350 mm 1996	PIPELINE 2 1,950 mm 1996	PIPELINE 3 2,750 mm 1995	PIPELINE 4 1,800 mm 1992
SOIL CONDITIONS	POOR (WEAK SILTY SANDS & CLAY)	POOR (WEAK SILTY SANDS & CLAY)	POOR (WEAK CLAYS & SILTY SANDS)	POOR (WEAK SILTY SAND & CLAYS)
GROUND-WATER	HIGH ABOVE PIPE ZONE	HIGH ABOVE PIPE ZONE	HIGH ABOVE PIPE ZONE	HIGH ABOVE PIPE ZONE
LIQUIFICATION POTENTIAL	HIGH	HIGH	HIGH	HIGH

3.2 PIPELINE & BEDDING DESIGNS

	PIPELINE 1 1,350 mm 1996	PIPELINE 2 1,950 mm 1996	PIPELINE 3 2,750 mm 1995	PIPELINE 4 1,800 mm 1992
PIPE & JOINT TYPE	PLRCP DOUBLE GASKET	PLRCP STEEL RING	PLRCP DOUBLE GASKET	PLRCP SINGLE GASKET
MIN. BEDDING REQUIREMENTS	0"	12"	24"	8"
ACPA BEDDING DESIGN	CLASS C SHAPED SUBGRADE	CLASS B GRANULAR BEDDING	CLASS B GRANULAR SUBGRADE	CLASS B GRANULAR SUBGRADE
BEDDING CONSTRUCTION REQUIREMENTS	NONE	MECHANICAL COMPACTION	NONE	MECHANICAL COMPACTION

3.3 PIPELINE FOUNDATION DESIGN

	PIPELINE 1 1,350 mm 1996	PIPELINE 2 1,950 mm 1996	PIPELINE 3 2,750 mm 1995	PIPELINE 4 1,800 mm 1992
DEWATERING REQUIREMENTS	5' BELOW SUBGRADE	3' BELOW SUBGRADE	2' BELOW SUBGRADE	DRY SUBGRADE
SHORING METHODS	TIGHT SHEETING TO BE REMOVED	TIGHT SHEETING & LAID BACK TRENCH	TIGHT SHEETING	TRENCH BOX
FOUNDATION CONSTRUCTION REQUIREMENTS	NONE	85% RELATIVE COMPACTION	NONE	85% RELATIVE COMPACTION
OVEREXCAVATION	12" MAX.	48" MAX	24" MAX	36" MAX
FOUNDATION INSPECTION	ENGINEER TO INSPECT	ENGINEER TO INSPECT	ENGINEER TO INSPECT	ENGINEER TO INSPECT

3.4 PIPELINE SETTLEMENT AND RESULTANT IMPACTS

	PIPELINE 1 1,350 mm	PIPELINE 2 1,950 mm	PIPELINE 3 2,750 mm	PIPELINE 4 1,800 mm
PIPELINE SETTLEMENT	16" PLUS	MINIMAL	±12"	LESS THAN 3"
JOINT DAMAGE	JOINT SEALS REQUIRED	NONE	JOINT SEALS REQUIRED	NONE
LEAKAGE	EXCESSIVE	MINIMAL	EXCESSIVE	MINIMAL

CONCLUSION

The design of large diameter gravity pipelines requires a rational approach that addresses the potential failure modes of the constructed environment. A rational design process is one where the design decisions are all based on engineering judgement, not just industry standards, and potential failure modes of the existing site conditions are addressed.

For pipeline projects in high ground water it is necessary to address the potential that it may not be feasible to dewater the soils. The designer must know what the alternatives are available and how to address the problem. In weak soils the designer must address the potential for excessive settlement and design a foundation and bedding system that will support the pipeline to be constructed.

It is apparent that a design that does not address the failure modes related to poor installation condition will not serve the owner as desired and will have a life expectancy much less than expected.

REFERENCES

American Concrete Pipe Association, *Concrete Pipe Design Manual*, 1992

American Society of Civil Engineers, *Gravity Sanitary Sewer Design and Construction*, 1982

American Water Works Association, *M9, Concrete Pressure Pipe*, 1995

American Water Works Association, *M23, PVC Pipe – Design and Installation*, 1988

American Water Works Association, *M11, Steel Pipe- A Guide for Design and Installation*, 1989

Bureau of Reclamation, *Geotechnical Branch Training Manual No. 7, Pipe Bedding and Backfill*, 1981

Amster Howard, *Pipeline Installation*, 1996

A. P. Moser, *Buried Pipe Design*, 1990

National Clay Pipe Institute, *Clay Pipe Engineering Manual*, 1990

O. C. Young and J. J. Trott, *Buried Rigid Pipes, Structural Design of Pipelines*, 1984

EVALUATION OF NEW INSTALLATIONS FOR CONCRETE PIPE

JAMES J. HILL, PE, MEMBER ASCE [1]
JOHN M. KURDZIEL, PE, MEMBER ASCE [2]
CHARLES R. NELSON, PhD, PE, MEMBER ASCE [3]
JAMES A. NYSTROM, PE, MEMBER ASCE [4]

Introduction

The Minnesota Department of Transportation is investigating the use of recent technological developments to change their traditional methods used to design and install concrete pipe. New methods of soil structure analysis, installation details, and the ability to directly measure soil properties with modern methods have made this advancement practical. The need for change is being forced by economic issues. New pipe installations must be reliable, less labor intense, safer during construction, and use native soil materials as much as possible for support of the pipe. At the same time, there has been a shift of the responsibility for the quality of construction from the owner to the contractor. Certification of conformance to specifications will become more common as on-site inspection becomes less common. This idea recognizes that when the owner sets performance criteria, it is more efficient for the contractor to determine how to meet those criteria.

This paper is the first of a series to provide an analytical means for comparing projected design loads and reaction forces for new concrete pipe standard installations with actual construction practices.

---

1 Civil Engineer, Minnesota Department of Transportation
2 VP Engineering, Condux
3 Principal, CNA Consulting Engineers
4 Engineer, Cretex

Background

Concrete pipes have been used for drainage structures in the United States for 150 years. These early pipe installations were probably not designed, but proportioned by intuitive or empirical processes. In the early part of the century, researchers at the Iowa Engineering Experiment Station in Ames published *The Theory of Loads on Pipes in Ditches and Tests of Cement and Clay Drain Tile and Sewer Pipe*. The design and installation methods developed at Iowa State University have been successfully used up to the present time. The method, which considered separately the tested structural strength of the pipe and the subjective quality of the surrounding soil envelope, was called Marston-Spangler (M-S), or standard design.

The American Concrete Pipe Association (ACPA) undertook a long-range research program in the 1970's to provide pipe designers a better understanding of the behavior and interaction of buried concrete pipe and the surrounding soil. The results of the research concluded with a new method of concrete pipe design and installation which was named Standard Installation Direct Design (SIDD). The new installations were developed through actual field performance evaluations of soils and finite element soil-structure modeling with Soil-Pipe Interaction Design and Analysis (SPIDA). The new installations are detailed for ease of construction as well as providing haunch support that reduces flexure in the invert of the pipe. The supporting stiffness of bedding material is based on objective criteria such as soil classification, placement, and measured density.

SPIDA was used to calculate the coefficients for moments, thrusts, and shears around the perimeter of the pipe wall. These coefficients became the basis for SIDD and were incorporated in the ACPA's concrete pipe design program PIPECAR. ACPA has compiled the results of their research in the *Concrete Pipe Technology Handbook*. Design, manufacturing, and installation methods have been published in the American Society of Civil Engineers (ASCE) Standard 15-93 *Direct Design of Buried Precast Concrete Pipe Using Standard Installations (SIDD)*. SIDD beddings have been adopted by the American Association of Highway and Transportation Officials (AASHTO) *Standard Design for Highway Bridges* and fill height tables may be calculated with ACPA's program 3EB.

Research

In the spring of 1997, the Minnesota Department of Transportation (MnDOT) entered into agreement with CNA Consulting Engineers, a Minneapolis consulting engineering firm, to analyze the effectiveness and constructability of the SIDD pipe installation. The performance of the SIDD installations was to be compared to the Marston-Spangler installations.

The comparison of the two types of concrete pipe installations were a part of a larger geotechnical study that involved evaluation of advanced instruments used to measure soil properties. According to current engineering practice, the stiffness of soil is defined by the combination of the Proctor Density and the soil classification. A recently developed electromechanical instrument directly measures the elastic modulus of the soil. The soil stiffness, as expressed by elastic modulus, provides the key soil parameter required for soil-pipe design methods.

Test Conditions and Pipe Design

The 1200 mm round and 1485 mm arch reinforced concrete pipe were installed in a subtrench with a surcharge load placed as overfill to model the effects of 6.1 meters of fill height on each pipeline. These two pipelines were laid in parallel with greater than two pipe diameters (4.4-meters) separating them to mitigate any effects of soil arching between the two structures. Figure 1 illustrates the geometry of the installation and the configuration of its surcharge loading.

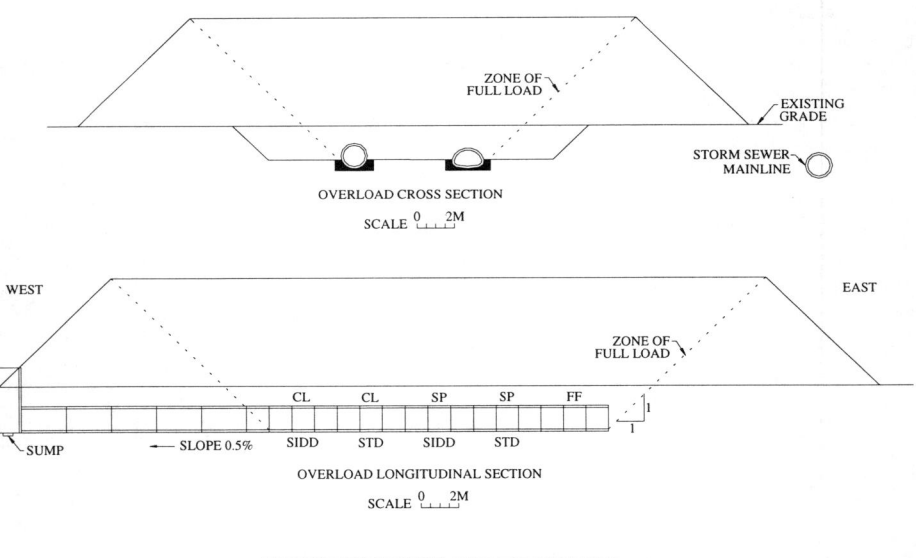

Figure 1. Overload Test Schematic

The test installation consisted of five standard bedding conditions. Each of these installations had three 1.2 meter long butt end pipes installed with a specified bedding. Three pipe sections were used for each bedding to insure, as a minimum, the mid-pipe performance was unique to the specific bedding being tested and not influenced by the adjoining test sections. Butt end sections were also specified to mitigate the possibility of any transfer of loading or shear forces from pipe to pipe through the joint.

The five standard installations which were specified for this test included a SIDD Type 2 (Sand), SIDD Type 3 (Clay), MnDOT Standard Sand Installation, MnDOT Standard Clay Installation, and foamed fly ash flowable fill installation. Use of these multiple installation details was intended to provide data in which to evaluated the performance of the new SIDD beddings to the existing MnDOT beddings. Figure 2 illustrates the bedding thickness, trench widths, bedding angle and compaction requirements necessary for one of these installations. Although only circular pipe bedding details are illustrated, the same details were used for the arch pipe.

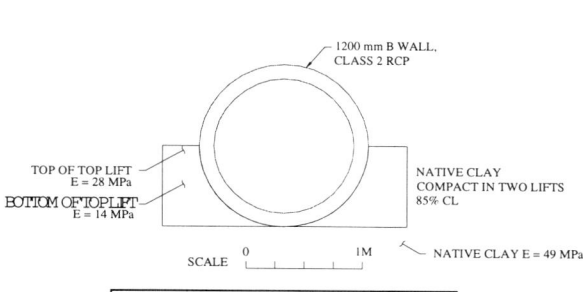

Figure 2. SIDD Clay Bedding Detail

The dimensions of the various bedding types were chosen based upon soil measurements taken during the installation of the storm sewer mainline. This data was used to estimate the stiffnesses that would be obtained using different soils and bedding types during the overload test. Figure 2 illustrates the cross-section of the clay Type 4 SIDD bedding, and the respective stiffness estimates.

In order to evaluate the performance of these installations relative to each other, the 1200 mm round and 1485 mm arch reinforced concrete pipe were designed to exhibit flexural, diagonal tension or radial tension strength limit. Obviously, if no distress was present in the pipe, one could not determine the benefit derived from the installations. The pipe was designed to show signs of distress at approximately 3.3 meters, with the level of distress worsening as the fill height or load increased. All distress was designed to be in a ductile failure mode and not a brittle, catastrophic failure. This safety provision required specifying a minimum fabric size to prevent failure of the steel wire.

The steel design for each of these structures is provided in Table A. In both cases, the pipe classifications represent the standard Class II ASTM C76 and ASTM C506 steel areas for the round and arch pipe, respectively. These pipe classifications have a D-Load capacity of 50 and 75 N/M/mm for their 0.3 mm crack and ultimate load criteria, respectively. In order to predict the pipe performance without any additional safety factors, the material design parameters represent the actual strengths of the concrete and steel used to manufacture the pipe.

TABLE A: PIPE MANUFACTURING DETAILS

Material Properties	1200 mm Circular Pipe	1485 mm X 915 mm Arch Pipe
Wall Thickness	125 mm	125 mm
Concrete Strength	44.9 MPa	44.9 MPa
Steel Yield Strength Reinforcing Type Concrete Cover	552 MPa Smooth Wire Fabric 25 mm	552 MPa Smooth Wire Fabric 25 mm
Steel Areas A_{si} A_{so}	350 mm²/M 220 mm²/M	590 mm²/M 470 mm²/M
D-Load Capacity Ultimate 0.3 mm Crack Diagonal Tension Radial Tension	114 N/M/mm 69 N/M/mm 121 N/M/mm 197 N/M/mm	120 N/M/mm 69 N/M/mm 117 N/M/mm 159 N/M/mm

The performance of the pipelines was modeled at various fill heights using both indirect and direct design analyses. Tables B and C represent the design required for each of these structures at the various applied loading and bedding conditions.

TABLE B: SIDD STEEL AREA REQUIREMENTS

1200 mm Circular Reinforced Concrete Pipe								
Installation Conditions Fill Height (M) Installation Type Arching Factor Soil Density (kg/M³)	7.3 1 1.35 1920	7.3 2 1.40 1920	7.3 3 1.40 1920	7.3 4 1.45 1920	6.1 1 1.35 1920	6.1 2 1.40 1920	6.1 3 1.40 1920	6.1 4 1.45 1920
Material Properties Steel Yield Strength (MPa) Concrete Strength (MPa)	552 44.9	552 44.9	552 44.9	552 44.9	552 44.9	552 44.9	552 44.9	552 44.9
Steel Areas A_{si} (Invert) (mm²/M) A_{so} (Springline) (mm²/M) A_{si} (Crown) (mm²/M) *Stirrups Required	431 228 362	653 303 444	1068 344 237	723* 246 602	346 183 289	492 240 352	692 291 393	1127* 405 472

The fill heights presented in Table B represent the design target with an additional analysis for possible overfill. The SIDD analysis was only conducted for the 1200 mm circular pipe because the original research used to develop the SIDD beddings was restricted to circular pipe. To ascertain the type of SIDD bedding obtained in the field, multiple installations were analyzed from a Type 1 through a Type 4. Given the steel area in the manufactured pipe, if little or no cracking is present in the pipe, a Type 1 or better installation was achieved. If large cracks or shear cracks are present, the expected Type 3 or Type 4 installation was obtained.

Table C presents the D-Load capacities of each pipe as the depth of fill increases. Given the depths of the final fill heights, the D-Loads were twice as high as the D-Load capacity for a Class II pipe.

TABLE C: D-LOAD REQUIREMENTS

Depth (M)	1200 mm Circular Pipe (N/M/mm)		1485 mm Arch Pipe (N/M/mm)	
	D-Load 0.3 mm	D-Load Ultimate	D-Load 0.3 mm	D-Load Ultimate
3.1	50	76	47	71
4.0	66	98	61	92
5.0	82	124	77	115
6.1	101	151	94	141
7.3	121	166	113	160

Pipe Measurement and Instrumentation

Buried concrete pipe is only one component of a soil-pipe composite structure. The structural design of the pipe and the quality and care in placing the bedding and haunching material determines the relative contribution of each component to the strength of the composite structure. The measurement of the density and the elastic modulus of the soil were of primary interest for this test. Data collected from the pipe was limited to measuring the changes in the vertical and horizontal diameters and noting crack width, cause and location. Inspection of the pipes was made without load, under service loads, at maximum loads and during unloading.

Four button head nails, used as measuring points, were installed ninety degrees apart at mid-length of each test pipe before construction started. The pipes were aligned during installation so the vertical and horizontal axis could be measured at each phase of loading. Measurements were taken with a micrometer.

Soil pressure cells were positioned (as illustrated in Figure 3.) around the pipe in critical areas identified by the SPIDA finite element model. The horizontal and vertical soil pressure was measured at the springline of the pipe. The ratio of horizontal pressure to vertical pressure, known as the horizontal arching factor (HAF), is an important element in SIDD design because it represents the side support on the pipe. Additional cells were installed at the pipe invert and at a point between the invert and pipe haunch. This is an area where SPIDA assumes there is little or no support from the soil.

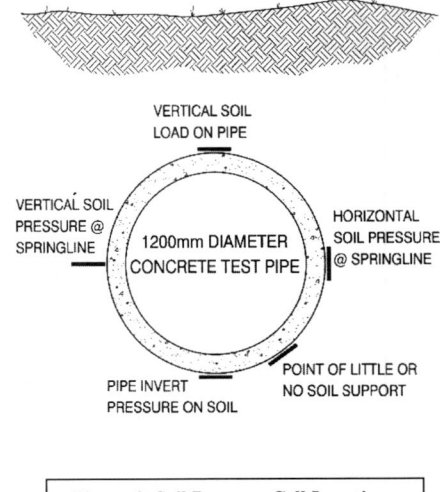

Figure 3. Soil Pressure Cell Locations

Installation

The test was designed to compare the constructability and performance of existing standard Marston-Spangler (M-S) and new SIDD installations of concrete pipe using native soils for the bedding and haunching material. Under special provisions of MnDOT's contract, the test pipes were intended to be included in pipelines required for drainage of the right-of-way of Minnesota State Highway 610 in Brooklyn Center, about 12 km northwest of downtown Minneapolis. The special provisions required that the contractor install, test, remove, and replace the test pipes within the pipeline. The contractor felt the testing process would excessively delay the progress of the pipe installation crew during the short construction season. It was then agreed that the contractor would construct an embankment suitable for the test on the project sites at the end of the construction season.

Discussions between the contractor, project engineers, and the pipe design task group revealed that details of Standard M-S and SIDD installations were not being followed in practice. Early photographs from Marston's work at Iowa State University showed small diameter pipe installed in a narrow, relatively shallow hand-dug trench. All modern concrete pipe installation specifications have been extrapolated from that beginning, but construction practices have changed dramatically. The task group decided to modify both the standard M-S and SIDD installations to reflect contemporary pipe installation practice.

Two lines containing fifteen 1.2 M long test pipes were installed in the embankment; one line of 1200 mm diameter circular pipe and one of 1200 mm equivalent span arch pipe. Five groups of three pipes each were installed in either sand (SP), clay (CH) or flyash slurry foundation and haunching material (See Figure 1 and Table D). Modified standard M-S Class C or SIDD Type 4 installation methods were used.

TABLE D: SCHEMATIC OF INSTALLATION CONDITIONS

	CLAY		SAND		FLOWABLE FILL
	SIDD	STD	SIDD	STD	
CIRCULAR	▭▭▭	▭▭▭	▭▭▭	▭▭▭	▭▭▭
ARCH	▭▭▭	▭▭▭	▭▭▭	▭▭▭	▭▭▭

▭▭▭ REPRESENTS THREE SAMPLE PIPE 1.2 M LONG

Constructing an embankment provided several advantages for the pipe test:

- The soil in the bedding and haunching areas could be selected and placed according to test parameters,
- Embankments are more common in highway construction so more useful knowledge may be obtained,
- Embankment installations can be used to model wide trench installations,
- An embankment was used for the SPIDA finite element model so information obtained may be compared to original assumption,
- The overload could be left on to evaluate creep.

Installation Characterization

During pipe and backfill installation and removal the soil properties were evaluated for density, moisture, stiffness and certain strength properties.

Figure 4 is a cross-section showing the locations of soil stiffness and nuclear density and moisture testing for the SIDD Sand installation. The testing locations for the other installations were similar. Locations were chosen so as to collect a data profile which could allow for back-analysis using properties of the surrounding native material as well as the fill material.

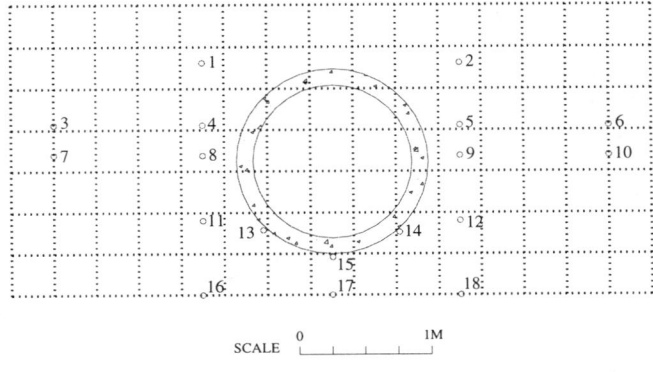

Figure 4. Soil Testing Locations

Photo 1. Soil Pressure Cell Location

Photo 1 shows a pressure cell positioned beneath the invert and the quarter point of a round pipe in a clay installation. Voids between the soil and the quarter point of the pipe can be observed in the photograph, and were a common occurrence in the clay installations.

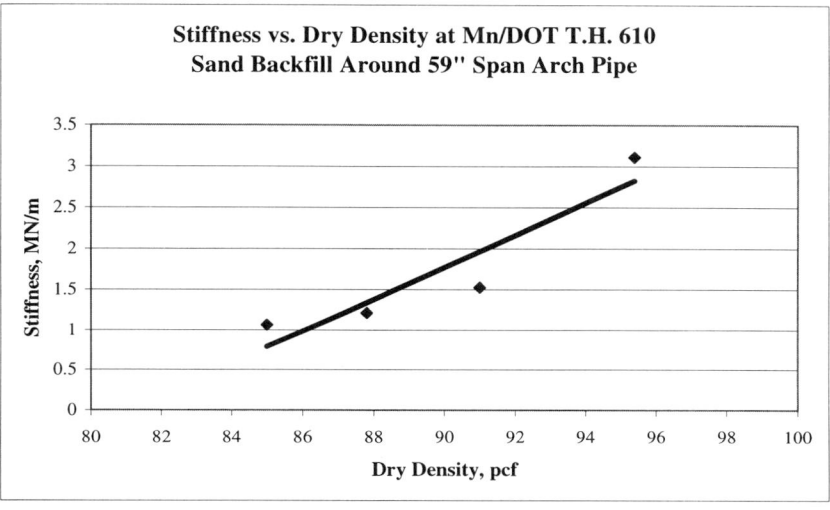

Figure 5. Variation in Soil Stiffness with Depth and Density

Figure 5 shows the variation observed in soil stiffness with depth and density in a sand backfill around arch pipe.

Conclusions

Full analysis of the test program will be presented in later papers. Initial conclusions and observations are:

- The reinforced concrete pipes performed well at the design load,
- There was extensive cracking in the pipe, but no total failure at the overload,
- Normal field installation deviates significantly from the idealized design conditions,
- Performance of the SIDD Type 3 and Type 4 design bedding installations indicated a reduction of the magnitude of the reaction force under the invert and a better distribution of the loads in the haunch than design assumptions,
- The use of higher class installations will require education, quality control measures and modified methods and equipment,
- The flowable fill installation provided excellent bedding support,
- The soil density and stiffness was found to be highly variable around and along the pipe.

TRWD Experience with Prestressed Concrete Pipe

David H. Marshall, P.E., Member, ASCE [1]

Abstract

The Tarrant Regional Water District, a wholesale water supplier in North Central Texas, operates two 120 km long (75 mile) large diameter prestressed cylinder concrete pipelines and is constructing a third pipeline. Solutions to problems encountered with the first two pipelines were incorporated into the specifications of the third pipeline. Problems still unresolved with the first two pipelines will lead to additional changes in specifying new pipelines, and the concepts of those changes are presented.

Introduction

The Tarrant Regional Water District is a political subdivision of the State of Texas whose role is to supply raw water for municipal, industrial, mining and irrigation use, and to provide for flood control in Fort Worth. The District owns four reservoirs and has water rights in a fifth reservoir, owned by the Corps of Engineers. Total safe yield for the system is about 532,790,000 cubic meters (432,000 acre feet) per year. The District serves 54 wholesale customers with raw water, with a total population in the service area of about 1.2 million people. Demands are growing consistently at about 3% per year.

The District supply utilizes two delivery systems. The West Fork Trinity River (the West Fork System) is located northwest of Fort Worth. The District owns two reservoirs on the West Fork System that feed to a small reservoir owned by the City of Fort Worth. Water flows by gravity into Fort Worth. The East Texas System delivers water from two reservoirs located southeast of Dallas using two pipelines. The pipelines supply three water plants in the Fort Worth vicinity, the City of Arlington's Lake Arlington and the Corps of Engineer's Benbrook Reservoir. Both

---

[1] Engineering Services Manager, Tarrant Regional Water District, 1022A North Calhoun, Fort Worth, Texas 76106

Lake Arlington and Lake Benbrook serve as terminal storage points. Last year the District delivered 92,251,000 cubic meters (74,800 acre feet) from the West Fork System and 180,600,000 cubic meters (146,000 acre feet) from the East Texas System.

Pipeline System

The East Texas System uses two pipelines. The first pipeline, from Cedar Creek Reservoir to Fort Worth, was finished in 1972. It consists of 109 km (68 miles) of 183-cm (72-inch) ID prestressed concrete cylinder pipe (PCCP), conforming to the AWWA Standard C301 of the time, and 10 km (6 miles) of 213 cm (84-inch) ID PCCP. Pressure classes of the pipe range from 15.521 dynes/cm^2 (225 psi) at the pumpstations to 3.45 dynes cm^2 (50 psi) at the termination of the lines. There are three pump stations on the line. Maximum delivery is 6.44 m^3/sec (147 mgd). Thrust restraint was accomplished using thrust blocks. When constructed, the pipe segments were not bonded for electrical continuity. The design called for a brass shim to be inserted between the bell and spigot to provide for electrical continuity. This shim proved ineffective in making the pipeline electrically continuous.

The design of the pipeline varied, differing in joint length and the use of shorting straps. A shorting strap is a thin, one-inch wide strip of steel that runs longitudinally down the pipe between the concrete core and the prestressing wire. Shorting straps are used to limit current drop for current used in cathodic protection. In areas where the soil resistivity was high and corrosion rate considered low, the 183 cm (72-inch), 7.3 meter (24-foot) segments were constructed without shorting straps. In areas with low soil resistivity, where corrosion was considered a possibility, the segments were constructed with shorting straps. The 213 cm (84-inch), 4.9 meter (16-foot) segments were constructed without shorting straps. The Cedar Creek pipeline had a single impressed current system constructed as part of the project to study cathodic protection of the line. The impressed current system ran from 1974 to 1992. Maximum on potentials from the rectifier was 1.2 volts.

The second pipeline, the Richland Chambers Pipeline, runs from Richland Chambers Reservoir to Fort Worth. Beginning near Ennis, Texas, the Richland line parallels the Cedar Creek line in the same right-of-way. The pipeline was finished in 1988 and consists of 116 km (72-miles) of 229 cm (90-inch) ID PCCP and 9.7 km (6-miles) of 274 cm (108-inch) ID PCCP. Pressure classes are the same as the Cedar Creek pipeline. There are two pumpstations on the pipeline. Total maximum flow is 6.49 m^3/sec (148-mgd). The segments are bonded to be electrically continuous using a single bolted strap between adjacent bells and spigots. All segments have two shorting straps under the prestressing wire, located 180 degrees apart. Thrust restraint was accomplished using welded joints.

Problems

The Cedar Creek line had its first failure in December 1981, after almost ten years of service. The failure was a catastrophic rupture due to corrosion of the prestressing wire. Since then there have been six other prestressing wire corrosion failures. There has been one failure, the latest one, caused by hydrogen embrittlement on the segment connected to the rectifier. Figure 1 is a graph of the timing of the failures, plotted cumulatively over time.

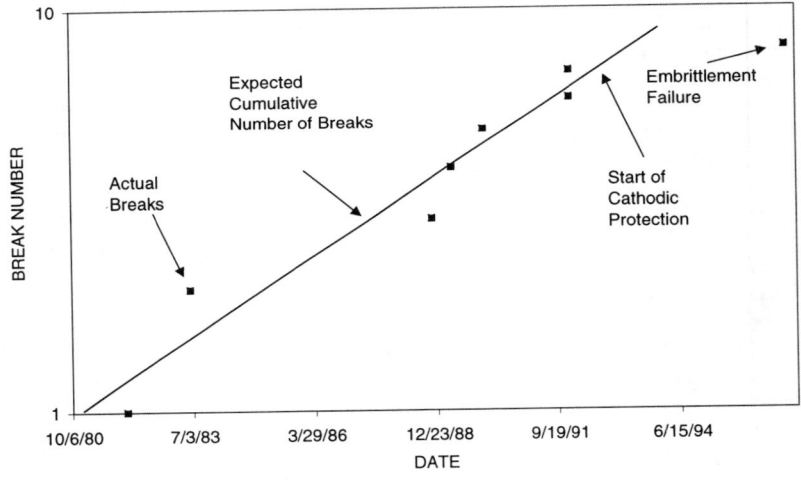

Figure 1. Timing of Cedar Creek Pipeline Failures

The Richland Line experienced its first failure during hydrostatic testing in October 1988. A segment began leaking immediately adjacent to a large (greater than 15 degrees) angle bend. The defect was thought to have been a failed weld between the bell and cylinder, and was repaired externally. Two months later (December 1988), during the first pump test, a second segment failed, again immediately adjacent to a large angle bend. The failure was a circumferential tear in the cylinder of the pipe. This time the pipeline was dewatered, and at every large angle bend inspected, segments immediately adjacent to the bends had circumferential cracks, indicating movement at the bends. The length of welded joints was doubled from the original design to mitigate the situation, and the pipe put back ready for service. The line began to be used in 1990.

Richland experienced its first corrosion failure in 1992. The first section of pipeline laid had a defect in the welding that was discovered during construction, when groundwater was leaking into the pipeline. A weld had penetrated the cylinder, permitting water to leak into the line. All pipes laid and all pipes manufactured to that point were tested for leakage, and some segments were found to be defective and repaired. The corrosion break in 1992 was in a defective segment that was not found during inspection. Product water had leaked onto the wire and created the failure.

Richland has experienced six additional failures, one due to operator error, two more due to thrust restraint problems, and three corrosion related failures. Two corrosion failures have been attributed to a minor amount of corrosion activity permitting hydrogen to be formed, and embrittling the wire. The embrittlement problem is still under investigation. Figure 2 is a graph of the timing of the failures in relation to the cumulative number of failures.

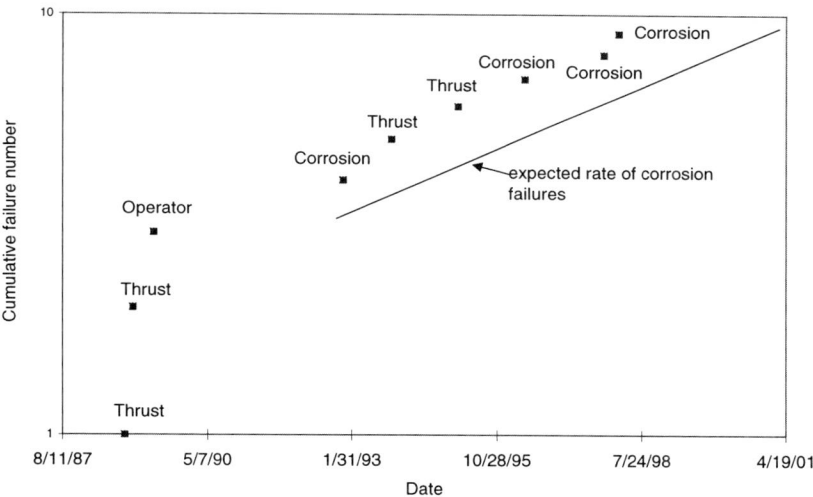

Figure 2. Timing of the Richland Chambers Pipeline Failures

The corrosion failure not attributed to hydrogen embrittlement, in early 1997, was due to extensive corrosion of the prestressing wires. The pipe had been inspected a year before the failure and longitudinal cracks on the floor and ceiling were found. This indicated the pipe had been compressed, possibly due to poor compaction on the sides and equipment running over it during backfilling. Inspection of that section of pipe revealed about 120 other segments had similar cracking. It appears during construction, backfill was improperly compacted and the pipe lacked haunch and sidewall support.

Mitigation

In 1989 the District began monitoring corrosion on the pipelines. Pipe-to-soil and cell-to-cell monitoring was started over both pipelines. Soil moisture limits the time the potential surveys can be run, so only a small portion of the pipeline can be covered each year. The Richland line's potential readings were completed in 1995. Since Cedar Creek was not electrically continuous, pipe-to-soil readings could not be done at the same time. As Cedar Creek is being bonded, the readings are taken.

Using potential measurements proved ineffective in determining active corrosion. Cell-to-cell measurements were sometimes corrupted by debris in the trench. Pipe-to-soil measurements were better than cell-to-cell, but still there were numerous unexplained spikes in the readings. Soil moisture influences the readings, limiting the accuracy. Figure 3 is a graph of four sets of pipe-to-soil readings over the same area, showing how differing soil moisture conditions affect the readings.

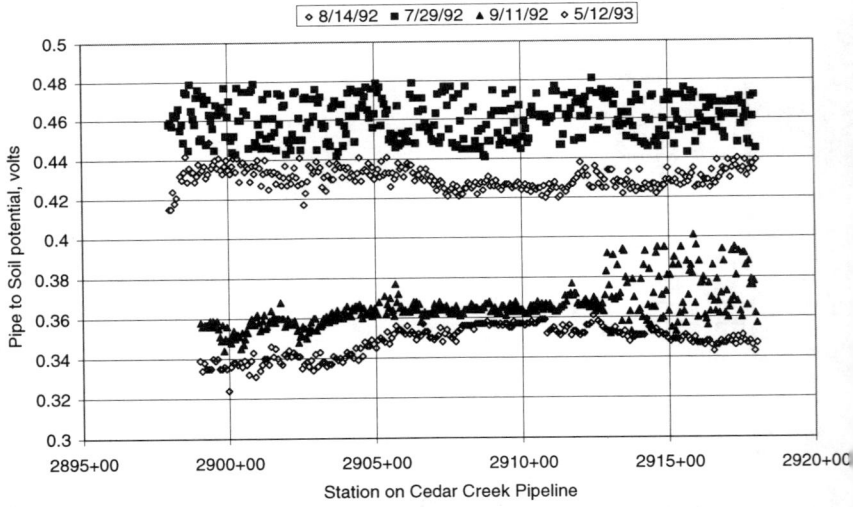

Figure 3. Comparison of Pipe to Soil Readings over time

The readings show that as soil moisture varies, so does the natural pipe-to-soil potential. The variability in readings creates problems in identifying areas of active corrosion and in determining effectiveness of a cathodic protection system. Pipe-to-soil potentials are still being taken to serve as a baseline in determining cathodic protection system effectiveness. The absolute values are viewed as an indication of activity rather than an exact measurement due to the variability that is observed.

Corrosion mitigation began in 1992, with the study of a 4.8 km (3-mile) section of the Cedar Creek pipeline that had been bonded internally so it would be electrically continuous. Testing included pipe-to-soil potentials, soil resistivities, coating analysis, and electrical resistance. Hydrogen embrittlement was a concern in the cathodic protection. Zinc sacrificial anodes were chosen to develop the protective current since they provided the most protection from hydrogen formation.

Cathodic protection of the Cedar Creek line began in 1993, with installation of sacrificial zinc anodes over the test section. The test section included an area where three corrosion failures had occurred. This section was then evaluated after the anodes were installed and the pipe polarized to determine the effectiveness of the design. Based on those results and specific tests in all areas, the design was done for both pipelines. Currently, about half the pipelines are under protection and work continues to complete the entire system within the next two years.

The design for the cathodic protection system varies from six 27 kg (60-pound) prepackaged zinc anodes per anode bed to over twenty anodes. Anodes are placed at springline of the pipe, with an equal number on each side of the pipe. Anode bed spacing varies from 61 meters (200-feet) to 152 meters (500-feet), depending upon soil conditions. Details of the design may be found in the May, 1997 Materials Performance (2). Design for the areas without a shorting strap in still under way. Protection on those segments may be require different anode configurations due to the drop in current running through the almost two miles of wire per segment.

Through the study of failures on the Cedar Creek pipeline, it was concluded that damage to the external mortar coating over the prestressing wire resulted in a pathway for water to collect on the prestressing wire, creating corrosion. Damage of the mortar coating was attributed to construction activities and to waterhammer damage. Waterhammer control on the system consisted of slowly closing the pump control valves during shutdown, thus keeping the flowline open. This procedure had limited success in minimizing waterhammer pressures. Many times the valve timing of the hydraulic system was too fast on closing, creating high pressures. In a effort to control waterhammer, the District modified the valve control system at all pumpstations using a PLC to keep the valve fully opened until the high pressure return wave passed the pumpstation, thus reducing pressure increases to within operating ranges. A paper presented Texas Water '97 highlighted the control system for the pump control valves (3).

On the Richland line, a thrust restraint failure occurred during the hydrostatic test. The failure was attributed to a bad weld between the cylinder and the bell at the time it occurred. A second thrust restraint failure occurred during the pump test. This time the pipe was dewatered and inspected. The inspection revealed the pipe had moved at every large angle bend. To mitigate the problem, the pipe joints were welded to twice the length called for in the original design. Another thrust restraint failure occurred in November 1993. Internal inspection of the line revealed a large number of circumferential cracks had developed due to continued movement of the line at the large angle bends. To try and mitigate the problem, segments were welded on each side of the bend to the theoretical length required to resist the thrust assuming no skin friction (as if the pipe were laid on top of the ground). All damage was carefully mapped and crack widths measured.

Fifteen months later, February 1995, another thrust failure occurred, in an area that had been welded with the additional length. The District, along with the pipe supplier and the design firm for the project consulted with Simpson, Gumpertz and Hagar (SGH) on the problem. SGH recommended a finite element analysis of the pipe and surrounding soil be made. Results showed that highly plastic soils did not offer sufficient side support. SGH redesigned the pipe adjacent to the bends with thicker cylinders that could resist tensile forces associated with the movement.

The new design called for cylinder thicknesses from 3.4 mm (10 gauge) to 7.9 mm (5/16 inches). Forty-six pipe segments were replaced with new pipe at seven bends. Some segments from the original construction that had minimal damage were reused in new locations as part of the design. Repairs were accomplished from January 1997 through March 1997. Five other bends remain to be repaired; all in areas where current operating pressure is much lower than anticipated future operations.

The District has tried several methods of repairing of the pipeline. For all emergency repairs due to a catastrophic failure, a three-piece closure section is used. The closure section consists of three short sections of PCCP pipe and a wide two piece butt-strap. The failed joint is removed, and the new short sections lowered in place. The lower half of the butt-strap is tack welded into one of the short sections prior to placing it. The new sections are aligned and the both lower and upper butt-straps welded on to make the closure watertight.

The District has also used 12.7 mm (½-inch) thick steel bands to surround corroded areas. These bands are bolted on so as to compress the pipe, then covered in grout to limit corrosion. These have been used for areas as wide as 1.2 meters (4-feet).

The latest repair method the District has tried has been using tendons to recompress the core. This method appears to have some advantages. During investigation of possible problem areas, if a bad area is found, the tendon repairs could be accomplished with little additional time and minimal additional disturbance to the line. The District will be moving forward to develop this repair methodology in-house.

Internal inspections are also done whenever the opportunity is available. Three water plants take directly from the line, so downtime is limited to repair of failures or maintenance during the winter when demands are low. With both lines having questionable integrity, downtimes are kept to a minimum. During internal inspections, the lining is visually checked for cracks. Each segment is also hammer tested. A disbonded area will produce a dull thud compared to the ring of a solid pipe. Tests have been done with a known bad joint for hammer testing. It was found that the defect has to be large (greater than .6 meters (2 feet) longitudinally) to be discerned.

The District has also conducted acoustic monitoring of a portion of the test section of the Cedar Creek pipe and echo impact monitoring. Both tests were conducted in the same area, and were part of an AWWARF study for nondestructive testing of pipe. The AWWARF report will go into some depth concerning the results of the study. One bad segment was found and replaced as a result of the testing. Both technologies show great promise in finding bad segments long before a failure would occur.

New Construction

The District's latest project, the Benbrook Connection Project, is a 229 cm (90-inch) pipeline to connect Benbrook Reservoir for terminal storage with the two east Texas Pipelines. The project runs through developed areas in south Fort Worth. The project was bid with the contractor's choice of PCCP or steel pipe. The low bid was won with PCCP. During the design process, the District tried to ensure that every problem faced in the past was addressed. Specifications dealing in thrust restraint, cathodic protection, backfill, and cylinder thickness were developed specifically for the project.

The District investigated the use of an impressed current system to protect the Benbrook line. Using the rectifier from the Cedar Creek line, it was attached to the Richland line, since the Richland line was electrically continuous, and the pipe-to-soil potential measured on both the Richland and Cedar lines (paralleling each other, 7.65 meters (25 feet) apart). The on-potential was limited to 0.85 volts to prevent hydrogen embrittlement. Instant off readings revealed that stray current was produced on the Cedar line when protecting the Richland line, with shifts of 10 to 20 millivolts on the Cedar Creek line during the instant off. With the potential for stray current to other utility lines in the vicinity of our pipeline and possibility of rectifier failure resulting in higher currents and hydrogen embrittlement, it was decided to construct the line with sacrificial anodes for cathodic protection.

The specifications used for the design built upon the AWWA C301-92 and AWWA C304-92 specifications (1). The District required that the steel cylinders be thick enough to resist internal pressure with no prestressing wire present. Cylinder thickness ran from 2.6 mm (0.1046-inches) to 4.7 mm (0.1875 inches) using 3033 dynes/cm^2 (44 ksi) steel. Thrust restraint was accomplished using welded joints, with a finite element analysis completed by SGH to determine length of restraint and cylinder thickness. Two shorting straps were required for each segment, with each end of the straps welded to the bell or spigot. To be sure the segments were electrically continuous, four horseshoe shaped clips were welded at each joint. The District required mechanically compacted select backfill bedding in most areas and flowable fill in areas of high risk.

Conclusions and Recommendations

The District's experience with large diameter concrete pressure pipe has definitely not been trouble free. The Cedar Creek Pipeline has experienced corrosion that went unchecked for almost twenty years, resulting in damage and a lowering of the integrity of the line. The District now believes that with the cathodic protection system we are installing and the investigative work we are doing, the Cedar Creek line will provide us with a reliable line for decades to come.

The Richland line has had problems since the hydrostatic test. The specifications for thrust restraint did not meet the requirements based on the soils the District encountered. It is hoped that the latest repair will prove successful. The District still faces a numbers of problems with the Richland line. In all, the District has identified about 350 segments with some damage. The damaged pipes will be ranked according to risk and repaired as time and budget permits. The problem with the hydrogen embrittling wire is being explored, but the extent of the problem and the best mitigation is yet to be developed and may change the repair schedule.

Sophisticated nondestructive testing to find bad segments is just now being developed. The acoustic monitoring offers the advantage of keeping the line in service while testing. As these technologies are further refined, the problem now faced of identifying bad segments in a timely fashion will be greatly diminished.

The specifications the District used for the Benbrook line should provide a reliable pipeline. There are some additional improvements in the specifications that I would like to make for our next line. The first improvement will focus on additional wire tests to examine the potential for hydrogen embrittlement. Improving the quality of wire so minor corrosion will not create a problem will improve reliability. The second improvement would focus on using fibers in the concrete core and exterior coating to improve the tensile strength. Making the pipe less susceptible to cracking may lower the risk of corrosion.

A better working relationship between the users and manufacturers of large diameter pipe could improve mitigation of current problems and future designs. Based on the bids for the Benbrook project, concrete pipe is still the most economical material in north central Texas. The changes the District made in the specifications to improve performance still kept the concrete pipe competitive with steel pipe. Time will tell if the life cycle cost of the line is the best economy for our rate payers. I am sure there could be other minor improvements in design and in quality control that would benefit the public without making concrete pipe uncompetitive. Owners and manufacturers need to work together to develop better specifications and a better pipe. The goal of both groups should be to provide the best service possible to the public at the lowest cost.

References

(1) AWWA Standard for Prestressed Concrete Pressure Pipe, Steel-Cylinder type, For Water and Other Liquids, ANSI/AWWA C301-92

(2) Benedict, Risque L., Jerald G. Ott, David H. Marshall, Dale White, "Cathodic Protection of Prestressed Concrete Cylinder Pipe Utilizing Zinc Anodes", *Materials Performance* May, 1997 Volume 36, Number 5, Pages 12-17

(3) Dang, Bill and Ed Weaver, "Implementation of Electric Controls to Improve Hydraulic Pump Control Valve Operation", presented at Texas Water '97

SOAP LAKE SIPHON RECEIVES REHABILITATION

by James E. Wolfe[1], F. ASCE
M. Wayne Cardwell[2]

ABSTRACT

Advancement in the state-of-the-art of the application of cement mortar to the surface of large diameter circular steel surfaces had progressed considerably in the past few years. The process described in this article gives confidence once again for large diameter applications.

The Soap Lake Siphon is one of the most important features of the West Canal, part of the Columbia Basin Project in central Washington. It is being rehabilitated in the winter of 1997-1998. This project is unique and challenging for several reasons, as follows.

First, with a 22' 4" internal diameter, this was the largest steel-lined concrete pipe ever constructed when the siphon was built in 1947-1948. It is quite possibly still the largest of its kind even today. Second, the siphon is 8250' long. Also the alignment and profile of the pipeline varies along its length. It reaches its lowest point, 250' below the intake, in the middle. Finally, there are two short, steep slopes which achieve an 18% grade.

The first part of the rehabilitation consists of removing all remnants of the original, hand-dobbed, coal-tar enamel coating. Some of this had deteriorated and exposed the steel liner plate, causing tuberculation. A cement mortar lining 3/4" thick will provide the new protection to the steel surface.

The challenge lies in applying the new lining under such unique conditions. The only project remotely similar in terms of the distance the mortar needs to be projected and trowelled was one completed 25 years ago on a pipe diameter almost as large with a less-difficult profile.

This project uniquely demonstrates the capabilities of the in-place mortar lining process. Cement mortar is being applied to a surface never before attempted, in unprecedented conditions. This paper fully details the role of this application process in the siphon's overall rehabilitation.

---

[1] Retired -- Consultant, La Verne, Calif., 91750
[2] Western Manager, Spiniello Construction Co., Santa Fe Springs, Calif., 90670

THE PROJECT

The Grand Coulee Dam provides the headwater for the massive Soap Lake siphon in the West Canal, which is a significant part of the Columbia Basin Project.

Called the most phenomenal structure on the West Canal by one of its original engineers, the Soap Lake siphon was indeed the largest structure of its kind in the world when it was constructed nearly 50 years ago. Even today it is still considered one of the largest such structures. Built in 1948 through 1950, the 3.91 km (2.43 mile) long inverted siphon traverses a U-shaped alignment through the Lower Grand Coulee around the north end of Soap Lake. The entrance to the siphon is located on the basalt cliff along the east side of Soap Lake, about 67.1 m (220 feet) above Highway 7. From there it drops 72.9 m (239 feet) to the Coulee floor and under the highway. It then rises again to terminate in the West Canal on the basalt bench west of Soap Lake.

Reinforced concrete pipe 7.62 m (25 feet) in diameter was used for the two ends of the siphon. However, in the length that dips sharply to the Coulee floor, where high pressure is induced by a large change in elevation, it was necessary to use steel lined concrete pipe 6.81 m (22 feet 4 inches) in diameter for a total distance of 2519 m (8,264 feet). (See Figure 1.)

Fig. 1: Last Steel Section

When the Soap Lake siphon began to show serious signs of aging experts were quickly called in. The diagnosis was similar to hardening of the arteries: the protective coal tar enamel lining was deteriorating, permitting water to come into contact with the ferrous steel liner and causing corrosion. Corrosion products already covered large areas of the inside of the line, producing flow losses and warning of future serious problems. In some areas there were porous mineral deposits over an inch thick adding to the problem. Continued dependable use of this critical siphon relied upon a rehabilitation and life-extension effort unequaled in the world both in size and scope.

The U.S. Bureau of Reclamation recommended rehabilitation of the siphon by employing in-place cement-mortar relining. A method described in the American Water Works Association Standard ANSI/AWWA C602-89 as "centrifugally placed" cement-mortar lining was selected for use.

The Quincy-Columbia Basin Irrigation District accepted bids on July 18, 1996, to rehabilitate this siphon as per U.S.B.R. recommendations. Four proposals were received and a the lowest bidder was selected to perform the work.

The steel liner used in this method was 0.635 cm (1/4 inch) plate, rolled to the 6.81 m (22 foot 4 inch) diameter with a coaltar enamel place on the surface by "hand-daubing." During the long period of the rehabilitation procedure, the enamel began to deteriorate in spots, leaving large areas of good, tight undisturbed bed coal-tar. Under these conditions the cleaning was considered to be exhaustive.

These cylinders were installed in an open trench and 53.34 cm (21 inches) of concrete was placed around their entire circumference. The encasement was well-crafted, leaving the pipe very true to round. There were several 60.96 cm (24 inch) man-holes spaced periodically as well as a blow-off placed at the low point in the line.

The profile of the pipe was relatively gentle except for two locations. Each of these areas was approximately 61.0 m - 91.4 m (200' to 300') long with a 18% slope. The first was at the center of the alignment, and the other was at the end near the discharge.

In addition to the steep slope, the 6.81 m (22 foot 4 inch) diameter pipe was the largest pipe ever considered for rehabilitation. Twenty-five years before approximately the same length of 6.25 m (20 foot 6 inch) I.D. steel tunnel liner was rehabilitated using this method of coating with cement-mortar. But in that case there was no previous coating to remove and the profile was flat.

Moreover, there was a 120-day down time restriction on the pipeline, making it necessary to split the job into two successive seasons. The down-stream half of the line was cleaned and in-place lined during the '96-'97 winter. The balance of the work was completed in the winter of '97-'98.

INSTALLATION

Considerable preparation was necessary to be able to apply the mortar. The application machine (the liner) needed to be improved to project mortar almost twice the distance as any before in the past 25 years. (See Fig. 1 and Fig. 2.) The trowelling, conveyor system, mortar batching, and mortar mixing equipment all had to be adapted for the larger pipe. To ensure success, a mock-up of the 6.81 m (22 foot 4 inch) pipe was erected and all the equipment was tested in the Southern California shop of the successful bidder. Then it was disassembled and sent to the site where it was reassembled prior to the shutdown.

The necessary material also had to be assembled and stored along the job site. This consisted of 40 truck loads of each sand and cement.

Quincy-Columbia Basin Irrigation District (owner) achieved a dry pipe by blowing off and pumping the water and provided access to the pipe through the intake and outlet ends, as well as the 60.96 cm (24 inch) manholes.

The first thing to do after the pipe was drained was to begin the cleaning operation. Considerable pipe had to be cleaned and ready for mortar application before the lining equipment could be placed in the pipe and tested. Being unprepared for the condition of the coal-tar enamel and determining its various stages of deterioration consumed more time than was anticipated and delayed the cleaning operation. This cut into the time determined necessary for the setup and trial period of the lining application. As everything fell into place the lining operation also proved greater than expected.

To commence the placement of the mortar, the equipment was placed in the pipe and walked to the center and lowest point of the siphon. To get the lining operation underway, all elements of the machinery had to be adjusted to work together perfectly.

The specifications required placing 1.91 cm (3/4 inch) of mortar on a surface 3.35 m (11 feet) away from the point where the material was released from the revolving application head. Not only did the material have to travel a considerable distance, it had to stick to a relatively flat, slightly curved surface. The surface receiving the continuous coating of cement-mortar completely surrounded the discharge point. In other words, mortar was being projected from the application horizontally to the right and to the left, as well as vertically, 3.40 m (11 feet, 2 inches) up and 3.40 m (11 feet, 2 inches) down. The vertical projection is the most difficult because gravity has an effect on the projection velocity.

Due to the weight of the applied mortar at the surface, the adhesion is insufficient to hold up the total 1.91 cm (3/4 inch) as specified. Therefore, to get the desired thickness two passes are required. Each course is 0.952 cm (3/8 inch) thick with the first pass being untrowelled.

To be assured the first pass will adhere to the prepared surface, which is mechanically- cleaned steel, a material is painted on the pipe to accelerate the setting of the mortar at the point and time when they come in contact. The second pass does not need this painting to help with the adhesion because the untrowelled mortar surface provides sufficient assistance for adhesion.

In order to apply an exact thickness of 3/8" continuously, many features of the operation must work together without fail. The extrusion of the application head is determined by measurement in cubic feet per minute, which is constant. The most exacting adjustment is the travel speed of the "lining machine." This speed is adjusted with variable speed transmissions measured in inches per minute. In this manner the desired thickness is applied at the pipe surface.

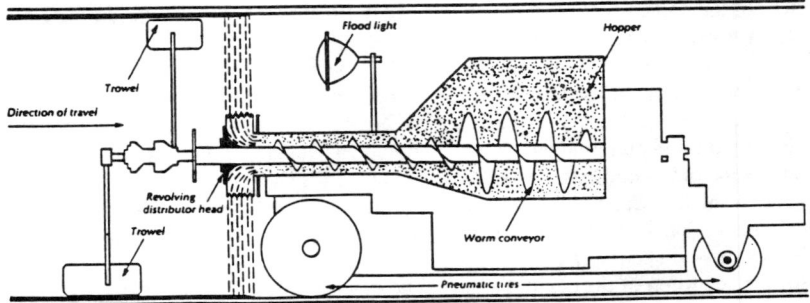

Fig. 2: Longitudinal Section of Lining Machine

Another exacting feature of the application is the quality of the finished surface. This is described in the specifications as a steel trowelled surface without any imperfections. Under the conditions of the application all trowelling must take place within one minute of the placement of the surface to be trowelled. To accomplish this, the surface texture, mortar consistency, speed and width of the trowels all must be adjusted and remain constant. The adjustment of the trowels can be done by varying the widths and tension of the steel finishers. The adjustments must take into consideration the applied pressure of finishers at four locations around the surface of the pipe: the upstroke on one side at the spring line, the down stroke on the other side at the spring line, the horizontal stroke at the invert and the horizontal stroke at the crown.

Any malfunction of the trowels that engage the fresh mortar would cause a disbandment and a large portion of the lining would come down.

Once these exacting adjustments to elements of the application have been made, the balance of the operation consists of providing the desired mortar in the correct consistency and amount at the lining machine.

To consistently providing the proper amount to the applicator, the mortar must be transported from the ground level, which is outside the pipe, in one-yard batches, and then transported distances up to 152 m - 183 m (500 or 600 feet). (See Fig. 3.) This is done by a machine called the "travel buggy." Despite its quaint name, this machine is substantial, especially when you realize it can carry 1361 kg (3000 pounds) of mortar 152 m (500 feet) in one minute. From the travel buggy, the mortar is then discharged into a feeder which deposits mortar at a consistent rate onto an elevator belt. The belt elevates the mortar approximately 3.66 m (12 feet) where it is released into a dispensing hopper of the lining machine. From there the mortar is screwed into the applicator head which is rotating at about 1800 rpm. The mortar is then projected to the pipe wall.

Fig. 3: Adjusting the Trowels

It takes approximately 20 minutes from the time the mortar leaves the mixer to the time the mortar is on the pipe surface. During this time the mortar receives two additional mixings. This additional working of the material improves the plasticity and dispels some of the heat of hydration, all of which prevent curing shrinkage in the applied and trowelled mortar.

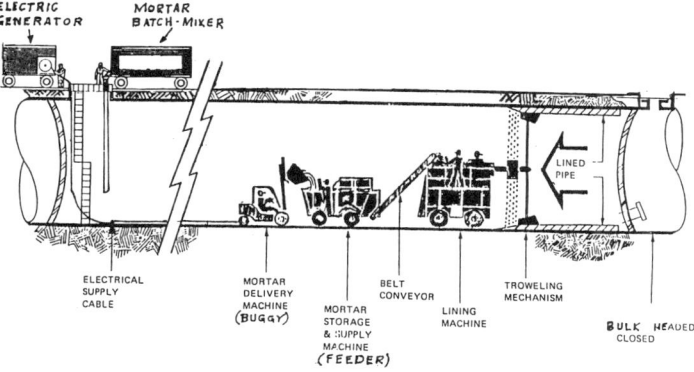

Fig. 4: The Continuous Operation

The mixing of the mortar is just as critical a process as all the others. The 1:1 mixture of sand and cement is measured by volume -- one cubic foot of cement to one cubic foot of sand. The water is added and the plasticity is adjusted through water control by the mixer man on instruction from the lining machine operator. The water-cement ratio is very close to .48. The actual amount of water used in the mix will vary during the day, due to atmospheric conditions inside and outside the pipe. If a miscount is made during the batching of the sacks of sand and cement the operator at the lining machine can usually detect the difference.

The mix is very rich in cement, which means there is a high tendency for heat of hydration and shrinkage. Therefore, the ambient temperature and humidity inside the pipe during application should be kept constant and within set limits. In this case air circulation should be kept to a minimum so as to prevent the air from absorbing detrimental amounts of the mixing water.

After an application has been placed to a stretch of pipe, and again at the end of the day, the pipe should be closed to a point where most of the water used in the mix stays in the pipe and therefore in the applied material.

Watching this application of the cement-mortar lining, one gains a real sense of appreciation for operation. First, to see the mortar adhering to the ceiling of the pipe some 5.49 m (18 feet) overhead gives no small degree of wonder. Then, immediately after adhesion it is trowelled to a smooth finish. There is no other method that can project mortar much more than one or two feet, let alone make it stick, apply it to a uniform thickness in both the vertical and horizontal directions simultaneously, and then trowell it to a smooth surface.

However marvelous the procedure, one can't help but ask: Will this material stay as placed for a long period of time, say, another 50 years? The answer: Yes, it can be expected to stay. This application procedure has been used on all sizes of pipe for over 50 years, and to the best knowledge of the industry all instances of pipes where cement-mortar lining has been applied in-place have remained operational as long as they were required to do so. But there is more convincing proof of this lining's serviceability than just an unfailing track record.

MORTAR CHARACTERISTICS

Each aspect of the process encompasses a complex set of principles. Therefore, in order for all the elements to work together successfully, a basic understanding of the actions and reactions of each element in the process must be achieved. Only then can the process become a system of checks and balances, where each step accommodates deficiencies in, or stabilizes variances of some other step.

The Portland cement has the most complex set of characteristics. Some of these can be detrimental to the overall quality of the mixture that originally intended to use Portland cement as an ingredient for a binder. Portland cement first of all is a very fine particle material. When it chemically reacts with water it can cause such things as heat or shrinkage. These reactions in turn effect plasticity, workability and curing, at various times from the moment the mortar is mixed through the period when it has reached its ultimate hardness.

This mortar-lining process recognizes and accommodate these special characteristics, using them to full advantage. With a very rich 1:1 sand to cement mix, these characteristics become acute.

Originally this 1:1 sand and cement mix was for adhesion to the pipe wall. But, this rich a mix caused problems such as high heat of hydration, resulting in severe curing shrinkage and false and flash sets. Changing the grind of the cement was the answer. These problems were overcome by periodic churning of the mortar. The sand necessary for the mix had to be well graded and completely pass through the #16 screen. If not well graded it does not perform as desired when adhesion is needed, and the trowelling finish is affected. Also, coarser grains cause rebound of the larger particles when they come in contact with the pipe surface.

The equipment used to mix this material is another crucial factor. A <u>concrete mixer</u> is not efficient because proper mixing depends upon coarse aggregate. A <u>plaster mixer</u> can properly mix two fine materials. But a <u>turbine mixer</u> is the only machine that can properly mix large amounts of fine material in a short time. It has been determined a concrete mixer takes at least three minutes to effectively stir a concrete mix, but a turbine mixer can correctly mix the ingredients in only one minute. A continuous mixer has been tried extensively, but has not yet become a favored method of mixing.

In the past 50 years there have been many admixtures for Portland cement mixes. Every one has its own advantageous contribution depending on the desires of the user. The main desire, though, is for additional strength of the Portland cement mixture. Also, to improve the mix for workability, for proper and easy placement and to make a better concrete at a lower cost.

All admixtures, as they came on the market, have been tried by the lining contractors. Several lasted for a number of years, usually until they were no longer available on the market. None of the additives' improvements carried through to the end product, though. Several added to the workability and adhesiveness, but none improved the cost of the end product.

This "rich" 1:1 mix has proven over the years to effectively protect a steel pipe's internal surface from deteriorating. To determine the thickness of the lining, consider how thick the lining must be within a cross sectional area in order to transmit the expansive forces to all parts of the circumference. The present thicknesses are determined by experience, and should take into account the practicality of placement and cost.

Two features of a cement-mortar lining that are not generally known by all people are: "A rich cement mortar, under water and pressure will expand over and above its original placed volume," and "when a pipeline is drained and any air is circulated through a pipeline that has been cement-mortar lined the lining will shrink in all directions substantially." (See Ref. 1.) These two principles are able to take place because there is no permanent adhesion of the mortar lining with the steel surface. This expansion and shrinkage are volumetric, and have nothing to do with the curing or chemical reactions of the Portland cement.

The volumetric expansion and contraction of the mortar will continue to take place over the life of the material. Similar to the compressive strength, the rate of the expansion and shrinkage will diminish during the years of continuous exposure to water under pressure. The basic reason for the reduction of expansion and shrinkage and the increase of compressive strength is because the free calcium liberated in the chemical reaction will become calcium carbonate. This calcium carbonate fills minute voids and cracks and hardens, sealing the porosity and any open areas between cracks that have swollen shut. This action of using calcium carbonate to fill in voids is referred to as "autogenious healing." (See Ref. 2.)

Therefore, armed with the longtime reliability of post-application over many years, and with a basic understanding of the actions and reactions of a relatively thin application of very rich mortar coating on the inside of a circular surface that transports potable water under pressure, you can explain many of the unusual features of this project.

With an understanding of these unusual and varied conditions, confidence about the ultimate serviceability of the coating protecting the steel pipe's internal surface can be obtained.

CONCLUSION

This project is a great example of placing a cement-mortar lining considering the extreme conditions of the application process. It also demonstrates that basic chemical and physical principles which have been observed for many years in much smaller pipelines still hold true even in pipelines of very large structures.

REFERENCES

(1) Lea, F.M. "The Chemistry of Cement and Concrete," p. 406, 3rd ed.

(2) Pekworth, H.F. Concrete Pipe Handbook, p. 435, 1959.

Evaluation of Prestressed Concrete Cylinder Pipe
in a High Chloride Environment
after 19 Years of Service

Jose L. Villalobos, P.E.[1]

ABSTRACT

A 60-Inch Diameter Prestressed Concrete Cylinder Pipe was evaluated in terms of corrosion after 19 years of service in a high chloride environment. The soils around the pipe were tested and found to contain over 330 parts per million (ppm) of chloride. The pipe was excavated at three locations and a 12-inch by 12-inch window was cut into the mortar coating at each location to observe the prestressing wire. Mortar samples were tested to determine the chloride concentration of the mortar at the surface, middle, and at the wire surface. The prestressing wires were found to be free of corrosion.

The paper presents the results of the investigation and conclusions relative to the lack of corrosion on the prestressing wires.

INTRODUCTION

On October 27, 1994, the author witnessed removal of the mortar coating by Gifford Hill American (GHA) on the existing Coastal Water Authority (CWA), 60-inch, A1, Prestressed Concrete Cylinder Pipe (PCCP). The purpose of this investigation was to determine if the PCCP prestressing wires had been affected by elevated chloride levels measured in the soil during the preparation of the Corrosion Investigation Report, dated April, 1994. The examination of the condition of the existing pipeline would then be used as a basis for design of the proposed pipeline segments and for determining the corrosion control requirements for the proposed A2 North and South Pipelines.

The existing CWA 60-inch A1 Pipeline has been in place for approximately 19 years. Excavation and testing was completed of the pipe to determine whether elevated chloride levels, found in the soils at various locations along the alignment, have had any deleterious effects on the mortar coated steel prestressing wires.

---
[1] Villalobos, Jose L. President. V & A Consulting Engineers. 1999 Harrison Street, Suite 975, Oakland, CA, 94612.

Stations 111+00, 179+00, and 199+00 were selected for excavation, based upon the elevated chloride levels found in these areas during previous testing.

From the results of the testing, it has been determined that corrosion activity has not caused degradation of the existing pipeline in the areas examined during this study. However, the A1 Pipeline was cathodically protected until 1990. In addition, since the de-energization of the cathodic protection system, elevated potentials have existed along segments of the A1 Pipeline. The elevated potentials are believed to have provided continued protection to some sections of the line and have exposed other areas to corrosion from stray current.

SAMPLE COLLECTION

CWA personnel excavated the 60-inch pipe at the three selected locations. Groundwater was located at approximately the crown of the pipe. The native soil was relatively dry and consisted of brown and dark grey to black clay soil. The pipe was found to be embedded in a sand envelope. Both mortar and soil samples were collected at each site. In addition to the samples obtained from destructive testing of the mortar coating on the pipe, one cast mortar section was fabricated by GHA and cut into samples for inclusion in the testing program. One sample was sent to the author and one sample was retained by GHA. The objective of testing the fabricated sample was to compare the test results on the 19-year old mortar with that of the fabricated sample in terms of meeting updated standards for mortar coating on prestressed concrete pipe. Two mortar samples from each excavation site were tested for absorption and compressive strength by Terra-Mar Inc. of Fort Worth, Texas, who was retained by GHA. Mortar samples from each site were also tested for water soluble chlorides, cement content, and petrographic examination by Schwein-Christensen Laboratories (SCL) of Lafayette, California. Two mortar samples from each site were cut into three layers and each layer was analyzed for a total of 18 water soluble chloride tests. Three mortar samples taken from the pipe were tested for cement content, along with the fabricated mortar, and petrographic analysis was completed on three samples removed during excavation as well as on the fabricated sample. The results of the testing are shown in Table 1. The GHA fabricated sample sent to the author is labeled, "Reference 1" in Table 1. It is recognized that the cement content testing resulted in very conservative numbers. This was done in order to ascertain the worst case results.

Three soil samples were collected at each excavation site. One sample was native material, another sample was taken from the sand envelope located on the west side of the pipe, and the other sample was sand taken from directly over the pipe. Of the nine soil samples collected, three were selected for chemical analysis. The analysis included determination of concentrations of chlorides, sulfates, pH, bicarbonates and resistivity. The testing was performed by Sequoia Analytical Laboratories of Redwood City, California. The data is shown in Table 2.

Electrical potential measurements were taken at each of the three excavation sites to determine the potential at the pipe surface.

TESTING PROCEDURES AND OBSERVATIONS

Station 199+00

After the soil samples were collected, GHA removed the mortar from a 12-inch by 12-inch "window" to expose the pipe at the level of the prestressing wires. At this location, 11 fragments of mortar were collected. The mortar was removed using a 10-pound chipping hammer. After collecting the 11 fragments of mortar, labeled Samples No. 1 through No. 11, the window was cleaned to facilitate more detailed visual inspection of the prestressing wires and the surrounding mortar. The mortar in immediate contact with the prestressing wire and the concrete core did not show any signs of corrosion. Observations made of the cleaned window area revealed clean wires with no evidence of corrosion of the wires. A detailed observation of the wire with the unaided eye did not reveal any cracks, splits or other damage to the wires.

Potential measurements were made using a copper-copper sulfate reference cell, a Fluke Model 8062A voltmeter, and wire connectors. A potential of -747 millivolts was measured, versus a copper copper sulfate reference electrode. This value is consistent with potential measurements taken in this area during previous testing.

At the completion of the testing, the window area was rinsed with fresh water and the 12-inch by 12-inch window area was repaired by GHA. GHA used Cel Crete to replace the mortar removed during observation.

A similar procedure was followed for the excavations at Stations 179+00 and 111+00. The following summarizes the findings at the excavation sites:

Station 179+00

Groundwater was located to a depth approximately equal to the crown of the pipe.

Three soil samples were collected as described for Station 199+00.

Eleven fragments of mortar were collected from the 12-inch by 12-inch window area and labeled Samples No. 1 through No. 11.

The pipe-to-soil potential measured at this site was -859 millivolts.

GHA replaced the mortar at the observation window area with Cel Crete.

Station 111+00

Groundwater was located at a depth approximately equal to the crown of the pipe.

Three soil samples were collected as described for Station 199+00.

Ten fragments of mortar were collected from the 12-inch by 12-inch window area and labeled Samples No. 1 through No. 10.

The pipe-to-soil potential measured at this site was -877 millivolts.

GHA replaced the mortar at the observation window area with Cel Crete.

ANALYSIS OF MORTAR SAMPLES:

From all of the mortar samples obtained during the investigation, samples were selected for compression and absorption testing, to be performed by GHA, and the remainder of the samples were included in the petrographic and chloride analysis. The following samples were given to GHA for the compression and absorption tests:

Station No.	199+00	Samples No. 7 and No. 9
Station No.	179+00	Samples No. 7 and No. 6
Station No.	111+00	Samples No. 3 and No. 4

As discussed previously, a mortar sample, which was intended to be representative of the mortar quality which would be expected in new pipe, was also fabricated by GHA. The sample was tested in accordance with the procedures used for those samples removed from the pipe. The mortar samples were cut into 0.5 inch square units and tested for compressive strength. The results of the analysis are shown in Table 1.

ANALYSIS OF SOIL SAMPLES:

The nine soil samples obtained at the excavation sites were analyzed for resistivity in both the as-received and saturated form for determination of the minimum resistivity. Three samples with the lowest resistivity values at each excavation site were sent to Sequoia Analytical Laboratory for chemical analysis. The results of the testing are shown in Table 2.

EFFECT OF ELEVATED CHLORIDE LEVELS ON PRESTRESSING WIRE

Based on the results of the chloride testing on the mortar samples, it appears that the prestressing wires at Stations 111+00 and 179+00 are exposed to chloride levels ranging from 42 to 639 ppm or 0.021% to 0.320% by weight of cement.

The chloride concentrations vary in terms of location from the surface of the pipe exposed to the soil, to a midpoint on the mortar coating, and at a point where the mortar coating meets the prestressing wire, as shown in Table 3. Corrosion of prestressing wire can occur under certain conditions based on these chloride levels. In order for corrosion to occur, the pipe must be subject to wetting and drying cycles. This condition does not exist along the proposed A2 alignment because the pipe is under the water table the majority of the time. However, penetration of chlorides into a mortar matrix does not necessarily destroy the passivity of steel in concrete. Pourbaix[2] prepared a diagram which illustrates the behavior of steel exposed to chlorides as shown on Figure 1.

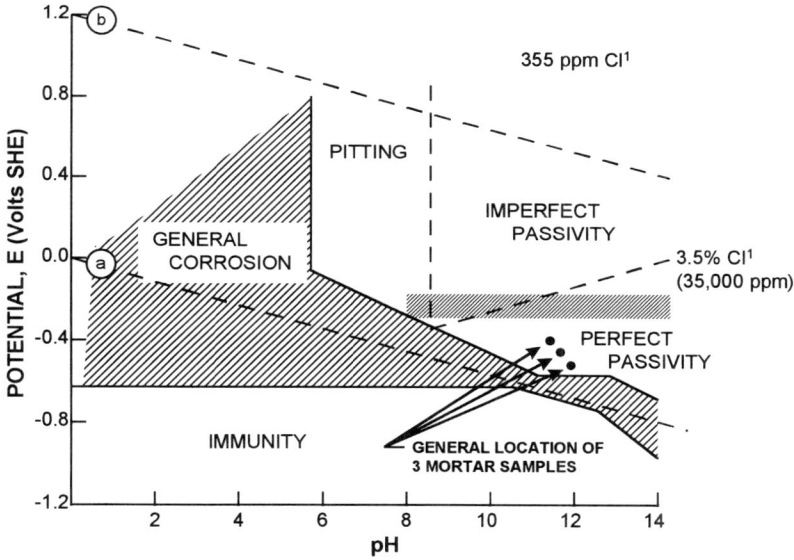

FIGURE 1.   INFLUENCE OF CHLORIDE CONCENTRATION ON CORROSION AND PASSIVATION OF IRON
(FROM M. POURBAIX, CORROSION SCIENCE, PAGE 14, JANUARY 25, 1994)

Pourbaix[2] found that even at high chloride concentrations, a zone of perfect passivity remains if steel in concrete is cathodically protected to a polarized potential more negative than -600 millivolts verses a Standard Hydrogen Electrode (SHE). It must be pointed out that the work of Pourbaix[2] is referenced for illustrative purposes, as the work of Pourbaix[2] was carried out in a laboratory in a controlled aqueous solution. For this paper, an extrapolation was made from the laboratory testing in an aqueous solution to a buried soil condition. The mortar samples collected at station 111+00, 179+00, and 199+00, would roughly fall at the locations shown in Figure 1. It is important to note that chlorides are not aggressive to steel unless oxygen is present. For example, chlorides in contact with the steel, in the absence of oxygen, will not lead to the types of failures which result from hydrogen evolution from high strength steel. The use of properly operated cathodic protection systems for buried pipe applications will provide protection of the steel in the presence of both chlorides and oxygen.

In a study completed by Edward Escalante and Satoshi Ito, entitled "Measuring the Rate of Corrosion of Steel in Concrete," published by ASTM in <u>Corrosion Rates of Steel in Concrete STP 1065</u> in 1990[3], the relationship between chlorides, pH and oxygen were studied to determine their impact on corrosion. The investigators found that cement mortar coated steel, which was continuously submerged, did not corrode when compared to cement mortar coated steel which was subject to cyclic wetting and drying.

In additional testing performed by Escalante[3], no corrosion was observed when cement mortar coated steel samples were immersed in a calcium hydroxide solution to which various amounts of sodium chloride were added. Oxygen levels and pH were also tested at various levels. When the pH was above 12, the oxygen concentration was below 6 parts per million, and the chloride concentration was less than approximately 360 mg/l, no corrosion occurred.

SUMMARY OF FINDINGS:

1. Chlorides have penetrated the mortar to the level of the prestressing wires in the areas where chloride concentrations in the soil samples are in the range of 53 ppm to 333 ppm.

2. Based upon review of the recommendations provided in AWWA C301-92 and ACI 318-83, the threshold value for corrosion of prestressing wire in concrete, or 0.06% water soluble chlorides by weight of cement, is exceeded at two of the three excavation sites.

3. Prestressing wires were found to be at potential values much more negative than the -200 millivolts common for steel in concrete. Because the cathodic protection system for the prestressed concrete pipe was de-energized in 1990, any polarization of the A1 Pipeline should have decreased to native potentials during the 4-year period since the rectifiers were turned off. Although some amount of protection has been achieved along the A2 Pipeline in the areas where stray current is being picked up along the line, at some point the stray currents must leave the A1 Pipeline to return to the point of origin, as dictated by Ohms Law. The point of origin will most likely be one of the many cathodically protected pipelines along the A1 right-of-way.

4. No corrosion was observed at the three excavation sites on the prestressing wires or on the mortar in contact with the wires.

5. No cracks, splits or other damage were evident on the prestressing wire when viewed in the field with the unaided eye.

6. The crown of the A1 Pipeline is at or below the groundwater level. The pipe is believed to be submerged in groundwater throughout the alignment.

7. The existing prestressing wire, No. 6 class three, will be similar to the prestressing wire to be used on the new PCCP.

TABLE 1
MORTAR ANALYSIS RESULTS

SAMPLE IDENTIFICATION	WATER SOLUBLE CHLORIDE (percent by wgt.)	WATER SOLUBLE CHLORIDE (ppm)	CEMENT CONTENT (sacks/yd^3)	COMPRESSIVE STRENGTH (avg. psi)	ABSORPTION (percent)
STATION 111+00 #1 SURFACE	.235	469			
STATION 111+00 #1 MIDDLE	.320	639			
STATION 111+00 #1 INTERIOR	.261	521			
STATION 111+00 #7 SURFACE	.252	503			
STATION 111+00 #7 MIDDLE	.280	559			
STATION 111+00 #7 INTERIOR	.243	485			
STATION 111+00 #3					6.8
STATION 111+00 #4				10,450	
STATION 111+00 #7			10.3		
STATION 179+00 #6 SURFACE	.073	146			
STATION 179+00 #6 MIDDLE	.075	150			
STATION 179+00 #6 INTERIOR	.065	130			
STATION 179+00 #10 SURFACE	.053	106			
STATION 179+00 #10 MIDDLE	.029	58			
STATION 179+00 #10 INTERIOR	.021	42			
STATION 179+00 #7					7.6
STATION 179+00 #9				5,880	
STATION 179+00 #10			8.6		
STATION 199+00 #6 SURFACE	.020	40			
STATION 199+00 #6 MIDDLE	.020	40			
STATION 199+00 #6 INTERIOR	.012	24			
STATION 199+00 #8 SURFACE	.022	44			
STATION 199+00 #8 MIDDLE	.015	30			
STATION 199+00 #8 INTERIOR	.007	14			
STATION 199+00 #5					7.9
STATION 199+00 #7				4,410	
STATION 199+00 #8			8.3		
REFERENCE #1			11.8	7,510	9.0

TABLE 2
SOIL RESISTIVITY DATA & CHEMICAL DATA

No.	Station	Location	Soil Resistivity As Received ohm-cm	Minimum. ohm-cm	pH	Sulfate (ppm)	Chloride (ppm)	Alkalinity: Bicarbonate (ppm)
1	111+00	Native	3,400	1,500	8.9	16	53	1,300
2	111+00	Top of Pipe	6,200	6,200				
3	111+00	Side of Pipe	4,600	4,600				
4	179+00	Native	750	620	8.3	27	250	3,400
5	179+00	Top of Pipe	3,200	3,200				
6	179+00	Side of Pipe	2,200	2,200				
7	199+00	Native	590	590	8.6	11	120	13,000
8	199+00	Top of Pipe	2,600	2,600				
9	199+00	Side of Pipe	1,200	1,200				
AVE			2,749	2,523	8.6	18	141	5,900
MIN			590	590	8.3	11	53	1,300
MAX			6,200	6,200	8.9	27	250	13,000

TABLE 3
CHLORIDE CONCENTRATION (PPM) vs. LOCATION ON MORTAR COATING

Mortar Location	Station 111 + 00		Station 179 + 00		Station 199 + 00	
	Sample 1	Sample 2	Sample 1	Sample 2	Sample 1	Sample 2
Surface	469	503	146	106	40	44
Middle	639	559	150	58	40	30
Wire	521	485	130	42	24	14
Interface						
Soil Sample 1	333		171		76	
Soil Sample 2	53		250		120	

NOTE: SOIL DATA FROM SOIL BORINGS AND EXCAVATION SITES

List of References

[2] Pourbaix, M. "Influence of Chloride Concentration on Corrosion and Passivation of Iron." Corrosion Science. Page 14. January 26, 1994.

[3] Escalante, Edward and Ito, Satoshi. "Measuring the Rate of Corrosion of Steel in Concrete." Corrosion Rates of Steel in Concrete STP 1065. 1990.

Effects of Environment on the
Durability of Prestressed Concrete Cylinder Pipe

Robert E. Price, P.E.[1]
Richard A. Lewis, P.E.[2]
Bernard Erlin[3]

Abstract

Durability of prestressed concrete cylinder pipe is dependent on the ability to resist deleterious environmental conditions without prestress wire degradation. Mechanisms involved in prestress wire deterioration and failure often include a combination of corrosion initiated by depassivating agents such as chloride ions, hydrogen embrittlement, and alteration of the protective portland cement mortar coating through carbonation or other means. In order for the prestress wire to degrade, the protective environment provided by the mortar coating must first be penetrated or compromised by aggressive exposure conditions. The design or evaluation of prestressed pipelines must consider the effect of environment on an individual basis.

Introduction

Although only a small percentage of the 28,000 or more miles of prestressed concrete cylinder pipes in service have failed, many of the failures have been sudden and catastrophic resulting in damage as loss of service, and in one case causing the death of several people. Rupture of the pipes usually results from external corrosion and progressive breakage of prestress wires until instability of the pipe under the internal pressure and imposed loads is reached. The burst is often explosive with a large flow of water that can carry shards of concrete and overburden materials hundreds of feet away from the failure site. The cause of these ruptures can frequently be traced to the effect of the environment on the materials used in manufacture of the pipe.

---

[1]President, Openaka Corporation, 565 Openaki Road, Denville, NJ 07834-9642

[2]Vice-President, Openaka Corporation, 565 Openaki Road, Denville, NJ 07834-9642

[3]President, The Erlin Company, 1693 Clearview Drive, Latrobe, PA 15650

Hydrogen Embrittlement

The conclusions reached by many failure investigators have emphasized the susceptibility of the high strength prestressing wires to corrosion and embrittlement. It is commonly noted that embrittlement had reduced the strength of the wires to the point where they failed at less than the design tensile strength without ductile reduction in area or loss of cross section by dissolution. Although the embrittled prestressing wires fail at loads that are only 60% to 65% of their rated minimum ultimate strength, the nature and processes involved in embrittlement are not completely understood. Most authorities, however, believe the loss in strength is caused by the absorption of atomic hydrogen into the steel.

There are several known poisons, which can induce hydrogen into high strength steels. Among these are hydrogen sulfide and arsenic compounds. The presence of these compounds at failure sites is not common, although hydrogen sulfide may be encountered in semitropical areas. The more usual source of hydrogen in buried pipelines, however, is thought to be the corrosion process itself. The corrosion of prestress wire in a mortar coating is under cathodic control. The oxygen available to combine with hydrogen at the cathode governs the rate and continuity of the corrosion cell. When the supply of oxygen is exhausted polarization takes place. The hydrogen produced at the anode plates out on the wire thereby insulating it and stopping the flow of corrosion current. The corrosion of steel in the absence of oxygen produces magnetite (Fe_3O_4) which is often observed to be present in the vicinity of freshly uncovered brittle failures.

Attempts have been made to correlate the quantity and rate of hydrogen absorption with the time to failure of the wires. The most commonly used test method is to expose the wire to Ammonium Thiocyanate. While this test indicates the relative susceptibility of the wire to hydrogen it does not differentiate between absorbed, adsorbed, or trapped hydrogen. Furthermore, the test environment bears little if any relation to the actual conditions surrounding a buried pipe. It is not surprising, therefore, that although the test results show greater susceptibility with higher strengths, failures involving embrittlement have been observed in all classes of wire used in prestressing pipes.

Prestressing Wire Environment

The environment, which surrounds the prestressing wires, consists of a portland cement mortar that provides protection from corrosion or embrittlement by virtue of the alkalinity of the cement paste and the ability of the mortar matrix to avert intrusion of depassivating agents and poisons. This protection is entirely dependent on the efficacy of the mortar coating to maintain a sufficiently stable environment around the wires to inhibit corrosion. Petrographic examinations have shown, however, that the wires may not be completely enveloped even though a twofold manufacturing procedure applies a neat portland cement slurry prior to application of the mortar. Inadequately encased wires with textural deficiencies in the mortar can lead to corrosion when the pipes are exposed to otherwise acceptable environments. The extent and rapidity of the corrosion are a function the quality of the mortar coating with regard to alteration and the volume of permeable voids as well as the aggressivity of the environment.

Examinations of failed pipes have shown that the most common "natural" aggressive materials from the outside environment are acids (such as carbonic and humic), chloride, sulfate, and pure water with low dissolved solids. Their effects can result in the gradual deterioration of the mortar coating through chemical transformations that disrupt the cementitious matrix and/or lower the pH of the coating around the wires. In the case of chlorides, the action is a breaching of the passivating film that normally inhibits corrosion of the wires. Industrial wastes and fertilizers can contribute a large variety of other aggressive chemicals.

The Corrosion Process

Steel embedded in concrete or mortar is passivated by encasement in a highly alkaline cement paste. Passivation is attributed to the formation of a thin film of gamma ferric oxide on the metal surface by the hydroxyl ions produced in the hydrating portland cement (Scott, 1965). It is generally accepted that local corrosion, including pitting, crevice, and stress corrosion, occurs only on passivated metals when the film is damaged or compromised in some manner (Szklarska-Smialowska, 1977). When the metal is exposed, either dissolution takes place or the surface is again passivated by reestablishment of the film (Lewelyn, 1977). Whether the metal corrodes or retains its passivation depends on the competitive reaction with hydroxyl ions and aggressive anions.

Weakening or disruption of the passivating film in a metal under tensile stress takes place either at inhomogeneities inherent in the matrix of the metal or at slip steps, which form in the metal under tensile loads. It is known that the presence of aggressive ions such as chloride destroys the film and initiates localized corrosion, but how this occurs is not agreed upon and may involve several mechanisms.

Stress corrosion cracking and hydrogen embrittlement are generally defined as environmentally induced degradations of strength resulting in brittle fracture of metals under stress. It is thought that stress corrosion cracking is caused by the dissolution of slip steps, which nucleate cracks. The process of stress corrosion cracking has been hypothesized as crack propagation in anodic regions which may be assisted by hydrogen evolution in acid conditions at the crack tip (Gulbransen, 1977). While the presence of hydrogen at the crack tip is accepted, it is generally not considered significant relative to the cracking (Staehle, 1977). Hydrogen embrittlement on the other hand is considered as a result of hydrogen absorption from the electrolyte in cathodic areas. Figure 1 shows the Pourbaix diagram for iron. The ordinate shows a range of steel potentials in volts with respect to a standard hydrogen electrode while the abscissa consists of a range of pH values. From Figure 1 a determination can be made of whether corrosion of the steel will occur dependent upon within which of the regions defined by the solid heavy lines the pH and electrical potential falls. Below the lower heavy broken line hydrogen evolution accompanies corrosion. In order for hydrogen to be evolved the pH must be lowered to within the hatched region labeled 'A'. A lowering of pH takes place during corrosive attack at micro cracks, crevices, and pits which develop on the metal surface as a result of dissolution initiated by chloride or other similar acting agents (Kreijer, 1977). Neither the process of stress corrosion cracking nor that of hydrogen embrittlement is completely understood and the distinction between them is far from clear.

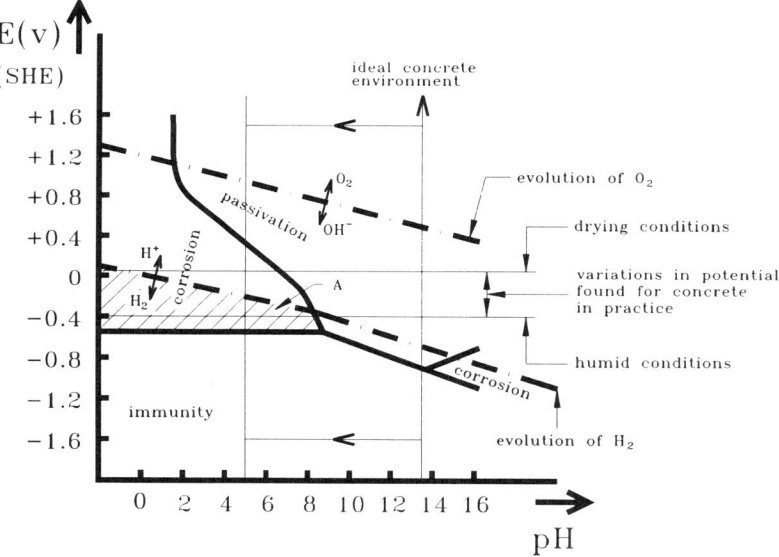

Figure 1. E-pH Diagram for Iron/Water at 25° C, According to Pourbaix.

The basic principles and complexities associated with prestress wire corrosion in mortar are illustrated in Figure 2, a schematic diagram showing the electrochemical reactions involved in a corrosion cell (Scott, 1962). It can be seen that an essential element in the corrosion process is the availability of oxygen at the cathodic area. Porosity of the mortar, which permits infiltration of air or diffusion of dissolved oxygen in water, is an important factor necessary for corrosion to initiate and continue.

In cathodic areas at potentials associated with moist concrete, reduction of hydrogen ions takes place at the cathode resulting in either atomic hydrogen that can be absorbed into the metal matrix or the evolution of hydrogen gas. These reactions occur only under acidic conditions. On the other hand, in an alkaline environment the cathodic

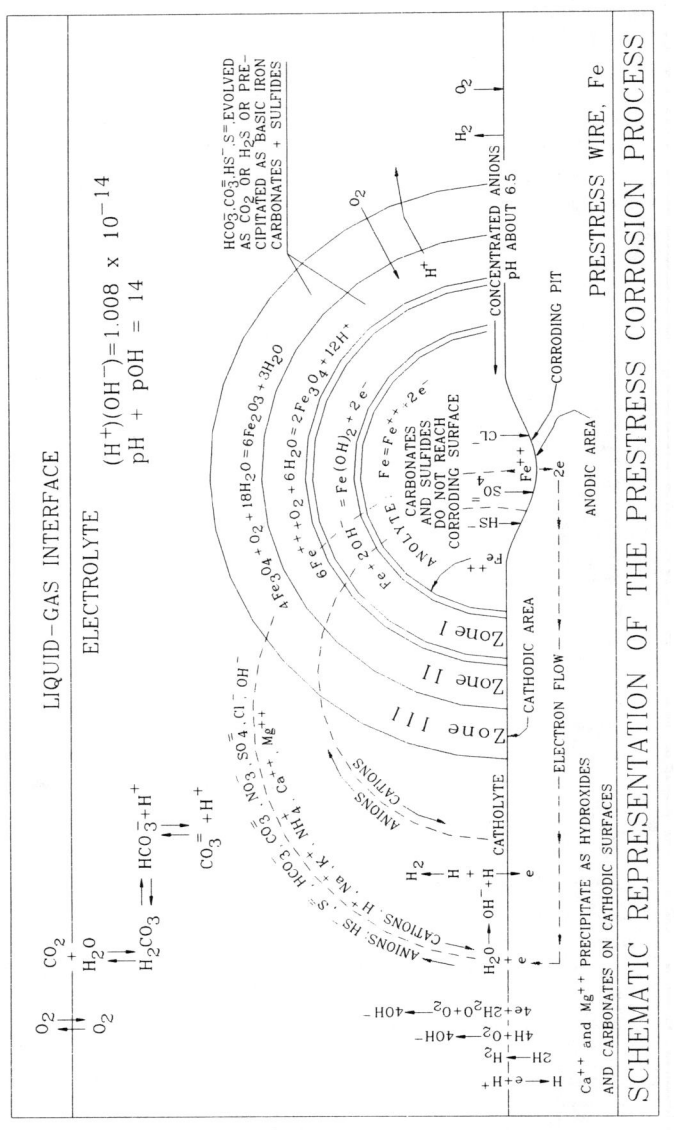

Figure 2. Prestress Wire Corrosion in Portland Cement Mortar

reaction is the reduction of oxygen producing hydroxyl ions. In view of the cathodic reactions involved, hydrogen embrittlement is unlikely unless there has been some local alteration of the alkaline environment.

In both stress corrosion cracking and hydrogen embrittlement the source of the hydrogen is corrosion activity. It can be concluded, therefore, that in the absence of specific poisons corrosion must precede embrittlement. In either case, corrosion cells, once initiated, can result in degradation of wire strength and subsequent brittle failure before significant amounts of dissolution take place.

Depassivating Agents

Prestress wire surrounded by portland cement paste is normally in an environment (pH≈13) where the corrosion rate is essentially zero even in the presence of moisture and oxygen while the passivating film remains intact. Chloride and sulfate, however, act as depassivating agents to initiate and accelerate corrosion when present in sufficient amounts. Unlike the available oxygen and water, which enter into the corrosion reactions, the chloride and sulfate ions are not consumed, but serve as initiators and intermediate products in the corrosion processes.

Chloride ions are pervasively mobile and probably the most prevalent depassivating agent affecting embedded steel in concrete or mortar. Although a small amount of the chloride ion will react with constituents of the portland cement paste to form chloroaluminates, significant amounts often remain available to participate in the corrosion process (Diamond, 1986). It is generally accepted that there is a minimum chloride content necessary to initiate corrosion in concrete or mortar. This minimum has been determined to be a function of the Cl^-/OH^- activity ratio. Hausmann established this ratio to be approximately 0.6 in a calcium hydroxide solution (pH 12.5) with free oxygen (Hausmann, 1967). The pore solutions in a concrete or mortar, however, usually have a somewhat higher pH due to the presence of sodium and potassium hydroxides. Investigations by Gouda and analysis by Diamond indicated that the onset of depassivation varies from a Cl^-/OH^- ratio of 0.57 at pH 11.8 to 0.30 at pH 13.3 (Diamond, 1986), (Gouda, 1970). These values suggest that at the pH level found in cement pastes the criteria determined by Hausmann might not be conservative with respect to depassivation of steel. Nevertheless, there is approximate agreement between the tests showing that the concentration of chloride necessary to initiate corrosion is much reduced in concretes and mortars that have been in service due to a lower pH. Older concretes tend to have pH values lower than the 13.0 to 13.5 typical of new concrete. The effect of pH reduction in the mortar on the susceptibility of embedded steel to chloride initiated corrosion is illustrated in Figure 3. It is apparent that corrosion can be initiated by even trace amounts of chloride when the pH is reduced as low as 9 to 10.

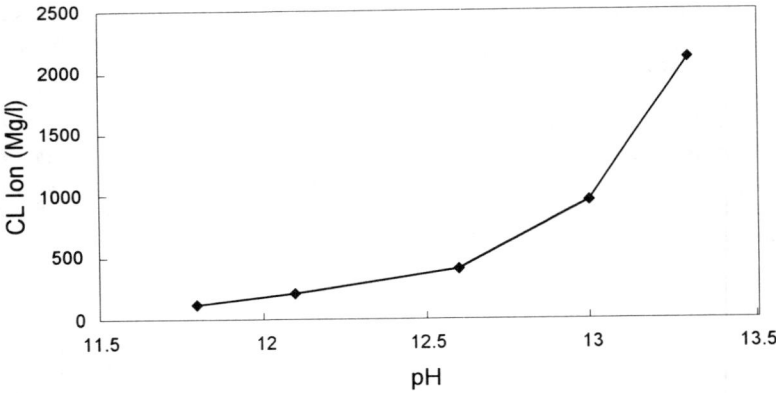

Figure 3. Cl⁻ ion concentrations required to initiate steel corrosion versus pH.

Sulfates, which can also act as depassivating agents, react with the cement paste to form insoluble calcium sulfoaluminates and are generally not available in concrete and mortar to affect the corrosion process. The formation of calcium sulfoaluminate compounds is often expansive when high amounts of sulfate are present and can result in disruption of the portland cement mortar coating.

Carbonation of the Mortar Coating

Carbonation of the constituents of cement paste, including calcium silicates, calcium hydroxide, and the alkalis, occurs in concrete or mortar when they are exposed to CO_2 in the presence of moisture. Common carbonation reactions with the calcium hydroxide produced by the hydration of portland cement are expressed as follows:

$$Ca(OH)_2 + CO_2 \rightarrow CaCO_3 + H_2O$$
$$Ca(OH)_2 + Ca(HCO_3)_2 \rightarrow 2CaCO_3 + 2H_2O$$

The reactions with the sodium and potassium hydroxides are similar but those involving the calcium silicates are more complex.

The overall effect of carbonation is to reduce the pH, particularly in the permeable voids prevalent at the haunches of the prestress wire and in other porous areas of the concrete or mortar. It is known the degree of passivation is reduced and corrosion rates are increased as the pH is lowered independently of the presence of a depassivating agent (Uhlig, 1948). It has been variously reported that steel can no longer be considered passive when the pH of the surrounding environment falls to between 8 and 10. As can be seen in Figure 3, the reduction in pH by carbonation can initiate corrosion of embedded steel when even small amounts of chloride are present.

Corrosion of prestress wire is most likely to occur where permeable voids are directly adjacent to the steel and where there is access to oxygen, CO_2, moisture, and chloride. The size of the voids is inconsequential as long as they can be permeated.

The process of carbonation of the cement paste surrounding the voids takes place either by reaction with atmospheric CO_2 in the presence of moisture, dissolved CO_2 in rain and ground water (carbonic acid), or by bicarbonates with calcium hydroxide, alkalis, and calcium silicates. Complete carbonation of concrete or mortar to equilibrium by atmospheric CO_2 at normal air pressure results in a pH of slightly above 8 (Verbeck, 1958). If CO_2 exposure is continued in the presence of moisture, the $CaCO_3$ in solution is converted to the bicarbonate according to the reversible equation:

$$CaCO_3 + CO_2 + H_2O \rightleftharpoons Ca(HCO_3)_2$$

The pH of the resultant solution is likely to be in the range of 6 to 7 which is sufficient to completely compromise the passivity of the steel. Subsequent partial drying redeposits calcium carbonate but allows the ingress of air or replacement water either of which can contain additional CO_2 and oxygen. Thus movement of moisture and air and the resultant wetting and drying is conducive to corrosion of steel in a porous mortar matrix. The drying process in a portland cement mortar, however, can be protracted, particularly in a buried pipe coating, and is likely to require months or years. As a result, corrosion usually proceeds slowly, concentrating around the larger voids which, in the typical mortar coating, tend to be located on the underside of the prestress wires.

Not only does carbonation act to reduce the pH of the environment around the wire and make it more susceptible to corrosion, but it can possibly induce stress corrosion cracking of steels by the bicarbonate solutions produced (Corrosion Basics, 1984). As can be seen from the schematic diagram of the corrosion process, Figure 3, carbonic acid dissociates in two stages, first to form bicarbonate ions and then carbonate ions. Below a pH of about 8.5, however, the first stage producing bicarbonate ions predominates and the amount of carbonate ions is negligibly small. In the presence of weak acids such as H_2CO_3 (carbonic acid) hydrogen evolves at a pH of 6 (Uhlig, 1948). Brittle fractures of high strength steels have occurred under stress in carbonate/bicarbonate solutions in laboratory tests (Fessler, 1977).

Mortar Coating Quality

The availability of moisture and oxygen, the amount of depassivating agent present, and the alkalinity of the surrounding environment are governed by the ability of the mortar coating to restrict the ingress of deleterious elements. If the mortar is porous enough to permit the movement of liquids and gases to the metal surface corrosion can be eventually initiated.

Investigations of prestressed pipe failures have revealed various characteristics in the mortar coating which can facilitate the ingress of deleterious substances. Petrographic examination shows mortar coating porosity may vary considerably throughout samples. Areas of high porosity observed between wires at the mid-height were attributed to over spray which filled the space between the wires before the main stream of mortar was applied in the brush mortar process. Over spray can also cause the formation of low angle, relatively porous layers in the coating extending from the exterior to the wire level which have been noted to have facilitated the passage of deleterious substances in a number of failures.

Interface mortar between two adjacent wires of vertically coated pipes have shown high porosity on one side and low to moderate porosity on the other side. Gravitational force acting on the slurry applied prior to the mortar was found to contribute to the formation of voids at the wire haunches on vertically coated pipe. This condition can result from slurry flowing from under the lower haunch of the wire.

Where coating separations are found there is usually a scarcity of aggregate bond and insufficient paste to fill the interstices in the vicinity of the wires. Although in many cases it may appear that the separations caused by expansion of corrosion products from the wire, there is often evidence that considerable porosity may have existed within the mortar constituting radial separation of the outer portion of the coating along the weak plane of the wires.

Summary

Although only a exceedingly small percentage of prestressed pipes fail, the environment which includes the mortar coating is usually involved. It is axiomatic that failures rarely occur due to a single cause; a porous poor quality coating in a benign environment may provide satisfactory protection for the prestressing wires, whereas a good quality dense mortar coating in a highly aggressive environment may not. Successful construction of a prestressed concrete pipeline depends upon detailed investigation of the environment for deleterious conditions as well as adequate quality control of the exterior mortar coating.

In conclusion it is incumbent upon the design engineer to thoroughly investigate environmental conditions to guard against alteration or penetration of the mortar coating which might lead to prestressing wire failures. When investigating the cause of prestressed concrete cylinder pipe deterioration or failures, the engineer should look beyond the prestressing wires to the environment provided by the coating and how it has been affected by external conditions.

References

1. Corrosion Basics, National Association of Corrosion Engineers, Houston, Texas, 1984.
2. Diamond, S., "Chloride Concentrations in Concrete Pore Solutions Resulting from Calcium and Sodium Chloride Admixtures," Cement, Concrete, and Aggregates, CCAGDP, Vol. 8, No. 2, Winter 1986, pp. 97-107.
3. Fessler, R.R., Groeneveld, T.P., and Elsea, A.R., "Stress-Corrosion and Hydrogen-Stress Cracking in Buried Pipelines," Stress Corrosion Cracking and Hydrogen Embrittlement of Iron Base Alloys, NACE-5, National Association of Corrosion Engineers, 1977.
4. Gouda, V.K., "Corrosion and Corrosion Inhibition of Reinforcing Steel," British Corrosion Journal, Vol. 5, Sept., 1970, p.198.
5. Gulbransen, E.A., "A Model for the Stress Corrosion Cracking of Metals Based on the Morphology and Structure of Localized Corrosions Processes," Nace-5, Stress Corrosion Cracking and Hydrogen Embrittlement of Iron Based Alloys, National Association of Corrosion Engineers, 1977.
6. Hausmann, D. A., "Steel Corrosion in Concrete," Materials Protection, Vol. 6, No. 19, Nov., 1967.
7. Kreijer, P.C., Sluijter, W.L., Bergsma, F., Etienne, C.F., and Boon, J.W., "Stress Corrosion in Prestressing Steel," Heron, Vol. 22, No.1, 1977.
8. Lewelyn Leach, J.S., "The Possible Role of Surface Films in Stress Corrosion Cracking," Stress Corrosion Cracking and Hydrogen Embrittlement of Iron Base Alloys, NACE-5, National Association of Corrosion Engineers, 1977.
9. Scott, G.N.,"The Corrosion Inhibitive Properties of Cement Mortar Coatings," NACE Annual Conference, March, 1962.
10. Scott, G. N., "Corrosion Protection Properties of Portland Cement Concrete," Journal American Water Works Association, Vol 57, No. 8, August, 1965.
11. Staehle, R.W., "Predictions and Experimental Verification of the Slip Dissolution Model for Stress Corrosion Cracking of Low Strength Alloys," Stress Corrosion Cracking and Hydrogen Embrittlement of Iron Base Alloys, Nace-5, National Association of Corrosion Engineers, 1977.
12. Szklarska-Smialowska, Z., "Various forms of Localized Corrosion, Common Features and Differences," Stress Corrosion Cracking and Hydrogen Embrittlement of Iron Base Alloys, National Association of Corrosion Engineers, NACE-5, 1977.
13. Uhlig, H.H., The Corrosion Handbook, The Electrochemical Society, Inc., New York, 1948.
14. Verbeck, G.H., "Carbonation of Hydrated Portland Cement," Research Bulletin 87, Portland Cement Association, February, 1958.

REHABILITATION OF 183-cm PCCP WITH STEEL PLATE LINERS

Gary P. Stine, P.E., Member ASCE[1]
Michael T. Stift, P.E., Member ASCE[2]

Abstract

The solution for rehabilitating long reaches of deteriorated, large diameter, prestressed concrete cylinder pipe (PCCP) in the early 1980's, was replacement with another pipeline. Time, right-of-way issues, environmental considerations, cost, and other constraints prompted the San Diego County Water Authority (Authority) to develop a construction system to reline long reaches of large diameter PCCP in place with steel plate liners rather than constructing a new pipeline. Using this technology, the Authority and other agencies, have relined various deteriorating PCCP pipelines since the early 1980's. The most recent PCCP pipeline the Authority has rehabilitated using this method is 2.4 km (1.5 miles) of the City of San Diego's 183-cm (72-inch) PCCP Shepherd Canyon Pipeline. American Society for Testing and Materials (A.S.T.M.) A-572, Grade 42 steel coil, 5 m (16 feet) long with 8-mm (5/16-inch) plate thickness, was wound circumferentially, secured with steel bands, transported into the corroded PCCP using a special forklift and welded longitudinally and at each steel plate liner joint. The annular space between the new steel plate liner and PCCP was grouted and the interior was cement mortar lined using conventional lining placement methods to finish the inside of the rehabilitated pipe. Construction was completed in 1989 at a fraction of the time and cost to replace the deteriorating pipeline. The pipeline has operated without problems for the past nine years.

Introduction

In the early 1950's, a new rigid wall, large diameter pipe design called prestressed concrete cylinder pipe (PCCP) was introduced in the western United

---

[1] Principal Civil Engineer, San Diego County Water Authority, 610 W 5th Avenue, Escondido, CA 92025
[2] Assistant Director of Engineering, San Diego County Water Authority, 610 W 5th Avenue, Escondido, CA 92025

States. This design was a cost effective alternative to the traditional large diameter, flexible wall, steel and concrete pipelines. The design of PCCP combines the compressive strength of Portland cement concrete and the tensile strength of steel to make a rigid pressure pipe for transmission and distribution water lines which is easy and quick to install and has a high external load-carrying capacity. The introduction of PCCP resulted in significant construction cost savings and made it the pipe of choice for many water agencies throughout Southern California. Over the next 20 years, the Authority installed 132 km (82 mile) of PCCP pipelines.

In the early 1980's, the Authority had several catastrophic failures of PCCP along one of its seventeen year old pipelines. It was determined that, although the design of PCCP resulted in an extremely strong pipeline on paper, the pipe sections were susceptible to the highly corrosive soils (high chlorides and very low soil resistivities), occurring throughout San Diego County and were not functioning to their expected design life. This left the Authority and other San Diego water agencies with the task of having to replace deteriorating PCCP pipelines. However, schedule, right-of-way issues, environmental considerations, cost and other constraints prompted the Authority in 1982 to develop a rehabilitation system to reline long reaches of PCCP with steel plate liners rather than construct new pipelines to replace the deteriorating PCCP pipelines. Since 1982, the Authority has used this method to reline over 10.5 km (6.5 miles) of PCCP. The most recent PCCP pipeline the Authority has rehabilitated using this method was 2.4 km (1.5 mile) of 183-cm (72-inch) diameter PCCP for the City of San Diego's Shepherd Canyon Pipeline. This paper discusses the design and installation of the steel plate liners for this pipeline.

Design Criteria

The steel plate liners to be placed inside the existing PCCP pipeline were designed to withstand the maximum operating design pressure of the existing PCCP pipeline with an allowable stress not to exceed 50 percent of the yield strength of the steel. No pressure credit was given to the ability of the existing PCCP to withstand internal pressure in the design of the steel plate thickness due to the deteriorated condition of the existing PCCP. To minimize the required steel plate thickness while still providing a ductile and conventionally weldable steel plate liner, steel conforming to American Society for Testing and Materials (A.S.T.M.) A-572, Grade 42 was selected for fabrication of the steel plate liners. After reviewing the high level of compaction achieved around the PCCP during the initial construction of the pipeline and the adjacent trench wall soil stability (combining to result in a high modulus of soil reaction, E'), the steel plate liners were assumed to experience minimal exterior loading. Therefore, no exterior loading requirements were considered in the design of the steel plate liners. To minimize the loss of flow capacity in the existing 183-cm (72-inch) diameter PCCP due to the relining process, an outer diameter for the steel plate liners of 179 cm (70.5 inch) was specified. To accommodate the pipe joint pulls and bevels within the existing PCCP pipeline due to the terrain of the project, each steel plate liner was manufactured to the same nominal length, 5 m (16 feet) as the PCCP into which it was to be placed.

Each steel plate liner section was fabricated with one end conventionally belled to make a bell and spigot joint. Each steel plate liner was belled with a bevel identical to the existing bevel in the original PCCP section up to a maximum bevel of five degrees. Angles greater than five degrees required fabrication of special steel liner plate sections with dimensions such that the special section could be transported within the PCCP and fitted in place at the PCCP pipeline angle point.

After fabricating the steel plate liners to the specified length and diameter, each steel plate liner was cut the entire length of the pipe in such a manner that the longitudinal seam would be located 50 mm (2-inch) from the vertical centerline at the bottom of the pipe, with the long seam located alternately on the right and left throughout the length of the project. Each long seam was provided with a backing bar to facilitate full penetration field butt welding following installation of the steel plate liner into the PCCP pipeline. The diameter of the steel plate liner was reduced to 163 cm (64-inch) by rolling the steel plate liner inside itself. The reduced diameter was necessary to accommodate the degree of grade and alignment changes throughout the length of the PCCP pipeline while incurring as little diameter loss and resulting carrying capacity loss as possible as shown in Figure 1.

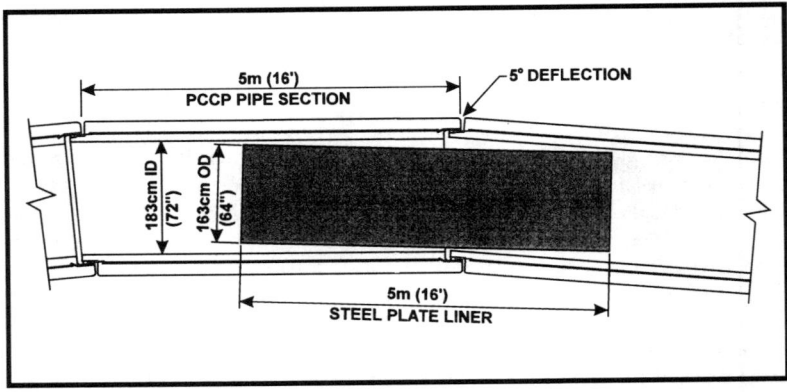

Figure 1- Typical Steel Plate Liner Installation

The reduced diameter of the steel plate liner was maintained by internal tack welds and external steel straps. The external steel straps were located close to the pipe ends so that they could be easily cut and removed following transport of the steel plate liner through the PCCP pipeline prior to fit-up with the adjacent steel plate liner.

The area between the new steel plate liners and the existing PCCP was pressure grouted and the interior of the pipe received a 13-mm (1/2-inch) coating of cement mortar lining to provide a smooth interior and protect the steel plate liners from internal corrosion. Included in the fabrication of the steel plate liners were 26-cm (1-inch) diameter steel couplings located at the top and bottom in the center of each steel plate liner. Spacer pads, 50-mm (2-inch) square, by 10-mm (3/8-inch) thick, were attached to the outside of the steel plate liners to ensure space was maintained between the existing PCCP and the new steel plate liners to fully distribute the grout throughout this space. Figure 2 depicts a typical PCCP with the steel plate liner installed.

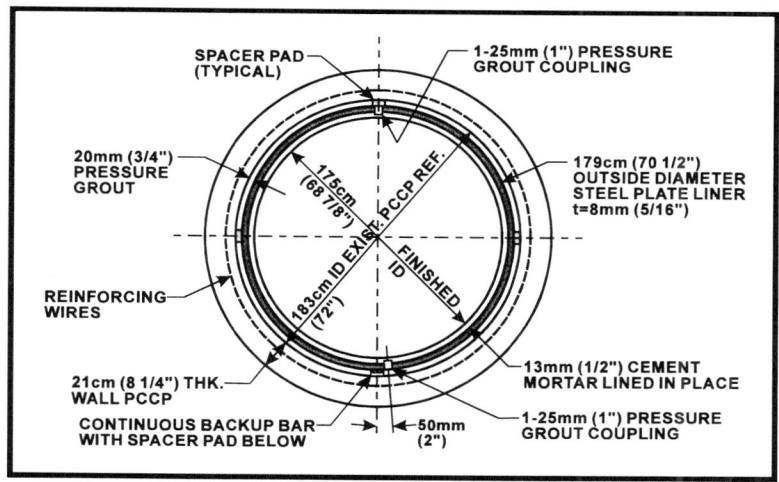

Figure 2 - Typical Relining Section

Access points to insert the steel plate liners into the PCCP were located throughout the length of the project to minimize the impacts of the terrain and the cycle time to transport each steel plate liner through the pipeline. One 5 m (16 feet) long PCCP section was removed at these locations and, following completion of the relining operation, replaced with a specially fabricated steel cement mortar lined and coated closure section designed to account for both internal pressure and external loading. A steel cement mortar lined and coated closure section was selected to provide a similar material environment to ensure that the closure section did not create a dissimilar material corrosion cell between the closure section and the adjacent PCCP section.

The criteria to determine the number and location of access points into the existing PCCP was:
- site ingress and egress;
- available right-of-way for the delivery, unloading, and temporary storage of the steel plate liners and other materials; and
- cost to excavate the access point site, fabricate and install the closure sections and backfill the access point site;

as compared to:
- the cost associated with the cycle time to transport a steel plate liner into position and return the carriage unit back to the access point to transport the next steel plate liner into position and possible schedule impacts;
- slope of the existing PCCP and anticipated degree of difficulty and safety consequences of transporting the steel plate liners through the PCCP; and
- grouting and field lining access constraints.

Construction

Work began during the winter of 1988/89 and was completed in a three month period from dewatering the existing PCCP pipeline to returning the rehabilitated pipeline into service. Following dewatering of the existing pipeline, the PCCP at the three access locations was excavated and removed at the access points. To ensure the integrity of the adjacent PCCP sections, it was essential that the bell and spigot of each adjacent PCCP not be damaged during any of the relining operations. In order to transport the steel plate liners through the PCCP pipeline, a pipe carriage using an electric forklift power unit as shown in Figure 3, was designed and built by the Authority. A double belled steel plate liner was loaded onto the pipe carriage and transported through the pipeline as the first liner plate section to be installed. Each subsequent steel plate liner was inserted into the pipeline, spigot end first, to allow the spigot to be inserted into the bell of the previously installed section.

After arriving at the assigned location the steel bands were cut and removed while still on the pipe carriage. The steel pipe liner was then placed into its approximate final location, removed from the pipe carriage, and the tack welds broken. After the tack welds were broken, the steel plate liner sprang open returning to its original diameter of 179-cm (70 1/2-inch) as the rolling process to reduce the steel plate liner diameter from 179-cm (70 1/2-inch) to 163-cm (64-inch) did not result in a permanent set of the steel at the reduced diameter. This rebound aided in fit-up of the steel plate liners.

Figure 3 - Modified Forklift Developed and Fabricated by the Authority to Transport Steel Plate Liners into Existing PCCP Pipelines

Following final fit-up with the adjacent steel plate liner, the longitudinal seam was field welded using a full penetration butt weld. After the longitudinal seam was completed, the standard bell and spigot lap joint was completed using a full fillet field weld. This operation was repeated until a steel plate liner had been placed in each PCCP section. Outlets from the mainline pipeline for air vacuum valves, blowoffs, manways, and turnouts were fitted with collars and welded to the original PCCP outlet, to ensure a water tight connection at each location.

After the installation of the steel plate liners was completed, the annular space between the existing PCCP and the steel plate liner was grouted using a sand cement slurry through the upper grout coupling. After grout started to flow out of the lower grout coupling, confirming complete grout distribution throughout the annular space, a grout plug was installed and seal welded to ensure water tightness of the steel plate liner. Close monitoring of the grout pressure at the coupling was necessary to prevent buckling the 8-mm (5/16-inch) thick steel liner plates. The target grouting pressure was established at 48 kpa (7 psi) with a maximum allowable pressure of 68 kpa (10 psi). Although these grouting pressure requirements were established, equipment failure and operator error resulted in the buckling of some of the liner plates during the grouting operation. The buckled areas were cut and removed, a replacement steel plate butt welded in place, and the repaired steel plate liner re-grouted. Following completion of the grouting operation, the interior of the pipeline was field lined with a 13 mm (1/2 inch) thick mixture of sand cement mortar. The field lining equipment was inserted into the relined pipeline at the access points and material delivered to the equipment through the existing manway structures. Following completion of the cement mortar lining operation, the closure steel pipe sections were installed at each access location and the entire relined pipeline hydrotested and returned to service.

Cost

The 1989 cost of this PCCP steel plate liner rehabilitation was $984 per meter ($300 per foot). This price included three access points and thirteen outlets over the 2.4 km (1.5 mile) of the project. The 1989 estimated cost to remove and replace the existing PCCP pipeline with a steel pipeline was $2625 per meter ($800 per foot). This resulted in a cost benefit ratio between the rehabilitation and replacement of 2.7 to 1. The pipeline replacement cost does not include any additional right-of-way requirements or environmental mitigation costs. The pipeline rehabilitation was completed in a three month period from draining the existing PCCP to returning the rehabilitated pipeline into service without significant community disruption. This is compared to an estimated ten to twelve month period for pipeline replacement at this same location with much greater community impacts.

Conclusion

Steel plate relining is a cost effective method to rehabilitate deteriorating large diameter PCCP pipelines up to operating design pressure in the intermediate thickness range to 8-mm (5/16-inch). Designs above this range require further investigation to ensure that the rolling process to temporarily reduce the steel plate liner diameter for ease of transport through the pipe to be relined, will not result in permanent set of the steel plate liner. Some degree of permanent set may be acceptable, provided the steel pipe liner can be easily expanded to the final diameter once the steel plate liner has been placed into position within the PCCP pipeline. Pipelines with internal operating pressures requiring steel thickness greater than 8-mm (5/16-inch) could be relined if maintaining the maximum diameter and, therefore, the capacity of the pipeline is not of great concern by eliminating the reduction in diameter of the steel plate liners to facilitate transport through the PCCP pipeline.

The rehabilitation of PCCP pipelines using this method should be considered where the terrain and PCCP pipeline horizontal and vertical alignments are not too varied and access points can be strategically located to optimize the cost, time, and difficulty to transport the pipe sections through the pipeline. Consideration of the long term ability of the deteriorating PCCP sections to withstand both dead and live external loads and allowing the steel plate liners only to withstand the internal pressure, requires additional investigation to better determine expected rehabilitated increases in overall PCCP pipeline service life. This evaluation is currently being conducted by the Authority on the original PCCP pipelines rehabilitated with steel plate liners. The results of this study will be made available when completed.

Overcoming the Challenges of Replacing 20 km of
Defective 1524 mm Diameter PCCP

Terry L. Walsh, P. E., M. ASCE[1]
David S. Hodge, P.E.[2]

Abstract

After repeated ruptures of its largest water supply main, Pinellas County, Florida undertook the replacement of 20 km (12 mi) of this defective 1524 mm (60 in) diameter prestressed concrete cylinder pipe (PCCP) transmission main. The existing pipe was made with dynamically strain-aged, hydrogen embrittled, Class IV prestressing wire and porous mortar coating. The replacement of this critical pipeline required overcoming the challenges of: determining a new route that would minimize public inconvenience, adverse environmental impacts and cost; assuring long-term, reliable service life for the new pipeline; complying with stringent environmental permit requirements; preventing additional breaks in the existing fragile main while working in very close proximity to it; and maintaining the contractor's strict adherence to the plans and specifications. These challenges were successfully met through: extensive alternative route studies; unusually comprehensive pipe materials and construction contract documents; continuous in-plant inspection of pipe manufacturing; continuous on-site resident engineering; and extremely thorough construction monitoring.

Introduction

Pinellas County is located on the west central coast of Florida. It is largely an urban area with a population of approximately 950,000. Pinellas County Utilities provides water supply service to the majority of the County from its own well fields and a regional shared system of primarily groundwater sources. Treated water is conveyed into Pinellas County through three major transmission mains. The largest of these is a 20 km (12 mi) long, 1524 mm (60 in) diameter PCCP (embedded cylinder)

---

[1]Partner, Greeley and Hansen, 1715 N. Westshore Blvd., Tampa, Florida 33607

[2]Regional Construction Engineering Manager, Greeley and Hansen, Tampa, Florida

main that carries about 1.75 m³/s (40 mgd). The pipeline was constructed in 1977 using pipe manufactured by the now-defunct Interpace Corporation.

In 1979, this transmission main ruptured and was taken out of service to replace the damaged pipe sections. Subsequently, the main ruptured four more times. Exhaustive investigations determined the pipe to be defective due to the use of Class IV prestressing wire and poor mortar coating in its manufacture. The consequences of the premature failure of this important transmission main were years of costly litigation and replacement of the entire pipeline.

Greeley and Hansen was retained by Pinellas County to provide engineering services for the studies, design and construction of the replacement transmission main.

Defective PCCP Manufactured by Interpace Corporation

Until the mid-1960's, prestressing wire utilized in the manufacture of PCCP under AWWA Standard C301 was Class I or II tensile strength. By 1968, a higher strength wire, Class III, emerged and became accepted in the industry. Interpace Corporation, however, pushed the tensile strength class yet higher and produced wire at tensile strengths in excess of any then existing, or current AWWA or ASTM prestressing wire specification. This wire, in 8, 6 and 1/4-inch gages, was designated Class IV by Interpace, and was manufactured by Interpace from 1972 to 1979.

Prestressing wire is made by taking hot-rolled, mill-produced rod, and pulling it through a succession of dies. This is a type of cold working that increases strength and some other mechanical properties, while reducing the wire diameter. The important manufacturing variables are control of the amount of diameter reduction at each die and control and dissipation of the considerable heat generated by the drawing process. Interpace Corporation purchased a wire mill in the mid-1960's and began drawing its own prestressing wire. The evidence obtained during later investigations and testing of Class IV prestressing wire manufactured by Interpace Corporation indicates that the wire was drawn at excessively high temperatures. Wire drawing temperatures exceeding about 200° C (400° F), while increasing tensile strength, compromises ductility, causes the wire to become dynamically strain aged and makes the wire susceptible to a form of stress corrosion cracking called hydrogen induced delayed brittle fracture, or hydrogen embrittlement. This condition can lead to failure of the wire.

The problems associated with defective Class IV wire have been exacerbated in many cases by poor mortar coatings. Permeable voids in the coating can facilitate the movement of ground water and gases through the coating to the steel elements. This can permit the entrance of deleterious compounds such as chlorides, which in sufficient concentration can initiate corrosion, or carbon dioxide in one form or another, which can result in carbonation of the cement paste, thereby reducing the pH and compromising the alkaline protection of the coating.

Challenges of Replacing the Transmission Main

The principal challenges encountered in designing and constructing the replacement transmission main included:

- Determining a route for the new pipeline which would minimize public inconvenience, adverse environmental impacts and cost;
- Assuring long-term, reliable service life for the replacement pipeline;
- Complying with stringent permit requirements while constructing the pipeline through protected forested wetlands, including a nature preserve;
- Avoiding additional ruptures of the fragile existing transmission main while working in very close proximity to it; and
- Maintaining the contractor's strict adherence to critical elements of the plans and specifications necessary to maximize the quality and service life of the replacement transmission main.

Determining the Best Pipeline Route

The existing transmission main runs from its connection to the regional water supply transmission system near the northern boundary of Pinellas County to the County's North Booster Pumping Station, where the water is pumped into the distribution system. All possible routes for the replacement pipeline must cross the Lake Tarpon Outfall Canal at a fixed location where the existing 1524 mm (60 in) diameter steel pipe elevated crossing is located. The County had already replaced a segment of the original pipeline adjacent to this crossing. Consequently, the project was divided into North and South Divisions for alternative route investigations. Furthermore, two construction contracts were preferable, since each division could be placed in service independently of the other and a substantial portion of the existing defective main could be removed from service without waiting for the entire project to be completed.

Separate alternative route studies were conducted for each contract division. In the South Division, where the area through which the pipeline would pass is almost entirely occupied by residential developments, the primary concerns were public inconvenience and cost. Six alternative routes were developed, evaluated and compared. A route evaluation matrix was used in the comparison. Each route was evaluated and scored against an array of thirteen priority-weighted, non-cost-identifiable criteria, as well as compared in terms of estimated project cost. The route with the best combination of high evaluation matrix score and low project cost was chosen. More than 65 percent of the 5.8 km (3.6 mi) long selected route for the South Division utilized an electric power transmission corridor. The precise alignment of the proposed pipeline in the power transmission corridor and the minimum distances from power transmission tower supports were agreed upon with power company representatives during the final design phase. About 20 percent of the route was located in very narrow easements requiring special construction provisions to protect existing structures. The use of the power transmission corridor and easements resulted in the least public inconvenience and lowest construction cost. Only about 15 percent of the route was located in residential streets.

Much of the North Division route would pass through pristine wetlands, including a forested wetland nature preserve, which is under strict regulatory control. The principal concern was protecting these environmentally sensitive lands from adverse impacts associated with major pipeline construction. Nine alternative routes were evaluated and compared utilizing the same procedures employed for the South Division. About 25 percent of the 13.7 km (8.5 mi) long selected route utilized an electric power transmission corridor, which continued from the South Division. A small portion of the route could follow in rural streets, while some of the route could be located in the County's well field. Much of the route, however, would have to pass through the wetlands. Extensive wetlands inventory and mitigation design activities were required to secure the required construction permits. The maximum allowable width of the construction corridor through the nature preserve and other wetlands was 15 m (50 ft).

The selected pipeline routes for both divisions of the project are shown on Figure 1.

Assuring Long-Term Reliable Service Life

After the problems resulting from the ruptures of its largest water transmission main, Pinellas County wanted the replacement pipeline to have a reliable, long-service life. A key element in achieving this goal was the development of comprehensive, conservative specifications for pipe materials, manufacturing, testing, protection, installation and inspection. PCCP, ductile iron and steel pipe were allowed, and technical specifications were based on equivalent performance to meet the various design criteria. Hydraulic studies indicated the pipe would need to withstand a maximum sustained operating pressure of 1240 kPa (180 psi) and a transient pressure of 2070 kPa (300 psi). The pipe could also experience total vacuum. Earth cover ranged from 1.2 m (4 ft) to 6 m (20 ft). At highway crossings, HS20 truck loads would be encountered.

The County was understandably concerned that the specifications for PCCP be especially thorough and require extensive testing of materials and finished pipe. Since the project was designed just prior to the publication of AWWA C301-92 and AWWA C304-92, the specifications for PCCP used in this project were based on AWWA C301-84 with extensive revisions and additions. Very stringent requirements for materials, testing and continuous pipe production inspection were incorporated. This resulted in a specification that was 45 pages long and included most of the requirements of the subsequently published AWWA C304-92.

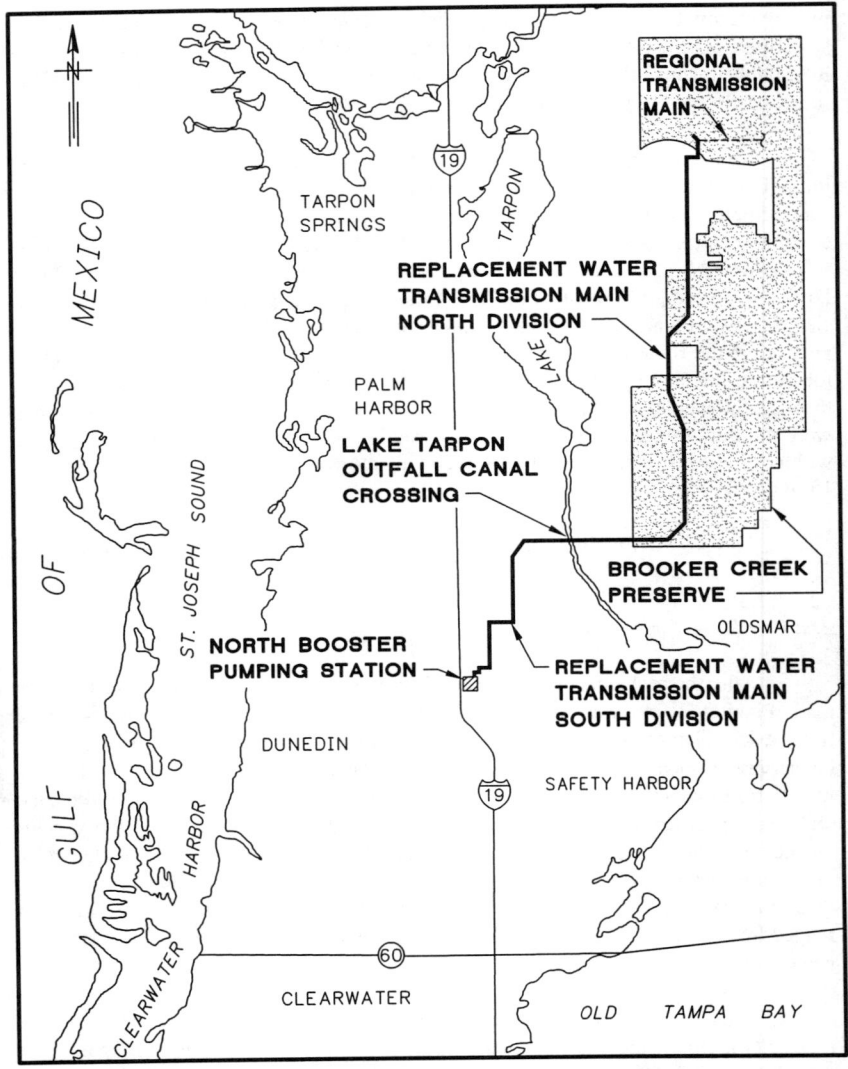

FIGURE 1   ROUTE OF REPLACEMENT WATER TRANSMISSION MAIN

Geotechnical investigations, including soil borings at 90 m (300 ft) intervals along the entire pipeline, yielded soil densities, corrosion protection requirements and thrust restraint friction factors for each type pipe material. Groundwater to grade would be encountered at nearly all locations. Soil pressure diagrams were developed for sheeting and shoring necessary in deep cuts, at tunnel shafts and adjacent to ponds. The contractors were required to furnish sheeting and shoring designs prepared by a Florida registered professional engineer. For the majority of the pipeline, the native sandy soils were determined to be adequate, if well compacted, for pipe bedding and backfill. In deep cuts, mostly adjacent to proposed tunnels, crushed stone was required. Crushed stone bedding was enveloped in filter fabric to prevent migration of sand and stone due to ground water movement. The project specifications required careful placement and compaction of pipe bedding in 15 cm (6 in) layers in a dry trench. In the generally low wet areas in which much of the pipeline would be installed, especially in wetlands and adjacent to ponds, dewatering was a major, carefully planned and monitored activity. The contractors were required to submit dewatering plans employing adequate capacity, primary and standby dewatering equipment. Corrosion protection would comprise achieving electrical continuity for PCCP to facilitate future corrosion monitoring, polyethylene encasement for ductile iron pipe and a triple tape wrap coating system for steel pipe. In addition, ductile iron and steel pipe would require passive cathodic protection systems with anodes placed at approximately 150 m (500 ft) intervals. The cathodic protection systems would also serve to ground the pipe in the power transmission corridors. Thrust restraint would be achieved by use of restrained joints. Internal circumferential welding would be employed for steel pipe, and the use of proprietary restrained joints would be used for ductile iron pipe and PCCP. Geotechnical investigations were also undertaken to determine the minimum distance that the replacement main could be constructed from the existing main and from the bases of power transmission towers without adverse structural impacts.

The project plans were developed from aerial and ground surveys. A construction survey baseline was established, and a detailed horizontal and vertical alignment design was performed. Each deflection and change in elevation of the proposed pipeline was defined and stationed. Each contractor was required to conduct his own alignment survey and was permitted to propose minor deviations from the alignment design for the engineer's consideration. The contractors and pipe supplier were required to collaborate in the development of detailed pipe laying schedules from which the pipe production schedules would be prepared. The laying schedules were to be submitted to the engineer for review. Each piece of pipe and fitting had its own identification number which could be used to trace the materials from which it was produced and all testing associated with its fabrication.

In order to achieve the highest level of quality assurance in the manufacture of the pipe and fittings for the replacement pipeline, a separate pipe purchase contract between the County and the pipe supplier was employed, rather than the traditional furnish-and-install contract. The County would obtain the pipe and provide it to the installation contractors. Although this imposed a responsibility on the County to arrange the manufacture and delivery of the pipe materials to the job site in a timely manner so as not to delay construction, it had the benefit of allowing the County to have direct and forceful oversight of the quality of the pipe materials and fabrication. In addition, this procurement method resulted in lower costs, since the County did not have to pay tax on the pipe and there was not cost markup by the installatiom contractor.

Competitive bidding documents were prepared for the purchase and delivery of PCCP, ductile iron and steel pipe with substantial penalties for failure to deliver the proper pipe to the job site to meet the contractor's installation schedule. Thompson Steel Pipe Company was the successful bidder and was awarded a contract to furnish all of the pipe and fittings used in the project.

Full-time, in-plant inspection of every phase of the pipe fabrication and coating was provided, as well as review of all shop drawings and materials testing. Prior to initiating fabrication of any pipe, engineers inspected the plant facilities and reviewed materials and fabrication testing equipment and procedures for strict compliance with the contract requirements. Careful scrutiny was especially given to the metallurgical chemistry of the steel coils and the shop welding. A complete paper trail of the pedigree of every length of pipe and fitting was developed in order to trace back to the source of any problem that might be experienced during or after installation. During periods of intense pipe manufacturing, inspectors were at the pipe plant round-the-clock. Extensive daily inspection reports and weekly summaries were prepared.

Complying with Stringent Environmental Permit Requirements

The Southwest Florida Water Management District imposed strict limitations on the use of environmentally sensitive lands for pipeline construction. The permit applications were extensive and definitive in the treatment of wetlands that would be disturbed. The primary effort was channeled toward impact minimization and avoidance. Where disturbance could not be avoided, on-site restoration and off-site forested wetlands mitigation were required. The narrow 15 m (50 ft) wide construction corridor meandered somewhat along the alignment of the pipeline to minimize adverse impacts and corresponding restoration and mitigation. Detailed on-site restoration and off-site wetlands construction designs were prepared as part the permit application process. A previously impacted upland was located and used for the contractor's staging area. Temporary low berms and culverts were constructed along the construction corridor where overland sheet flow of water was encountered. Throughout the construction, a representative of the Pinellas County Department of Environmental Management monitored the contractor's efforts to protect the wetlands in the Brooker Creek Preserve, conducted water quality testing and directed corrective actions as required.

Avoiding Ruptures in the Fragile Existing Main While Working in Close Proximity

The condition of the existing transmission main and its close proximity to the construction to of the replacement main required special attention. For considerable distances, the replacement main was constructed parallel to and within 6 m (20 ft) of the existing main. At many locations construction equipment had to pass over the existing pipeline. At these sites temporary bridges were built using timbers and steel plates. The new main had to cross under the old main at seven locations. At each of these sites, the crossing was accomplished by installing the replacement main in a liner plate tunnel. Tunnels were used to cross under the existing main in order to minimize potential disturbance of the soils under and around the existing pipe. Subsidence of these soils could lead to a catastrophic break in the existing main, which was considered to be in fragile condition. The tunnel shafts were either circular or rectangular depending on the contractor's needs to facilitate installation of the new pipe into the tunnels. Because of the soil and ground water conditions, three of the tunnels had to be constructed under compressed air. This required special physical exams for the workers and inspectors. A doctor and decompression facilities were on standby, if needed. Entrance to these tunnels was restricted to air locks, and close communications between workers inside the tunnel and support staff outside the air lock was necessary. At one tunnel, a void in the soil above the limerock was encountered which, combined with local high volumes of migrating ground water, forced the abandonment of the tunnel shaft. After additional soil borings and geotechnical analysis, the tunnel was moved approximately 45 m (150 ft) south of its original location. Grouting above and ahead of the tunnel face was necessary at some locations to control loss of air pressure.

Maintaining the Contractor's Strict Adherence to the Plans and Specifications

The planned service life of even the best-designed pipeline can be seriously compromised if the facility is not constructed in accordance with the critical elements of the design. In order to assure the contractor's continual compliance with the plans and specifications, the County retained Greeley and Hansen to provide continuous, on-site resident engineering and construction monitoring. No significant construction operations were permitted in the absence of an inspector.

The critical elements of design which were given special attention in the field included: the condition of the pipe and coating; placement and compaction of pipe bedding; welding; and the corrosion protection system. Each length of pipe and every fitting were carefully inspected by representatives of the pipe supplier, the contractor and the engineer. All repairs to damaged tape coating on the steel pipe were made by the pipe manufacturer's representative, who was required under the pipe purchase contract to remain on-site throughout the entire pipe installation operation.

Installing 1524 mm Steel Pipe in Power Transmission Corridor
South Division

Installing 1524 mm Steel Pipe in Forested Wetland Preserve
North Division

No placement of pipe bedding was permitted unless the trench had been dewatered to a dry condition. A test program was conducted to demonstrate the contractor's means and methods of placing and compacting the pipe bedding. Compaction densities were taken at various depths in the trench to assure proper compaction. When the engineer was satisfied that the contractor's methods would consistently produce the compaction results, the contractor was required to continue following the those methods throughout pipe laying operations. Density testing of the pipe bedding was conducted on a daily basis. Where densities were determined not to meet the specifications, the contractor was required to re-excavate and correct the compaction. If this became a reoccurring problem, the contractor was required to modify his methods until uniform compliance with the specified densities was achieved.

All welding was thoroughly examined by both the engineer and an independent welding inspection consultant. Generally, field welding was required only for thrust restraint and was performed inside the pipe. External joint welding was not required for the pressures that would be experienced by the replacement transmission main. Field welding procedures were reviewed for compliance with applicable standards. In addition to visual inspection, magnetic particle testing was conducted on all field welds. Defective welds were removed and replaced.

Each cathodic corrosion protection station was carefully inspected during installation. Upon completion of the work, an independent corrosion consultant was retained to test the system for compliance with the plans and specifications. In addition, operation and maintenance manuals for the cathodic protection systems were prepared.

Successful Completion of the Project

The South Division of the replacement water transmission main was constructed by Westra Construction Corporation and placed in service after final pressure testing and disinfection in March 1995. The North Division was constructed by Kimmins Contracting Corporation and placed in service in October 1997. The old, defective pipeline was decommissioned and filled in a number of locations with flowable fill. The pipe purchase and installation costs for the entire project totaled approximately $17,508,000. The thorough attention devoted to the critical design, fabrication and construction elements of this project has provided Pinellas County with a high degree of confidence that its primary water supply transmission main will have a long, reliable service life.

IN MEMORIAM
Victory W. Formby, P.E.
1950 - 1998
Director of Engineering
Pinellas County Utilities

DESIGN AND CONSTRUCTION ASPECTS OF THE PCCP WATER MAIN REHABILITATION
By:
Sufian A. Khondker, Ph.D., P.E., F. ASCE[1], and John R. Mitchell, Jr.[2]

Abstract

After considering several alternative methods of correcting the defects, Washington Suburban Sanitary Commission (WSSC) decided to rehabilitate its 2440mm (96-inch) PCCP (Prestressed Concrete Cylinder Pipe) water main by relining with 13mm (1/2-inch) thick steel liner. In developing the optimum rehabilitation schemes, the factors considered included the determination of an optimum diameter steel liner that can be inserted into the existing PCCP; structural, geotechnical, and hydraulic analyses; corrosion control system design; constructibility; and sensitivity to community impacts. The construction of the project was broken into three phases. Construction of first two phases using two separate contractors began in 1997 and the third is scheduled to start in early 1998. The paper presents the salient design features and the construction aspects of the rehabilitation project.

Introduction

Between 1977 and 1983 WSSC installed approximately 12,500m (41,000 ft.) of 2440mm (96-inch) PCCP water main under six separate contracts. At that time, WSSC began to experience failures of other PCCP mains in the supply system and discontinued the use of PCCP. Other factors also delayed the immediate use of 2440mm (96-inch) PCCP water main. During that delay, WSSC began to understand the problems associated with PCCP and launced extensive investigations into the condition of the 2440mm (96-inch) PCCP water main.

---

[1] Project Manager, Raytheon Infrastructure Incorporated
Two World Trade Center, New York, NY 10048-0752

[2] Project Manager, Washington Suburban Sanitary Commission
14501 Sweitzer Lane, Laurel, Maryland 20707-5902

As a result of its analysis, WSSC concluded that due to material defects the water main will experience premature failure and to ensure future reliability of the water supply, WSSC decided to rehabilitate the water main. In 1993, WSSC retained Raytheon Infrastructure Incorporated to develop optimum rehabilitation schemes. The rehabilitation schemes consisted of relining the existing PCCP with 13mm (1/2-inch) thick steel liner as well as replacement of sections with new steel pipe installed by cut and cover method. The choice of liner method was primarily dictated by recent developments in the area around the water main that made simple open cut replacement of the existing PCCP difficult. The rehabilitation schemes also included the evaluation, upgrading and/or replacement of the in-line valves, valve vaults, blow-offs, and entry ports.

The alignment of the existing PCCP water main, which follows the ground profile, contains numerous horizontal and vertical compound curves as well as elbows making the use of a mock-up test impractical and constructibility issue a paramount consideration. As a result, a detailed survey of the pipe interior was conducted and the data used in a computer model to determine the optimum diameter for insertion. In addition to the actual survey data, allowances for transportation equipment, corrosion protection devices, and survey tolerances were factored into the computerized clearance study. Based on the clearance study; structural, hydraulic, and cost analyses; a 2030mm (80-inch) OD steel liner was selected. The annular space between the new liner and the existing PCCP was filled with a low strength concrete/fly ash grout or a cellular concrete grout. Corrosion control system consisted of tape coating, mortar lining, sacrificial anodes and a selective deep-anode impressed current system.

The rehabilitation of 2440mm (96-inch) PCCP was broken into three phases and currently being constructed under three separate contracts. Contract- I included the rehabilitation of the existing PCCP along the east side of I-95 from north of Baltimore-Washington Parkway to Good Luck Road, Contract- J from Good Luck Road to Maryland State Route 450 and Contract- K included the selective rehabilitation of existing PCCP along the east side of I-95 from Baltimore-Washington Parkway to the intersection of I-95 and I-495 in Prince George's County, Maryland. The scope of Contract-I, Contact-J, and Contract-K included approximately 1730m (5,680 ft.), 2350m (7,700 ft.), and 2990m (9,800 ft.) of relining and 100m (320 ft.), 180m (600 ft.), and 1370m (4,500 ft.) of replacement, respectively. The design allowed for flexibility in the contractor's method of transportation and installation of liners and the contractors have taken different approaches. In subsequent sections, the paper discusses the design considerations outlined above in detail and the implementation of that design in construction.

Design Aspects

The salient design features consisted of optimum diameter and clearance study; hydraulic analyses; selection of pipe/liner material and structural design; selection of in-line valves, blow-off valves, air/vacuum valves; corrosion control system design; and assessment of permit requirements.

Optimum Diameter Study and Clearance Study

The purpose of the diameter study was to determine the largest diameter liner that could be inserted into the existing 2440mm (96-inch) PCCP. This required the study of various parameters such as the extent of open cut and liner sections; clearance requirements during and after construction; method of construction; accessibility; flow and hydraulic requirements; and overall cost of the rehabilitation schemes.

The extent of the open cut sections was determined based on data from aerial mapping and topographic survey, walkdown of the entire pipe line and photographic records of all surface features. The clearance study was predicated by the alignment of the existing PCCP, which consisted of numerous horizontal and vertical compound curves and bends. Due to the nature of this alignment, mock-up tests to establish the required clearance were considered impractical, especially if more than one diameter and length of liner were to be studied. Therefore, it was decided that the clearance study would be performed using a computer simulated model, which would facilitate the selection of the optimum diameter and length of the liner section. A computer program known as "Tunnel System" specially developed to deal with the problems encountered with a curvilinear alignment and widely used in the design of rapid transit tunnel system was used for the study. The program was modified to simulate a circular cross section liner inside a circular cross section tunnel (PCCP).

The success of the clearance study depended on the accurate delineation of the interior of the existing PCCP. Consequently, a detailed survey of the interior of the existing PCCP, which identified and located four quadrants, namely the crown, invert and spring lines of each 6m (20 ft.) PCCP section, was conducted. Based upon the survey data, the radius center of each measured cross section at 6m (20 ft.) intervals was developed. Mathematically controlled baselines for both the horizontal and vertical alignments of the existing PCCP were established within 25mm (1-inch) tolerance, so that any point on the 2440mm (96-inch) PCCP would not be more than 25mm (1-inch) off from the circumference.

The methodology used in the clearance study was similar to what is expected during the transportation of the liner. The selected liner diameter and length ("can") was inserted into the model PCCP. The ends of the "can" were assumed to follow the control baseline (centerline), as shown in Figure 1. At any given section, where the clearance was desired, the "can" was placed at three different locations. First, the leading end was placed at the desired station and the clear distance or interference between the PCCP and the liner was measured at thirty-six locations along the ring at 10 degree intervals (Figure 1); next the midpoint of the "can" was placed at the desired station followed by the trailing end and the clearance/interference data was obtained. The worst clearance/interference results were obtained by combining the data computed for the three locations.

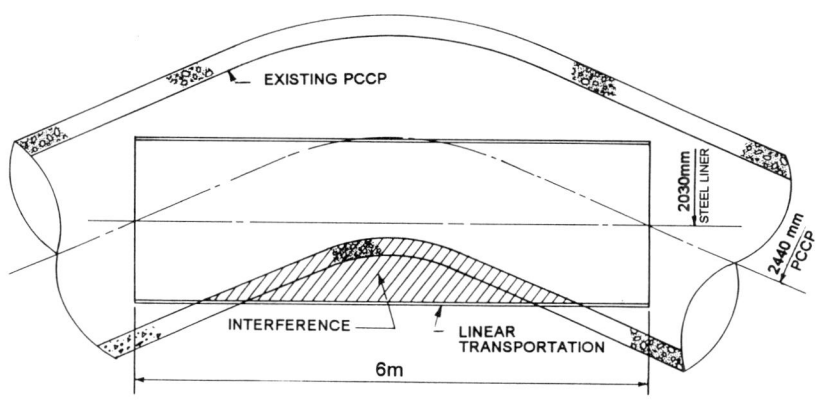

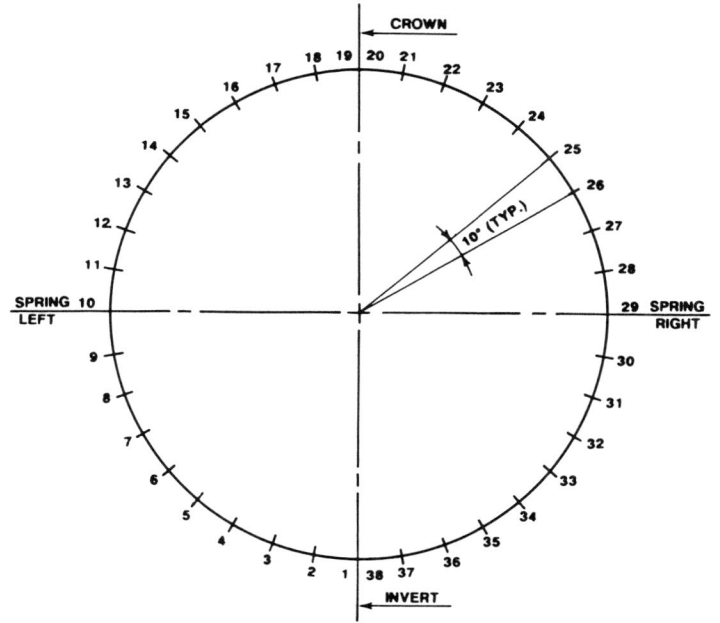

FIGURE 1 : CLEARANCE STUDY

The first trial section was the maximum possible diameter of 2160mm (85-inch) OD based on survey and fabrication tolerances and allowance for corrosion monitoring devices, while the smallest liner diameter of 1850mm (73-inch) OD was selected on the basis of minimum flow requirements for year 2020 established by WSSC. Initially the 2160mm liner with three lengths (6m/9m/12m) were studied thoroughly. The study showed numerous interferences at curves and bends as well as at straight portion of the pipeline, hence it was concluded that the 2160mm (85-inch) OD liner could not be inserted and transported. The next trial run was with the 1850mm (73-inch) OD liner and clearances were in excess of 150mm (6 inches) practically everywhere, except at few sharp bends. Three additional liner diameters, 1930mm (76-inch), 1980mm (78-inch), and 2030mm (80-inch) OD with 6m (20 ft.) lengths were subsequently studied. Based on the results of diameter and clearance studies as summarized in Table 1, a 2030mm (80-inch) OD (1980mm flow diameter) liner and pipe were selected for the rehabilitation of the existing PCCP.

TABLE 1: OPTIMUM LINER DIAMETER

Liner Outside Diameter (mm)	1850	1930	1980	**2030**	2160
Maximum Capacity (m^3/s)	5.61	6.04	6.35	**6.70**	7.40
Total Installed Cost (million)	$41.7	$43.0	$43.9	**$44.5**	$47.0

Hydraulic Analyses

A steady state analysis of the entire system was performed for the 1980mm (78-inch) flow diameter(2030mm OD) steel liner/pipe and maximum capacity of 6.70 m^3/s (153 MGD), Table 1. The results of the steady state analysis established the operating pressure to be 1344 kN/m^2 (195 psi).

Next the transient (water hammer) analysis was performed utilizing LIQT-386, a widely used PC based computer program co-developed by Stoner Associates, Inc. and Professors Wylie & Streeter. As a result of the transient analysis, the maximum and minimum water hammer pressures; the opening/closing time for the in-line valves; and location and size of the air/vacuum valves were established.

Selection of Pipe/Liner Material and Structural Design

The selection of the steel pipe/liner material was based on the intended service and the fracture toughness, weldability and fabrication considerations as well as experiences with two recently completed steel water mains. Additional information received through questionnaires from fabricators regarding the consistent, reproducible quality and predicted chemical requirements from qualified steel mills was also considered. Based on thorough evaluations, ASTM A 572 Gr. 50 or A 607 Gr. 50 were selected for pipe/liner material. The pipe/liner cylinders were fabricated by double groove butt welding, spiral seam using an automatic welding process conforming to AWWA C200. The pipe/liner ends were plain, fitted with back-up bars to facilitate full penetration butt welding from inside in accordance with AWWA C206. Grout couplings were provided for grout backfill and for anti-floatation supports.

Structural steel elements including steel liners, steel pipes and special pieces were designed by the Working Stress Method to withstand all loading combinations, without any contributions from the existing PCCP. The rehabilitation scheme comprised of two different methods. In the areas where an open-cut was not feasible, direct liners were inserted inside the existing PCCP, and the annular space was backfilled with grout. In the areas where open-cut excavations were feasible, steel pipes were installed to replace the existing PCCP. The design criteria for these two methods were different. The steel liners inside the PCCP were supported by undisturbed soils whereas the direct buried pipes were placed on compacted bedding inside a trench. The liner sections thus rested on a more rigid support than the pipe sections resulting in differential settlements that required special design of transition.

Selection of In-Line Valves

In-line valves are required for the isolation of pipeline segments for repair, inspection and maintenance. It is a common practice to use butterfly valves as isolation valves, since the butterfly valves are relatively inexpensive and require less space than other type of valves such as gate, cone, and plug valves. However, because of specific difficulties experienced with other large diameter butterfly valves, WSSC adopted a stringent testing requirements for all new butterfly valves which increased the cost of the butterfly valve to be comparable to the costs of other types of valves. In addition to costs, the valve closure time, the change of flow area and surge mitigation, and operation and maintenance characteristics played an equally important role in selecting the in-line valves.

Based on a parametric study and evaluation, double disc gate valves were selected for use as in-line isolation valves. In order to reduce costs, downsized 1520mm (60-inch) gate valves with reducer/expansion pieces were selected. The effect of the increased head loss due to downsized 1520mm (60-inch) gate valves at five locations was limited to the reduction in flow from 6.70 m^3/s (153 MGD) to 6.57 m^3/s (150 MGD), which was considered insignificant.

Blow - Offs

All low points of the rehabilitated pipeline were provided with blow-offs to facilitate draining of the pipeline. The existing blow-offs had discharge pipes tangentially connected to the 2440mm (96-inch) PCCP. Since the diameter of the rehabilitated water main was 1980mm (78-inch), the tangential connection from the PCCP could not be used. Besides, all existing blow-off vaults were leaky and under sized. Hence it was decided to replace all existing blow-off valves and vaults.

The selection of type and size of blow-off valves depended on velocity of blow-off jet, throttling characteristics, and the duration of draining the water main. In controlling the adverse downstream effects, such as possible flooding and erosion damage, the control of blow-off jet and energy dissipation were the most important factors. Based on thorough evaluations, resilient seat gate valves were selected as blow-offs.

Air/Vacuum Valves

Air/vacuum valves allow the escape of large volume of air during filling and admit air into the pipeline during emptying of the line. In addition, properly designed air/vacuum valves reduce the surge pressure and release the small amounts of air that accumulate at high points during normal operation. The preliminary size of air/vacuum valves were determined based on the following criteria:

- Air release differential pressure = 7 kN/m^2 (1.0 psi).
- Air intake differential pressure = 35 kN/m^2 (5.0 psi).
- Filling rate velocity = 0.3 m/s (1.0 ft/sec).

The actual size, type and location of air/vaccum valves, however, were finalized through transient (water hammer) analysis.

Corrosion Control System Design

Soil studies were performed as part of the corrosion control evaluation. The studies included the measurement of soil resistivity using the Wenner 4-pin method, calculations of resistivity using the Barnes Layer Method, pH and other chemical analyses performed on soil and water samples taken from the test borings. Analyses of soil and water samples did not reveal the need for special corrosion control design measures based on soil/water chemistry alone.

Another series of tests were performed to determine qualitatively the extent of stray current activity in the soils adjacent to the route of the 2440mm (96-inch) PCCP water main. These tests consisted of measuring and recording of stray current voltage gradients at various locations along the pipe route. Indications of stray current activity were found at all test locations. The voltage gradients with the greatest magnitude were found adjacent to a new WMATA (Washington Metropolitan Transit Authority) storage and maintenance facility.

Based on the initial investigations, the corrosion control system design was developed as follows:

- The exterior of the steel liner within the PCCP was tape coated and additionally protected by the grout backfill placed in the annulus between the steel liner and the PCCP. The small bare portion of the liner at every field weld was protected with three strands of zinc ribbons.

- Internal corrosion control of the steel pipe and liner was by a 13mm (1/2 inch) thick cement mortar lining. Any interior surfaces that cannot be mortar lined were coated with two coats of a chemically cured self-priming epoxy/amine coating designed for continuous immersion service.

- The cathodic protection system design was a unique site specific design, which consisted of deep anode ground bed installations. Each installation consisted of a rectifier and two deep anodes between 60m (200 ft.) to 90m (300 ft.) deep. Additionally, zinc ribbon anodes were also installed at each pipe end at every welded joint. Special installation technique was required for very difficult installation and transportation of liner sections inside the existing PCCP, while corrosion protection devices were in place.

Assessment of Permit Requirements

The type of construction operations dictated the type and number of permits necessary for the work. The project involved three major and distinct construction operations. These operations comprised of: installation of steel liner for a major portion of the existing PCCP, excavation and removal of portions of the existing PCCP and replacement with new steel pipe sections, and addition of various appurtenances to the pipeline.

Portions of the project which involved the installation of a steel liner required the least number of permits, since surface disturbances were limited to the entry points. Those portion of the project that involved excavation and removal of the existing PCCP and installation of new pipe sections required most of the permits, since the disturbed areas included the entire right-of-way plus off-site spoil areas. The permits required for the addition and replacement of various appurtenances depended on their locations. On the whole, the project required approximately fifteen permits including State Highway permit, a joint permit from Maryland Department of Natural Resources for waterway construction and wetland permits, Maryland Forest Conservation permit, Tree permit, National Park Service permit, Sediment Control permit, and Railroad permit.

Construction Aspects

The most interesting and challenging construction aspects included the liner installation/ transportation inside the existing PCCP; Joint Fit-up (field welding) from inside; and backfilling annular space with grout.

Liner Installation/Transportation

Based on the diameter study conducted during design phase, it was guaranteed that a 6m (20-foot) section liner would negotiate all curves and bends inside the existing PCCP. Both Contractors, however, elected to exercise the option to maximize the length of the liner and in most cases used 12m (40-foot) sections. The limitation was set at this length due to shipping constraints. When shorter sections were required, typically 6m (20-foot) lengths were used. This was again primarily due to shipping costs that made it more cost effective to ship two 6m (20-foot) sections on one trailer than a larger length on two separate trucks.

Each of the Contractors elected to transport the liner sections using a different method. One contractor (Contract I) who had relatively straight segment of the pipeline chose a more labor intensive method where the liner section was set on two carts. The carts were connected by cables allowing removal of the carts from underneath the liner section when finally transported to its final destination. The liner was moved through the PCCP with a bobcat and only adjustment could be made at the "free" (bobcat) end of the liner. Once the liner was transported to its "final resting place", it was jacked into place using a pair of come-alongs and some hooks that were welded to the liner interior. Once the liner was set to the proper orientation, the liner joint was tack welded in place, liner supports at the free end were placed and the next piece was installed. This method averaged installation of about three liner sections per day.

The second Contractor (Contract J) fabricated a special, very sophisticated transporter for moving and setting the individual liner section in place. This method proved extremely valuable since this contract contained a greater number of both vertical and horizontal curves to negotiate. The transporter permitted movement of both the ends of the liner and simplified the transportation of the liner over greater distances. When the liner was placed at its final location, the transporter had the flexibility to place the liner in its proper orientation and even permitted rotation of the liner should such changes be necessary with relative ease within the PCCP itself. There was no need to tack weld and then remove hooks for come-alongs within the liner because the transporter itself forced the liner home. When going through relatively straight lengths of the PCCP, this method allowed installation of up to seven liner sections in one day.

Joint Fit-up

For field welding of liner segments, the design required full penetration butt welds and to facilitate the welding, the liner sections were fitted with back-up bars. During the initial days of the installation, it was theorized that if the root gap was correct, the liner would automatically follow its proper course. However, it did not take long to realize that this approach was only leading to problems. Especially, with 12m (40-foot) lengths, a small deviation at one end could have devastating effects one, two or three liner lengths down the road. The more prudent course of action was to make sure that the free end was properly oriented while trying to maintain the allowable tolerances at the root gap of 6mm (1/4-inch) plus 6mm (1/4-inch), or minus 1.6mm (1/16-inch). When necessary, grinding was permitted in areas where root gap was less than 4.8mm (3/16-inch) and weld areas where the root gap exceeded 13mm (1/2-inch) were built up in accordance with AWS D1.1. Due to the additional welding efforts and additional testing requirements for these repair procedures, the Contractors had built in incentives to bring the joints as close to the specified tolerances as possible.

Backfill Grout

The deign recommended the use of temporary bulkheads to facilitate the placement of backfill grout in the annular space between the new liner and the existing PCCP. Both Contractors, however, proposed the use of permanent bulkheads and the placement of the backfill grout after a certain section of liner was installed. Because of the use of 12m (40-foot) lengths by both Contractors, dimpling of the liner at the supports and excessive sagging at the center during backfill operations, were major concerns. To alleviate these concerns, both Contractors placed the backfill grout in lifts and one Contractor used light weight cellular concrete. Since the backfill grout consisted of a low strength flowable mix design, little pressure was needed during grout placement.

Conclusions

Although not yet complete, the Washington Suburban Sanitary Commission has every indication that the rehabilitation of the 2440mm (96-inch) PCCP water main is a successful project. Careful analysis during the design phase, accurate design documents and survey data along with conscientious construction by the Contractors have all led to a successful and enjoyable construction project. In addition, although this work is being accomplished in a highly populated areas, community impacts have been kept to a minimum.

The Lake Gaston Pipeline: 76 Miles of Controversy

Joe Bivins, P.E., Member, ASCE[1]
Tom Leahy, P.E. Member, ASCE[2]
Jim Richards, P.E., Member, ASCE[3]

Abstract

The Lake Gaston Water Supply Project is a 227 Million Liters Per Day (60 MGD), 122 Kilometers (76 miles), 1.52 Meter (60 inch) Pipeline project designed to provide a new source of raw water to the citizens of Virginia Beach and Chesapeake, Virginia. The project is owned and operated by the City with Chesapeake being a one sixth partner. The project is relatively simple in concept, but became very complex in reality. It has been extremely controversial and lengthy (15 years), and has involved the City, the Commonwealth of Virginia, the State of North Carolina, and numerous local, state and federal regulatory agencies. This project had many engineering, construction, environmental and legal issues that were somewhat unique. Environmental issues included fish habitat mitigation, wetlands mitigation and completion of a full Environmental Impact Statement. Some of the legal and regulatory issues included review for Coastal Zone Management Act compliance, Federal Energy Regulatory Commission review, 9 separate lawsuits and appeals filed against the project and ultimate submission to the Supreme Court of the United States.

Many of the engineering issues required unusual solutions. The project was designed and bid with three alternate pipe materials. Soil and right-of-way conditions varied significantly. The project included six overhead river/swamp crossings that were utilized to minimize environmental consequences as well as numerous large diameter tunneled road/railroad crossings. Construction issues were also somewhat unique, but solvable. Due to the delays caused by the legal issues the construction schedule was accelerated from 39-months to a 24-month schedule utilizing eight separate construction contracts.

---

[1]Project Manager, Michael Baker Jr., Inc., 770 Lynnhaven Pkwy, Suite 120, Va. Beach, VA 23452
[2]Water Resources Director, Public Utilities, City, Municipal Center, Va. Beach, Virginia 23456
[3]Southeast Regional Manager, Michael Baker Jr., Inc., 770 Lynnhaven Pkwy, Suite. 120, Va. Beach, VA 23452

Other construction issues included: (1) changing pipe materials based on soil conditions to achieve construction economies without importing large amounts of fill, (2) streamlining construction methods for grouting, diapering and backfill operations, (3) unique requirements for coordination between construction contractors, and (4) factoring in pipe material suppliers production and delivery capabilities. In spite of all the obstacles, the construction work was completed ahead of schedule, and the project was placed in service in the fall of 1997.

Introduction

The Lake Gaston Water Supply Project, one of the largest utility projects in the history of the Commonwealth of Virginia, has ended the decades-long threat of a crippling water shortage for the City by providing a new source of raw water to meet present and future needs. The project originated after many years of severe droughts, especially the drought of 1980 when strict water allocation forced residents and businesses in Virginia Beach, Norfolk, Chesapeake and Portsmouth to cut back to 75% of normal water usage. Although conservation measures implemented since then, when the population was about half its current 460,000, have enabled the city to grow and prosper, City decision makers recognized that another drought would prove devastating without an additional water supply.

South Hampton Roads, Virginia is a major population center located on the mid-Atlantic coast of the United States. It is bounded by the estuary water of the Atlantic Ocean, Chesapeake Bay and James River which combine to form a deep, warm-water seaport. The region includes one of the country's largest port facilities and is home to the largest military complex in the world. Since 1945, South Hampton Roads has experienced rapid and steady population growth which has strained local water supplies. Unfortunately, the same flat, coastal plane, estuary geographical features which give rise to the massive port and military facilities, and in turn the population growth, also limit the ability to develop significant fresh water supplies in the region.

After an in-depth analysis of different options, the City Council settled on the Lake Gaston Water Supply Project to provide a long-term solution for this shortage. The major elements of this $150-million project are: (1), a 122 KM (76 miles), 1.52 Meter (60 inch) diameter pipeline carrying 227 million liters a day (60 MGD) of water from Lake Gaston, a manmade reservoir on the Virginia/North Carolina border about 160 KMs (100 miles) west, to a reservoir in the Hampton Roads area; (2), a pump station on the shores of Lake Gaston; and (3), three pressure control structures that regulate the flow in the pipeline. (See Figure 1)

In addition to the large size and length, this project had numerous engineering, construction, and environmental and legal issues that were unique. Many of the engineering issues required unusual solutions. The project was designed and bid with three alternate pipe materials: Prestressed Concrete Cylinder Pipe (PCCP),

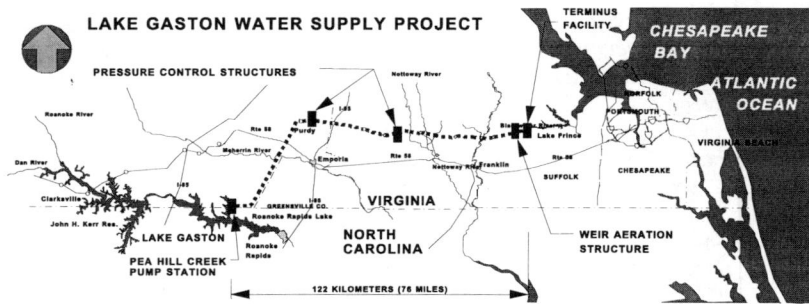

Figure 1: Location and Major Features

Ductile Iron Pipe (DIP), and Welded Steel Pipe (WSP). Each pipe material required separate bedding and construction requirements and all three pipe materials were used in final construction. The Right-Of-Way selection provided major cost savings and included heavily forested, previously unused land, 500 KV power line Right-Of-Way, both active and abandoned railroad right-of-way, and paralleling of some existing pipeline right-of-way. The project included six overhead river/swamp crossings that were utilized to minimize environmental consequences. Some of these crossings were as long as 274 meters (900 ft.). There were eight tunneled road/railroad crossings 2.44 meters (96 inches) in diameter with the longest tunnel being over 114 meters (375 ft.). Pipe pressures ranged from 1724 kPa (250 psi) down to 275 kPa (40 psi). The topography of the route allowed the use of only one pump station for the complete project. Due to the delays caused by the legal issues the construction schedule was accelerated from 39-months to a 24-month schedule utilizing eight separate construction contracts.

Michael Baker Corporation of Virginia Beach, VA was the prime consultant, managing the planning and engineering for the overall project, and bore primary responsibility for engineering and construction administration of the pipeline. Malcolm Pirnie, Inc., of Newport News, VA, was the prime subconsultant and responsible for engineering and construction administration of the pump station and pressure control facilities.

Engineering and Design

The Pump Station

The Lake Gaston pump station has a 650 square meter (7000 sq. ft.) footprint, and contains six vertical-turbine centrifugal pumps and a flooded wetwell that extends down 11 meters (35 ft.) below the lake surface. Water flows through intake screens located offshore into two 1.52 meter (60 inch) diameter pipelines which fill the wetwell. Because of the pump station's location near an upscale lakeside development, noise and aesthetics were identified early on as significant concerns. The facility was designed and constructed to contain noise and maintain a low visual impact. The submerged pumps contain most noise in an underground concrete vault, and the underground wetwell design creates a low visual profile from the lake. Because noise requirements were specified in the construction permits, maximum noise levels were included in motor specifications and special sound-reducing and sound absorbing materials were incorporated, making pump and motor noise undetectable from 30 meters (100 ft.) away. Five of the six pumps are dual-speed, 507/1267 horsepower pump and motor combinations operating at 4,160 volts AC (VAC). The nominal capacity of each of these five pumps is approximately 38 million liters per day (MLD) (10 MGD) at low speed (900 rpm) and 57 MLD (15 MGD) at high speed (1200 rpm). The sixth pump is a 253 horsepower, 440 VAC pump and motor combination that can deliver 15 to 30 MLD (4 to 8 MGD). This pump is used for filling the line and for maintaining small flows when larger flows are not necessary.

The City also made an extensive effort to insure that the pump station and its associated facilities would be as attractive as they were quiet. Because the development around Pea Hill Creek involves small clusters of homes on the peninsulas which extend into the lake, the City laid the facilities out in the same manner. Both the profile and the footprint of the pump station were kept as low as possible. In fact, the pump station was designed to look very much like an upscale lake home.

The Pressure Sustaining Control Structures

The three pressure control structures, a critical feature of the pipeline, are quite unique in their design and operation. From an elevation of 61 meters (200ft) MSL at Lake Gaston, the pipeline climbs to 101 meters (330 ft) MSL before it descends to its terminus 122 KMs (76 miles) away at elevation 15 meters (50 ft) MSL. At flows below about 170 MLD (45 MGD), the hydraulic head required to move the water is less than the net elevation difference between the high point in the pipeline and the terminus structure. Under this scenario, water would flow through the pipeline too fast causing pressure surges, water column separation, and water hammer which could damage the pipeline To address this issue, there are three pressure control structures located on the pipeline. These structures create a back pressure in the pipeline which throttles the water flow when flows of 170 MLD (45 MGD) or less are being pumped and assures that the pipeline remains full. Two of the structures are buried vaults which contain an array of automatic, pressure controlling valves. These valves sense the upstream pressure in the pipeline and close or open to maintain the necessary back pressure at low flows. The two valve vaults are located at roughly the one-third and two-thirds mark along the pipeline route. (See Figure 2) The third structure is located near the end of the pipeline. It is an elevated weir which creates a back pressure in the pipeline by forcing the water to flow over the top of the weir which is approximately 9 meters (30ft) above the pipeline grade. This is combined with a set of cascading steps which begin at the top of the weir. From the weir, the water flows over the steps where it mixes with air and is re-oxygenated. From the bottom of the weir/aeration structure, the water flows by gravity through two more miles of pipeline where it discharges into a stream channel. Hydraulic controls enable the system to operate safely without interruption even during power interruptions, a real possibility given the facilities' remote locations. Other important safety features include various air and surge relief valves located along the pipeline where needed.

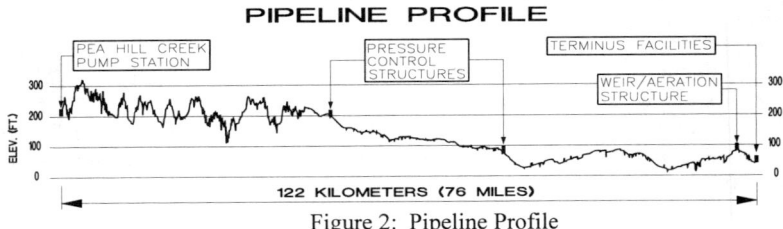

Figure 2: Pipeline Profile

The Pipeline

The pipeline itself consists of approximately 20,000 sections of either prestressed concrete cylinder-pipe (PCCP), ductile iron pipe (DIP) or welded steel pipe (WSP), each 6 meters (20 ft) long and 1.52 meters (5 ft) in diameter. The sections varied in weight from 3,628 kg to 10,886 kg (8,000 to 24,000 lbs). Over 18,000 junkyard autos were recycled and used to make the DIP. The three pipe materials were specified for all five pipeline contracts in order to achieve maximum competition. DIP was the low bid for two of the five pipeline contracts and PCCP was the low bid for the other three. WSP was utilized for the 609 meters (2000 ft) of overhead river/swamp crossings. Overall, the line is 57% PCCP and 42% DIP. Pipe segments with working pressures ranging from 1,724 kPa to 1,034 kPa (250 to 150 psi) were used in the 119 KM (74 miles) of pipe from the pump station to the weir/aeration structure. The 3.2 KM(2 mile) gravity flow section of pipe from the weir/aeration structure to the terminus has a working pressure of 275 kPa (40 psi).

Approximately 101 KMs (63 miles) of pipeline is in relatively firm soils which do not require special foundation preparation or special backfill placement. Since these conditions prevailed over the majority of the pipeline, they were designated as the "normal" trench construction case for developing cross sections and trench construction criteria. An additional 11.3 KMs (7 miles) of the pipeline is in areas where rock, or weathered rock was expected to be encountered within the depth of the trench excavation. Most of these sections required pre-blasting. Approximately 10 KMs (6 miles) of the pipeline were areas underlain by very soft silty and clayey soils. Very soft soils were encountered primarily in the marsh, and in several low embankment areas of the abandoned railroad segment. Foundation stabilization measures were required in these areas to improve support conditions for the pipeline.

In order to facilitate pipe cleaning in the future, pig launching and retrieval assemblies were designed into the pipeline at each end and at the third points. These assemblies consisted of capped 1.52 meter wyes. A pig launching wye was located directly downstream of the Pea Hill Creek Pump station and retrieving and launching wyes were located directly upstream and downstream of each Pressure Control Structure. The weir structure was also designed to retrieve a pig near the end of the pipeline. Land was purchased at each of the pig retrieving locations to use as settlement basins in the future, if needed.

Hydrostatic Testing

Hydrostatic testing of all five pipeline contracts, the overhead river crossing contract, the pump station contract, and the pressure control structure contract was performed with a coordinated effort. To reduce costs associated with the provision of temporary dead ends needed only for the hydrostatic testing, it was decided to test the pipeline after all of the contracts had connected. Temporary internal bulkheads were used in most cases to segment the test sections. Test sections ranged in length from 3.2 KM (2 miles) to as long as 24 KMs (15 miles) depending on the terrain and the pressure requirements. The pipeline was tested to approximately 120% of design working pressure. The leakage allowance for the testing was 9 lpd/(cm-km) (10 gpd/(inch-mile)) for the gasketed joints.

Overhead River Crossings

As previously noted, 69 KMs (43 miles) of the pipeline follows an abandoned railroad ROW. When the railroad abandoned the right-of-way, it removed the tracks, ballast, and railroad ties, but left behind concrete piers and abutments where the railroad crossed five major rivers/swamps. These piers and abutments were inspected as part of the Geotechnical Investigations phase of the Project. The results of the inspection generally indicated that the existing piers and abutments were adequate to support the proposed pipeline after some remedial work. Virginia Beach staff approved the use of these existing railroad piers for the five river crossings to eliminate any need to disturb the river bottom or the adjacent wetlands. This would help with permitting, and the Corps of Engineers agreed, authorizing the City to construct these crossings using overhead methods. The construction concept for the overhead stream crossings is generally identical for all six crossings (five along the railroad ROW and one across another major river). The construction method utilizes concrete piers/bents, either existing or new, with ring girders to support the nominal 1.52 meter (60-inch) diameter WSP. These crossings ranged in length from 43 meters (140 feet) to over 274 meters (900 feet). The use of WSP for the crossings had many advantages, the most important being that it makes a superb beam all by itself. It can easily span the distances between the existing bents. Ballistics tests were even performed on various steel thicknesses to ensure that hunting rifles would not significantly damage the pipeline.

Environmental Issues

True environmental impacts were minimal, however, due to the opposition and controversy surrounding this project, these issues and impacts were analyzed in great detail. Therefore, extra effort was taken during the design to further minimize any potential environmental impacts. The project included original environmental solutions to fish habitat migration and wetlands mitigation. To minimize impacts on fish life in Lake Gaston, the stainless steel intake screens were designed to keep normal water velocity at the intake below 3 cm/second (0.1 ft/sec.), with screen openings less than the thickness of a dime, protecting fish larvae. In an unusual engineering solution, a one-third acre fish spawning structure was created in a nearby cove to provide enhanced fish spawning habitat.

The reduced environmental consequences of utilizing 34 KMs (21 miles) of power line ROW and 69 KMs (43 miles) of abandoned ROW was a factor in pipeline routing. Because the power line ROW and railroad ROW were already cleared, the total acreage of trees that had to be cut and permanently removed was small. Additionally, the railroad ROW included five of the six major river crossings and most of the wetland areas. Although these wetland areas were extensive, the railroad embankment formed a large, earthen berm constructed almost a century ago. In these areas, pipeline construction was restricted to the railroad embankment with little or no impact to the wetlands or the environment. The five overhead river and swamp crossings utilized existing abandoned railroad track piers as the support structure for the overhead pipe. This reduced those potential environmental impacts to almost zero and helped significantly in obtaining necessary permits.

Legal/Regulatory Issues

There were extreme external legal and regulatory issues associated with this project. These issues were compounded by the fact that the logical choice of water for the area involved an inter-basin transfer and this basin was partially in another state. Because of the extensive hydroelectric and flood control development in the Roanoke Basin, the drought capacity of the lower Roanoke River dwarfs all other river systems in either Virginia or North Carolina. The maximum withdrawal of 227 MLD (60 MGD) is about one percent of the average discharge from the impoundments and about three percent during a major drought. Although 75% of all the water in this reservoir system originates in Virginia, it all flows downstream to North Carolina. Therein lies the controversy. Any water diverted by Virginia Beach, no matter how little, is water that will not ultimately flow downstream, to North Carolina. As a result, North Carolina and the Roanoke River Basin Association (RRBA) have vigorously opposed the project in almost every federal agency proceeding. They have also filed numerous lawsuits in various federal courts challenging the validity of each federal permit, license, or approval issued for the project. As the chronology shown in Table A demonstrates, the Lake Gaston Project has survived a fifteen-year legal and regulatory gauntlet that is among the most extensive ever leveled at any project. To summarize the above, since 1983 the project has been the subject of six environmental reviews by three independent federal agencies, and five lawsuits in four different federal courts challenging the validity of those environmental reviews. All eleven of those proceedings have been resolved in favor of the City.

Construction

After all the problems and delays associated with the legal and regulatory issues related to the project, the actual construction phases proceeded very smoothly and actually finished ahead of schedule. The original construction schedule for the complete project required 39 months from award of the first contract to completion of the project. While the litigation was ongoing, the City was able to obtain permission from the courts to construct the overhead river crossings and the foundation and wetwell for the pump station. This work was started in 1992 and was completed in 1994, but a court restraining order to had to be resolved before any other construction could commence. Originally the pipeline work was scheduled to be completed over 36 months, being completed three months ahead of the pump station in order to allow testing of the pump station utilizing the pipeline. This 36-month schedule for pipeline work provided several advantages. It required less that 914 meters (3000 feet) of pipe per week to be produced which would not put a strain on any one pipe manufacturer. It also allowed a contractor to be able to bid on more than one project, which was expected to provide additional cost savings to the City.

In establishing the final construction schedule after the delays were reconciled, it was determined that the pump station construction would be the critical path. The remaining portion of the pump station above the wet well would require approximately 24 months (including testing) to complete. This required that all the pipeline work be completed in 21 months from notice of award. This was accomplished by reducing the number of construction contracts to attract contractors with large scale capabilities, and assuming that two or more pipe materials would be utilized in order to expect production from the pipe plants to peak at over 2,286 meters (7500 feet) per week while still not impacting cost significantly.

This accelerated schedule worked well. Price Brothers Corporation provided 69,494 meters (228,000 ft.) of PCCP from their Perryman, MD, plant in a little under 14 months, averaging almost 1219 meters (4000 feet) per week of pipe. American Ductile Iron Pipe Co. provided 52,120 meters (171,000 ft) of DIP from their Birmingham, AL plant over a period of nine months. The three contractors who installed the pipe [J.D. Stephens, Stone Mountain, GA-80 KMs (50 miles), Garney

TABLE A

REGULATORY AND JUDICIAL REVIEWS OF THE LAKE GASTON PROJECT AS OF 2/98

12/83 - The Corps of Engineers published an Environmental Assessment (EA) that determined the project would have no significant impacts.

12/84 - COE Final Environmental Impact Statement (EIS) determined there was a need and would be no significant impacts.

6/87 - Federal District Court, Raleigh, ruling in response to a lawsuit filed by North Carolina and the Roanoke River Basin Association (RRBA) claiming that the December 1983 EA was flawed, dismissed 38 of 40 complaints and asked for more study on two issues.

12/88 - COE Supplemental Environmental Assessment (SEA) resolved in favor of Virginia Beach the two issues remanded by the Federal Court.

2/90 - Federal District Court in Raleigh, ruled in response to North Carolina lawsuit alleging that the 1983 EA and 1988 SEA were flawed. The Court dismissed all of the complaints and ruled that the Corps had done a careful and thorough environmental review.

7/91 - Fourth Circuit Court of Appeals in Richmond, VA rejected N.C. appeal and upheld the Federal District Court's ruling that the 1983 EA and 1988 SEA were valid. (North Carolina and RRBA subsequently petitioned the Supreme Court to certify an appeal but were turned down).

5/94 - The Department of Commerce and National Oceanic Atmospheric Administration published a Coastal Zone Management Act (CZMA) Environmental Review that determined the project was needed, would have no significant impacts, and that there were no other practical alternatives.

6/94 - Federal Energy Regulatory Commission (FERC) published a Final Environmental Assessment (FEA) which determined the project would not have significant impacts on the environment. However, the agency decided to prepare an Environmental Impact Statement to update the environmental data.

7/95 - FERC published a Final Environmental Impact Statement approving the project.

9/95 - FERC rejected a petition by North Carolina to reconsider and/or stay the Order approving the project.

9/95 - Federal District Court, District of Columbia ruling in response to North Carolina's lawsuit claiming that the CZMA Decision was flawed. The court upheld the original decision.

5/97 - Federal Appellate Court, District of Columbia in response to lawsuit by N.C. claiming that the FERC ruling was flawed ruled against N.C. and upheld the FERC decision.

8/97 - Federal Appellate Court, District of Columbia denied rehearing by N.C. with respect to the FERC decision.

2/98 - Supreme Court refused to hear N.C. appeal of Federal Appellate Court ruling on FERC Decision.

Construction, Kansas City, MO-27 KMs (17 miles), and Rockdale Construction Co., Conyers, GA-14 KMs (9 miles)], all completed their respective contracts ahead of schedule. Several contractors utilized unusual equipment and techniques that are not usually seen on smaller pipe projects. For example, compaction was assisted by using metal wheels mounted to a hydraulic boom which compacted both sides of the pipe simultaneously. The grouting of the PCCP joints was accomplished by utilizing sand, cement, mortar bins and mixing equipment mounted on a track vehicle that hydraulically boomed down to inject the grout into the diaper. This was essentially a one man operation. During peak laying some contractors installed as much as 488 meters (1600 feet) of the pipe per day. In fact one contractor installed (excluding testing and cleanup) 80 KMs (50 miles) of pipe in 12 months. All three contractors utilized the Komatsu 1000 as their primary excavation and pipelaying machine. The accelerated schedule did not impact on quality — pressure testing of the entire 122 KM (76-mile) pipeline found only one improperly installed joint gasket.

In addition, the right of way selection, which included extensive use of existing power line and railroad right of way and existing pipeline right of way, provided major economies by minimizing utility crossings, traffic problems, and allowing contractors to operate without space concerns. This translated into significant cost/time savings for the City.

Conclusion

The engineering team was challenged to think on a much larger scale than typical pipeline work. Lessons learned: non-traditional bid package preparation, overcoming and accommodating significant delays mandated by challenges from federal, state and local agencies, complex hydraulic operating conditions of the long route with more than 91 meters (300 ft) range in elevation, and the utilization of complex coordination requirements between the various construction contractors.
The project was successful because the engineering, legal, regulatory and environmental issues were all molded together and resolved by one team. The engineers, lawyers, environmentalists, and managers coordinated their actions at all times, which enabled the team to make unified decisions that would not have a negative impact on other project efforts.

The project's success assures the City and other Hampton Roads communities adequate drinking water to enable them to continue to grow and flourish into the foreseeable future without threat of water restrictions The City had specific needs regarding the project's successful completion and startup to lift long standing water restrictions and meet obligations to nearby cities whose water systems interconnected. The project team met the City's needs and improved on their scheduling commitments -the facilities were placed into service in the winter of 1997, approximately six months ahead of planned completion.

DESIGN AND CONSTRUCTION OF THE VALLEY CENTER PIPELINE

Mark Butier[1], W. Jeffery Moncrief[2], Gary P. Stine[3], and Richard Trembath[4]

Abstract

This paper describes a number of the challenges encountered during the design and construction of the Valley Center Pipeline. Relative to design, the following issues will be addressed: public involvement in alignment selection, single and double welding for thrust restraint, and geotechnical issues. During construction, special consideration was given to the following issues: coordination with the public, the Keys Creek crossing, and the tunnel crossing of Interstate 15. Partnering between the owner, the engineer, the construction manager, and the contractor assisted in successful construction of the project.

Introduction

The San Diego County Water Authority (Authority) currently operates two aqueduct systems importing water from Lake Skinner in Riverside County into San Diego County. The Authority supplies water on a wholesale basis to approximately 2.8 million people in San Diego County. The proposed Valley Center Pipeline will provide a 1680 mm diameter 6,520 meter long crossover pipeline between Pipeline 4 on the Second Aqueduct and Pipelines 1 and 2 on the First Aqueduct as shown on Figure 1. This pipeline will improve the operational flexibility of the system. The pipeline crosses Interstate 15 (I-15), a busy eight lane freeway. The pipeline was constructed for a total construction cost of $14,969,583.

---

[1] Senior Project Manager, ALBA Construction Management, 101 Pacifica, Suite 150, Irvine, CA 92718, (714) 753-7333, Fax (714) 753-7320
[2] Principal Civil Engineer, San Diego County Water Authority, 3211 Fifth Avenue, San Diego, CA 92103, (619) 682-4100, Fax (619) 692-9356
[3] Principal Civil Engineer, San Diego County Water Authority, 3211 Fifth Avenue, San Diego, CA 92103, (619) 682-4100, Fax (619) 692-9356
[4] Senior Project Manager, Parsons Engineering Science, Inc., 9404 Genesee Avenue, Suite 140, San Diego, CA 92037, (619) 453-9650, Fax (619) 453-9652

Design

Public Involvement in Alignment Selection

The public was involved in the project's alignment selection as a result of the Authority's sensitivity to the issue of property and right of way acquisition. Initially, 27 alignments were developed and screened, considering the issues of engineering feasibility, cost, environmental impacts, and property acquisition. These were reduced to three alignments presented in the draft Environmental Impact Report and evaluated in detail in the Alignment Study Report. The public was given notice and invited to attend the Authority's Board Meetings. The engineering design team met with property owners along the alignments with the goal of minimizing impacts to the properties to the extent possible. A goal of the team was to select an alignment which would not require taking a residence. After considering all aspects of each alignment, a final alignment was selected which had the least interruption to residential areas.

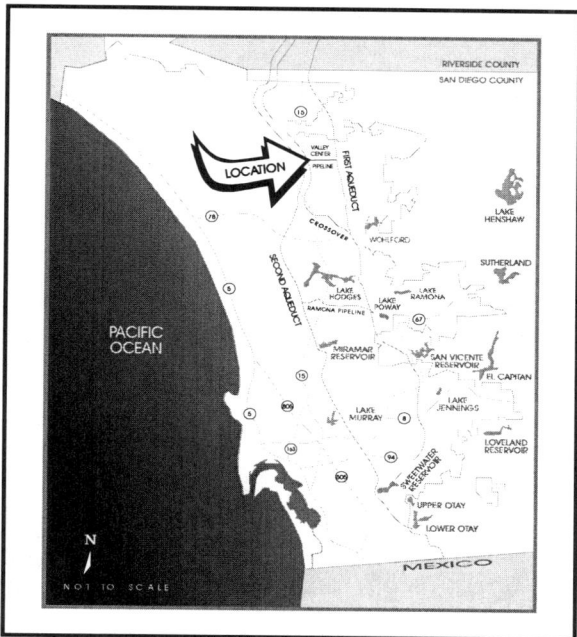

Figure 1. Location Map

Single and Double Welding for Thrust Restraint

The pipeline design specified both single and double lap welded pipe joints. Double welds were specified at locations where longitudinal thrust was being carried by the pipe. At locations with minimal longitudinal thrust, single welds were specified. These areas were in the reaches where soil pipe interface friction dissipated the longitudinal forces.

During construction, hydrostatic testing dictated that a bulkhead be installed in one area of single welded joints. This required an evaluation by the design engineer of the capacity of single welded joints for carrying longitudinal forces. A concern arose because, unlike butt joints, axial forces cause bending stresses in the joint region because of the geometric eccentricity. Strength of Bell-and-Spigot Joints (Roger Brockenbrough, 1990, Ref. 1) concludes under certain conditions that a single full thickness fillet weld can provide adequate strength. In Buried Pipe Design (A.P. Moser, 1990, Ref. 4), it is concluded that the eccentric loading can produce bending stresses as much as seven times greater than the longitudinal stress levels. Considering the uncertainties associated with joint geometry, and pipe and weld material properties, including ductility, it was decided to adopt a conservative approach and double weld the joints carrying longitudinal loads.

Geotechnical Issues

The pipeline alignment was generally characterized by variably weathered intrusive granitic rocks associated with the Southern California Batholith that lie within the Peninsula Range Geomorphic Province in California. The Maximum Credible Design Earthquake corresponded to a mean peak ground acceleration of 0.36 on the Elsinore fault zone with a design earthquake of Magnitude 7.4.

The geotechnical investigation (Woodward Clyde, 1992, Ref. 3) consisted of 25 borings ranging in depth from 1 meter to 15 meters for the pipeline, and 4 borings drilled to a depth of 12 to 23 meters for a tunnel crossing under Interstate 15 (Moncrief & Trembath, 1997, Ref. 2). Additionally, 18 seismic refraction surveys were performed to assess rippability of the granitic rocks and to identify subsurface rock characteristics for the tunnel. Geotechnical material property and environmental laboratory investigations were also completed to provide baseline material characteristics for design recommendations and construction.

The main geotechnical issues were limited to possible hazards over limited lengths of the alignment. These hazards included: traversing ancient landslides, high groundwater areas where crossing streams, creek or canyons, and liquefaction of saturated alluvial valleys.

Pipeline Coating System

The extremely rugged terrain in San Diego County is one of the primary reasons the Authority's Standard Specifications for construction projects includes a 25.5 mm thick cement mortar coating over the three layer cold-applied plastic tape coating system. Several reaches of the pipeline alignment had slopes in excess of 50 percent as shown on Figure 2. The construction equipment necessary to haul, lift, place, and backfill large diameter steel pipe on steep slopes can damage the tape coating system. The mortar coating over the tape provided an armor coating to protect the tape during handling and installation operations.

The mortar coating was applied at the pipe manufacturing plant at the same station as the tape coating operation. This prevented handling damage to the tape coating during the manufacturing process.

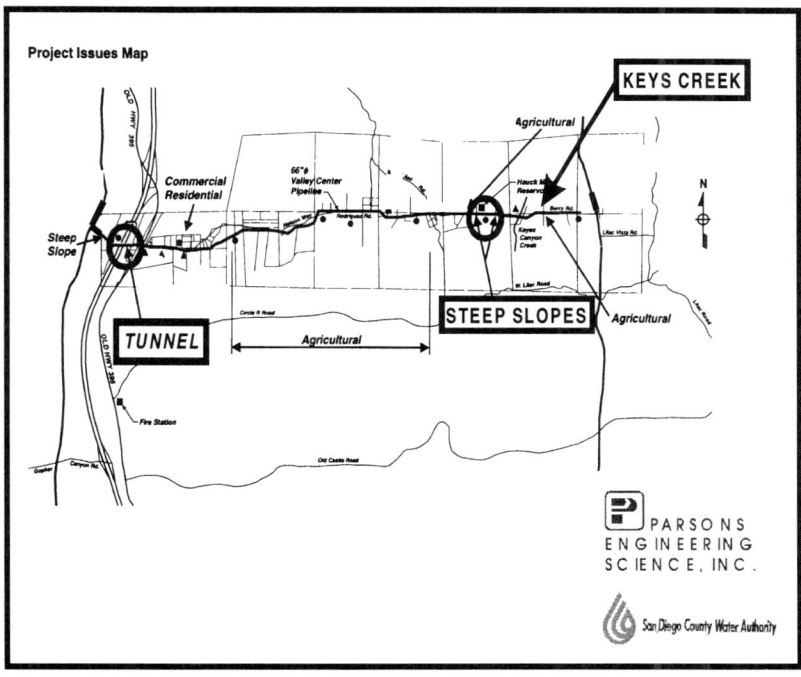

Figure 2. Pipeline Alignment

Construction

Coordination with the Public during Construction

Coordination with the public continued during the construction phase. A toll free telephone number was provided to the public to provide information and to enable the public to voice their concerns. Bulletins were issued regularly providing status information on the project construction. The construction manager was proactive in updating the property owners on the schedule and in ensuring any inconvenience to the property owners was minimized.

Keys Creek Crossing

Thick accumulations of alluvium were encountered in three major drainages crossed by the alignment. The most significant area was considered to be the South Fork of Keys Canyon, an alluvial reach that was approximately 91 meters long and approximately 7.9 meters deep at its deepest point. The creek is an ephemeral stream and during the dry season, flow is limited to minor nuisance flows. An old 6-meter high rock and concrete gravity dam was located approximately 152 meters downstream of the pipeline crossing. The dam was completely filled with sediment. Because of the high potential for liquefaction in this reach, the pipe was designed to be constructed in the granitic bedrock below the alluvial material and encased in reinforced concrete. The geotechnical investigation (Woodward Clyde, 1992, Ref. 3) included estimates for dewatering in this area of approximately 187 cubic meters per day, based on tests conducted during the design.

The contractor constructed a deep well dewatering system in Keys Creek to dewater the trench zone. Ten, 150 mm deep wells were installed to the top of the granitic bedrock along the alignment at depths from 5.5 meters to 8.5 meters. Supplementary pumps were also used in the trench to remove nuisance water. Water was discharged downstream of the excavation immediately north of the construction zone. Dewatering for a five week period did not result in significant water level reductions. The groundwater discharge was recycling back into the trench excavation due to the presence of the dam downstream of the site. The contractor extended the discharge pipe downstream of the dam and sufficiently dewatered the construction site for work to proceed. The presence of the dam was determined to be an unknown site condition which was not disclosed in the contract documents and could not have been reasonably seen during pre-bid site investigations. The contractor received an additional $82,000 for the additional dewatering and the extension of the discharge pipe.

Partnering

The Authority made the decision to include a partnering specification in the construction documents which invited the contractor, the design engineer, and construction manager to partner during the construction contract. The successful bidder accepted the invitation. One initial and two subsequent partnering sessions were held. The goal of the initial partnering session was to define responsibilities, establish lines of communication, identify areas of significant concern for each participant, and to establish some procedures for dispute resolution. Communication throughout the construction was very good and is considered to have contributed to completion of construction on time, within budget, and with no claims at the end of the job. The subsequent partnering sessions were useful in smoothing out and resolving some minor coordination issues.

References

[1] Brockenbrough, R.L., Strength of Bell-and-Spigot Joints, Journal of Structural Engineering, July 1990.

[2] Moncrief W.J. and Trembath R.J., Valley Center Pipeline Tunnel, ASCE, Trenchless Pipeline Projects, Practical Applications, 1997.

[3] Woodward-Clyde, Geotechnical Investigation Summary Report for San Diego County Water Authority, July 1992.

[4] Moser, A. P., Buried Pipe Design, McGraw & Hill, 1990.

Designing and Building a Major Transmission Main through a Constructed Environment, The North County Distribution Pipeline

San Diego County Water Authority
Escondido, California

Steve Tedesco[1]
Edward Stewart[2]

Abstract

The North County Distribution Pipeline (NCDP) was placed in service in January 1997, and can serve up to 0.4×10^6 m^3 (105 mgd) of water to four member agencies of the San Diego County Water Authority. The project included the design and construction of:

- 5.5-km (3.4 mi) of 1,830-mm (72-in) pipeline
- 0.004×10^6 m^3 (1.0 million gallon) regulator structure
- A chlorine contact tank
- Three flow control and metering structures
- A 213-mm (700-ft) rock tunnel beneath three existing SDCWA aqueducts.

The total project cost was $30 million including planning, public information, design, construction management, construction, acquisition of right-of-way and environmental mitigation.

---

[1]Director of Water and Wastewater, ASL Consulting Engineers, 11770 Bernardo Plaza Court, Suite 116, San Diego, CA 92128-2423

[2]Principal Civil Engineer, San Diego County Water Authority, 610 W. 5th Avenue, Escondido, CA 92025

This paper will outline the design and construction challenges which were met in order to successfully complete this project. The project was constructed through all types of conditions including:

- Cross Country Terrain
- Sensitive Environmental Areas
- Crop Land
- Rural Residential Communities
- Country Club Golf Course
- Rock Tunnel

Extensive coordination, communications and public relations were required during the design and construction phases.

Introduction

The San Diego County Water Authority (SDCWA) is a wholesale water supplier that supplies water to 23 retail member agencies in San Diego County, California. SDCWA encompasses 367,000 ha (907,000 acres) and provides water to more than 2.6 million residents.

The North County Distribution Pipeline (NCDP) was placed into service in January 1997. It provides up to 0.4×10^6 m³/d (105 mgd) of potable water to four agencies in northern San Diego County (City of Oceanside, Vista Irrigation District, Vallecitos Water District and Rainbow Municipal Water District). The project consisted of 5.5-km (3.4 mi) of 1,830-mm (72-in) diameter steel pipeline, three flow control and metering facilities, a 0.004×10^6-m³ (1-mil-gal) regulatory structure, a chlorine contact tank, and a 213-m (700-ft) hard-rock tunnel under three large existing SDCWA aqueducts that transport most of the region's water supply.

The total project cost was $30 million, including planning, public information, design, construction management, right-of-way acquisition and environmental mitigation. The project was constructed through an existing rural/residential area including extensive work at an operating water treatment plant and construction through an operating golf course and country club.

Design. The design of the NCDP posed a number of significant challenges. The project provides potable water from two separate sources. Treated water is provided from either the City of Oceanside's Weese Water Treatment Plant or from SDCWA treated water aqueduct. The City of Oceanside had previously provided water to its city using two 24-inch diameter pipelines from the treatment plant to the city which is approximately 8.1-km (5 mi) away. Treated water was disinfected using chloramines and the pipelines were used to achieve required chlorine contact time (CT).

NCDP design would change the overall operating conditions. The first customer on the new pipeline would now be 8.1-km (5 mi) closer to the treatment plant.. A new chlorine contact tank would have to be added to the pipeline in order to provide the required chlorine contact time.

Figure 1 shows the schematic of the overall NCDP system. The chlorine contact tank was designed and constructed below grade at the existing plant. It is connected to the pipeline through a 213-M (700-foot) hard-rock tunnel. The second source, the connection to the SDCWA aqueduct, is regulated through the NCDP1 flow control and metering structure. The two sources are connected to a 0.004×10^6-m^3(1.0 million gallon) regulating structure. This structure is used to balance flows and allow regulating storage in case a member agency is unable to accept ordered flows.

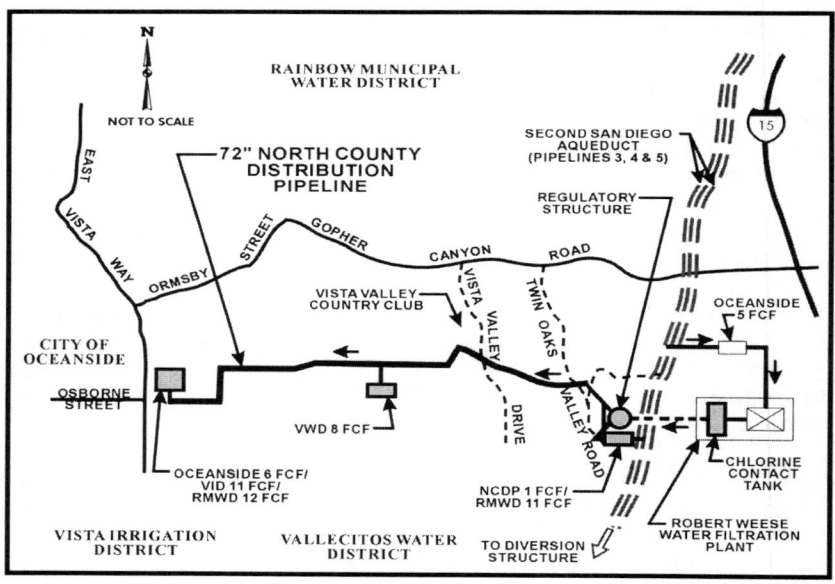

Figure 1 - Schematic Diagram of NCDP

Two additional flow control structures were also constructed to serve the member agencies. These facilities consist of above ground buildings which house meters, flow control valves, instrumentation and electrical equipment required to operate the system.

Because of highly corrosive soil known to exist in the San Diego County area it was decided to use a steel pipe with a dielectric coating and an armor rock shield. Figure 2 is a diagram showing the pipe coating system. Pipe thicknesses ranged from a minimum of 9.53- mm (3/8-inch) to 14.30-mm (9/16-inch). The dielectric coating used was a three layer tape coat. A 25.42-mm (1-inch) rock shield was designed to protect the pipeline tape coat during transportation and installation. It was extremely valuable on the project due to the rocky soil conditions.

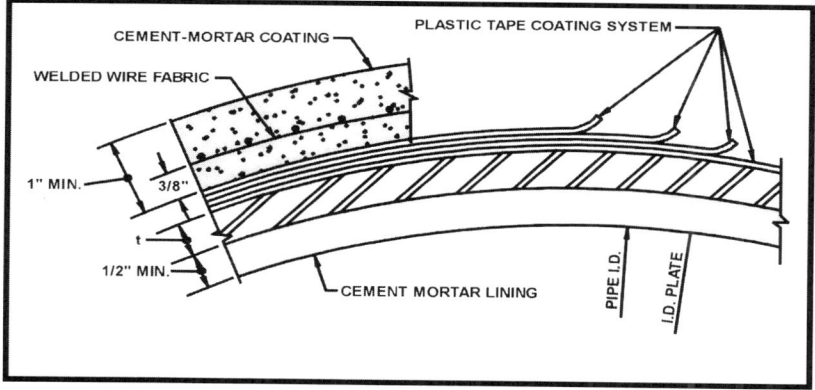

Figure 2 - Right-of-Way

One of the early design challenges was to find a way to provide adequate working room for construction. The alignment traverses very steep and rocky terrain. No roadways existed which followed the alignment. Many of the areas were accessible only by foot.

After extensive review, it was decided to obtain a 24.3-m (80-ft) to 30.4-m (100-ft) wide right-of-way along the alignment and to use a series of existing roadways to provide access to the site. Figure 3 shows the pipeline alignment along with the access road system.

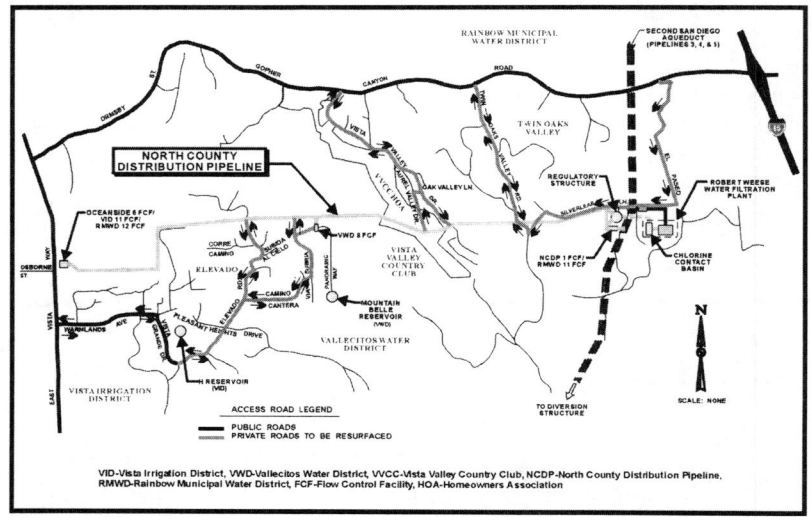

Figure 3 - Access Road System

During the design phase the access roads were evaluated. It was decided to require the contractor to adequately sign each access road, to only use the designated access roads and to repave the roads at the completion of the project. This approach, along with an extensive public relations campaign was very well excepted by the local community.

Probably the most difficult challenge on the project was how to traverse the Vista Valley Country Club and golf course. The pipeline would have to traverse the club and golf course and would affect at least three of the course's 18 holes including a high profile elevated green. After analyzing several alternatives it was decided to traverse perpendicular through one fairway and then traverse directly down the fairway on the 16th hole.

After extensive negotiations with the Country Club it was decided that the way to construct the project with the least amount of disruption was to close the courses for a nine week period. All construction including complete course restoration was required to be completed in this period.

Contact documents were prepared which specifically outlined requirements in the following areas:

Construction Packaging - The project was bid in four packages; the $13.3 million pipeline; the $7.7 million chlorine contact tank, regulatory structure, hard-rock, tunnel and flow control facility; the $1.7 million three member agency flow control facility; and the $320,000 single member agency flow control facility. All contracts had specific requirements to minimize community impacts.

Work Hours - Construction was limited to 7:00 a.m. - 5:00 p.m., Monday - Friday. Overtime would only be allowed with prior authorization by SDCWA. The nature and type of work to be performed along with how it would affect the community would have to be carefully considered.

Dust Control - Strict dust control requirements were placed in the contract documents. A large portion of the alignment traversed through fields of expensive fruit crops. Local farmers had expressed concerns that dust would lower crop yields and cause a hardship. Contractors were required to have a water truck on site at all times. Ground surfaces were wet down before and after trenching and grading to reduce potential dust problems.

Truck Access - A major concern of many local homeowners was high speed truck traffic and traffic on private roads. Approved haul routes were clearly shown on the project plans. SDCWA posted 24-km/h (15-mph) speed limit signs as well as signs designating haul routes. Meetings were held with subcontractors and haul road maps were provided. Truck drivers were required to have the maps on hand when using private roads. Additionally, homeowners were encouraged to notify SDCWA if trucks were seen speeding or using unauthorized roads.

Blasting - The geology in the project area included a large amount of hard rock which required blasting. Contractors were required to give seven days written notification to all residents located within 91-m (300-ft) of the blast area. Specifications required the contractor to perform pre-blast surveys to document the condition of all structures prior to the start of blasting.

In addition, mats or thick soil blankets were required to be placed over all blast sites to prevent flying rock. In areas with steep slopes, steel plates were used to prevent rocks from rolling down hills.

Golf Course Construction - During right-of-way negotiations a stipulation was placed on SDCWA that construction through the golf course must be completed and the course fully restored in a nine-week period. In order to meet this condition SDCWA hired a golf course architect to work closely with the golf course management on construction and restoration strategies.

The general contractor was required to subcontract with one of three experienced restoration subcontractors. Work could not start until May 1st and the course needed to be opened by July 7th in time for a tournament on July 8th.

Several large oak trees were required to be removed. Some were more than 100 years old. Existing trees were side-boxed six months prior to construction in preparation for the move. One tree required a 3,658-mm (144-in) box and a 90.7-metric ton (100-ton) crane.

Once the pipe was installed, the subcontractor immediately came in and replaced the irrigation system. Excess trench spoil was used to re-grade several areas of the course and reduced the contractor's hauling costs.

Environmental Considerations - The major environmental considerations were related to truck traffic, emergency access and disposal areas. The pipeline crossed a private road which was the only access into the neighborhood. The contractor was required to build a detour on a steep slope and to limit any traffic delays to a maximum of 10 minutes.

In order to reduce truck traffic the contractors were encouraged to set up a rock crushing operation. The rock crusher would not only reduce the material to be hauled from the site but it also reduced the need to import pipe zone material. An estimated 4,000 truck trips were saved by using the crusher. To further reduce truck traffic, SDCWA set aside a designated disposal site. The site was located on a sensitive coastal sage habitat which required special revegetation. This restoration needed to comply with the CEQA and USDFW recommendations.

Public Affairs

The planning and design of NCDP was conducted to provide a project which not only served its purpose but which also could be constructed without excessive disruption to the local community. However, SDCWA went one step further prior to and during construction. SDCWA embarked on an extensive communications and public affairs campaign.

SDCWA focused on an extensive public affairs plan early in the project. Community meetings were held more than a year prior to ground breaking. Large and small meetings were held to obtain input from concerned citizens along the alignment.

The SDCWA Project Manager corresponded with more than 300 residents describing the project's scope and schedule. A "hotline" telephone number was provided. Six detailed newsletters were published throughout the construction. Press releases, media coverage and door hangers were also provided.

As a result of this extensive public affairs campaign, very few complaints were received on this major project. Fewer than 50 calls were received during construction.

Summary

The NCDP project was constructed for a total cost of approximately $30 million. The golf course sections were completed within the nine week period and on July 8, 1996 Vista Valley Country Club held their tournament as planned.

Fewer than 50 calls were received by the Hotline during the construction period. Very few complaints were received and in fact, many residents have complemented SDCWA on how well they handled the project.

The facilities are operating and additional connections to the system are being planned.

PCCP DESIGN CHECK USING C304-92 IN A SPREADSHEET

Dr. Jey K. Jeyapalan, P.E. and Dr. Sri K. Rajah, Ph.D., P.E.
Pipeline Engineering Consultants
21 Meetinghouse Terrace, New Milford, CT, USA 06776
Phone 1-860-354-7299
email: jkjeyapalan@snet.net

Abstract

The manufacture of prestressed concrete cylinder pipe(PCCP) has been governed by AWWA C-301 for several decades. The pipe suppliers used either Method A or Method B to help the design engineer and the owner design and specify PCCP pipe for major water transmission and sewer force main projects. Numerous owners and design engineers began to question the validity of these over-simplified design curves in the appendices of AWWA C-301, particularly when the pipe intended to last forever or long time were falling apart. American Concrete Pressure Pipe Association(ACPPA) and its members responded to this criticism by developing a pair of new standards, namely AWWA C301 and C304, to address material properties and design methodology. The design steps involved in C304 are so complicated that ACPPA had an elaborate computer software developed to implement C304 for designers of this pipe. In most cases, the software was purchased primarily by the pipe manufacturers who belong to the ACPPA and either they would run design analyses for interested parties or ACPPA would sell a copy of the software for an enormous fee. The authors of this paper were experts on a number of lawsuits involving PCCP failures and came across the above problems. As part of the discovery phases for the members in the legal profession, the authors developed a simple spreadsheet based on C304-92 to check the structural adequacy of PCCP lines designed to the old C301-72 to C301-84 design standards rather efficiently. This paper provides an overview of this approach and a sample application of this spreadsheet for specific PCCP projects to illustrate the point that an adequate design methodology does not necessarily have to be complicated.

Introduction

For several decades, the design and manufacturing of Prestressed Concrete Cylinder Pipe (PCCP) has been in accordance with ANSI/AWWA C301 (American Water Works Association, 1972, 1984). Two distinct procedures designated as Method A and Method

B governed the design of PCCP and were described in Appendices A and B of ANSI/AWWA C301. These methods are also known as Cubic Parabola Design Method and Stress Analysis Method, respectively. These procedures only limited concrete core tensile stress and used generic design values for all component material characteristics, and thus did not address material and environmental variations. The existence of two methods for the PCCP design caused confusion and the extremely simplified design curves raised questions on the validity of the design procedures. These procedures were replaced with the new PCCP standard C304-92. This new standard presents a unified design method for both types of PCCP, namely lined cylinder pipe (LCP) and Embedded Cylinder Pipe (ECP).

The methodology presented in AWWA C304-92 is based on a literature review of the state of art of prestressed concrete design, theoretical work by Zarghamee and Fok (1990), and on the analysis of hundreds of PCCP performance tests made over several years. The new standard presents simplified practical design equations developed by evaluating time related stress variations using computerized numerical integration procedures. The resulting design steps in the PCCP design methodology are so complex that an elaborate computer program was developed by ACPPA for use in the design of PCCP. As experts providing litigation assistance to clients, the authors have encountered a number of projects where the failed PCCP pipes were designed using the old C301-72 to C302-84 standards. To help the discovery process, the authors developed a simple spreadsheet to check the structural adequacy of the PCCP lines efficiently using the current design standard C304-92. An overview of this spreadsheet is presented in this paper.

Review of PCCP Design Methodologies

In a prestressed concrete pipe, the concrete core and the steel cylinder act as primary load carrying components. The purpose of prestressing the core is to maintain compressive stresses in the concrete core under normal working conditions, as concrete is weak in tension.

Method A

The Cubic Parabola Design Method specified in the Appendix A of the C301 standard is

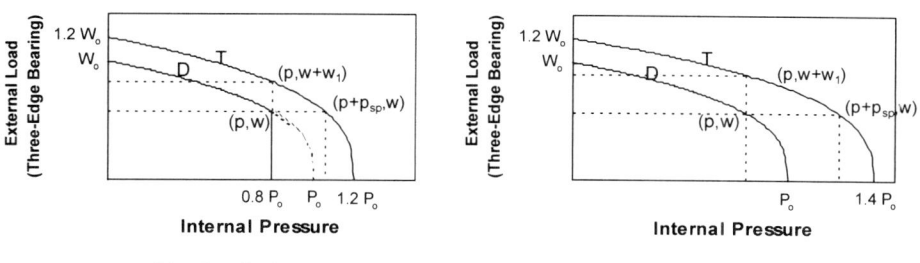

a) Lined Cylinder Pipe　　　　　　　　　　b) Embedded Cylinder Pipe

Figure 1: Design and Transient Capacity Curves for Lined and Embedded Cylinder Pipe using Cubic Parabola Design Method (Appendix A, AWWA C301-84)

generally known as "Method A". This method is based on: 1) W_o, which is nine tenths of the three-edge bearing test load that causes incipient cracking; and 2) the theoretical hydrostatic pressure P_o that relieves the residual compression in the concrete core due to prestressing. The allowable combinations of three-edge bearing load and internal pressure were determined by a cubic parabola, passing through W_o and P_o, as defined by the following equation:

$$w = W_o \left(\frac{P_o - p}{P_o} \right)^{\frac{1}{3}}.$$

Where

$$W_o = 0.9 W_{3-edge}$$

$$P_o = \frac{f_{cr}}{6D_y} \left[A_c + n_r \left(A_s + A_y \right) \right]$$

$$f_{cr} = f_{ci} \left[\frac{A_c + n_r \left(A_s + A_y \right)}{A_c + n_r \left(A_s + A_y \right)\left(1 + C_r \right)} \right]$$

$$f_{ci} = \frac{A_s f_{sg} \left(1 - R_1 - R_2 \right)}{A_c + n_i \left(A_s + A_y \right)}$$

where, R_1 and R_2 are relaxation factors; C_r is shrinkage strain; f_{cr} is final prestress in the core; f_{ci} is the initial prestress in the core; P_o is the internal pressure required to overcome all compression in the core concrete, exclusive of the effect of external load; W_o is nine tenths of the three-edge bearing load producing incipient cracking in the core, with no internal pressure; p is the maximum design pressure in combination with three-edge bearing load w; and w is the maximum three-edge bearing load, equivalent to dead load, in combination with design pressure p. All other variables are as given in AWWA C-304. Design (D) and transient (T) capacity curves resulting from the Cubic Parabola Design Method is shown in Figure 1 for both embedded-cylinder and lined-cylinder pipes.

Method B

The Stress Analysis Design Method outlined in the Appendix B of AWWA Standard C301 is also known as 'Method B'. This procedure is based on limiting the maximum combined net tensile stress in pipe under static external load and internal pressure to a value equal to $7.5\sqrt{f_c'}$, where f_c' is the specified 28-day compressive strength of the concrete. This concept is illustrated in Figure 2 and the resulting design curve is defined by the following equation:

$$p = \left(f_{cr} + 7.5\sqrt{f_c'} \pm \frac{M}{S} \pm \frac{F}{A_t} \right) \frac{A_t}{12 R_y}$$

where, p is maximum design pressure in combination with field dead-load; w, not to exceed 0.8 P_o for lined-cylinder pipe or P_o for embedded-cylinder pipe; f_{cr} is resultant induced compression; M is total moment in the pipe section due to pipe weight, water weight, and external static load; and F is total thrust in the pipe section due to pipe weight, water weight, and external static load; S is section modulus of the control pipe section based on the total pipe wall at the crown and invert sections and on the prestressing wire and core only at the side section; A_t = transformed cross-sectional area of the control section based on the total pipe wall at the crown and invert sections and on the prestressing wire and core only at the side section; and R_y = outside radius of the steel cylinder. The design (D) and transient (T) capacity curves resulting from the Stress Analysis Design Method are shown in Figure 2 for lined-cylinder and embedded-cylinder pipes.

PCCP Design using C304-92 Standard

The current design methods for buried PCCP under internal pressure defined in the C304-92 standard includes the effects of working, transient, and field-test load, and internal pressure combinations. The new standard uses appropriate design values for component material characteristics as summarized in Table 1 and uses limit states design criteria. The limit state design criteria limit circumferential thrust and bending moment resulting from internal pressure, external loads, pipe weight, and fluid weight. Also, the design criteria assures that the state of prestress and safety margins for adequate strength will be maintained even if the pipe is subjected to abnormal conditions that may cause visible cracking. The design procedure consists of the following three limit-states criteria: 1) Serviceability Limit State; 2) Elastic Limit State; and 3) Strength Limit State.

Serviceability limit-states criteria ensure the performance of the pipe under service loads by precluding the microcracking from occurring in the core and by controlling microcracking in the coating under working loads and pressures. Also it is intended that the criteria preclude visible cracking from occurring in the core and the coating under working plus transient loads and pressures. The serviceability limit state criteria include 1) Core-crack control; 2) Radial tension control; 3) Coating-crack control; 4) Core-compression control; and 5) Maximum pressure.

Elastic limit states criteria define the onset of material non-linearity. The criteria limit combined working plus transient loads and pressures so that if cracks develop in a prestressed pipe under the transient condition, the pipe response will be elastic and damage or loss of prestress will not occur. The elastic limit state criteria include 1) Wire-

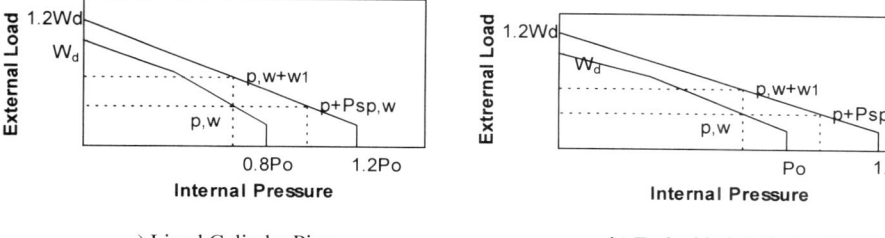

a) Lined Cylinder Pipe   b) Embedded Cylinder Pipe

Figure 2: Design and Transient Capacity Curves for Lined and Embedded Cylinder Pipe using Stress Analysis Design Method (Appendix B, AWWA C301-84)

Table 1: Essential differences in component properties between new and old PCCP design standards

Old PCCP Design Methods (ANSI/AWWA C301-84)	New PCCP Design Standard (AWWA C304-92)
Stress/Strain Diagram for Mortar/Concrete (linear elastic in tension and compression)	Stress/Strain Diagram for Mortar/Concrete (with visible cracking branch in tension)
Creep and Shrinkage factor for Cast Pipe = 2.0 Factors for Coating or Inner and Outer Core were not separated	Creep Factor Formula is: $$\phi = \frac{(h_{co} + h_m)\phi_{com} - h_m\phi_m + h_{ci}\phi_{ci}}{h_{ci} + h_{co}}$$ Shrinkage Factor Formula is: $$s = \frac{(h_{co} + h_m)s_{com} - h_m s_m + h_{ci} s_{ci}}{h_{ci} + h_{co}}$$ The effect of volume-to-surface ratios and relative humidity exposures is included
Limits stresses/strains in the concrete core	Limits stresses/strains in all component materials (concrete core, mortar coating, steel cylinder, and prestressing wire to preclude cracking or yielding
Steel Modulus Wire Modulus, $E_s = 28 \times 10^6$ psi Cylinder Modulus, $E_y = 28 \times 10^6$ psi	Steel Modulus Wire Modulus, $E_s = 28 \times 10^6$ psi Cylinder Modulus, $E_y = 30 \times 10^6$ psi
Cast concrete initial modular ratio, $n_i = 7$	Cast concrete initial modular ratio $$n_i = 109(f_c')^{-0.3}$$ $$n_i' = 117(f_c')^{-0.3}$$ (Prime indicates ratio to cylinder steel)
Resultant Modular ratio, $n_r = 6$	Resultant modular ratio $$n_r = 93(f_c')^{-0.3}$$ $$n_r' = 99(f_c')^{-0.3}$$ (Prime indicates ratio to cylinder steel)
Wire relaxation loss = 5 %	Wire relaxation loss Cast concrete : $R = 0.111 - 3.5\dfrac{A_s}{A_c}$ Spun concrete : $R = 0.13 - 3.1\dfrac{A_s}{A_c}$
Pipes were designed using generic material characteristics without requiring material testing	All component materials used in each pipe factory are tested. If a characteristic of available material does not meet or exceed the default characteristic assumed in design, then the actual material characteristics are used in design.

stress control; and 2) Steel-cylinder stress control.

Strength limit states are defined to provide safety and to protect the pipe under extreme loads. The criteria protect the pipe against yielding of the prestressing wire, crushing of the concrete core under external load, and tensile failure of the wire under internal pressure. Safety factors are applied to loads and pressures that produce the strength limit states. The strength limit state criteria include 1) Wire yield-strength control; 2) Core compressive-strength control; 3) Burst-pressure control; and 4) Coating bond-strength control.

The three limit state criteria described above determine the allowable combinations of internal pressure, external loads, pipe weight, and fluid weight to assure adequate serviceability of the pipe under working plus transient design loads and pressures. The load factors to be used in the design for differing loading conditions are summarized in Table 2 for both embedded and lined cylinder pipes. Table 3 provides a summary of different limit state conditions and associated loading conditions for both type of pipes.

Table 2: Load and pressure factors for Lined and Embedded Cylinder Pipes

Loading Conditions	Loads and Pressures (ECP)							Loads and Pressures (LCP)						
	We	Wp	Wf	Wt	Pw	Pt	Pft	We	Wp	Wf	Wt	Pw	Pt	Pft
Working Loads and Pressure Combinations														
W1	1.00	1.00	1.00	-	1.00	-	-	1.00	1.00	1.00	-	1.00	-	-
W2	1.00	1.00	1.00	-	-	-	-	1.00	1.00	1.00	-	-	-	-
FW1	1.25	1.00	1.00	-	-	-	-	-	-	-	-	-	-	-
Working Load Plus Transient Load and Pressure Combinations														
WT1	1.00	1.00	1.00	-	1.00	1.00	-	1.00	1.00	1.00	-	1.00	1.00	-
WT2	1.00	1.00	1.00	1.00	1.00	-	-	1.00	1.00	1.00	1.00	1.00	-	-
WT3	1.00	1.00	1.00	1.00	-	-	-	1.00	1.00	1.00	1.00	-	-	-
FWT1	1.10	1.10	1.10	-	1.10	1.10	-	1.20	1.20	1.20	-	1.20	1.20	-
FWT2	1.10	1.10	1.10	1.10	1.10	-	-	1.20	1.20	1.20	1.20	1.20	-	-
FWT3	1.30	1.30	1.30	-	1.30	1.30	-	1.40	1.40	1.40	-	1.40	1.40	-
FWT4	1.30	1.30	1.30	1.30	1.30	-	-	1.40	1.40	1.40	1.40	1.40	-	-
FWT5	1.60	1.60	1.60	2.00	-	-	-	1.60	1.60	1.60	2.00	-	-	-
FWT6	-	-	-	-	1.60	2.00	-	-	-	-	-	1.60	2.00	-
Field-Test Condition														
FT1	1.10	1.10	1.10	-	-	-	1.10	1.10	1.10	1.10	-	-	-	1.10
FT2	1.21	1.21	1.21	-	-	-	1.21	1.32	1.32	1.32	-	-	-	1.32

Spreadsheet Implementation

With increasing availability of faster computers and powerful spreadsheet software, problems with increasing complexity are being solved utilizing the spreadsheets. Although, the need for programming the complex equations are not eliminated, the ease with which the problems are modeled and differing model situations are analyzed is remarkable. One of the strengths of this approach is that the computations are being performed even while the data is being entered and the model is being manipulated. Overall, this distributed computing resulting in a higher apparent speed, with which the calculations are performed, better than that of a conventional stand-alone program.

Table 3: Limit States and load combinations for Embedded and Lined cylinder pipes

Location	Purpose (To Preclude)	Load Combination	Criteria - ECP	Criteria - LCP
Serviceability Limit State				
Full Pipe	core decompression	W1	Internal working pressure	Internal working pressure
Circumference	coating cracking	WT1	Internal working pressure	Internal working pressure
Invert/Crown	onset of core microcracking	W1	Inside core tensile strain	Inside core tensile strain
		FW1	Inner core-to-cylinder radial tension	-
	onset of core visible cracking	WT1, WT2, FT1	Inside core tensile strain	Inside core tensile strain
		WT3	Inside core tensile strain	-
Springline	onset of core microcracking & to control microcracking of coating	WT1	Outer core tensile Strain & outer coating Tensile strain	Outer core tensile Strain & outer coating Tensile strain
	coating visible cracking	WT1, WT2, FT1	Outer core tensile strain & outer coating tensile strain	Outer core tensile strain & outer coating tensile strain
	core compression	W2	Inner core compression	Inner core compression
		WT3	Inner core compression	Inner core compression
Elastic Limit State				
Invert/Crown	exceeding limit stress in steel cylinder	WT1, WT2, FT1	Cylinder stress Reaching yield	-
		WT3	Onset of tension in Cylinder	-
Springline	exceeding wire limit stress & maintain core compression stress below 0.75 fc'	FWT1, FWT2, FT2	Wire stress limit & Core compression	Wire stress limit & Core compression
Strength Limit State				
Springline	wire yielding	FWT3, FWT4	Wire stress limit	Wire stress limit
	core crushing	FWT5	Ultimate moment	Ultimate moment
Burst pressure	to prevent burst failure	FWT6	Internal pressure is less than burst pressure	Internal pressure is less than burst pressure

With the ease of post-processing, report generation within a spreadsheet and the flexibility of modeling, the spreadsheet programming is attractive for problems that can be programmed within a spreadsheet. With programming languages and numerical tools linked with spreadsheet programs, the range of problems that can be analyzed within a spreadsheet is very wide.

The PCCP design problem presented here has been implemented in an EXCEL spreadsheet. Use of macros, written in Visual Basic has been kept to an absolute minimum with the equations embedded in the cells. By protecting the cells that contain equations, the interface between the user is kept only to the necessary user input. A sample input-sheet is shown in the Figure 3.

Figure 3: Input Screen for the PCCP Design spreadsheet

Type of Pipe: LCP
Units: English

Data From Purchaser		Data from Pipe Manufacturer		Design Values	
Inside diameter of the Pipe, in.	72	Outside diameter of the steel cylinder, in.	75.5	Design modulus of elasticity of prestressing wire, psi	28000000
Fluid Unit Weight, lb/ft^3	62.4	Thickness of the steel cylinder, in.	0.0598	Unit weight of prestressing wire, lb/ft^3	489
External Dead Load, lb/ft	6000	Diameter of the prestressing wire, in.	0.192	Unit weight of concrete, lb/ft^3	145
External surcharge load, lb/ft	0	Class of Prestressing wire (II/III)	252000	Design 28-day compressive strength of core concrete, psi	5500
External Transient Load, lb/ft	0	Number of layers of prestressing wire (1/2/3)	1	Design 28-day compressive strength of mortar, psi	5500
Internal Working Pressure, psi	150	Coating thickness over the prestressing wire, in.	0	Unit weight of mortar, lb/ft^3	140
Internal Transient Pressure, psi	60	Coating thickness between layers of prestressing wire, in.	0	Design modulus of elasticity of cylinder, psi.	30000000
Internal Field-test Pressure, psi	180	Concrete 28-day compressive strength, psi	5500	Tensile yield strength of steel cylinder, psi	33000
Time Period of Exposure to outdoor environment (days) if more than 270 days	270	Concrete modulus of elasticity multiplier (if less than 0.9)	0.9	Design tensile strength of steel cylinder at pipe burst, psi	45000
Time Period of Exposure of Pipe to Burial environment before water filling (days) if more than 90 days	90	Concrete creep-factor multiplier (if greater than 1.1)	1.1	minimum tensile strength of wire, psi	40000
Relative humidity of the outdoor environment	70	Concrete shrinkage strain multiplier (if greater than 1.1)	1.1	Height of Earth Cover, ft	5.5
		Prestressing wire intrinsic relaxation multiplier (if greater than 1.1)	1.1		

The present implementation offers a larger window on the computations performed each step has been reproduced in the spreadsheet. However, this increased visibility of the data (i.e., results) is often confusing and overwhelming to an inexperienced user. In such a situation, increasing the variables in the Visual Basic macros can reduce the visibility of the data or select sheets can be hidden from the user keeping the similar programming style. Table 4 shows the computed design variables for the above example. Also, Table 5 shows a typical calculation summary for the serviceability conditions at invert/crown and the results show that the design satisfies inside core micro-cracking, inside core visible cracking, and inside core-to-cylinder radial tension.

Summary

The paper presents a novel and an efficient spreadsheet program to check the design of PCCP using AWWA C304-92 standard. Owners of PCCP pipelines would find this approach extremely useful, as the material properties of the pipe, ground conditions, and loading on their pipelines vary during the life of the project. As part of the periodic inspection and maintenance program, this design check could be done effectively to determine the structural health of the pipeline, rather than let the pipeline blow up with absolutely no warning. The approach of using spreadsheet programming to perform PCCP design check in a computer is flexible, simpler, and efficient.

Table 4: Computed design variables for a sample problem

Core		Cylinder		Wire		Concrete		Mortar	
D_i	72	t_y	0.0598	d_s	0.192	f_c'	5500	f_m'	5500
D_y	75.5	f_{yy}	33000	f_{su}	252000	γ_c	145	γ_m	140
h_c	5.5	$f_{yy}*$	45000	f_{sg}	189000	E_c	3840887	E_m	4E+06
Steel cylinder and concrete		E_y	3E+07	E_s	28000000	n	7.28998	m	0.9484
A_y	0.718	h_{ci}	1.69	f_{sy}	214200	n'	7.81069	f_{tm}'	519.13
A_c	65.28	d_y	1.72	λ_s	0.017455	f_t'	519.134	ε_{tm}'	0.0001
Coating:		λ_y	0.31	ε_{sg}	0.00675	ε_t'	0.00014	ε_{km}'	0.0011
h_m	0.94			ε_{sy}	0.00765	ε_k'	0.00149		
λ_m	0.0856			γ_s	489				
R	39.22								
Environment		PH	70%	t_1 270 days		t_2 90 days			
Pressures:		Earth Load and Fluid Weight:		Earth Load (bedding: 90° Olander):		Pipe weight (bedding: 15° Olander)		Fluid weight (bedding:90° Olander)	
P_w	150	γ_f	62.4	C_{m1e}	0.1247	C_{m1p}	0.2157	C_{m1f}	0.1208
P_t	60	W_e	6000	C_{m2e}	0.0885	C_{m2p}	0.1016	C_{m2f}	0.0878
P_{ft}	180	W_t	0	C_{n1e}	0.3255	C_{n1p}	0.1029	C_{n1f}	-0.2703
		W_f	1764.3	C_{n2e}	0.5386	C_{n2p}	0.3026	C_{n2f}	-0.0617

References

American Water Works Association, (1992), "AWWA Standard for Design of Prestressed Concrete Cylinder Pipe- AWWA C304-92", AWWA.

American Water Works Association, (1992), "AWWA Standard for Prestressed Concrete Pressure Pipe, Steel-Cylinder Type, for water and other liquids- AWWA C301-92", AWWA.

American Water Works Association, (1984), "AWWA Standard for Prestressed Concrete Pressure Pipe, Steel-Cylinder Type, for water and other liquids- AWWA C301-84", AWWA.

Zarghamee, M. S. and Fok, F. L.(1990), "Analysis of Prestressed Concrete Pipe Under Combined Loads," ASCE Journal of Structural Engineering, Vol. 116, No. 7, p 2022-2039.

Table 5: Typical calculation summary for Serviceability conditions at Invert/Crown

	WT1	WT2	FT1	W1	FW1	WT3
P	210.0	150.0	198.0	150.0	0.0	0.0
M_1	51693.7	51693.7	56863.1	51693.7	59030.0	51693.7
N_1	93483.7	66303.7	87883.1	66303.7	-2134.5	-1646.3
v	10.0	10.0	10.0	10.0	10.0	10.0
v_2	1.7	-0.1	1.2	-0.1	-1.7	-1.7
k	0.7	0.4	0.6	0.4	-0.3	-0.3
t_t	1.4	2.4	1.6	2.4	2.4	2.5
t_e	2.4	-0.3	1.9	-0.3	-3.9	-4.4
λ	0.7	-5.6	0.9	-5.6	-0.4	-0.4
"=>" m	0.3655	0.5926	0.5926	0.5926	0.9484	0.9484
	2.9959	3.0857	3.0098	3.0857	3.2160	3.2160
ε_{mm}	-0.0003	-0.0002	-0.0003	-0.0002	0.0000	0.0000
"=>" m	0.3655	0.5926	0.5926	0.5926	0.9484	0.9484
			Strains			
ε_{ci}	0.00036	0.00012	0.00030	0.00012	-0.00009	-0.00010
Δ_{ey}	0.00020	0.00002	0.00015	0.00002	-0.00019	-0.00019
eco	0.00015	0.00019	0.00017	0.00019	0.00040	0.00039
Des	0.00016	0.00020	0.00018	0.00020	0.00041	0.00040
emm	-0.00007	-0.00005	-0.00006	-0.00005	0.00016	0.00015
emo	-0.00003	-0.00002	-0.00002	-0.00002	0.00019	0.00018
			Stresses			
"=>" m	0.9	0.9	0.9	0.9	0.9	0.9
fci	431.8	452.9	456.4	452.9	-339.6	-384.9
Dfy	6044.2	620.6	4589.7	620.6	-5614.9	-5755.5
fcy	493.7	79.5	512.3	79.5	-718.9	-736.9
fco	586.0	741.3	641.4	741.2	1552.3	1510.3
Dfs	4524.1	5555.7	4903.4	5555.6	11470.4	11153.2
fms	-386.8	-252.5	-337.4	-252.6	516.9	475.7
fmm	-258.8	-175.3	-221.8	-175.3	595.3	548.4
fmo	-98.1	-78.3	-76.5	-78.3	693.8	639.8
			Internal Forces			
F'ci	-49440.4	-81919.5	-54704.3	-81921.8	-80672.9	-86929.8
F"ci	31092.6	65805.2	38430.8	65804.4	-55915.3	-68800.9
Fci	-18347.8	-5669.2	-16273.5	-5668.8	-3138.8	-4344.8
Fy	-3983.1	-388.3	-2925.9	-388.3	3513.4	3601.4
Fco	5727.9	15183.7	7591.8	15183.6	65571.3	66887.1
Fs	2774.6	3281.7	2961.1	3281.6	6188.7	6032.8
F'm	-2371.1	-1539.1	-2074.3	-1539.1	2808.2	2583.2
F"m	-554.5	-442.7	-432.5	-442.8	3921.5	3616.4
Fm	-2925.5	-1981.8	-2506.8	-1981.9	6729.7	6199.6
ΣF	0.0	0.0	0.0	0.0	0.0	0.0
			Internal Moments			
Mci	72266.3	24409.5	67675.1	24408.1	13514.5	18707.1
My	15437.9	1505.2	11340.6	1505.0	-13617.5	-13958.6
Mco	-3660.1	-18736.4	-5720.9	-18736.4	-160176.4	-170990.7
Mm	-811.9	-571.0	-682.3	-571.1	2698.4	2487.1
ΣM	nearly zero	nearly zero	nearly zero	nearly zero	nearly zero	nearly zero

Assuring Top Quality Prestressing Wire in PCCP

Ralph T. Rundle, P.E. [1], John Olden, P.E. Member ASCE [2] and Will Worthington, P.E. Member ASCE [3]

1. Abstract

The City of Houston, Texas, and the Coastal Water Authority determined the need for a 2.44 meter (96-inch) raw water supply pipeline of 11 km (7 miles) in length, to augment the raw water supply to the City of Houston's Southeast Water Purification Plant. The two utility owners have used prestressed concrete cylinder pipe, (PCCP) in the past, however the otherwise excellent performance of PCCP has been marred by several ruptures which have been attributed to the "hydrogen embrittlement" phenomenon. The recent improvements in PCCP which have been incorporated into AWWA Standard C301-92 were noted, however additional measures were identified which would further assure the quality of PCCP, particularly with regard to the prestressing wire quality. These measures were summarized in a "Supplement to the Requirements of C301-92," satisfying these owners that inclusion of PCCP as an allowable pipe material in their specification would be advisable. These measures included the pre-qualification of wire manufactures by requiring that the manufacturers demonstrate the performance of their wire products in a hydrogen embrittlement sensitivity test, and certain other quality assurance measures. Bidders were allowed to bid either welded steel pipe or PCCP. The successful bidders in fact did include PCCP. Implementation of the AWWA Standard C301 supplementary wire quality assurance measures during wire and pipe manufacturing operations will be discussed. This paper will be of great interest to pipeline engineers and owners who have encountered the hydrogen embrittlement phenomenon in their PCCP pipelines or are considering PCCP for future work.

1. Ralph T. Rundle, Project Engineer, Executive Director, Coastal Water Authority, 1200 Smith Street, Suite 2260, Houston, TX 77002, phone 713-658-9020, fax 713-658-9429

2. John Olden, Project Engineer, Brown & Root, Inc., P.O. Box 3; Houston, TX 77001-0003; phone 713-260-3172; fax 713 260 3225

3. Will Worthington, President, Pipeline Technologies, Inc. 1435 North Hayden Road, Scottsdale, AZ 85257-3773; phone 602-451-3500; fax 602-970-6355

2. Background

Coastal Water Authority

The Coastal Water Authority (CWA), created by the Texas Legislature in 1967, is located in a three-county area encompassing all of Harris County and parts of Chambers and Liberty Counties, Texas. Acquisition and construction of facilities to transport water from the Trinity River to the Houston area was the fundamental, but not limiting, purpose for CWA's creation.

The Coastal Water Authority is governed by a Board of seven Directors, three appointed by the Governor of Texas and four by the Mayor of the City of Houston. The Coastal Water Authority and the City of Houston have entered into a contract wherein CWA agreed to finance, construct, maintain, and operate facilities to transport the City's untreated water from the Trinity River to the Houston Metropolitan Area.

The Coastal Water Authority project, designed by the Authority's consulting engineers, Brown & Root, Inc., was built in two segments which are referred to as the Conveyance System and the Distribution System.

The Conveyance System facilities include the Trinity River Pump Station, the 35 km (22 mile) Main Canal, 23 km (14 miles) of lateral canals and the 81 ha (200 acres) Lynchburg Reservoir at the termination of the Main Canal. In 1996, CWA contracted with the City of Houston to operate and maintain the City's Lake Houston Facilities to transport raw water from Lake Houston to the City's East Water Purification Plant. The Lake Houston Facilities include the Lake Houston Pump Station and the 23 km (14-mile) West Canal.

At the south end of the Lynchburg Reservoir, the Distribution System begins with the Lynchburg Pump Station. The pump station has eleven pumps with the capability to pump 29.5 cubic meters per second (673 million gallons per day) with one stand-by pump on each system (A, B, & C) and an ultimate design capacity of 44 cubic meters per second (1 billion gallons per day). The Distribution System provides water for two City of Houston water purification plants as well as Houston's industrial areas along the Houston Ship Channel and in the Bayport Industrial Complex. The Distribution System downstream from the Lynchburg Pump Station contains over 81 km (50 miles) of pressure pipelines varying in size from 2.74 m (108-inch) diameter to 1.22 m (48-inch).

Need for the 96-Inch Pipeline

The City's growth in water demand and conversion from ground water to surface water dictated the need for a new raw water pipeline to supply the City's Southeast Water Purification Plant. The water plant and the Bayport Industrial complex were supplied by a 1.52 m (60-inch) raw water supply pipeline flowing at capacity, 4.8 cubic meters per second (110 mgd). Further industrial growth was anticipated as well as expansion of the water plant.

Previous Experience with PCCP

The CWA System includes over 81 km (50 miles) of large diameter pipelines including 6 km (4 miles) of steel and 74 km (46 miles) of PCCP. CWA experienced two failures of 2.74 m (108-inch) PCCP manufactured in the early 1970's. The failures occurred in 1990 and 1991. Wire from the 1990 failed pipe as well as wire from adjacent intact pipe was examined by Battelle Laboratory (Kajumdar & Cialone, 1991). The failed wire had exceedingly high tensile and yield strength as compared to the intact wire. It was concluded that the failed wire had been fully strain aged prior to being placed into service and thus was most likely dynamically strain aged during the wire drawing operation. The intact wire had excellent resistance to hydrogen stress cracking in spite of being in service for the past 16 years in a cathodic protection environment. The resistance of the intact wire was comparable to the best Class III wire tested previously at Battelle.

3. Improving PCCP Performance

Prestressed Concrete Cylinder Pipe (PCCP)

Prestressed Concrete Cylinder Pipe (PCCP) was developed and used in the United States in the mid 1940's as a means to provide an economical means of conveying water under pressure. (Prosser, 1996) The nature of the steel alloy used and the method of applying the steel wire to the concrete pipe core under tension, made very efficient use of the concrete and steel materials which constitute PCCP. The competitive cost of PCCP, its durability, and its anticipated long life led to increasing acceptance and use of the product. By the late 1960's, PCCP was the preferred pipe product for large pressure water pipelines in the United States. (Prosser, 1996) The product is widely used in other parts of the world as well, including Europe, the Middle East, Australia, South America, and China - where the world's newest PCCP plant was opened in 1994. (ACPPA Concrete Pressure Pipe Digest, 1996)

PCCP is designed to utilize the compressive strength of concrete and the tensile strength of steel. The principal components of this type of pipe are a core consisting of concrete and a light gage steel cylinder to which steel joint rings are welded, high tensile steel wire and a cement mortar coating. The design and manufacture of this type of pipe has been standardized under American Water Works Association Standard C301-Prestressed Concrete Pressure Pipe-Steel Cylinder Type, for Water and Other Liquids. PCCP is available in sizes from 400 mm (16") in diameter and greater.

The light gage steel cylinder functions as a water-tight membrane insuring the impermeability of the finished pipe. The concrete core is placed either centrifugally in a horizontal mold or vibrating into a vertical form, and then steam cured. High tensile wire is helically wound around the core under measured tension and at uniform spacing to induce the compressive strength required for the design considerations. The pipe is completed by placing a machine applied cement mortar coating on the exterior of the wrapped core.

The core is compressed so that at the highest operating pressure for which the pipe is designed - no tension is induced in the core. All of the tension is instead taken in the prestressed wire. When subjected to both internal operating pressure and external loads, the compression must be sufficient to prevent any cracking of the core. When the pipe is subjected to transient loads such as water hammer or live loads such as vehicle loads, in addition to the operating loads, the induced compression must insure elastic behavior of the pipe structure.

When corrosion occurs, it generally starts on the exterior of the pipe with a breakdown of the protective cement mortar coating (typically 1.9 cm (3/4") in thickness). Once the mortar coating has been compromised, the prestressed wire is subjected to corrosion and deterioration of the wires, which causes the wire to loose its strength and break or snap. After a wire breaks, the Poisson effect causes the severed ends of the wire to immediately re-anchor into the mortar coating after slipping only a very short distance. The disturbance of corroding wires tends to further stress and crack the mortar coating, which in turn leads to corrosion on adjacently wound prestressed wires. As corrosion and breakage of wires occurs in a localized area, the strength of the compressed concrete cylinder is diminished - eventually leading to instantaneous rupture. (Worthington, 1992)

A Typical PCCP Failure Investigation

Due to a wide variety of causes, there are over five hundred (500) recorded failures of PCCP in the world. (Price, 1995) The causes of failure are related to improper design, manufacturing flaws, improper construction, improper operation, and damage by outside agents. Generally, there is a relatively even spread of these causes, both geographically and temporally. (Clift, 1991) There was, however, a period of time in the mid-1970's when manufacturing practices resulted in a product which has suffered unusual number of ruptures, and which affected PCCP's reputation and statistical performance figures significantly. These practices were the result of well-intended efforts to improve the efficiency of use of the prestressed steel wire, and entailed drawing the wire at higher and higher temperatures. In the course of increasing steel strength, a point was reached when the steel quality and durability was sacrificed. This practice has come to be known as "dynamic strain aging". The effect of dynamic strain aging has become more obvious as a number of pipelines have failed in the first or second decade of service, leaving pipeline owners to cope with the consequences. Investigations and/or litigations have followed, notably in Newport News, VA; San Francisco; San Bernardino, CA; Oklahoma City; Tampa; Phoenix; San Diego; El Paso; Wichita Falls, TX; Washington D.C.; New Jersey and many other locations. (Denver, 1990)

A comprehensive investigation of the cause of a PCCP failure will explore every aspect of pipeline performance. It will address each stage of the pipeline's production - from pre-design data collection through manufacturing, construction, and operation. It will look at each component of the PCCP section itself and everything in contact with the pipe - the concrete, the steel, the mortar, the gaskets, the bedding, and the water. The investigations consume a great deal of time and money.

Dynamic Strain-Aged Wire Tests

One of the more telling tests which has been frequently performed on the prestressing wire seeks to identify the presence of dynamically strain aged prestressing wire, and the resulting susceptibility to hydrogen embrittlement. The test is not a standard test in the US or Canada, but rather is borrowed from Europe where it was developed. It is known variously as the "ammonium thiocyanate test", or the "FIP test", (for the French federation involved in its development), or the "direct charging hydrogen embrittlement susceptibility test". We will refer to it as the Hydrogen Embrittlement Sensitivity Test, or HEST. One of the few labs in the US with the capability of performing the HEST is that of Lewis Engineering in Gainesville, Florida.

The HEST determines the time to fracture of a sample maintained at a constant tensile force and immersed in a solution of ammonium thiocyanate, NH_4SCN, at a constant temperature. The ammonium thiocyanate causes embrittlement of the wire by atomic hydrogen, mimicking the hydrogen embrittlement phenomenon on PCCP. The test is performed by applying a tensile force of 70% ultimate tensile strength on the prestressing wire test sample in a closed still frame equipped with a time- recording force indicator. An ammonium thiocyanate cell is placed around the test sample and the solution of at least 150 mm is introduced and continuously circulated. The sample is then tested in the ammonium thiocyanate solution for the determination of time to failure. The test is completed either on fracture of the sample or upon the test time reaching 200 hours, and the time to fracture is recorded on the chronometer.

The Pass/fail Criterion for the test is established by the laboratory based on extensive experience and on the concentration of solutions used. Lewis Engineering suggests a 10-sample test, and for ASTM A648- 90A six gage and 6.35 mm (1/4 inch) wire, that the mean time to failure minus one standard deviation for ten samples tested exceeds 75 hours. Provisions are advised for statistical integrity of the test.

Some of us have had the experience of receiving the results of this testing with the metallurgist's conclusion that the wire which was used in the manufacture of your pipeline is inherently susceptible to hydrogen embrittlement. This author vividly recalls the consultant's conclusion that no matter what else was done or not done, this pipeline was doomed to premature failure.

The idea presented in this paper was born at that moment.

The test for hydrogen embrittlement susceptibility should be done before the pipe is manufactured, not ten years after!

A Supplement to AWWA C-301

So that is what we did for the 2.44 cm (96-inch) raw water supply pipeline for the City of Houston, Texas, and the Coastal Water Authority. The construction specification included a "SUPPLEMENT TO THE REQUIREMENTS OF AWWA C301-92", which was developed with the cooperation and consultation of Gifford-Hill-American Inc. Bidders for the 2.44 cm (96-inch) pipeline project were permitted to use either steel or PCCP for the pipe material, but the spec required that if the bidder elected to use PCCP then the prestressing wire manufacturer was required to pre-qualify. The Supplement also required certain other testing and record-keeping aimed at achieving and documenting the use of the highest quality prestressing steel in PCCP.

The project was constructed in two segments with identical prestressing steel testing requirements. Both successful bidders included PCCP and as a result, both were required to pre-qualify the manufacturer of the prestressing steel. The tests were performed for two prospective suppliers of prestressing wire. One supplier consistently passed the test. One did not. The prestressing wire which was used in the manufacture of this PCCP appeared to be of very high quality as measured by the HEST and other tests required.

The engineer included PCCP plant inspection and prestressing wire manufacturer plant inspection in services provided to the owner.

4. Summary and Conclusions

In summary, the City of Houston, Texas, and the Coastal Water Authority constructed seven-mile 96" water conveyance pipeline of PCCP. The project was awarded in two specifications, both of which allowed the use of steel or PCCP pipe materials. The specifications included a supplement to the requirements of AWWA Standard C-301 requiring that the prestressing wire manufacturer be pre- qualified by the Hydrogen Embrittlement Sensitivity Test. Both successful bidders did in fact use PCCP, and the PCCP manufacturer did pre-qualify the wire manufacturer.

The authors are left with these conclusions:

- The quality of prestressing wire is a significant factor in the premature failure of PCCP.

- The Hydrogen Embrittlement Sensitivity Test appears to be a valid test to measure the resistance of prestressing wire to hydrogen embrittlement

- The use of this test aided in assuring top quality prestressing wire in the PCCP for this project.

- This test did not adversely affect the relative competitiveness of PCCP with other pipe products.

- This test did not interfere with the manufacturers or contractors in any significant way.

- This test did not add materially to the cost of the contract.

5. References

American Concrete Pressure Pipe Assn., "Concrete Pressure Pipe Digest," 1996.

Clift, James S., "PCCP – A Perspective on Performance, "American Water Works Assn. Conf. Proceedings, 1991.

Denver Water Department letter dated Oct 23 90: to USBR (Uyeda); with comments and summary of the first meeting of the PCCP Users' Group.

Kajumdar, B.S. & H.S. Cialone; "Evaluation of the HSC Resistance and Strain Aging Condition of Prestressing Wires from a Field failure"; Batelle, Columbus, Ohio; June 19, 1991 (Houston, TX, failure of June 1990).

Price, Robert; "Evaluation of Concrete Pressure Pipelines and Prevention of Failures", ASCE Conference; October 1995.

Prosser, David, "Research, Product Improvement, New AWWA Standards ... Contribute to the Increased Usage of PCCP," American Concrete Pressure Pipe Assn, 1996.

Worthington, Will; "Prestressed Concrete Pipe Inspection and Monitoring Methods", Proceedings of the Conference of Nondestructive Evaluation of Civil Structures and Materials, Boulder, CO; May 1992.

ADVANCEMENTS IN DESIGN AND INSTALLATION OF PRESTRESSED CONCRETE CYLINDER PIPE

David P. Prosser, P.E., Member, ASCE[1]

ABSTRACT

The American Concrete Pressure Pipe Association (ACPPA) and the concrete pressure pipe industry which it represents undertook a program designed to make significant advances in design, manufacture and installation of Prestressed Concrete Cylinder Pipe (PCCP) through materials research, application of modern structural engineering principles, improvements in manufacturing practices, and upgrading of AWWA standards. Practices for monitoring and protection of the pipe were researched. An independent certification program was also initiated. Implementation of those improvements was closely coordinated with the results of market studies which identified customer responses.

Introduction

Prestressed Concrete Cylinder Pipe, a widely used and trusted water transmission pipe product, continues to improve its performance and service to the water industry. In recent years a number of significant improvements have been brought about by the American Concrete Pressure Pipe Association and the industry which it represents, through:

- A strong focus upon quality control and technology development;
- A water industry landmark compliance audit and certification program and;
- Development of rigorous standards.

---

[1] President, American Concrete Pressure Pipe Association, 11800 Sunrise Valley Drive, Reston, VA 20191

ACPPA presented a comprehensive report covering the performance of PCCP for the period 1943-1990 at the June 1991 National Conference of the American Water Works Association (AWWA) held in Philadelphia, Pennsylvania, (Clift, 1991). ACPPA also challenged all other piping materials covered by AWWA standards to issue a similar report. To date, no other pressure pipe producer has accepted the challenge.

RESEARCH

Mortar Coating - For 50 years, the producers of PCCP have understood the importance of the dense, cement-rich mortar coating that protects the steel wire and cylinder in PCCP. Considerable work was done to improve the quality of the applied mortar coating.

In the 1990's, new research was initiated. A study was conducted first at several manufacturing plants and later by Wiss, Janney, Elstner Associates, Inc. of Northbrook, Illinois, (Krauss, 1992) and by Ameron International Corp. (Hall, 1996), to measure the effects on mortar performance of moisture content, silica fume additive and various corrosion inhibitors. This research generally concluded that:

- As mix water was increased from 5.9 percent to 8.1 percent, chloride permeability was reduced by 84 percent, based upon 90-day salt water ponding tests.
- Similar results were observed with the use of silica fume without addition of extra mix water.
- Little or no improvement was experienced in the relevant mortar properties as a direct result of the use of corrosion inhibiting additives.

The results of this research have been put in place. Today, AWWA C301-92 requires increased water content, a maximum limit on absorption, and a minimum compressive strength for mortar coating. Production equipment and processes have been altered to run with a wetter mix.

More recently, G.M. Idorn of Denmark analyzed samples of mortar coating taken from old and new pipe (Palbøl, 1998). Idorn said:

"Mortar coating samples taken from in-service Concrete Pressure Pipes of varying ages have shown excellent durability as well as having provided effective corrosion protection for the steel elements of the pipe. These coatings have functioned well even in cases where they are exposed to high chloride concentrations or low pH levels. Improvements to the mortar coatings for new Concrete Pressure Pipe should offer even better protection. Based upon the evidence of corrosion prevention performance of mortar coatings on existing concrete pipe, and the significant improvements delineated above, it would be reasonable to project even longer service life for the mortar coatings currently produced."

Monitoring - The condition assessment of buried structures is a major concern for some utilities because each pipe material requires different assessment and repair techniques. Large diameter water transmission lines represent a major investment and a vital artery serving the needs of major populations. Unexpected interruption of flow is not just inconvenient, it represents a threat to life and property. Therefore, the objective of condition monitoring programs is to identify potential problems in sufficient time to permit repair or provide for alternate sources of flow.

The ACPPA, industry, and independent experts have researched several methods for monitoring PCCP. The four technologies showing the most promise are: Over-the-line potential survey, acoustic emissions monitoring, impact echo and remote-field eddy-current.

Over-the-line potential survey was the subject of an ACPPA test program (Hall, 1995). That technique starts with measuring and recording a base line electrical potential survey shortly after the pipeline is installed. Then, periodic surveys are conducted to monitor changes in electric potential. Active corrosion will set up a unique electrical pattern that can be identified.

ACPPA established a test site in California that allowed monitoring of pipe lengths that could be electrically bonded or unbonded to each other. The testing was done on pipe with the standard mortar coating and pipe with supplemental exterior protection over the mortar coating. The conclusions of the test program were:

- Moist soil is necessary for the potential monitoring technique to be meaningful.
- Bonded and unbonded PCCP can be effectively monitored for active corrosion.
- Bonded pipe is substantially more sensitive to potential monitoring than unbonded pipe.
- Identifying corrosion on the bottom of the pipe is less sensitive. However, experience has shown that in rare cases where corrosion of PCCP occurs, it generally occurs at the springline where the mortar coating is subject to the greatest stress.
- PCCP, with supplemental protection over the mortar coating, can be effectively monitored using these techniques.

The overall conclusion was that over-the-line potential surveys were useful and indicative, but more refinement of the methods is still needed.

Acoustic Emissions Monitoring has been used to monitor PCCP. Acoustic Emissions Monitoring, as applied to PCCP, involves inserting hydrophones into a pipeline, while in use, to detect sound waves created by breaking wire or other related movement. As a passive sonar application, it was adopted from the U.S. Navy. A similar application is being marketed for parking structures.

In 1997, ACPPA entered into a project agreement with the National Research Council of Canada, Institute for Research in Construction (IRC) to evaluate the effectiveness of the hydrophone technology in identifying wire break sounds and location. The IRC has included two venders and a water authority, the Halifax Regional Water Commission, in the project. Test results will be documented. Performance parameters will be described and parameters for effective monitoring with the acoustic emission technology will be developed.

Impact-Echo is a sonic echo for measuring concrete thickness, evaluating concrete quality and detecting flaws from one surface. The method was researched and developed at the National Institute of Standards and Technology in the early 1980's.

ACPPA, through Wiss, Janney and Elstner, evaluated impact-echo and identified its effectiveness at detecting anomalies in embedded cylinder pipe, but recognized that considerable work would be involved to adopt the technology to pipeline monitoring. Additionally, as conceived, impact-echo would require draining of the pipeline to be monitored.

Olson Engineering, Inc., is currently monitoring garages, slabs and tunnels using impact-echo. It developed a four-head scanner, which required a clean, level surface, and subsequently a hand held scanner which an individual could use. Washington Suburban Sanitary Commission has done developmental work to adopt impact-echo to monitoring embedded cylinder pipe and has applied for a patent for the software applications.

Remote-field eddy-current monitoring developed from eddy-current monitoring technology used for examining oil and gas well casings. That technology, used by Russell Technologies, was adapted into remote-field eddy-current monitoring, with the assistance of Dr. David L. Atherton, Queens University, Kingston, Ontario, Canada. Dr. Atherton's speciality is nondestructive evaluation.

Dr. Atherton believed that remote-field eddy-current technology can be applied to embedded cylinder pipe. In November, 1997, a successful test was witnessed. While in its infancy, remote-field eddy-current technology is an emerging PCCP monitoring technology which shows promise in locating prestressing wire disconnections.

Cathodic Protection of PCCP - The exterior mortar coating on PCCP provides an alkaline environment and physical barrier to protect the steel elements of the pipe.

In most soil conditions, mortar coating is adequate to prevent deterioration of the structural steel elements and to assure the intended service life. Some soil or environmental conditions can disrupt the passivating effects of the mortar coating. In those cases, supplemental exterior protection and/or cathodic protection (CP) have been used to protect the pipe.

Two independent studies on the frequency of use of CP systems indicated that only 0.1 percent to 0.5 percent of all PCCP installed had active CP systems in use. Those were predominately in the western United States (Clift, 1991). Because of relatively scarce application of cathodic protection of PCCP, little design and operating criteria existed.

Therefore, beginning in 1991, the ACPPA and its members undertook a multi-year study to determine the threshold potential above which CP systems can damage the steel prestressing wire and also to determine the correct current level to achieve the required protection. In the first phase of the study, it was shown that the hydrogen content of wire increases at potentials more negative than -1000 mV versus a copper-copper sulfate electrode (CSE). Increases in hydrogen content may cause embrittlement of the wire, which can lead to sudden and unpredicted failure of the wire. Wire tested at the normal residual tension of wrapped pipe (60% of wire ultimate tensile strength) at potentials of -1200 mV (CSE) failed within 9 to 41 months. In an on-going program, wire tested at -1000 mV (CSE), or less negative, showed no failures after over five years of exposure. (Hall et al, 1996)

The above clearly indicates a hydrogen embrittlement threshold where problems may occur somewhere between -1000 mV and -1200 mV (CSE). That result is further substantiated by failures of two lines which had CP systems operating at -1200 mV (CSE) or more negative (Houston, Texas and Salt Lake City, Utah). The obvious conclusion is that if CP systems are used, maximum negative potential should be -1000 mV (CSE).

The second phase focused on the amount of current required to protect PCCP. While it has been shown by Hall, that potential shifts as small as 20 mV can provide adequate protection, the more traditional 100 mV polarization shift, supported by NACE International, was used for this research.

The results of the research showed the following current densities required to achieve a 100 mV polarization shift in potential:

Mortar-coated PCCP 12 $\mu A/ft2$
Mortar-coated PCCP with 3 $\mu A/ft2$
Supplemental Protection

Using these research results and comparing the cost of CP systems for PCCP to organically coated steel pipe, it was found that the cost of CP for mortar-coated PCCP is approximately the same as for dielectrically coated steel pipe; and CP for mortar-coated PCCP with supplemental protection is considerably less than for dielectrically coated steel pipe.

The ACPPA continues to advise that CP systems for any pipeline be developed and supervised only by qualified corrosion engineers.

Prestressing Wire - In 1993, the ACPPA assembled experts from Europe, Japan and the United States in Denver, Colorado, to discuss the state of research and manufacturing advances for prestressing wire as used on pipe. According to the participants, that was the first time such a group has been assembled. It was agreed that strain-aging and hydrogen embrittlement of wire had been a worldwide problem in the 1960's and early 1970's and that each geographical area had independently discovered and taken essentially the same steps to correct the problem. The factors common to worldwide improvement in wire quality and performance were:

- Selection of good raw material (drawing rod);
- Proper cleaning of the drawing rod;
- Control of temperature during wire drawing;
- Restrictions on upper tensile strength limits; and
- Enhanced mechanical testing procedures.

A follow-up study was made, but not published, comparing test methods used in Europe and North America. The mechanical tests, while somewhat different, proved equally effective. The European hydrogen embrittlement sensitivity test did not prove to be a discriminating test for U.S. purposes, nor as reproducible and accurate as the FIP research tests used by forensic engineers. It was generally concluded that the additional wire testing requirements included in AWWA C304-92 and AWWA C301-92, and in ASTM A648, adequately prevented the use of unsuitable wire.

COMPLIANCE AUDIT AND CERTIFICATION PROGRAM

Every ACPPA member company has its own detailed quality assurance program. In 1993, the concrete pressure pipe industry conducted a major survey of users and specifiers of PCCP, and assembled regional focus groups to refine its understanding of customer needs and perspectives. The customers were asked to define quality. This is what they said, "Quality is consistently meeting the requirements." They were asked to define the requirement. They said, "AWWA standards."

The industry reviewed typical, broad-based quality assurance systems used around the world, including ISO 9000, and concluded that the best program would have the following features:

- The same standard of quality must be used by every supplier of PCCP.
- The standard of quality should be consistent compliance with AWWA standards.
- Periodic audits must be conducted of every plant by an independent, knowledgeable and respected party.

In 1994, ACPPA developed the program and contracted with Lloyd's Register, the most well-known, independent, not-for-profit quality assessment company in the world. Today, every ACPPA member, U.S. and Canadian alike, suppliers of PCCP and other concrete pressure pipe products, has successfully completed four separate audits by Lloyd's. The certification provides users and specifiers the assurance that all PCCP suppliers have the required systems, training and equipment to comply with the AWWA and project specific standards. Today, to be a member of ACPPA, requires that each company's plants be in full quality compliance.

The Compliance Audit and Certification Program has become so popular that consulting engineers and owners have requested additional concrete pressure pipe products be certified. It must be said that this quality system is unique in the North American waterworks industry. No other pipe materials have this industry-wide, consistent quality audit program tied to the national standards.

AWWA C301-92 AND C304-92 STANDARDS

In 1992, the first version of a totally new PCCP design standard was introduced as AWWA C304-92. Also a separate, thoroughly revised and updated manufacturing standard, AWWA C301-92, was approved.

Despite the fact that design methodology for PCCP was not a problem, the industry realized that the existing methods were not sufficiently detailed, and out of date relative to the latest design technology. Using more than 40 years of experience, new research, a worldwide review of current technical knowledge and computer modeling, the new design standard provides a highly accurate, science-based design method that incorporates state-of-the-art structural engineering principles and yet maintains a traditionally conservative design philosophy.

Unlike changes in design methods for most other piping materials, C304 generally results in designs that are more conservative when compared to the previous design methods. Widely accepted by customers, the transition to AWWA C304 has been smooth and without any adverse experience.

Also in 1992, AWWA C301, the manufacturing standard for PCCP, was completely revised to incorporate new and more detailed testing of virtually every aspect of the pipe manufacturing process. For example, more controls were imposed on wire manufacture, such as upper limits for wire drawing temperature, and tougher mechanical property tests. Mortar coating mixing, application, performance and testing were also upgraded.

The net effect of these two new standards is that PCCP designed and manufactured today is better than at any time in its over 50-year history.

THE PERFORMANCE OF PCCP

In 1991, the industry provided a statistical report card on the performance of PCCP since 1943, the year PCCP was introduced to the U.S. pipe market. That data was updated, using the same criteria and categories as the previous report, and presented in Toronto, Ontario, Canada, during the AWWA annual convention (Prosser, 1996).

The statistics indicate the continuing and excellent performance of PCCP that was reported in 1991. They are also confirmed by a National Research Council of Canada study commissioned by Uni-Bell, covering 21 Canadian cities. That report documents the performance of 17,554 km (10,900 miles) of pipelines and shows that PCCP had the lowest number of breaks/100 km of any product.

COMPARATIVE MAINTENANCE COSTS OF PIPING MATERIALS

The aspect of pipeline costs that has the most public attention and scrutiny is the cost of new construction. While that is an important figure, it is, in general, far less each year than the maintenance cost to keep the system in operation. To date, too little attention has been focused on what happens after a line is installed. One major obstacle in this evaluation is that water departments either have little segregation of costs for pipeline maintenance. ACPPA is not aware of any independent, broad-based U.S. study of these costs and has recommended that AWWA set up standard cost formats and conduct annual surveys to identify costs by material used and type of problem.

In 1997, ACPPA executed a cost of pipeline operations and maintenance study covering 1991 through 1995, (ACPPA, 1997). The results of that study revealed that operations and maintenance of Concrete Pressure Pipe were one-half of what ductile iron costs, and less than the one-third of what cast iron or steel cost. Furthermore, according to the U.S. Army Corps of Engineers, Manual 1110-2-2902, concrete pipe will last twice as long as any other pipe.

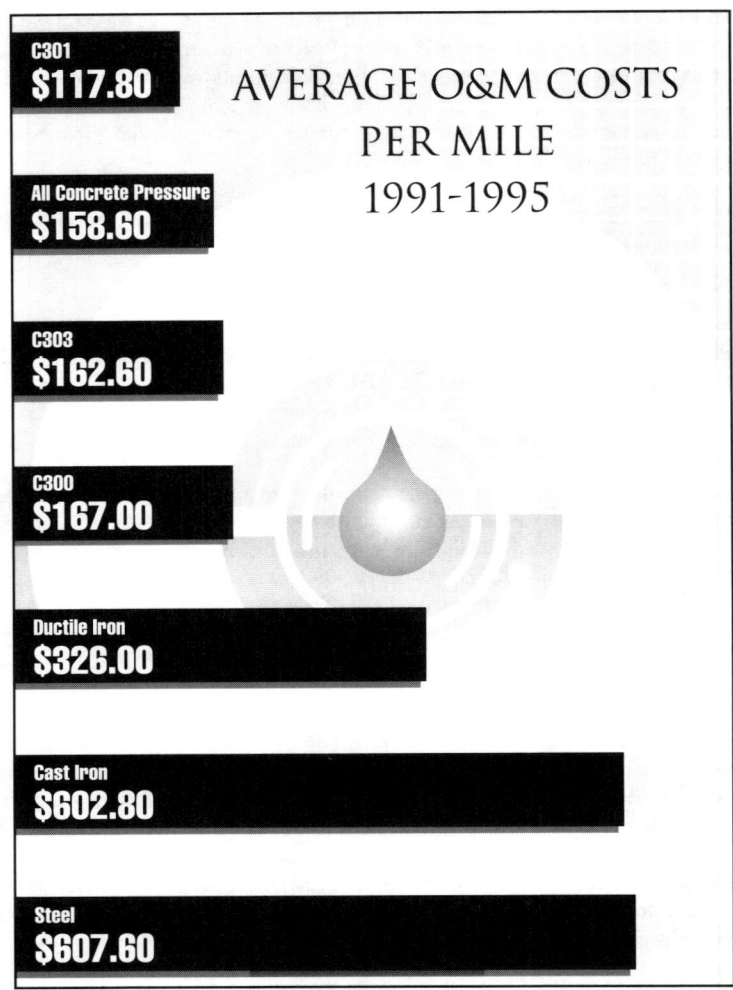

ACPPA PIPELINE OPERATIONS STUDY

INCREASED USAGE OF PCCP IN OTHER COUNTRIES

Today, the U.S. and Canada are the envy of the developing world for many reasons. In addition to stable political structure and industrial experience, it is the well-developed infrastructure of roads, communication and piping systems that provide an unmatched advantage and strength. Countries like China, India, Indonesia, and even Korea are struggling to get enough water to support their fast-growing cities and industry. As they seek to replace corroded steel pipe, or older unreinforced concrete pipe, they are increasingly turning to PCCP. For example, Korea is currently writing its first PCCP standard, modeled after AWWA Standards. China completed that task within the last two years. And India recently bid its first project with PCCP.

Today, 90 of the 100 largest U.S. metropolitan areas use PCCP in their water systems. Before long, other countries will have a similar percentage of PCCP serving their growing communities.

REFERENCES

American Concrete Pressure Pipe Association. (1997) "1997 Pipeline Operations and Maintenance Cost Study," Reston, Virginia.
Clift, James S. (1991) "PCCP - A Perspective on Performance," reprinted from 1991 ACE Proceedings, American Water Works Association National Conference, June 25, 1991, Philadelphia, Pennsylvania.
Hall, Sylvia C. and Matthew, Ivan. (1995) "Cathodic Protection Requirements of Prestressed Concrete Cylinder Pipe," reprinted from Advances in Underground Pipeline Engineering, Proceedings, Second International Conference, Pipeline Division and TCLEE/ASCE, June 25-28, 1995, Bellevue, Washington.
Hall, Sylvia C., Matthew, Ivan and Sheng, Qizhong, Ph.D. (1996) "Prestressed Concrete Pipe Corrosion Research - A Summary of a Decade of Activities," of Ameron International, Corrosion '96, the NACE International Annual Conference and Exposition, Houston, Texas.
Krauss, Paul D. and Pfeifer, Donald W. (1992) "Corrosion Testing of Embedded Steel Cylinder Pipes for the American Concrete Pressure Pipe Association (ACPPA)," WJE No. 910597, Wiss, Janney, Elstner Associates, Inc., July 24, 1992, Northbrook, Illinois.
Palbøl, Lars, Jakobsen, Ulla Hjorth and Thaulow, Niels. (1998) "Petrographic Examination of Mortar Coatings," *ACI's Concrete International*, pp. 47-53.
Prosser, David P., P.E. (1996) "Research, Product Improvement, New AWWA Standards, Compliance Certification & Performance - Contribute to the Increased Usage of Prestressed Concrete Cylinder Pipe," 1996 ACE Proceedings, American Water Works Association National Conference, June 25, 1996, Toronto, Ontario, Canada.
U.S. Army Corps of Engineers. (1997) "Conduits, Culverts, and Pipes," Engineering Manual 1110-2-2902, Washington, D.C.

North City Tunnel Connects to New Water Reclamation Plant

Greg Arakaki, P.E., M. ASCE [1], Duane Larson, P.E., M. ASCE [1], John Kinneen M. ASCE [1], and Alan Redmon, P.E., M. ASCE [2]

Abstract

Construction materials, regulatory restrictions, and environmental issues affected the design of the City of San Diego's Metropolitan Wastewater Department's (City) North City Tunnel Connector (NCTC) that was recently completed. These issues forced the project team to develop innovative approaches to meet the City's objectives. The NCTC contains four pipelines and is a 5.5-meter (18-foot) high by 570-meter (1,870-foot) long tunnel connecting the North City Water Reclamation Plant (NCWRP) with the Rose Canyon Trunk Sewer and other utilities.

Introduction

The purpose of the NCTC is to connect the NCWRP, located on the east side of I-805 with the Rose Canyon Trunk Sewer and other infrastructure located on the west side of I-805. Figure 1 shows the relationship between these facilities. The NCWRP is a skimming plant with an ultimate capacity of 170,000 m^3/day (45 million gallons per day). During the initial planning for the project, it was decided that four pipelines were needed to service the plant - an influent sewer line, an effluent line, a waste line, and a reclaimed water line, The influent sewer is a 2,100 mm (84-inch) pipeline that conveys flow from the northern portion of the plant's service area extending from Del Mar in the north to as far east as Poway. A diversion chamber allows the plant to redirect sewage that would normally flow to the Point Loma Wastewater Treatment Plant (PLWTP) into the NCWRP. Treated sewage is conveyed to the City's ocean outfall bypassing the PLWTP in a 1,350 mm (54-inch) effluent pipeline. The 900 mm (36-inch) waste pipeline return sidestreams from the treatment process to the Rose Canyon Trunk

---

[1]Metcalf & Eddy, Inc. 701 B Street, Suite 1100, San Diego, California 92101
[2]City of San Diego Metropolitan Wastewater Department, 600 B Street, MS 905, San Diego, California 92101

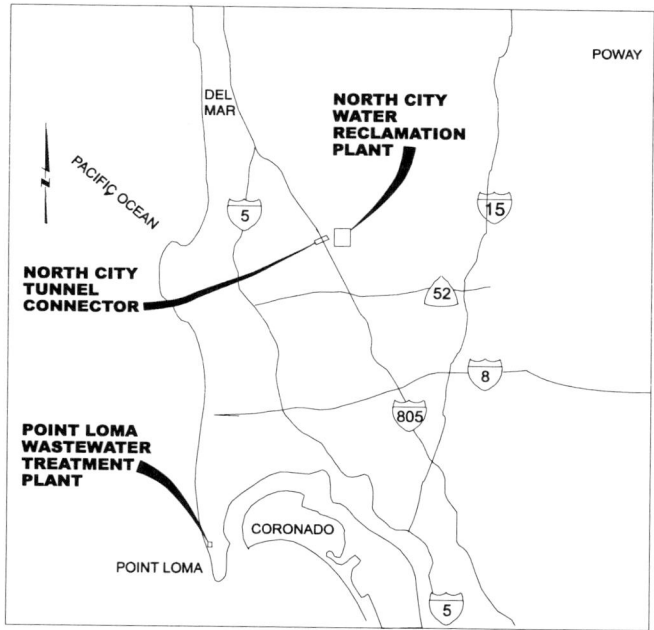

Figure 1. Location Map

Sewer and ultimately to the PLWTP. Finally, a 900 mm (36-inch) reclaimed water pipeline provides a connection to the City's reclaimed water distribution system. and a - under Interstate 805 (I-805). With the advent of the City's Repurification Project, the effluent line was routed back to the Rose Canyon Trunk Sewer, thereby eliminating a separate downstream line to the ocean outfall.

Regulatory Requirements

Five agencies played major roles in shaping the final design of the NCTC. These are the California Department of Health Services (DHS), the California Department of Transportation (CalTrans), the Regional Water Quality Control Board (RWQCB), the Army Corps of Engineers (ACE), and the U.S. Fish and Wildlife Service (USFWS). The DHS has strict regulations governing separation distances between sewage and treated (reclaimed) water pipelines. The limited cross-sectional area of the tunnel, however, precluded such separation. In lieu of this, the DHS required that rather than completely grouting the tunnel upon completion, that the tunnel would be partially grouted to cover the influent, effluent, and waste pipelines and that the reclaimed water pipeline would remain exposed. This would allow an air gap separation between the reclaimed water line and the other pipelines. CalTrans, imposed strict settlement limitations to minimize any adverse impacts on Interstate 805, one of the regions major transportation corridors. Environmental agencies such as the RWQCB, the ACE, and the USFWS indirectly influenced the design. The RWQCB required implementation of Best Management Practices to control storm water pollution during construction. ACE approval was required to divert the flow of a small intermittent stream around the construction site. The USFWS imposed restrictions to avoid disruption to the habitat of the endangered California Gnat Catcher.

Tunnel and Pipeline Design

The development of the NCTC was undertaken with the intent of providing the City with a highly functional design, while remaining cost-effective. Several alternatives were considered during the design of the NCTC. Value engineering and numerous City and internal reviews helped pare down the alternatives to the ultimate final design, a 19-foot diameter circular tunnel. A horseshoe shaped cross-section was also detailed as an alternative and the contract specifications were written to allow the contractor to propose alternative methods that the engineer would consider during the shop drawing phase.

In addition to the influent, effluent, reclaimed water, and waste pipelines, a 200 mm (8-inch) drain line to conduct condensate from the tunnel and four 100 mm (4-inch) PVC fiber optic ducts for current and future communication systems were added. The tunnel was partially grouted to comply with DHS requirements, but this feature also offered the City the opportunity to inspect the tunnel as well as the opportunity to route other utilities through the tunnel in the future.

It was anticipated that the tunnel would be bored using a tunnel boring machine and shield from a launching pit on the west side of I-805 to a receiving pit on the NCWRP site where connections to existing pipes would be made. However, a conflict in the construction schedules of the NCTC and the NCWRP, resulted in a situation where the receiving pit area on the NCWRP site was regraded to its final elevation roughly 18 meters (60-feet) above the crown of the tunnel. A modification was made that required the contractor to extend the tunnel past the CalTrans right-of-way to the existing pipes so that the connections could be made inside the tunnel.

Access to the tunnel was accommodated on two fronts. First, the tunnel is located in a small canyon that required the construction of a steep access road for both construction and maintenance. The design and construction of this road required coordination with the City's Water Department which was concurrently constructing the Rose Canyon Trunk Sewer Upgrade adjacent to the tunnel site. In addition, physical entry into the tunnel was accommodated by two shafts - one at either end of the tunnel. The shafts are available for entry to the tunnel, but ventilation and lighting within the tunnel were not provided because of the perceived infrequent need for access.

The pipelines in the tunnel were designed based upon anticipated hydraulics from the NCWRP. Flexibility in the choice of pipeline materials (ductile iron or steel) was written into the contract documents to encourage competition among the suppliers for the effluent, waste, and reclaimed water lines. Ultimately, the contractor supplied each of these lines in ductile iron pipe. The influent line was specified as a reinforced concrete pipe.

Cathodic protection systems were installed to safeguard the tunnel ribs and the pipelines and to insure their 50-year service life. The tunnel rib cathodic protection system consisted of anodes and reference electrodes placed external to the tunnel at roughly 90 meter (300-foot) intervals. The pipeline cathodic protection system consisted of reference electrodes and anodes embedded in the tunnel grout. Both systems are controlled by a rectifier panel located outside of the tunnel.

Another aspect of the tunnel design involved acquisition of easements. The NCTC passed through six parcels zoned for commercial development. A total of 15 easements were required to address construction staging, site access, and the tunnel routing. An innovative, three-dimensional underground easement for the tunnel routing, modeled after an aerial-type easement for a girder bridge, reduced property acquisition costs. All of the tunnel components were designed with allowances to account for loading from future construction.

Construction Approach

Upon award of the project, the contractor proposed an alternative construction method utilizing a variation of the Belgian Method. The designer worked with the contractor to insure that all of the design criteria were met and eventually approved this approach. Figure 2 shows a cross section of the constructed tunnel. The construction method consisted of a top heading and bench approach. It was selected because of its past success in similar "soft rock" soil conditions. By employing this method, the exposed soil height was minimized at the face of the tunnel where most of the workers were located. A rotary road heading machine was used instead of a tunnel boring machine and shield to advance the top bench with a semi-circular section. The bottom bench was then excavated using conventional equipment, after steel arch support ribs and reinforced concrete wall plates were installed. Innovative use was made of expanded metal mesh as a permanent form for poured-in-place concrete tunnel lining. The contractor had previously used mesh as a permanent form for lining shafts, but not for a tunnel. Trials were conducted to find the optimum concrete mix consistency which could be pump-placed to fill the mesh up to the tunnel crown without excess bleeding or segregation and still meet the strength requirements. The tunneling and installation methods used resulted in significant cost and time savings to the City.

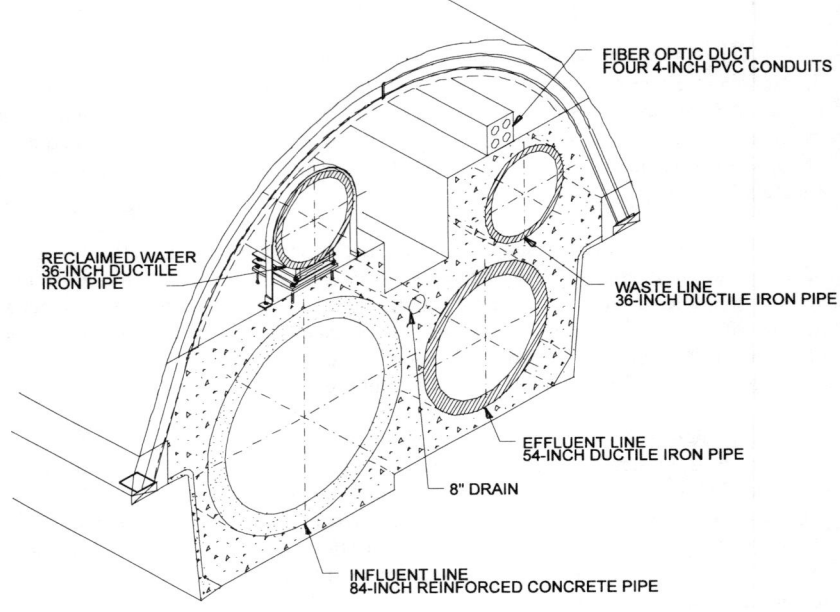

Figure 2. Tunnel Cross-Section

The differing diameters, elevations and slopes of the four pipelines in the tunnel required each pipeline to be constructed individually, in sequence. Pipe was loaded through the single portal one length at a time using a hydraulic pipe carrying and loading machine and assembled from the tunnel face back to the portal as shown in Figure 3. When each pipeline was completed, it was tested and then grouted in place using lightweight (wet density = 0.77 kg/m^3 (60 lbs/ft^3)), cellular grout.

Although cellular grout is more expensive than mass concrete, the contractor hoped to recover the additional cost through economies in labor, formwork and time. However, problems were experienced with heat of hydration resulting from placing large volumes of grout in the tunnel. As a result, the initial grout placement experienced significant cracking and delamination and had to be removed. Subsequent grouting was approached using smaller lifts to avoid this problem. The use of the cellular grout was not a cost saving measure.

Figure 3. Pipe Loading Machine

The final task in the tunnel construction was completion of landscaping to mitigate the impacts of construction. Figures 4, 5, and 6 show the general site before, during, and after completion of construction. Disruption of California Gnat Catcher habitat was avoided and the slopes of the access road were successfully hydroseeded with native grasses

Conclusion

Design of the North City Tunnel Connector began in 1992 and was completed in 1994. Construction commenced in 1995 and was completed in 1997 at a total cost of $8.5 million dollars. As a result of the cooperative efforts between the designer, the contractor, and the City, the project was completed on schedule and within the allocated budget. Impacts to the environment were minimized and the tunnel's successful completion enabled the timely start-up of the North City Water Reclamation Plant.

Figure 4. Site Prior to Tunnel Construction

Figure 5. Site During Tunnel Construction

Figure 6. Site After Tunnel Completion

Construction Challenges for Soft Ground Tunneling

Gregory W. McBain[1], Luciano Meiorin[1], Jon Y. Kaneshiro[1], Stephen J. Navin[1], and Rolf H. Lee[2]

Abstract

For more than 60 years, raw sewage has flowed down the hills of Tijuana, Mexico into the Tijuana River, crossing the U.S.-Mexican border. The sewage contaminates the low-lying river valley in San Diego and then flows into the Pacific Ocean, creating a health hazard for beaches at Border Field State Park, Silver Strand Beach in Coronado, and the most severely affected, Imperial Beach. To alleviate these problems, the South Bay Ocean Outfall (SBOO) will convey effluent down a 11 m diameter by 58 m deep drop shaft, through a 3.35 m diameter by 5.8 km long tunnel, and up a 46 m high by 2.7 m diameter riser shaft connecting to a 1.5 km seafloor pipeline and 1.2 km of diffusers on the ocean floor.

Project Description

The South Bay Ocean Outfall (SBOO) will convey up to a total of 14.6 m^3/sec/day of treated wastewater effluent originating in both Tijuana (8.8 m^3/sec/day) and the City of San Diego (5.8 m^3/sec/day) (Figure 1). Design flows are 7.6 m^3/sec/day (average dry weather flow) and 14.6 m^3/sec/day (peak flow). A peak flow of 11.3 m^3/sec/day will be conveyed through the SBOO by gravity, and the capacity can be increased to 14.6 m^3/sec/day in the future by addition of a pump station. The construction of the South Bay International Waste Water Treatment Plant (SBIWTP) is proceeding in parallel with the South Bay Ocean Outfall (SBOO) project (Figures 1 and 2). The South Bay Land Outfall (SBLO) is a 3.66 m diameter by 3,750 m long pipeline constructed in 1991-1994, which connects the SBIWTP with the SBOO. The SBOO is designed for a service life of 75 years. The cost of the SBOO is $138 million dollars, with a 39 month construction schedule. The construction of the SBOO was divided into three contract packages as follows.

Parsons Engineering Science, La Jolla, CA[1] and City of San Diego, San Diego, CA[2]

Contract Package 1 covers the work required for construction of pipelines and special structures through which treated effluent flows from the SBIWTP through an Effluent Distribution Structure (to co-mingle and distribute effluents from the treatment plants to conveyance structures), an Entrained Air Vent Structure (to help remove entrained air); and an Energy Dissipation Structure (to dissipate head occurring at low flow) before entering the existing SBLO. Notice to proceed was given in April 1996 to Colich & Sons (Gardena, CA) who was low bidder at $9,950,000. The pipe was supplied by Ameron (Rancho Cucamonga, CA). The special structures were completed in September 1997.

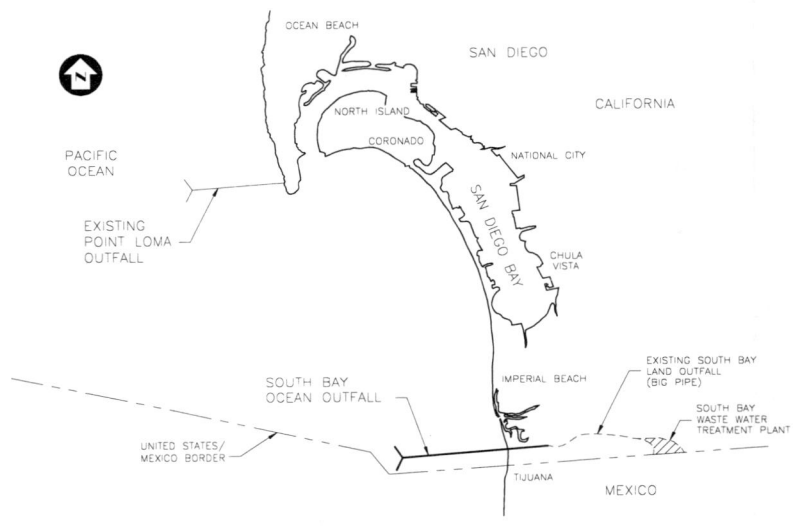

Figure 1. Site Location and General Project Arrangement

Contract Package 2 is for the construction of a drop shaft tunnel and riser which carries effluent from the end of the SBLO through an anti-intrusion structure (to prevent seawater from backing up at high tide and low start-up flows) down an 11 m diameter drop shaft to an elevation of about -50 m below sea level. At that point, a tunnel is being constructed with an internal diameter of 3.35 m and an overall length of approximately 5.8 km. The tunnel will terminate about 4.27 km offshore where it will connect to a 2.8 m diameter by a 47 m long riser, where effluent will be conveyed vertically to a seabed pipeline part of Contract Package 3. Notice to proceed was given in September of 1995 to Traylor Brothers/Obayashi, a J.V. (Evansville, IN / San Francisco, CA) who were low bidder at $88,285,000. Major subcontractors include Case Foundation (Chicago, IL) for the Riser and Layne Northwest (Pewaukee, WI) for the drop shaft freeze wall and dewatering.

Major suppliers include Boretec/Mitsubishi J.V. (Solon, OH / Kobe, Japan) for the earth pressure balance tunnel boring machine, Sehulster Tunnels / Pre-Con (Rancho Cucamonga, CA) for the precast concrete segmented one-pass liner system, and NESCO Fabricators (Mare Island, CA) for the Riser. Completion of Contract Package 2 is scheduled for the end of summer 1998.

Contract Package 3 included construction of a 3.05 m by 1,433 m long sea floor pipeline connecting to a wye where the treated effluent is discharged along two, 610 m diffuser legs on the seabed. Notice to proceed was given in January 1996 to Fletcher General (Seattle, WA) who was low bidder at $36,442,000. The pipe is supplied by Hydro Conduit (Corona, CA). Contract Package 3 was completed in 1997.

The owners are the International Boundary and Water Commission and the City of San Diego. Parsons Engineering Science performed the design, provides shop drawing review and provides design services during construction. Funding is being provided by the Environmental Protection Agency, with project oversight being provided by the EPA, IBWC, U.S. Army Corps of Engineers and the State of California.

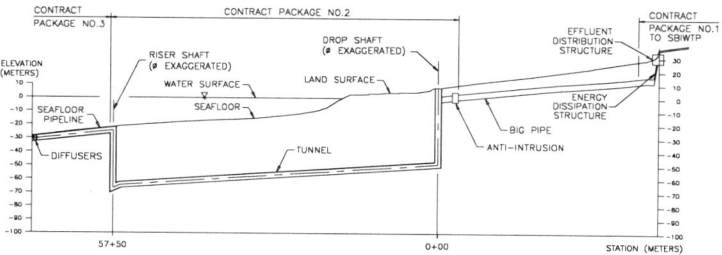

Figure 2. Generalized Project Profile

Geotechnical Considerations

Three geologic units are found along the tunnel route: Fill, Alluvium and Alluvial Gravels, and the San Diego Formation (as shown in Figure 3). The tunnel is approximately 61 m below sea level. Cover over the tunnel ranges from about 55 m onshore to about 43 m offshore. The San Diego Formation is an indistinctly bedded, poorly indurated, fossiliferous marine siltstone, and is relatively flat lying to gently westerly dipping. It consists of very stiff to hard clays; very dense silts; very dense, fine silty sands; clean sands; clayey, silty and sandy gravels/cobbles/boulders, (with boulders possibly up to 1 m diameter); and well-cemented concretions. From a tunneling standpoint, the material is classified as "soft ground".

Numerous faults, both active and dormant, many of which cross the SBOO alignment, are located within the South Bay of San Diego County. The key seismic sources are the Rose Canyon fault and the Coronado Banks fault, which is about 10 km west of the Riser. These faults are capable of producing a Magnitude of 7-1/4 and 7-1/2 earthquake, respectively. A probabilistic seismic hazard analysis and a probabilistic fault displacement analysis were performed, which gave a peak ground acceleration of 0.63 g and a nominal fault displacement of 76 mm with a probability of exceedance of 10 percent in 75 years. This was used to design the underground structures for seismic shaking and fault displacement.

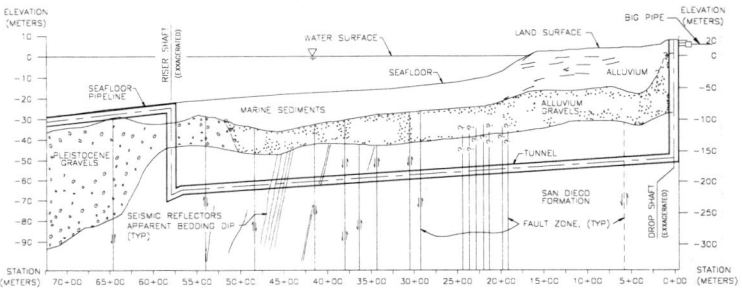

Figure 3. Generalized Geological Profile of Tunnel Alignment

Contract Package 1 - Special Structures

Key features of Contract Package 1 are the Energy Dissipation Structure and two 3.6 m knife gate valves, the largest in the world.

Energy dissipation is required during low flows, to dissipate about 40 feet of gap difference in elevation between the liquid levels in the effluent Distribution Structure and the outfall pipe. Several means of dissipating energy were explored during the design phase; from sleeve valves, to helicoidal ramps drop shaft. The preferred solution was a section of pipe with artificially increased friction. Multiple finger type weirs were built into a section of the pipeline, built on a slope of 1 to 10.

A physical model of the energy dissipation structure was built and tested to provide necessary design information. Modeling was conducted in the W.M. Keck Laboratory of Hydraulics at the California Institute of Technology. A scale ratio of 14.4 to 1 was selected, resulting in the 3.6 m Energy Dissipation structure being modeled with a 25 cm diameter, clear acrylic tube. Three types of baffles were investigated.

The modeling indicated that the best results were achieved with the baffles arranged as shown in Figure 4. Modeling identified some flow behavior instabilities that related to certain combinations of flow and baffle spacing. The best results were obtained at a baffle spacing of 8 feet. Modeling confirmed that the total head losses in the structure are maximum for low flows and are reduced to low valves at high flows.

Knife Gate Valves

Two large knife gate valves were provided on Contract Package 1. The valves were specially designed and built for this application. The 3.6 m valves were built in Vancouver, Canada, under tight specifications. The 8.75 cm thick, stainless steel disks were machined to very tight dimensional tolerances and passed all the leakage tests with no visible leaks. To avoid water hammer surges, the minimum closing and opening times were calculated and specially designed operators were provided.

Figure 5 shows one of the valves during manufacturing.

Figure 4. Energy Dissipating Structure   Figure 5. 3.6 m Knife Gate Valve

Contract Package 2 - Drop Shaft and Tunnel

Drilling of over 40 freeze holes to a depth of 90 m around the perimeter of the drop shaft began in December of 1995. The freeze subcontractor monitored the freeze wall with temperature profiling of the freeze holes and temperature probes. In May of 1996, however, during excavation at about 24 m below the ground surface, leakage was observed in the bottom of the shaft. With the realization that there was a hole in the freeze wall, the freeze subcontractor turned off the sump pumps in the shaft. Before backfilling of the shaft could begin, a sinkhole developed adjacent to the shaft. The sinkhole was backfilled with shaft muck and the shaft itself was backfilled with imported sand. Freezing resumed and continuous temperature profiling and monitoring ensued. A warm spot was detected as a high groundwater velocity zone on the side of the shaft adjacent to the Tijuana River Valley and was ruled as a differing site condition. Four additional Nitrogen freeze holes were drilled in the warm spot zone. Within three weeks, the closure was achieved as indicated by a continuous rise of cold water within the monitoring well inside the shaft. The concrete lining for the shaft was completed in November of 1996. There was a five month delay in the construction of the drop shaft due to the hole in the freeze wall during excavation.

The 214 metric ton, 3.98 m diameter by 10.978 m long TBM was shipped from Japan to Long Beach, CA where it was trucked on a 93-wheeled trailer. The route from Long Beach to the construction site was a 250 km journey, which required careful planning. Even so, the TBM was delayed one week before being allowed to cross a city bridge, so that the bridge design could be checked. After onsite assembly of the TBM and over 90 m of backup trailing gear (including a 30 m screw conveyor) from July to September 1996, a Manitowoc 360 metric ton crane was required for picking up the TBM was lowered into the Shaft in November (Figure 6 and 7).

The tunnel is unique in that it will carry effluent under internal operating heads of 1.1 to 2.7 bar (in excess of external pressure), and it will be constructed under the ocean with potential for high external groundwater pressure up to 7 bars. The tunnel boring machine (TBM) must withstand the high external head, while the liner must withstand the internal and high external heads and overburden stress. The contractor selected an earth pressure balance machine, which maintains muck in a pressure chamber (earth plenum), and in front of the cutterhead via a screw conveyor.

Figure 6. TBM Pick (note armor plating)   Figure 7. TBM Lowering (note freeze pipes)

The tunnel liner incorporates a five-piece segmented, single-pass liner with continuous hoop steel to carry the internal pressure (Figure 8). The segment rings are left and right tapered for steering adjustments. Segments are 1.2 m long, and 3.35 m ID by 229 mm thick and provide a minimum 38 mm cover on each face for corrosion protection of the reinforcement.

The tunnel liner is designed for full overburden and maximum external groundwater pressure for construction/ inspection. For operation, the liner is designed for a differential water pressure of 27 m head, resulting in hoop tension reinforcement within each segment and tensioned radial bolts. The reinforcement steel meets ACI-224 crack control design with an 0.2 mm crack limit in direct tension for 12 m head (normal operation load condition) and working stress at 60 percent of yield for 27 m head (short term flushing load condition).

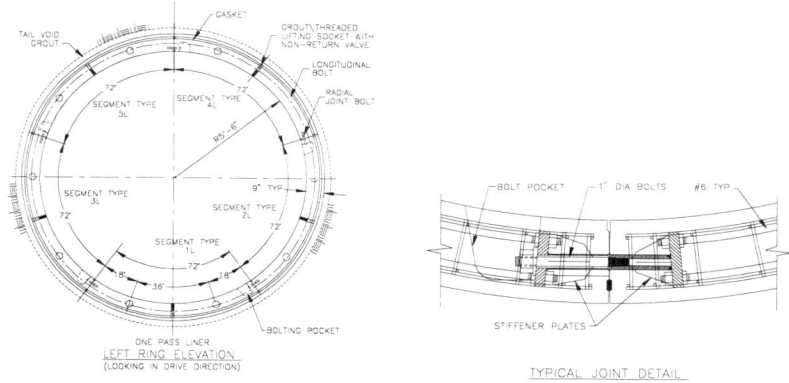

Figure 8. Precast Segment Layout

Circumferential bolts are spaced at 36 degrees, two per segment. The circumferential joint is designed for 25 mm diameter (No. 8) high strength longitudinal bolts at all fault crossings (a total of 966 m of tunnel) to accommodate movement at the numerous fault crossings. At all other locations, the Contractor used smaller 19 mm diameter (No. 6) size (60 Ksi grade) bolts. Also, all longitudinal joints are staggered in order to avoid cross joints at segment intersections.

Figure 9. Erected Tunnel Liner

Immediate grouting of the tail void is required to control liner ovalization. The grout mix must be sufficiently stiff and fast setting so as to provide for the development of immediate passive reaction, to limit liner ovalization/float, in response to the external loads that develop after the ring leaves the tail shield.

Initial (or startup) operations lasted to about the middle of May 1997. These startup activities included scheduled stops for tunnel eye construction; installation of trailing gear, screw conveyor, umbilical support and hydraulic lines approaching 200 m length, shaft elevator, train switches, muck car shaft guides, and other miscellaneous structures; as well as delays due to hydraulic and operational problems, and other learning curve related issues. As of the end of May 1998, the contractor has installed over 3,936 rings or 4.81 km (83 percent of total). Best production rate has been 30 rings (36.6 m) per day for three 8 hour shifts. For a 6 day a week mining cycle, the best weekly advance has been 60 m. Monthly progress is shown in Figure 10.

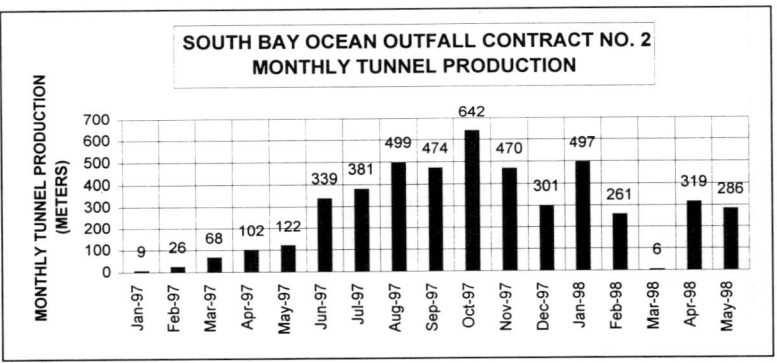

Figure 10. Monthly Production

Contract Package 2 - Offshore Riser

The offshore riser is a pipe encased by tremie concrete within a vertical shaft (Figure 11). The riser connects to the tunnel with a concrete-lined steel elbow and connects to the seabed pipe with a horizontal outlet in the riser head. Figure 12 shows the Riser arriving in San Diego. Offshore riser shaft construction began in October 1996 and included excavating the upper 7.6 m of marine sediments and driving a two-stage casing. The 3.96 m ID upper stage casing was driven through Pleistocene gravels into approximately 1.5 m of the San Diego Formation. A full length 3.65 m ID lower stage casing was driven through the upper casing to within 9 m above the planned tunnel crown. The void between the casing was grouted. The offshore platform for installation of the riser is shown in Figure 13. The platform, was constructed specifically for this project by the subcontractor for the riser. Drilling, riser installation and grouting were performed from the platform.

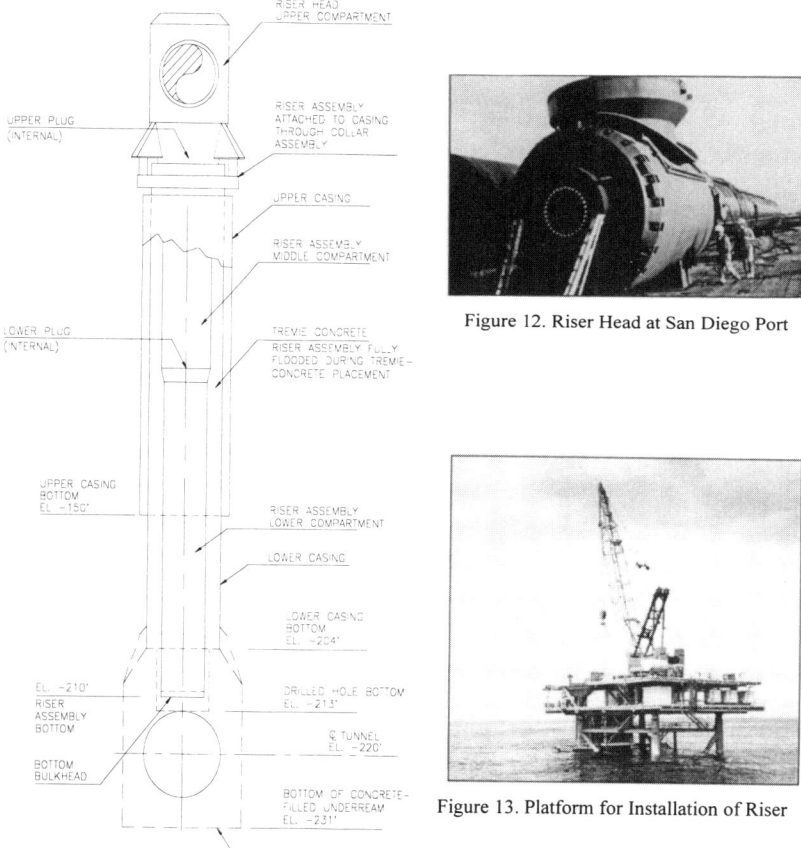

Figure 12. Riser Head at San Diego Port

Figure 13. Platform for Installation of Riser

Figure 11. Riser Shaft Assembly

Underreaming of the area below the lower shaft casing was needed to allow replacement of soil with concrete in the tunnel-to-riser connection zone. The concrete plug provides the target for the tunnel boring machine to pass through. The Offshore Riser was completed in July 1997. The TBM bores through the concrete plug where the TBM shield is parked/abandoned, stripped of internal machinery, and backfilled with cementatious materials. The connection from the tunnel to riser is made from the tunnel by cutting a hole in the shield and excavating out the concrete plug materials between the shield and the bottom of the riser. The concrete plug materials provide a homogenous material to allow connection to the riser under stable conditions. The riser is dewatered from the tunnel and a prefabricated elbow is lowered from the riser where the final connection of mitered pipes are brought in from the tunnel and connected to the elbow.

Contract Package 3 - Seabed Pipeline

The seabed pipeline is 1,433 m long and consists of 3.05 ID by 6.1 m long reinforced concrete pipe that connects the riser head at the 22.9 m water depth to the diffuser wye at the 29 m water depth. The pipeline was trenched and backfilled with graded stone. Armor stone will provide stability during the 100 year design wave and protects the pipe from dragging anchors. Pipe joints are bell and spigot with double gasket o-ring seals. Joints are pressure-tested during installation. The joints can deflect approximately 1.5 degrees from center position, and can be extended axially 127 mm from the fully closed position and still maintain a seal. Construction of the seabed pipeline is complete. Best pipe laying production amounted to 10 sections of pipe per day, and the average was about 5 sections of pipe per day. The seabed pipe laying barge is shown in Figure 14.

Contract Package 3 - Diffuser System

The diffusers were installed in water approximately 29 m deep. The diffuser system consists of a diffuser wye (Figure 15) that diverts effluent flow from the 3.05 m seabed pipeline into two 610 m-long diffuser legs, each consisting of approximately equal lengths of 2.13, 1.83, and 1.37 m ID pipe segments, and a termination structure with flap gate that allows for flushing of each leg. Slide gates in the wye permit flow diversion to one leg during flushing operations. A 3.05 m end gate in the wye also allows for possible future extension of the pipeline. The diffuser legs are oriented parallel to the seabed contours. This configuration minimizes salt water and sediments intrusion during low flows.

Figure 14. Barge for Installation of Seabed Pipe

The diffuser pipe top is flush with the seabed bottom, and is covered with a 1 m thickness of armor stone. Diffuser risers, spaced 7.3 m on center, project 30 cm above armor stone. Each diffuser riser incorporates four ports and is made of high density polyethylene because of its toughness, flexibility, and desirable hydraulic characteristics (Figure 16). Steel shields protect the risers from anchors, and the riser heads are of low profile making them less likely to be hooked by an anchor, fishing net, or wire.

Figure 15. Wye Structure Figure 16. Diffuser Pipe with Riser

Conclusions

The South Bay Ocean Outfall is expected to be completed by end of 1998 with change orders, due to changed conditions, presently amounting to less than three percent for all three contracts. Beaches that have had restricted access is the past, will once again be usable. In addition to these benefits, for oceanographic, geotechnical, tunnel, and marine engineering, state-of-the-art studies were performed.

San Diego's Conveyance Tunnels,
A Historical Perspective

Gregory L. Raines[1]
Rick Wright[2]

Abstract:

San Diego's dependence on water as a resource has necessitated a long history of conveyance tunnel construction to alleviate the threat of drought. The subsequent supply of water coupled with San Diego's natural attractiveness, has led to significant population increases, and the need for the construction of even more conveyance pipelines. This paper will provide a historical summary of the progression of tunnel and trenchless technology construction projects that contributed to the advance of current technology, feasibility and constructability. A bibliography of relevant reports and data is included.

Introduction:

Tunnel construction began in San Diego in 1947 with the installation of the original San Diego Aqueduct by the San Diego County Water Authority. Necessitated by the influx of military personnel into San Diego during WWII, it was considered by President Roosevelt, a matter of national concern. The aqueduct originates at the San Jacinto River / Reservoir and proceeds through Hemet, Temecula, Escondido and finally to Poway, where it became the first link to a continuous water supply east of the desert. A second barrel was added to the first aqueduct in 1954, along a parallel route but utilizing the same tunnels. The Second San Diego Aqueduct began construction in 1960 and was added to in 1973 and 1997. The pipeline route originates from the Colorado River and proceeds to the Lower Otay Lake Reservoir, crossing through the heart of San Diego. These Aqueducts have the largest pipelines constructed in San Diego County and their cross-country nature called for significant tunnel construction projects within them.

---

[1] Senior Engineer, Tunnels Office, Haley & Aldrich Inc., 2918 Fifth Avenue, Suite 202, San Diego, CA 92103, 619.296.7181
[2] Civil Engineer, Tunnels Office, Haley & Aldrich Inc., 2918 Fifth Avenue, Suite 202, San Diego, CA 92103, 619.296.7181

The Metropolitan Wastewater Department, with their 1.1 billion dollar Clean Water Capital Improvement Project, is another major, yet recent contributor to the history of tunnels in San Diego. Four world class tunnel construction projects originated from this program: the South Bay Ocean Outfall paralleling the Mexican border into the Pacific Ocean, the North Metro Interceptor Phases I and II, which collect and redirect flows to the Point Loma Treatment Plant, and the North City Tunnel Connector which directs various flows under the 805 freeway to the North City Water Reclamation Plant. All of these large tunnel construction projects are being completed on a fast track schedule.

San Diego Gas and Electric also contributed to San Diego's diverse tunnel history with the gas distribution pipeline known as Pipeline 2000 which so far has included 4 medium sized tunnels, they also recently completed a 2000-foot directionally drilled crossing under the San Luis Rey River. Two of these projects set records for the strongest rock ever excavated in North America using microtunneling and directional drilling.

San Diego Water Utilities performed a variety of installations including horizontal directional drilling, microtunneling, and conventional jack and bore tunneling. The City of San Diego has installed at least 4 microtunnel projects, including a 5000-foot installation as part of the North Mission Valley Interceptor. The Port of San Diego added microtunnels under the Lindbergh Field Runway, and the Old Pacific Highway.

San Diego's tunnel construction history has also been contributed to by: the Sweetwater Authority, who constructed a tunnel associated with the Sweetwater Dam in 1911, the City of Oceanside which completed three medium sized tunnels in north San Diego County, the San Marcos Water District who installed an 800-foot crossing under Lake San Marcos, and the International Boundary Water Commission who constructed a 1500-foot horizontal directionally drilled crossing under the Tijuana River.

A summary table is attached at the end of this paper that summarizes the vast majority of the tunneling projects that were constructed in San Diego County.

Highlighted Tunnel Projects:

The following projects are some of the more high profile tunneling projects constructed in San Diego County, and deserve to be mentioned in more detail:

<u>The First San Diego Aqueducts:</u> The First San Diego Aqueduct program consisted of 71 miles of pipeline extending from the San Jacinto Tunnel at the Colorado River Aqueduct to the San Vicente Reservoir. The high priority project was rapidly conceived and designed in the early 1940's due to the large influx of military personnel into San Diego prior to World War II threatening to deplete the City's

water supply. Construction began in September 1945 and was completed in November 1947.

The pipeline consisted of reinforced concrete pipe ranging in diameter from 48-inch to 96-inch, however all tunneled sections of the pipeline were designed for the ultimate maximum capacity of 165 cubic feet per second, and were 72 inches in diameter. The seven tunnels that were constructed for the project, had a mean length of almost 3600 feet, and were given the names as follows: Rainbow, Lilac, Red Mountain, Oak Hills, Poway, Fire Hill, San Vicente. The tunnels were bored through hard rock consisting of the Basement Complex and the Poway or Stadium Conglomerate and finally lined with concrete. Limited documentation suggests that the tunnels were often unsupported during excavation, and steel ribs or timbers were used as temporary support in the faulted fractured areas.

Cowels Mountain Tunnel: The City of San Diego Water Authority's Cowles Mountain Water Transport Tunnel was constructed with a Construction and Tunneling Services Tunnel Boring Machine. The 11' 3" tunnel was 6800 linear feet in length and was completed in January 1992. The excavation took place in prolonged runs of granodiorities of up to 64,000 pounds per square inch and currently holds the record for the strongest rock excavated in North America with a tunnel boring machine.

North Metro Interceptor Phase I: The North Metro Interceptor Phase I was conceived as part of San Diego's Clean Water Program to increase the capacity of, and allow for the rehabilitation of the original North Metro Interceptor Pipeline. The 3.5 mile, 108" diameter pipeline began construction in mid 1995 and was completed in early 1997, and consisted of 10,000 feet of open trench installation and 3000 linear feet of directly jacked and two pass tunneling through soft ground. The most notable of the jacked pipe segments was the Taylor Street drive which was 1900 linear feet; the longest directly jacked drive on the West Coast to date.

North Metro Interceptor Phase II: The North Metro Interceptor Phase II was constructed as a two-pass tunnel, 4500 feet in length, under the property of the Marine Corps Recruiting Depot located adjacent to San Diego's Lindbergh Field International Airport. The initial supports were concrete segments that were cast on site, and assembled in the tail shield of the Lovat Tunnel Boring Machine. The 114" reinforced concrete pipe segments were set into place with a pipe carrier and locomotive.

North City Tunnel Connector: The North City Tunnel Connector was designed to carry various pipelines including; sewer, water, reclaimed water and fiber optics to and from the North City Water Reclamation Plant under the 805 Freeway at La Jolla Village Drive and Miramar Road. The 2000 linear foot tunnel was constructed in soft ground using a road header and scooptram to excavate the heading. The initial support was steel ribs and timber blocking, covered with a wire mesh, which was later filled with stiff cement, and finally covered over with shotcrete.

South Bay Ocean Outfall: The South Bay Ocean Outfall was constructed to relieve San Diego's additional supply of treated sewer water for discharge into the Pacific Ocean. The 19000 linear foot, 132" tunnel is being bored at a depth of 150 feet below the seabed at 7 bars of atmospheric pressure, and was constructed with a Lovat Tunnel Boring Machine that was 220-foot in length with trailing gear attached. The one pass tunnel is being supported with concrete segments that were assembled in the tail shield.

The San Diego Gas and Electric Projects: In 1995 SDG&E launched the Pipeline 2000 campaign, a 36" gas transmission pipeline, which included the crossing of 4 major thoroughfares. SDG&E crossed under Interstate 8 and Fletcher Parkway using a digger shield in Stadium Conglomerate, they used a rotary cutter head to cut through decomposed granite under Hwy 94, and microtunneled their way under State Route 125 in Santiago Peak Volcanics. SDG&E also projects completed a 2000-foot horizontal directional drilling project under the San Luis Rey River directly installing a 30" steel gas transmission line.

The City of San Diego Projects: In 1992 the City of San Diego began the Kearney Mesa Trunk Sewer project, attempting to microtunnel a 51" steel casing but had to finish the installation using hand excavation due to the presence of a large cobble formation. The Middletown Trunk Sewer, in mid 1993, was successful in directly jacking a 1500-foot length of 18" clay pipe. The North Mission Valley Interceptor Sewer project, in 1997, successfully installed 5000 ft of 24" concrete pipe by mean of direct jacking behind a microtunnel machine, and was awarded the "1997 Trenchless Technology Project of the Year Award."

The San Diego Water Utility Projects: San Diego Water Utilities successfully pulled a 2200-foot, 16" HDPE run under the San Diego River Jetty from Mission Bay to Ocean Beach using horizontal directional drilling techniques in 1995. They also microtunneled an 1100-foot run of 20" clay pipe to Shelter Island across Mission Bay in 1997. The Water Utilities Department also completed a 1200-ft rib and lagging tunnel through Friars Conglomerate as part of their North Mission Valley Interceptor project, 78" concrete pipe was later installed in the initial support.

Geologic Setting:

The following is provided to introduce the reader to the basic formations encountered in San Diego County as mentioned in the subject paper:

Alluvium and Slope Wash: Alluvium in the area consists primarily of poorly consolidated stream deposits of silt, sand and cobble sized particles derived from bedrock sources that lie within and to the east of the study area. Alluvium and slope wash have not been differentiated on most geologic maps. (Kennedy, 1975)

San Diego Formation: These exposures, attaining a maximum of 30 meters, are composed of yellowish-brown, fine to medium-grained, poorly indurated sandstone. Cobble conglomerate, thin beds of bentonite, marl, and brown mudstone further characterize this section. (Kennedy, 1975)

Mission Valley Formation: Marine, lagoonal, and non-marine sandstones reaching a maximum thickness of 60 meters. The sandstone is soft and friable, light olive gray, and fine to medium grained. It is locally interstratified with cabonate cemented beds. Cobble conglomerate tounges within the Mission Valley Formation, which are identical to the Stadium Conglomerate in lithology, comprise up to 30 percent of the strata. (Kennedy, 1975)

Friars' Formation: Non-marine and lagoonal sandstones with interbeds of claystone reaching a maximum thickness of 150 meters. The sandstone is massive yellowish gray, medium grained, poorly indurated, and caliche-rich. The claystone is dark greenish gray, well indurated, and expansible. Landslides are common in the clay-rich part of the formation. (Kennedy, 1975)

Poway Conglomerate: One of the most widespread and distinctive rock units in southern California. The rock is mostly non-marine sandstone with course cobble conglomerate. Three formations are recognized within it the Stadium Conglomerate, the Mission Valley Formation, and the Pomerado Conglomerate. (Kennedy, 1975)

Stadium Conglomerate: The formation consists of a massive cobble conglomerate with a dark yellowish-brown coarse-grained sandstone matrix. The Stadium Conglomerate is moderately well sorted with an average clast size in the cobble range. Fine-grained rocks within the Stadium Conglomerate constitute less than 20 percent of the unit. (Kennedy, 1975)

Santiago Peak Volcanics: The Santiago Peak Volcanics comprise an elongate belt of mildly metamorphosed volcanic, volcanoclastic, and sedimentary rocks. They are hard and extremely resistant to erosion and form topographic highs. Most of the volcanic rocks are dark greenish gray where fresh, but weather to grayish red or dark reddish brown. (Kennedy, 1975)

Santiago Formation: Marine and non-marine, greenish-gray and yellowish gray, medium to course grained sandstone with interbedded fine grained clayey sandstones. The friable, medium density, sandstone has a maximum thickness of 200 meters. (CDMG, 1976)

Ardath Shale: Ardath Shale is predominantly weakly fissile, olive-gray shale. Concretionary beds containing mooluscan fossils are common. Expansible claystones locally comprise as much as 25 percent of the unit and landslides are commonly associated with those areas. Sieve analyses indicate that the particle size distribution is 81 percent silt, 16 percent clay, and 3 percent sand. (Kennedy, 1975)

Crystalline Basement Complex: The Crystalline Basement Complex consists of two principal rock units: (1) the Upper Jurassic Santiago Peak Volcanics, a succession of deformed and metamorphosed volcanic, volcaniclastic, and sedimentary rocks; and (2) mid-Cretaceous plutonic rocks of the Southern California batholith, which intrude the Santiago Peak Volcanics. (Kennedy, 1975)

Methodology:

The information gathered to write this paper originated from the best possible resources available at the time including plans, geologic investigations, geologic reports, previously written papers, correspondence with local government organizations and municipalities, and project experience. The authors will continue to update this paper and would greatly appreciate any additions or corrections that can be found. A location map of the tunneling projects contained in the table, will be available at the conference, or can be obtained from the authors upon request.

Summary and Conclusions:

San Diego has a rich and diverse history of tunnel construction. Emergency programs such as the First San Diego Aqueduct, providing drought insurance for the citizens, and security for the nation, environmentally mandated programs such as Metropolitan Wastewater's Clean Water Capital Improvement Project, and expansive programs such as the SDG&E Pipeline 2000 Project have all contributed. These projects have not only contributed to San Diego's tunneling history but also to the knowledge needed to perform these projects successfully. As agencies find that they can perform these jobs reliably, and at a lower cost or with less surface disruption, they will turn to these solutions more often, and San Diego's tunneling history will lie within the future.

References:

Autobee, R., 1993, "Research on Historic Reclamation Projects", Bureau of Reclamation History Program, Denver, Colorado.

Black and Veach, 1996, Construction Plans for the Second San Diego Aqueduct Pipeline 5 Extension Phase II, San Diego County, California.

California Division of Mines and Geology, 1976, "Geology and Engineering Geologic Aspects of the Laguna Beach Quadrangle" Orange County, California.

James M. Montgomery, Consulting Engineers, Construction of the 96" Pipeline No. 4 Extension Phase I, San Diego County, California.

Kennedy, M.P., 1975, "Geology of the San Diego Metropolitan Area, California", California Division of Mines and Geology, Sacramento, California.

Swallow, W., 1998, "Case Studies for 5 City of San Diego Microtunneling Projects. Moving up the learning curve to Provide Efficient Microtunnel Design projects", North American No-Dig '98 Conference Paper.

PIPELINES IN THE CONSTRUCTED ENVIRONMENT 699

Project	Location	Owner, Purpose	Date Final	Length	Initial Support	Final Size	Geologic Conditions	Construction Methods
First San Diego Aqueduct	Rainbow	SDCWA, Water	1947	4860	Hard Rock	72 in	Basement Complex, Poway Conglomerate	Drill and Blast
	Lilac			500				
	Red Mtn.			4980				
	Oat Hills			3750				
	Poway			3350				
	Fire Hill			5300				
	San Vicente			2450				
Cowles Mountain Tunnel	Cowles Mountain	SDCWA, Water	Early 1992	6800	Concrete Segments	135"	Granodiorites and sand conglomerate	TBM
Las Posas Tunnel	Questhaven / S12	SDCWA, Water	1992	1685	Steel Ribs - Mod HS	108" Steel	Santiago Peak Volcanics	Drill and Blast
Pipeline 4 Extension Phase I	Jackson Drive Tunnel	SDCWA, Water	1996	340	Steel Casing	96" Steel	Stadium Conglomerate	Boring and Jacking / Shield Tunneling
	Interstate 8 Tunnel			466	Steel Rib or Casing	108" Steel		
	Lemon Grove RR crossing			235	Steel Casing	96" Steel		
	South Bay Parking Tunnel			327	Steel Casing	96" Steel		
Pipeline 2A	Interstate 15 Tunnel	SDCWA, Water	Mid 1997	620	Steel Ribs and Lagging	66" Steel	Granatic Rock	Blasted / Hand Mined
Second San Diego Aqueduct Pipeline 5 Extension Phase II	Mercy Road Tunnel	SDCWA, Water	Mid 1997	615	Steel Casing	108" Steel	Santiago Peak Volcanics	Drill and Blast
	Del Dios Hwy Tunnel			70				
	Carmel Mtn Road Tunnel			450			Mission Valley Formation	
San Marcos Tunnel	Mission Rd / Hwy 78	SDCWA, Water	1992	390, 170	Ribs and Boards	108"	Santiago Peak Volcanics	Digger Shield
South Metro Interceptor	Harbor Drive	MWWD, Sewer	1960's	2500	Ribs and Boards	144"	Cretacious Sedement	Digger Shield
Point Loma Tunnel	Point Loma	MWWD, Sewer	1960's	5000	Steel Ribs	6' x 8'	Cretacious Sedement	Drill and Shoot
North City Tunnel Connector	Miramar / Frwy 805	MWWD, Various	Early 1996	2000	Steel Ribs	18' Dia. Horse-shoe	Weak Sandstone	Road Header
North Metro Interceptor Phase I	Hwy 8 / Hwy 5	MWWD, Sewer	Late 1996	6000	Direct Jack	108" RCP	Bay Deposits	TBM

SAN DIEGO COUNTY - TUNNELING SUMMARY TABLE								
Project	Location	Owner, Purpose	Date Final	Length	Initial Support	Final Size	Geologic Conditions	Construction Methods
North Metro Interceptor Phase II	Lindbergh Field / Harbor Dr.	MWWD, Sewer	Mid 1998	4500	Concrete Segments	114" RCP	Bay Deposits	TBM
Fiesta Island Repl. Project Tunnel	Hwy 52 / Convoy	MWWD, Sludge	Early 1997	600	Liner Plates	Multiple Pipe Types	Weak Sandstone	Open Face Digger Shield
Penasquitos Pump Station Tunnel	Hwy 15 / Mercy Rd.	MWWD, Sewer	Mid 1997	600	Steel Casing	48" DIP	Santiago Peak Volcanics	Blast
The South Bay Ocean Outfall	US border / Pacific Ocean	MWWD, Rec. H2O	Late 1998	17500	Concrete Segments	132"	San Diego Formation	EPBM
Pipeline 2000	Interstate 8 Tunnel	SDG&E, Gas	1995	350	Liner Plates	30" Gas	Stadium Conglomerate	Digger Shield
	Fletcher Parkway Tunnel	SDG&E, Gas	1995	300	Liner Plates	36" Gas	Stadium Conglomerate	Digger Shield
	Highway 94 Tunnel	SDG&E, Gas	1995	250	48" Steel Casing	36" Gas	Decomposed Granite	Rotary Cutter Head
	State Route 125 Tunnel	SDG&E, Gas	1996	250	48" Steel Casing	36" Gas	Santiago Peak Volcanics	Microtunneling
San Luis Rey River Crossing		SDG&E, Gas		2000	None	30" Gas	Alluvium	Horizontal Directional Drilling
Mission Valley River Crossing		SDG&E, Gas		500	48" Steel Casing	36" Gas	Alluvium	Rotary Cutter Head
Middletown Trunk Sewer		City of SD, Sewer	July-93	1500	Direct Jack	18" Clay	Alluvium	Microtunnel
North Mission Valley Interceptor Phase II		City of SD, Sewer	Nov. 1997	5000	Direct Jack	24" Sewer	Alluvial Sands / Hard Clay / Cobble	Microtunnel
Group Job #506		City of SD, Water	Mar-97	570	30" Steel Casing	16" Water	Ardath Shale	Microtunnel
Kearney Mesa Trunk Sewer		City of SD, Sewer	Mar-92	250	51" Steel Casing	27" Sewer	Large Cobble Formation	Microtunnel / Hand excavation

SAN DIEGO COUNTY - TUNNELING SUMMARY TABLE

Project	Location	Owner, Purpose	Date Final	Length	Initial Support	Final Size	Geologic Conditions	Construction Methods
Shelter Island Tunnel	Mission Bay / Shelter Is.	SDWU, Water	Early 1997	1100	Direct Jack	20" Clay	Dredged Bay Deposits	Microtunnel
Mission Bay / San Diego River	Mission Bay / San Diego River	SDWU, Water	1995	2200	HDPE	16"	Dredged Bay Deposits	Horizontal Directional Drilling
North Mission Valley Interceptor	Mission Rd / Fairmont	SDWU, Water	1992	250	Jacked Pipe	84" RCP	Alluvium	Hand Excavation
	Stardust / Mission Valley	SDWU, Water	Mid 1997	1200	Steel Ribs and Wood Lagging	78" RCP	Friars Conglomerate	Jack and Bore
	Hwy 163 / Hazard Center Road	SDWU, Water	Mid 1997	300	Steel Rib Steel Liner Plate	78" RCP	Alluvium	Jack and Bore
Lindbergh Field Runway Crossing	Lindbergh Field	Port of SD	1995-1996	1100	Steel Casing	10"Jet Fuel, Elec.	Fill	Microtunnel
Old Pacific Highway Crossing	Hwy 8 / Pacific Hwy	Port of SD	1995-1996	420	Steel Casing	16" Water	Fill	Microtunnel
Old Utility Tunnel		Oceanside	1950's	2000	6' x 8' Shotcrete	24"	Existing Tunnel	Place and Backfill
El Camino Real / Mesa Upgrade Tunnels		Oceanside	1994	200	Jacked Pipe	48"	Santiago Formation	Rotary Cutter Head / Jacked
		Oceanside	1994	600	Jacked Pipe	48"	Santiago Formation	Rotary Cutter Head / Jacked
Sweetwater Dam Tunnel	Sweetwater Dam	Sweetwater Authority, Water	1911	250	Drill and Blast	2-24", 2-30"	Hard Rock	Drill and Blast
Lake San Marcos Crossing	Lake San Marcos	Valacitos Water District	1995	800	Direct Jack	20"	Alluvium	Microtunnel
Tijuana River Crossing	Dairy Mart Rd. / TJ River	IBWC, Water	July-96	1500	HDPE	20"	Very Dense Sands	Horizontal Directional Drill

CONDITION ASSESSMENT AND REPAIR OF PRESTRESSED CONCRETE PIPELINE

Mehdi S. Zarghamee[1], F. ASCE, Rasko P. Ojdrovic[1], M. ASCE, and Roger Fongemie[2], M. ASCE

Abstract

The water reclamation supply system of the Palo Verde Nuclear Power Generating Station is a 36.5 mile long pipeline consisting of 114, 96, and 66 inch diameter prestressed concrete cylinder pipe (PCCP). The corrosion of prestressing wires resulted in three ruptures, and an unacceptable risk of emergency shutdown. A program of risk management was adopted that involved placing the pipeline under cathodic protection, and conducting a condition assessment program to identify and repair corroded pipe segments. Condition assessment was based on internal visual inspection and sounding, nondestructive testing, and external inspection. The accuracy of internal inspection and nondestructive testing methods in locating corroded pipe was determined by excavation and external inspection. Over 276 pipe segments were exposed for external inspection.

Seventy corroded pipe segments were repaired by post-tensioning using a circumferential tendon system. The post-tensioning repair design criteria and stress and strain limit states under the combined effects of internal pressure and external loads are discussed. Tests were performed to evaluate the effect of pipe curvature on sheathing and on the coupler friction. Special hardware was developed to alleviate the problems encountered, and to allow for cathodic protection of the tendons. The post-tensioned pipe was coated using a 3 in. thick polyolefin fiber-reinforced shotcrete.

Background

The water reclamation supply system of the Palo Verde Nuclear Generating Station consists of approximately 36.5 miles of prestressed concrete cylinder pipe (PCCP) extending from the Wastewater Treatment Plant in Phoenix to the Water Reclamation Plant at Palo Verde Nuclear Generating Station, with internal diameters of 114 in., 96 in., and 66 in. The pipe wall consists of a concrete core with a 16

---

[1] Principal and Senior Project Manager, respectively, Simpson Gumpertz & Heger Inc., Consulting Engineers, 297 Broadway, Arlington, MA 02174
[2] Senior Project Manager, Arizona Public Service, Tonopah, AZ 85354

gage embedded steel cylinder with a yield strength of 33 ksi. Prestressing wire, 8 gage – Class 2, is helically wrapped around the concrete core, and a 3/4 in. thick mortar coating is applied on the wrapped core. The design characteristics of the pipe are as follows:

Core thickness	$h_c = 5.25$ for 66 in. diameter pipe
	$h_c = 7.25$ or 9.0 for 96 in. diameter pipe
	$h_c = 8.75$ or 12.25 in. for 114 in. diameter pipe
Core compressive strength	$f'_c = 4500$ psi (most pipe designs)
Pipe segment length, typical	$L = 24$ ft for 66 and 96 in. diameter pipe
	$L = 16$ or 20 ft for 114 in. diameter pipe

The pipeline was constructed with bonded joints and was monitored.

A total of three pipe segments ruptured in service: one 66 in. diameter pipe segment ruptured in 1994, and a 66 in. and a 96 in. diameter pipe segments ruptured in January 1997. The back-to-back failure of the pipeline created a high risk of an emergency shutdown of the Nuclear Generating Station.

A program of risk management was adopted that involved placing the pipeline under cathodic protection with impressed current, and conducting a condition assessment program to identify and repair corroded pipe segments. Condition assessment was based on internal visual inspection and sounding, nondestructive testing, and external inspection. The corroded pipe segments were repaired.

Rupture Scenario

The typical rupture scenario, in absence of a wire break caused by hydrogen embrittlement, involves three stages. The first stage involves cracking and/or debonding of concrete coating, causing corrosion of prestressing wire. Alternatively, corrosion of prestressing wire can result from permeation of chloride ions through the coating, causing cracking and debonding of the coating. In the second stage, cracking of the concrete core occurs after loss of prestress from the external loads, and causes debonding of inner concrete core. In the third stage, rupture of the steel cylinder occurs when pressure exceeds the capacity of the uncorroded steel cylinder, or corrosion of the steel cylinder causes the rupture after loss of thickness to corrosion.

The above failure scenario indicates that pipe does not rupture after a partial loss of prestress. For rupture to occur, the prestress must be lost, the core must be cracked allowing pressure to be applied to the steel cylinder, and the pressure must exceed the ultimate strength of the steel cylinder in either the uncorroded or the corroded state.

Condition Assessment

The goal of condition assessment was to identify corroded prestressed concrete pipe for repair through an appropriate inspection program in such a way that power production is not affected.

All pipe segments selected for external inspection based on the results of internal inspection and nondestructive testing were excavated, externally inspected, and repaired if corroded. The results of this correlation study was used to evaluate the effectiveness of internal inspection and testing for locating corroded pipe. (Note that the excavations performed do not provide information on the condition of the pipe in areas where internal inspection has not shown major anomalies.)

The schematic of the condition assessment program is shown in Fig. 1.

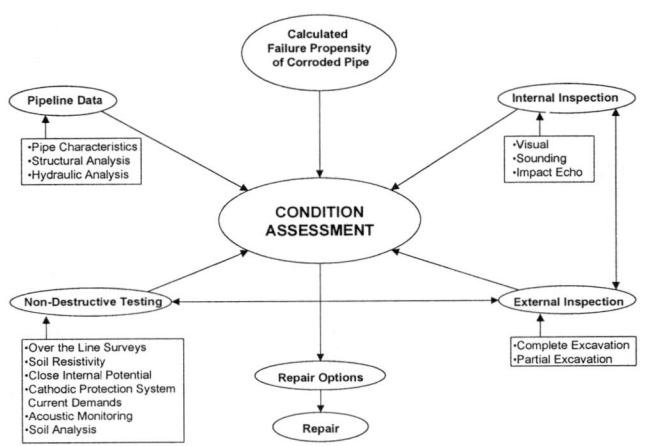

Fig. 1. Condition Assessment and Repair Schematic

Database. A database was constructed from the files with pipe layout data and inspection and test results, and pipe design data. The database contains the following data:

Basic Pipe Data:

- Pipe segment number
- Pipe station
- Pipe design data
- Design pressures and cover depths
- Operating and surge pressures

Data from Different Assessment Techniques:

- Internal inspection results and observed anomalies
- External inspection results
- Acoustic monitoring data
- Soil resistivity survey data

- Close interval potential survey data
- Cathodic protection current demand data
- Soil test results from pipeline design surveys and from tests performed on samples retrieved from excavation sites

Structural Strength Data:

- Adequacy of design
- Internal pressure causing the steel cylinder to yield after complete loss of pipe prestress
- Internal pressure causing the steel cylinder to rupture after complete loss of pipe prestress

External Inspection. External inspection is a direct method for detecting mortar coating cracking and debonding, extensive corrosion of prestressing wires, concrete core cracking, and steel cylinder corrosion condition if outer concrete core is removed over a limited area at a core crack. It, however, is expensive and time consuming to perform, and must be performed during planned outages.

Through external inspection of pipe segments with major internal distress or their adjacent pipe segments, a total of 70 pipe segments were found to be corroded and were repaired (Table 1). From the 70 repaired pipe, 25 were located in the 66 in. diameter section (at an average corrosion rate of 3.1 pipe per mile), 45 were located in the 96 in. diameter section (at an average corrosion rate of 2.0 pipe per mile), and none in the 114 in. diameter section. On several corroded pipe segments in the 66 in. and 96 in. diameter sections, the outer concrete core was removed at cracks varying in size up to 1/8 in. in width and the condition of steel cylinder was inspected for evidence of corrosion. At most, a very light surface rust was observed on the exposed part of the steel cylinder without any loss of cross section.

Table 1. Summary of Pipe Inspection and Repair Results

Pipe Diameter	Miles of Pipe	Total No. of Pipe	Number of Pipe with Internal Distress*		No. of Pipe Repaired	Pipe Repair Rate Per Mile
			Medium/ Minor	Major		
66	8.0	1,797	75	8	25	3.1
96	22.5	5,000	83	15	45/32**	2.0
114	6.0	1,836	1	0	0	0.0

*   Pipe with hollow sound and/or longitudinal cracking.
** Of the 45 repaired 96 in. diameter pipe, 32 have documented internal inspection results; the other 13 repairs were based on undocumented internal inspections performed before March 1997, and are not used in our correlation of internal and external inspection.

Internal Inspection. The anomalies observed are indirect indications of pipe condition. Typical anomalies observed in internal inspection included:

- Hollowness in inner core, which is an indication of prestress loss and cracking of the outer core, thus allowing pressure to separate the steel cylinder from the inner core. Extensive hollowness is typically indicative of severe damage. Hollowness in inner core may occur in uncorroded prestressed concrete pipe over a limited area, due to radial shrinkage of the inner core and large bending moment, especially near the invert.

- Longitudinal cracking, which is an indication of pipe overload with earth load (or live load) and internal pressure. Propensity of a pipe to develop longitudinal cracking increases significantly after partial or full loss of prestress.

- Circumferential cracks, which occur typically on PCCP within about 1 ft from the joint, or with equal spacing away from the joint due to shrinkage or differential settlement. Such cracks, if hairline, are not significant.

- Other anomalies, such as efflorescence at cracks, joint opening, map cracking and stains, which are not likely indications of loss of prestress unless accompanied by hollowness or longitudinal cracking.

The severity of the anomalies observed on the inside of the pipe may be an indication of the extent of prestress loss. We subdivided the hollowness and longitudinal cracking records in the database into two categories: (1) major anomalies – if 3 ft or larger in dimension (2) minor anomalies – if less than 3 ft in dimension or of undocumented size.

A correlation study was performed between the results of internal inspection and the results of the external inspection of those pipe segments that were excavated during the 1997 outages.

In total, 133 pipe segments were fully exposed and are included in this correlation study; 65 pipe segments were located on the 66 in. diameter and 68 on the 96 in. diameter section of the pipeline. Only one 114 in. diameter pipe segment was excavated and it was found to be uncorroded. In addition, the excavation provided an opportunity for partial inspection of adjacent pipe segments, which increased the total number of fully and partially inspected pipe segments to 276. Some adjacent pipe segments were partially excavated, while others were fully excavated due to either an error in locating the pipe or the existence of coating cracks and wire corrosion on the exposed part of the pipe.

To account for the uncertainty in the location of internally inspected pipe relative to the excavated pipe, we allowed for a possible location mismatch by one pipe segment in relating the external and the internal inspection results to each excavation site. The comparison showed that such a mismatch occurs only rarely.

A summary of the results obtained from the internal and external inspections of the pipe is shown in Table 2. In this table the number of excavated pipe includes those pipe that were adjacent to the fully excavated pipe but were only partially excavated. The exterior surface of partially excavated pipe segments was inspected to the extent exposed.

Based on the results of this correlation study, the following conclusions were derived:

- Pipe segments with major internal distress, i.e., major hollow sound and major longitudinal cracking in the internal inspection have a 91% probability of being corroded.

- Pipe segments with medium/minor internal distress in the internal inspection have a 29% probability of being corroded. For such pipe segments, longitudinal cracking is a better indicator of corroded pipe than hollow sound.

- Pipe segments with no internal distress located adjacent to corroded pipe segments have an 8% probability of being corroded.

- Overall, 21% of fully and partially excavated pipe were found to be corroded and repaired and 37% of the repaired pipe had major internal distress.

- Internal inspection can identify pipe in advanced stages of distress. It cannot, however, identify pipe segments in earlier stages of distress.

Table 2. **Correlation of Internal and External Inspection Results Based on Fully and Partially Excavated Pipe**

Pipe Condition from Internal Inspection	66 in. Pipe		96 in. Pipe		Total	
	Repaired/ Excavated	% Repaired	Repaired/ Excavated	% Repaired	Repaired/ Excavated	% Repaired
Major Distress	8/8	100%	13/15	87%	21/23	91%
Medium\ Minor Distress	10/46	22%	13/34	38%	23/81*	29%
No Distress	7/83	8%	6/90	7%	13/173	8%
TOTAL	25/137	18%	32/139	23%	57/277	0.21

* includes one 114 in. diameter pipe that was excavated and found to be good.

The probability of corrosion of a pipe occurring within three segments of a corroded pipe was calculated to determine if the pipe deterioration occurs in clusters. The results show that approximately two-thirds of corroded pipe are in clusters.

Soil Resistivity. Field soil resistivity, measured using the Werner four-pin method, was found to be low over the entire pipeline due to the presence of electrolytic salt ions such as chlorides and sulfates. We denote soils with resistivity below 1,000 ohm-cm as very corrosive, between 1,000 and 2,500 ohm-cm as corrosive, between 2,500 and 5,000 ohm-cm as fairly corrosive, and above 5,000 ohm-cm as mildly corrosive. Correlation of soil resistivity measurements and internal inspection results show the following:

- Approximately three quarters of the pipeline is installed in very corrosive and corrosive soils, and the remainder is in fairly corrosive and mildly corrosive soils.

- All of excavated 96 in. diameter pipe are located in very corrosive and corrosive soils.

- For 66 in. diameter pipe, all except three adjacent repaired pipe segments are located in corrosive and very corrosive soils. These three pipe segments are in mildly corrosive soils located in the area of Luke Wash, with significant seasonal variations in available water. Other excavated pipe segments in fairly and mildly corrosive soils did not have externally observed distress.

- For 114 in. diameter pipe, most of the pipe is in very corrosive and corrosive soils. However, only one pipe segment was excavated and found to be not corroded.

The results show that nearly all corroded pipe are in very low and low resistivity soils, and that resistivity cannot be used to pin-point a corroded pipe, but it can be used to identify areas where more corrosion can occur.

Close Interval Potential Survey. A correlation study of close interval potential survey and pipe distress as determined from external excavation showed no apparent correlation. Based on these results, one can conclude that close interval potential survey cannot be used to locate corroded pipe.

Cathodic Protection System Current. Correlation of cathodic protection system current demands and the external excavation inspection results show no apparent correlation.

Protected Pipeline Crossings. Seven cathodically protected gas and petroleum pipelines cross the pipeline. No significant correlation was found between the crossing location and observed distress in the pipe at such crossings.

Acoustic Monitoring. In acoustic monitoring, several hydrophones are installed at manholes and acoustic events are recorded during a monitoring period. The location of acoustic events can be determined from cross reference of time of arrivals recorded by two or more hydrophones. Acoustic data was collected by monitoring a section of the 66 in. pipeline between Sta. 199+50 and Sta. 263+50 during the time period from 25 June to 4 August 1997.

There were no known corroded pipe in this interval. During this period, 7 possible wire breaks, and 3 locatable events were recorded. From the total of 10 pipe segments that produced acoustic events, our database shows that one pipe was excavated after internal inspection showed distress and found to be uncorroded, two pipe segments were encased during the monitoring period and apparently continue to corrode, and one pipe segment was a replacement pipe adjacent to a failure site.

Acoustic monitoring performed under steady loads can detect breaks of prestressing wires during the monitoring period, but it does not show the existing damage to the pipe, and it cannot be used to identify corroded pipe. Note that lack of wire breakage during a recording period does not mean that the wire is not already corroded.

Laboratory Soil Analysis. A correlation study was performed between the results of laboratory chloride content and minimum soil resistivity tests of samples retrieved from the soils adjacent to the excavated pipe and pipe corrosion condition. The results of this study show that:

- Chloride content and minimum soil resistivity are highly variable along the pipeline and around the pipe at a station.

- Corrosion of pipe has not been observed where chloride content is less than 250 ppm and where minimum soil resistivity exceeds 1500 ohm-cm.

- For the case where data is available, the variability of local chloride content and minimum soil resistivity is in good agreement with the occurrence of local corrosion around the pipe.

Corroded Pipe Safety Priority. For monitoring and repair of pipeline, corroded pipe safety priorities were assigned to each pipe, assuming that prestressing wire is lost to corrosion, based on (1) internal pressure effect on pipe failure after loss of prestress and (2) soil resistivity measurements.

Priority 1. Loss of prestress causes the uncorroded steel cylinder to rupture under working pressure. All pipe segments in this priority were encased.

Priority 2. Loss of prestress causes the uncorroded steel cylinder to rupture under surge pressure. Internal inspection of such pipe should be performed annually. All pipe with internal distress should be excavated and repaired immediately in the next planned outage.

Priority 3. Loss of prestress does not result in the rupture of the uncorroded steel cylinder under surge pressure, and the pipe is in corrosive or highly corrosive soils. Internal inspection of such pipe in this priority should be performed every 2 years, and distressed pipe should be repaired in the next planned outage.

Priority 4. Loss of prestress does not result in rupture of the uncorroded steel cylinder under surge pressure and the pipe is not in corrosive or highly corrosive soils. Internal inspection of such pipe should be performed every 5 years and distressed pipe should be repaired in planned outages.

Priority 5. Loss of prestress cannot result in rupture, e.g., repaired, encased or lined pipe. No action is required.

Repair Design

This section summarizes the methodology used in the design of the repairs of the prestressed concrete cylinder pipe using post-tensioning tendons. The repair design assumptions, criteria, and analyses performed are described below.

Repair Design Assumptions. The design of the post-tensioning repairs is based on the current AWWA C304-92 Standard for Design of Prestressed Concrete Cylinder Pipe with the following assumptions and modifications:

- Prestressing is lost over a certain length of the pipe to be repaired. Post-tensioning may be applied over unprestressed as well as prestressed portions of a pipe.

- Concrete core of the pipe is cracked and cannot take any tension.

- The 28-day design compressive strength of concrete core is 4500 psi, the same as in the original design.

- Earth load for a given cover depth is computed from Marston embankment installation for granular backfill and a unit weight of 120 lb/ft^3, accounting also for the weight of soil in the pipe shoulders.

- The maximum working pressure and the maximum transient pressure are based on the results of hydraulic analysis performed during the initial pipeline design.

Repair Design Criteria. The repair pipe was designed so that when post-tensioning tendons are applied to the pipe, both the unprestressed and the prestressed parts of the core satisfy all serviceability and strength design limit states and criteria in AWWA C304-92 with the following modifications:

- Under working conditions, limit the tensile strain in the core to about 100 to 150 micro strains at the inside surface of the core at invert and at the outside surface of the core at springline. This strain limit is small enough and the neutral axis is close enough to the exterior surface of the pipe that no water would permeate into the core and reach the steel cylinder from the outer surface.

- Under working conditions, limit the compression in the core to 0.6 f'c.

- Loss of prestressing is calculated assuming that the pipe is mature and does not creep as much as a newly cast concrete pipe, shrinkage has already taken place, and post-tensioning tendons are made of low relaxation wires.

Repair Design Procedure. An existing computer program developed initially for design of prestressed concrete pipe was modified to enable us to calculate the stresses and strains in the pipe wall under the assumptions described above. The analysis was performed for post-tensioning on both the unprestressed and the prestressed cores. The stresses and strains for different tendon spacings were calculated; the spacing that satisfies best the criteria stated above was selected.

To observe the impact of creep, shrinkage and wire relaxation factors, we repeated the design calculations with full creep, shrinkage, and wire relaxation. The results show that for the working loads, the strain in the concrete core remains small.

Core compression was calculated for the condition of post-tensioning the prestressed part of the core. The maximum compression under working conditions was limited to about 0.6 f'c.

Concentrated Anchorage Load Effects. A finite-element analysis was performed on the effect of concentrated loads resulting from an angular offset in the tendons at the anchor joints. The results show that a sizable force can develop on the pipe depending on anchorage geometry and pipe diameter, and cause high stresses in the pipe wall. To ensure that the resulting stresses are small, design changes were made in the anchor block, and the circumferential spacing and stagger of anchor blocks were determined.

Longitudinal Stresses During Post-tensioning. Since hoop stresses during post-tensioning process can produce longitudinal stresses and crack the core, a finite-element model of post-tensioning process was developed for the pipe with partial prestress of the core. The results show that the prestressed concrete core has a quarter wave length attenuation distance of about 24 in. for the 96 in. diameter pipe and 16.5 in. for the 66 in. pipe. As the post-tensioning tendons are applied, the cracked core becomes an integral wall capable of withstanding longitudinal stresses. Outside of the prestressed part of the core, the longitudinal stresses for intact concrete core is nearly equal, but with the outside surface of the pipe in tension. If the concrete core is not intact, the stress in the steel cylinder is small. Since the longitudinal stresses in the pipe are quite large, the tendons were tensioned in two steps, in the first step every other tendon is placed and tensioned and in the second step the intermediate tendons are placed and tensioned.

Longitudinal Stresses after Post-tensioning. The longitudinal stresses in the pipe wall after post-tensioning may result from differential prestress between a part of the core that is post-tensioned with tendons and a part that is post-tensioned by the tendons and the existing prestress. The difference of prestress along the length of the pipe is less than the post-tensioning effect of the tendons on an unprestressed pipe discussed above.

Longitudinal Stresses Resulting from the Added Weight of New Coating. This calculation was performed on finite-element model of the pipeline as a beam-on-elastic foundation. The soil spring was selected at 100 lbs/in.^{3}, corresponding to the stiff foundation that exists under the pipeline. The results calculated for an added weight on the pipeline of 400 psf (2.8 psi) show that the longitudinal stresses are small.

Repair Hardware

The repair hardware is shown in Fig. 2. A photo of the repaired pipe with installed hardware is shown in Fig. 3.

Post-tensioning Tendons. The tendons used were 0.6 in. diameter 7-wire strand, 270 ksi minimum specified ultimate strength, with 0.08 in. thick polypropylene sheathing. The tendons and anchorage were supplied by Lang Tendons, Inc. The high curvature of pipe results in large radial forces between the strand and the sheathing that tend to embed the tendon into the sheathing and puncture it. A laboratory and a field pipe test was performed to determine whether such a radial load can cause puncture in the sheathing. It was determined that the polypropylene performed significantly better than the polyethylene sheathing.

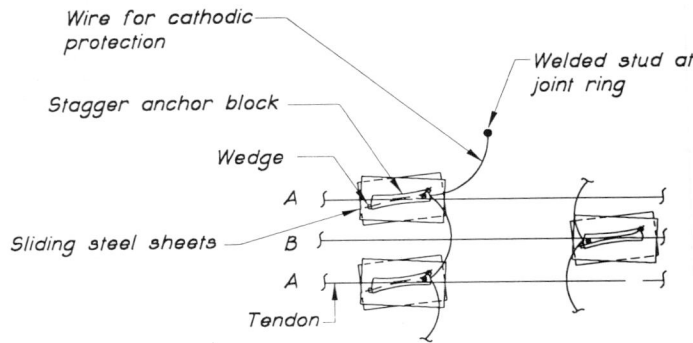

Fig. 2. Repair Hardware

Fig. 3. Installed Tendons

Anchor Block. The initial anchor blocks required tendon installation at a sizable offset from the pipe to eliminate the frictional force between the anchor block and the concrete pipe surface. This offset would result in high stresses and possible cracking of the core. Working with Lang Tendons, a compact anchor block was developed along with an installation procedure that would allow the anchor blocks to rest on the pipe, thus eliminating the offset. The procedure eliminated the frictional force between the anchor block and the pipe surface by placing two galvanized steel sheets with grease in between under the anchor blocks.

Electric Continuity of Tendons. Since the pipeline was to be placed under cathodic protection, the tendons had to be made electrically connected to the pipe. The electric continuity was achieved by exposing the bell ring and welding studs to the bell ring and connecting the stud to each anchor block.

Shotcrete Coating of Post-tensioned Pipe. The coating of the post-tensioned pipe with cracked core must be capable of bridging over the existing cracks in the core and prevent the cracks from extending into the coating, as the width of core cracks may change with internal pressure. The coating must also withstand shrinkage and thermal strains without cracking. The highly corrosive soil environment may corrode the steel reinforcement, such as welded wire fabric that is typically used to reinforce the coating, and the corrosion products can cause delamination of the coating. Therefore, a polymeric fiber reinforced shotcrete was used for its increased tensile strain capacity. Different types of fibers and concentrations were tested to determine the shotcrete mix with sufficient volume of fibers for ductility and adequate placement properties. A 3M polyolefin fiber, 0.015 in. in diameter, and 1 in. in length were selected for this application. An appropriate mix consisting of 20 lb/yd^3 of fibers was used for the shotcrete. The fiber reinforced shotcrete was applied with no difficulty. A 3 in. thick shotcrete coating was applied over post-tensioning tendons. Special care was exercised to prepare the pipe surface for coating application and to cure the coating.

Conclusions

The program of risk management adopted by PWNGS involved condition assessment and repair with a goal to identify corroded prestressed concrete pipe before their rupture and to repair them in the normal outages so that power production is not affected. The accuracy of inspection and test procedures used for identifying corroded pipe was evaluated. For pipe segments with working pressures less than the ultimate strength of steel cylinder, the corrosion of prestressing wire does not result in immediate pipe failure. For such pipe segments periodic internal inspection and sounding is still the most effective method for identifying pipe in the final stages of its failure scenario. A repair procedure, based on post-tensioning with tendons was developed and successfully applied to the pipe segments found to be corroded through excavation and external inspection.

In-line Electromagnetic Inspection of PCCP

Dr. Brian J. Mergelas[*] and Professor David L Atherton[**]

Abstract
Inspection of embedded-cylinder pipe is possible using an in-line inspection tool. The tool operates using a electromagnetic technique called remote field eddy current/ transmission coupling. The inspection system, consisting of an internal pair of coaxial coils with instrumentation, can be passed through a man hole. Field tests in both 1200 mm (48") and 2100 mm (84") embedded cylinder pipes have been conducted and sensitivity to a single broken wire anywhere along the length or circumference of the pipe, has been demonstrated.

Introduction
There are more than 11, 000 km of Concrete Pressure Pipe in use by nearly every major water utility in North America. At an estimated replacement cost between $2 million to $15 million/ km, this infrastructure is valued at over $40 billion. These pipes are used mostly for pressurized water supply and distribution lines and for cooling systems in power plants. A rupture of one of these lines represents millions of dollars in repair costs and lost service.

Concrete pressure pipe is constructed with an expected life time of 50-100 yr. Many of the original concrete pressure pipelines are already over 50 years old and it is estimated that one half of the existing lines will be replaced in the next 20 years. To minimize the enormous capital cost of replacing an entire pipeline, it will become increasingly more important to conduct selective maintenance. By identifying and replacing specific pipes which are at the highest risk for failure, the life expectancy of a pipeline can be extended at significant savings to the operator.

---

[*] The Pressure Pipe Inspection Company Ltd. 481 Mallorytown Ave, Mississauga, ON, Canada, L4Z 2M7; Phone: (905) 712-9711; Fax: (905) 712-0058

[**] Queen's University; Department of Physics; Kingston, ON, Canada , K7L 3N6 Phone: (613) 545-2701; Fax: (905) 545-6463

The two most common configurations of concrete pressure pipe are prestressed concrete embedded-cylinder pipe (ECP or SP-12) and prestressed concrete lined-cylinder pipe(LCP or SP-5). Both products consist of a thin steel cylinder in which a layer of concrete is cast. For lined cylinder pipe, prestressing wire is wrapped directly onto this steel cylinder. For embedded cylinder pipe a second layer of concrete is cast outside the steel cylinder before the prestressing wires are wound on the pipe. A layer of protective mortar is placed on the outside of these pipes.

Pipes are designed according to AWWA C301 and C304 standards but job by job variations are common. Differences in the gauge of the steel cylinder, thickness of the concrete layers, and in the diameter and level of pretension of the steel wires reflect different operating pressures for a range of pipeline diameters. In all cases the pre-stress in the steel windings is sufficient to guarantee that the concrete portion of the pipe remains in compression over the range of operating pressures.

Problems can arise if the prestressing steel windings begin to corrode. If a sufficient number of these wires break, the concrete will no longer remain in compression and a catastrophic failure will occur. Thus, detection of broken wires is of concern to operators of concrete pressure pipe.

Pipeline inspection techniques which are typically used in the oil and gas industry are not appropriate for concrete pressure pipe. The inner concrete layer provides a large non ferromagnetic standoff which makes the Magnetic Flux Leakage (MFL) techniques ineffectual. The multiple layered structure makes inspection by an Ultrasonic Tool (UT) impractical. Two further consideration which make conventional pipeline inspection difficult are the diameters of some of the concrete pressure pipelines, which can be as large as 5300 mm, and the fact that, with few exceptions, no facilities exist for launching and receiving inspection tools. Typically, access to the pipelines is through 450mm (18") manholes.

In response to the need for a technology which could be used to detect broken prestressing wires in concrete pressure pipe, an investigation was launched nearly 5 years ago. The starting point, was the remote field eddy current technique but subsequent discoveries has lead to the development of a new method called the remote field eddy current / transformer coupling (RFEC/TC) technique.

Remote Field Eddy Current/ Transformer Coupling

The remote field eddy current technique is an electromagnetic technique which is useful for inspecting both ferromagnetic and non-ferromagnetic pipes and tubes [1,2]. The technique has being used to inspect well casings, heat exchanger tubes and most recently ductile and cast iron water distribution systems. A remote field system operates with an internal pair of coils. The exciter coil is driven by a low frequency a.c. signal. Two distinct coupling paths exist between the exciter and the remotely spaced detector. The direct path, inside the tube or pipe, is attenuated rapidly by circumferential eddy currents induced in the conducting pipe wall. The indirect coupling path originates in the exciter fields which diffuse radially outward through the wall. At the outer wall, the field spreads rapidly along the pipe with little attenuation. At remote field spacing, these fields re-diffuse back through the pipe wall and are the dominant field inside the pipe. Any disturbance in this indirect path causes a change in the magnitude and phase of the received signal.

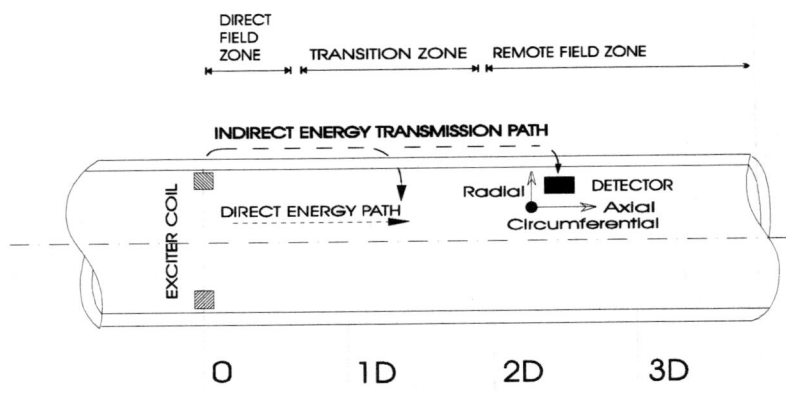

Figure 1. A pair of internal coil are used in remote field testing. The signal from the transmitter is attenuated and slowed as it passes through the pipe wall. Wall thinning will allow the signal to be picked up by the detector sooner and stronger.

Transformer coupling can be understood as an interaction between the indirect transmission path and the external winding in the concrete pressure pipe. The external winding is excited and will behave as a solenoid coupling flux between the exciter and detector. Thus, the signal received in a remote field eddy current transformer coupling system has two components. The remote field component will show a phase shift and attenuation consistent with the double through wall transmission of the field through the steel cylinder. The larger transformer component will dominate the response, however will be reduced in the presence of broken wires.

Field Tests

Field tests of a remote field eddy current / transmission coupling system were conducted at Lafarge Canada Ltd. The inspection system was pulled through a 30 m test line of 1200 mm embedded-cylinder pipe. None of the pipes had any shorting straps, however in all cases the prestressing wire was anchored to the inner steel cylinder. Tests were conducted with and without electrical connections between adjacent pipes. Field tests were conducted on four different occasions offering the

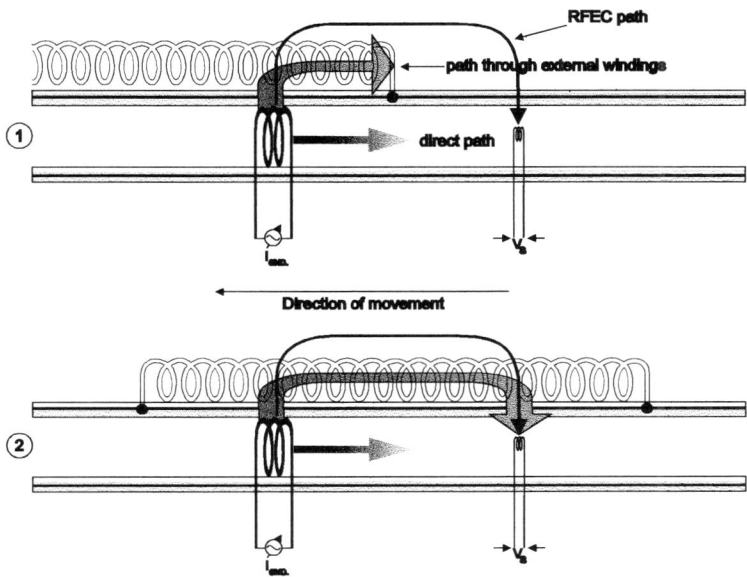

Figure 2. When both transmitter and detector are underneath the same set of prestressing windings, transformer coupling results in a very large amplification of the detector signal. Disrupting the coupling by breaking a wire results in a decrease in the detector signal.

opportunity to demonstrate the system with wet and dry concrete as well at temperatures from -5°C to 30°C. No significant variation in the system response could be seen at different temperatures, with wet or dry concrete or with or without electrical contact between the pipes.

The mortar coating was removed from various sections of the test line to expose the prestressing windings. Several wires were broken at various positions along the length and around the circumference of a pipe. The RFEC/TC technique offers the unique opportunity to 'turn on' or 'turn off' the defect by short circuiting the gap across a given break. In this way multiple tests were performed to demonstrate sensitivity of the system to a single broken wire at any position along the length of the pipe. The method is equally sensitive to breaks anywhere around the circumference of the pipe. Notably, this means that breaks at the top and bottom of the pipe are indistinguishable at present.

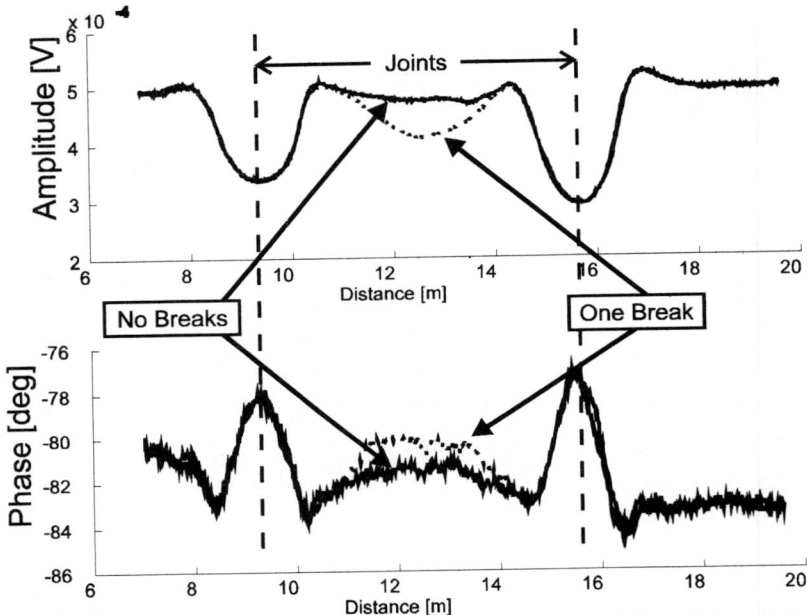

Figure 3. The signal from a single broken wire is clearly distinguishable from the defect-free signal

Multiple breaks in the prestressing wire were examined at individual locations along the length of the pipe. Using voltage plane polar plot analysis[3], an interpretation techniques developed specifically for examining remote field eddy current data, it is possible to see small, but distinct, correlation between the defect signal trace and the number of broken wires.

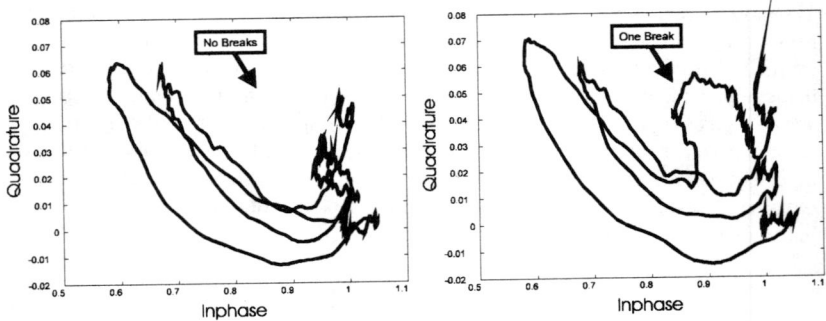

Figure 4. Voltage plane analysis can be used to show very distinct differences in the signal in the presence of a broken wire.

Tests were performed at different operating conditions in an attempt to optimize the sensitivity of the system to broken wires. Often in non-destructive testing, it is acceptable to use calibration test samples to set operating conditions for an inspection tool. The huge size of concrete pressure pipe makes calibration samples impractical. In practice the operating parameters for a given inspection are chosen by performing a series of setup measurements inside the pipe of interest.

The diameter of the inspection tool is less than 350 mm; the length of the tool is variable. These points are significant because the system is designed to be inserted into a pipeline through a 350 mm man hole. Once inside the line some assembly of the tool is required. For large diameter pipes an inspection crew can walk the self contained system through the pipeline. For smaller diameter tools, a wire-line tool which transmits data back to an external recording device can be used.

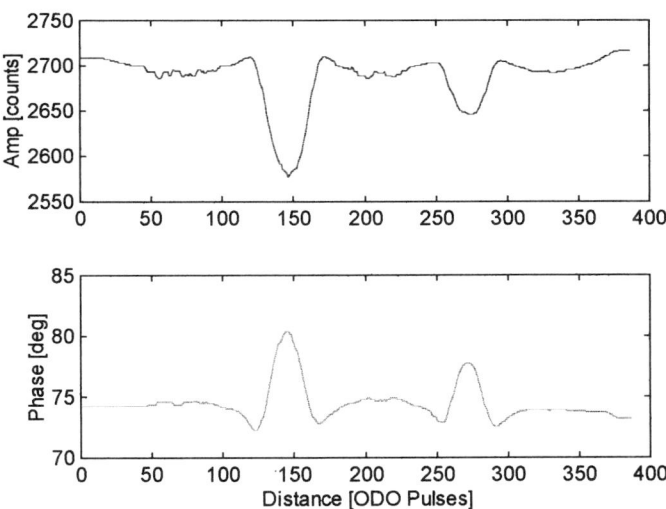

Figure 5. Defect free result. in 2100 mm pipe. Both 1200 and 2100 mm pipes showed similar behaviours at different operating conditions

Tests were also conducted in a 18 m test line of 2100 mm embedded-cylinder pipe. While no broken wires were made in the larger diameter pipes, the system response was examined at different operating conditions. Good signals were received in the large pipe; the defect-free behaviour of the system was similar for both diameters.

Conclusion

A new electromagnetic technique for the nondestructive testing of concrete pressure pipe has been demonstrated. The system operates based on remote field eddy current/ transformer coupling. Tests conducted in sections of 1200 mm embedded-cylinder pipe have demonstrated sensitivity to single breaks in prestressing wires anywhere along the length or around the circumference of a pipe. Voltage plane analysis can be used to perform semi-quantitative predication of how many broken wires are present in a given pipe section. While no broken wires were examined in 2100 mm embedded-cylinder pipes, the defect-free response in both diameter pipes is very similar.

The inspection technique, which is currently commercially offered by the Pressure Pipe Inspection Company, can only be applied to embedded-cylinder pipe at the present time. Lab tests in lined-cylinder pipe are promising however several additional complications must be overcome before a commercial inspection service for LCP can be provided.

References

1. Schmidt T.R., "The remote field eddy current inspection technique", *Materials Evaluation*, Vol 42, pp 225-230, 1984.

2. Atherton D.L., "Remote Field Eddy Current Inspection", *IEEE Trans. on Magnetics*, Vol.31, No.6, pp 4142-4147, Nov. 1995.

3. Atherton D.L., Mackintosh D.D., Sullivan S.P., Dubois J.M.S. and Schmidt T.R.`Remote Field Eddy Current Representation,' *Materials Evaluation*, July 1993, pp. 782-789.

Internal Inspection And Database Development Of PCCP

John J. Galleher Jr., P.E.[1], Michael T. Stift, P.E.[2]

Abstract

The San Diego County Water Authority (Authority) has over 400 kilometers (km) of large diameter pipelines that provide up to 95 percent of the water to San Diego County's 2.7 million people. Included in this 400 km is 133 km of prestressed concrete cylinder pipe (PCCP) manufactured between 1958 and 1982. In order to ensure that the Authority fulfills its mission statement, "to provide a safe and reliable supply of water to San Diego County", the Authority implemented the Aqueduct Protection Program (APP) in 1992 which includes the condition assessment of the Authority's PCCP. Over the past six years, the Authority has developed detailed investigative procedures and methodology used to locate distressed sections of PCCP prior to failure. The program includes developing non-invasive/non-destructive inspection procedures to determine the condition of existing PCCP, gathering data on each section of pipe relative to design, construction, manufacturing, and installation, developing a condition assessment database allowing access to and the ability to cross reference this information, monitor deterioration, and determine how the distressed sections should be rehabilitated.

Introduction

San Diego County California has some of the most corrosive soils in the United States due to high chlorides combined with low soil resisitivities. Within these aggressive soils resides 133 km of the Authority's PCCP. In February 1979, the Authority had its first failure along one of its PCCP pipelines. The 17 year old, 1.75 meter pipeline ruptured with a "roar like a tornado" a local resident said. Fortunately, no one was injured. However, there was significant damage to two homes from the flooding. There were two more PCCP ruptures; one in January 1981 and one in October 1990. At the time, this was the only pipeline serving the southern part of San Diego County and the Authority needed to take a pro-active approach to determine the nature and extent of PCCP corrosion problems in their aqueduct system to avoid emergency repair situations like the section that failed in 1979.

---

[1] Engineer II, San Diego County Water Authority, 610 West 5th Ave., Escondido, Ca 92025
[2] Assistant Director of Engineering, San Diego County Water Authority, 610 West 5th Ave., Escondido, Ca 92025

FAMILIES ESCAPE INJURY
Huge Pipeline Bursts, Two Homes Flooded

A major pipeline of the San Diego Aqueduct near the Sweetwater Reservoir burst yesterday morning and flooded the homes of two Spring Valley residents.

The underground pipeline, 69 inches in diameter and carrying water from the Colorado River to the reservoir, ruptured at about 9:30 a.m. with a "roar like a tornado," a resident said.

Roberto C. Stanley of 166 Lakeview St. said he and his family were in bed when the huge pipe burst, cascading water high into the air. Authorities said no one was injured. "It sounded like a tidal wave or tornado," said Stanley, a computer systems analyst.

He said the water flooded his four bedroom home, located about 180 feet from the break, and he and his family were forced to leave. Water in some of the rooms reached a level of more than two feet. Stanley's next door neighbor, Albert Gomes of 168 Lakeview St., also reported flood damage to his home.

Pete Rios, public information officer for the San Diego County Water Authority, said the force of the water from the pipeline excavated a hole 20 feet across, 30 feet long and at least six feet deep.

Rios said no one was injured when the waterline burst and that water flow was brought under control in a matter of minutes through the use of valves along the line and at the reservoir. Repair crews, operating with cranes, digging equipment and bulldozers, were removing a 20 foot section of damaged line and will replace it with a new one.

Rios estimated it would take four to five days to repair the line. He said water will be supplied through alternate routes.

Dave Jones, left, and Bob Gonzales of the Otay Water District help dig out a section of a main pipeline of the San Diego Aqueduct that burst yesterday morning near the Sweetwater Reservoir in Spring Valley. Two homes were damaged by water when the 69-inch water line ruptured.

Figure 1 - February 23, 1979 San Diego Union Article on PCCP Failure

In 1990, there was a limited amount of knowledge on performing condition assessment of PCCP. In addition, the Authority's pipelines could only be taken out of service for a limited time, during which investigations could be performed, without adversely effecting the water supply to San Diego County. Taking a pro-active approach, the Authority began to develop techniques to inspect PCCP along with database management to analyze and monitor the condition in order to determine when and if a section of PCCP would fail.

History of PCCP

The development of PCCP began in 1937 with the first commercially manufactured PCCP installed in 1942. The first PCCP used in the western United States was installed in the early 1950's.

There are two basic types of PCCP; embedded cylinder and lined cylinder. Embedded cylinder PCCP is constructed of a thin, typically 16 gauge, steel cylinder embedded within a high strength concrete core with high-tensile prestressing wire helically wrapped around the concrete core with a dense cement mortar coating as shown in Figure 2. Lined cylinder PCCP is essentially constructed the same way except that the high tensile prestressing wires are wrapped right on the steel surface as shown in Figure 3.

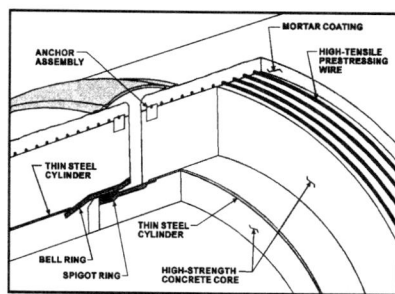
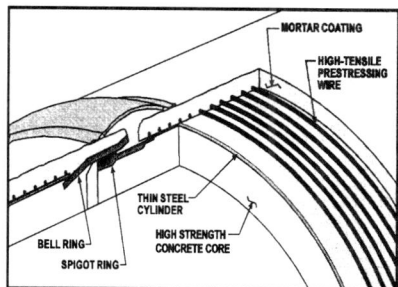

Figure 2 - Embedded Cylinder PCCP   Figure 3 - Lined Cylinder PCCP

Protection against corrosion of the steel elements in PCCP is provided by the high alkaline environment of the concrete core and mortar coating (minimum pH of 12.5). However, any cracking of the mortar coating during pipe manufacture, handling, or installation can lead to corrosive elements, such as chloride ions, present in the soil, to penetrate the coating. These chloride ions negate the beneficial effects of the high pH cement mortar coating on the steel and may cause corrosion and failure of the prestressing wires. Breakdown or failure of this protective coating can lead to corrosion related failures that can be catastrophic and endanger both life and property in the vicinity of the failure.

Beginning in the late 1970's, a thin steel shorting strap was placed under the prestressing wires on embedded cylinder PCCP so that cathodic protection could be applied as shown in Figure 4. However, 88% (116 km) of the Authority's pipelines were constructed prior to this and do not have the shorting strap installed. In addition, 34 km of PCCP constructed between 1959 and 1962 do not have bonded joints. For this reason, the Authority's PCCP can not be economically cathodically protected and had to develop alternative methods to extend the service life of the PCCP.

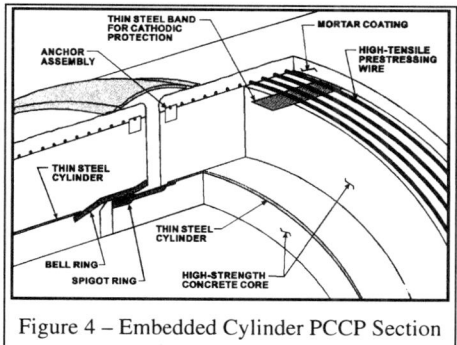

Figure 4 – Embedded Cylinder PCCP Section With Shorting Strap

Aqueduct Protection Program (APP)

The objective of the APP is to determine and monitor the condition and, if feasible, implement methods to extend the service life of existing pipelines along the

Authority's aqueducts. To achieve this objective, the APP has been approached as a three phase program.

Phase 1 - Data Collection and Correlation

Phase 1 involves data collection and correlation to determine, and when possible correct, factors contributing to corrosion and loss of pipeline service life. This involves conducting extensive literature research of past pipeline design, fabrication, and construction records and correlating information to each pipe section along each pipeline. Corrosion and soil resistivity field surveys along all the Authority's pipeline rights-of-way are conducted and correlated to each pipe section. Pipelines are internally inspected to locate potentially deteriorating pipe. External pipe inspections are conducted when internal inspections indicate a pipe section may be near failure. Forensic investigations are conducted to determine the physical and chemical properties of the pipe materials and soils surrounding these pipes. Finally, all information is assembled into a database to establish the baseline condition of the pipeline, and to facilitate monitoring of subsequent deterioration during future inspections.

Soil resistivity is generally recognized as the most significant soil characteristic with regard to the corrosivity of the soil. Although no standard has been accepted by such organizations as the National Association of Corrosion Engineers (NACE) or the American Society for Testing and Materials (ASTM), most corrosion engineers use a table similar to Table 1 to classify soil resistivity. Table 1 and Figure 5 show soil resistivity measurements taken along 10.6 km of the Authority's 2.13 km PCCP Pomerado Pipeline during Phase 1.

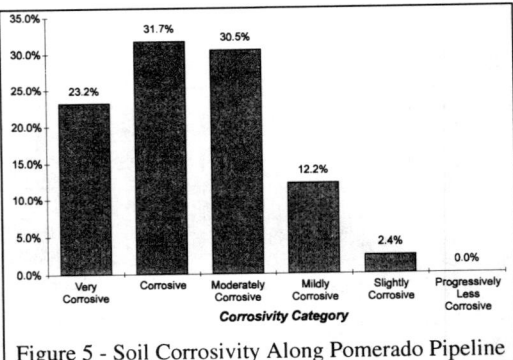

Figure 5 - Soil Corrosivity Along Pomerado Pipeline

Table 1 - Soil Corrosivity Along the Pomerado Pipeline

Corrosivity Category	Resistivity (ohm-cm)			No.	%	Cumulative %
Very Corrosive	0	to	1,000	38	23.2	23.2
Corrosive	1,000	to	2,000	52	31.7	54.9
Moderately Corrosive	2,000	to	5,000	50	30.5	85.4
Mildly Corrosive	5,000	to	10,000	20	12.2	97.6
Slightly Corrosive	10,000	to	25,000	4	2.4	100
Progressively Less Corrosive	Above 25,000			0	0	100

Internal inspections of PCCP require the interior sections of the pipeline to be numbered on the pipe's lining per the original pipe laying diagrams for easy identification, physically sounded, and visually inspected. The Authority's PCCP was sounded with a hollow aluminum rod with steel caps. The sounding is performed in a spiral back and forth motion slowly along the pipe (see Figure 6). Moderate and severe hollow sounds were correlated from failed PCCP that was previously removed. Generally, hollow areas greater than 0.24 but less than 0.56 square meters with no cracking are classified moderate and areas greater than 0.56 square meters with cracking are classified severe. However, PCCP sounding is more of an art than a science and requires practice to obtain consistent results. It is not just the size but the level of disbondment that is important.

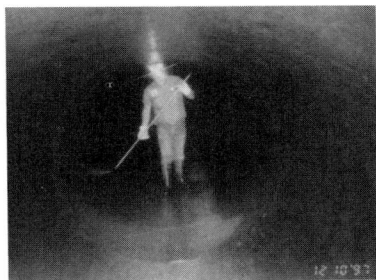
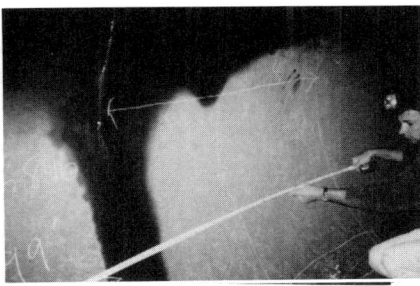

Figure 6 - Internal Sounding of PCCP   Figure 7 - Measuring Distressed Area

When a hollow was found, it was measured (Figure 7), recorded on the internal inspection data sheet (Figure 8) and later input into the database.

Internal Inspection Data Sheet

Spec: _____
Date: _____   Recorded By: _____

Pipe #	Class			Hollow					Type		Cracking					
	Slight	Mod	Severe	Dist. from Joint		Location on Pipe					Dist. from Joint		Location on Pipe			
				Up Stream	Down Stream	From	To	Length	Circum.	Longit.	Up Stream	Down Stream	From	To	Length
1278	X					4.5 feet	10:00	2:00	5 feet	X		6 feet		8:00	4:00	

Figure 8 - Internal Inspection Data Logging Sheet

Presently there are two databases used for the assessment of the Authority's pipelines. A Corrosion Database (Figure 9) and an Internal Inspection Database (Figure 10).• The Corrosion Database contains information pertaining to the surrounding environment, type of pipe, repair information, and any laboratory tests performed on the pipe (i.e., chlorides, conductivity, sulfates, & pH). The Internal Inspection Database contains information found during internal inspections (i.e., distressed PCCP sections, hollow areas, hollow size & location, etc.). These databases provide the needed information to easily go back and locate distressed areas and provides a history so that when the pipeline is reinspected, an assessment can be made to determined if the suspect area is increasing in size. If the distressed area has increased, the pipe will be excavated and inspected on the outside to determine if the pipe is in danger of rupturing.

San Diego County Water Authority Corrosion Data, Spec 224, *"POMERADO PIPELINE"*

SDCWA			PIPE DATA						SOIL RESITIVITY					SOIL ANALYSIS			
Survey Station	Cover TOP	#	Manufacturer	Type	Class	Joint	ID (in.)	Year	Hollow Data Date Class Area (in²)	Date	Clns. Rd (ohm-cm)	Barnes Layer Analysis (ohm-cm) 0-2.5' 2.5-5' 5-10' 10-15' 15-20' 20-25'		Date	Chloride (ppm)	Sulfate (ppm)	pH	Conductivity (µ-mhos)

San Diego County Water Authority Corrosion Data, Spec 224, *"POMERADO PIPELINE"*

MORTAR ANALYSIS			POTENTIAL DATA		WORK/REPAIR DATA
Date	Chloride (ppm)	Sulfate (ppm)	pH	Date	Vg (CSE) (volts)	Date	

Figure 9 - Corrosion Database Fields

Pipeline 4 - Spec 221 - 90" PCCP   Internal Inspection Data Analysis

Date	Shutdown No.	Area No.	Pipe Diameter (in)	Pipe No.	End Stations of Pipeline Segment	Distance from Joint (ft) South North	Location o'clock From To	Width (hours)	Width W1 (arc)	Hollow Dimensions Width W2 (estimated)	Average Width (ft)	Length (ft)	Hollow Area (sq ft)	Hollow Classification

Pipeline 4 - Spec 221 - 90" PCCP   Internal Inspection Data Analysis

Sector 1 Top Sector Hollow Long Crack	Sector 2 Bottom Sector Hollow Long Crack	Sector 2B Bearing Sector Hollow Hollow Arrea Long Crack	Sector 3 East Side Hollow Long Crack	Sector 4 West Side Hollow Long Crack	Circum. Ctrack	Comments

Figure 10 - Internal Inspection Database Fields

Before the pipeline is repressurized, severe hollow areas are located from the surface and excavated. Resistivity tests, chloride tests, and sulfate tests are taken on the soil samples surrounding the pipe. Mortar samples are taken from the coating and tested for chloride and sulfate ion concentrations and recorded in the database. The exterior of the pipe is then visually inspected adjacent to the internal hollow for signs of corrosion products or cracking. Figure 11 shows a section of PCCP that was replaced due to a large severe hollow found during a scheduled internal inspection. The hollow area was located between 8:00 and 10:00, approximately 1.5 square meters in area, and had a longitudinal crack in the center of the hollow.

Figure 11 - Pipe Forensics

During internal inspections of PCCP, it is extremely important to document all hollow areas and any cracking. The hollow areas, depending on the severity, usually indicate the loss of prestressing on the outside of the pipe and cracking of the concrete core. Longitudinal cracking is usually indicative of pipe distress due to some type of loading on the pipe. This could be excessive earth loads, live loads, or internal pressure exceeding the residual prestress after corrosion of the prestressing wires.

Circumferential cracking is generally caused from concrete core shrinkage, differential settlement, or longitudinal pipe bending due to poor bedding conditions. Generally, a severe hollow combined with longitudinal and circumferential cracking would cause pipe repair or replacement.

Phase 2 - Data Analysis and Condition Assessment

Phase 2 of the APP involves analyzing all information collected in Phase 1, performing pipeline condition assessment studies, and estimating the remaining pipeline service life. Work also includes determining interim and long-term rehabilitation, repair and replacement projects and priorities, and developing budget estimates and schedules.

Presently the Authority has completed the pipeline condition assessment for 46% of the PCCP within the Authority's Aqueduct system. During this condition assessment, twenty suspected distressed pipe sections were identified during the course of the internal pipeline inspections. Of the 20 pipe sections, 14 required replacement with steel pipe, one was rehabilitated by installing a steel liner, three required removal of large boulders from the pipe zone backfill, and two needed no further work as the pipe was found to be in good condition.

Since inception, the APP internal inspection techniques have yielded a 90% success rate in locating severely distressed PCCP. The Authority has never had a catastrophic pipe failure in any pipeline that has been internally inspected.

Phase 3 - Implementation

Phase 3 is the implementation phase and consists of preventative maintenance, rehabilitation and repair, and pipeline replacement. Preventative maintenance involves development and implementation of pipeline internal inspection schedules and a corrosion monitoring and control procedures manual to extend pipeline service life. Specific remediation and repair work may include installation of corrosion monitoring test stations and cathodic protection (where applicable), internal joint repairs, concrete coating and lining rehabilitation, pipeline replacement and steel plate relining. Pipeline replacement or steel plate relining will be done when a pipeline nears the end of its service life.

Common Characteristics of PCCP Sections That Were Replaced

Common characteristics of PCCP sections that were replaced are as follows:

- Pipes were located within a environment with resistivities less than 2,000 ohm-cm.
- Pipes were located in areas subject to wet and dry cycles allowing chlorides to concentrate on the mortar coating.
- Elevated concentrations of chlorides were present in the backfill material (greater than 1,400 ppm).

- Pipe bedding and backfill was not as dense as the unexcavated surrounding native soil.

Prioritization of Remediation Measures

Information from the following areas are used to prioritize the remediation measures:

Internal Inspection and Corrosion Database

The information from the database is analyzed using a matrix developed from comparing the database information with similar data from the pipe sections that had been removed from the pipeline. This information is used to prioritize pipelines as to their potential for failure. This data included pipe manufacturer, year manufactured, pipe cover, soil resistivity, soil analysis for chlorides, sulfates, pH and conductivity, mortar analysis for chlorides, sulfates and pH, pipe-to-soil electrical potential data, backfill data, hollow location and size, cracking, and other information from the construction and manufacturing inspection records.

Pipeline Operation

The critical hydraulic bottleneck areas with no redundancy are identified. Areas of transient conditions different from the design pipeline operating conditions are located to determine if the PCCP could have the potential to operate at pressures greater than the original design pressures.

Risk to Life and Property

Since the aqueducts were constructed, considerable development has taken place adjacent to the pipelines and in the natural drainage areas of the pipelines. The design criteria has also evolved to include testing for corrosiveness of the soil. This new information has uncovered risks that could not have been anticipated by the original designers and demonstrates the need for an annual corrosion monitoring program.

Methods of Extending the Life of PCCP

The life of PCCP can be extended by relining, repairing, replacing, and monitoring. These options are described below.

Relining PCCP - The Authority has relined with steel plate, 8 km of 1.68 m PCCP to a finished inside diameter of 1.61 m in 1982, 1983, and 1984. In addition, the Authority relined 2.42 km of 1.83 m PCCP to a finished inside diameter of 1.75 m for the City of San Diego's Shepherd Canyon Pipeline in 1989. The relining cost was $1,000 per linear meter of pipeline in 1989. Relining is cost effective in environmentally sensitive and urban areas. In areas of sharp horizontal and vertical curves, additional entry points can increase the cost of the relining to the point where replacement becomes a better economic alternative. The disadvantage of relining is that it reduces the capacity of the pipeline.

Replacing PCCP - The Authority has replaced twenty-two 6.1 to 7.3 meter long sections of PCCP with a steel repair section encased in reinforced concrete. Six were replaced due to failure and sixteen were located by internal investigations. In 1997, the cost of replacing a single 6.1 meter long section of PCCP in San Diego County was $90,000 when done on an emergency (around-the-clock) basis.

Monitoring - The Authority has a very aggressive annual inspection and monitoring program. The goal of this program is to eliminate the need for relining or replacing all the PCCP pipelines in non-critical areas. Monitoring should be considered when the annual cost of monitoring the PCCP and replacing only severely corroded pipes is less than the annual debt amortization cost of relining or replacing the same section of PCCP.

The advantage of annual monitoring and maintaining the database is that trends and rates of deterioration can be established. Replacement, repair, or relining can then be planned and budgeted in a cost effective manner. By stretching out the repair costs, changes in pipeline repair technology can also be used to the pipeline owner's advantage.

Conclusions

The Authority is presently investigating alternative internal inspection techniques and composite linings for PCCP. However, with the limited amount of time that the Authority has to internally inspect the PCCP and the success rate that has been achieved, internal inspection procedures by sounding the pipe with the specially fabricated hollow steel rod still appears to be the quickest method for identifying distressed PCCP. In addition, the Authority has not had a catastrophic failure on any of the PCCP that has been inspected.

The APP has been of great interest to other agencies due to new technologies pioneered by the Authority. Relining of PCCP with steel plate technology was developed and first installed by the Authority. This new rehabilitation method has been utilized by the Bureau of Reclamation, City of San Diego, and Seattle Water Department. A method for spot repair, when there is only limited corrosion of prestressing wires, has also been used by the Authority and has performed well in extending PCCP service life. Additional conclusions are as follows:

1. In general, severe hollow areas (greater than 0.56 square meters) combined with longitudinal cracking indicated that the prestressing wires had corroded and the pipe section needed to be replaced.

2. Specific attention should be directed to areas where there are wet/dry cycles affecting the pipeline. These are areas where minor amounts of chlorides from the surrounding environment can settle and concentrate on the cement mortar coating

causing damage to the prestressing wires. These areas also exhibited low resistivity.

3. A typical inspection and remediation program can be summarized in three phases. Phase I involved a corrosion survey, internal/external pipeline inspections, and database development. Phase II consists of analysis of the database to determine the methods for extending the life of the PCCP and design of any remediation requirements. Decisions on future monitoring versus repair or replacement are also made during this phase. Phase III is the physical construction of any remediation measures.

4. The key to a cost effective repair program is to collect enough information (database) to identify all problems. Once the extent of the problem is known, remediation measures can be customized to provide solutions that maximize benefits to all concerned. Over a period of time, the data base can be used to determine rates of deterioration, as well as estimates of useful remaining life of the PCCP.

5. Annual inspection is required to continually update the database and to identify new problem areas.

6. Monitoring versus repair decisions can be made when the operating and design requirements are used in conjunction with the database information. These decisions are more difficult when the PCCP cannot be cathodically protected.

7. Internal inspections give clues as to corrosion of the prestressing wires but the corrosion must be confirmed by external excavation.

References

1. Kennison, Hugh F., (1950) "Design of Concrete Cylinder Pipe," Reprinted from Journal American Water Works Association Vol 42, No. 11, November 1950

SOURCES OF FUNDING FOR REPLACEMENT AND/OR REPAIR OF DEFECTIVE PCCP

Geoffrey Johnson, Esq.[1]

ABSTRACT

Any large construction project, particularly one utilizing complex equipment or materials such as prestressed concrete pressure pipe (PCCP), risks that latent defects will surface long after the project has been put into service. If the defect is pervasive, the owner will probably want to replace the structure or perform extensive repairs. However, given the construction costs, owners typically are unwilling to do so unless they can be assured of recouping a major portion of the expense from those responsible for the original construction, namely the engineer, the contractor and the material supplier. Construction litigation, however, is expensive and time consuming. Therefore, any owner contemplating action against one or more of these parties should first examine carefully the various potential liabilities of the parties, defenses available to them and the likelihood that the parties will be able to satisfy any award either directly or through insurance or bonds. Conversely, an owner or engineer designing such a project should be aware of the legal effect of the specifications they are drafting and the various means by which the owner can achieve greater accountability.

INTRODUCTION

Prestressed concrete pressure pipe has been in use for almost 50 years and over 18,000 miles have been put into service successfully in North America. However, there have been numerous failures due to manufacturing defects and/or improper design of the pipe for the anticipated conditions. While the great majority of these cases have involved some of the six million feet of PCCP with Class IV wire manufactured by Interpace Corporation in the 1970's, most of the manufacturers in the United States and Canada have been or are currently involved in disputes concerning the manufacture of their pipe. The remedy sought by the owners of the pipelines has varied from ignoring the problem to complete replacement of entire pipelines. Understandably, given the very large cost of replacing these pipelines, a

---

[1] Partner, Lewis & McKenna, 82 East Allendale Road, Saddle River, NJ 07458. Email: gmjohnson@ibm.net. Since 1987, as part of his construction law practice, Mr. Johnson has handled four cases involving defective Interpace pipe.

major factor in determining the owner's response has been the availability of funding. In many instances, owners have recovered substantial sums in litigation against one or more of the various parties involved in the construction. However, the likelihood of a successful outcome for the owner depends on many variables which this paper will explore. By examining the general principles involved in large construction projects, it may also give some guidance to those responsible for drafting specifications and contracts on future projects.

RESPONSIBILITIES OF THE VARIOUS PARTIES

Every construction project of any size involves the same core players: The engineer, the contractor and the manufacturer. Usually, when a project ends in litigation, the owner looks to these three participants for recompense (or as Claude Rains said famously in *Casablanca*: "... round up the usual suspects").

1. The Engineer

In a case involving a highly engineered composite product like PCCP (as compared with materials fabricated on-site), the engineer's potential liability can derive from three sources: 1) contractual agreements between the engineer and the owner or between the contractor and the owner, 2) statutory or common law (the "standard of care"), or 3) applicable national standards.

With respect to contractual obligations, the specified scope of work in the engineering services agreement should be determinative, in theory. However, this is not the case in practice because the engineer is viewed by judges and juries as the protector of the owner (and public). Practically, unless the engineer states specifically in its agreement with the owner that it bears no responsibility for checking materials used on the project, the engineer is not entitled to rely on designs and material test reports, or affidavit of compliance of the manufacturer. This is particularly true where the material is on the leading edge of available technology, as was the case with Class IV wire in the 1970's. In these instances, the engineer will be required to conduct an independent investigation of the manufacturer's submittals. The result has been that even though no engineer in the early 1970's challenged Interpace's use of Class IV wire (which would seem to establish the "standard of care used by similar professionals in the community"), several prominent engineering firms that permitted its use have had to pay millions of dollars to the owners of the pipelines as a result of settlements or judgments.

A second source of contractual liability for the engineer is the contract between the owner and the contractor (although at least one court has held that the engineer's obligations cannot be governed by a contract to which it was not a party). For example, some specifications have required the engineer to approve the design of the PCCP and to observe a series of "proof of design" tests. By doing so, the engineer may not assure a more reliable product (because the tests are not properly conceived or monitored) but may, paradoxically, augment its own liability by creating additional responsibilities for itself. Moreover, in at least one of the Interpace cases the engineer sought to protect the owner (and presumably itself) by specifying a testing laboratory for the wire. Even though the contractor retained the testing laboratory, a jury held the engineer liable for the use of Class IV wire.

Additionally, if the engineer has contractually assumed resident inspection services, some owners have argued that the engineer assumes responsibility for detecting latent manufacturing defects that could be detected only through expert observation at the pipe plants (an option under C301).

Recently, PCCP specifications (AWWA C301) have also tended to enlarge the engineer's exposure. Whereas an engineer probably could have avoided liability under the C301-72 specification, which is at the heart of much of the recent Interpace litigation, had the materials fallen unarguably within the specification (i.e. the pipe was not made with wire having a higher tensile strength than Class III) and the pipe otherwise had a proven track record, the current C301-92 specification involves the engineer more intimately with the pipe's design. C304-92 gives the owner (and therefore the engineer) the option of approving the pipe's design. As the history of the Interpace cases teaches us, courts will probably interpret this "option" as a requirement if the pipe fails. Additionally, in its forward, the current standard explicitly puts the overall design responsibility for the pipe on the engineer:

> Purchasers are advised that, while this standard presents information on materials and procedures for the manufacture of the pipe, it does not contain all of the engineering information needed to prepare a complete specification for a particular pipeline installation. A specific installation may require provisions more restrictive than those in the standard and most certainly will require additional design and installation features.

While some engineers might seek to avoid this added liability by reverting to C301-84, courts and juries would almost certainly not look favorably on an engineer who consciously refused to use the latest standard.

2. The Contractor

The threshold issue regarding the contractor's liability for defectively manufactured pipe is whether the contractor furnished it. If the owner furnished the pipe to the contractor for installation, then the contractor will almost certainly not be held responsible for latent manufacturing defects. On the other hand, when the contractor purchases the pipe, AWWA C301, Section 1.2.3 puts it in the shoes of the manufacturer. This is so even though contractors lack the expertise to understand PCCP design and no contractor has ever been known to inspect its manufacture. (C301-92, Section 1.8.3 introduced ambiguity into this requirement by stating only that the constructor is not relieved of its liability to furnish materials in accordance with the standard.) However, Section 1.2.3, which defines that liability, was removed following C301-84.)

The contractor's legal liability though can be attenuated by certain legal doctrines. First, if the manufacturing defect is design-related then the contractor will not be held responsible under the so-called "*Spearin*" doctrine." A specific example has been Interpace's calling out the use of Class IV wire on its submittal sheets, which in turn are reviewed and approved by the engineer. The engineer is deemed to have made a conscious design decision, thus relieving the contractor of liability. If, on the other hand, the defect arises from poor quality control in manufacturing (which would not generally be evident to the engineer reviewing the submittals), the question becomes whether the resultant defect is patent or latent. If patent (e.g. soft or damaged mortar coating), then the contractor would probably be held accountable. Yet, if the engineer had resident inspection duties and observed the patent defect, but failed to reject it, the contractor might successfully argue waiver by the owner. (The courts are divided on the enforceability of "no waiver" provisions if the owner's agent actually sees the defect.) If the defects are latent (e.g. split prestressing wire), then the contractor may escape liability - even an express warranty of materials - if it can show that the owner specified the manufacturer. Arguably, this occurs when the manufacturer in question is named in the specifications as a standard of quality and the location of the project makes use of alternate suppliers a practical impossibility. This is often the case with PCCP because of the paucity of suppliers and the cost of shipping.

Finally, while C301 makes the contractor's liability coextensive with that of the manufacturer, in practice, courts and juries have tended to minimize the contractor's liability. The author's study of or participation in a dozen cases involving Interpace pipe has revealed no instance where a contractor has had to pay more than the cost to the contractor to defend the case. In many instances, the owner even elected not to join the contractor in the case. The reasoning for this "leniency" is a reluctance by courts and juries to hold responsible an organization that is not equipped to evaluate the complex design and materials issues at the heart of systemic failures of PCCP.

3. The Manufacturer

If the owner purchased the pipe directly from the manufacturer, then its claim would be based on breach of contract, a theory of recovery that may be limited by the manufacturer's warranty limitations (e.g. cost of recovery limited to the cost of the pipe). If the owner did not purchase the pipe directly, then it can resort to tort theories, which has the added advantage of permitting the owner to seek punitive damages (which happened in two Interpace cases.) In its defense of the various cases filed against it, Interpace sought to take advantage of perceived ambiguities in the 1972 specification to argue that the PCCP supplied met the letter of the national standard. The approach, however, was unsuccessful, specifically because of the material and workmanship standard in C301. If the pipe failed, the courts concluded, then Interpace must have breached the workmanship standard. Furthermore, inspection of the pipe by the purchaser, engineer or contractor did not relieve Interpace of liability, but served only to establish the liability of the entity doing the inspection because it missed the defects. Probably, the only defense that would have been successful (other than establishing some other cause for the failure such as contractor damage or surge), would have a *Spearin* defense based on the owner's improper specification of internal pressures, burial depth and the like.

THRESHOLD LEGAL IMPEDIMENTS

1. Statutes of Limitation and Repose

Discovery of defective conditions in a pipeline may start the running of the time to bring suit known as the statute of limitations. Typically statutes of limitation, which vary from state to state in their duration and application, impose time limits of four to ten years from when the owner knew or should have known of the defect which results in failure. (The statute of limitations for actions brought by the federal government is six years.) This, of course, presents a problem for owners seeking to assert a claim based on a national problem such as Interpace's use of Class IV wire, which was reported nationally as a probable cause of failure beginning in 1982. Defendants argue that responsible owners should have begun investigations in the 1980's. The response has been that Class IV wire by itself does not necessarily doom the pipe; other factors, such as a porous coating, must also be present.

A companion to the statute of limitations is the statute of repose which has the effect of blocking litigation within a certain period after completion of the project in question (as opposed to discovery of the defect). These time limitations, therefore, commence earlier than their sister statutes of limitation, but generally run longer (ten to fifteen years). Statutes of repose are absolute; they commence on a date certain and do not depend on when the defect should have been discovered.

Despite the seemingly draconian effect of theses statutes on claims by owners for their defective pipelines, the author is not aware of any case involving Interpace pipe where such a defense has been asserted successfully. Typically courts are loath to find against a public body unless it is very clear that it knew or should have known of the defect before the statutes of limitations or repose expired.

2. **Limited Warranties**

In addition to statutory time limitations, the specifications themselves will often have a contractual time limit applicable to actions against the contractor. In at least two Interpace cases, the contractors have successfully claimed that their liability for defective pipe ended with the expiration of the one year warranty.

Interpace sought to limit its warranty obligations to the cost of the pipe through conditions printed on its invoices. While such a limitation is theoretically effective among the contracting parties, it is not binding on an owner who did not furnish the pipe. As a practical matter, Interpace's limited warranty has not played any role in resolution of the cases involving its pipe.

Finally, because the engineer does not warrant its work, it has no need to limit its warranty obligations. On the other hand many specifications will require the contractor (and possibly even the manufacturer) to warrant the work and materials to the engineer as well as to the owner, thereby providing the engineer with an additional remedy in the event of litigation. Typically these warranty claims are very difficult for a contractor to defeat unless there is a time limitation as discussed above. And even then, there is substantial authority that a warranty limited in duration merely shortens the owner's ability to demand repair, but does not lessen its time to seek damages.

PAYMENT OF JUDGMENTS

Should an owner obtain a judgment against one or more of these parties, its next concern is finding the money to satisfy the judgment. Typically engineers, as service organizations, have few assets other than goodwill. Contractors are frequently merged or dissolved or have sought to insulate themselves through project specific corporations. Manufacturers, like Interpace, have been known to divest themselves of their assets and go into bankruptcy. (This is not necessarily fatal to the owner's claim as the recent recovery of tens of millions of dollars by water authorities and one engineer in the Interpace bankruptcy demonstrates). Therefore, it may be necessary for owners to look to secondary sources of collection such as insurers and bonding companies.

1. Insurance

Engineers carry professional liability insurance which would cover claims related to design. However, such policies are "claims made" policies meaning that the policy in effect when the claim is made and not the policy in effect when the damaged occurred, will provide coverage. In cases involving latent defects and extensive subsequent investigation, the lapse between the damage and the claim can be many years. Furthermore, the damage can take several years to manifest itself. Thus, an owner potentially may have to look to a policy purchased by the original engineer 10 to 20 years after completion of the project. By then, any required level of coverage in the engineering services agreement for the project will have been long forgotten. Furthermore, the engineer's defense costs generally fall within the policy limits, meaning that every dollar spent by the engineer in its defense will reduce the coverage by a dollar. This points toward early resolution of the case before the costs of typically expensive litigation make any victory Pyrrhic, unless it can be determined early that the engineer has sufficient coverage.

In contrast, contractors typically purchase only comprehensive general liability policies which are designed to insure against third party damage, but which exclude damage arising out of the contractor's own work. If the pipeline ruptures and causes personal injury or property damage, there will be coverage for the third party damage, but not for replacement of the pipeline. Manufacturers, too, can purchase general liability insurance, but it too would exclude coverage to the manufacturer's own products. Furthermore, Interpace was not known to have purchased liability insurance.

2. Performance Bonds

While professional liability insurance is the most likely source of funding in claims against engineers, performance bonds may be the best alternative against contractors and manufacturers. Owners commonly look to performance bonds to insure completion of the work of a defaulting contractor. The courts, however, have consistently ruled that that sureties can be made to respond to claims of latent defects in the work long after it has been completed (assuming that the bond contains no enforceable time limit). While rare, manufacturers too can be required to submit material bonds, although their duration is usually limited to the length of any express warranty. Performance bonds, therefore, have become the owner's insurance against latent defects.

CASE HISTORIES

A review of past Interpace cases will demonstrate the application of these principles.

Western Lake Superior Sanitary District - This project involved the installation of 42-inch Interpace PCCP near Duluth, Minnesota in 1974. The District had entered into an engineering agreement with the Engineer for design of the interceptor sewer system and with the Contractor to furnish and install the necessary pipe. The District accepted the Contractor's work in 1976 and the system was put into service in 1978. The pipeline ruptured in 1980 and 1981. An independent testing laboratory hired by the District reported that the tensile strength of the prestressing wire exceeded the maximum allowed by C301. As was its practice, Interpace also tested the pipe and unsurprisingly reported that surge caused the problem and also reported that the pipe met specification. The pipeline ruptured again in 1982 and 1985. The District tested the wire again and learned that it was defective because it contained seams. In 1986, the District sued Interpace and the Contractor, but not the Engineer. However, the surrounding property owners did sue the Engineer and this suit was later consolidated with that of the District.

The court denied motions for summary judgment based on the statute of limitations for various reasons. First, it held that mere rupture of the pipe did not mean that the District knew or should have known that Interpace pipe was defective; there could have been other causes. Second, the District alleged that the Engineer fraudulently concealed the defective nature of the pipe from its client. However, the court did grant summary judgment to the Contractor based on the one year warranty provision in its contract, which read: "All work shall be and is guaranteed by the contractor for a period of one year from and after the date of final acceptance of all work by the owner."

Subsequently, the case proceeded to trial against the Engineer which was found liable by the jury.

Pinellas and Pasco Counties, Florida - These cases involved three PCCP transmission mains, installed between 1973 and 1980. The original pipeline, a 84-inch and 66-inch transmission main known as the Cypress Creek Transmission Main, was built in 1973 to 1975. It was constructed with roughly equal amounts of Interpace and Price Brothers pipe furnished by the Contractor, a joint venture formed by three pipe installation contractors. The second pipeline, built in 1976 to 1977, connected to the southern end of the first and was constructed entirely of Interpace 60-inch and 54-inch PCCP. Both pipelines were designed by the same Engineer. The Pinellas County Water Authority purchased this pipe directly from Interpace because it was learned that pipe furnished through the Contractor would be subject to state sales tax. In addition, the owner used its own resident inspectors in order to reduce engineering fees. The third pipeline, known as the Cross Bar Pipeline, was connected to the first two through a pumping station. It was constructed in 1979 and 1980 from Interpace 66-inch PCCP. This pipeline was designed by a second national engineering firm. The owner purchased this pipe.

The 60-inch line ruptured in 1979 and 1980, causing its owner, Pinellas County, to launch an extensive investigation. Ultimately, the county concluded that the pipe was defective in nearly every component (particularly the Class IV wire and the mortar) and commenced suit against the Engineer, Interpace and the Contractor. However, the Contractor settled early in the case for a nominal amount (reportedly $50,000) because it merely installed, but did not furnish, the pipe. After extensive investigation and discovery the case was tried in 1989. The trial before a Pinellas County judge lasted a year and resulted in a judgment against the Engineer and Interpace for the replacement cost of the pipe. However, the judgment was reversed because of the judge's bias against the Engineer. It was retried before a jury in 1996 against only the Engineer; Interpace had filed for bankruptcy protection in 1991. The jury returned a $10,000,000 verdict against the Engineer. On appeal, the appellate court found sufficient evidence to conclude that the Engineer had breached its "design duties" when it failed to make recommendations to the owner concerning the suitability of Class IV wire and to require additional testing or samples of the wire.

In the case involving the 84-inch line, the owner, West Coast Regional Water Supply Authority, reached a settlement with Price Brothers prior to instituting suit, thereby removing its half of the line from litigation. The Contractor then obtained a partial summary judgment from the court on the ground that the Contractor should not be held responsible for the design of the pipe and because its liability was limited to the one year warranty of materials. The owner continued to allege, however, that the Contractor's installation practices contributed to the problems with the pipeline. Subsequently, the case settled following mediation with the Engineer paying 95 percent of the total settlement amount.

The Cross Bar Pipeline case settled subsequently for a reported $7,500,000 from the Engineer (of which a reported $2,000,000 was obtained from Interpace successors in the pending bankruptcy proceeding). Pinellas County and West Coast Regional Water Supply Authority also obtained an additional $9,000,000 from the Interpace bankruptcy estate directly. West Coast Regional Water Supply Authority, continues to pursue its claim against the non-settling Contractor, which is in bankruptcy.

Richmond, Virginia - In 1979, the City of Richmond awarded a contract to construct a water transmission main. The contract required the Contractor to furnish and install pipe, which it purchased from Interpace. The pipeline ruptured twice in 1987 and again in 1988, causing the city to bring suit against Interpace, the Engineer and the Contractor. The Engineer settled with the city and agreed to cooperate in a case against the remaining defendants. The case was tried before a jury which found that the Contractor did not breach its contract with the city by furnishing Interpace pipe, but awarded $10,000,000 in compensatory damages against Interpace and its successors Madison Management and GHA Lockjoint (which had purchased Interpace's pipe manufacturing business). It also found these defendants to have committed fraud and awarded an additional $500,000 in punitive damages. An appellate court affirmed the jury award. With respect to the Contractor, it found that the jury could have reasonably concluded that its express warranty of the pipe was limited to one year.

Other cases - Other Interpace cases have been: *Washington Suburban Sanitary Commission* (Interpace paid in settlement an estimated $5,000,000; the Contractor paid less than $100,000); *Oklahoma City Municipal Improvement Authority* (Interpace paid a confidential amount; the Engineer settled by offering future services; the claim against the Contractor was dismissed voluntarily); *Hampton Roads Sanitation District* and *City of Norfolk* (this authority pursued its claim in the Interpace bankruptcy proceeding; it decided not to sue the installation contractors); *Howard County, MD* (ultimately the Contractor's surety paid $120,000 to settle the $10,000,000 claim against the Contractor); *City of Ft. Lauderdale* (the Engineer settled for $250,000 when it was discovered that city engineers had also approved the use of Class IV wire.)

CONCLUSION

An owner can maximize its protection by requiring the engineer to participate not only in the design, but in the approval of materials and inspection of construction, and also by requiring it to provide insurance commensurate with the risk. In addition, the owner should require the contractor to purchase all materials and should require a material bond from the manufacturer in addition to the contractor's performance bond. Of course, this will increase the cost of the project through higher bids or proposals from the engineer, contractor and manufacturer. The owner will have to determine if the higher cost is offset by cost savings from use of advanced materials or designs. Owners who purchase the materials directly in order to save sales tax should be aware that the immediate savings could be offset by much greater long-term exposure should the materials prove to be defective.

Biography of
Geoffrey Johnson, Esq.

Mr. Johnson is a 1973 graduate of Princeton University and a 1976 graduate of Georgetown University Law Center. He has been in private practice for twenty-two years and has specialized in construction law and litigation for the past fourteen years. For the past decade Mr. Johnson has been actively involved in four cases arising out of the installation of Interpace prestressed concrete pipe. Mr. Johnson is a partner in the law firm of Lewis & McKenna with offices in New Jersey, New York and Florida.

Practical Repair Procedures for Concrete Pressure Pipe

Richard I. Mueller, P.E., F.ASCE

Abstract

This paper describes various procedures used and taught by Gifford-Hill-American, Inc's field service crews for the repair of reinforced and prestressed concrete pressure pipe.

Introduction

Concrete pressure pipelines have been installed throughout North America to move raw and treated water and sewage. Concrete pressure pipe was and is typically specified because it is bottle-tight, relatively inexpensive, and has a reputation for durability. For most installations, concrete pipe performance meets these expectations of durability, so many water utilities are unprepared to repair concrete pressure pipe. The repair procedures described herein can be used as a guide for utility operators as they consider how they should prepare for the possibility of repairing their concrete pressure pipelines.

Making a Durable Repair

Completed repairs on concrete pressure pipe should provide the strength to withstand external loads and internal pressures and protect the pipe and repair from corrosion.

The reinforcement in the concrete pressure pipe wall is usually designed to provide the strength to withstand the external loads on the pipe. With the exception of American Water Works Association (AWWA) standard C303 Bar-Wrapped Cylinder Concrete Pipe, all concrete pressure pipe types are designed as "rigid". Excessive damage to the reinforcement of rigid pipe will significantly reduce the pipe's load bearing capacity.

Vice President, Engineering/Marketing, Gifford-Hill-American, Inc., 1003 N. MacArthur Blvd., Grand Prairie, TX 75050

Consequently, damaged pipe should be evaluated to assure that external loads will not be excessive for the repaired pipe. If the load-bearing capacity of the repaired pipe is questionable, well-consolidated sand or crushed rock, soil cement, or concrete encasement can typically provide the needed strength for the pipe to withstand the external load. If a soil cement or concrete encasement is used, the encasement should stop at flexible joints in the pipeline, if possible, to minimize potential problems from shear caused by differential settlement.

Pressure containment strength is generally established by assuring adequate gasket compression, by welding, by installing additional circumferential reinforcement on existing pipe, or by pipe replacement.

Corrosion protection of concrete pipe repairs is generally provided by coating all exposed steel with 25-mm (1-in) thick portland cement mortar. If the pipeline being repaired is bonded for corrosion monitoring or cathodic protection, the repair steel should also be electrically connected to the pipeline steel.

Interior corrosion of concrete pressure pipe has been the cause of some failures of "force mains" which were not actually flowing full, and of a pipeline carrying corrosive wastes in an industrial facility. In such cases, interior concrete or mortar was insufficiently protected from the extremely acidic liquids inside the pipe. To provide a permanent repair for these failures, the inside pipe mortar or concrete must be separated from the acidic liquids by buffering the liquid or covering the mortar or concrete with an inert material. The reader is cautioned to research the performance of any proposed inert liners - several types of inert liners or paints have a history of delamination from the pipe wall. However, ceramic epoxies and anchored liners such as T-Lock have generally performed very well.

Pipe Identification

Pipe identification is a key step in determining a proper procedure for repair of a particular concrete pressure pipe. By determining the type of pipe damaged, one also determines whether the pipe has a cylinder, how deeply the cylinder is located in the pipe wall, what type of joints the pipe has, and whether the pipe is prestressed. If the type of pipe is known, current concrete pressure pipe manufacturers can typically assist in the determination of the above data regarding the pipe. Depending on the type of damage to the pipe, each of these bits of data can be very helpful toward expediting the pipe repair.

Various types of concrete pipe have been produced across the country. Typical cross sections of pipe made to American Water Works Association standards AWWA C300, AWWA C301, AWWA C302, and AWWA C303 are shown in Chapter 2 of the AWWA Manual M9, Concrete Pressure Pipe. Some types of noncylinder prestressed pipe have also been produced.

Typical attributes of each of these types of pipe are as follows.

AWWA C300 –
Steel joint rings, a steel cylinder from 1.6 mm (1/16 in) through 10 mm (3/8 in) or greater thickness (depending on pipe pressure and diameter), either separate cage reinforcement or bar wrap on cylinder (or both), smooth concrete exterior, usually no circumferential reinforcement inside the cylinder, diameters generally from 910 mm (36 in) and larger, though some pipe as small as 610 mm (24 in) were manufactured, pipe wall thickness typically ranged from ID/8 to ID/12 in thickness.

AWWA C301 –
Lined Cylinder Type: Steel joint rings, steel cylinder usually 1.6 mm (1/16 in) thick, with prestressing wire wrapped directly on the steel cylinder, 25 mm (1 in) spray-applied mortar coating over cylinder, concrete lining approximately ID/16 thick, pipe diameters from 410 mm (16 in) to 1520 mm (60 in), this pipe type generally found from the eastern US and Canada west through Texas.
Embedded Cylinder Type: Steel joint rings, steel cylinder usually 1.6 mm (1/16 in) thick embedded in concrete core of thickness ID/16 to ID/12, prestressing wire wrapped over concrete core and coated with 25 mm (1 in) spray-applied mortar (pre-1970 installations may have smooth concrete coating 38 mm (1.5 in) thick), pipe diameters generally from 1220 mm (48 in) and larger, though pipe as small as 910 mm (36 in) were made, pipe installed across the US and Canada.

AWWA C302 –
Either steel, concrete, or a combination of steel and concrete joint rings, no cylinder, one or more reinforcement cages, smooth concrete exterior, pipe diameter range from 300 mm (12 in) and larger, pipe wall thickness at least ID/12 in thickness.

AWWA C303 –
Steel joint rings, steel cylinder from 1.6 mm (1/16 in) through 10 mm (3/8 in) or greater thickness (depending on pipe pressure and diameter), mild steel reinforcement bar wrapped directly on cylinder, 25+ mm (1+ in) spray-applied mortar coating over cylinder and bar, mortar or concrete lining approximately 19 mm (3/4 in) thick, pipe diameters from 250 mm (10 in) through 1520 mm (60 in) installed throughout the western US and Canada, including Texas.

Noncylinder Prestressed Pipe –
Attributes typically the same as embedded cylinder AWWA C301 pipe except the joint rings may be either concrete, steel, or both, and the pipe has no cylinder. Very few pressure pipe projects were supplied with this type of pipe.

Repairing Mortar or Concrete

In most circumstances, portland cement mortar or concrete in direct, intimate contact with steel will passivate the steel from corrosion. The high pH of the portland cement causes an iron oxide to form on the surface of the steel which makes the steel approximately as resistant to corrosion as copper or bronze. This built-in corrosion protection is lost if the mortar or concrete is delaminated from the steel. Consequently, repairs to concrete pressure pipe should include coating or recoating all steel with a structurally-sound mortar or concrete that is in intimate contact with the steel.

To repair pop-outs, breaks, or large cracks in mortar (or concrete), all broken or delaminated mortar should be chipped out. The borders of the remaining, sound material around the repair area should be slightly undercut to key the repair mortar in place. Care must be taken during chipping to avoid breaking any steel, especially any cylinder or prestressing wires.

The repair area should then be cleaned, wetted with fresh water, and painted with a thick portland cement slurry. Epoxy bonding agents should **not** be used! Most such bonding agents are barrier materials which will shield the steel from both the high pH of the mortar and from any cathodic protection which might be applied. Bonding agents are typically not needed. If one is used, it should be a permeable, non-dielectric type.

While the cement slurry is still wet, the repair mortar should be rammed and compacted into the repair area. The mortar should be worked under the chipped-out borders of the repair area and under or around any exposed reinforcement or prestressing wire. Wire mesh may be useful for supporting the mortar over large repair areas. The surface of the repair mortar should be shaped to match the contour of the surface while providing a minimum 19 mm (3/4 in) coverage over any pipe steel.

The repair mortar should be a mixture of one part cement to not more than three parts clean sand, mixed with as little water as possible so the mortar will be very stiff but workable. The cement should be a **portland** cement – some other "cements" do not have the high pH that is essential for passivating the steel surface. Repairs made with regular portland cement should be covered with wet burlap, a plastic sheet, a curing compound, or otherwise kept moist for 24 hours or so to allow the repair to cure. Alternatively, a quick-setting, portland-cement based mortar can be used so the repair will harden quickly and not require special curing procedures. In any case, the repair material must be protected from freezing until it is cured.

Except for longitudinal cracks in circumferentially-prestressed pipe, cracks on the interior of steel cylinder pipe which will remain full when in operation typically can be ignored. (Longitudinal cracks in prestressed pipe are an indication of structural distress.) Even if the pipe is to carry seawater or brine, lining shrinkage cracks up to 5 mm (3/16 in) will close as the lining reabsorbs water and autogenous healing occurs. Filling such cracks will typically cause flaking of the lining at the crack edge as the lining swells when reabsorbing water.

If exterior mortar or concrete repairs are needed on pipe which has already been installed, the broken mortar or concrete should be removed and the repair area cleaned as described above. The repair mortar can then be placed as described above, or the pipe exterior can be encased in mortar or concrete. This can be accomplished by placing a cloth "joint wrapper" to straddle the repair area and then filling the wrapper with mortar, or by backfilling the trench pipe zone in the repair area with concrete or mortar.

These same procedures should be used for coating any steel repair bands or patches which are used to repair the pipe.

Repairing Joint Rings on New Pipe

On occasion, new pipe are bumped in such a manner that the steel bell or spigot ring is deformed. Such damage can often be repaired before the pipe is laid by heating the joint ring and hammering it back to its original configuration. Roundness and gasket groove confinement for the repaired joint must be carefully checked, and any damaged mortar or concrete properly repaired. Similar "reforming" repairs can be made on pipe with concrete joint rings by using an epoxy mortar to replace broken concrete and reshaping the gasket-sealing surfaces; however, judgment must be used to assure the repair will withstand the force from the gasket being compressed.

After each segment of pipe is laid, the tightness of any rubber gasket joint should be checked by forcing (or trying to force) a 0.3 mm (0.01 in) thick feeler gauge past the gasket in the joint. If the gasket is not snug in the gasket groove, the joint should be relaid. However, sometimes this condition is not discovered until many additional pipe segments have been laid. Such joints are sometimes identified by infiltration of either groundwater or water used to jet backfill materials.

Suspected loose joints should be rechecked with a feeler gauge before any interior joint mortar is applied. If the looseness is extensive or the gasket is damaged or out of place, the joint should be welded. If the looseness is localized and the gasket is in place, the joint can often be sealed by pounding the spigot against the bell with a sledgehammer. To accomplish this, a 75 mm (3 in) width of mortar or concrete should be removed from the interior of the spigot at the loose length of the joint. Caution should be used to assure the pipe cylinder is not perforated with a chipping gun or similar tool. After the loose area of spigot is exposed, it can be pressed against the mating bell by hammering on the back of the spigot gasket groove. When the feeler gauge shows the joint to be tight, the mortar or concrete should be repaired.

Other new joints that are not sealed can be repaired by using full-circumferential, watertight welds on either the interior or exterior of the pipe. See Chapter 9 of the AWWA Manual M9, <u>Concrete Pressure Pipe</u>, for typical welded joint details. For exterior welds, an 8 mm (5/16 in) or larger diameter filler rod should be placed under the bell flare. The filler rod is then fillet-welded to both the adjacent bell and spigot.

For interior welds with a gasket in the spigot groove, the first welding pass between the spigot and the bell should be "downhill", placed as quickly and with as little heat as possible. Skip welding and cooling with a damp rag may be necessary to prevent excessive gasket smoke in the pipe. As additional welding passes are placed, the heat from the welding will further burn the gasket. The downhill "sealing pass" will help keep additional gasket fumes away from the welders inside the pipe.

Repair of Leaking Joints on Pipe Already in Service

On rare occasion, a slow joint leak in a new pipeline will not be discovered before the pipeline is accepted and placed in service. Some such leaks become worse over time, such that a significant joint leak must be repaired in a pipeline that can not or should not be taken out of service.

Such joints in AWWA C303 and lined cylinder AWWA C301 pipe can be welded shut if the leakage is not excessive and the water spray can be kept off the welder. If the leakage is too great, the joint can still be indirectly welded shut by using a special joint repair saddle, (see Fig. 1). The joint repair saddle cross section is U-shaped, with the legs of the U on the inside of the saddle. The saddle is built as two half sections which will fit together to cover and encircle the entire leaking joint. A steel coupling and plug are built into both the top and bottom halves of the saddle.

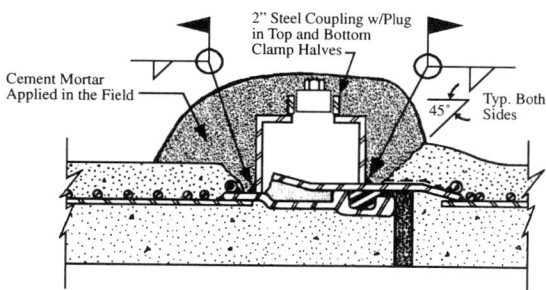

Fig. 1 – Joint Repair Saddle

To install the joint repair saddle, exterior mortar is chipped off the leaking bell and spigot. The plugs are removed from the repair saddle, and the repair saddle is positioned over the joint with one open coupling at the bottom and one at the top. This arrangement will allow the leaking water to drain from the bottom of the saddle while the saddle is welded into place. The saddle halves are tightened together with a come-along and welded to the bell, spigot, and to each other. The plugs are then installed, (bottom plug first), the repair inspected for leaks and touched up if needed, and then exterior mortar is applied.

Repair of Holes in Pipe

An easy way to repair holes in concrete pressure pipe is to use a tapping saddle designed for the type of pipe damaged. The saddle can be placed on the damaged pipe and centered around the hole. The saddle gland, with or without a small valve attached, can be positioned over the hole. If the pipe is under pressure, the valve can be left open to allow leaking water to escape. Once the gland is secured into its final position, the valve can be closed and the repair steel coated with mortar.

Another method to repair holes in non-prestressed concrete cylinder pipe, (AWWA C300 or AWWA C303 pipe), is to clean the area around the hole in the cylinder and weld a patch directly on the cylinder. **Caution – this type of repair requires welding to a cylinder which may be only 1.6 mm (1/16 in) thick. Only welders experienced in making watertight welds on thin steel should attempt this procedure.** After the cylinder is patched, any broken rebar should be rewelded and all exposed steel coated with mortar.

Holes in any type of concrete pressure pipe can be permanently sealed by squeezing a rubber pad over the hole, (see Fig. 2). On cylinder pipe, the area around the hole in the cylinder should be cleaned so the rubber pad can seal against the cylinder. On non-cylinder pipe, the rubber pad can be sealed against the smooth concrete surface of either the exterior of the pipe or the pipe core, whichever is applicable. In any case, layers of rubber are stacked until the outside of the rubber is at least 13 mm (1/2 in) above the exterior of the pipe. A steel clamp is then bolted around the circumference of the pipe and over the rubber to squeeze it tightly against the pipe. The pipe should be repressurized to assure the repair is sufficiently tight. All exposed steel is then encased in mortar or concrete to complete the repair.

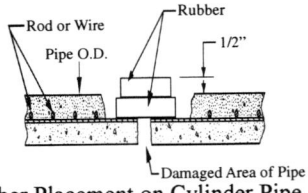

a. Rubber Placement on Cylinder Pipe

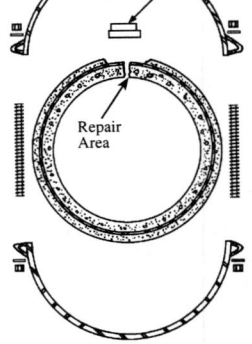

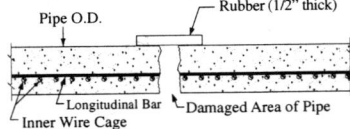

b. Rubber Placement on Noncylinder Pipe   c. Component Parts - Assembly

Fig. 2 – Gasket Clamp Secures Rubber Over Hole in Pipe

If the pipeline is in operation and can not be turned off, holes in the barrel of cylinder pipe can often be repaired in a manner similar to that previously presented for repair of leaking joints. A weld-on repair saddle, (see Fig. 3), is fabricated in two halves to be clamped over the length of pipe with the hole. Each half of the saddle has a coupling with a plug. Each half of the repair saddle is somewhat U-shaped in cross section so the legs of the U will fit against the pipe cylinder but the center of the saddle will bridge over the concrete and reinforcement on the outside of the cylinder. Two strips of exterior concrete and reinforcement are removed from the cylinder so the ends of the saddle can be secured to the cylinder. The plugs are removed from the saddle, the saddle halves are positioned and tightened with a come-along, and the saddle halves are welded to the cylinder and to each other. Again, only welders experienced in making watertight welds to thin cylinders should attempt this repair. The plugs are then installed, the repair checked for leaks, exterior reinforcement on either side of the saddle is attached to the saddle, and all exposed steel is coated with mortar or concrete.

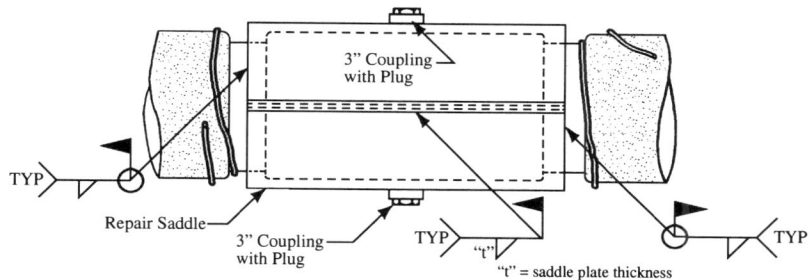

Fig. 3 – Weld-On Repair Saddle

Repair of Cylinder Pipe When External Reinforcement is Damaged but Cylinder is Intact

Three methods are available for repair of external reinforcement on cylinder pipe when the cylinder is still intact. The simplest procedure is to remove any unsound concrete or mortar, repair the exterior of the pipe to match the original contour, and then squeeze a full-circumference steel wrapper around the damaged area with a come-along. Pounding the wrapper with a hammer as it is being squeezed with the come-along will help assure tight contact and good stress transfer between the pipe and the wrapper. When the wrapper is tight against the pipe, the hammer sound will be distinctively solid, and the wrapper may be welded in place. Alternatively, a wrapper with gaskets on each end can be secured to the outside of the pipe, and the void underneath the wrapper can be filled with nonshrink grout. This repair wrapper is known as a bolt-on reinforcing clamp, (see Fig. 4). In either case, direct stress transfer

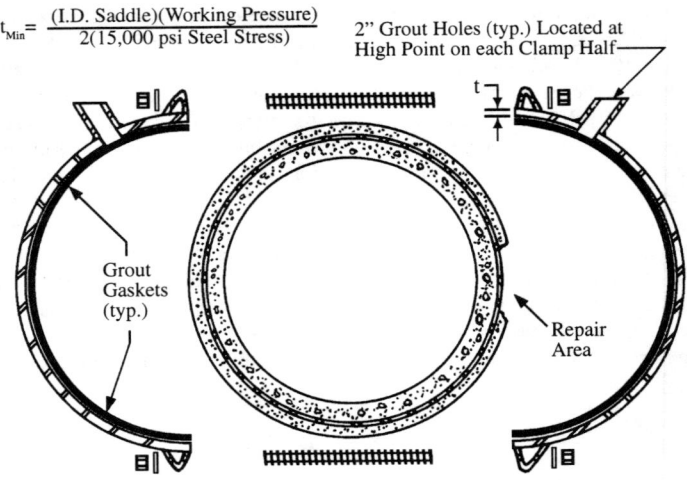

$$t_{Min} = \frac{(I.D.\ Saddle)(Working\ Pressure)}{2(15{,}000\ psi\ Steel\ Stress)}$$

Fig. 4 – Bolt-On Repair Clamp

between the cylinder, exterior concrete or mortar, and the saddle must be achieved. The exposed steel can then be coated with mortar to complete the repair.

A newer method to repair external reinforcement on cylinder pipe when the cylinder is still intact is to wrap the pipe with sleeved prestressing tendons. For large-diameter pipe, this technique has the advantage that each repair "piece", (the tendon), can be lifted manually, without the need for mechanical equipment. For prestressed pipe, the tendons can be wrapped at a stress sufficient to replace the effects of any broken prestressing wire. For reinforced pipe, the tendons can be snugged into place with sufficient stress to assure good stress transfer, or come-alongs can be used to snug rebar around the damaged pipe so the rebar can be welded to itself. In either case, after the reinforcement is installed, all exposed steel must be encased in mortar or concrete.

Segment Replacement

If the damage to the pipe is sufficiently extensive, it may be most economical to simply replace the damaged piece. For all types of concrete pressure pipe, this can be accomplished by cutting and lifting out a short piece of the damaged pipe, and then working the remaining segments gently to disengage the adjacent joints without damage. A new short pipe and closure section can then be installed.

A section of AWWA C303 Bar-Wrapped Cylinder Concrete Pipe can be replaced without using a closure section. For such a repair, the top half of the bell of the adjacent original pipe and the bottom half of the bell of the replacement pipe are cut and removed. The replacement pipe is then installed so the remaining bottom half of the adjacent pipe bell cradles the replacement pipe spigot, and the top half of the replacement pipe bell laps over the other adjacent pipe's spigot. The two bell halves which were removed are then replaced and welded in place, and the joints at each end of the replacement pipe are welded for watertightness.

Conclusion

Many utility operators are inexperienced in repairing concrete pressure pipe, and may be apprehensive regarding attempting such a repair. Concrete pressure pipe manufacturers are always available to assist in determining the best repair procedure for any particular problem. However, utility operators can successfully accomplish such repairs by assuring the repair results in a sound structure, with adequate circumferential reinforcement and properly-applied mortar or concrete for continuing corrosion protection.

When making repairs, utility operators should be cautious about attempting welds on thin pipe cylinders, or cutting uncoated, stressed prestressing wire. Bolt-on repair saddles, or saddles which are tightened around the outside of the pipe and welded to themselves, can frequently be applied to provide needed circumferential strength. Coating these repairs with mortar can result in a repair as long-lasting as the pipe.

Note

Interested parties may contact the author to receive a copy of Gifford-Hill-American, Inc.'s *Concrete Pressure Pipe Repair Manual*. This manual shows details and presents step-by-step procedures for various repair methods.

References

AWWA (1992) *Standard for Prestressed Concrete Pressure Pipe, Steel-Cylinder Type, for Water and Other Liquids*, AWWA Standard C301, American Water Works Association.
AWWA (1995) *Standard for Reinforced Concrete Pressure Pipe, Noncylinder Type*, AWWA Standard C302, American Water Works Association.
AWWA (1995) *Standard for Concrete Pressure Pipe, Bar-Wrapped, Steel-Cylinder Type*, AWWA Standard C303, American Water Works Association.
AWWA (1995) Concrete Pressure Pipe, AWWA Manual M9, American Water Works Association.
AWWA (1997) *Standard for Reinforced Concrete Pressure Pipe, Steel-Cylinder Type*, AWWA Standard C300, American Water Works Association.
Gifford-Hill-American, Inc. (undated) Concrete Pressure Pipe Repair Manual, Gifford-Hill-American, Inc.

LESSONS LEARNED ABOUT CURED-IN-PLACE PIPE DURING CONSTRUCTION

Mark W. Hutchinson, PE. Member ASCE[1]

Abstract

Design and construction of cured in place pipe (CIPP) in the City of Portland, Oregon was once thought to be a trouble-free pipe rehabilitation method requiring little or no design engineering or inspection know-how. It was as simple as writing a purchase order to the local installer and asking them to call us when they were done. Unfortunately, after several installation failures we no longer feel that way.

Consider the following scenarios: a 914 mm (36") sewer pipe with leaking joints under the runway at a major airport, plugged up with a CIPP installation gone bad; a large CIPP installation with a 30% collapse in a 1219 mm (48") combination storm/sanitary sewer; talking to a maintenance supervisor who just spent three days jackhammering resin out of a downstream pump station and diversion manhole that had caused sewage to spill into the Willamette River; a liner that is not completely set under a major arterial to the Portland Airport.

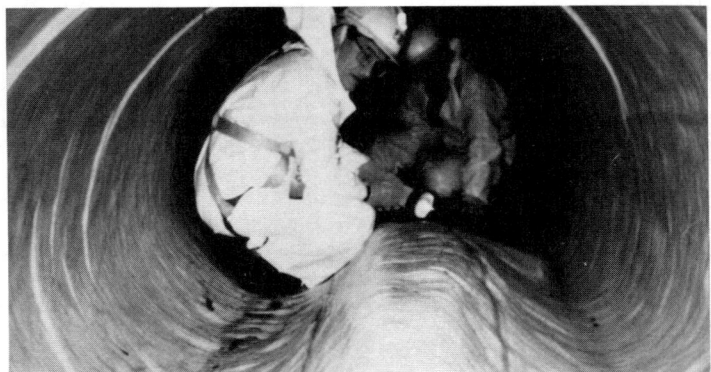

Figure 1: A CIPP liner which has failed due to excessive ground water pressure

---

Engineer, Construction Services Division, Bureau of Environmental Services, City of Portland, 1120 SW 5th Ave., Rm. 1100, Portland, OR 97204

Introduction

From the lessons learned over the last ten years, the City has created new specifications and has increased scrutiny of CIPP design, design parameters and installers. In this article we will share what the City has learned.

Lesson 1: The Design Phase

The City needed to replace a 1219 mm (48") brick pipe built in 1890 that was cracked and had sections of concrete missing. The project was initially designed for conventional construction methods because there were no conflicting utilities and the existing sewer was 4.5 m to 8.5m (15' to 25') deep. After meeting with residents along the pipe route, we discovered that difficulty with access to the apartments along the street as well as the potential for damage to old utilities were issues, but that they could be resolved by using trenchless construction methods. The existing pipe had adequate capacity so cured-in-place pipe was chosen as a repair alternative. The consultant selected to design the concrete pipe replacement was not experienced in designing cured in place pipe, so he contacted a local vendor. Using the vendor's design program and literature, a CIPP construction option was designed .

The design information came in the form of a computer program and design guide. The design guide provided information on the ranges of design factors for installations all over the country. Individual owners needed to apply appropriate design criteria to fit their particular needs and site conditions.

The City did not expect the CIPP option to be the lowest bid. However, a CIPP option was awarded, based on the lowest bid. Unfortunately, we did not review the CIPP option thoroughly enough. After the first CIPP installation was completed, water began seeping out of the street through cracks. We suspected a leaking 19 mm (3/4") water service which could no longer drain to the sewer. The water from the leak caused an artificial groundwater elevation and cause the liner to deform into the shape shown in Figure 2.

Although the installers questioned the design, they could not tell us the nature of the problem. They stated numerous times that large diameter liners need to have a thicker wall than we had designed. They also maintained that the liner deformed after the resin had been cured. Their assertion was confirmed when the holes were drilled in the liner to relieve the water pressure and the liner went back to within inches of its original shape. The liners rely on their round shape for structural integrity - once the shape is deformed, the liner may continue to deform over time. Also, deterioration of the material will be accelerated if it is under stress.

Downstream on this same project, the liner began to collapse as the internal curing water was pumped from inside the liner. Cracking and popping was heard from the liner as it failed. The liner had hardened and it deformed as the internal water pressure was decreased. Inspection of the liner revealed a serious failure. The storm/sanitary sewage was flowing under the liner and back into the pipe through a 60mm (24") rip in the liner. The liner was deformed as shown below:

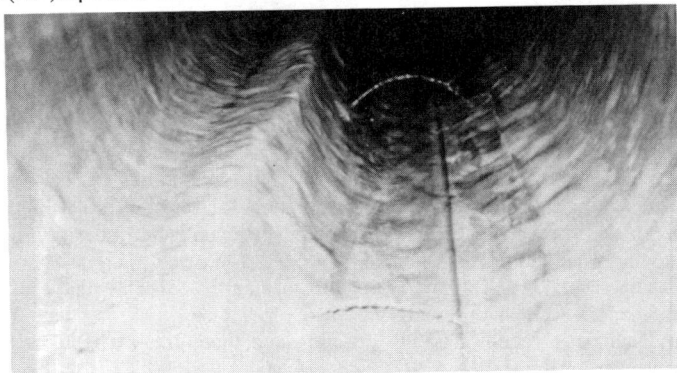

Figure 2: Water seeped out under pressure when holes were drilled in the boat in the CIPP

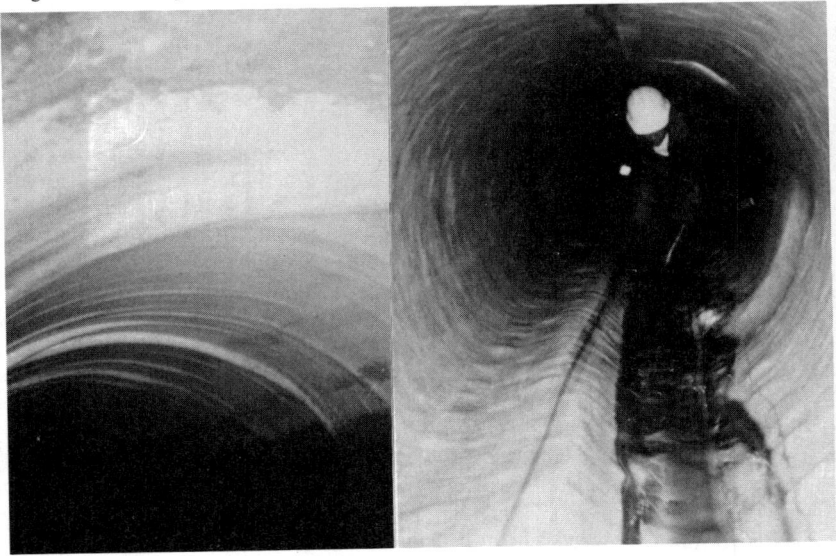

Figure 3: Wrinkles in liner on the left, a boat in the liner on the right

Our first assumption was that these deformations were caused by a faulty installation. Samples of material were taken to our testing lab and tested for strength, the water column height to overcome the head created by branch lines was investigated, the heat and cure log times and temperatures were checked and the surrounding waterlines and service connections were checked before it was determined that this was not the installer's problem.

In this case, no external water source was evident. Inspection of the liner revealed evidence of failure after installation and curing. Brick marks were found imprinted in the outside of the pieces removed and the layers of the liner that had delaminated, showing that the liner had been in contact with the existing brick sewer when it cured. After investigating the installation, structural integrity and connecting services, the external water pressure here, too, became the prime suspect. The two possible causes were a leaking 203 mm (8") branch or the sewer trench acting as a drainage path for groundwater in this area. During the installation of the liner, a 203 mm (8") branch line required little flow diversion pumping even though it was running 1/3 full upstream. The flow was exiting the pipe and flowing along the outside and back into the sewer through the cracks in the pipe wall. The other source of ground water contributing to the failure was an upstream project which was backfilled with sand. Rain water filtered down into the sand backfill and moved through the sand faster than the surrounding soil and eventually entered the cracks in the sewer. Once the sewer was sealed, the water created an artificial high groundwater pressure on the liner.

Once the nature of the problems was identified, the external groundwater elevation was assumed to be the top of the pavement because it was likely that the 1914 and 1922 vintage waterlines had developed other leaks and the upstream trench and branch would continue creating artificially high groundwater levels. A liner was designed taking into account the external water pressures using an 18 foot ground water depth to invert. Using these parameters, the liner thickness nearly doubled from 10 mm (0.4 inches) to a the new liner thickness of 20 mm (0.78 inches).

The deformed portion of the liner was removed and another liner installed in the area of the failure over the top of the first liner. Methods of attaching a liner to another liner will be described later in this paper in Lesson 6: Repairs.

Lesson Learned:
1. Assume ground water levels at the road surface when working around old water lines, areas prone to flooding and sewer pipe trenches that are transporting water.
2. Other design factors to consider are: ovality of pipe, resin type versus cost and resin strength and flexibility, time of year for installation and sewage diversion requirements.

Lesson 2: Installer Prequalifications

For years the City has encouraged competition for local CIPP installers. The City felt that the price for this type of repair was nearly double what it should be. In 1991, there was finally some competition. At this same time, the City had taken over maintenance responsibility for the sewer lines under the Portland Airport, next to the Columbia River. The concrete pipes were in good shape but the joints leaked because of poor construction, being 1.5 m to 3 m (5 to 10 feet) under the groundwater table. Over half the sewage flow in this pipe was attributed to infiltration of ground water.

Due to the location of the pipe under the airport runways, a major highway and airport parking, adequate capacity in the pipe, and, the need to perform the repairs quickly, CIPP was the chosen repair method. A CIPP liner was designed and bid. A contract was awarded for $1.1 million to line the pipes. An installer who had just purchased a CIPP franchise was the low bidder. The installer was new to the sewer lining method of repair and had only lined a 203 mm (8") pipe prior to this project. The franchise parent company was to be on site to help them. The contract was awarded, basic submittals were approved and the contractor was directed to begin construction. Since inspection of CIPP installation had always been similar to watching paint dry, the City thought there was no reason to worry. Not this time. On the night of the first installation, a visit to the warehouse where the resin was being impregnated into the felt sock revealed the following: the installer was mixing the resin and catalyst by hand in a 50 gallon drum with no personal protective gear and minimal ventilation. Because the resin was not flowing, the feed apparatus was being heated with a torch, despite the warnings on the side of the containers which read: "Warning Flammable and Catalyst May Self Ignite If Temperature Exceeds 75° F". The resin was being poured from the 50 gallon drum into five gallon buckets, then from the buckets into the sock. Resin was all over the floor, creating quite a mess. After about 254 kg (560 pounds) of resin was poured into the sock in globs, the sock was run through rollers to try to distribute the resin evenly throughout the sock. The entire operation fell off the conveyor, requiring a boom truck to lift it back up. City staff decided to leave the site, partly due to the chemical fumes, the warning labels on the containers and the fear of being seen anywhere near the mess.

When the impregnated sock arrived at the installation site twenty-four hours later, the contractor was asked about sewage diversion. Several services had not been plugged and they had partially filled the pipe to be lined. The contractor was told to pump it dry. The contractor sent for a pump to dry up the line. Two hours later, he returned with a pump more suited for use with a hot tub. After a colorful conversation, the contractor sent for a sewage pump and dried up the pipe. The sock was pulled into place, only to find that it was too short. The sock was then dragged down the pipe to a manhole run where it would fit, filled with water, and heated and cured for approximately forty-eight hours. The contractor's workers began to get fatigued after sixty hours of continuous work. Some even tried to leave the job site. However, the contractor was not finished yet. The liner still needed to be cooled down and the ends cut out. This work was finally finished after seventy-two hours of continuous labor. Workmen were "walking zombies"and could be found sleeping behind the trucks.

Later, on this same project, a section of liner was not heated long enough and did not set completely. The contractor requested permission to re-heat the liner. After discussion with a chemist and some university professors familiar with the process, it was determined to be possible, although they had never observed it before. The liner was re-heated. A man-entry into the sewer was requested to inspect the twice-heated liner. However, when entry into the pipe was attempted, the installer said they could not reduce the flow. Upon further inspection, it was found that the flow was backed up purposely to prevent inspection of the liner. Inspection of the liner revealed that the liner was cracked top and bottom for about 80 feet and had "boats" (see Lesson 5) the rest of the way. The cured liner could be broken by hand. The entire liner had to be removed and replaced. Re-heating the partially set liners didn't work, because the resin is made to be heated once. During the curing process, portions of the liner pass through the exothermic portion of the chemical reaction. Re-heating makes them brittle and produces cracks(see picture). Partially set liners must be removed and replaced.

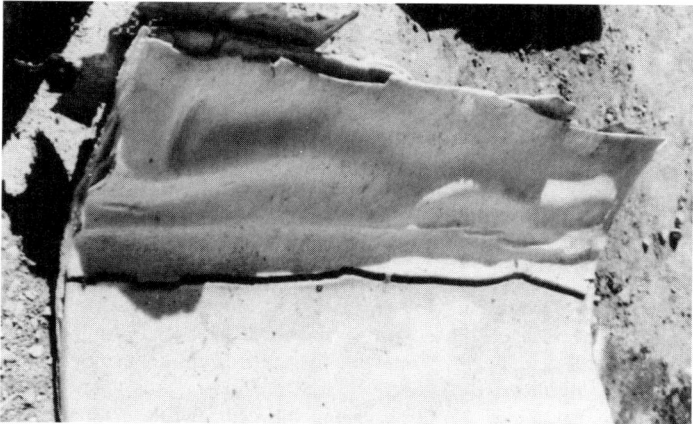

Figure 4: The split liner with cracks shown above resulted from the re-heating process

This same contractor tried to install a liner that was 1.8 m (6') too short. There were holes in the liner that were drilled to observe what was behind the bulges in the liner. The flow into these holes was too great to allow a plug to be installed.

Water was pouring out of the space between the liner and the concrete pipe at the ends of the liner. While the contractor was trying to seal this water off an inflatable plug exploded. The area between the the liner and the pipe was eventually sealed by first using a chemical to stop the water and then applying the approved sealant.

A post inspection of the liner found that the sock was not completely impregnated with resin. It was found later that pigment mixed in with the resin will allow the inspector to determine if the sock is completely full of resin.

Lesson Learned:

1. Address the risk of failure and prequalify bidders based on that, check the experience of the installers, request references from other projects and resumes of key personnel and check them.
2. Require through specifications the resumes of key personnel and require one or two years experience installing the same diameter CIPP and similar experience for sewage diversion
3. Require through specifications that blue pigment be added to the resin.
4. Require that an hour by hour procedure plan be submitted prior to the installation, enabling the inspector to check that the liner was heated for the minimum time and cooled down for the minimum time.
5. Require through specifications that thermal couplers be attached to the sock part way between manholes and at either end. Require the installer to record the temperature of the CIPP in a cure log every half hour.

Lesson 3: Runaway Resin

This lesson was learned while installing a 50 mm (2 inch) thick, 1372 mm (54") diameter CIPP pipe in an 1829 mm (72") brick sewer. When CIPP is installed by inverting the liner, the wet resin side comes in contact with the pipe as the liner inverts under the pressure of water. In front of this inverting liner is a mixture of resin and whatever else is in the host pipe. The liner pushes standing water and debris out of its way. Although heat is required to cure the liner, this resin will eventually set up downstream. In this case, we learned this when a maintenance crew had to remove a blockage in a diversion manhole and pump station. The diversion manhole had plugged up with the slug of resin, resulting in sewage unknowingly being diverted to the river. The maintenance crew spent two days jackhammering the 76 mm (3") thick resin mixture into two foot squares to take it out of the manhole.

Lesson Learned:

1. When reconstructing a pipe, add a requirement to the specifications to place dams down stream to catch runaway debris and check the dams daily. This time it was resin, other times it has been tools, pieces of pipe, gloves and wood. Depending on the flows, the dams can be made of sand bags or expanded metal, wedged or bolted to the pipe.

Lesson 4: Preventing Flow Around the CIPP

A watertight liner is only watertight until you put a hole in it. CIPP liners are installed such that the resin comes into contact with the host pipe and they allegedly adhere to the host pipe. However they do not adhere to the host pipe everywhere. In fact, most of the liner does not adhere to the host pipe. Water or sewage travels between the host pipe and the liner. Every installation requires two holes to be cut in the liner, one upstream to let the flow in and one downstream to let the flow out. Each one of these holes needs to be sealed as soon as possible. This is usually a two step process consisting of stopping the water with a water setting compound and then a glue sealant compatible with the resin.

In sewer applications, more holes need to be put in the liner to provide for services or inlet leads. Usually the services are drilled with a 13 mm (1/2") hole to relieve pressure immediately after cutting out the ends. Occasionally, holes are drilled where there are no services and another place for infiltration into the pipe is created. It is important to know the location of services by station and clock position. Fortunately, the laterals usually cause dimples in the main. After the laterals are relieved, the holes are opened up completely by man-entry into large pipe or with robots using TV cameras on small diameter pipes.

It should be noted that after the services are opened up, the area around them needs to be sealed or the flow from the service will travel between the host pipe and the liner, exiting at another service or worse yet, ground water will infiltrate at each one of these holes. Without sealing the services to the liner, water will come through these holes, defeating the purpose of the liner.

Lesson Learned:

1. Require through specifications that the installer seal between the host pipe and the CIPP at manholes.
2. Require through specifications that the installer seal between the host pipe and liner at each hole cut for a service lateral.

Lesson 5: Wrinkles, Boats and Deformations

The City has learned from the hundred plus liner installations that CIPP liners very seldom completely fit the host pipe without wrinkles. The City of Portland limits deformation sizes to 5%. One of the reasons for this is that the CIPP pipe is a flexible pipe and it relies on its circular shape for strength. A circular pipe when subjected to conditions of of internal vacuum or groundwater pressures in excess of internal pressures buckles when the external load exceeds the strength of the pipe material. A circular pipe line with CIPP with a 5% deformation is only 64% as strong as it should be, a 10% deformation results in a liner that is only 41% as strong as it should be (Insituform Engineering Design Guide). ASTM 1216-91 does not provide limits on deformations. The only requirement found in ASTM is that the liner fit tight to the pipe wall when installed.

Deformations in CIPP liners take two forms: wrinkles and boats. Wrinkles form when the liner does not exactly match the diameter of the host pipe or the pipe is on a curve. The wrinkles can be ground off or left in place depending on the location and size. Wrinkles in the top of the pipe below 5% can probably be left (gravity pipe), larger wrinkles and wrinkles in the bottom half of the pipe should be removed. Normally, wrinkles are just thicker places in the wall of the liner.

Boats are a different story. These are locations where the liner has separated from the host pipe. Boats in the liner are caused by trapped water, external water pressure in excess of internal water pressure during cure and trapped debris between the host pipe and the CIPP. Boats are a structural problem that need to be removed and repaired if they exceed the 5% rule. Additionally, the cause of the deformation needs to be identified and dealt with.

Lesson Learned:
1. Expect wrinkles in the liner, and grind or cut them out if they are in the invert or exceed the 5% requirement.
2. Determine if deformations are wrinkles or boats and determine the cause.
3. Remove boats and repair them if they exceed 5% to maintain the structural integrity of the liner.
4. ASTM Specifications do not address liner deformations. The designer needs to determine the acceptable deformation based on calculations and specify that in the design.

Lesson 6: Repairs

Once it has been determined that the liner needs to be repaired, the type of repair needs to be identified. Liners need to be repaired when they are not structurally sound, when they are too short, when they collapse and when they have not been completely cured. Jacking and bolting liners in place, grouting behind them and attaching steel plates does not provide a long term structural fix. Repairs need to be accomplished with similar materials so that the CIPP retains its circular shape, and reacts to loads as a whole.

Boats should be removed as discussed previously. If the liner has the correct shape, the remaining liner can be left attached to the host pipe and lined over if the pipe capacity allows. If the boat is the result of a structurally deficient design, a liner of adequate thickness will need to be installed over the remaining liner. To get the first liner to stick to the new liner, the area of overlap will need to be roughened up by grinding or sanding. After the new liner is in place and the ends removed, it will need to be sealed over with resin or other adhesives compatible with the lining system.

When portions of the liner are removed for deficiencies or when the CIPP is not long enough, this section can be relined with a short sock. The procedure is to taper and roughen the mating surface between the new liner and short sock at each end. The short sock can be dragged into place and cured as the large one was.

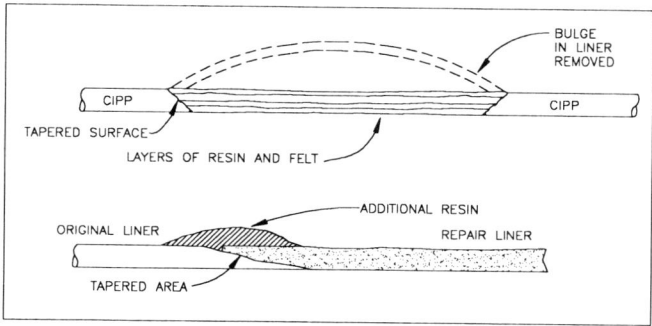

Figure 5: Drawing of liner overlap area with repair sock (bottom) and small repair using laminate (top).

Small boats can be repaired with a laminate of ambient-cure resin applied to the sock material. The procedure is to remove the bulge then grind or sand the edges around the hole. Then apply several layers of laminate until the thickness matches that of the surrounding CIPP.

Lesson Learned:
1. Repair long boats by removing the boat and/or liner and installing an overlapping CIPP.
2. Roughen the end area where the liners make contact to improve the bond between them and then place resin over these areas.

3. Repair small boats in the liner (less than 1 square meter) by roughening the mating area and applying layers of felt and ambient cure resin until the desired thickness is achieved.

Conclusion

In conclusion, CIPP properly designed and installed will extend the life of pressure mains and sewer pipes for at least 25 years. The repair can be accomplished in days rather than weeks or months, with little disruption above ground. Well thought-out design parameters, bidder prequalification requirements, specifications and inspection will help to ensure a good product for the owner. Poorly designed and poorly installed CIPP liners will cause problems and provide opportunities to learn lessons.

References

Insituform Technologies, Inc. (1992). "Insituform Engineering Design Guide"

Uni-Bell PVC Pipe Association (1983). "Handbook of PVC Pipe Design and Construction"

PIPELINE MARKET - 20 BILLION DOLLARS FOR 1998 WOULD YOU LIKE SOME OF THIS WORK?

By

Dr. Jey K. Jeyapalan, P.E.
Pipeline Engineering Consultant
21 Meetinghouse Terrace, New Milford, CT, USA 06776
phone 1-860-354-7299
email: jkjeyapalan@snet.net

Abstract

The market volume for fiscal year 1998 in America for new pipeline construction and renovation of aging pipelines would exceed 20 billion dollars. At one time, new pipelines for sanitary sewers were constructed of unlined concrete pipe. Storm sewers were of either concrete pipe or corrugated metal pipe. For sanitary sewer force mains and water transmission pressure lines, we used primarily ductile iron and pre-stressed concrete cylinder pipe. In 1998 however, a major segment of sanitary sewers would be constructed of plastics, either with a smooth wall or with profiles, cores, and ribs. In storm market, a major share would be of corrugated polyethylene. The pressure market would go to PVC and HDPE in smaller sizes and to welded steel in larger diameters in major volumes. By year 2000, we Americans will build over 50 % of our new pipelines with plastics, mostly with PVC and HDPE. The above trends are very different from most countries around the globe. In the renovation market, 10 years ago, Insituform, was the only game in town offering Cast In Place Pipe(CIPP). Now, there are numerous players in the CIPP technology alone in America, and the winner is chosen primarily on price and in some cases, based on superior track record and prior established relationships. The pipeline renovation market in America in 1998 will spend over 4 billion US dollars in all market segments, with some 25 % of this volume being done by trench-less methods. There are over 1,000 players in the pipe renovation business alone offering either a product or service at some level to cities, counties, water authorities, and private utility companies. Needless to say that the pipelines construction and renovation is a very competitive business, with some winners and some losers in the next 10 years. For those aggressive companies who want to be part of the Great American Dream or those aggressive City Officials who are looking to compare notes with their counterparts will find this paper useful. One needs to become aware of the various market entry barriers, what criteria pipe users set to select new pipe materials, our standardization efforts and even local guidelines such as the Southern California Green Book or the City of Houston's Standard Wastewater Products Committee. The aggressive companies and City Officials also need to be aware of what technologies have been tried and rejected by the American market and why.

Background

There are over 60,000 community water supply systems in America servicing over 230 million people. In 1998 dollars, $ 138 billion needed to meet clean water requirements for the next 10 years, and out of this $ 77 billion would be for improving pipelines involved in transmission and distribution of drinking water. There are over 18,000 wastewater treatment systems in America serving over 200 million people. In 1998 dollars, the overall market size is $ 47 billion for combined sewer overflow corrections, $ 45 billion for other sewer related work, another $ 15 billion for non-point pollution controls for the next 10 years. The U.S. EPA predicts another 2,000 treatment plants to be built. The amount of money involved in sewer renovation work in America is on the order of $ 2 billion per year. The industrial wastewater market is growing some 30 % each year to $ 15 billion in year 2005, but the funding for industrial sewer renovation work is privately controlled and would not become available in large volumes until some time in the future. The Gas industry in America installs approximately 25,000 miles(40,000 km) of pipe and repairs over 700,000 leaks every year either as new or by renovation. Only 4 percent of this work is done by trench-less methods. The pipe renovation market could be on the order of $ 300 million per year for distribution piping and another $ 300 to 900 million for transmission pipeline work. The Crude Oil, Products Pipeline industry in America probably does about $ 300 million worth of pipe renovation work each year.

Obviously, the above figures indicate a nation with a mega market in all sectors of the pipeline industry; but, is this market easy to penetrate for anyone who wants a piece of the play? Probably not. Then who gets to enjoy the great American Dream. The remaining sections of this paper provide much food for thought for entrepreneurs, pipe manufacturers, and the engineers from the private and public sectors.

Pipeline Technology for Open Cut-Past, Present, and Future

One hundred years ago, most pipelines in sanitary sewers, storm drains, and water supply were constructed of concrete, clay, and wood. Some fifty years ago, corrugated metal pipe took some share of this market only to lose during the last 10 years to plastics. In 1977, plastics took over 30 percent of the total length of the pipeline market with predictions of over 50 percent for year 2000. Steel would be the next dominant material taking over 18 percent of the length in 2000. Copper would be used for about 15 percent, concrete would do about 5 percent of the length, ductile iron another 2 percent, and clay less than 0.2 percent. The total expenditure for pipelines installed by open cut in America this year alone would exceed 18 billion dollars with an anticipated increase of about 5 to 10 percent per year.

Why do we Americans so much in love with plastics and steel? We are a trendy culture turning to lowest possible initial cost of construction with new and fashionable materials such as plastics sometimes without a major consideration for materials which last forever on one hand while on the other steel had given great service in our pipeline infrastructure regardless of the market segment. So, why not continue to use steel for pipelines where pressure carrying capacity and larger size pipe are in demand.

Technology for Trench-less Renovation-Past, Present, and Future

Municipal Sewer Market

Cleaning

In many municipal sewer systems and their components, proper cleaning alone could return some of the lost capacity back to the line. Lack of sufficient flow capacity in old pipelines have generated many new pipeline construction projects. In this regard, jet cleaning is most commonly used to renovate sewer collection systems. A CCTV video is prepared before any cleaning is commenced as a baseline measurement of the current condition of the system and after jet cleaning, a second video is prepared to determine the level of rejuvenation done to the collection systems. If a lining method using a paint mixed into an epoxy hardener would enhance the collection system capacity, this is always undertaken following the jet cleaning operations. It is important to recognize that the skill level and past experience of the operators employed by the city would determine the degree of success of cleaning operations. Although chemical cleaning is being tried with some good results for water distribution systems, use of such technology for sewer collection system cleaning would have to ensure that the chemicals utilized do not interfere with the treatment functions of the sewage treatment plants. The best tool for verifying the effectiveness of cleaning efforts is CCTV.

Root Control and Removal

Chemical root control is commonly used to kill root growth into the sewer system and to inhibit re-growth without causing significant damage to trees and plants, the ambient environment and the wastewater treatment process. It is necessary to recognize that the tools used in root control could also cause damage to the walls of the collection pipes, if proper care is not taken. The usual ingredient for killing roots is a special herbicide that kills roots at low concentrations. Common materials include sodium methyldithiocarbonate, diclobenil and other. Rootpru is one such fumigant gel, and others use foam for delivery of materials provided by Avanti, Tobys, Airrigation in North America Although some attempts were made to include formation of copper sulfate solution in situ by providing copper wires to react with other chemicals around pipes with root intrusion problems, due to the contamination caused, this approach is no longer accepted by the US EPA.

Mainline Renovation

There are numerous technologies which have been in use for some time for rehabilitation of sanitary sewers in North America. These methods could be broadly classified into two types namely, "trench-less" and "less-trench". In the trench-less methods, the process needs to be applied from one existing manhole to the next one with the laterals needing reconnection using robotic cutters. It is most common not to have any open-cut excavation at all in or around the job site. Examples of such methods are Cured in Place Pipes, Fold and Formed Pipes, Pipes by Directional Drilling, Robotic Repairs, and Fill & Drain Technologies. In the less-trench methods, some excavation is always required. These are used to introduce the new pipe into the system and for reactivating the laterals. Some processes require pits, auger holes, while others need sloping trenches leading into the old pipe. Few examples are Slip-linings, Swaged & Rolled Down Pipes, Spiral Wound Pipes, Segmented Linings, Pipe Bursting, Micro-tunneling, and Pipe Ramming.

Coatings

Reinforced shotcrete and cast-in-place concrete are possible coatings only when corrosion is not a problem. In the presence of sulfide corrosion, one would not use either of these two coatings. Other engineering materials and resins also could be used with or without fibrous reinforcements for coatings.

Point Repairs

Most of the spot repair systems currently available involve the application of chemical grout to fill cracks, a cured-in-place sleeve or mat, or an epoxy-based resin. Differences lie in the method of application, with some processes using robotic devices, a winched-into-place sleeve, or a combination of methods. The latest advances in point repair include performance liner, link-pipe, amkrete, econoliner, fibers in gunite or shotcrete, and chemical grouts. In these methods either a pre-formed segment of a new pipe is inserted into the right location where the existing sewer system needs renovation and the new pipe is expanded or cured to form a new structural and/or hydraulic liner to provide the point repair. The most effective point repair is to use a grout either cementitious or chemical using an inflatable packer guided by a close circuit television camera to the location of the sewer system needing the repair.

Chemical Grouting

Although the earliest form of sewer collection rehabilitation technology involved some form of chemical grouting, this method is still seeing many new developments. Use of many new chemical and cementitious grouts with varying properties, set times, and functions are emerging in the North American market place. The most common type include epoxy, gels, acrylamide grout, acrylic grout, acrylate grout, urethane grout, and urethane foam. Cementitious grouts continue to see several enhancements such as the use of centrifugally acting spraying machines which travel along the pipeline and provide new coatings and linings completely unmanned, use of admixtures, and fibrous reinforcements to increase tensile and corrosion characteristics of such grouts. The internal grouting is usually applied using a remotely inflatable packer guided by CCTV, the external grouting process involves a detailed study of soil conditions to establish which grout would flow into the subsurface around the pipe effectively. Chemical grouting has been successfully used for over 30 years. The success rates of grouting was based upon the level of competence of the contractor and the ability of the engineer to design and specify the project out in a manner which would guarantee good results. Those who are inexperienced in working in varying soil and pipe conditions and capabilities of different grouts are quick to claim that "grouting doesn't work". The reality was grouting was looked upon as a cure for all pipe problems and if grouting didn't on a given project, those who tried never researched to learn the causes of its failure. In summary, currently there are two primary methods/procedures exist for this type of work. Chemical grouts, which use a packer to first air test a joint to test its water tightness. If the joint fails the air test then the joint is grouted using a grout with a viscosity of less than 10 cps. This process allows for a pipeline system installed in the 1940's to be brought up to 1990 standards. Chemical grouting as a form of maintenance for pipelines is increasing in popularity. Many managers of aging systems are seizing upon this technology to ensure additional longevity of their pipes. This philosophy is much the same as a home owner painting his dwelling, the work is looked upon as a preventive effort for maintaining the integrity of a house. The process must be repeated after a period of time. Epoxy placement is performed with robotic tools, again with the aid of CCTV. Some of these processes utilize routing bits to allow the epoxy to bond to a clean surface.

Cured in Place Pipes

Cured in Place Pipe (CIPP) systems enable sewer pipelines to be repaired from within by insertion of a lining material through existing manholes. The liner is composed of a fabric reconstruction tube which is impregnated with a thermosetting resin that hardens into a structurally sound joint-less pipe when exposed to hot circulating water or steam. Once cured the pipe is allowed to gradually cool to prevent thermal shock and then the laterals are reconnected. The rehabilitation liner not only serves to repair the deteriorated structure of the existing pipe, but reduces infiltration of unwanted ground water. The CIPP process was introduced in the United States in 1977 and millions of feet of CIPP have since been installed. Until recently, competition was almost non-existent from other trench-less rehabilitation products and the primary competitors with CIPP were open-cut construction and slip-lining. In the mid-80's several new CIPP products were introduced. Now, the North American market is flooded with numerous CIPP systems, such as Insituform, Inliner, Spinello-KM Liner, Masterliner, Nationaliner, etc, without having to worry about patent infringement lawsuits by Insituform), the winner is based on cost and cost alone. Particularly, when Applied Felts Ltd from UK and those similar to them are prepared to deliver either the dry liner or the resin-soaked liner to the job site for any installer, the CIPP process has dropped in price significantly the past year in North America. The fact that multiple options exist for owners to consider for pipe renovation has its strengths and weaknesses. For example, on one hand there is more competition among many technologies, while on the other, many of the new products in the market place do not have well-trained staff or research results comparable to those collected by the original technology provider. Cured in Place Pipes are either winched into place or inverted in place using air or water pressure. The curing process is done either using steam or hot water. All thermoset resins commonly used for RPM and FRP pipes are used for Cured in Place Pipes. Among these, polyester is the most common resin used. In situ cutters guided by CCTV provide means for opening the laterals. Insituform Technologies has a new Insitu Launcher to do the lateral reopening. The liner pipe is designed either to perform just the hydraulic function or both hydraulic and structural functions depending on whether the existing pipe has structural capacity intact or not. Even here, there is continuing confusion and lack of uniform design guidance on how to design the thickness for liner. The most significant advantage of Cured in Place Pipe is that it could be used for old pipes of any size and shape. New innovations include using reinforced felt to handle internal pressures of substantially high values. The design of Cured in Place Pipes would be quite similar to slip-liner pipes with the exception of the absence of any pulling or pushing forces and annular grouting.

Fold and Formed Pipes

Fold and Form Pipe (FFP) allows for pipelines to be repaired through existing manholes as the CIPP process. The FFP system uses a thermoplastic material which have been deformed from a circular shape, i.e., folded, to result in a smaller cross-section that can be easily fed into an existing sewer through an open manhole and pulled from the next manhole at the end of the reach. These products utilize either extruded polyvinyl chloride (PVC) filled or unfilled, alloyed or unalloyed, or high density polyethylene (HDPE) pipe that is flattened and folded longitudinally. The plastic pipe is fed from a spool into an existing pipe where hot water or steam is applied until the liner reaches its temperature for rounding. After rounding the materials are allowed to cool and then the laterals are reactivated. Softer PVC cell class resins and HDPE folded pipe require minimal heating to form the pipe back to its initial shape. For folded pipes, it is common to use a rounding device and apply the technology effectively when the old pipe is not surrounded by excess amounts of groundwater. In the presence of groundwater the heat input from the source would not be sufficient to keep up with the loss of heat into the ambient ground, unless a heat containment tube is utilized. It is important to recognize that the fold and formed pipes using HDPE resins could never come close to those pipes using PVC resins in the pipe stiffness property particularly to meet long term buckling capacity requirements. The HDPE fold and formed pipe has a mere 10 % of equally thick PVC fold formed pipe when it comes to long term buckling strength. Thus, HDPE in essence would never be the material of choice in comparison to those pipes made of PVC resins, if the long term buckling strength would govern the choice of the pipe material.

Directional Drilling

In directional drilling methods, a pilot bore is made and the product pipe is pulled right behind the back-reaming operation on the return trip from the exit point to the starting point. This method cannot be used for minimum grade gravity sewer collection systems. The drift control is within inches using electromagnetic tracking systems. Cobbles and boulders cause serious problems and may prevent finishing of the line. Usually, fluid additions reduce friction on the outer wall of the pipe, assist in cutting of the bore hole, and prevent the bore hole from collapsing. Most projects use either steel or HDPE pipe materials for renovation or new line installation but copper pipes and cables also could be installed by this technology. The capability is subdivided into mini, midi, and maxi where pipe sizes up to 1,500 mm for a distance of up to 1,500 m to a depth of 30 m with pull forces up to 300 tons could be handled. The most successful story in America is for the use of horizontal directional boring technology for new pipe and pressure pipe replacement. The first machine was sold in 1985 and we have exceeded 3,000 machines this year. The growth will continue till we reach over 8,000 machines by year 2000. It is predicted that 2 or 3 major players will manufacture some 70 to 80 % of these machines. This business will expand into Europe and Pacific Rim, but never anywhere close to the density we will see in North America.

Robotic Repairs

Intelligent robots which could brush away the dirt in the collection systems, jet clean the pipe walls, cut down the root growth back to the pipe wall, fill holes with proper grouts, mill away damaged and badly fitted pipes and laterals, perform point structural or hydraulic repairs, provide continuous video record of the internal condition of the collection systems are becoming more and more popular. The most common are those by Sika, Ka-Te, and American Robotics in North America.

Fill and Drain Methods

In this method, two chemical solutions which react with one another when they are brought into contact are filled and pumped out one after another into the collection system to form a third material which provide structural repair to all components of the system in a monolithic fashion. Although the same method and materials could be used for any existing pipe material and for all components of the collection systems, there are some drawbacks in this method of repair such as clogging of the main waterway in the smaller sized collection systems and when these products come into contact with roots. The only company of this type of technology which has tried to do work in North America is Sanipor.

Segmented Slip Linings

In this process in its most recent form, a new pipe is inserted either by pulling or pushing in continuous length or in short discrete lengths. It is very common to have the annulus grouted with proper care once the slip-liner is inserted, if no compression fit is present. Without the grout the slip-liner will not be able to withstand most of the external water pressure and other buckling loads. The most preferred material for slip-lining is HDPE due to its superior characteristics in corrosion resistance, abrasion, impact strength, and strain tolerance. So, Chevron's Plexco Spirolite made to the German License "Bauku,", Hobas made to the Swiss License, Lamson-Session's Vylon made to the Greek License "AG Petzatakis," and Meyer Polycrete made and brought all the way from Germany are good examples of such technology already in the North American market. Short pipes using either gasketed or mechanical joints are inserted in situ using hydraulic jacks. In summary, for discrete pipes, smooth wall HDPE, profiled walled HDPE, profiles walled PVC, FRP, RPM, Ductile Iron, Steel, Clayware are all used quite effectively to slip-line collection systems. The design checks should lead to the selection of proper wall thickness for the liner pipe. In addition to ensuring that the hydraulic capacity does not diminish excessively, the structural strength and stiffness and the liner pipe and its composite action with the existing old pipe needs to be evaluated. The liner pipe should be able to withstand grouting pressures during construction in addition to the pull or push force in the axial direction. Also, the capacity of the liner-existing pipe composite to withstand the soil loads, live loads, and groundwater loads need to be checked. In many situations, use of simple ASTM or AWWA equations for structural checks could lead to either under-design or over-design of the liner pipe. Detailed structural analysis calculations using tools such as the finite element method or German ATV Calculation System is always a more cost-effective and reliable means for evaluating such composite structures or any renovation liners.

Continuous Pipes

A nose cone is attached to a butt-fusion welded HDPE for continuous pulling or pushing into a sewer requiring rehabilitation. It is common to clean the line before slip-lining operations start. There are many players in the North American market and the three largest are: Philips Drisco Pipe, Chevron Plexco, and CSR-Polypipe.

Swaged or Rolled Down Pipes

Because HDPE is a soft resin, this permits swaging or rolling down of this pipe to be able to insert into smaller ID old pipe. Use of swaging die, gradually sized down rollers, etc. are used to insert HDPE pipes into old pipe. The design of swaged pipe would be similar to that of slip-liner pipe. Examples of technology in North America already are: Swagelining by British Gas and Titeliner by Insituform's wholly owned subsidiary named United Pipeline Systems.

Pipe Bursting

Pipe bursting falls as a matter of fact within the generic category of slip-lining when up sizing of a system is required. Pipe bursting is generally not used if congestion underground is a question or if the existing pipeline is not of a brittle nature. In smaller sized pipes, pipe bursting becomes a viable tool where, a hydraulically or pneumatically activated cutting head breaks the old pipe and pushes into the native ground making way for the new pipe to take its place up to almost twice its size. This method however, has major noise and vibration problems, on some cases, somewhat uneconomical if too many laterals have to be reconnected. The examples of players in the North American market are: Impipe by CSR, Pipebursting by British Gas, Con-Split by Consolidated Edison Gas Company, Expandit by Clearline Technologies-Miller Pipeline. There are recent attempts using micro-tunneling to break and replace pipe in-situ.

Spiral Wound Pipes

In the spiral wound process, the seam is made using the ribs located at the edges of the extruded strip and the pipe is wound into an insertable form at the bottom of the manhole in small sizes under the names Danby, Ribloc, etc.. In larger sizes, either segmented spiral pipe or continuous spiral pipe is unwound inside the broken pipe while grouting the annulus for strength and stiffness in the Danby process and with or without the need to grout in the Ribloc process. In essence the spiral pipe acts only as a form work to make the grout liner and whatever grouting one does is the main structural help this process can give to the old pipe. Some in situ testing of the effectiveness of the grouting is very essential. In the Ribloc process, steel reinforcing could be included so that the liner could act as a stand-alone structural liner without the need to depend on the grout curtain in the annulus.

Manhole Renovation

Over the past ten years, pipelines have received the greatest attention. Recently, however, manholes, lift stations and house laterals are gaining recognition as a significant part of the sewer system requiring serious rehabilitation. Both cities and industry are concluding that all segments are interrelated and to concentrate on only one segment and ignore the others is both short sighted and inefficient. Those contractors who can provide a full range of solutions for all segments are the ones that are in the best position to profit from this expanding sewer rehabilitation market. Cementitious Liners, Protective Coatings, Structural Repair Systems, Flexible Seals & Full Inserts, and Spot Repairs are the primary methods available for manholes.

Cementitious Liners

Low pressure hand spray application of mortar grade material which are then troweled for compaction and appearance. The major players are: Strong-Seal, Quadex, Master Builders, Shotcrete, Fosroc, Innerguard, Mainstay, Monoform, Permacast, Sewpercoat, SewerGuard, Southwest, Strong-Seal.

Protective Coatings

Various chemical formulations sprayed like paint onto the existing interior after the surface has been specially prepared; some are sand filled and troweled. The major players in the North American market are Raven/Aquatapoxy, Spay-Wall, Spray-Seal, Tenemic, Allseal, Carylon, Cor+Guard, IPA, Parson Lining Systems, Renderoc SP-15, Sauereisen, and Sancon

Structural Repair Systems

Mixture of fiberglass fabric, felt or similar industrial cloth saturated with epoxy, urethanes, polyesters or vinylesters which are triggered to harden by chemical reaction or by elevated temperatures once set in place. Referred to as cured-in-place liners. The major suppliers of this type of technology are, Insituform, Suncoast Environmental's Polytriplex Liner System, A-Lok, CSR Polypipe, Econoliner, Fiberline, Foam Seal, Monoform, PA Glazier, Permaform, Protective Liner System, and Timesaver.

Flexible Seals and Full Inserts

In this type of technology, pipe sections or rigid material such as fiberglass or polyethylene slipped within existing manhole cylinder and then annulus is grouted. The manufacturers in this category are L&F Fiberglass, Associated Fiberglass, Phillips DriscoPipe, Hancor, ADS,

Cretex, Encapseal Safety-T-Seal, Grappler, Infi-Shield, Press-Seal, Rainstopper, StomaSeal, and Wrapid Seal.

Spot Repairs

There are not too many named players in this market and everyone in other methods are doing some work using spot repair technologies with the exception of Hydra-Plug, Avanti, Buchan, and DeNeff.

Lateral Renovation

Again there are two types of problems with laterals: one, water entering the collection system through the laterals and the other, structural repair needs. The industry is starting to use grouting, cured-in-place liners, and pipe bursting combined with new plastic liners for lateral renovation. Many agencies are beginning to require airtesting of the laterals. Logiball lateral packers are used by some agencies for grouting. Cues also provides technology in this market. Even longer lengths are grouted by Video Pipe Services. ABC Services is beginning to install electro-cure liners. CSR offers the product called "IHC Liner." The others in the lateral renovation markets are: for CIPP, Advanced Trench-less Rehabilitation Systems, Insituform, Performance Liner, Roboliner, and Superliner; for HDPE Liner, Trench-less Replacement Systems; for PVC Liner, Exmethod; for robotic lateral connection, Ka-Te, Sika, and American Robotics Corporation; for Grouting, Avanti and DeNeef.

Potable Water Market

In the potable water market, there are two major sub-markets namely: renovation of transmission lines and distribution piping. The technology used in North America for renovation of large diameter transmission pipelines whether they carry raw water or potable water are usually either for steel or ductile iron pipes requiring relining with cement mortar or pre-stressed concrete cylinder pipe that is falling apart in many locations due to premature corrosion of the tendons. The two methods commonly used for pre-stressed concrete cylinder pipe are: adding pre-stressing high tension tendons and shotcreting or slip-lining inside with welded steel pipe. For distribution pipe, we have jet cleaning are epoxy lining using TotoSakanka-Tokan Engineering from Japan, Chemical Cleaning by HERC, cement mortar lining, Paltem and Thermapipe by Insituform, and conventional slip-lining using continuous HDPE.

Gas Market

The gas markets are subdivided into three sectors namely: gathering lines, transmission lines, and feeder main & distribution lines. The renovation market for gathering market almost non-existent due to wells out of their supply about the same time or sooner than the useful life of the gathering pipelines. The main transmission lines are pigged, cleaned on the inside, and re-coated on the outside for corrosion protection in a periodic fashion. The primary technology used where structural or corrosion protection on the side are needed are using slip-lining with HDPE. In North America, therefore the pipe renovation market in the gas industry is mainly in the distribution section. The technology providers are Swagelining by British Gas, Pipe Insertion Method(PIM) by Miller Pipeline Company, ConSplit by Consolidated Edison, U-Liner by CSR, Titeliner(under different name Ultrapipe abroad) by United Pipeline Systems, a wholly owned subsidiary of Insituform, and Paltem by Insituform. Some projects are being done by conventional sliplining companies again using HDPE. In addition, Gas Research Institute(GRI) plans to tests a number of new systems for possible adoption among their member Gas Companies and systems considered are those like Reverse Lining Method, High Speed Internal Resin Lining Technique(HIT), and No Excavation Resin Lining Technique(Next) from Tokyo Gas Company, British Gas, AMEX GmbH using epoxy resins, flexible woven liners, and polyethylene liners.

Crude Oil Market

In the crude oil market, the primary renovation work is on the outside of the pipe. The pipe is taken out of the ground and re-coated to ensure that the corrosion protection is adequate and sometimes new cathodic protection systems are installed. On the side of the pipe, the pipe is cleaned, scrubbed, and scraped in a preventive maintenance manner, using pigs to improve the flow of oil. Any damaged areas are replaced using either a patch of steel tubing welded on both ends or an outside sleeve is used to reinforce the structural capability of the pipe. Robots are used to grind down the welds or corroded sections of the pipeline and new coatings are applied on the inside. U-Liner from CSR has been tried on an experimental basis for relining crude oil pipelines.

Industrial Market

Industrial plants generate two types of effluents namely, municipal wastes by those working at the plant and industrial wastes generated as part of the production process. The treatment systems on site and the collection systems they use inside the industrial compounds require renovation work just like municipal waste systems run by Cities and Counties. The only major difference in a market study such as this is that public agencies are funded mostly by local taxes and are under higher level of scrutiny to generate the renovation market volume. It is not that there is no need to renovate pipelines in the industrial sector. It is simply that the funds have to come from annual operating budgets of private companies and the priority for renovating sewers in the industrial sector normally is of lower importance than any other expenditures within the company. Therefore, the market is pursued by very few players. Paltem, Titeliner, and Pressure CIPP systems offered by Insituform have done some work in this market along with U-Liner from CSR, and conventional slip-lining with HDPE.

Market Volumes in Pipe Renovation

Municipal Sewer Market

There are multiple sources of information for Municipal Sewer Market namely: Associated Construction Publications(ACP), U.S. EPA Reports to the Congress on Needs Assessment, Association of Metropolitan Sewer Agencies, and articles in the public domain. All these sources were searched, to the extent possible given the time limitations, and the following are some market projections for sewer renovation works in the coming years in the US. The Constructor magazine indicates that there are some 800,000 miles of corroded and leaking sewer pipes in the US and some 25 % of these pipes need immediate replacement or renovation. With some 3% of the existing system of pipe networks being added on an annual basis. The United States has over 18,000 sewerage collection systems serving 200 million people or about 75% of its total population. These collection systems have more than 20 million manholes of which 4,000,000 are fifty (50) years old or older and another 5,000,000 manholes are thirty (30) to fifty (50) years old. EPA estimates that fully 50% of the former and 30% of the latter (approximately 3.5 million combined) are suffering from serious structural decay and are in need of immediate replacement. Additionally, there are another 11,000,000 manholes that require some lesser degree of rehabilitation in order to restore them to a serviceable condition or to correct specific minor repairs. The potential revenues for manhole rehabilitation as in most of the sewer infrastructure market are not determined by the size of the need but rather by the amount of the funding that is available. The need far outweighs any historical expenditures and funding commitments would have to increase substantially to even begin to have an impact on this market. Current annual expenditures for manhole rehabilitation in the USA is $100 million dollars. A portion of this amount is publicly bid and a portion is spent under annual maintenance budgets through independent contractors or through in-house personnel at city and district levels. The amount of funding committed to any particular rehabilitation method is not proportional to the need within that category since costs can vary significantly from one method to another. More specifically, costs for structural replacement such as with PERMAFORM are significantly higher than costs' for moderate repairs. In units alone, this results in fewer replacements than those which are repaired. About 51% of all manholes (72% of those needing repair) require some minor rehabilitation, but this segment receives about 80-85% of the available revenues. Estimates show that while $50 to $60 million is spent annually on minor repairs another $ 40 to $ 50 million is spent on structural replacement of manholes and lift stations.

Potable Water Market

Every four years, U.S. EPA prepares a report to the Congress advising the Congress on the Infrastructure funding needs for Drinking Water in the United States. Last such report was compiled in 1996 and released to the public in January 1997. The total 20-year need for Transmission and Distribution of Drinking Water involving mostly pipeline work is about $ 77 billion to avoid breaks. There are over 500,000 miles of distribution piping in the potable water market and it is expected that the water market will spend some $ 3 billion per year on the water supply network for the next 20 years to meet the requirements of US EPA's Clean Water Act and Safe Drinking Water Act.

Gas Market

Transmission Pipelines

There are 531, 000 miles of transmission pipelines in the U.S. and some 240,000 miles total are for gas. The age of these pipelines are as follows: 27 % over 40 yrs or older, 24 % in the range of 26 to 40 yrs of age, the remaining 49 % are under 26 yrs of age. Sizewise, some 19 % in 6 inch in diameter, 21 % are 8 inch, 25 % in the range of 10-14 inch, 8 % in 16-18 inch, 8 % in 20-24 inch, and 19 % are 26 inch and larger. According to Underground Construction magazine, in 1996 alone 2,800 miles of gas transmission pipelines were built in the US at a cost of $ 3 billion. Again in 1998, 3,300 miles at a cost of $ 3.7 billion would be constructed. Industry estimates are that at least 10 % or as high as 30 % of this money is spent on renewal of existing transmission pipelines to meet US DOT safety regulations. This will put the gas transmission market volume to about $ 300 to 900 million.

Distribution Pipelines

There are over 800,000 miles of gas distribution piping in America according to a GRI Study completed in May 1997. Pacific Gas and Electric company alone has plans to spend some $ 1.6 billion in the next 15 years for distribution pipe renewal either by renovation or replacement for their network of 5,000 miles of transmission lines of size 16 to 30 inch and pressure of 150 to 450 psig, and 33,000 miles of distribution mainlines. Overall, 3 % of metal main piping or some 3,000 miles are renewed annually.

Crude Oil and Product Market

Some 185,000 miles are for crude oil transmission and 106,000 miles are for transmission of chemicals and other industrial products. The age of these pipelines are as follows: 27 % over 40 yrs or older, 24 % in the range of 26 to 40 yrs of age, the remaining 49 % are under 26 yrs of age. According to Underground Construction magazine, in 1996, crude oil/chemicals/products segment of the market saw 1,500 miles of new pipeline construction for a total cost of $ 2 billion and in 1998, only about 400 miles would be constructed for a total of $ 460 million. Petroleum industry estimates 10 to 30 % of this budget for pipeline renovation and this would be about $ 300 million per year.

Industrial Market

No doubt there is a large need for pipeline renovation in the industrial sector but the private companies focus their attention on the end product, which is a chemical for Rohm and Hass, Photofilm from Kodak, and a ·pentium chip from Intel. Sewers carrying toxic fluids if they leak, they leak and companies delay doing repair work. In the municipal market, the end product for the City Mayor is her or his ability to have a working sewer that removes wastes from his people and his commercial clients. Because of this major difference, the need far exceeds the money available for industrial sewers while, according to AMSA survey in 1996, most large public agencies feel that they have adequate funding for meeting their critical needs in sewer renovation.

Cost of Renovation Processes

The cost of renovation process is affected by many factors namely, production rate, environment, location, brevity of the engineering specifications, etc. For example,

Production Rate: length of the pipeline, is the project continuous or random lines to be done at several locations, number of house connections or laterals to be made, amount of cleaning needed, availability of qualified personnel.
Environment: amount of bypassing needed, competition, the level of construction risk, weather.
Location: traffic control, distance from the main office of the contractor.
Specifications: did the contractor provide good input to the development of the specifications on materials and methods and is there a prior-established relationship between contractors' staff and the city engineering staff.

Note: The author has a large database of unit costs of various technologies. And these are updated on a monthly basis as they change with market conditions. However, because of the rapidly changing prices in this market, printing specific numbers would be of no value to the readers, due to such data becoming out of date by the time the reader gets to this paper.

What Market Entry Barriers Are There in America?

There are numerous barriers in America toward either a new material or technology in the pipeline industry entering the market place. To name a few in a summary form:

1. has the new material or technology been codified into ASTM, AWWA, ASME, AGA, PPI, GRI, API, AASHTO, AGC, APWA, etc, where task groups are at work developing new or updating existing consensus standards?
2. has the new material or technology been used in sufficient number of demonstration projects to produce long enough a track record?
3. has the new material or technology been tested by independent laboratories for validation of the vendors' claims?
4. has the new material or technology faced and passed the lengthy scrutiny of local new material & methods committees' protocols?
5. has the material or technology vendor working with qualified installers or contractors who are either licensed or pre-qualified to perform work within the jurisdiction of the agency or private company?
6. Does the new material or technology have any resemblance what so ever to another material or technology that had been tried in the prior years or decades and found to be problematic?
7. And most importantly, does the new material or technology vendor aware of the capabilities and limitations of their offering in the market place?

Future Outlook

1. We have tried pre-stressed concrete and we do not like it due to its problems in certain site conditions. For gravity sanitary sewers, we have had major problems with unlined concrete pipe, and if U.S. EPA were ever to require zero-leakage requirements, then concrete pipe will become even more of a loser. Welded steel, we have been in love with at least 100 years ago, 50 years ago, 10 years ago, and even now. So, we probably will continue to love welded steel for most pipeline market segments because it is the most cost effective pipeline material, if it is properly engineered. Then how about plastics. Indeed, we are very committed to plastics for new open cut pipelines, new trench-less pipelines, and for pipe renovation. We will continue to make most of them in PVC and more and more in HDPE, MDPE, and thermoset resins.
2. Trench-less pipeline renovation business in America has matured significantly in the past 10 years. Although more cities are willing to consider this construction option, only 25 % of the pipe renovation work needed is done using trench-less methods. The city officials, engineers, the American public, and the businesses want the least disruption and inconvenience in their daily lives.
3. The pipe renovation business in the municipal market is a highly competitive industry with decreasing prices for some technologies. Ten years ago, we had names such as Insituform and Paltem, as dominant players using similar CIPP technology in two different continents, former growing its business in municipal sewer market, while the latter in gas market. Now, both are forced to compete against numerous others like them or against other pipe renovation technologies.
4. The factors which will continue to provide momentum for the market are:
    - aging underground infrastructure
    - doing more work with less funds
    - protecting the environment
    - increasing congestion in urban and suburban centers
    - faster rate of technology transfer and information
    - privatization of utility companies
    - growing needs for water, energy, and waste handling
5. However, the engineering and contracting environment would be more complex and different. More parts of the underground would be managed and maintained for the public by private "total solution" companies. It is reasonable to think that, many players in business now

will not be around in 10 years due to market consolidation. Without a doubt, there will be losers and winners.
6. The present system of licensing of new technology and the relationship of licensor-licensee will turn sour in many cases in America and will go completely out-of-date due to contractors unwilling to pay license fees when they face stiffer competition in price wars. We already are seeing plenty of broken marriages of this type happening all over America and some all the way into the courtroom. But the market is there in America, exceeding $ 5 billion per year in all sectors of the pipeline infrastructure for renovation and replacement.
7. More emphasis will be placed on evaluating the current condition of the entire pipeline network for cost effective maintenance and renovation strategies. More localized repair work would be the preferred option of private companies managing the network for public entities. The use of intelligent robots for inspection, documentation, and renovation will grow.
8. We no longer are an American economy; we are part of a global economy, where many parts of the world are growing even faster then we do. So, any of you who want to be part of the American Dream has to think in terms of this global economy, but provide tremendous flexibility and decision-making agility to your local centers of business activity, so that your business leaders could act locally fast enough to react to rapidly changing demands of the American(oops, Global) customers. This could only be done when you have the best possible technology, the best people, the financial resources, and a strategic plan with a long term vision to serve as the road map to make it happen in America and the World.

Miami Beach Infiltration/Inflow Reduction Program—
A Project that Pays for Itself

Russell Barnes, P.E.[1]

INTRODUCTION

Ahhh, Miami Beach—18.4 km^2 of sparkling blue waters, soft, sandy beaches, and a cultural wonderland. With the refurbishment of South Beach's Art-Deco District came an economic boom that has kept the City on the fast track. With this incredible economic boom came people—lots of them. Unfortunately, the City's aging sanitary sewer infrastructure was ill-equipped to handle the demands imposed by continued growth.

In 1993, because of continuing maintenance problems experienced by the sanitary sewer department, the City of Miami Beach initiated its sanitary sewer rehabilitation program. The problems included collapsed lines, blockages, and overall unacceptable service to the residents and tourists in Miami Beach.
In addition, the City buys potable water and sanitary sewer treatment from Miami-Dade County, and the volume of water purchased compared to the volume of sanitary sewage that was treated was also a concern. At the time the rehabilitation program started, the City was purchasing several million liters *more* of sanitary sewer treatment than it was of potable water—an indication of a serious infiltration problem.

BACKGROUND
The City of Miami Beach has more than 228,600 m (750,000 linear feet) of sanitary sewer line, most of it made of clay. As this clay sanitary sewer pipe ages, cracks and separation of pipe segments allow infiltration and inflow (I/I) to enter the sanitary sewer system. Miami Beach is located on a barrier island and the groundwater elevation is therefore influenced by the tides. The groundwater

---

[1]Project Manager, Kimley-Horn and Associates, Inc., 5100 NW 33rd Avenue, Suite 157, Ft. Lauderdale, FL 33309

fluctuates between 0.6 and 0.9 m (2 and 3 feet) daily, bringing a majority of the gravity sanitary sewer system below the groundwater level, thus significantly increasing the opportunity for I/I to enter the system. This, coupled with the majority of the system being over 50 years old, was a key cause of the infiltration.

The Environmental Protection Agency had also lodged a Consent Decree against Miami-Dade County, stipulating that I/I must be addressed by each high volume municipal customer within the county. I/I levels must be reduced to below 5,000 GPD/IDM (gallons per day/inch diameter mile). Recognizing the high cost of treating I/I and wanting to comply with the Consent Decree, the City of Miami Beach hired Kimley-Horn and Associates, Inc. to develop a sanitary sewer I/I reduction program.

The five-phase approach began with broad program development and proceeded with successively narrowing targets of further study to identify and ultimately construct only those portions of the sanitary sewer system that would provide substantial, cost-effective I/I elimination—meaning less cost to repair the system than continuing to treat the I/I. The design team's approach to address the greatest I/I reductions as early in the program as possible resulted in immediate savings in sanitary sewer services and higher long-term savings for the City.

This paper details the unique I/I program developed for Miami Beach, including discussions on developing the program, prioritizing the work, analyzing the data, choosing a rehabilitation method, designing a solution, and monitoring the construction effort.

Developing the Program
The first element of the rehabilitation program assessed the overall system based on existing information. Sanitary sewer flows were compared to water consumption, time, tidal effects, and rainfall influences. The entire system was divided into basins based on pump station service areas. The basins were then prioritized based on the volume of infiltration. An overall program budget was also developed in the early stages of the program.

Prioritizing the Work
Phase 2 of the program involved measuring sewage flow on a basin-wide basis and isolating 378 clusters of sanitary sewer pipe to quantify sewage flows for individual portions of the sanitary system. Each cluster contained approximately 914 to 1,524 m (3,000 to 5,000 linear feet) of sanitary sewer pipe. Late night flows were established for each cluster using hand-held weirs and these flows were evaluated for infiltration by subtracting the estimated volume of actual usage based on land use in the area. This phase prioritized the clusters by volume of infiltration with the assumption that the clusters in the worst physical condition

would have the largest volume of infiltration. Another factor in prioritizing the clusters was other construction work already scheduled. If the streets in a particular area were already scheduled to be opened up for other construction work, then these areas were given a high priority so that the sewer rehabilitation work could be completed at the same time, thereby saving the City time and money by not duplicating work.

A cost–benefit analysis was performed on every cluster to determine if it was cost effective to further investigate (less expensive to repair the system than to continue to treat the I/I). The cost–benefit analysis was based on the present worth cost of treating infiltration over the 20-year pay-off period of the bond funds required for the project. The treatment costs were conservatively estimated to increase 3% per year. The average treatment costs (two years since the beginning of the project) were actually $1.65 per 3,785.4 L (1000 gallons), which is an increase of approximately 20%. The cost analysis resulted in a present worth of 20 years of treatment cost of $2.24 L per day (LPD) [$8.47 per gallon per day (GPD)]. This present worth analysis provided the justification to continue full field investigations on 171 of the 378 clusters and limited investigation on 158 clusters. The analysis also determined that it was not cost effective to further investigate 49 clusters. Data from this phase of the study indicated that approximately 24.6 million LPD (6.5 million GPD) of infiltration could be removed cost-effectively from the sanitary sewer system.

Analyzing the Clusters
Cluster studies involved the actual television, smoke, visual, and dye tests conducted on the clusters identified in Phase 2 to have significant infiltration. Each preceding phase of the program considered the cost/benefit ratio of the subsequent step. Phase 3 identified pipe and manhole defects, system cross-connections, and other sources of non-use related flows entering the system. These sources were identified by means of television inspection, smoke testing, and dye testing of the line segments, as well as visual inspection of the manholes. Each defect observed during the inspections was assigned an estimated amount of infiltration. Only those sources that were cost effective were recommended for rehabilitation. Estimated rehabilitation costs were divided by the amount of infiltration observed. This computation provided a rehabilitation cost per liter of removable infiltration. This cost was compared with the benefit value of $2.24/LPD ($8.47/GPD) determined in the Phase 2 study. If the rehabilitation cost per liter of removable infiltration was less than the $2.24/LPD ($8.47/GPD) value benefit, then the individual sewer segment was considered cost effective for rehabilitation and was moved into the design phase.

During this element of the program, the engineering team maintained close coordination with the City staff. When voids in the system such as missing sections of pipe, large holes, etc., were identified by the consultant, the City was immediately contacted and could correct the problem areas quickly.

Economic feasibility was an integral part of the Miami Beach I/I program, as it focused on rehabilitating only those portions of the sanitary sewer system that would be cost effective. The City owns and maintains approximately 228,600 m (750,000 linear feet) of sanitary sewer line and 2,700 manholes. During the Phase 2 assessment, approximately 42,672 m (140,000 linear feet) of sewer line were eliminated from further consideration because continued study would cost more than continued treatment of the I/I. Nearly 39,624 m (130,000 linear feet) more were eliminated from rehabilitation during the field investigations phase of the project because repair would cost more than continued treatment of the I/I. The following is a summary of the field investigation element of the program:

- 148,623 m (487,608 linear feet) of sanitary sewer investigated
- 94,511 m (310,075 linear feet) recommended for rehabilitation by lining
- 15,371 m (50,428 linear feet) recommended for rehabilitation by alternative methods (i.e., spot repairs, joint grouting)
- 2,197 manholes inspected
- 2,125 manholes recommended for rehabilitation

Choosing a Rehabilitation Method
Several rehabilitation options are available for this type of work, including lining, sliplining, sani-pour rehabilitation, joint grouting, and line replacement. Because of traffic densities in the area, as well as the fact that the majority of the gravity sewer system is 0.45-m (18-inch) pipe and smaller, trenchless technology lining (meaning little or no digging) was the method chosen for Miami Beach.

Two main methods are used for lining sanitary sewers: cured-in-place (CIP) and fold-to-form (FTF). Both methods were evaluated based on performance, availability, and cost, and both were used on this project. CIP technology was the preferred method because of its ability to cure in the shape of the host pipe and therefore reduce the ability of infiltration to track between the host pipe and the liner. Reconnecting lateral service connections was also easier with the CIP method because the liner "dimples" at service connections, which allowed the service connection to be visible from inside the line. However, the FTF method was considerably less expensive for 0.20-m and 0.25-m (8-inch and 10-inch) lines. The considerable cost savings was a definite benefit of FTF technology. Insituform Technologies was the lowest bidder for this program.

Designing the Fix
Plans and specifications preparation began immediately for those priority clusters determined to be cost effective in accordance with the Phase 3 analysis. Because

corrections of substantial I/I sources would begin to pay for themselves as soon as they were constructed, this fourth phase was initiated for each identified cluster of rehabilitation prior to completion of the Phase 3 analysis of other clusters. The extensive effort that went into the plans and specifications preparation resulted in unit prices that were the lowest in the country for the recommended rehabilitation method. The contract unit prices for the City's contract were nearly 25% lower than the Miami-Dade County contract unit prices.

Following the Phase 4 design process, rehabilitation and reconstruction activities began. Again, some clusters were under construction or rehabilitation while lower priority clusters were under Phase 3 analysis and/or Phase 4 design. This flexible approach addressed the greatest I/I reductions as early in the program as possible, which resulted in more long-term savings for the City.

Monitoring the Work
While continuing to rehabilitate the sanitary sewer system in Miami Beach, the consultant developed a database that allowed detailed tracking of the program. The database contained information (length, depth, cluster number, amount of observed infiltration, recommended rehabilitation method, probable cost of rehabilitation, etc.) for each of the thousands of sewer lines and manholes in the City. The information was linked so that complete status reports could be generated very quickly—also allowing the consultant to track progress and status of individual lines or the project as a whole—all in just a few minutes. The database proved to be a very effective tool for tracking and scheduling the inspection, design, and rehabilitation of all lines and manholes throughout the project area.

A Program That Works
As a result of the pipe rehabilitations to date, daily sewage flows from the City have decreased by approximately 18.9 million LPD (5 million GPD), saving the City more than $3 million in direct treatment costs annually. The reduction in flow volumes also resulted in savings in the operating and maintenance costs of the pump and lift stations located in the City.

Pump station flow records provided by the City have been compared from 1993, 1996, and 1997. To date, approximately 50% of the recommended pipe rehabilitation has been completed, with the greatest sources of I/I being addressed first. This rehabilitation has resulted in eliminating approximately 75% of the estimated 24.6 million LPD (6.5 million GPD) of I/I from the sanitary sewer system.

Pipeline Drainage Discharge Analysis

Timothy M. Smith[1]

ABSTRACT

The unsteady flow condition during draining of pipelines is the topic of this report. A model was developed for determining various parameters including discharge, velocity, volume, and time to drain. Both time and space parameters are used to accurately model the cross-sectional area in terms of head by describing the pipeline in stations and elevations (x and y coordinates) at changes in slope. The model computes head, discharge, and velocity at each time step. The total time to drain the system containing water is also computed. The effect of selecting the number of coordinate points and time intervals of the system was examined. A finer resolution gives results very much closer to the solution of the differential equation. Initial results using a simplified method are compared with the results by using a model established in this paper. When calculating the time to drain, the model developed, or a similar multiple coordinate description of the pipeline is recommended.

INTRODUCTION

Water is expelled at various low points through discharge facilities when emptying a pipeline that traverses an undulating terrain. These facilities, usually referred to as blowoffs, typically have discharge valves, discharge piping, and some form of energy dissipator. The head forces the water column as it is expelled through the blowoffs. Opening of the valve over a period of time can reduce any harmful water hammer

---

Associate Member, American Society of Civil Engineers, Black & Veatch, 9665 Chesapeake Drive, Suite 450, San Diego, CA 92123

caused by transient flow. The unsteady flow condition during draining of pipelines is the topic of this report. Initial results using a simplified method are compared with the results by using a model established in this paper.

BACKGROUND

Research
A review of the literature of similar work on the topic and applicable equations was performed. Several texts [1, 2, 3, 7, 8, 10, 11, 12, 13] describe unsteady flow in closed conduits. However, the focus of these texts was transients caused by stopping flow in a closed conduit. A significant amount of work has been discussed in the literature on this topic. For the problem discussed in this report, a very short opening time (approximately 15 seconds) can alleviate any transients caused by opening of the discharge valves. One recent paper described filling pipelines with undulating profiles [5] and presented a model that simulates the velocity, pressure, and length as a water column advances downstream from a supply reservoir and fills an empty pipeline. The focus was on air release at high points, and examination of potential water hammer during filling of a pipeline rather than examination of unsteady flow during pipeline drainage. Although similar in nature, the equations and focus vary from this paper.

Project
The project driving this research was the San Diego County Water Authority's Rancho Penasquitos Pipeline (also known as Pipeline 5 Extension, Phase II). The project is a 15.6 kilometer (9.7 mile), 2.74-meter (9-foot) diameter pipeline with a maximum of pool head of over 213 meters (700 feet). One phase of the project was to design discharge facilities that would adequately protect downstream residential property from significant erosion during pipeline drainage.

Examination of discharge from pipelines during drainage is becoming more prevalent in southwestern United States as population and communities continue to grow and new regulations are established. Discharge did little damage to property in uninhabited area, but as communities expanded, property damage has become an issue. Currently, National Pollution Discharge Elimination System (NPDES) regulations are being drafted for water and utility company discharges to water courses. Their primary concern is treatment prior to discharge of chlorinated water to receiving bodies of water. In this case, the utility's primary concern is the damage to residential properties and potential lawsuits.

Equations of Flow
Several equations are available to describe the flow condition. By use of the continuity equation $Q = VA$ for steady flow where Q is flow in cubic meters per second, V is velocity in meters per second, and A is area in square meters, the flow equations can be described as [2, 4, 7]:

PIPELINES IN THE CONSTRUCTED ENVIRONMENT 789

Manning's	$Q = A R^{2/3} S^{1/2} / n$	(1)
Hazen-Williams	$Q = 0.35 A C_h R^{0.63} S^{0.54}$	(2)
Chezy	$Q = CA(RS)^{0.5}$	(3)
Darcy-Weisbach	$Q = A (8g/f)^{0.5} (RS)^{0.5}$	(4)

Where R is the hydraulic radius in meters (R = A/P where P is the wetted perimeter in meters); S is the slope (dimensionless); n is the Manning's friction coefficient, C_h is the Hazen-Williams friction coefficient, C is the Chezy friction coefficient, g is gravity, and f is the roughness coefficient.

Another flow equation for this specific problem is the orifice equation:

Orifice $\qquad Q = C_d A_o (2gH)^{0.5}$ (5)

Where C_d is the coefficient of discharge, A_o is the orifice area, g is gravity, and H is the head.

The Manning's equation was selected for the analysis presented because of its continued use in the water utility industry. The Hazen-Williams formula may yield erroneous results for pipe diameters smaller than 0.05 meters (2 inches) and larger than 1.83 meters (6 feet) [7]. The Chezy formula is used for open channel flow. The Darcy-Weisbach equation requires the use of the Moody diagram to determine the roughness coefficient "f". Explicit equations have been developed for f and Q [4] but they are complex. The solution to this problem using the Darcy-Weisbach equation is left for further study. The orifice equation alone will not describe the additional friction and minor head losses of the discharge facilities. However, it will be used to equate an equivalent length of pipe to be used in the Manning's equation.

ANALYTICAL MODEL

The governing equation of discharge for a system containing water with varying cross section is [4]:

$$Q \, dt = -A_t \, dh \qquad (5)$$

Where Q is the discharge, dt is the time derivative, A_t is the cross-sectional area, and dh is the head derivative. An expression for the cross-sectional area, A_t, as a function of h must be determined. The time, t, to empty the system containing water from height h_1 to a lower height h_2 is:

$$t = \int_{h_1}^{h_2} A_t \, dh / Q \qquad (6)$$

Now the discharge equation (1) is operated on for pipe full flow resulting in:

$$Q = 0.3117 \, D^{8/3} \, S^{1/2} / n \tag{7}$$

Where D is the pipe diameter in meters. The slope, S, can be described as the slope of the hydraulic grade line for pipe flow as the change of head over the length, L:

$$S = \Delta h / L \tag{8}$$

Combining equation (6), (7) and (8) and using the law of derivatives, we obtain:

$$t = A_t \, \Delta h \, n / 0.3117 \, D^{8/3} \, (\Delta h / L)^{1/2} \tag{9}$$

This equation will determine the time to drain the system containing water. The model developed uses a system of stations (x-coordinates) and elevations (y-coordinates) to express A_t in terms of head. The discharge facilities head losses can be converted to equivalent length of pipe for use as the length term. Figure 1 describes the system.

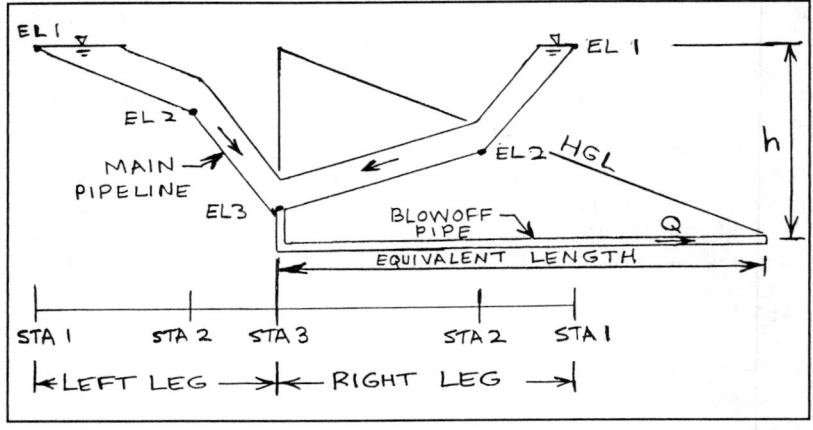

Figure 1 Definition Sketch of the System

The initial conditions of head, discharge, and velocity are also important for design. Initial head is simply the pool head. Pool head refers to the static water surface elevation in the pipeline if upstream flow is stopped and downstream flow is allowed to drain by gravity. A pool of water forms between topographic features. Initial discharge can be determined using Equation (7). Initial velocity can be determined using the continuity equation Q = V A. In addition, head, discharge, and velocity can be determined for each time step during drainage.

Instantaneous values of discharge are expressed in the same manner as for steady flow. This is not strictly correct, since for unsteady flow the energy equation should also include an acceleration head [3]. In cases where the head does not vary rapidly, no appreciable error will result if the acceleration term is disregarded [3].

APPLICATION

Numerical Experiments

An initial sample problem is presented herein to describe the model application. A single static pool is examined with identical left and right legs of 914 meters (3000 feet) length and initial head of 61 meters (200 feet). Discharge piping size is selected to limit the initial velocity to 10.7 meters (35 feet) per second, the limit for ball valves as described in AWWA C507-91. Manufacturers indicate a similar limit for plug valves and sleeve valves for intermittent use. A Manning's n of 0.013 is selected for steel pipe.

The effect of selecting the number of coordinate points of the system was examined. A station interval, or Δx, is selected. This will determine the number of points entered into the model. It also defines the resolution of the problem. Figure 2 shows the results of using a Δx of 914, 457, 308, 152, and 76 meters. A finer resolution gives results very much closer to the solution of the differential equation. The more coordinate points required, the more data intensive the problem will become. However, if the slope of the pipeline changes along a given leg, the system can only be accurately described by entering coordinate points at changes in slope. This is due to the fact the A_t as a function h must be described to obtain the time to drain in equation (9).

The effect of selecting the time interval for the system is examined. A time interval, or Δt, is selected. The model establishes coordinate points for each time interval between points entered into the model. It also defines the resolution of the problem. Figure 3 shows the results of using a Δt of 3, 1, 0.5, 0.25, 0.10 and 0.05 hours. A finer resolution gives results very much closer to the solution of the differential equation. Coordinate points must be entered at changes of slope or the model will assume a straight line between points and truncation errors will result. For a simplified three point system, a minimum of data entry can be accomplished be using Δt to discretize the physical system.

For both station interval (Δx) and time interval (Δt), a broader resolution results in lack of convergence due to truncation errors but good stability. Convergence refers to the ability of the scheme to reproduce the terms of the differential equation with sufficient accuracy [6]. Stability refers to the ability of the numerical scheme to march in time without generating unbounded error growth [6]. For both Δx and Δt, a finer resolution results in convergence but lack of stability due to round off errors.

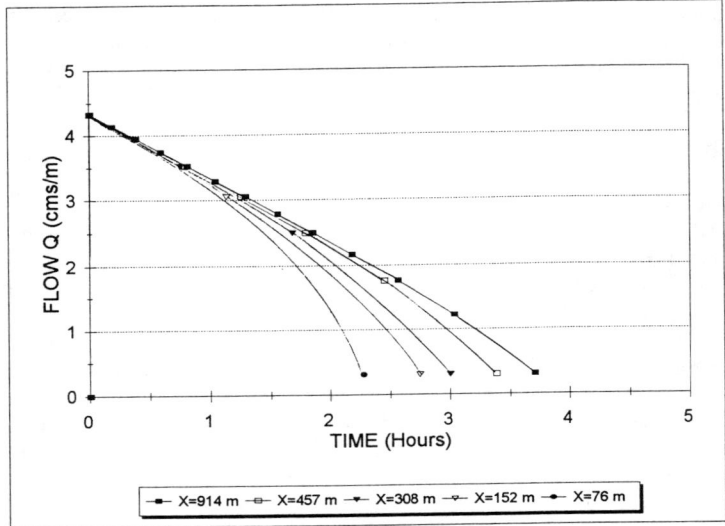

Figure 2 Variation in Delta X

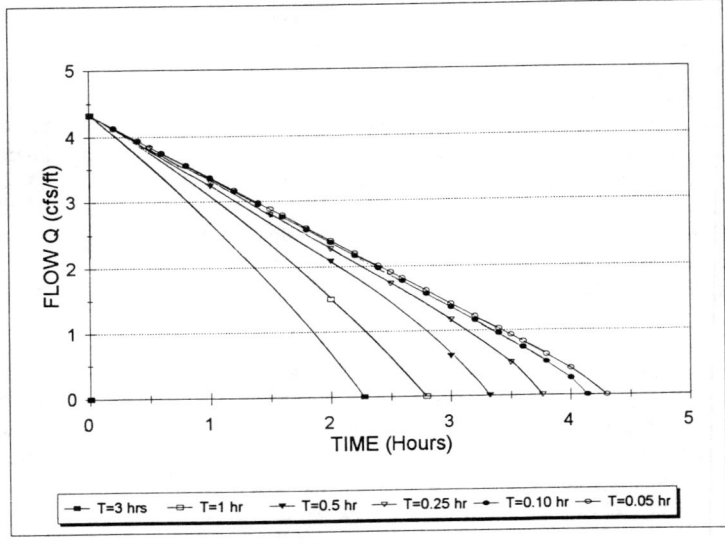

Figure 3 Variation in Delta T

Verification of Physical Problem

The San Diego County Water Authority is currently constructing the Rancho Penasquitos Pipeline, a 15.6 kilometer (9.7 mile) long, 2.74 meter (9-foot) diameter pipeline. One phase of the project was to design discharge facilities that would adequately protect downstream properties from significant erosion during pipeline drainage. There are fifteen blowoff facilities at low points along the pipeline. About half the blowoffs are in developed areas and the other half are in currently undeveloped areas. The maximum pool head is over 213 meters (700 feet). Initial discharge, velocities, head, and drain times calculated using a simplified method are presented in Table 1. Effective pool head refers to the pool head less head broken with a throttling valve. For blowoffs B4 and B15, the pool heads are 222 meters (728 feet) and 163.5 meters (536.5 feet) respectively, however, sleeve valves are used to break the majority

Table 1 Simplified Method Results

Blowoff	Effective Pool Head (meters (ft))	Equivalent Length (meters (ft))	Initial Discharge (cms (cfs))	Initial Velocity (mps (fps))	Time to Drain (Hours)
B1	69.8 (229)	167.6 (550)	1.40 (49.51)	10.81 (35.46)	2.32
B2	29.9 (98)	62.2 (204)	0.43 (15.19)	8.49 (27.84)	4.12
B3	38.1 (125)	62.2 (204)	0.49 (17.15)	9.59 (31.45)	6.49
B4	22.6 (74)	55.5 (182)	1.39 (48.93)	10.68 (35.04)	6.67
B5	11.3 (37)	62.2 (204)	0.26 (9.33)	5.22 (17.11)	6.14
B6	42.7 (140)	167.6 (550)	0.73 (25.85)	10.03 (32.91)	4.57
B7	69.8 (229)	41.8 (137)	1.40 (49.51)	10.81 (35.46)	2.62
B8	9.8 (32)	41.8 (137)	0.17 (5.84)	5.10 (16.73)	4.95
B9	21.6 (71)	41.8 (137)	0.25 (8.70)	7.60 (24.92)	5.9
B10	39.6 (130)	41.8 (137)	0.33 (11.77)	10.28 (33.73)	6.09
B11	36.6 (120)	41.8 (137)	0.32 (11.31)	9.88 (32.40)	5.01
B12	19.5 (64)	41.8 (137)	0.24 (8.26)	7.21 (23.66)	5.85
B13	20.1 (66)	41.8 (137)	0.24 (8.39)	7.32 (24.03)	4.95
B14	11.0 (36)	62.2 (204)	0.26 (9.20)	5.14 (16.88)	6.2
B15	23.0 (75.5)	55.5 (182)	1.40 (49.42)	10.79 (35.40)	4.35

of the head. Drain times were calculated based on the average head. These results will be compared with the results using the model established in this paper.

A model was developed for determining various parameters including discharge, velocity, volume, and time to drain. A computer model was written in Fortran for the simulation. Both time and space parameters to accurately model the cross-sectional area (A_t) in terms of head by describing the pipeline in stations and elevations (x and y coordinates) at changes in slope. The model computes head, discharge, and velocity at each time step. The total time to drain the system containing water is also computed.

The program was run for fifteen blowoffs on the Rancho Penasquitos Pipeline and compared with a simplified method using only three coordinate points to describe the pipeline. Initial values of discharge, velocity, and head were the same for the simplified method and for the model because initial conditions are based on the pool head. For initial values of discharge, velocity, and head refer to Table 1. The time to drain, however, varied significantly. A comparison of the time to drain computed by the simplified method and the model are shown in Table 2. The model used a $\Delta t = 0.05$ hours to increase convergence. The results indicate that the simplified method calculated a time to drain of approximately 32 percent less than the model. This amount of error is not considered satisfactory because operations and maintenance personnel must operate and monitor the blowoff facilities during drainage. The length of the planned total shut down time is also based on the calculated time to drain. When calculating the time to drain, the model developed or a similar multiple coordinate description of the pipeline is recommended. In some cases (B2 and B3), the time to drain calculated by the simplified method was more than 50 percent less than that calculated using the model. For blowoffs B2 and B3 this results in an additional time to drain of 6.28 and 8.99 hours respectively. The higher error is due to the shape of the draining pipeline with a majority of the volume at lower heads.

Table 2 Time to Drain Comparison			
Blowoff	Simplified Method (Hours)	Model (Hours)	Error
B1	2.32	3.34	-31%
B2	4.12	10.40	-60%
B3	6.49	15.48	-58%
B4	6.67	8.86	-25%
B5	6.14	8.77	-30%
B6	4.57	6.19	-26%

Table 2 Time to Drain Comparison (Continued)			
Blowoff	Simplified Method (Hours)	Model (Hours)	Error
B7	2.62	3.16	-17%
B8	4.95	6.70	-26%
B9	5.90	7.26	-19%
B10	6.09	9.51	-36%
B11	5.01	7.21	-31%
B12	5.85	8.25	-29%
B13	4.95	6.85	-28%
B14	6.20	8.45	-27%
B15	4.35	5.85	-26%
		AVERAGE	-32%

Hydrology
Currently, the discharge velocities are reduced to below the maximum permissible velocity of the downstream drainage channel based on existing soil conditions. Hydrologic routing of the time-discharge curve to the downstream drainage channel would assist in determining the erosion and flooding effects in the downstream drainage channel. Routing of the flood wave is left for an area of additional research.

SUMMARY AND CONCLUSION
The unsteady flow condition during drainage of pipelines was analyzed. In emptying a pipeline, the head drives the water column as it is expelled at low points. A model was developed for determining various parameters including discharge, velocity, and time to drain. The Manning's equation was selected for the analysis presented because of its continued use in the water utility industry. Various values of space and time were examined to study the sensitivity of the model to change in model parameters. Values of discharge, velocity, volume, and time to drain were computed for a sample problem. The model was then used to compare with a simplified model previously developed to determine the accuracy of the simplified method. The error of the simplified method is not satisfactory considering the manpower and planning effort of a pipeline drainage event.

The following recommendations are offered for further research:
1. The solution of this problem using the Darcy-Weisbach equation.
2. Verification of actual times to drain (as well as other parameters) during a pipeline shutdown and drainage event. Laboratory experiments using physical models could also assist in the verification of the model presented.
3. Hydrologic routing of the time-discharge curve to the downstream drainage channel to determine erosion and flooding effects.
4. Artificial flooding has been accomplished as seen in the recent artificial flooding of the Colorado River [9]. If the drainage parameters are known at the blowoff facilities, the discharge at blowoff facilities could be controlled and would be beneficial for planning artificial flooding of rivers.

REFERENCES

[1] Benedict, R.P., *Fundamentals of Pipe Flow*, New York: John Wiley & Sons, Inc., 1980.
[2] Chow, V.T., *Open Channel Hydraulics*, New York: McGraw-Hill, Inc., 1959.
[3] Daugherly, R.L. and Franzini, J.B., *Fluid Mechanics with Engineering Applications, Seventh Edition*, New York: McGraw-Hill, Inc., 1977.
[4] Lindeburg, M.R., *Civil Engineering Reference Manual, Sixth Edition*, Belmont, CA: Professional Publications, Inc., 1989.
[5] Liou, C.P. and Hunt, W.A., "Filling of Pipelines with Undulating Profiles", *Hydraulics of Pipelines*, Proceedings of International Conference held at Phoenix, AZ on June 12-15, 1994, New York: American Society of Civil Engineers, 1994.
[6] Ponce, V.M., Indlekofer, H., and Simons, D.B., "Convergence of Four-Point Implicit Water Wave Models", Journal of Hydraulics Division, July 1978, pp 947-958.
[7] Roberson, J.A., Cassidy, J.J., and Chaudhry, M.H., *Hydraulic Engineering*, Boston: Houghton Mufflin Company, 1988.
[8] Roberson, J.A. and Crowe, C.T., *Engineering Fluid Mechanics*, Third Edition, Boston: Houghton Mufflin Company, 1985.
[9] San Diego Union-Tribune, "Artificial Grand Canyon flood is declared a rousing success", April 13, 1996.
[10] Stephenson, D., *Pipeline Design for Water Engineers*, Amsterdam: Elsevier Scientific Publishing Company, 1981.
[11] Tullis, J.P., *Control of Flow in Closed Conduits*, Proceedings of the Institute Held at Colorado State University on August 9-14, 1970, Fort Collins, CO, 1971.
[12] Watters, G.Z., *Modern Analysis and Control of Unsteady Flow in Pipelines*, Ann Arbor, MI: Ann Arbor Science Publishers, Inc., 1979.
[13] Wylie, E.B. and Streeter, V.L., *Fluid Transients*, New York: McGraw-Hill, Inc, 1978.

Design and Construction of the
Sweetwater Reservoir Urban Runoff Diversion System 48-Inch Pipeline

Richard Bottcher[1], Claud Seal[2], Tucker[3], Jim Smyth[4]

Abstract

Degradation of water quality in the Sweetwater Reservoir has been occurring as a result of uncontrolled inflow of poor quality runoff from within the watershed. One of the solutions to this issue was to construct the Sweetwater Urban Runoff Diversion System (URDS), a diversion system to intercept and divert poor quality runoff upstream of the reservoir and convey it downstream of the reservoir. Phase II of this system (URDS-II) consists of approximately two miles of 48-inch gravity pipeline placed in a highly environmentally sensitive area within rocky terrain of the north embankment of the Sweetwater River. This paper describes the numerous challenges that were encountered in the planning, design and construction of this pipeline.

Introduction

The Sweetwater Authority (Authority), a retail water agency serving approximately 165,000 people in southern San Diego County, owns and operates the Sweetwater Reservoir which stores local and imported water prior to treatment. Degradation of water quality in this reservoir has been occurring as a result of uncontrolled inflow of poor quality runoff resulting from urbanization within the watershed. One of the solutions was to construct a diversion system to

---

[1] Senior Civil Engineer, Boyle Engineering Corporation, 7807 Convoy Ct., San Diego, CA 92111, (619) 268-8080,
[2] Associate Civil Engineer, Boyle Engineering Corporation, 7807 Convoy Ct., San Diego, CA 92111
[3] Construction Manager, Sweetwater Authority, 505 Garrett Ave., Chula Vista, CA 91912-2328, (619) 420-1413
[4] Chief Engineer, Sweetwater Authority, 505 Garrett Ave., Chula Vista, CA 91912-2328, (619) 420-1413

intercept and divert poor quality runoff upstream of the reservoir and convey it downstream of the reservoir.

In 1991, the Authority completed the first phase of its Sweetwater Reservoir Urban Runoff Diversion System (URDS-I). The Phase I system consists of a series of three collection ponds, pipelines, and instrumentation and control systems that automatically divert poor quality water around the reservoir **(see Exhibit 1)**. Three detention ponds also allow settleable solids containing much of the contaminants to settle out before the diverted runoff is conveyed around the reservoir. The runoff is diverted to the Lower Sweetwater River Basin, a relatively shallow alluvial basin with groundwater of similar salinity to the diverted runoff.

The diversion point for the URDS-II system is located at a very narrow portion of the Sweetwater River, approximately two miles upstream of the reservoir. Here, rock outcroppings cause groundwater to surface. A low flow diversion barrier crossing the river diverts the poor quality flow from the river to a 2-mile long, 48-inch pipeline that outlets into a 32 acre-ft detention basin. The system will then connect to the URDS-I to divert the captured flows the last distance around the reservoir. The system will be fully automated with telemetry allowing remote control of all diversion gates and instantaneous sampling of the flows and salinity of flow being diverted.

Exhibit 1: One of the diversion ponds of the Sweetwater Urban Water Diversion System, Phase 1. Sweetwater Reservoir is in the background. The system diverts poor quality urban runoff around the reservoir.

Currently under construction is Phase II of the diversion system (URDS-II). The system is located upstream of the Sweetwater Reservoir in the main stem of the Sweetwater River. It is designed to divert low volume dry season flows dominated by poor quality urban runoff and highly saline surfacing groundwater. The system also offers protection against the potential for sewage spills from upstream sewer pump stations and a sewer/reclamation treatment plant.

Numerous challenges were encountered in the planning, design and construction of the URDS-II pipeline which include:

- **Unusual hydrologic and hydraulic criteria** used in the pipeline sizing and alignment. This included the need to intercept a variety of flow conditions such as "first flush" urban runoff, poor quality dry season flows and potential sewage spills and the need for self cleansing flow velocities for gravity flow in a highly depositional environment.

- **Difficult environmental compliance issues** requiring proactive environmental coordination during the planning, design and construction processes. This included developing an alignment and construction corridor to minimize impact to three federally endangered local species; seasonal construction to avoid breeding and nesting seasons; and coordination with numerous environmental and regulatory agencies.

- **Difficult construction** in the rocky terrain along the north embankment of the Sweetwater River with limited accessibility. This included benching, blasting and trenching in fractured bedrock and construction scheduling to avoid work in the river during the wet season and during the breeding and nesting seasons of federally endangered species.

Planning and Design Issues

Flow Diversion Concept

The concept of urban runoff diversion includes the capture and delivery past the reservoir of "first flush" storm flows that occur immediately after a dry period, and the diversion of low volume dry season stream flows. "First flush" storm flows are contaminated with disproportionately high levels of salts, bacteria, and other contaminants including fertilizers, insecticides, herbicides, rubber tire dust, and petrochemical compounds. Studies conducted for the Authority by the lead author's firm have shown that 90 percent of the contaminant load within the watershed is flushed out by the first 0.5 inches of runoff (requiring approximately 1.4 inches of rainfall in the Sweetwater watershed). Likewise, dry season flows from tributaries adjacent to the reservoir contain salinity on the order of 3,000 mg/l.

Due to the highly depositional environment of the Middle Sweetwater River, the URDS-II gravity pipeline was designed to allow for self-cleansing flow velocities during full-pipe flow conditions. In this section of the river, the maximum obtainable slope was 0.0014 ft/ft. Such a low slope necessitated full time inspection of pipeline bedding, backfill and associated compaction during construction to prevent future subsidence at joints, which could create low points along the pipeline profile.

Design in Harmony with Habitat

The URDS-II pipeline traverses a portion of the Middle Sweetwater River that houses three federally endangered species – the Arroyo Toad, the Least bells Vireo and the California Gnatcatcher and traverses property owned by the U.S. Department of Fish and Wildlife. The listing of these endangered species has greatly altered the original plans for the Urban Runoff Diversion System, Phase II. The initial URDS-II concept included three additional ponds and several additional miles of pipeline into the heart of the Middle Basin of the Sweetwater River (approximately 5 miles upstream of the Sweetwater Reservoir). Because of the environmental impacts of such a concept, the final configuration of the project was changed to include the construction of a low flow barrier for diversion where the river narrows (only 2 miles upstream of the reservoir), thus omitting 3 miles of pipeline and three detention basins. This revision, however, has its consequences in that not all "first flush" urban runoff can be captured from the urbanized Middle Basin. It does, however, enable capture of saline dry season flows and sewage spills along with a limited amount of urban runoff. The design change was a compromise to avoid environmental impacts.

During the design and construction phases of the URDS-II, the design team worked with the biologists and regulatory agencies to minimize disturbances to the sensitive habitat. This included changes in pipeline alignment, minimization of the construction corridor (requiring one-way pipeline construction with staging areas allowing construction traffic to pass), a comprehensive revegetation plan and vegetation removal plan within the construction corridor and inclusion of seasonal construction constraints preventing construction during the breeding and nesting seasons for the endangered species. Exhibit 2 shows the seasonal construction constraints imposed on the contractor.

As a result of early input from regulatory agency staff, the design of URDS-II incorporates a method to enhance the habitat downstream of the diversion point in the Sweetwater River. Through the inclusion of two groundwater wells located at the reservoir, good quality reservoir water is pumped back to the low flow barrier in the Sweetwater River to feed the habitat in the river. The groundwater pipeline also incorporates several take-off points where portable irrigation systems can be connected on the way to the barrier. Currently, flows in the Sweetwater River are intermittent. This system will allow for a continual flow in the river for habitat enhancement. Like the other portions of URDS, the groundwater well operation is fully automated and is connected to the telemetry system to monitor flows. The diverted flows are also monitored for flow rate and salinity levels.

PIPELINES IN THE CONSTRUCTED ENVIRONMENT 801

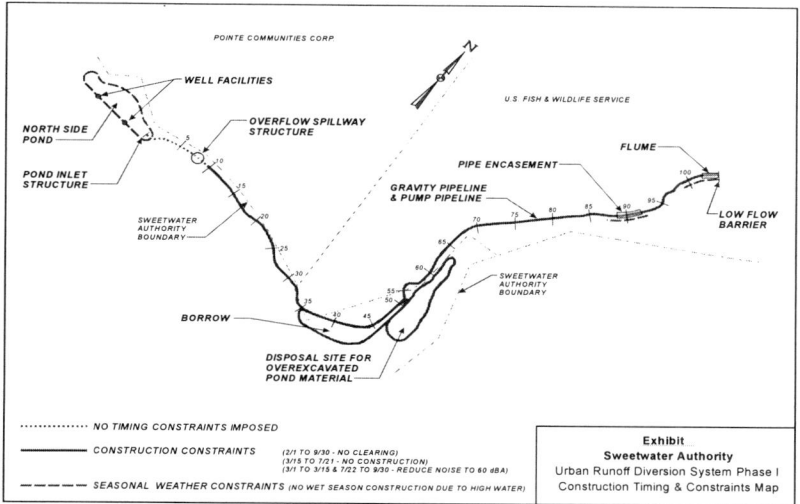

Exhibit
Sweetwater Authority
Urban Runoff Diversion System Phase I
Construction Timing & Constraints Map

Regulatory Agency Involvement

Throughout the planning and design phases of the Urban Runoff Diversion System, the Authority has worked hand in hand with numerous environmental and regulatory agencies to develop an overall scheme to the mutual benefit of the Authority and the environment. Such agencies as the San Diego Regional Water Quality Control Board, the California State Department of Fish and Game, the U.S. Fish and Wildlife Service, the San Diego County Department of Environmental Health, the California State Department of Health Services, the U.S. Army Corps of Engineers, the U.S.E.P.A., the San Diego County Water Authority, the Sierra Club, and the U.S. Bureau of Reclamation have played an integral part in the conceptualization, design and construction of key components of the Authority's system.

Construction Issues

Accessibility

Of the six large reinforced concrete structures required on the URDS-II project, three of them were within the environmentally restricted area shown on Exhibit 2. Pipeline construction required 3800 feet of pioneering, blasting, and bench development in order to gain access to its construction site. At the upstream end of the project, structures allowing diversion of the river's low flows into the 48-inch pipe required an additional 1400 feet of similar work through near vertical rock walls. The only access to the site was at one location, located at the downstream portion of the project. This limited accessibility coupled with the narrow corridor widths along the highly environmentally sensitive portions of the project required the contractor to set up several staging areas along the construction corridor.

Benching and Blasting

Benching was accomplished initially using a Hitachi 450 Excavator. Caterpillar D-9's and D-8's were used after an initial bench was cut to move the larger rocks and cut the bench levels closer to grade. Drilling for blast holes was done by either one or two Ingersoll Rand ECM 370 hydraulic drills. Hole patterns varied in size from 17 to 300 holes, depending upon physical constraints and production needs. Until wet holes were encountered, dry fertilizer grade material mixed with fuel oil was used for the explosive. Toward the end of the blasting when water was encountered in the lower levels of the drill holes, dynamite was used.

The permitted construction corridor, varying from 40 to 80 feet in width, required close control of the charges, firing delays, and direction of explosive energy release. The contractor was subject to permit revocation and possible monetary fines for allowing exploded materials to cross the permit limits, uphill or downhill. Blasting was required for initial bench elevation construction and again for loosening the trench excavation route. Generally, trench patterns were three rows wide, with 4 feet to 6 feet center to center spacing while the development drill holes were usually placed on 6 feet by 6 feet centers. Blasting was conducted in October through December, 1997, and in January, 1998. Blasting, in a horizontal distance of about 5200 feet loosened some 25,000 cubic yards of rock.

Pipe Encasement

Due to the narrowing of the river channel under an existing utility bridge and to the near-vertical face of the bedrock embankment at this location, we designed a 250-ft reinforced concrete encasement (pinned into the embankment) for the 48-inch pipe. Loose rock had to be chipped and firm surfaces exposed with hydraulic hoe-rams mounted on excavators and hand operated air hammers (jack hammers). Over 300 anchor pins, 1-inch diameter by 3-ft long, were epoxy grouted in the holes bored into the rock faces. Five 48-ft long sections of base and upper encasement concrete were alternately placed. Nearly 800 cubic yards of concrete were used in this structure. To finish the exterior, the stream sided walls were formed by molds giving a convincing river run rock appearance. The horizontal surfaces received a stamped flag stone texture. Hardeners containing various pigments were used later to color the "stone." The texturing and colors blend with the elements of the preserve.

48-Inch Pipe Placement

Pipe placement began at the upstream flume structure in December 1997. Initial pipe placement started very slowly due to the discovery of highly fractured rock bedding. The unanticipated degree of fractured rock had resulted from over blasting. In some locations, loose fractured rock to depths of 5 feet was over-excavated and replaced with a screened and sized 8-inch x 2-inch rock bedding material. Due to high runoffs in the adjacent river, the water table was encountered in the bottom of the first several hundred feet of trench. We found that without clean rock (i.e. rock with a minimum particle size of 2-inches), the water would be absorbed into the finer aggregate sizes and would not allow full contact between all the larger rock surfaces (resulting in uneven compaction which would not properly support the 7,000 to 11,000 lb. pieces of pipe). A filter fabric was placed on the surface of the rock bedding followed by 12-inches of well-graded sand compacted to 90% relative compaction. The pipe was then set and aligned.

Delays were encountered when it was found that the special beveled pipe pieces were not cast with the required bevel angles. In some cases, pipe was rejected upon delivery to the job by on-site inspectors due to non-compliance with contract requirements. Quality control of the delivered pipe products remained a problem throughout the project.

Environmental Constraints

Perhaps the biggest surprise, and the area of greatest difficulty to enforce, was construction crew compliance of the environmental permit conditions. Every employee that worked on the project was required to attend an environmental briefing given by the Authority's Resident Biologist. Even with the briefing, infractions were witnessed daily:

- **Trash**: No trash of any kind was to be left on the job. The contractor provided trash receptacles. <u>Result</u>: Old habits are hard to break. Soda cans, candy and sandwich wrappers were seen at every work activity center. Trash items were seen in the pipe trenches prior to final fill and compaction. No one of course knew where these items came from. Having the crews stop work and physically remove the trash did little to deter future trash throwers.

- **Spilled petroleum products**: The USFWS permit was specific about fueling conditions inside and outside the sensitive habitat work areas. Approved sites for refueling were identified. The contract detailed specific requirements for handling spilled fuel and lubricating products. Result: Spilled fuel or petroleum products were observed. The contractor was told to clean it up and dispose of the contaminated soil and the contractor's personnel did so. However, equipment was fueled wherever needed, regardless of whether it was at an approved site. When the contractor was caught, then they would resume fueling at an approved fueling site.

- **Cigarettes**: After nearly catching the fuel storage area on fire with a carelessly tossed cigarette, the General Superintendent placed a job wide "no smoking" policy into effect. Result: In less than one month, smokers were back (when inspectors were gone) as evidenced by cigarettes found throughout the work area.

- **Parking outside the construction corridor**: The corridor boundaries were marked with lath and "hot pink" flagging. Result: Contractor's employees who drove their own vehicles to the work areas often, at least initially, parked and turned their vehicles around in "out-of-bounds" areas. On-site inspectors noted these infractions. The USFWS then mitigated penalties for the infractions against the contractor. The contractor was ultimately penalized because the hourly personnel were not properly informed of the boundary violation impact, or supervisory enforcement was not used.

Conclusion

Numerous challenges were encountered in the planning, design and construction of the URDS-II 48-inch pipeline. Most challenges were due to the environmental sensitivity of the habitat. Proactive coordination with environmental and regulatory agencies was accomplished during the planning, design and construction phases of this project. Through early discussions with regulatory personnel, the URDS-II concept was redesigned. Through ongoing consultation with project biologists, the alignment was changed to minimize environmental disturbance.

Project environmental constraints were further emphasized by incorporating a special environmental section in the bidding documents (with permits in the appendices), presenting construction limits on the planning documents while staking construction limits in the field, establishing a seasonal construction constraints map limiting construction periods, incorporating a comprehensive revegetation plan into the contract documents, hiring a full-time biologist to monitor activities during construction, and requiring all contractor field personnel to attend an environmental briefing prior to activities on the job site.

Even with the environmental documentation and briefings, environmental disturbances still occurred (though more may have occurred without the environmental documentation in place). In working with the regulatory agencies proactively from the onset of the project and by keeping them apprised of any environmental disturbances, a good working relationship between agency personnel and the Authority was developed. This relationship helped minimize the required mitigation for the environmental disturbances, which could otherwise have been more severe. The violations that did occur show that even with the environmental documentation in place, full-time inspection during construction in this type of habitat is recommended.

Subject Index

Page number refers to the first page of paper

Acoustic detection, 250, 468, 477
Aerial photographs, 113
Alignment, 172, 298
Alkalinity, 345
Aqueducts, 405, 413, 423, 721

Beaches, 782
Bedding, 536
Bids, 154
Budgets, 241
Bureau of Reclamation, 80
Buried pipes, 28, 180, 241, 277, 485, 516

California, 41, 66, 71, 88, 144, 195, 203, 232, 291, 308, 692
Canals, 566
Case reports, 47, 377, 423
Cast iron, 187
Cathodic protection, 187, 356, 367, 664
Certification, 546, 664
Chlorides, 575, 584
Coastal environment, 144
Coating, 566
Cogeneration, 387
Community support, 210, 221, 232
Compliance, 260
Computer programs, 28
Computer software, 162
Concrete pipes, 66, 250, 268, 318, 345, 356, 367, 413, 433, 468, 477, 528, 536, 546, 556, 566, 575, 584, 594, 602, 612, 646, 656, 664, 702, 714, 721, 731, 742, 763
Concrete, precast, 250

Concrete, prestressed, 345, 356, 367, 468, 477, 528, 556, 575, 584, 594, 602, 612, 646, 656, 664, 702, 714, 721, 731, 742
Concrete, reinforced, 66, 536, 742
Conduits, 180
Constraints, 241
Construction, 1, 57, 71, 88, 113, 133, 144, 210, 221, 260, 291, 308, 405, 556, 612, 622, 632, 638, 752, 763, 797
Construction materials, 674
Construction methods, 277, 328, 516
Contaminants, 681
Contracts, 260
Conveyance structures, 692
Corrosion, 468, 575, 584, 702
Corrosion control, 172, 180, 187, 356, 367, 377
Cost effectiveness, 172
Cost estimates, 291
Cost minimization, 284
Cost savings, 66, 71, 195
Costs, 80, 638, 782
Crack propagation, 656
Crossings, 57, 494
Curing, 485, 752
Cylinders, 250, 345, 367, 468, 528, 575, 584, 594, 602, 646, 664, 714, 721

Defects, 731
Deflection, 1, 14
Design, 47, 57, 71, 88, 103, 124, 162, 172, 277, 423, 528, 546
Design criteria, 1, 268, 451
Design modifications, 195

Dewatering, 124, 356, 397
Differential equations, 787
Discharge, 334, 787
Diversion structures, 797
Drainage, 787
Drilling, 47, 57, 162, 494
Droughts, 692
Ductility, 277, 494, 506
Durability, 584

Effluents, 681
Encasements, 180
Engineering, 284
Environmental factors, 268
Environmental issues, 622, 638, 674
Environmental planning, 298
Environmental quality, 308
Error analysis, 97
Evaluation, 318, 546, 575
Excavation, 187

Failures, 433, 468, 528, 731
Finite element method, 413
Flexible pipes, 1, 14, 28
Florida, 782
Flow, 334
Foundations, 268, 536
France, 241
Funding allocations, 731

Gas pipelines, 387
Geographic information systems, 80
Geology, 113
Gravity sewers, 66
Ground water, 133, 536
Ground-water depletion, 203

Hong Kong, 57
Hydrogen, 656
Hydrostatic tests, 433

In situ tests, 714
Infiltration, 782
Inspection, 260, 318, 356, 387, 397, 433, 602, 714, 721
Installation, 66, 494, 536, 546, 664
Intercepting sewers, 66, 71, 154
Iron, 277, 494

Jet grouting, 133

Liners, 752
Litigation, 622, 731

Maintenance, 298, 334
Mapping, 113
Marketing, 763
Materials, 1, 154, 277, 451
Methodology, 172, 451
Mexico, 47, 681
Microtunneling, 113, 124, 133, 516
Missouri, 162
Models, 28, 787
Monitoring, 468, 721
Mortars, 566, 584
Municipal wastes, 506

Natural gas, 387
Networks, 80
Nondestructive tests, 250, 477, 702, 714, 721

Outfall sewers, 681

Partnering, 632
pH, 345
Pigs, 334
Pipe design, 28, 284, 594, 612, 646, 664, 752
Pipe jacking, 516
Pipe joints, 433
Pipe lining, 566
Pipe tests, 782

Pipeline design, 97, 232, 284, 291, 632, 638, 797
Pipelines, 113, 124, 133, 241, 260, 268, 277, 298, 334, 356, 451, 485, 506, 681, 702, 742, 763, 787
Pipes, 187, 494, 506
Planning, 71, 80, 124, 291, 797
Plastic pipes, 180, 763
Polyethylene, 180, 328, 763
Potable water, 57, 328, 405, 423
Power loss, 528
Predictions, 14
Pressure pipes, 97, 268, 433, 731, 742
Pressure responses, 88, 103
Programs, 664
Projects, 71, 154, 195, 203, 210, 232, 268, 284, 308, 622, 638
Public opinion, 221
Public participation, 210, 232, 632
Puerto Rico, 413
Pumps, 103

Quality assurance, 656

Regulations, 674
Rehabilitation, 260, 298, 308, 328, 377, 485, 566, 594, 602, 612, 752, 782
Reinforcing steels, 367
Reliability, 298
Remote control, 397
Renovation, 451, 763
Repairing, 318, 468, 485, 506, 702, 731, 742
Research, 1, 210
Reservoirs, 797
Restraint systems, 97
Retrofitting, 405
Revegetation, 144
Right-of-way, 345
River crossings, 41, 47, 162, 632

River flow, 41
Rivers, 681
Robotics, 397

Safety analysis, 241
San Francisco, 113
Sanitary sewers, 318, 763, 782
Scour, 41, 47
Seismic design, 405, 413
Seismic hazard, 413
Selection, 1
Service life, 377, 594
Sewage disposal, 681
Sewage treatment, 782
Sewer design, 66
Sewers, 154, 241, 298, 308, 318, 674, 752
Shafts, 133
Sheet piles, 405
Sheet piling, 124
Simulation, 88
Siphons, 566
Soil properties, 345, 546
Soil strength, 14, 28
Soil tests, 575
Specifications, 433, 546, 556
Splitting, 328
Spreadsheets, 646
Standards, 284, 451, 742
Steel, 566, 594, 612
Steel pipes, 41, 221, 328
Stiffness, 14, 28
Strain, 14
Structural analysis, 318
Substitutes, 154
Subsurface investigations, 516
Sulfates, 268
Surge, 88, 103, 528

Testing, 172, 477
Tests, 187, 656
Texas, 556

Thrust, 97
Topsoil, 144
Trenchless technology, 328, 485, 506, 692, 752
Tunnel construction, 692
Tunneling, 423, 516, 602, 681
Tunnels, 210, 397, 674

Underground conduits, 485
United States, 377
Unsteady flow, 787
Urban runoff, 797
Utilities, 133, 387, 423

Value engineering, 195
Valves, 103
Virginia, 622

Wastewater disposal, 124, 334, 506, 674
Wastewater treatment, 47
Water demand, 162
Water distribution, 377, 405, 638
Water hammer, 787
Water pipelines, 41, 47, 57, 80, 88, 103, 144, 154, 162, 172, 195, 203, 210, 221, 232, 250, 291, 328, 377, 397, 413, 477, 556, 602, 612, 622, 632, 638, 646, 656, 797
Water quality control, 797
Water reclamation, 80, 674, 702
Water reuse, 80
Water storage, 203
Water supply, 41, 203, 221, 232, 405, 423, 556, 612, 622, 692, 721
Water table, 536
Water treatment plants, 423

Author Index

Page number refers to the first page of paper

Ahinga, Zachary, 210
Arakaki, Greg, P.E., 674
Arzamendi, Moi, P.E., 47
Ash, George L., 180
Atherton, David L., 714

Barden, Peter J., 308
Barnes, John, 367
Barnes, Russell, P.E., 782
Barrett, Stephen V. L., P.E., 57
Batta, Jamal, P.E., 66
Bell, G. E. C., 172
Benedict, Risque L., 345
Biery, James, P.E., 318
Bivins, Joe, P.E., 622
Bottcher, Richard, 797
Bradish, Bryan M., P.E., 377
Bramwell, Dave, 80
Brovold, Frederick N., 405
Brunzell, Wayne R., P.E., 268
Buchanan, Doug, P.E., 468
Butier, Mark, 632

Camacho, Anibal, 413
Capossela, Theodore A., P.E., 14
Cardwell, M. Wayne, 566
Carpenter, Ralph, 506
Cass, Tim, 144
Chamberlain, Dave, 144
Cimbora, Roger M., Sr., 334
Collins, Frank X., 291
Collins, Janice, 210
Conner, Michael E., 308
Conner, Michael E., P.E., 298
Conner, Randall C., 494

Davis, R. Ted, 423
de Leon, Carlos, P.E., 195
Deering, Stephen L., P.E., 298
Diab, Youssef Georges, 241
Dodge, Christopher F., 405
Doniguian, Ted, P.E., 367
Duke, Steve, 133

Erlin, Bernard, 584

Ferguson, Ken, 203
Fongemie, Roger, 702
Foster, R. Scott, 88, 103
Fowles, Deon T., P.E., 277

Galleher, J. J., Jr., 172
Galleher, John J., Jr., P.E., 721
Garcia, Felix, 413
Giandoni, Mark, P.E., 66, 154, 536
Gifford, John S., P.E., 328
Gill, Jesse, 124
Goodwin, John, 124
Gwaltney, Tim, P.E., 485

Hale, Marnel, P.E., 298
Hale, Marnell L., 308
Hall, Sylvia, P.E., 318
Hall, Sylvia C., P.E., 356
Harris, John, P.E., 66
Heffron, Ronald E., P.E., 397
Henry, Karen Larson, P.E., 154
Hill, James J., P.E., 546
Hodge, David S., P.E., 602
Holley, Mark, 468
Hutchinson, Mark W., P.E., 752

Irias, Nicholas J., 187

Jeyapalan, Jey K., P.E., 451, 528, 646, 763
Johnson, Geoffrey, 731
Johnson, Paul W., P.E., 195

Kalkman, Thomas, 71
Kaneshiro, Jon Y., 681
Khondker, Sufian A., P.E., 612
Kienow, Kenneth K., P.E., 433
Kinneen, John, 674
Kipps, Harry, P.E., 367
Klein, Steve, 133
Kramer, Steven R., P.E., 57, 162
Kurdziel, John M., P.E., 546
Kwong, James, 133

Larson, Duane, P.E., 674
Leahy, Tom, P.E., 622
Lee, Rolf H., 681
Lewis, Richard A., P.E., 584
Lynch, Scott, P.E., 80

Marks, Michael, 41
Marshall, David H., P.E., 556
Martin, Kim, 80
Martinez, Hector, 41
Masnada, Dan, 203
Mathy, David, 124
Mathy, David C., 113
McBain, Gregory W., 681
McGrath, Timothy J., 28
Meinhart, Thomas J., P.E., 162
Meiorin, Luciano, 681
Melton, James R., 221
Melton, Lyndel, 203
Mergelas, Brian J., 714
Metts, D. Michael, P.E., 298
Miles, Robert W., 1
Miller, Marilyn L., 187
Mitchell, John R., Jr., 612
Moncrief, W. Jeffery, 632

Motley, Edward M., 413
Mueller, Richard I., P.E., 742

Nagle, Galen, 133
Navin, Stephen J., 681
Nelson, Charles R., P.E., 546
Nichols, Daniel J., P.E., 162
Nielson, Dru R., 113
Nystrom, James A., P.E., 546

Ojdrovic, Rasko P., 702
Okita, Glen, 133
Olden, John, P.E., 656
Olson, Larry D., P.E., 250
O'Malley, Denis M., 308
Ostrander, Barbara B., P.E., 387

Pannell, R. A., 172
Perez, Antonio J., P.E., 195
Pflum, Timothy, P.E., 154
Pratt, David L., 405
Price, Robert E., P.E., 584
Prosser, David P., P.E., 664

Raines, Gregory L., 516, 692
Rajah, Sri K., P.E., 451, 528, 646
Rao, Rajesh S., 413
Ratliff, Alison, 318
Redmon, Alan, P.E., 674
Richards, Jim, P.E., 622
Romer, Andrew E., P.E., 97, 284
Ruffin, Larry J., 328
Rundle, Ralph T., P.E., 656
Ryan, Philip K., 423

Sack, Dennis A., 250
Schluter, James C., P.E., 14
Schrock, B. Jay, 1
Seal, Claud, 797
Smith, Terry, P.E., 47
Smith, Timothy M., 787
Smyth, James, 41

Smyth, Jim, 797
Stewart, Edward, 638
Stift, Michael T., P.E., 594, 721
Stine, Gary P., 632
Stine, Gary P., P.E., 594
Szeliga, Michael J., P.E., 377

Tedesco, Steve, 638
Tenbusch, Al, 506
Trembath, Richard, 632
Tucker, 797

Villalobos, Jose L., P.E., 575

Wallace, Steven W., 291
Walsh, Terry L., P.E., 602
Watson, Michael, 71
Weinberger, Marc R., 232
Williams, Paul J., P.E., 260
Wilson, Howard O., 405
Wolfe, James E., 566
Worthington, Will, P.E., 477, 656
Wright, Rick, 692

Yako, Michael A., 413
Yu, Burt K., P.E., 195

Zarghamee, Mehdi S., 413, 702
Zoumaras, Dave, P.E., 154